THE
SOCIAL
BIOLOGY
OF WASPS

14

34.95
60u

D1383069

MAY 2 3 1993

OCT   5 1994

OCT 17 1994
APR 06 1995

APR 2 9 1996

DEC 1 3 1996

FEB1 0 1997

APR- 3 1997

MAY 2 1 1997

DEC 0 8 1997

FEB 2 7 2004

DEC 2 1 2004

# THE SOCIAL BIOLOGY OF WASPS

*Edited by*

## Kenneth G. Ross
## Robert W. Matthews

*Department of Entomology, University of Georgia*

Comstock Publishing Associates

A DIVISION OF CORNELL UNIVERSITY PRESS
ITHACA AND LONDON

ERINDALE
COLLEGE
LIBRARY

Copyright © 1991 by Cornell University

The copyright does not apply to the Introduction, which is in the public domain.

Figures 2.7, 2.9, 3.9, 4.7, 4.9, 5.2, 6.3, 6.6, 6.7, 7.8, 8.9, 15.2, 16.4, and 17.3 by Amy Bartlett Wright.

All rights reserved. Except for brief quotations in a review, this book, or parts thereof, must not be reproduced in any form without permission in writing from the publisher. For information, address Cornell University Press, 124 Roberts Place, Ithaca, New York 14850.

First published 1991 by Cornell University Press.

Printed in the United States of America

∞ The paper in this book meets the minimum requirements of the American National Standard for Information Sciences—Permanence of Paper for Printed Library Materials, ANSI Z39.48-1984.

Library of Congress Cataloging-in-Publication Data

The Social biology of wasps / edited by Kenneth G. Ross and Robert W.
   Matthews.
      p.    cm.
   Includes bibliographical references and index.
   ISBN 0-8014-2035-0 (cloth : alk. paper)
   ISBN 0-8014-9906-2 (pbk. : alk. paper)
   1. Vespidae.  2. Sphecidae.  3. Insect societies.  I. Ross,
Kenneth G., 1955–   .  II. Matthews, Robert W., 1942–   .
QL568.V5S58  1991
595.79′8—dc20                   90-44178

# Contents

*Preface* *xiii*

*Contributors* *xv*

**Introduction** *1*
MARY JANE WEST-EBERHARD


PART I  *The Social Biology of the Vespidae*

**1**

**Phylogenetic Relationships and the Origin of
Social Behavior in the Vespidae** *7*
JAMES M. CARPENTER

**Phylogenetic Relationships** *8*
  Subfamily Relationships *8* Generic Relationships *9*
    Stenogastrinae *9* Vespinae *12* Polistinae *14*
**The Polygynous Family Model for Wasp Social Evolution** *21*
  Description of the Model *21* Test of the Model *25*
**Concluding Remarks** *29*

## 2

# The Solitary and Presocial Vespidae 33
DAVID P. COWAN

**Nesting** 36
  Nest Construction 36
    Construction materials 40 Burrowers 41 Renters and tube dwellers 42 Builders 42
    Chimneys 43
  Dispersion of Nests 43 Oviposition 44 Development 45 Foraging and
  Hunting 46 Mass Provisioning by Solitary Wasps 47 Delayed- and Progressive-
  provisioning, Subsocial Wasps 49 Communal and Quasisocial Wasps 52
  Immature Mortality 55
    Brood parasites and predators 55 Nest usurpation 57 Prey quality 58
**Mating Behavior** 58
  Male-Male Competition 59 Female Choice 63 Courtship and Copulation 64
**Population Structure and Sex Ratio** 66 **Concluding Remarks** 69


## 3

# The Stenogastrinae 74
STEFANO TURILLAZZI

**Systematics** 75 **Habitat, Adult and Larval Morphology** 75 **The Abdominal
Substance and Rearing of the Brood** 77
  Oviposition 78 Development of the Larvae and Larva-Adult Interactions 79
  Larva-Adult Trophallaxis 80 Storage of Liquid and Solid Food 80 Ant
  Guards 81 Other Functions of the Abdominal Substance 82
**Colony Development and Social Organization** 82
  Nest Foundation 82 Colony Size 83 Female Reproductive Status 84 Social
  Interactions, Dominance Hierarchies, and Division of Labor 85
**Male Behavior and Mating** 88 **Nest Architecture** 90
  Evolutionary Trends in Nest Architecture 90
**Predators, Parasites, and Nest Defense** 93 **Concluding Remarks** 95


## 4

# *Polistes* 99
HUDSON K. REEVE

**Systematics and Distribution** 101 **Nesting Biology** 103 **Colony Cycle** 104
**Founding Phase** 106
  Foundress Strategies: Founding, Joining, and Usurpation 106 Foundress
  Strategies: A Synthesis 118 Foundress Interactions 120
**Worker Phase** 127
  Fates of Subordinate Foundresses 127 Worker Dominance and Reproduction 128
  Social Integration and Colony Growth 132 Early Males 135 Seasonal Changes in
  Characteristics of Emerging Adults 135

Contents

**Reproductive Phase** *137*
 Sex Investment Ratios and Timing of Male and Gyne Production *137*
**Intermediate Phase** *139*
 Mating *139* Winter Diapause *143*
**Concluding Remarks** *143* **Appendix 4.1** Renesting Model *145* **Appendix 4.2**
Survivorship Insurance Model *146* **Appendix 4.3** Foundress Three-strategy
Model *147*


## 5

### *Belonogaster, Mischocyttarus, Parapolybia,* and Independent-founding *Ropalidia* 149

RAGHAVENDRA GADAGKAR

**Distribution and Systematics** *151* **Nest Architecture** *153* **Nesting Cycle** *158*
 Seasonal Nesting Cycle *161* Aseasonal Determinate Nesting Cycle *162*
 Indeterminate Nesting Cycle *164*
**Enemies and Colony Defense** *165* **Food and Feeding Habits** *170* **Mating
Behavior** *171* **Social Organization** *172*
 Dominance Hierarchies *173* Role of the Egg Layer *174* Behavioral Caste
 Differentiation *175*
**Caste** *179* **Origin of Social Life: A Perspective from Studying Independent-
founding Polistines** *181* **A Possible Route to Eusociality** *185* **Concluding
Remarks** *186* **Appendix 5.1** Common Parasites of *Belonogaster, Mischocyttarus,
Parapolybia,* and Independent-founding *Ropalidia* *188* **Appendix 5.2** Common
Predators of *Belonogaster, Mischocyttarus, Parapolybia,* and Independent-founding
*Ropalidia* *190*


## 6

### The Swarm-founding Polistinae 191

ROBERT L. JEANNE

**What Are the Swarm-founding Wasps?** *191* **The Success of the Swarm-founding
Polistinae** *193* **Biology** *199*
 Colony Founding *199* Polygyny and the Fluctuation of Queen Number *200*
 Colony Growth and Reproduction *204* Types of Swarms *206* Independence of
 the Nesting and Colony Cycles *207*
**Causes of Success** *210*
 The Socialization of Dispersal and Founding *210* Colony Size *212*
**Consequences of Large Colony Size** *214*
 Specialization *215*
  Queen-worker specialization *215* Specialization among workers *216*
 Communication *216* Defense *220* Homeostasis *223*
**Ecological Influences on Colony Size and Cycle** *224*
 Latitudinal Patterns in Colony Size *225* Latitudinal Patterns in Seasonal
 Synchrony *227* Indeterminate Nesting Cycles *229*
**Concluding Remarks** *231*

## 7

## *Vespa* and *Provespa* 232
MAKOTO MATSUURA

**Life History of *Vespa*** 233
  Life History Strategies 233 Usurpation of Nests 235
**Nesting Habits** 236
  Nest Materials and Nest Sites 236 Structure of the Nest 237 Relocation of the
  Nest 240
**Feeding Habits** 242
  Food Sources 242 Group Predation 243
**Social Structure** 245
  Caste Differentiation and Division of Labor 245 Queen Control and
  Communication 248
**Population Dynamics** 250
  Survivorship Curves 250 Colony Size 251 Nest Survival 252 Sympatry and
  Mechanisms of Coexistence of *Vespa* Species 252
**Life Cycle Characteristics in the Tropics** 254
  Asynchronous Colony Cycles 255 Polygyny and Nest Size 255 Nesting
  Period 256 Structural Characteristics of Tropical Nests 257
**Life History of *Provespa*** 257
  Nocturnal Habits 257 Nest Foundation 258 Nest Structure and Colony
  Cycle 258 Interactions among Individuals in a Nest 260
**Concluding Remarks** 261

## 8

## *Dolichovespula* and *Vespula* 263
ALBERT GREENE

**General Life History Patterns** 265 **Colony Initiation and Queen
Competition** 274 **Worker and Reproductive Production** 281 **Colony
Maintenance** 284
  Nest Site Modification 284 Sanitation 285 Construction 285
  Thermoregulation 287 Food Transfer 288 Defense 289
**Queen Control and Reproductive Competition** 290
  Reproductive Dominance 293
    Nutritional Aspects 293 Physical Aspects 293 Pheromonal Aspects 296
  Colony Cohesion 299 Queen Production 300
**Mating** 302 **Concluding Remarks** 304

# PART II   *Special Topics in the Social Biology of Wasps*

## 9

## Reproductive Competition during Colony Establishment 309
PETER-FRANK RÖSELER

**Occurrence and Distribution** 311
  Nest Usurpation and Social Parasitism 312 Competition among Foundresses of

Different Nests *313* Competition among Associates on Multiple-foundress
Nests *315*
**The Dominance Hierarchy** *318*
Social Dominance *318*
  Characteristics *318* Dominant-Subordinate Interactions *320*
Reproductive Dominance *322*
  Characteristics *322* Inhibition of Egg Formation *324*
**Establishment of a Dominance Hierarchy** *325* **Maintenance of Dominance** *330*
**Reproductive Success** *331* **Concluding Remarks** *333*

# 10

# Evolution of Queen Number and Queen Control *336*

J. PHILIP SPRADBERY

**Queen Control in *Polistes*** *337*
Dominance Hierarchies: A Case Study *337* Differential Oophagy *342* Cell
Building and Other Activities *342* Physiological Basis *344*
**Queen Number and Queen Control in the Stenogastrinae and Polistinae** *344*
Stenogastrinae *345* Old World Polistinae *349*
  *Ropalidia 349 Belonogaster 354 Polybioides 357 Parapolybia 358*
New World Polistinae: *Mischocyttarus* *358* New World Polistinae:
Swarm-founding Genera *359*
**Queen Number in Vespinae** *360*
Pleometrosis *360* Secondary Polygyny *362*
**Queen Control in the Vespinae** *364*
*Vespa* and *Provespa* *364* *Dolichovespula* *366* *Vespula rufa* Species Group *367*
*Vespula vulgaris* Species Group *368*
**Evolution of Queen Number** *371*
Predisposing Features *371* Benefits from Social Nesting *372* Possible Routes to
Eusociality *376* Genetic Basis for the Evolution of Eusociality *376* The
Polygynous Route to Eusociality *380*
**Evolution of Queen Control** *382*
Confrontation, Competition, and Compromise *382* From Physical to Chemical
Dominance *384*
**Concluding Remarks** *387*

# 11

# Polyethism *389*

ROBERT L. JEANNE

**Patterns of Polyethism** *390*
Reproductive Division of Labor *391* Age Polyethism among Workers *396* Task
Partitioning *402* Individual Variability *403*
**Polyethism and Reproductive Strategies** *407*
Options for Co-foundresses *408* Options for Offspring *412* The Role of
Dominance *414* Releasers of Worker Behavior *420* The Regulation of Colony
Cycle and Size *421*
**Concluding Remarks** *423*

**12**

## Nourishment and the Evolution of the Social Vespidae  *426*

JAMES H. HUNT

Foundations of the Nutritional Perspective *428* "Nutritional Castration" and the
Two Kinds of Workers *431* Ontogenetic Workers in the Evolution of
Eusociality *433*
  The Nutritional Scenario *433* Elements of the Scenario *434*
      The eusocial lineages *434* Parasitization and predation *435* Pleometrosis and division of
      labor *435* Large provisions *436* Larval provisions and adult nourishment *436*
      Cannibalization of larvae *437* Larval saliva *437* Saliva nutrient concentration *438*
      Nourishment and saliva retention *439* Poor nourishment *440* Emergence date *441*
      Inclusive fitness and kin selection *442* Selection for behavioral repertoires *443*
Social Wasps as Annuals *444* Generalization of the Model *447* Concluding
Remarks *448*

**13**

## Population Genetic Structure, Relatedness, and Breeding Systems  *451*

KENNETH G. ROSS AND JAMES M. CARPENTER

Primitively Eusocial Vespidae *456*
  Stenogastrinae *456* Independent-founding Polistinae *457*
      *Ropalidia, Parapolybia, Belonogaster,* and *Mischocyttarus 458 Polistes 465*
Highly Eusocial Vespidae *469*
  Swarm-founding Polistinae *469* Vespinae *471*
Social Sphecidae *475* Concluding Remarks *476*

**14**

## Evolution of Nest Architecture  *480*

JOHN W. WENZEL

Stenogastrinae *483*
  Building Materials *483* Nest Design *483*
Diversification and Convergence in Independent-founding Polistinae *485*
  Variation within Species *487* Variation among Species *488*
      Eight Elements in *Mischocyttarus 489* Conclusions from *Mischocyttarus 491*
Variation among Genera of Vespinae and Polistinae *491*
  Structural Variation *491*
      Building Materials *492* Pedicels *495* Sessile Combs *496* The Envelope *498*
      Engineering with Paper *500*
  Ontogenetic Variation in Architecture *502*
      Vespinae *503* Polistinae *505*
Concluding Remarks *507* Appendix 14.1 Definitions *518* Appendix 14.2
Abbreviations of Collections Containing Nest Specimens Cited in the Text *519*

## 15

## The Nest as the Locus of Social Life 520
CHRISTOPHER K. STARR

Functions of the Nest 521 The Elements of Nest Structure 525 The Boundary between Colony and Noncolony 529 Communication and Nest Structure 532 Conflicts of Genetic Interest 535 Concluding Remarks 538


## 16

## The Function and Evolution of Exocrine Glands 540
HOLLY A. DOWNING

**Description of Exocrine Glands** 540
  Gland Morphology 540 Gland Locations and Functions 542
    Glands of the Head 542
      Ectal mandibular glands 542 Mesal mandibular glands 543 Hypopharyngeal glands 544
      Labial palp glands 544 Maxillary-labial glands 545 Maxillary-hypopharyngeal
      glands 545 Sublingual glands 545 Endostipital glands 545
    Thoracic Gland 545 Glands of the Abdomen 546
      Sternal glands 546 Tergal gland 551 Poison gland 551 Dufour's gland 552
**Pheromones and Gland Evolution** 553
  Dominance and Queen Control 553 Social Parasitism 557 Recruitment 558
    To an Aggregation 558 To a Food Source 559 To a New Nest Site 559 To a Site of
    Defense 561
  Nest Construction 562 Recognition of Nestmates and Nests 564 Brood
  Care 564 Nest Defense 565 Reproductive Behavior 566
**Concluding Remarks** 568


## 17

## Evolution of Social Behavior in Sphecid Wasps 570
ROBERT W. MATTHEWS

**An Overview of Sphecid Social Behavior** 573 **Factors Favoring Social Behaviors** 577
  Pressure from Parasites and Predators 578 Pressure from Conspecifics:
  Usurpation 580 Philopatry and Construction of Persistent Nests 581 Nest Site
  and Type: The Opportunity to Be Useful 582 Female Longevity, Multivoltinism,
  and Adult Overwintering 586 Maternal Control over Sex Ratio, Size, and
  Fecundity 587 Provisioning with Multiple Small Prey 589 Nestmate or Kin
  Recognition 591
*Microstigmus comes:* **Case Study of a Eusocial Sphecid** 592
  Nest Construction 593 Nest Establishment 594 Parasitism, Predation, and
  Usurpation 594 Genetic Relatedness 597 Sex Ratio 598 Reproductive Division of
  Labor 600 Evolution of Eusociality in *Microstigmus comes* 600
**Concluding Remarks** 601

*Literature Cited*  603

*Subject Index*  667

*Taxonomic Index*  675

# Preface

This book is an introduction to and overview of our present knowledge of wasp social biology. Social wasps are important model organisms for the study of the evolution of social behavior, yet no comprehensive survey of wasp social biology has appeared since Evans and West-Eberhard (1970) and Spradbery (1973a) published their major reviews, a period during which our understanding of these organisms has advanced considerably. The publication of this volume not only fills a perceived need for a review of these advances but also complements the recent appearance of major reviews of the social biology of bees (Seeley 1985, Winston 1987, Roubik 1989) and ants (Hölldobler and Wilson 1990), the two other groups of social Hymenoptera.

The authors of the chapters in this edited collection have made fundamental contributions in the primary literature to the topics on which they write. Their papers are organized into two groups, those dealing with the basic natural history of specific vespid taxa as it pertains to social behavior (or lack thereof) and those dealing with selected topics of special interest to students of social evolution. Chapter 1 sets the stage for what follows by providing the broad outlines of the taxonomy and phylogenetic relationships of social vespid wasps. The remaining chapters in Part I serve as primers on the ecology, behavior, and social organization of solitary, presocial, and eusocial vespids; these aspects of wasp biology are interpreted largely in the context of how they bear on the emergence and development of eusocial behavior. The themes introduced in Part I are developed in greater detail in Part II, which is a sampler rather than an exhaustive compendium of conceptual topics of interest to sociobiologists. All eight chapters in Part I and eight of the nine chapters in Part II focus exclusively or primarily on vespids, a focus that reflects the taxonomic distribution of social behaviors in wasps: most social wasp species are members of the family Vespidae.

The final chapter combines both the natural history and theoretical themes of the book in considering social evolution in the family Sphecidae. Social behavior is not as well developed or as well represented in the Sphecidae as in the Vespidae, but comparative studies of sphecids may provide fresh insights into our understanding of the evolution of eusociality in winged hymenopteran predators that construct nests.

The book is intended for specialists as well as for those who are broadly interested in sociobiology yet somewhat unfamiliar with entomology or wasp biology. To make the material more accessible to the nonspecialist, the authors usually define technical terms where they first appear in each chapter. A subject index and a taxonomic index are provided to point the reader to topics of particular interest. Because of their broad relevance, the same topics often are treated in more than one chapter; the indexes can be especially useful as guides to a diversity of viewpoints on such topics. The literature cited throughout the volume is presented in a single section at the end of the book. Papers by one or two authors are listed alphabetically by the authors' names; papers by more than two authors are listed alphabetically by the first author's name and then ordered chronologically.

Many individuals helped us put together this volume. First and foremost, we thank the contributors for their papers. For critical scientific reviews of the contents we thank Michael E. Archer, H. Jane Brockmann, George C. Eickwort, George J. Gamboa, Michael H. Hansell, Robert W. Longair, Robin E. Owen, Justin O. Schmidt, Joan E. Strassmann, Philip S. Ward, and Seiki Yamane, as well as all the contributors to the volume. We especially appreciate the assistance of James M. Carpenter, Christopher K. Starr, and John W. Wenzel, who provided expert advice in all matters technical and scientific, and of Janice R. Matthews, who undertook the indexing. We also thank Tonia R. Jones, Amanda M. Lambert, Daniel J. Noles, and Charles R. Santa Maria for help with the graphics and word processing. Finally, we are grateful to Robb Reavill of Cornell University Press, who has encouraged us in this project from its inception.

KENNETH G. ROSS
ROBERT W. MATTHEWS

*Athens, Georgia*

# Contributors

**James M. Carpenter**
Museum of Comparative Zoology
Harvard University
Cambridge, Massachusetts 02138 USA

**David P. Cowan**
Department of Biological Sciences
Western Michigan University
Kalamazoo, Michigan 49008 USA

**Holly A. Downing**
Department of Biology
University of Wisconsin–Whitewater
Whitewater, Wisconsin 53190 USA

**Raghavendra Gadagkar**
Centre for Ecological Sciences and
    Centre for Theoretical Studies
Indian Institute of Science
Bangalore 560012 India

**Albert Greene**
Government Services Administration
WPBO-B, Room 7719
7th and D Streets, S.W.
Washington, D.C. 20407 USA

**James H. Hunt**
Department of Biology
University of Missouri–St. Louis
St. Louis, Missouri 63121 USA

**Robert L. Jeanne**
Department of Entomology
University of Wisconsin
Madison, Wisconsin 53706 USA

**Makoto Matsuura**
Faculty of Bioresources
Mie University
1515 Kamihama
Tsu 514 Japan

**Robert W. Matthews**
Department of Entomology
University of Georgia
Athens, Georgia 30602 USA

**Hudson K. Reeve**
Section of Neurobiology and Behavior
Cornell University
Ithaca, New York 14853 USA

**Peter-Frank Röseler**
Zoologisches Institut (II) der Universität Würzburg
Röntgenring 10, D-8700 Würzburg, Germany

**Kenneth G. Ross**
Department of Entomology
University of Georgia
Athens, Georgia 30602 USA

**J. Philip Spradbery**
Division of Entomology
CSIRO
Black Mountain, Canberra, ACT 2601 Australia

**Christopher K. Starr**
Biosystematics Research Center
Research Branch, Agriculture Canada
Ottawa, Ontario, Canada K1A 0C6

**Stefano Turillazzi**
Dipartimento di Biologia Animale e Genetica
Centro di Studio per la Faunistica ed
  Ecologia Tropicali del C.N.R.
Università degli Studi di Firenze
50125 Firenze, Italy

**John W. Wenzel**
Department of Entomology
University of Georgia
Athens, GA 30602 USA

**Mary Jane West-Eberhard**
Escuela de Biología
Universidad de Costa Rica
Ciudad Universitaria, Costa Rica

# Introduction

MARY JANE WEST-EBERHARD

The title of this volume makes no secret of the fact that the entire book is about a single aspect of a single group of organisms: the social biology of wasps (Vespidae and Sphecidae). Specialists will be delighted, and nonspecialists may be incredulous, that such a fat book has been written on such an esoteric subject. Why, indeed, a whole book on the social biology of wasps?

The answer to that question invites reflection on the relationship between specialized knowledge and conceptual progress in science. Progress clearly depends on both depth and breadth of understanding. On the one hand, it requires extremely detailed, meticulous comprehension of particular cases; on the other hand, it requires a search for patterns and the "laws of nature." Neither endeavor can prosper without the other. Detailed data without concepts have little significance; and concepts without information can become vacuous fantasies. So there has to be a continuous interweaving of fact and theory, both within a discipline and in the minds of individual scientists.

Consequently, the deep understanding of particular groups of organisms—such as the wasps (for students of sociality) or the genus *Drosophila* (for students of genetics) or the Galapagos finches (for students of ecology and speciation)—plays a special role in the advancement of knowledge in biology. To test generalizations with facts about particular organisms, one must be able to move among levels of explanation and information without running into walls of ignorance at any level. Consider the example at hand—wasps and social life. In recent years, we have seen the creation of hypotheses regarding the evolution of

1

social life that aspire to a high order of generality. Kin-selection theory, for example, pretends to apply to all kinds of organisms and even to the origins of multicellular life itself. The wasps have been a prime group in testing kin-selection theory; and, in turn, questions raised by the theory have tested our knowledge of the wasps. The kin-selection generalization, like Darwin's theory of evolution by natural selection, is so simple that it may appear trivial. It says that helping relatives (or allelic co-carriers) is favored if the benefit (in terms of allelic replication) is greater than the cost of aid. But testing that generalization, and confronting the predictions and confusions it has raised, has led researchers through labyrinthine paths of hypothesis and observation. As test organisms, the wasps have been scrutinized in a surprising variety of ways, most of them discussed in this book. There are now data on causes of success and mortality in newly founded nests; longevity of immatures, workers, and queens; sex ratios of progeny; behavior and reproduction in the solitary and rudimentarily social relatives of highly specialized social wasps; genetic relatedness among nestmates as revealed by protein electrophoresis; mating behavior and the contents of female spermathecae; ovarian condition (and, hence, reproductive role) of individual females; division of tasks among workers; social dominance in relation to reproductive success; hormonal correlates of behavior and ovarian condition; microscopic structure of exocrine glands and the chemical nature of their pheromonal products; and the phylogeny of higher groups as revealed by their nests, morphology, and behavior.

The absence of any element in this vast array of kinds of information would represent a roadblock to understanding the evolution of sociality. Furthermore, as such information accumulates, the wasps become more and more valuable as a testing ground; it becomes increasingly easy to augment the assembled facts with new ones, filling in the ever smaller holes alongside existing information, so that the realm of understanding increases at an accelerating rate.

The scientific study of wasp sociality began with the works of René Antoine Ferchault de Réaumur (1683–1757). Réaumur's 1719 article "Histoire des guêpes" and his beautifully illustrated quarto volumes *Mémoires pour servir à l'histoire des insectes* (1740) were among the first published works to challenge the partly imaginary accounts of insect behavior promulgated by Aristotle and Pliny nearly two thousand years earlier. The wasps began moving toward their present status as a key group for understanding social behavior at the beginning of the twentieth century, when the French entomologist Etienne Roubaud joined a mission to study sleeping sickness in the African Congo. There he took up research on wasps as a pastime. Roubaud made unprecedented observations of solitary and primitively group-living wasps, especially of the genera *Synagris* (Eumeninae) and *Belonogaster* (Polistinae), groups that seemed to be at the brink of true sociality (eusociality), and he developed a theory for the evolution of insect sociality that even today remains insightful and modern in tone. Roubaud's classic papers (1911, 1916) (see Cowan, Gadagkar, this volume), and the work of Ducke (1905–1914) in Brazil (see Jeanne, this volume: Chap. 6; Wenzel, this volume), were publicized in lectures by William Morton Wheeler, first in Boston (1922) and later in Paris (1925), and in Wheeler's influential books *Social Life among the Insects* (1923) and *The Social Insects* (1928). These early studies of tropical species drew attention to the kinds of transitions likely to have taken place in the origins of a sterile worker caste, and established the wasps, with their easily observed and socially diverse colonies, as prime material for comparative study.

The first half of the twentieth century was marked by a steady, if slow, accumulation of information on the natural history and behavior of social wasps, especially in the work of É.-P. Deleurance, C. D. Duncan, L. Pardi, P. Rau, O. W. Richards, and J. van der Vecht.

With the "New Age" of evolutionary biology ushered in by kin-selection theory in the 1960s, the wasps came once again into prominence, being haplodiploid Hymenoptera singled out for special mention by Hamilton (1964a,b), who had observed *Polistes* in Brazil. If progress in theory then benefited the study of wasps, the reverse was also true. All of the earliest (pre-1970) applications of kin-selection theory to field studies of animal behavior were on social insects (Haskins and Whelden 1965; West-Eberhard 1967, 1969; Eickwort 1969), and all but one of them (Haskins and Whelden 1965) were on wasps (early general applications were discussed by Haskins [1966], Williams [1966], and Wilson [1966]). But the wasps have been important in sociobiology and behavioral ecology not only because they were first on the scene, or even because they are easily observed and socially diverse. In addition to these qualities, they display, perhaps more obviously than any of the other social insects, a striking individuality of social role, shown by Pardi and others to be partly associated with degree of aggressive dominance and ovarian state. The transparently selfish, status-related and kin-directed nature of social behavior in *Polistes* and other wasps invited reflection on exactly the principles that were being emphasized in evolutionary analyses of sociality—selfish cooperation and discriminating aid among kin. In wasp colonies, the drama of Darwinian competition and its sometimes surprising consequences—ritual combat, measured beneficence, group integration, and even idleness—are played out in fascinating detail. It is hard to imagine a group of organisms better suited to inspire and inform the study of social evolution.

Research on wasps, and on the evolution of social behavior, is now at another turning point. With increasing interest in development and phenotypic plasticity in evolutionary biology, the social insects are likely to take on an even greater role as model organisms for studies of behavioral evolution. The enormous literature on hormonal "caste determination," perhaps the most highly advanced traditional area of research on the social insects, places these organisms in a favored position for testing ideas relating development and evolution. In this field, too, the wasps are promising material for research. The regulatory origins of the worker-queen dichotomy can be sought in the hormonal cycles of solitary species; and closer examination of the genetics and ontogeny of "individuality" in behavior, for which wasps are famous, could provide valuable ideas regarding the evolutionary history of division of labor. There are already some data on the behavioral endocrinology of primitively eusocial species (see Röseler, this volume), and these data offer unique insights into the evolution of caste regulation, as they begin to unravel the mechanism of the association between social status (dominance-subordinance) and reproductive success (ovarian state)—a correlation that is fundamental to group organization in all eusocial organisms.

It has been prophesied (Wilson 1989) that in the future taxon-centered research such as that represented by this book will become increasingly important in biology. With this compendium, the wasps stand poised for continued leadership in teaching humans about the evolution of sociality.

# PART I

*The Social Biology of the Vespidae*

# 1

# Phylogenetic Relationships and the Origin of Social Behavior in the Vespidae

JAMES M. CARPENTER

As Wilson (1971:1) remarked, the social insects, "together with man, hummingbirds, and the bristlecone pine . . . are among the great achievements of organic evolution. Their social organization . . . in respect to cohesion, caste specialization, and individual altruism—is nonpareil." Because they exhibit social behavior, wasps in the family Vespidae have long fascinated evolutionists, but significantly, not all wasps in the family are social. *Eusociality* is defined as the presence of overlapping adult generations, cooperative brood care, and extreme reproductive division of labor—namely, an essentially sterile caste. The origin and maintenance of sterile castes has been one of the major problems in selection theory, and because wasps and bees exhibit a graded series of behaviors, ranging from solitary to eusocial, they are the principal insect groups allowing testing of evolutionary models for the origin of eusocial behavior.

A widely accepted model for the origin of eusociality in wasps is the *polygynous family hypothesis* elaborated by West-Eberhard (1978a). After surveying life history data for parasocial and primitively eusocial wasps (see Cowan, this volume, for definitions of types of nesting behaviors), she suggested a number of possible transitions in the evolution of a worker caste and the development of different types of eusocial behavior. The model envisions polygynous groups of related females as the setting for the evolution of workerlike behavior and holds that caste formation preceded matrifilial monogyny. Several routes to long-term polygyny (multiple female reproductives in a nest) and long-term monogyny (a single female reproductive) are outlined.

This model has never received a critical test, for the critical test of an evolutionary model is a phylogenetic one and well-corroborated phylogenetic hypotheses for vespids have not previously been available. Evolutionary models allow predictions about particular results, or evolutionary patterns, which may be critically compared with independently generated phylogenetic hypotheses. Simply put, models (process theories) can be tested by being mapped onto cladograms (phylogenetic theories), a point particularly well discussed by Eldredge and Cracraft (1980). I have recently analyzed all of the genera of vespid social wasps cladistically, and in this chapter I use these results to propose a test of West-Eberhard's model.

## PHYLOGENETIC RELATIONSHIPS

The family Vespidae has traditionally been recognized as a well-defined, but isolated group. It is monophyletic, as is evidenced by the autapomorphies (unique derived characters) of an elongate discal cell (at least equal to the submedian cell in length), spined parameres, and oviposition into an empty cell (Carpenter 1982). However, the identity of the sister-group of the vespids is not clear. Brothers (1975) suggested Scoliidae as the closest relative of Vespidae, with prothoracic characters constituting the strongest evidence. These included a produced posterolateral angle, concave posteroventral margin and acute ventral angle of the pronotum, immobile union of the prothorax and mesothorax, and a sunken prosternum. Brothers (1976) added lateral expansion of the third phragma as another synapomorphy (shared derived character) for these two families. Königsmann (1978), Day et al. (1981), and I (1982) followed Brothers's hypothesis, but Rasnitsyn (1980) disputed it, suggesting Formicidae instead as the sister-group of Vespidae. Although Rasnitsyn failed to employ cladistic methods or to restrict consideration to groundplan (ancestral) features, Gibson (1985) has shown that the immobile prothorax must have arisen independently in scoliids and vespids, and there may yet be evidence for monophyly of the traditional taxon Scolioidea (Rasnitsyn 1980) or a relationship of Scoliidae to part of Scolioidea (Osten 1982). A comprehensive reevaluation is necessary, but at present Rasnitsyn's suggestion is less well supported than Brothers's scheme. Recently, Piek's (1987) review of venom chemistry identified wasp kinins as a further possible synapomorphy of Scoliidae + Vespidae (and see Carpenter 1990).

### Subfamily Relationships

In the first application of cladistic analysis to the Vespidae, I investigated the monophyly and interrelationships of all the suprageneric taxa

(Carpenter 1982). Hitherto the vespids were usually arranged in the three families Masaridae, Eumenidae, and Vespidae (Richards 1962), each with three subfamilies. I showed that the Masaridae is a paraphyletic group and proposed a strictly cladistic classification with one family and six subfamilies. The interrelationships of the subfamilies are shown in Fig. 1.1.; Table 1.1 lists the approximate taxonomic diversity and distribution of these groups. Eusocial behavior is found in three of the subfamilies, the Stenogastrinae, Polistinae, and Vespinae, which in this system are all descendants of a recent common ancestor not shared with solitary Vespidae. Vecht's suggestion (1977a) that Stenogastrinae is more closely related to a subgroup of Eumeninae ("Zethinae") cannot be supported (Carpenter 1982, 1988a).

My 1982 study referred primarily to morphological data, but behavior has been subsequently treated in more detail (Carpenter 1988a). Although there are specific behavioral differences in the three eusocial subfamilies, a considerable number of behavioral traits can be interpreted as part of the groundplan of these groups. Morphological synapomorphies uniting the three groups include a distally pointed forewing marginal cell and narrowed larval labrum (the latter is widened in various Polistinae, a reversal to the primitive condition). Derived behavioral traits shared by social vespids include simultaneous progressive provisioning with a masticated paste from a broad range of adult and immature arthropod prey, probably only after egg hatch; care extended until adult eclosion; construction of complete nests that hang free from the substrate, probably primitively of plant material; cell reuse; nest sharing among adults; adult-adult trophallaxis; and, probably, cooperative brood care and temporary reproductive division of labor. Moreover, the Eumeninae is a monophyletic group, with the autapomorphies of a parategula, hindcoxal carina, bifid claws, and bisinuate larval labrum. Because Stenogastrinae has none of these traits, it cannot be a subgroup of Eumeninae, as Vecht suggested.

## Generic Relationships

### Stenogastrinae

Stenogastrinae (hover wasps) is the most poorly known social group. The approximately 50 described species are endemic to the Oriental tropics. Although hover wasps are notable for great diversity in nest architecture, their behavior has received very little study until recently (reviewed in Carpenter 1988a and Turillazzi, this volume). Vecht (1977a) arranged the species in seven genera but did not treat phylogeny. I recently investigated the monophyly and phylogenetic relationships of these taxa (Carpenter 1988a) and proposed a reclassification into six genera, synonymizing Vecht's genus *Holischnogaster* with *Parischno-*

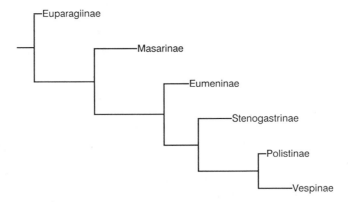

**Fig. 1.1.** Cladogram for the subfamilies of Vespidae. (After Carpenter 1982.)

**Table 1.1.** Diversity and distribution of the extant subfamilies of Vespidae

| Subfamily | Common name | Diversity | Distribution |
|---|---|---|---|
| Euparagiinae | | 9 species in 1 genus | Southwestern Nearctic |
| Masarinae | Pollen wasps | ca. 250 species in 19 genera | Australia, Southern Africa, Neotropics, USA, Mediterranean |
| Eumeninae | Potter wasps | ca. 3000 species in 184 genera | Cosmopolitan |
| Stenogastrinae | Hover wasps | ca. 50 species in 6 genera | Oriental tropics |
| Polistinae | Paper wasps | ca. 800 species in 29 genera | Cosmopolitan |
| Vespinae | Hornets and yellowjackets | ca. 60 species in 4 genera | Holarctic and Oriental tropics |

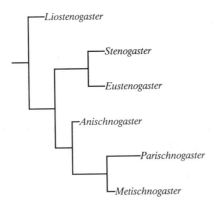

**Fig. 1.2.** Cladogram for the genera of Stenogastrinae. (After Carpenter 1988a.)

10

*gaster.* The interrelationships of the genera are shown in Fig. 1.2. This cladogram is based on analysis of 40 characters and is supported by 10 informative morphological characters and 3 possibly informative behavioral ones. The grouping of all the genera apart from *Liostenogaster* is established by fusion of the occipital and hypostomal carinae. *Eustenogaster* and *Stenogaster* are shown to be sister-groups by possession of elongate palpi, the male with the clypeus ventrally rounded and mandibular teeth reduced, and, possibly, nests campanulate in shape and cells not completely closed by adult females. The remaining three genera are grouped by a petiolate second metasomal segment, and, possibly, nests with comb elongated (see also Turillazzi, this volume: Fig. 3.9). Finally, *Metischnogaster* and *Parischnogaster* are sister-groups based on a narrow propodeal valvula and dilated aedeagus.

Stenogastrines have usually been characterized as showing extreme diversity in social behavior, ranging from solitary to eusocial (e.g., West-Eberhard 1978a). However, the females of all studied species care for their young until adult eclosion, behavior that is, strictly speaking, at least subsocial according to the usual definition (cf. Wilson 1971, see Cowan, this volume: Table 2.1), a point also made by Iwata (1967, 1976). The earlier characterization of stenogastrines as extremely diverse in social behavior is no doubt partly due to the relatively few studies of the subfamily available at the time. Some species with large colonies have been shown to have a dominance hierarchy such as occurs in the eusocial paper wasp *Polistes* (e.g., Yoshikawa et al. 1969, Pardi and Turillazzi 1981, Turillazzi and Pardi 1982), and permanent sterility has even been reported in one of these species (*Liostenogaster flavolineata*: Samuel, cited in Hansell 1987a). Species in the genera *Stenogaster* and *Eustenogaster*, which have small colonies and have been less frequently studied, do not show such complex social organization. Although there are still considerable gaps in our knowledge, the recently published information on hover wasp social biology is sufficient to permit inferences, however tentative, of the groundplan, or ancestral condition, in several behavioral traits of interest.

The most pertinent traits to assess are nest sharing and reproductive division of labor. Both are widespread in the subfamily, but until recently (Carpenter 1988a) they were not considered ancestral. Regarding nest sharing, multiple-adult colonies have been reported for every species that has been subjected to long-term study (reviewed in Carpenter 1988a), but in many of these species nests with only a single female also occur. Thus, although nest sharing is universal in these species, it does not necessarily characterize the entire colony cycle (as is also the case in many Polistinae and Vespinae) or, at the very least, it is facultative. Haplometrosis (solitary nest initiation) is on present evidence the general mode of colony foundation in the subfamily, and so the single-

11

ported may have been recently founded. Other adults
t commonly either by the emergence of offspring that
natal nest for some period or by joining behavior, which
panied by usurpation. As for reproductive division of
wly emerged adults assist in brood care for a period of
time ⌐ ⌐ attempting to initiate their own nests, even in species
where colonies are always small and nest sharing persists over a relatively short part of the colony cycle (cf. Krombein 1976 and Hansell
1987b on *Eustenogaster*). Short-term nest sharing and a temporary reproductive division of labor are presumably primitive states relative to
the linear dominance hierarchies and large colonies with multiple egg
layers of some species, and so these traits are parsimoniously inferred
to be ancestral in the subfamily as a whole (see Carpenter 1988a for
further details).

## Vespinae

In contrast to the preceding subfamily, the subfamily Vespinae (yellowjackets and hornets) is perhaps the best known wasp group. About
60 described species are distributed throughout the Holarctic Region
and the Oriental tropics. Since they are broadly sympatric with temperate-zone biologists and their colonies are readily noticed, they have
received more study than other wasps. The taxonomy of the group has
been quite unstable, however, and phylogenetic relationships unclear.
The taxonomic history of the genera has been most recently summarized by Edwards (1980, but see Carpenter 1987a), and I have reviewed
phylogenetic treatments (Carpenter 1987a). Of the previous phylogenetic studies, Yamane (1976) did not include *Provespa,* whereas Matsuura
and Yamane (1984) studied all of the genera but did not resolve the
relationships completely. Varvio-Aho et al. (1984) presented an allozyme data set for eight species of European yellowjackets, and phene-
tic and cladistic analyses that they interpreted as showing that *Dolichovespula* is not a monophyletic group. However, their analyses failed to
find the correct trees for their data (Carpenter 1987b), data that appear
in any case to be uninformative at this level. I recently published a
comprehensive analysis of the genera, subgenera, and species groups
of Vespinae (Carpenter 1987a), including a reanalysis of the data of
Yamane (1976) and Matsuura and Yamane (1984). The results at the
generic level are shown in Fig. 1.3. Figure 1.4 shows additionally the
interrelationships of species groups in *Vespula*. These species groups
have frequently been treated as subgenera or even genera, following
Blüthgen (1938), but I do not consider that any of these groups merit
formal recognition (Carpenter 1987a).

The cladograms in Figs. 1.3 and 1.4 are based on analysis of 46 characters, in addition to the 42 characters listed by Matsuura and Yamane

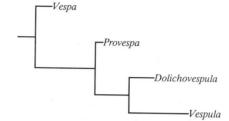

**Fig. 1.3.** Cladogram for the genera of Vespinae. (After Carpenter 1987a.)

(1984). The relationships depicted in Fig. 1.4 are established by 39 informative characters. The grouping of all genera apart from *Vespa* is supported by four synapomorphies (forewing with basally truncate second submarginal cell, apical clustering of hamuli, reduction of the pronotal carina, and loss of the pretegular carina). *Dolichovespula* and *Vespula* are shown to be sister-groups by possession of four synapomorphies (labial palpi without strong seta on segment 2, loss of the scutal lamella, and embryo nests with a twisted pedicel and first sheet bonded to this). No fewer than twelve derived traits support the monophyly of *Dolichovespula*. An elongate malar space is only the most obvious one. *Vespula* is established as a group by at least six synapomorphies (tyloides lost on male antennae, loss of the pronotal carina, aedeagal rods fused apically, nest site enclosed, and embryo nest with disc and uncoated pedicel). Within *Vespula*, the *V. koreensis* and *V. vulgaris* groups share three synapomorphies (female metasomal sternum VI with a dorsolateral process, male with metasomal tergum VII depressed and with sternum VII transverse). Finally, the *Vespula squamosa* and *V. rufa* groups are sister-taxa, established by three synapomor-

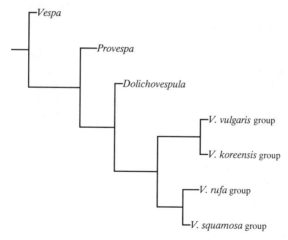

**Fig. 1.4.** Cladogram for the genera of Vespinae showing also the interrelationships of the species groups in *Vespula* recognized by Carpenter (1987a).

phies (occipital carina ventrally effaced, volsella shortened, and digitus slender and fingerlike). This last grouping contradicts the view of Mac-Donald and Matthews (1975, 1984), who considered the *V. squamosa* group to be closely related to the *V. vulgaris* group. However, they did not analyze the characters they adduced as evidence, and I have shown that these consist mostly of symplesiomorphies (shared primitive characters) and autapomorphies (Carpenter 1987a), which are uninformative on phylogenetic relationship (Hennig 1966).

## Polistinae

The subfamily Polistinae (paper wasps) is the most diverse group of social wasps, both in terms of species richness (about 800 described species) and morphological and behavioral diversity. Polistines are cosmopolitan, although concentrated primarily in the tropics, especially the New World tropics. Presently, 29 genera are recognized, although some European authors place the socially parasitic *Polistes* species in a separate genus (*Sulcopolistes*). Such placement is completely unjustified; it renders *Polistes* paraphyletic because this genus can then be diagnosed only by the absence of the apomorphies of *Sulcopolistes*.

Bequaert (1918) arranged the polistine genera in three subfamilies: Polistinae (for *Polistes*), Ropalidiinae (for *Ropalidia*), and Epiponinae (for the remaining genera).* Bequaert's subfamilies were later reduced to tribes in Richards's (1962) Polistinae. Charnley (1973) questioned the diagnostic value of the chief distinguishing feature for the tribes (namely, a dorsally pointed propodeal orifice in *Polistes* and *Ropalidia*), and I pointed out (1982) that the Polybiini (properly Epiponini) is a paraphyletic group. Jeanne's (1980a) proposal that the Old World Polybiini (properly Epiponini) form a monophyletic group with *Ropalidia* is confirmed below. Thus, there is no basis for the recognition of Richards's tribes.

No cladistic analysis of the relationships of the polistine genera has yet been done, although there have been several traditional treatments (Ducke 1914, Richards 1978a, Raw 1985). Ducke (1914) represented the Polistinae as polyphyletic, having evolved independently from the Eumeninae, a view now known to be unwarranted (Richards and Richards 1951, Carpenter 1982). Richards presented a tentative dendrogram (1978a: fig. 40) and eleven characters (six of which were three-state) "used in assessing relationships." Inspection of his dendrogram shows that it is not actually based on analysis of these characters: his tree is almost completely resolved, yet his data are insufficient to resolve it,

---

*Bequaert (1922), in a footnote, proposed Polybiinae as a replacement name for Epiponinae, and this name has subsequently been used. However, this replacement, evidently done at the behest of some colleagues, is unjustified under the current rules of nomenclature, and the senior name must be retained for this taxon.

and numerous branches are either supported only by primitive characters or are completely unsupported. Raw (1985) contended that Richards's genus *Occipitalia* was not monophyletic, and he described a new genus, *Asteloeca*, for one of the two included species. He presented a table of eleven characters that varied among five polistine genera, and although he did not polarize these into derived and primitive states, he suggested that they showed a closer relationship of *Asteloeca* to *Metapolybia* than to *Occipitalia*. The characters he cited do not in fact establish this relationship, although the present analysis leaves open the possibility.

I present here the first cladistic analysis of the Polistinae. Because it is not a comprehensive study but, rather, represents preliminary results from work in progress, a reclassification is not attempted. However, it should be noted that not all of the genera appear to be monophyletic.* Furthermore, the recognition of other genera is indicative of the same sort of oversplitting that has rendered the classification of other vespid groups chaotic (e.g., Eumeninae: Carpenter and Cumming 1985, Carpenter 1986; Vespinae: Carpenter 1987a).

The characters used in this analysis are listed in Table 1.2. Only informative characters are included, that is, no autapomorphies are incorporated as such. The data matrix for the genera is given in Table 1.3. Character polarities have been inferred with reference to other Vespidae, especially Vespinae, the sister-group of the Polistinae. Where the characters vary within genera, the scorings in Table 1.3 represent the parsimonious inferences of the groundplans for the relevant genera based on preliminary resolution of the ingroup. Analysis was performed using the Hennig86 program by J. S. Farris. The "mhennig" algorithm, with extended branch swapping and tree buffer set to available memory, produced 304 trees of length 77 and consistency index of 0.48 (see Kluge and Farris 1969). The trees differed only in alternative resolutions of the component (= branch point: Nelson 1979) including *Pseudopolybia* and *Polybia* in Fig. 1.5, specifically in the positions of *Pseudopolybia*, *Charterginus*, *Synoeca*, *Clypearia*, *Asteloeca*, *Metapolybia*, and *Occipitalia*.

Successive approximations character weighting (Farris 1969) was employed to reduce the ambiguity indicated by the multiple trees by selecting those supported by the most consistent characters (Carpenter 1988b). The characters were weighted by a product of their best unit character consistency index (Farris 1969) and retention index on the 304 trees, new tree(s) were calculated, new unit character consistencies and retention indexes were calculated and used as weights for further

---

*For example, *Protopolybia* is apparently paraphyletic in terms of *Pseudochartergus*, and these two genera are synonymized elsewhere (Carpenter and Wenzel 1989).

**Table 1.2.**  Characters and inferred polarities for cladistic analysis of Polistinae

1. Antennal articles: 12 in ♀/13 in♂, 0. 11 in ♀/12 in ♂, 1.
2. Tyloides: present, 0. Absent, 1. Replaced by specialized flattened areas, 2.
3. Palpal formula: 6-segmented maxillary palpi: 4-segmented labial palpi, 0. 6:3, 1. 5:3, 2.
4. Labial palpomere 2: without strong bristle, 1. With strong bristle, 0.
5. Occipital carina: present, 0. Absent, 1.
6. Eyes: without bristles, 0. With bristles, 1.
7. Clypeal apex: sharply pointed, 0. Rounded to truncate, 1.
8. Clypeal lateral lobes: well-developed, 0. Reduced and rounded, 1.
9. Tempora: broader than eye, 0. Narrower than eye, 1.
10. Gena: curved, 0. Sinuous, 1.
11. Pretegular carina: present, 0. Absent, 1.
12. Pronotal fovea: present, 0. Absent, 1.
13. Pronotal carina: present, 0. Absent, 1.
14. Pronotal prominence: short, 0. Extended dorsally, carinate, 1.
15. Fore coxa: laterally produced, 0. Rounded, 1.
16. Dorsal groove: present, 0. Absent, 1.
17. Scrobal sulcus: present, 0. Absent, 1.
18. Epicnemium: carina present, 0. Absent, 1.
19. Scutal lamella: developed adjoining tegula, 0. Reduced, 1.
20. Propodeal carinae: absent, 0. Dorsal carinae present, 1.
21. Scutellum: rounded, 0. Angled, 1.
22. Metanotum: rounded, 0. Compressed, 1.
23. Metanotum with posterior lobe: absent, 0. Present, 1.
24. Metasomal petiole: absent, 0. Segment I < half the width of Segment II, 1. Linear, 2.
25. Thyridium: linear, transverse and basal, 0. Not transverse or basal, 1.
26. Van der Vecht's gland: external modified area present, 0. Absent, 1.
27. Volsella: cuspis and lamina not fused, 0. Fused, 1.
28. Larval mandible: tridentate, 0. Bidentate, 1.
29. Meconium extraction: not extracted before adult emergence, 0. Extracted through back of cell, 1. Cells repaired by adult secretion, 2.
30. Nest: envelope absent, 0. Present, 1. Phragmocyttarus, 2. Astelocyttarus, 3.
31. Swarm founding: absent, 0. Present, 1.

*Note*: The primitive state is listed first; states are linearly ordered except for character 30, which is treated as nonadditive.

analysis, and this procedure was repeated until the trees did not change between iterations. The output stabilized at 30 trees, with a length of 253 and a consistency index of 0.76 for the weighted data. The strict consensus tree (Nelson 1979) for these cladograms is shown as Fig. 1.5, and the cladistic diagnoses (Farris 1979) are listed in Table 1.4. Note that the placements of *Pseudopolybia* and *Charterginus* are resolved on this tree.

Although use of a consensus tree as a phylogenetic hypothesis should generally be avoided (Mickevich and Farris 1981, Miyamoto 1985, Carpenter 1988b), the length of the tree in Fig. 1.5 for the data in Table 1.3 is 80, a loss of only three steps from the best-fitting cladograms. The consensus tree is also a considerable improvement over previously available treatments. Richards's dendrogram (1978a: fig. 40) requires a length of 96 to account for the data as coded in Table 1.3, 16

**Table 1.3.** Data matrix for cladistic analysis of Polistinae.

| Genus | 1 | 2 | 3 | 4 | 5 | 6 | 7 | 8 | 9 | 10 | 11 | 12 | 13 | 14 | 15 | 16 | 17 | 18 | 19 | 20 | 21 | 22 | 23 | 24 | 25 | 26 | 27 | 28 | 29 | 30 | 31 |
|---|---|---|---|---|---|---|---|---|---|---|---|---|---|---|---|---|---|---|---|---|---|---|---|---|---|---|---|---|---|---|---|
| Polistes (= Sulcopolistes) | 0 | 0 | 0 | 1 | 0 | 0 | 0 | 0 | 0 | 0 | 0 | 0 | 0 | 0 | 0 | 0 | 0 | 0 | 0 | 0 | 0 | 0 | 0 | 0 | 0 | 0 | 0 | 0 | 0 | 0 | 0 |
| Ropalidia | 0 | 0 | 0 | 1 | 0 | 0 | 0 | 0 | 0 | 0 | 1 | 1 | 0 | 0 | 0 | 1 | 1 | 0 | 1 | 0 | 0 | 0 | 0 | 1 | 0 | 0 | 1 | 1 | 2 | 0 | 0 |
| Parapolybia | 0 | 0 | 0 | 1 | 0 | 0 | 0 | 0 | 0 | 0 | 0 | 0 | 0 | 0 | 0 | 0 | 0 | 0 | 1 | 0 | 0 | 0 | 0 | 1 | 0 | 0 | 1 | 1 | 2 | 1 | 0 |
| Polybioides | 1 | 0 | 2 | 1 | 0 | 0 | 0 | 0 | 0 | 0 | 0 | 0 | 1 | 0 | 0 | 0 | 0 | 1 | 1 | 0 | 0 | 0 | 0 | 2 | 1 | 0 | 1 | 1 | 1 | 1 | 1 |
| Belonogaster | 0 | 2 | 2 | 1 | 0 | 0 | 0 | 0 | 0 | 0 | 0 | 1 | 0 | 0 | 0 | 1 | 1 | 1 | 0 | 0 | 0 | 0 | 0 | 2 | 0 | 0 | 1 | 1 | 1 | 1 | 1 |
| Mischocyttarus | 0 | 0 | 0 | 1 | 0 | 0 | 0 | 0 | 0 | 0 | 0 | 0 | 1 | 0 | 0 | 0 | 0 | 1 | 1 | 0 | 0 | 0 | 0 | 1 | 0 | 0 | 1 | 1 | 0 | 0 | 0 |
| Apoica | 0 | 1 | 0 | 0 | 0 | 0 | 0 | 0 | 0 | 0 | 0 | 1 | 0 | 0 | 1 | 1 | 1 | 1 | 0 | 0 | 0 | 0 | 0 | 1 | 0 | 0 | 1 | 1 | 0 | 0 | 1 |
| Agelaia (= Stelopolybia)[b] | 0 | 1 | 0 | 1 | 0 | 0 | 0 | 0 | 0 | 0 | 0 | 0 | 0 | 0 | 0 | 0 | 0 | 1 | 1 | 0 | 0 | 0 | 0 | 1 | 1 | 0 | 1 | 1 | 0 | 1 | 1 |
| Angiopolybia | 0 | 1 | 0 | 0 | 1 | 0 | 0 | 0 | 0 | 0 | 0 | 0 | 0 | 0 | 0 | 0 | 0 | 1 | 1 | 0 | 0 | 0 | 0 | 0 | 0 | 0 | 1 | 1 | 0 | 1 | 1 |
| Pseudopolybia | 0 | 1 | 0 | 1 | 0 | 0 | 0 | 0 | 0 | 0 | 0 | 0 | 0 | 0 | 0 | 0 | 0 | 1 | 1 | 0 | 1 | 0 | 0 | 0 | 0 | 0 | 1 | 1 | 0 | 1 | 1 |
| Parachartergus | 0 | 1 | 1 | 0 | 1 | 0 | 0 | 0 | 1 | 0 | 1 | 0 | 0 | 0 | 1 | 1 | 0 | 1 | 1 | 0 | 0 | 0 | 0 | 1 | 0 | 1 | 1 | 1 | 0 | 1 | 1 |
| Chartergellus | 0 | 1 | 0 | 1 | 0 | 1 | 0 | 0 | 0 | 0 | 0 | 0 | 0 | 0 | 0 | 1 | 0 | 1 | 0 | 0 | 0 | 0 | 0 | 0 | 1 | 1 | 1 | 1 | 0 | 1 | 1 |
| Nectarinella | 0 | 2 | 2 | 0 | 0 | 0 | 0 | 0 | 0 | 0 | 0 | 0 | 0 | 0 | 0 | 1 | 0 | 1 | 1 | 0 | 0 | 1 | 0 | 1 | 1 | 0 | 1 | ? | 0 | 3 | 1 |
| Leipomeles | 0 | 1 | 2 | 1 | 0 | 0 | 0 | 0 | 0 | 0 | 0 | 1 | 1 | 0 | 1 | 1 | 0 | 1 | 1 | 0 | 0 | 0 | 0 | 1 | 1 | 1 | 1 | 1 | 0 | 1 | 1 |
| Marimbonda | 0 | 1 | 2 | 0 | 0 | 0 | 0 | 0 | 0 | 0 | 0 | 0 | 1 | 0 | 1 | 1 | 0 | 1 | 0 | 0 | 0 | 0 | 0 | 1 | 1 | 1 | 1 | ? | 0 | 3 | 1 |
| Synoecoides | 0 | ? | 0 | 1 | 1 | 0 | 0 | 0 | 0 | 0 | 0 | 0 | 1 | 1 | 1 | 1 | 0 | 1 | 1 | 0 | 0 | 0 | 0 | 0 | 0 | 1 | ? | ? | 0 | 2 | 1 |
| Epipona | 0 | 2 | 0 | 1 | 1 | 0 | 1 | 0 | 0 | 0 | 0 | 1 | 1 | 1 | 1 | 1 | 0 | 1 | 1 | 0 | 0 | 0 | 0 | 1 | 1 | 1 | 1 | 1 | 0 | 2 | 1 |
| Clypearia | 0 | 2 | 0 | 1 | 1 | 0 | 1 | 0 | 0 | 0 | 0 | 0 | 1 | 1 | 1 | 1 | 0 | 1 | 0 | 0 | 0 | 0 | 0 | 0 | 1 | 1 | 1 | 1 | 0 | 3 | 1 |
| Synoeca | 0 | 2 | 0 | 1 | 1 | 0 | 1 | 1 | 0 | 0 | 0 | 1 | 1 | 1 | 1 | 1 | 0 | 1 | 1 | 0 | 0 | 0 | 0 | 1 | 1 | 1 | 1 | 1 | 0 | 3 | 1 |
| Metapolybia | 0 | 2 | 1 | 1 | 1 | 0 | 1 | 1 | 1 | 0 | 0 | 1 | 1 | 1 | 1 | 1 | 0 | 1 | 0 | 0 | 0 | 0 | 0 | 1 | 1 | 0 | 1 | 1 | 0 | 3 | 1 |
| Occipitalia | 0 | ? | 0 | 1 | 0 | 1 | 1 | 1 | 0 | 0 | 0 | 1 | 1 | 1 | 1 | 1 | 0 | 1 | 1 | 0 | 0 | 0 | 0 | 1 | ? | 0 | ? | ? | 0 | 3 | 1 |
| Asteloeca | 0 | ? | 0 | 1 | 0 | 0 | 1 | 1 | 0 | 0 | 0 | 0 | 1 | 1 | 0 | 1 | 0 | 1 | 1 | 0 | 0 | 0 | 0 | 1 | 1 | 1 | 1 | 1 | 0 | 3 | 1 |
| Polybia | 0 | 1 | 0 | 1 | 0 | 0 | 1 | 0 | 0 | 0 | 0 | 0 | 0 | 1 | 1 | 0 | 0 | 1 | 1 | 0 | 0 | 0 | 0 | 0 | 0 | 1 | 1 | 1 | 0 | 3 | 1 |
| Protonectarina | 0 | 1 | 0 | 1 | 0 | 1 | 1 | 0 | 0 | 0 | 0 | 0 | 0 | 0 | 1 | 0 | 0 | 1 | 1 | 0 | 0 | 0 | 0 | 0 | 1 | 1 | 1 | 1 | 0 | 2 | 1 |
| Protopolybia | 0 | 1 | 0 | 0 | 0 | 0 | 1 | 1 | 0 | 0 | 0 | 0 | 0 | 0 | 1 | 0 | 0 | 1 | 1 | 0 | 0 | 0 | 1 | 0 | 0 | 1 | 1 | 1 | 0 | 2 | 1 |
| Pseudochartergus | 0 | 1 | 0 | 0 | 0 | 0 | 1 | 1 | 0 | 0 | 0 | 0 | 0 | 0 | 1 | 1 | 0 | 1 | 1 | 0 | 0 | 0 | 1 | 0 | 0 | 0 | 1 | ? | 0 | 2 | 1 |
| Charterginus | 0 | 1 | 0 | 1 | 0 | 0 | 1 | 1 | 1 | 1 | 0 | 1 | 0 | 0 | 0 | 1 | 1 | 1 | 0 | 1 | 0 | 1 | 1 | 0 | 1 | 1 | 1 | 1 | 0 | 1 | 1 |
| Chartergus | 0 | 1 | 0 | 0 | 1 | 1 | 1 | 1 | 0 | 1 | 1 | 0 | 0 | 0 | 0 | 1 | 1 | 1 | 1 | 1 | 1 | 1 | 0 | 0 | 0 | 1 | 1 | 1 | 0 | 2 | 1 |
| Brachygastra | 0 | 1 | 0 | 1 | 0 | 1 | 1 | 1 | 0 | 1 | 1 | 1 | 0 | 0 | 0 | 1 | 1 | 1 | 1 | 1 | 1 | 1 | 0 | 0 | 1 | 1 | 1 | 1 | 0 | 2 | 1 |

*Note:* A question mark indicates missing data.
[a] Numbers correspond to character numbers in Table 1.2.
[b] See Carpenter and Day 1988.

17

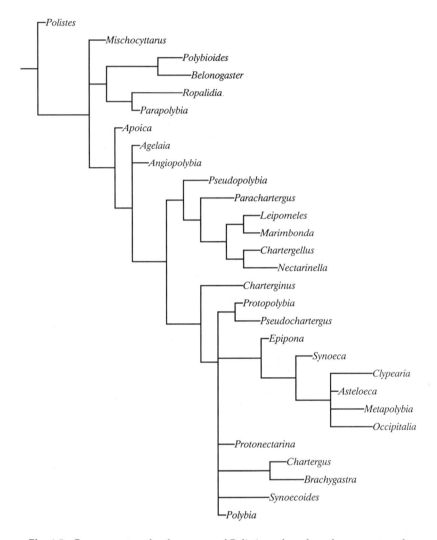

**Fig. 1.5.** Consensus tree for the genera of Polistinae, based on the present work.

steps more than are required for the tree in Fig. 1.5. By contrast, the latter tree accounts for Richards's data almost as well as his own dendrogram. It requires a length of 39 (consistency index of 0.44) to account for Richards's characters (with polymorphisms coded as ambiguous), whereas Richards's tree requires a length of 38. (Over 3,000 trees of length 32 exist for these same data; their consensus is largely unresolved.) Furthermore, inclusion of a preliminary data set from the

**Table 1.4.** Cladistic diagnoses for polistine consensus tree (Fig. 1.5)

*Mischocyttarus–Polybia* component
    18. Epicnemium not carinate (1).
    24. Metasoma petiolate (1).
    25. Thyridium not transverse or basal (1).
    27. Volsella with lamina volsellaris more-or-less desclerotized and fused with cuspis (1).
    28. Larval mandible bidentate (1).
*Mischocyttarus*
    15. Fore coxae rounded (1).
*Polybioides–Parapolybia* component
    29. Meconium-extracting behavior (1).
*Polybioides + Belonogaster*
    1. Antennae 11/12 (1).
    3. Palpal formula 5:3 (2).
    13. Pronotal carina lost (1).
    24. Petiole linear (2).
*Polybioides*
    30. Nest with envelope (1).
    31. Swarm founding (1).
*Belonogaster*
    12. Pronotal fovea lost (1).
    16. Dorsal groove lost (1).
    17. Scrobal sulcus lost (1).
*Ropalidia + Parapolybia*
    19. Scutal lamella reduced (1).
    29. Cells repaired after extraction of meconium (2).
*Ropalidia*
    11. Pretegular carina lost (1).
    12. Pronotal fovea lost (1).
    16. Dorsal groove lost (1).
    17. Scrobal sulcus lost (1).
    18. Epicnemium carinate (0; reversal).
*Apoica–Polybia* component
    2. Tyloides lost (1).
    31. Swarm founding (1).
*Agelaia–Polybia* component
    30. Nest with envelope (1).
*Angiopolybia*
    19. Scutal lamella reduced (1).
*Marimbonda–Brachygastra* component
    16. Dorsal groove lost (1)?
    24. Metasoma not petiolate (0; reversal).
    26. Van der Vecht's gland lost (1).
*Pseudopolybia–Polybia Component*
    4. Labial palpi setate (0).
    16. Dorsal groove lost (1)?
*Pseudopolybia*
    5. Occipital carina lost (1).
*Parachartergus–Nectarinella* component
    3. Palpal formula 6:3 (1).
    16. Dorsal goove lost (1).
*Parachartergus*
    6. Eyes bristled (1).
    11. Pretegular carina lost (1).
    17. Scrobal sulcus lost (1).

19

**Table 1.4**—*Continued*

*Leipomeles–Nectarinella* component
    3. Palpal formula 5:3 (2).
*Leipomeles + Marimbonda*
    24. Metasoma petiolate (1).
*Marimbonda*
    30. Nest astelocyttarus (3).
*Chartergellus + Nectarinella*
    5. Occipital carina lost (1).
*Nectarinella*
    21. Scutellum angled (1).
    22. Metanotum compressed (1).
    30. Nest astelocyttarus (3).
*Charterginus–Polybia* component
    7. Clypeal apex rounded (1).
    8. Clypeal lateral lobes rounded (1).
    16. Dorsal groove lost (1).
    19. Scutal lamella reduced (1).
*Charterginus*
    9. Tempora narrowed (1).
    11. Pretegular carina lost (1).
    22. Metanotum compressed (1).
    25. Thyridium transverse and basal (0; reversal).
*Protopolybia–Polybia* component
    30. Nest phragmocyttarus (2).
*Protopolybia + Pseudochartergus*
    23. Metanotum with posterior lobe (1).
*Pseudochartergus*
    9. Tempora narrowed (1).
*Epipona–Occipitalia* component
    2. Male antennae with specialized flattened areas (2).
    5. Occipital carina lost (1).
    12. Pronotal fovea absent (1).
    24. Metasoma petiolate (1).
*Synoeca–Occipitalia* component
    13. Pronotal carina lost (1).
    15. Fore coxa rounded (1)?
    30. Nest astelocyttarus (3).
*Synoeca*
    15. Fore coxa rounded (1).
*Clypearia–Occipitalia* component
    14. Pronotal prominence raised (1).
    15. Fore coxa rounded (1)?
*Clypearia*
    9. Tempora narrowed (1).
    15. Fore coxa rounded (1).
*Metapolybia*
    17. Scrobal sulcus lost (1).
*Occipitalia*
    9. Tempora narrowed (1).
    15. Fore coxa rounded (1).
*Protonectarina*
    6. Eyes bristled (1).
*Chartergus + Brachygastra*
    10. Gena sinuous (1).
    12. Pronotal fovea lost (1).

**Table 1.4**—*Continued*

---

  20. Propodeum with oblique dorsal carinae (1).
  21. Scutellum angled (1).
  22. Metanotum compressed (1).
*Chartergus*
  25. Thyridium transverse and basal (0; reversal).
*Brachygastra*
   6. Eyes bristled (1).
  11. Pretegular carina lost (1).
  17. Scrobal sulcus lost (1).
*Synoecoides*
   5. Occipital carina lost (1).
   9. Tempora narrowed (1).

---

*Note*: Apomorphies are listed for each component (branch point) and genus; ambiguity in characters 15 and 16 is indicated by question marks. The components are identified by the names of included genera that are furthest apart in the cladogram. The numerical codings for each apomorphy in Tables 1.2 and 1.3 are in parentheses.

more-detailed treatment of nest architecture by Wenzel (this volume) and reanalysis results in just five trees, the consensus of which differs from that in Fig. 1.5 by resolving the multifurcation among the genera related to *Polybia*, and by indicating a sister-group relationship between *Agelaia* and *Angiopolybia*. All the original cladograms are identical to the tree in Fig. 1.5 in its major points; instability exists primarily in the component including *Protopolybia* and *Polybia*. As a hypothesis of the phylogenetic relationships of the basal lineages of Polistinae, therefore, the consensus tree presented in Fig. 1.5 appears to be quite robust. Resolution of the multifurcations or modifications of the relationships in the component that includes *Pseudopolybia* and *Polybia* will not affect the conclusions reached below.

## THE POLYGYNOUS FAMILY MODEL FOR WASP SOCIAL EVOLUTION

### Description of the Model

West-Eberhard (1978a) considered data on nest sharing by a wide range of parasocial and primitively eusocial wasps in the formulation of her polygynous family model of social evolution. Her table 1 listed 29 species, and she had additional data for 10 others; however, only 12 of these species are vespids. The other wasps are sphecids and pompilids, which are not closely related to vespids (Brothers 1975, Königsmann 1978, Rasnitsyn 1980). Only one (*Microstigmus comes*) shows what is likely to be eusocial behavior (see Matthews, this volume; West-Eberhard did not consider it eusocial), and the nest-sharing behavior of the other species has clearly evolved independently from that in the vespids. Although these unrelated species may be informative for inferring possible selective forces involved in social evolution, they are unlikely to illustrate the actual stages in the evolution of eusocial behavior in wasps, which is restricted to Vespidae.

Of the vespids, three subfamilies (Masarinae, Eumeninae, and Stenogastrinae) were represented, with stenogastrines predominant among them. The masarine and eumenine species listed are not primitive in these respective groups (Carpenter, unpubl.), hence they cannot indicate groundplans nor can their behavior be regarded as homologous. The stenogastrines are crucial, however, since they represent the sister-group of Polistinae + Vespinae. The stenogastrine sample comprised all of the species (eight) for which any behavior was known to West-Eberhard. Seven were characterized as polygynous. Even the sole species designated as monogynous, *Stenogaster concinna*, was noted to have occasional nest sharing by multiple females. Adult wasps on the nests of half of the species were considered relatives, usually mother and daughter; in two cases they possibly were not related, and in two cases the relationship was unknown. Finally, generational overlap of adults was listed as occurring in three cases, possibly not occurring in two others, and unstudied in three species.

West-Eberhard synthesized these observations into her polygynous family hypothesis. Her summary figure is reproduced here in modified form as Fig. 1.6. In the initial stage (I), the *solitary stage*, a female provides a mass of food after ovipositing, and then seals the nest and provides no other care, as is seen in potter wasps (subfamily Eumeninae; see Cowan, this volume). The transition to the next stage (II), the *casteless polygynous family group*, is brought about by nest sharing by more-or-less closely related females, *including* daughters (cf. Pollock 1983). The selective force involved here is thought to be mutualism—for example, in response to habitat limitation caused by a shortage of suitable nesting sites. This is the *primitively social stage*. (Note that this term is not equivalent to *primitively eusocial*, as used by most authors.) In this stage, the rudiments of caste differentiation are considered to be present already in the form of differences in reproductive success among females sharing the nest. In an example given from facultatively aggregating species, adult body size is associated with different reproductive strategies, so that larger females are more aggressive and better provisioners, while smaller females are joiners and are more parasitic. Staggered attainment of reproductive maturity among females was also mentioned as leading to differential reproductive abilities. Such differences are expected to become exaggerated by intragroup competition, and incipient castes can thus be formed, as elaborated by West-Eberhard (1981). This means that in a group of related females, the individ-

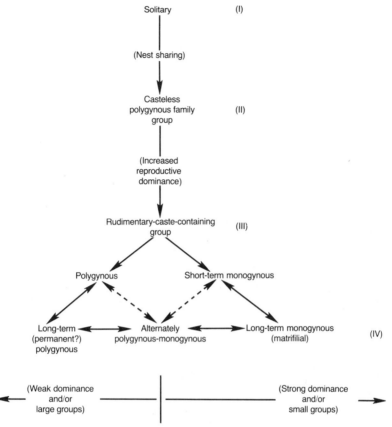

**Fig. 1.6.** Model for the evolution of eusociality in wasps, modified from West-Eberhard (1978a: fig. 2, courtesy of the Kansas Entomological Society). The original legend read: "Hypothetical evolutionary relationships of monogyny and polygyny in wasps. In region of double-headed arrows, weak dominance and/or large groups tend to produce poly-gyny (movement toward left side of diagram), and strong dominance and/or small groups tend to produce monogyny (movement toward right side of diagram). Double-headed arrows could be added to link polygyny and short-term monogyny with alternate polygyny-monogyny in caste-containing groups." The present figure reproduces her original, except that the arrows (dashed) have been added according to the latter sugges-tion.

uals losing in intragroup competition for, say, oviposition sites, can still enhance their inclusive fitness by contributing to the care of related offspring. As West-Eberhard (1981) put it, worker behavior is an alter-native reproductive strategy for losers in social competition.

This leads to stage III, that of the *rudimentary-caste-containing group*, in which reproductive division of labor is still only partial. It may be tem-porary, with all females eventually laying eggs, or incomplete, with some females being more "queenlike" (laying more eggs and foraging less) and some being more "workerlike" (laying fewer eggs). At this stage, worker behavior is facultative (West-Eberhard 1987a). Increasing reproductive and social dominance then brings about the final, *highly social stage* (IV) (distinct from the concept of *highly eusocial*, as used by most authors), where some "loser" females regularly fail to reproduce at all—that is, a true sterile worker caste is present. This is the eusocial stage.

West-Eberhard's model differs from the classical subsocial scenario championed by Wheeler (1923) in holding that incipient caste formation preceded permanent monogyny. In other words, daughters did not suddenly start rearing their sisters instead of their own offspring, but gradually lost the option of reproducing directly. It differs from the semisocial model (Lin and Michener 1972) in emphasizing that groups of relatives, not simply females of the same generation, formed incipient colonies. It also holds that haplodiploidy of the Hymenoptera is not the critical factor in the origin of eusociality (cf. Hamilton 1964b), because long-term monogyny would have to be present during the evolution of reproductive castes in order that the unusually high levels of relatedness between workers and brood made possible by haplodiploidy be fully realized (see Ross and Carpenter, this volume). The absence of long-term monogyny among primitively social species, as well as the presence of facultative sterility among nonrelatives, are also considered to vitiate the maternal control model of Alexander (1974).

Note that in Fig. 1.6 there are several different states of stage IV, eusociality, and multiple possible transformations between them and the rudimentary-caste-containing stage (III). West-Eberhard concluded that permanent monogyny, like that in vespines, evolved from short-term, behaviorally enforced monogyny, as in *Polistes*, as an adaptation for overwintering in the temperate zones, and, indeed, that behavior is relatively more characteristic of temperate-zone taxa. She suggested that the permanent polygyny of some swarm-founding paper wasps might also have evolved from short-term monogyny because of "the variety of intermediate forms" with temporary monogyny—namely, *Polistes, Mischocyttarus*, and various species of small-colony swarm-founding paper wasps (see West-Eberhard 1973, 1981). This suggestion envisions permanent polygyny as evolving as colony size increases. However, West-Eberhard also stated that it was possible that permanent polygyny evolved directly from the rudimentary-caste-containing stage without an intervening monogynous stage, citing the suggestion of Ducke (1914). She reasoned that such change could occur if a group of queens suppressed the reproduction of a group of workers, and the evolution of increasing group size kept pace with the evolution of increasingly effective reproductive dominance so that a single female was never able to gain complete control. As noted above, Ducke's notion of polyphyly of Polistinae is untenable, so this scenario would be possible only if the groundplan state of the paper wasps was the rudimentary-caste-containing group.

West-Eberhard did not attempt to identify the groundplan state of the relevant taxonomic groups. She characterized the Stenogastrinae as containing species that belong to each stage, whereas various paper wasp species (Polistinae) were placed in each category from primitively social (stage II) (*Polistes* foundress associations), to rudimentary-caste-containing (stage III) (*Belonogaster* and possibly some *Ropalidia*), to each of the types of eusocial behavior (stage IV). Hence it is unclear which of these hypothetical transitions, if any, might actually have occurred in the phylogenetic history of these wasps. To establish this it is necessary to characterize precisely the ancestral condition in each taxon. The phylogenetic system of these wasps does this, permitting specification of the possible transitions while testing the essential features of the model. That is the step I wish to take here.

**Test of the Model**

The uses of cladograms in testing evolutionary theories have been alluded to above, but the subject deserves elaboration in this particular context. Because cladograms are theories about general patterns resulting from genealogy, they provide summaries of the independent evidence bearing on patterns predicted by evolutionary process theories. A trivial application of this property is to a trait found in the descendants of a common ancestor: no process theory is required to account for the presence of inherited traits. This statement applies to the extended brood care shared by Stenogastrinae and Polistinae + Vespinae. Such behavioral similarities in the groundplans of the three subfamilies are most parsimoniously inferred to have been inherited from their common ancestor (Hennig 1966), so that arguments that such traits are convergently evolved are ad hoc (Farris 1983). A related example is West-Eberhard's hypothesis for the evolution of permanent polygyny directly from the rudimentary-caste-containing stage. If a well-supported phylogenetic hypothesis suggests that such a transformation did not occur (see below), then this explanation is unnecessary.

Phylogenetic systems also indicate the evidentiary value of data. For example, the nonvespid nest-sharing species listed by West-Eberhard (1978a) are of little value to the issue of the evolution of eusociality in Vespidae because they are so remotely related to vespids that their common ancestor was unquestionably solitary. Thus nest sharing has arisen independently and convergently in these groups. For study of, say, the adaptational basis of group formation, this series of less closely related species might be desirable. Apomorphic (derived) factors common to the ecology of all the species are valid candidates for promoting apomorphic function, that is, adaptation (Coddington 1988). But without a phylogenetic context the explanation is not necessarily even relevant (Coddington 1988). In investigations of which of the putative transitions in social behavior were followed, only the species closely related to the eusocial vespids can serve as phylogenetic intermediates, that is, as candidates for having retained actual steps to their eusociality.

Application of cladistics in tests of evolutionary theories is still relatively new, and the formulation of a test for a complex scenario can be difficult. But in this case, the separate stages of Fig. 1.6 can be defined in terms of the possession of several independent features—for example, nest sharing, reproductive division of labor, and number of queens. These features can then be treated as characters and mapped onto cladograms with ordinary character state optimization techniques (Farris 1970). A match between the resulting pattern and the hypothesized stages corroborates the model. A mismatch is evidence against the model. A simple example of this approach is given in Carpenter (1987a) from the Vespinae. MacDonald (1977) suggested that the evolution of a single worker-cell comb in the *Vespula rufa* group was due to that group's reliance on live prey. But the single worker-cell comb is uniquely derived in Vespinae; other vespines have multiple worker-cell combs. On the other hand, reliance on live prey is primitive and is shared with most other vespines—only the *V. vulgaris* group is primarily scavenging. The ancestral vespine had multiple worker-cell combs yet relied on live prey. Single worker-cell combs cannot be causally related to live prey, and the explanation must therefore be rejected.

In what follows, a test for West-Eberhard's model is developed. It should first be stressed, however, that the reference cladograms are developed independently of the behavioral features under discussion. Morphological data establish the cladograms of Figs. 1.1–1.4. The behavioral characters discussed in Carpenter (1982, 1987a, and 1988a) support the cladograms, and in the case of subfamily relationships seem more convincing than morphology. But the behavioral characters are not required to establish the patterns of relationship as they are for the paper wasp consensus tree (Fig. 1.5). However, even in the latter case the phylogenetically informative behavioral features of paper wasps have nothing to do with the social behavior of concern here (see diagnoses in Table 1.4). Character 31, swarm founding, is one of the features of interest in the model, but it is not required for construction of the tree in Fig. 1.5. It may be deleted and the same tree results. It is included for reasons that will become clear below.

The first point concerns the transition from the solitary to the eusocial stages (Fig. 1.6, stages I and IV). Vespidae is primitively solitary, but the common ancestor of Polistinae + Vespinae was evidently eusocial (Carpenter 1982, see also below). Their joint sister-group, the Stenogastrinae, thus becomes crucially important. If it shows a hypothesized intermediate stage in behavior, the model is corroborated, for that stage is then phylogenetically intermediate. I earlier considered subsocial behavior to be the ancestral state of the Stenogastrinae (Carpenter 1982), with simultaneous progressive provisioning the salient evolutionary novelty in the social behavior of the common ancestor of the three distal subfamilies. This state is important in the model of Evans (1958), but the common ancestor is then only at a transition between the solitary and casteless nest-sharing stages (I and II) in West-Eberhard's scheme. However, as discussed in Carpenter (1988a) and mentioned above, the more-detailed information on stenogastrine behavior published since 1982 suggests a different interpretation. The ancestral state of the Stenogastrinae is most parsimoniously inferred to be nest sharing (short-term or facultative) with temporary reproductive division of labor. In the state seen in *Stenogaster* and *Eustenogaster*, each newly emerged female may initially perform worker tasks but eventually has the opportunity to reproduce, either by founding or usurping a new colony or by contending for dominance on the natal nest or in another colony (e.g., Hansell 1987b). *Anischnogaster* is apparently similar (Spradbery 1989). *Liostenogaster* has larger colonies and dominance hierarchies (Iwata 1967, Hansell et al. 1982), as does *Parischnogaster* (e.g., Turillazzi and Pardi 1982). *Metischnogaster* is poorly known. From consideration of this information the rudimentary-caste-containing stage (III) is inferred to be ancestral for Stenogastrinae. Referring to Fig. 1.1, stage III then must also be inferred to be ancestral for the social Vespidae as a whole. In terms of the model, the major difference between this type of social behavior and that in paper wasps such as *Polistes* may be the lack of permanent sterility characterizing stage III.

West-Eberhard (1978a) therefore appears to be correct in suggesting that polygynous family groups have been important in the origin of eusociality in vespids, since they appear to be ancestral, rather than merely common, in the family. Based on the condition found in Stenogastrinae, the rudimentary castes were primitively temporary and group living itself was possibly facultative. Note, however, that casteless nest sharing is not exemplified in the social wasp clade under this interpretation. The only vespids known to have strictly casteless nest sharing are a few species of Masarinae and Eumeninae. Both of these subfamilies, as well as the basal subfamily Euparagiinae, are primitively solitary, and so the transition seen on the basis of the phylogeny presented in Fig. 1.1 is from solitary (stage I) to rudimentary-caste-containing (stage III) to eusocial (stage IV). This is not to say that the common ancestor of the social wasps could not have passed through a stage of casteless nest sharing, but that stage does not appear to have been distinct from stage III in the evolution of eusociality in Vespidae. Little distinction might be expected, however, if workerlike behavior originally arose as a conditional phenomenon that was expressed facultatively (West-Eberhard 1981, 1987a).

The next point concerns the transitions between different types of eusocial behavior: short-term monogyny (also termed serial polygyny), long-term or matrifilial monogyny, and long-term or swarm-founding polygyny. Beginning with the Vespinae, these wasps have until recently all been considered to exhibit matrifilial monogyny. Polygynous colonies are well known in species of the *Vespula vulgaris* group and in *V. squamosa* (reviewed in Spradbery 1986, Greene, this volume), but these are obviously secondary. New queens are recruited at the end of annual colony cycles in perennating, originally haplometrotic colonies in warm temperate regions. The other vespine genera do not show this trait. However, a type of swarm founding has now been observed in *Provespa* (Matsuura 1985), and polygynous foundress associations have been reported in *Vespa affinis* (Matsuura 1983a,b, Spradbery 1986). The swarm founding of *Provespa*—by a single queen and a swarm of workers —is not the same as the swarm founding of polistines, in which multiple queens participate. Thus, this trait is an autapomorphic development from matrifilial monogyny that is only superficially convergent to polistine swarm founding.

The foundress associations reported in *Vespa affinis* are a different matter. Since *Vespa* is the relatively basal vespine genus (Fig. 1.3), its foundress associations could represent the primitive condition in the subfamily—if such associations represent the groundplan of the genus. The most parsimonious optimization of the behavior would then infer it to be ancestral for the subfamily, considering the outgroup (Polistinae). However, the nature of this polygyny is not yet well understood. Colony foundation is haplometrotic and dominance interactions have not been observed in the species, but there is apparently some form of monopolization of oviposition (Matsuura 1983a, Spradbery 1986). The temperate-zone species of *Vespa* always have single functional queens (Matsuura 1984) and only two tropical species are known to have polygynous colonies (reviewed in Spradbery 1986, Matsuura, this volume), so polygyny may be a secondary phenomenon in this genus. The phylogenetic relationships of *Vespa* species must be analyzed to establish the groundplan and social organization in tropical colonies must be investigated further to settle this issue. The ancestral condition for the rest of the subfamily is matrifilial monogyny, but for *Vespa*, and so for the subfamily as a whole, the ancestral condition remains uncertain.

Clear inferences may be drawn for the Polistinae, however. Short-term, behaviorally enforced monogyny among multiple foundresses is the primitive state in the subfamily. Long-term, matrifilial monogyny and swarm-founding polygyny are both derived separately from this groundplan. Short-term monogyny is found in *Polistes, Ropalidia, Parapolybia*, and *Mischocyttarus* (reviewed in West-Eberhard 1978a, Jeanne 1980a, Akre 1982, Itô 1986a, Reeve, Gadagkar, this volume), as well as in *Belonogaster* (Keeping and Crewe 1987, Gadagkar, this volume). Matrifilial monogyny seems largely restricted to temperate-zone species of these genera (Itô 1986a) and is probably correctly interpreted as an adaptive syndrome related to overwintering. At least the North American species of *Polistes* and *Mischocyttarus* that exhibit this behavior are not primitive in their respective genera (Carpenter, unpubl.) and so probably do not pertain to the groundplan. The remaining polistine genera, along with some species of *Ropalidia*, are swarm founding (Richards 1978a). Periodical monogyny occurs in several of these genera, but they are primarily polygnous (West-Eberhard 1973, 1978b, 1981). Treating short-term monogyny and swarm-founding polygyny as alternative character states (scoring *Ropalidia* as missing because it has both states) and fitting the states to the tree in Fig. 1.5 results in short-term monogyny being inferred as primitive for the subfamily as a whole (including *Ropalidia*) (Table 1.3), with swarm-founding polygyny derived from it on several occasions (within *Ropalidia*, in *Polybioides*, and in the common ancestor of the component including *Apoica* and *Polybia*).

The behavior of *Belonogaster* accords with its phylogenetic position. Since Roubaud's (1916) studies this genus has been supposed to show a unique and primitive type of social behavior—namely, polygyny without caste formation. West-Eberhard (1978a) treated it as occupying the rudimentary-caste-containing stage. *Belonogaster* has since been shown to have a true sterile caste (Pardi and Marino Piccioli 1981, Keeping and Crewe 1987), and, furthermore, some of these workers accompany the inseminated females during colony foundation in *B. grisea* (Pardi and Marino Piccioli 1981). If *Belonogaster* is transitional between anything, it is transitional between short-term monogyny and swarm-founding polygyny in forms with already well-developed social behavior, not between presociality and eusociality.

## CONCLUDING REMARKS

The vespid cladograms presented here have numerous implications for the study of the evolution of social organization. Apropos of West-Eberhard's (1978a) model, the main conclusions of the present work are that the distinction between the primitively social (II) and rudimentary-caste-containing (III) stages is not clear-cut or else is unimportant in social Vespidae, and that long-term polygyny was always derived from short-term, behaviorally enforced monogyny. The first conclusion suggests that some form of reproductive division of labor originated simultaneously with nest sharing and accords with the view of worker behavior as initially facultative (West-Eberhard 1987a). The second conclusion resolves a long-standing question about the evolution of the unique social organization of the swarm-founding paper wasps (cf. Ducke 1914, West-Eberhard 1978a, Jeanne 1980a, Fletcher and Ross 1985). These conclusions corroborate several of the possible transitions proposed by West-Eberhard (1978a) but indicate that others probably did not occur.

There are several questions concerning the evolution of social behavior in Vespidae that might now be raised. What is different about Vespidae that permitted the evolution of eusocial behavior in this group and not in other wasps (certain wingless or hirsute groups aside)? What is different about nest sharing in the ancestor of social wasps (Stenogastrinae + Polistinae + Vespinae), compared with that evolved within Masarinae and within Eumeninae, that has resulted in the evolution of cooperative brood care in the former group and not in the latter two? And why did permanent sterile castes evolve in the ancestor of Polistinae + Vespinae and not in the ancestor of the Stenogastrinae? Convincing answers to these questions may prove elusive, but I conclude by pointing out some distinguishing features of the behavior of these groups that perhaps relate to these problems.

The outstanding behavioral autapomorphy of Vespidae is oviposition into an empty cell. This trait, found in very few other wasps, has been thought to be important in permitting the evolution of the extended brood care characteristic of social wasps (see discussion in Evans 1958). As for the nest sharing of the most recent common ancestor of social wasps, the distinguishing features are cell reuse, simultaneous progressive provisioning of macerated prey, and care extended until adult eclosion. This combination of traits does not occur in any of the Masarinae or Eumeninae that have convergently evolved nest sharing. Progressive provisioning by communal species occurs in *Zethus miniatus* (Krombein 1978), *Montezumia cortesioides* (Evans 1973a as *M. dimidiata*, M. J. West-Eberhard, unpubl.), and perhaps also in *Xenorhynchium nitidulum* (West-Eberhard 1987b). From the details published on *Z. miniatus* by Ducke (1914; misidentified as *Z. lobulatus*) and West-Eberhard (1978a, 1981, 1987a), females may remain on the natal nest, which can persist for several years, but each female usually tends her own independently built cells. Larvae are not tended simultaneously by single females. However, West-Eberhard (1987a) stated that females without brood sometimes adopt orphaned larvae and suggested that this workerlike behavior originated as misplaced continuing brood care. Progressive provisioning may thus have been of considerable significance in permitting the evolution of cooperative brood care, with simultaneous and extended care having further promoted it.

Regarding the permanent sterile castes of Polistinae + Vespinae, it may be remembered that females joining together to form colonies in the relatively primitive polistine genera are all presumed to be capable of founding their own nests, and thus of reproducing on their own. Nests are initiated by single females who are then joined by other females, but joining may not necessarily entail complete sterility (e.g., Metcalf and Whitt 1977a, Noonan 1981, Strassmann 1981a). Nests without joiners may have very low survival rates (e.g., Pardi and Marino Piccioli 1981, Itô 1985a, Keeping and Crewe 1987), and latitudinal variation in group size suggests that extrinsic factors may influence joining (West-Eberhard 1969, Strassmann 1983). The distinction between the temporary castes of Stenogastrinae and the permanently sterile workers of Polistinae + Vespinae may therefore not be as sharp as these categories imply. The ancestor of the latter group developed effective means for monopolization of oviposition, or queen control (West-Eberhard 1977a). Dominance is strongly enforced, primitively by direct physical attack in Polistinae (see Jeanne 1980a, Röseler, Spradbery, this volume). Similar traits also occur in those stenogastrines with clear dominance hierarchies (cf. Hansell et al. 1982, Turillazzi and Pardi 1982), although interactions have been characterized as relatively "less violent" in these wasps (Sakagami and Yamane 1983). Such hierarchies are evidently the result of convergent social evolution within Stenogastrinae (Carpenter 1988a and Test of the Model, above), and so perhaps are the mechanisms for queen control. Queen control merits much further study in hover wasps and independent-founding paper wasps, particularly with the application of genetic techniques to clearly delineate the number of functional queens (see Ross and Carpenter, this volume).

An obvious difference between Stenogastrinae and Polistinae + Vespinae is colony size. The colonies of hover wasps are usually far smaller than those of most paper wasps or yellowjackets (stenogastrines, 5–80 cells, and polistines or vespines, 50–>100,000 cells: Akre 1982). Vecht (1977b) suggested that this difference in colony size is related to a simple difference in nest architecture. These three subfamilies are alike in building complete nests that hang freely from the substrate. The primitive condition for the rest of the Vespidae is ground nesting, so free-standing nests are a shared derived feature of the three social subfamilies. Paper wasps and yellowjackets differ from hover wasps in that the nests are primitively begun with a petiolate cell, whereas in stenogastrines the nest is attached directly to the substrate. Hansell (1985, 1987a) showed that the nest paper of hover

32                                      James M. Carpenter

wasps is significantly more fragile than that of paper wasps and yellowjackets, and he speculated that selection for petiole construction led to the evolution of tough paper (the adaptive basis being wide choice of nest sites while maintaining protection against ants; see West-Eberhard 1969 and Jeanne 1975a). This tougher paper is viewed as an initial prerequisite for large colonies. Hansell further suggested that small colony size restricted the evolution of social organization in stenogastrines, but as Wenzel (this volume) points out, group size in itself should not affect the particular form of social organization. On the other hand, ancestral differences in nest architecture probably did facilitate the evolution of large colonies. Large colonies are the setting where the partial breakdown of queen control may have resulted in the evolution of secondary polygyny, which allowed the production of large worker forces necessary for swarm founding. The relatively simple synapomorphy of a petiolate first cell is thus perhaps the fundamentally most distinctive behavioral feature of Polistinae + Vespinae, one that ultimately allowed entry into several new adaptive zones.


*Acknowledgments*

I am grateful to George Eickwort, Howard Evans, Bob Jeanne, Junichi Kojima, Dave Queller, Ken Ross, Chris Starr, Joan Strassmann, Stefano Turillazzi, Phil Ward, and especially Mary Jane West-Eberhard for critiquing the initial draft of this paper. I also thank John Wenzel for sharing the results from his study of nest architecture. Support for computer analyses has been provided by NSF grant BSR-8508055 to the author.

# 2

# The Solitary and Presocial Vespidae

DAVID P. COWAN

Students of wasps and bees are particularly fortunate because there are close living relatives of the social species that provide insight into the possible pathways from ancestral solitary life histories to the threshold of eusociality. Thus it is logical to look to the noneusocial members of these groups to learn of the ecological, behavioral, and genetic factors that may have predisposed or preadapted some of their relatives to evolve eusocial societies. In this spirit, I review current knowledge of the solitary and presocial Vespidae, with a view toward their significance for understanding the evolution of eusociality.

*Solitary* vespid females nest individually and show considerable parental care, but there is no contact between parents and their developing young (Table 2.1). Most solitary species are *mass provisioners*: a female prepares a brood cell, lays an egg, then rapidly makes several foraging trips, and stores the food she has gathered with her egg. When enough food has been placed in the cell for the complete growth of the larva, she seals the cell and begins work on another. The females of a few solitary vespids may oviposit after they have stored the larval food.

The term *presocial* applies to diverse behaviors that include more than sexual interactions but do not fulfill all three criteria necessary for eusocial behavior (Table 2.1). *Subsocial* insects nest alone and have extended parental care such that parents interact with their developing offspring; however, this interaction generally ends before the offspring reach adulthood. Females of subsocial vespids practice *progressive provisioning*. They prepare a cell and oviposit, but may not start hunting for and providing food until the egg hatches. Food is then brought in slowly,

33

**Table 2.1.**  Categories of nesting behavior (levels of sociality) found among Vespidae

---

*Solitary*: Females nest alone and mass provision their nests. They do not interact with
   their developing young.
*Presocial*: Females exhibit social behavior beyond sexual interactions, yet short of eu-
   sociality.
  *Subsocial*: Females nest alone but interact with their developing larvae by progressive
     provisioning. Females that live sufficiently long may occur on the nest with their
     adult daughters.[a]
  *Parasocial*: Females of the same generation interact on the same nest. All known para-
     social (and eusocial) vespids practice progressive provisioning.
    *Communal*: Each female builds, oviposits in, and provisions her own cells.
    *Quasisocial*: All females cooperate in building and provisioning brood cells, and all
       females oviposit.
    *Semisocial*: Some females (reproductives) lay most or all of the eggs. Other females
       (workers) with limited egg-laying opportunities are relegated to foraging, nest
       building, and caring for the young.
*Eusocial*: Multiple females (1) cooperate in nesting and (2) exhibit reproductive division of
   labor (as in semisocial), but there is also (3) an overlap of generations, so that adult
   offspring assist their parents.
  *Primitively eusocial*: Colonies are relatively small and short-lived, and morphological
     differences between reproductive and nonreproductive females are minimal or non-
     existent.
  *Highly eusocial (advanced eusocial)*: Colonies are relatively large and complex and often
     are long-lived. Reproductive castes often are morphologically distinct from non-
     reproductive castes.

---

*Source*: Modified from Wilson 1971, Michener 1974, and Eickwort 1981.
[a]In bees and perhaps some wasps, progressive provisioning may not be required for subsociality.
Wilson (1971) recognized three levels of subsociality, but the terms are not in widespread usage.


and the cell may not be sealed until the larva is fully grown. Behaviors
intermediate to mass and progressive provisioning are often referred to
as *delayed provisioning* (see Nesting: Delayed- and Progressive-provisioning, Subsocial Wasps). *Communal* behavior occurs when females of
the same generation use the same composite nest but do not cooperate
in brood care. Each female builds her own cells and tends her own
offspring. In *quasisocial* societies females cooperate in nest building and
brood care, and all females oviposit. Whether quasisocial behavior actually occurs and whether it can be evolutionarily stable are open to
doubt (Eickwort 1981), because powerful females may be able to lay a
disproportionate share of the eggs, in which case there would be a
reproductive division of labor and the society would be *semisocial*.
West-Eberhard (1978a, 1987b,c) uses the term *primitively social* (cf. *primitively eusocial*) for wasps with more than one female sharing a nest but
without reproductive division of labor (communal and quasisocial).
This term is useful because it focuses attention on the evolution of sterility rather than overlap of generations as the trait of greatest significance.

   Those species of wasps in which multiple females cooperate on a
nest, reproductive division of labor occurs, and adult generations over-

lap are regarded as *eusocial*. Eusocial species whose reproductive and nonreproductive females are morphologically similar to one another and whose colonies are relatively small and short-lived are considered to be *primitively eusocial*, whereas *highly eusocial* wasps form relatively large and complex societies and often have castes that are morphologically dissimilar. The nesting biology of eusocial vespids is discussed in subsequent chapters in Part I of this volume.

Traditionally, taxa that exhibit different social behaviors have been arranged in an ethocline so that apparent ancestral and derived traits are connected by a logical sequence of intermediates. Thus, for wasps, ancestral solitary nesting is followed by subsocial behavior, communal behavior, and so on to eusociality (Wheeler 1923, Evans 1958), and it is reasonable to propose that ancestors of eusocial species actually proceeded through these behavioral stages over time. This technique is useful because by comparing populations with different life-styles it may be possible to understand the selective forces that led to each variant (Evans and Hook 1986). Michener (1985) cautions that attempts to place each taxon at a point along the continuum are sometimes ill-advised because at least some species of bees exhibit intraspecific variation that includes solitary and eusocial behavior but does not include presumptive intermediate stages. Also, to label a species as solitary, for example, suggests that its members make their living simply by building a nest, ovipositing, and foraging for prey. In fact, individuals in species that are formally classified as solitary may exhibit a wide range of nesting behaviors. For example, females of the solitary sphecid *Trypoxylon politum* may build mud nests, or they may occupy preexisting nest cavities, or two females may co-provision a nest until one drives the other away, or a female may break into the closed nest of another female, destroy the egg, and replace it with her own (Brockmann 1980, see also Matthews, this volume, for further examples in sphecid wasps). Intrapopulation variation as rich as this, though not yet documented for solitary vespids, forms the basis of social evolution. Thus, a primary goal of this chapter is to describe and attempt to understand variation in vespid nesting behavior.

The likelihood that eusocial behavior will evolve is influenced not only by nesting circumstances, but also by aspects of the breeding system and population structure that present females with opportunities to interact harmoniously with relatives. Populations may be fragmented into localized demes because suitable nest sites are highly localized, relatedness between relatives may increase if mating behavior results in inbreeding, and seasonal shifts in sex ratios may increase the opportunities for females to help rear sisters at their natal nest. These features of solitary and presocial vespids are also considered in this chapter.

The classification of the Vespidae proposed by Carpenter (1982) includes six subfamilies, three of which (Euparagiinae, Masarinae, and Eumeninae) are solitary or presocial in their habits. The subfamily Euparagiinae contains a single genus with nine described species and occurs only in the southwestern United States and Mexico (Richards 1962). Euparagiines nest in the ground and provision their young with curculionid beetle larvae (Williams 1927, Clement and Grissell 1968). The Masarinae (two tribes [Gayellini and Masarini], 19 genera, and 250 species) are found worldwide, primarily in warmer climates; exhibit a variety of nesting habits; and are distinctive in that they provision their nests with nectar and pollen rather than with insect prey. The majority of species in the Vespidae belong in the subfamily Eumeninae, with more than 180 genera and 3,000 species (Carpenter 1986, unpubl.) occurring throughout the world. Early workers (Bradley 1922, Bequaert 1928) divided the Eumeninae into three separate subfamilies: the Raphiglossinae (two genera), Zethinae (eight genera), and the Eumeninae, but recent workers argue that many of these groupings are unnatural (Carpenter 1982, 1986; Carpenter and Cumming 1985). A sound classification as well as good comprehensive manuals for identification of the Eumeninae are not yet available. The subfamily Stenogastrinae may be partly subsocial or communal, but is considered in detail elsewhere (Turillazzi, this volume). The vespids mentioned in this chapter are listed in Table 2.2 following the system of Carpenter (1982), with Bradley's and Bequaert's subfamilies treated as tribes within the Eumeninae. Earlier valuable reviews of solitary and presocial vespids include Olberg (1959), Malyshev (1968), Spradbery (1973a), and Iwata (1976).

## NESTING

The major behavioral advance of many aculeate Hymenoptera over their parasitoid ancestors was the evolution of nesting behavior, which plays an important role in reducing immature mortality (Evans 1977a). Among the noneusocial Vespidae, major aspects of nesting include the construction and form of the nest, the methods of providing nutrition to larvae (e.g., mass vs. progressive provisioning), and the social organization (e.g., whether females nest alone or with others).

### Nest Construction

One feature common to virtually all solitary and presocial vespids is that they employ a plastic material such as mud or chewed leaves to modify or build their nest cavity. Iwata (1976) classified the nesting habits of solitary Vespidae into three categories: (1) burrowers that ex-

**Table 2.2.** Classification of the vespid wasps discussed in the text

Euparagiinae
   *Euparagia*
Masarinae
  Gayellini
  Masarini
    *Ceramius*
    *Pseudomasaris*
    *Trimeria*
Eumeninae
  Eumenini

| | |
|---|---|
| *Abispa* | *Orancistrocerus* |
| *Allorhynchium* | *Oreumenes* |
| *Ancistroceroides* | *Pachodynerus* |
| *Ancistrocerus* | *Parachilus* |
| *Antepipona* | *Paralastor* |
| *Anterhynchium* | *Paraleptomenes* |
| *Delta* | *Parancistrocerus* |
| *Epsilon* | *Pararrhynchium* |
| *Eumenes* | *Pareumenes* |
| *Euodynerus* | *Pterocheilus* |
| *Leucodynerus* | *Stenodynerus* |
| *Monobia* | *Synagris* |
| *Montezumia* | *Xenorhynchium* |
| *Odynerus* | *Zeta* |

  Raphiglossini
  Zethini
    *Calligaster*
    *Zethus*
Stenogastrinae
Polistinae
   *Polistes*
Vespinae
   *Vespula*

cavate nests in the soil (Fig. 2.1), (2) renters that occupy and modify preexisting cavities (Figs. 2.2a, 2.3, 2.4), and (3) builders that construct their entire nest from materials such as mud (Figs. 2.2b,c, 2.5) or masticated plant material (Fig. 2.6). In such a large assemblage of species, there are many with habits that cannot be pigeonholed within this classification scheme. For example, *Leucodynerus russatus* obtains nest holes in the ground by usurping the burrows of tiger beetle larvae rather than digging its own (Knisley 1985), females of *Zethus laevinodus* (Bohart and Stange 1965) and *Z. otomitus* (Calmbacher 1977) excavate their nest cavities by chewing holes in rotting wood, and some species facultatively practice different nesting behaviors (Fig. 2.2) (Cooper 1979) or exhibit geographic variation in their nesting habits (Isely 1913).

    Nest construction is a significant aspect of parental behavior that functions to hold immature wasps in proximity with prey and to protect them from environmental insults, predators, and parasites. The

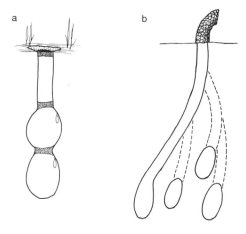

**Fig. 2.1.** Nests of two burrowing eumenines. (a) Nest of *Euodynerus crypticus* (redrawn from Isely 1913). (b) Nest of *E. annulatus* (redrawn from Evans 1956, courtesy of the author).

construction material and form of nests are influenced by the availability of nest sites and construction materials, as well as by the ability of particular designs to thwart nest parasites and predators. For example, nests of masticated plant material may be vulnerable to parasites that oviposit through cell walls, but they can be suspended by petioles that discourage ant attacks (Jeanne 1975a). Burrows in the soil are relatively immune to parasites that attack through the nest wall but are susceptible to ants and flooding. Also, because some building materials cannot be fashioned into large nests capable of physically supporting mul-

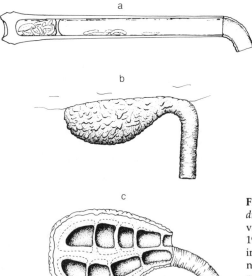

**Fig. 2.2.** Nests of *Orancistrocerus drewseni* illustrating intraspecific variability (redrawn from Iwata 1938b). (a) Cut-away view of a nest in bamboo. (b) Side view of mud nest built on the underside of a rock. (c) Interior of a double nest made from mud; the nest has been removed from the substrate.

**Fig. 2.3.** Cut-away view of a nest of *Ancistroceroides ambiguus* inside bamboo, showing individual mud cells. (Redrawn from Claude-Joseph 1930, courtesy of *Annales des Sciences Naturelles Zoologie et Biologie Animale*.)

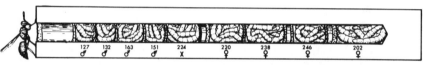

**Fig. 2.4.** Cut-away view of a nest of *Euodynerus foraminatus* in a trap nest. The number under each cell indicates the weight (in milligrams) of the caterpillar provisions. The symbols indicate the sex of the wasp that was reared from each cell. X indicates that the wasp died before reaching adulthood. (Redrawn from Cowan 1981, courtesy of Springer-Verlag.)

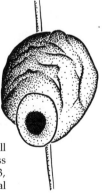

**Fig. 2.5.** Two views of a mud cell built by *Eumenes micado* on a grass stem. (Redrawn from Iwata 1953, courtesy of Fukuoka Entomological Society.)

**Fig. 2.6.** A nest of the communal eumenine *Calligaster cyanoptera* made from partly chewed leaves. (Redrawn from Forbes 1885.)

tiple females and their brood, building techniques may impose some limits on the development of social behaviors (Hansell 1987a,1989, but see Wenzel, this volume).

CONSTRUCTION MATERIALS   The great majority of mud-daubing vespids obtain their mortar for nest construction by drinking water at one place and then flying to a site of dry clayey soil, where they bite at the ground, regurgitate water, and mix the resulting mud to the proper consistency (Iwata 1938b, 1939). They then fly to the nest with the globule of mud and incorporate it into the nest structure. Because the exposed mud nests of some solitary vespids are quite durable and do not wash away in rain, it may be that saliva is added to the water and soil to strengthen the nest (Isely 1913), as has been suggested for a sphecid (Qureshi and Ahmad 1978). Nests often are located near a source of water (Isely 1913, Markin and Gittins 1967, Freeman and Jayasingh 1975a), and Gess (1981) has argued that some wasps may be limited geographically and seasonally by lack of water to make mud.

Because they depart from the typical method of making mud, a few exceptional cases are noteworthy. Some wasps make mud by robbing dried mortar from the nearby nests of other wasps (Brooke 1981); *Delta curvatum* collects premoistened mud rather than mixing its own (Williams 1919); and *Pachodynerus nasidens* is facultative in using preexisting mud or mixing water with dry soil (Jayasingh and Taffe 1982). Rather than making mud and transporting it back to the nest, females of *Stenodynerus canus* pick up dry pieces of clay, carry them back to the nest, moisten them there, and incorporate them into the nest (Clement 1972). The masarine *Pseudomasaris edwardsii* uses nectar collected at flowers to make mud (Torchio 1970), and the females of *Oreumenes decoratus* appear to use the hemolymph of prey caterpillars to knead mud pellets (Iwata 1953). These exceptions, plus the fact that most species have not been studied in detail, suggest that facultative variation in mud making may be common.

Some vespids cut pieces of leaves, chew the margins into a paste, and incorporate the pieces into their nests somewhat like shingles (Fig. 2.6) (Forbes 1885, Williams 1919). Some *Zethus* make cells by coiling ribbon mosses and cementing them together (Beebe et al. 1917) or completely macerate pieces of leaves and incorporate the paste into a nest that, when dry, becomes very tough and hard (Ducke 1914; Rau 1933, 1943).

There are a few reports of species using materials other than mud or masticated leaves for nest construction. The renting wasp *Odynerus vespoides* uses sticky tree gum to close its nest (Williams 1919). *Xenorhynchium nitidulum* builds a mud pot and then smears a sticky gum over the entire nest before provisioning (West-Eberhard 1987b). Other species waterproof their nests with weathered wood fibers, sand, pieces of paper, or a mixture of saliva and ashes (Iwata 1938a, 1953). Females of *Eumenes architectus* collect wood fibers in a manner similar to *Polistes* and make a paper cover over their nests (Iwata 1939). This habit seems to represent a nest-building stage that is intermediate between the use of mud and the exclusive use of plant fiber to make paper nests, as is seen typically in the eusocial vespids.

BURROWERS Females of burrowing species usually excavate nest holes in areas of firm or hard, clayey soil by regurgitating water and removing the earth as globules of mud. There is considerable variation in nest structure (Isely 1913, Evans 1956, Iwata 1976). Some simply excavate a tunnel and then fill it with a series of linearly arranged cells (Fig. 2.1a), whereas others build single cells at the end of diverticula (Fig. 2.1b). Partitions between cells as well as final closures are made with mud. Once the nest hole is excavated some species simply use the cavity as it is, but others modify the walls (with water and saliva?) so that they are harder and more resistant than the surrounding soil (Isely 1913), and some build separate mud cells within the cavities (Gess and Gess 1980).

A limited number of burrowing vespids, particularly in the genus *Pterocheilus*, dig in friable soil and do not regurgitate water as a soft-

ener; nor do they close their nests with mortar (Evans 1956). Though
nesting in friable soil is possibly ancestral among the hunting wasps,
this habit as seen in *Pterocheilus* is certainly a condition derived from a
mud-daubing ancestor (Carpenter and Cumming 1985). Rather than
possessing sand rakes on the legs as do the burrowing Pompilidae and
Sphecidae, members of this genus have a basket formed by long setae
on the labial palpi (Bohart 1940, Gess and Gess 1976, Gess 1981). When
digging, females bite at the dry sand, collect it in the basket, and remove it from the hole (Grissell 1975). When closing a nest they kick
loose, dry soil into the hole and tamp it down with their mandibles.

RENTERS AND TUBE DWELLERS    Renting vespids are opportunistic and
occupy a wide variety of preexisting cavities. Almost any type of nook
or cranny may be used, but the most common are tubes in the hollowed pith of twigs (Figs. 2.2a, 2.3) or vacant insect borings in dead
wood (xylophilous wasps). Also, vacant cells in nests made by building
species of wasps may be reused (Iwata 1976). The opportunistic hole-
nesting behavior of xylophilous species has made it possible to attract
nesting females to artificial nest sites (trap nests) that often are made
by drilling holes in blocks of wood (Fig. 2.4) (Krombein 1967) or are
simply bamboo sticks with open ends (Itino 1986). Although renters
remove loose material and patch cracks with mud, usually they do not
enlarge the cavity in pith or wood. Reports of renting species burrowing or modifying nest sites (Cooper 1979), or even building complete mud nests (Iwata 1938b), indicate that these wasps may be much
more flexible in their nesting behavior than is generally acknowledged.

BUILDERS    Building wasps construct entire cells by collecting raw
materials from the environment, processing them into a plastic material
that is transported to the nest site, and there incorporating and shaping the material into the nest structure (Figs. 2.2b,c, 2.5, 2.6). Nests are
usually built in sheltered places such as under the eaves of human
dwellings or rock overhangs, but some species build on relatively exposed plant stems or the upper surface of rocks. The cells produced by
building vespids vary from the elegant mud pots produced by members of the genus *Eumenes* (Fig. 2.5) (Iwata 1953) and the delicate parallel tubes of *Paraleptomenes* (Krombein 1978) to what seem to be crude
lumps of mud made by some *Euodynerus* (Clark and Sandhouse 1936)
and *Montezumia* (Evans 1973a).

In addition to variation in the basic shape of nests, the methods of
mud manipulation exhibit variability warranting further investigation.
Females of *Orancistrocerus drewseni* with a pellet of mud enter the nest
head first, turn around, and then apply the mortar from the inside
with their mandibles. The foretarsi are not used. This results in a nest
with a smooth interior but a wrinkled exterior. Species of *Eumenes* construct from the outside using mandibles and foretarsi as opposable

tools (Iwata 1938b, 1953; Jayakar and Spurway 1967, 1968; see also Smith 1978). After a nest is provisioned and sealed, some species add additional material to the outside surface (Williams 1919, Jayakar and Spurway 1965a). Jayakar and Spurway (1968) distinguish between lumps of mud daubed on the outside to thicken the walls and "crepissage," which is an irregular covering of ridges and closed chambers that may function to confuse parasites (Iwata 1938b, 1942) or to control microclimate within the cell.

CHIMNEYS  Mud chimneys or turrets are a common feature of the nests of many solitary vespids (Figs. 2.1a, 2.2). These structures may be found on the entrances of renting and building species but are most often associated with burrowing wasps. Most observers report that broken chimneys are rebuilt. There is no agreement about the function of the chimneys, and there have been no tests of the various hypotheses. Chimneys may discourage nest parasites (Isely 1913, Markin and Gittins 1967, Smith 1978, Gess 1981), but Iwata (1938b) and Bristowe (1948) considered them to be no barrier to invaders. These structures may help keep dust or water out of nests or help control internal microclimate. Clement and Grissell (1968) even speculate that chimneys serve as landmarks used in finding the nest. The fact that chimneys often are built by wasps that nest in places where foreign material, microclimate, and orientation are unlikely to be problems mitigates against these ideas.

## Dispersion of Nests

The dispersion of nests may be controlled by physical factors such as the availability of proper soil type for burrowers, sheltered areas for builders, or hollow stems for renters, as well as by biotic factors such as parasitism, predation, and competition (Wcislo 1984). By controlling the dispersion of nests, physical and biotic factors influence the likelihood and nature of social interactions among females. Thus, students of the aculeate Hymenoptera have tended to search for ecological factors that favor aggregation (Wcislo 1984) and communal nesting (Lin and Michener 1972, Abrams and Eickwort 1981, and Eickwort 1981 for bees; Evans and O'Neill 1988 for sphecid wasps). Among ground-nesting hymenopterans, localized soil of the proper consistency does not seem to be the causative factor of aggregation of nests (Andersson 1984, Wcislo 1984). However, seasonal hardening of soil may be a factor causing some females of some bees to join established nests (becoming communal) rather than attempting to dig their own (Abrams and Eickwort 1981). Ground-nesting vespids may not be affected by this limitation because they are able to dig in hard soil by softening it with water. The consensus of the authors cited is that aggregated nesting in sphecid wasps and bees reduces the risk of nest parasitism. Brooke (1981) found that, overall, immature survivorship in the eu-menine wasp *Delta alluaudi* is higher in aggregated than solitary nests and that females preferentially build nests near each other, suggesting that similar factors may favor aggregated nesting in vespid wasps.

Although biotic factors favor grouping by nesting females under some circumstances, the concentration of grouped wasps may attract large numbers of their parasites, and nearby conspecifics may interfere with each others' nesting. Infestations of the chalcid parasite *Melittobia* on the building wasp *Zeta abdominale* are severe in rock shelters where numerous females nest in close proximity, but because this parasite is a poor disperser mortality is lower in scattered isolated nests built on roots or stems (Freeman and Taffe 1974, Taffe and Ittyeipe 1976, Taffe 1983). Itino (1986) found that parasitism by sarcophagid flies increases when nests are clumped. Neighboring female wasps may steal prey (Eberhard 1974, Alexander 1986) or building materials (Brockmann 1980, Brooke 1981), replace a neighbor's egg with their own (Eickwort 1975, Brockmann 1980), or ursurp an entire nest (Markin and Gittins 1967, Eickwort 1975, Brockmann 1980, Cowan 1981, Alcock 1982). To-gether, selection pressures associated with the concentration of para-sites at groups of nests and interference among conspecifics may ac-count for the occurrence of many species that nest in isolation.

## Oviposition

Solitary and presocial vespids finish building their cells then ovi-posit into the empty cell; oviposition into an incomplete cell is known only for *Zethus miniatus* (Ducke 1914). Reports of provisioning before ovipositing (Piel 1935, Clement and Grissell 1968, Zucchi et al. 1976) generally have not been substantiated (Moore 1975) or are based on circumstantial evidence (Carpenter and Cumming 1985), but Iwata (1964a) confirms Piel's (1935) account that *Pareumenes quadrispinosus* completes provisioning before ovipositing. The majority of solitary ves-pids suspend their egg from the ceiling of the cell by a short filament, although a few affix the egg directly to the cell wall without a filament (Iwata 1942). For mass provisioners, the filament may keep the egg away from excessive moisture or partially paralyzed prey that could crush it (Taffe 1983). Species that provision progressively (so that young larvae are not crowded into cells with semi-active prey) tend to deposit their eggs loosely on the bottom of the cell (Roubaud 1911, Ducke 1914, Williams 1919, Jayakar and Spurway 1966a), and some masarines that mass provision with pollen also lay their egg loosely in the cell (Gess and Gess 1980, Houston 1984). Oviposition into an empty cell may be advantageous because it gives the wasp larva a developmental head start on parasites introduced later with the prey.

Occasionally individual females deposit more than one egg into a cell (Roubaud 1916, Jayakar and Spurway 1967, Brooke 1981). This behavior is not associated with the death of the first egg, but two adults never emerge from the same cell. Roubaud and Brooke speculate that females lay a second egg because of the physiological need to deposit an egg that has matured in the reproductive tract. Such physiological inefficiency could set the stage for cleptoparasitism (the parasite feeds on the food stored for the host larva) if wasps deposited these "excess" eggs in foreign nests. Yet cleptoparasitism has not evolved among solitary vespids, as it has among sphecid wasps (Bohart and Menke 1976, Wcislo 1987) and bees (Eickwort 1975).

Not infrequently, wasps seal cells without ovipositing or provisioning. Although some biologists have considered these empty cells the result of errors or adverse conditions, Tepedino et al. (1979) have argued that sealing empty cells is an adaptation that reduces immature mortality by discouraging parasites.

## Development

Typically eggs hatch in about two days. First-instar larvae have sets of lateral thoracic and abdominal teeth, which enable them to escape the chorion at the time of hatching by ripping the chorion on each side (Cooper 1966). There are six molts after hatching. The first four instars are feeding and growth stages. Among species that provision with intact prey, the early larval instars tend to simply bite a hole in the prey and suck out their fluids (Iwata 1953, Medler 1964). Not until the later instars do the wasp larvae consume the solid parts of the prey, leaving only the head capsules (Markin and Gittins 1967). Fully fed fifth-instar larvae produce a varnishlike secretion that covers the inside of the cell and then spin a delicate cocoon closely appressed to the inner cell wall. After cocooning, larvae pass the meconium (fecal pellet) and become inactive. This is the prepupal stage, which in temperate-zone species with a winter diapause is the dormant stage. Depending on the species and sex, the pupal stage can last 9–22 days, with females requiring more time (Krombein 1967). Before exiting the nest, teneral adults remain within their cells for several days while their cuticle hardens.

Usually, adult emergence is accomplished by breaking through the mud wall or partition of the nest. To this end, young adults have a supply of liquid that is regurgitated onto the mortar to soften it and ease their exit (Roubaud 1911; Cooper 1953, 1966). Even though they hatch first and may reach adulthood before their siblings in the outer cells, wasps derived from eggs laid in the innermost cells of nests within tubular cavities (e.g., Fig. 2.4) delay emergence until all wasps are mature and ready to exit (Krombein 1967).

## Foraging and Hunting

Virtually all eumenines prey on caterpillars that lack large numbers of setae, or on coleopterous larvae in the families Curculionidae or Chrysomelidae. Several reports suggest that some species of wasp specialize on particular prey (Iwata 1938a, 1939; Krombein 1967; Bohart et al. 1982), but many eumenines take a variety of prey (Medler 1964, Krombein 1978); some species even include lepidopterans and coleopterans in their diet (Evans 1956, Markin and Gittins 1967). One nest may contain several kinds of caterpillars, while an adjacent nest of the same species contains cells provisioned exclusively with one prey species. This suggests that females learn particular search habits that bring them into contact with a limited range of prey, or that hunting females return repeatedly to exploit concentrations of prey items, as Brooke (1981) concluded on the basis of differing rates of return with prey. However, Jayakar and Spurway (1967) concluded that the prey of *Delta campaniforme esuriens* are widely scattered.

Females of a few species attack exposed caterpillars that they simply pounce on and sting (Iwata 1953), but most search out concealed prey, especially leaf webbers and rollers (Isely 1913). These are taken by entering a caterpillar's retreat (Iwata 1938a) or chewing a hole in it (Williams 1919, Cooper 1953). Other species seek prey by digging into flowers (Grissell 1975) or pine buds (Lashomb and Steinhauer 1975). In a study of the stimuli used during hunting, Steiner (1984a) concluded that foraging females are generally attracted to the color green. The stimuli of rolled leaves, caterpillar silk, and droppings elicit an excited response of running quickly, chewing at leaves, and probing with the antennae. Contact with caterpillars elicits stinging, primarily on the underside of the head and thorax (Cooper 1953, Bonelli et al. 1980, Steiner 1983, Veenendaal and Piek 1988).

The condition of paralyzed caterpillars is important to the survival of wasp larvae within the nest. Normally caterpillars are stung so that they are capable only of uncoordinated twitching but remain alive (fresh) for a week or two. Caterpillars stung too many times die, shrivel, or rot and become inedible (Cowan, unpubl.). Caterpillars not stung often enough may become inedible because they pupate or even transform into adults (Jorgensen 1942, Medler 1964, Gess and Gess 1976).

Once females have subdued their prey by stinging, they often knead it by chewing it with their mandibles, working from one end of the caterpillar to the other without breaking its skin. Nielsen (1932) has suggested that this may help to distribute the venom throughout the prey. Isely (1913) observed a case where a particularly active caterpillar was difficult to place into the wasp's burrow; the wasp quieted it by kneading it further, suggesting that this behavior may serve to help incapacitate prey. Chewing prey in this manner may also provide a means for wasps to detect prey that have been attacked by parasitic hymenopterans or dipterans (Lashomb and Steinhauer 1975, Cowan 1981, Bohart et al. 1982).

Masarine wasps collect pollen and nectar rather than prey to provide nutriment for larvae. They collect pollen with their foretibiae and foretarsi (Neff and Simpson 1985), swollen antennae, or on the labrum and front of the clypeus (Gess and Gess 1980). Hicks (1929) observed that after visiting several flowers females clean pollen from their heads and swallow it. They then regurgitate the pollen and nectar mixture into their brood cells (Torchio 1970, Houston 1984). Most masarines are apparently oligolectic (using pollen from only a few plant species) rather than polylectic (foraging at a wide variety of plants) (Cooper 1952, Gess and Gess 1980, Neff and Simpson 1985), but this conclusion is disputed by Tepedino (1979).

## Mass Provisioning by Solitary Wasps

Mass provisioning, exhibited by the vast majority of Euparagiinae, Masarinae, and Eumeninae, is certainly the ancestral way of provisioning brood among the Vespidae. After she oviposits, the female takes provisions to the nest at a steady pace, so that a cell is filled and sealed before the egg hatches. The usual order of activities—cell preparation, oviposition, provisioning, and then cell closure—is quite stereotyped; however, some species are able to alter the sequence in response to novel events, for instance, the damaging of cells, which must then be repaired (Jayakar and Spurway 1967, 1968).

The burrowing wasp *Parachilus insignis* deviates from the typical mass-provisioning sequence by building a nest with cells divided into two compartments (Gess and Gess 1976). A female lays her egg and places several caterpillars with it at the bottom of a tunnel. She next builds a thin mud partition across the cell, walling off the small egg chamber. Most of the provisions are then placed above the partition in a larder chamber. After it hatches, the wasp larva consumes the two or three caterpillars in the egg chamber and then must break down the thin mud wall to gain access to the remaining provisions. This specialization may thwart some nest parasites.

The amount of food provided and the size of the larval chamber of mass-provisioning species can be quantified; such data give us a measure of the effort expended by a parent on each offspring. These aspects of wasp biology are best known for xylophilous species that occupy trap nests. When a nest contains individuals of both sexes, the inner cells nearly always contain female eggs and are provisioned with more food than the outer cells, which contain males (Fig. 2.4) (Medler 1964, Krombein 1967). Factors other than the sex of the offspring also are correlated with the amount of provisions. Offspring in large-diameter holes receive more food than those in small holes (Krombein 1967, Cowan 1981), apparently because the diameter of the nest hole limits offspring size. Also, multiple females nesting in close proximity tend to rear larger offspring than females nesting in isolation. One explanation is that there is increased competition for nest sites among females and for mates among males when populations are dense, and that large adults are better competitors in both respects (Cowan 1981).

The mass of food provided to an individual offspring apparently is not determined by the number of prey acquired or the length of time spent foraging. Rather there seems to be a mechanism that results in females providing enough food for their offspring to reach a particular size. Thus, females provide many prey items when they are capturing small larvae, but they provide only a few when the caterpillars are large (Brooke 1981). *Delta emarginatum conoideum* apparently judges whether provisions are adequate by how full the pot is. By removing caterpillars as they were brought to the nest, Hingston (1926, 1929) found that even though females acted "perturbed," enough caterpillars for several offspring would be placed in one pot. The final amount of provisions is apparently determined primarily by a memory of the sex of the egg that was deposited and the size of the nest chamber.

The sequence of production of the two sexes and the amount of food provided the larvae are not as well studied in building wasps as they are in trap-nesting species. *Delta alluaudi* provisions a series of mud pots adjacent to each other, usually producing a male egg in the first cell followed by a series of females (Brooke 1981). This species is unusual in that the males are larger than the females and presumably get more food. A similar pattern with male cells provisioned first in a cluster is found in *Delta campaniforme esuriens*, but there is some tendency to rear only one sex in a cluster of cells, and occasionally the order of sexes is reversed (Jayakar 1963, Jayakar and Spurway 1968). There is evidence that males of some building species are reared in smaller pots than females (Iwata 1964a), but Jayakar and Spurway (1965a) found no correlation between pot size and the sex of the offspring in *D. emarginatum*.

## Delayed- and Progressive-provisioning, Subsocial Wasps

The term *delayed provisioning* is applied to cases in which, because of inclement weather or prey scarcity, or perhaps innately, wasps that usually mass provision fail to supply the final prey to a cell before the egg hatches (Evans 1977b). Such provisioning is distinct from *truncated progressive provisioning*, in which wasps bring no prey until their egg is about to hatch, provision slowly at first while the larva is small, but then fill the cell quickly and seal it before the larva is fully grown (Roubaud 1911, 1916; Iwata 1938b; Evans 1966; Itino 1986). The fundamental difference between the two methods of provisioning is that with delayed provisioning the rate of storing prey is independent of the condition of the larva, but with progressive provisioning stimuli received from the offspring influence how rapidly prey are placed in a nest. *Fully progressive provisioning* occurs when prey are provided slowly until the time the larva is fully fed and ready to spin its cocoon (Roubaud 1911, Ducke 1914, Williams 1919, Jayakar and Spurway 1966a). Thus, delayed-provisioning females forage and supply prey as rapidly as possible, remaining in the nest only during inclement weather and spending long periods away from the nest hunting. The progressively provisioning vespids are generally distinctive in that, during periods of slow provisioning, females spend long periods sitting in their nest regardless of weather so that the nest is unguarded for only short periods (Jayakar and Spurway 1966a, Iwata 1976, Krombein 1978). Usage of these terms in the literature is sometimes not precise, and slow provisioning that may be an incidental effect of environmental vagrancies (scarce prey) may be confused with behaviors that are evolved adaptations (extended parental care in the form of nest guarding, see below).

Some eumenines show intraspecific variation between mass and progressive provisioning on a geographic basis (Roubaud 1916, Iwata 1964a) or seasonally ( Markin and Gittins 1967). Roubaud observed that *Synagris spiniventris* practices mass provisioning when prey is abundant but shifts to progressive provisioning during times of shortage, a feature that perhaps has confused thinking about the proximate and ultimate causes of progressive provisioning. It is not clear how prey scarcity (which would necessitate long periods away from the nest to hunt) could select for reduced foraging and intensified nest guarding. I suggest that changes in provisioning behavior of the latter sort are more likely to be correlated with changing predator and parasite pressure at the nest (that a guarding female can mitigate) than with prey availability.

Progressive-provisioning eumenines process prey items more extensively than do mass-provisioning species. Mass provisioners apparently sting prey so that they are only lightly paralyzed; it is important
that the prey remain fresh (alive) for the larva to eat during its entire
developmental period. Prey brought to nests by progressive provisioners are flaccid and completely immobile, as though severely
kneaded (Roubaud 1911; Iwata 1938b, 1976). This treatment may kill
parasites within the prey more effectively, and prey need not remain
fresh a long time for progressive provisioners. *Calligaster cyanoptera* kills
prey by simply biting off the head (Williams 1919), and *Synagris cornuta*
chews prey into a paste before presenting it to the larva (Roubaud
1911) (Fig. 2.7). When the sting is not needed to paralyze prey it may
evolve into a more effective weapon in nest defense—significantly,
mandibular prey capture and mastication are universal among the
powerfully stinging presocial and eusocial vespids (Evans and West-
Eberhard 1970). On the other hand, mass-provisioning eumenines that
successfully paralyze prey have stings as painful to humans as most
eusocial vespids (Starr 1985a, Waldbauer and Cowan 1985).

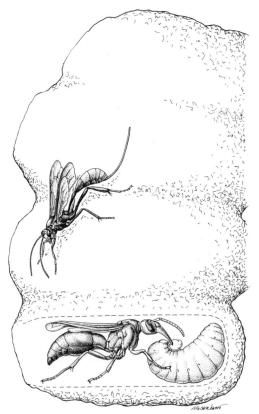

**Fig. 2.7.** Cut-away view of nest
showing a female *Synagris cornuta*
providing her developing larva with
a globule of masticated caterpiller.
The ichneumonid parasite *Osprynchotus violator* lurks on the outside of
the nest. (Based on descriptions in
Roubaud 1911 and Bequaert 1918.)

Because of the additional parental care lavished on each offspring, it is predicted that progressive provisioners should suffer lower larval mortality than mass provisioners. Williams (1919) speculated that because female *C. cyanoptera* remain in their open cell most of the time during progressive provisioning, fewer larvae are killed by ants and eusocial wasps. Yet this species still suffers attacks by parasitic chalcids with long ovipositors that pierce the nest wall. Iwata (1938a,b) raises similar questions regarding *Orancistrocerus drewseni* and *Pararrhynchium ornatum*. Although these species are host to a number of parasites, these parasites also gain access to brood cells by methods that circumvent the mother's defenses, such as ovipositing through the nest wall or entering the nest on the mother wasp as minute first-instar larvae.

To test the hypothesis that progressive provisioning reduces immature mortality, Itino (1986) compared immature mortality between the progressive provisioner *O. drewseni* and the mass provisioner *Anterhynchium flavomarginatum*. Mortality due to parasites that entered the nest as adults was lower in *O. drewseni*, the progressive provisioner. However, because it suffered high losses to other parasites, such as rhipiphorid beetles, which enter the nest as minute first-instar larvae, and to microbial pathogens or developmental abnormalities, overall nest mortality was similar for the two species.

Only limited information about mortality in other progressive provisioners is available. Jayakar and Spurway (1966a) have recorded that for *Paraleptomenes miniatus* 14 out of 17 (82%) eggs reached adulthood, and Krombein (1978) has reported that only 23 of 181 cells (13%) of the closely related *P. mephitis* were parasitized. These wasps do not suffer the 33—80% immature mortality suffered by some mass provisioners (Freeman and Taffe 1974; Taffe 1978, 1979). On the other hand, Rocha and Raw (1982) have reported only 18% immature mortality for the mass-provisioning *Zeta argillaceum*. Clearly, additional data comparing larval mortality in related species that practice mass as opposed to progressive provisioning are needed before the effect of the method of provisioning can be assessed.

Data on adult survivorship are virtually nonexistent. However, because progressive provisioners spend so much time sitting in nests, their death rates by accident or predation are likely to be lower than those of mass-provisioning species. This could mean slower senescence (Williams 1957) and increased longevity. Among sphecids, Evans (1966a) suggests that progressively provisioning *Bembix* lives two to four times as long as mass-provisioning *Gorytes*.

Progressive provisioners are estimated to have low lifetime fecundity, only four to eight eggs, compared with mass provisioners that have fecundities of 7–60 eggs (Iwata 1964b, Freeman and Jayasingh 1975b, Taffe 1979, Cowan 1981, Itino 1986). There are two techniques for measuring fecundity: (1) a life table is constructed from data on immature mortality, assuming a stable population, in order to calculate average fecundity (Freeman and Taffe 1974, Taffe 1979); or (2) the activities of marked individuals are followed and their nest cells are counted (Brooke 1981, Cowan 1981). Given that females are known to move from one nest site to another (Jayakar and Spurway 1965a, Taffe 1983), neither of these methods is entirely accurate because of the problem of distinguishing emigration from mortality. The lower fecundity of progressive provisioners is associated with slow egg maturation and an inability of females to replace lost eggs quickly. Thus, among subsocial species there may frequently be females that are temporarily non-reproductive, and although they cannot rear their own offspring at such times they may increase their inclusive fitness by assisting nearby relatives.

Another important evolutionary step toward eusociality is the ability to care for more than one cell and larva simultaneously (Jayakar and Spurway 1966a). The only noneusocial vespid reported to do this is the African *Antepipona tropicalis* (Roubaud 1916). Females excavate burrows into clay banks and progressively provision with intact caterpillars. When the development of one larva is well under way, females often dig a second adjacent burrow, lay a second egg, and tend both offspring. Nesting in this way would seem to forfeit some of the presumed advantages of progressive provisioning because one of the nests would always be untended. This problem may be ameliorated when nest cells diverge from a single common entrance or if multiple females occupy the same nest.

## Communal and Quasisocial Wasps

The vast majority of eumenines and masarines are strictly solitary or subsocial and are aggressive toward conspecific females that approach their nest, but the few notable exceptions represent an important behavioral link with the eusocial species. Bohart et al. (1982) report that occasionally two females of *Odynerus dilectus* actively provision the same nest but seem to be unaware of each other. When they happen to encounter each other in the nest, they fight, the smaller female is driven away, and the larger one closes the nest. After emerging, young adults of *Stenodynerus claremontensis* remain together on peaceful terms in the burrow of their natal nest for several days (Markin and Gittins 1967). The nest is reused by only one female, however, and the others leave to nest elsewhere by themselves.

Brauns (1910) reported communal nesting by the South African masarine *Ceramius lichtensteinii*, but Gess and Gess (1980) disputed this conclusion and argued that the occurrence of multiple females in a single

nest is temporary and does not involve provisioning. Also, this s
reuses nests from previous seasons, and, for a short period afte
females emerge, several sisters seek shelter together in their natal
As nesting begins, however, only one dominant individual remains
the natal nest, and the others disperse to establish nests elsewhere.
These temporary groups of wasps appear similar to prereproductive
assemblages known for some bees. Michener (1985) pointed out that
these groupings provide a context for selection to reduce mutual antag-
onisms and that, at least among bees, this behavior has led to the evolution of populations with both solitary and semisocial or eusocial
nests. The presumptive intermediate subsocial and communal stages
do not exist. On the basis of the occurrence of multiple females inside a
single active nest, only one actively tended cell, and open cells with
larvae and pollen but no closed cells with pollen, Zucchi et al. (1976)
concluded that the Brazilian masarine *Trimeria howardi* provisions progressively and forms communal or quasisocial groups in reused nests.

Among eumenines, Evans (1973a) described a nest of *Montezumia dimidiata* with three cells being progressively provisioned and three associated females. By marking individuals of this species, West-Eberhard
(1978a) confirmed that some wasps return to reproduce on their maternal nest, but as yet it is unclear whether females nest alongside their
adult daughters or whether communal groups comprise only sisters.
Similar nesting habits may also occur in *Xenorhynchium nitidulum* (West-
Eberhard 1987b), *Zethus laevinodus* (Bohart and Stange 1965), and *Calligaster cyanoptera* (Williams 1919).

The South American eumenine *Zethus miniatus*, in which 15 or more
females may occupy a common nest made from masticated leaves, displays a larger repertoire of alternative behaviors than their solitary and
subsocial relatives (Fig. 2.8) (Ducke 1914; West-Eberhard 1978a, 1987c).
Females with a mature egg in their ovary may build a new cell, reuse a
vacant cell, or usurp the cell of a less aggressive female. Independent
nests are initiated by females that have had cells usurped. Sometimes
females without mature oocytes remain on the nest and adopt orphaned larvae or wait on the nest surface and patrol (nest guarding?).
Although all individuals can readily engage in the complete spectrum
of behaviors, it is possible that by accident or because of the presence
of a particularly dominant individual some wasps fail to reproduce.
Such females may still have a fitness greater than zero by protecting or
feeding larvae of relatives that are likely to be present.

Increased competition and likelihood of nest usurpation are apparently the most important costs to communal and quasisocial behavior,
but we lack detailed information for vespids about possible offsetting
benefits. Communal sphecid wasps and bees enjoy reduced nest parasitism and predation (Evans and Hook 1982a, Andersson 1984) and, as

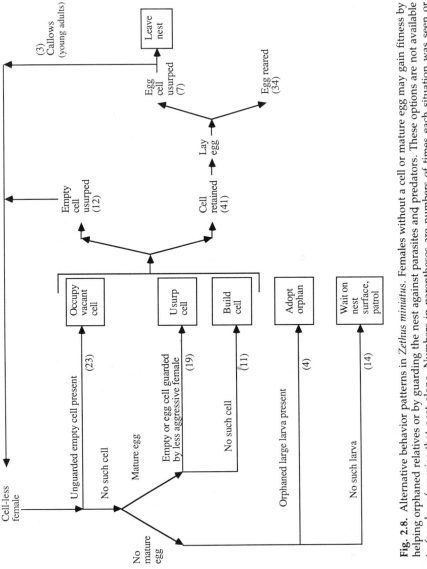

**Fig. 2.8.** Alternative behavior patterns in *Zethus miniatus*. Females without a cell or mature egg may gain fitness by helping orphaned relatives or by guarding the nest against parasites and predators. These options are not available to females of species that nest alone. Numbers in parentheses are numbers of times each situation was seen or could be deduced from records of 17 females resident on a nest for 115 days. The "wait on nest surface, patrol" alternative includes only cases that lasted one or more days. (Modified from West-Eberhard 1987c, courtesy of the author and Japan Scientific Societies Press.)

54

with *Zethus miniatus*, benefits may accrue by aiding orphaned relatives or by having one's own orphaned offspring reared to adulthood. Communal and quasisocial behaviors seem not to be as commonly represented as solitary and eusocial behavior. This observation may be due only to our lack of knowledge, but it may also be that these behaviors are evolutionarily unstable in most evironments with respect to eusocial or solitary life.

## Immature Mortality

Much of the present variation in nesting behavior among solitary and presocial vespids can be interpreted as responses to parasite and predator pressure and other sources of brood mortality. To provide insight into the problems faced by nesting wasps and how these influence nesting behavior, I conclude this section on nesting with a brief review of the causes of immature mortality.

BROOD PARASITES AND PREDATORS   Nests are especially vulnerable to attack when the female is away; this is when many enemies take the opportunity to despoil the nest. Minute adult phorids (Diptera) enter open nests, kill the wasp eggs by sucking out fluids, and then oviposit among the prey caterpillars (Iwata 1953, Itino 1986). The phorid larvae then feed on the provisions (Krombein 1967). The presence of phorid flies in the nest may cause the female to wall off a cell before it is fully provisioned (Itino 1986) or to clear out all of the last cell's contents, including the egg, and seal the nest (Iwata 1938a). Maggots of the sarcophagid tribe Miltogrammini (Diptera) are serious pests of solitary wasps (Krombein 1967). Female flies follow provisioning females back to their nests (Chapman 1959), sometimes causing the wasps to take evasive action or to fly away without depositing their prey item in the nest. When a wasp enters her nest, flies alight nearby, wait until the wasp leaves, and then enter the nest and deposit several small larvae. Maggots develop rapidly, and they may be completely fed and have destroyed the contents of a cell within two days (Markin and Gittins 1967). In multicellular wasp nests, the maggots may break down the partitions that separate the cells and destroy the entire nest (Krombein 1967).

Cuckoo wasps of the family Chrysididae (Hymenoptera) typically oviposit among the prey caterpillars while the nest is still open and being provisioned (Krombein 1967), but some species make a hole in the side of a mud pot and oviposit into sealed cells (Iwata 1953). Females of *Euodynerus foraminatus*, upon discovering adult *Chrysis coerulans* inside their nest, bite and sting at them while dragging and pushing them from the nest. Chrysidids do not seem to suffer from these attacks, as they curl up so that their hard integument deflects the attack. Females of *Delta campaniforme esuriens* do not attack chrysidids that are breaking into their nests, but parasitism will cause them to abandon the nest (Jayakar and Spurway 1968). Inside the nest, chrysidid larvae usually hatch and kill the host wasp larva and then feed on the caterpillars, but a few chrysidids delay development until the host wasp has reached the prepupal stage and then feed externally on the wasp larva. Adult chrysidids generally emerge in synchrony with the wasps from unparasitized cells, but sometimes chrysidids develop more rapidly than their hosts and emerge through unparasitized cells that contain larvae or pupae in delicate developmental stages. This invariably results in the death of the wasps in those cells.

Minute adult females of the eulophid genus *Melittobia* (Hymenoptera) enter a nest while it is open and being provisioned, or squeeze through tiny cracks, or chew an entrance through the mud wall to gain entrance to a cell. When the host wasp larva has reached the prepupal stage, they sting it and lay a large number of eggs (up to several hundred) on it (Krombein 1967). Because of the many offspring produced on a single host and the tendency for dispersing female *Melittobia* to crawl rather than fly in search of hosts, serious infestations of this parasitoid can build up in places where wasp populations persist. *Melittobia* is known to act as a density-dependent mortality factor on several wasp species (Freeman and Taffe 1974, Freeman and Jayasingh 1975a, Taffe 1978).

Two groups of parasites gain access to wasp nests not as adults but as tiny first-instar larvae that attach to nesting females (perhaps when they visit flowers) to be carried back to the nest. Once in a nest, larvae of Strepsiptera enter a wasp egg but do not complete development until their host reaches adulthood (Krombein 1967). At this time, the strepsipteran pupa protrudes through the intersegmental membranes of the host wasp. Females never leave the host, but males emerge to fly off in search of mates. Mating occurs on the host, and active larvae eventually crawl out of their mother and off the host as it flies about. Parasitized wasps are probably sterile. The effects of these parasites are easy to overlook because the first-instar larva is very small, and parasitized wasps show no external signs until they have emerged from the nest and have been active for several days (Iwata 1953). This means that wasps reared from nests must be kept alive for several days and then inspected for strepsipteran pupae. It seems that Strepsiptera populations are very localized. In some areas they seem nonexistent, but in other areas they have been known to parasitize a large fraction of the hosts (37% in a study by Freeman and Jayasingh [1975a]).

Beetles of the family Rhipiphoridae gain access to host wasp nests in the same manner as strepsipterans (Iwata 1953, Krombein 1967, Itino

1986). Rhipiphorid larvae feed internally for a period and then emerge to finish feeding externally when the host larva is full-grown. Adult females are free-living and deposit thousands of eggs about the environment.

Ants invade nests of solitary wasps and carry off the stored provisions and larvae (Jayakar and Spurway 1967). Species of *Crematogaster* are prone not only to steal nest contents (Freeman and Taffe 1974) but to then usurp the cavity as a colony home (Iwata 1953, Cowan, unpubl.). The presence of even a few ants around a nest may cause the mother wasp to desert. Females of *Parachilus insignis* seem helpless when ants raid their nests, and Gess and Gess (1976) suggested that persistent ant raids reduce population sizes. The full impact of ants on solitary wasps is probably underappreciated; unlike many other enemies, they may be in a nest for only a short time, and they leave no evidence (such as cocoons) of their visit. There is no question that ant predation has a major impact on populations of eusocial species of wasps (Jeanne 1975a, Hansell 1987a).

In addition to the enemies that gain access to nests through open entrances, a variety of parasitic hymenopterans attack immature vespids by piercing the nest wall with a long ovipositor (Krombein 1967, Krombein et al. 1979).

While the proportion of mortality caused by various parasites and predators has been determined for some populations of solitary vespids, we know almost nothing about the search strategies of different wasp enemies or how different nesting behaviors of host wasps increase or decrease vulnerability. For example, subsocial behavior with a female guarding her nest entrance for extended periods may discourage flies, chrysidids, and ants, but this behavior may actually facilitate the transfer of beetle larvae into the nest.

NEST USURPATION    Nest usurpation (supersedure) occurs frequently in species that occupy tunnels in wood or hollow stems (see Nest Construction). It is particularly easy to recognize when the inner part of the cavity is used by one species and the outer part contains the nest of a different species of wasp or bee (Krombein 1967). This arrangement can occur when the first occupant does not use the complete cavity or when the original occupant is displaced by a usurper. Generally, joint occupancy need not result in the death of the immature wasps in the inner part of the nest; they may still complete development and emerge. In some cases, however, the usurper cleans out the original contents of the nest and discards them. Markin and Gittins (1967) have reported that in some years megachilid bees usurp 30–50% of nests excavated by *Stenodynerus claremontensis*. Before beginning their own provisioning, the bees cast out the eggs, larvae, and provisions from any open cells; apparently the bees ignore closed cells. In this particular case, the harm to the wasp may be more from the loss to the female of time and effort rather than the actual destruction of immature wasps.

Intraspecific nest usurpation by solitary wasps is also significant. Markin and Gittins (1967) report fighting between females of *S. claremontensis* for ownership of nest burrows. When an intruding female wins the encounter, she discards the nest contents. *Euodynerus foraminatus* usurpers not only invade active nests but also break through the mud closures of completed nests and clean them out (Cowan 1981). Usurpation occurs only when all the nest sites in the immediate vicinity are already occupied. The threat of usurpation must constitute a major barrier to nest sharing among wasps.

PREY QUALITY    Prey caterpillars occasionally pupate and rarely even transform into adults in the cell of a wasp nest (Jorgensen 1942, Medler 1964, Gess and Gess 1976). Wasp larvae are unable to feed on the hard-bodied lepidopterous pupae. Caterpillars are commonly attacked by parasitic insects (Tachinidae, Braconidae, Ichneumonidae) that complete their development inside the wasp cells even though their host has been stung (Roubaud 1911; Iwata 1938a, 1953; Cowan, unpubl.). Such parasites may be consumed along with the prey caterpillar, but their activities can kill the wasp larva. For example, braconid larvae emerging from the prey spin cocoons that entangle and kill small wasp larvae. The adult braconids that emerge perish in turn because they are unable to escape from the nest. When only a few parasites emerge, the wasps may complete development, but their growth may be stunted owing to lack of food. Caterpillars that have been attacked by polyembryonic Encyrtidae (Chalcidoidea) become bloated, rigid, and inedible; if many of the provisions are so parasitized the wasp larva may starve.

## MATING BEHAVIOR

The mating behavior of wasps is best understood as it relates to the two major components of sexual selection, male-male competition and female choice (Thornhill and Alcock 1983). Because females invest heavily in the well-being of offspring but males do not, males are expected to compete among themselves to monopolize females and to fertilize their eggs, and females may choose among potential mates (Kirkpatrick 1987). Competition among males can take many forms depending on the distribution of receptive females (Emlen and Oring 1977). The distribution of females is, in turn, controlled by the distribution of the resources required for nesting and provisioning. But females are not sexually receptive at all times. The temporal aspects of female receptivity in combination with the females' spatial activity patterns determine male mating strategies. Female receptivity can be categorized according to whether females mate only once or repeatedly during their lives. When females mate once after emergence and then become unreceptive, males exhibit behaviors that bring them into contact with young unmated females. If females mate with more than one male, then sperm competition between the ejaculates of different males may occur within females' reproductive tracts, and the nature of this competition can have significant consequences for male behavior.

Most commonly in insects, when females are inseminated by multiple males the sperm from the last male to mate have an advantage in fertilizing eggs over the sperm of a female's earlier mates (Parker 1970, 1984; Gwynne 1984). In this case, there is little advantage to locating and mating with virgins, and males would be expected to compete to be the last to inseminate a female before she oviposits. The patterns of sperm utilization are not known for solitary vespids and, indeed, have been investigated for only a few hymenopterans (Starr 1984). Two parasitic wasps (Chalcidoidea) exhibit sperm precedence contrary to the typical insect pattern. Females normally mate only once, but when artificially manipulated they can be induced to mate a second time (Wilkes 1965, 1966; Holmes 1974), with the first male to mate fertilizing most of the eggs. Sperm utilization has also been explicitly investigated in honey bees (Page and Metcalf 1982) and eusocial vespids (Ross 1986). Both of those studies report that sperm from different males mix within the spermatheca; the sperm of one male are not used before those of another male. However, these data also seem to indicate that sperm from different males are not used in equal proportions. Some paternal genotypes are responsible for 60–80% of the fertilizations, while other genotypes fertilize only 20% or less of the eggs. Possibly, several males of the predominant genotype mated with the same female, but it is also possible that priority in fertilizations results from the order of mating. The behavior of solitary vespids is consistent with the hypothesis that when females mate with more than one male, sperm from the last male to mate have precedence.

## Male-Male Competition

If females mate only once, then males will compete for access to virgin females. In the extreme manifestation of this competition, males find nest sites where newly eclosed females are emerging. This may happen in two ways. Males of species that have isolated nests may "know" the location of their natal nest, where females (their sisters) are likely to emerge. For species that nest gregariously at the same place generation after generation, the presence of actively nesting females, as well as the fact that a male emerged there himself, may indicate that virgin females will be emerging in a particular area.

Females of *Paraleptomenes miniatus* construct and provision nests so that sons and daughters emerge simultaneously, and the siblings mate before dispersing (Jayakar and Spurway 1966a). This first incestuous mating requires no searching or interacting with rivals on a male's part. In the twig-nesting *Euodynerus foraminatus*, males in the outer cells emerge from the nest several days before their sisters and compete among themselves for access to females. The winning male (generally the largest brother) waits at the nest entrance and mates in succession with up to five of his sisters as they exit their nest. Once inseminated, females reject other males' advances (Cowan 1979, 1986). Because males wait at nests where virgin females emerge, sibling mating may also occur in *Ancistrocerus adiabatus* (Cowan 1986), *Epsilon* species (Smith and Alcock 1980, Thornhill and Alcock 1983), *Pseudomasaris coquilletti* (Richards 1963), *Synagris cornuta* (Fig. 2.9) (Roubaud 1911), and *Euodynerus crypticus* (Isely 1913). Males of *E. foraminatus, A. adiabatus*,

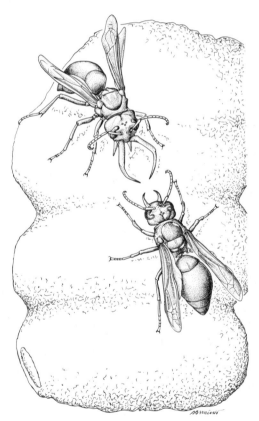

**Fig. 2.9.** Males of *Synagris cornuta* joust for position on a cell from which a receptive virgin female will eventually emerge. Males of this species exhibit polymorphism in body size and in the size of their mandibular horns.

and *Epsilon* species are active at nest sites primarily in the morning, when females emerge. Later in the day they leave the nests and search around plants for females that did not mate when they emerged. Subordinate males that were unable to obtain territories at nest sites devote all of their searching efforts to areas away from the nest.

Males of some colonial ground-nesting species do not locate and defend individual nests and are not territorial; they simply perch near or patrol nesting areas (Markin and Gittins 1967; Gess and Gess 1976, 1980). Males may pounce on nesting females in an apparent attempt to mate but break off contact quickly. Because male activity is conspicuous but matings with nesting females have not been observed, it seems likely that females mate only once early in life. At the densities observed in these studies, males were not territorial, perhaps because the high level of male activity around a colony makes defense costly and the low probability that receptive females will emerge from a small defendable area reduces the benefits.

If nests and emerging virgin females are scattered and difficult to locate, then males search for virgins at localities that contain resources that might attract females, including flowers and water. Clay and prey items typically are widely dispersed resources and generally are not known to concentrate females. However, some males of *Pseudomasaris zonalis* apparently wait for females at potential clay-gathering sites (Longair 1987). Flowers and especially water (Isely 1913; Longair 1984, 1985; Gess and Gess 1980) in arid habitats are often clumped resources and, not surprisingly, mating activity is commonly observed at these sites.

Longair (1984) described mating-related behavior of *Stenodynerus taos* around a watering trough, where females got water for making mud. Here, a few large males maintained territories and their smaller rivals patrolled vegetation. Over 90% of the females that arrived to drink were discovered and contacted by the males. Usually, males immediately separated from the females. In only 8% of cases did the males grasp the females. These pairs flew away from the trough to surrounding vegetation, where males attempted to initiate copulation. Because the water trough was small (defendable) and reliably visited by females, male territoriality appeared to be advantageous. Two Australian species of *Paralastor* also mate at pools of water in arid habitats (Smith and Alcock 1980). Here, males are nonterritorial; they patrol pools of water, pouncing on and mating with some of the females that land on the water to drink. However, after closely approaching other drinking females, males quickly depart. The behaviors of *S. taos* and *Paralastor* suggest that females mate only once and that males can distinguish virgins, which they court vigorously, from previously mated and unreceptive females, which they quickly abandon after contact. For these species searching for mates at widely scattered nest sites would be less economical than visiting the water sources that concentrate females.

Several species of *Pseudomasaris* have oligolectic pollen-foraging habits, and clumps of their preferred plants serve as centers of activity for males and for mating (Longair 1984). Males follow regular routes from clump to clump or perch near a clump, defending it as a small territory. Females mate more than once, but are receptive only during periods of cell construction (Longair 1987). Flowers provide not only food for the adults of these wasps, but also larval provisions and possibly the liquid used to make mud as well (Torchio 1970).

If females mate with several males and if sperm precedence or displacement occurs with advantage to the last male to mate, then males will compete to be the last to mate before oviposition, rather than competing to be the first to find a female. In the extreme, males may establish themselves in active nests to wait for females and copulate with them just before they oviposit. Although these behaviors are known for some Sphecidae (*Trypoxylon*, *Pison*, and *Oxybelus*; Bohart and Menke 1976, Krombein et al. 1979), they have not been confirmed for solitary or presocial vespids. However, Zucchi et al. (1976) excavated ground nests of *Trimeria howardi* and found adult males in eight out of 35 nests where females were still provisioning. Judging from the composition of brood cells, they concluded that most of these males had not been produced in the nests where they were found and thus were probably potential mates of the females.

Even if females mate multiply, males may not always find it profitable to locate nests and await the female's return. Males of the Australian *Abispa ephippium* patrol stream pools, where they intercept females visiting for water (Smith and Alcock 1980). Because females have never been observed to resist males' advances, it seems likely that females mate many times with a number of males. In temperate North America where water is widely available, males of *Ancistrocerus antilope* patrol flowers in search of mates. Although females of *A. antilope* do not copulate with all suitors, they do mate with several males (Cowan 1984, 1986; Cowan and Waldbauer 1984).

Members of a species may be so thinly dispersed that no particular resource effectively concentrates females. In an arid region of North America, nest sites and host plants that supply pollen and nectar (and perhaps nest-building liquid) for *Pseudomasaris maculifrons* are abundant and widely distributed. Males of this species have adopted hill-topping behavior to locate mates. They sit on perches along ridges and fly out to investigate passing insects approximately the size of conspecific females (Alcock 1985, 1987). Males are widely spaced and are not strongly territorial.

## Female Choice

Although much of male behavior makes sense in terms of competition for mates, the nature and effects of female choice are still quite obscure. Females of many kinds of insects discriminate among males on the basis of their ability to provide material resources that females require (Thornhill and Alcock 1983). Some males may be able to monopolize resources such as flowers or watering places, excluding rivals and limiting female choice, but in such cases it is difficult to separate the effects of male-male competition and female choice. Does a female mate with a particular male because she "chooses" the most vigorous male on the basis of his ability to subdue rivals, or do dominant males reduce a female's choice by driving off competitors? There is no evidence that vespid males contribute anything more than genetic material to their mates and offspring; thus, female choice, if it occurs, will be based on phenotypic traits of males. Whether females select male phenotypes that indicate good genes that will combine to make vigorous offspring, or whether choice is arbitrary and leads to maladaptive male structures, is currently an open debate (Kirkpatrick 1987).

That female choice does occur is evident from the fact that females mate with some males and reject others. Communication of unreceptivity may involve chemical cues that cause males to withdraw after only close approach or brief touching of a female (e.g., *Stenodynerus taos*: Longair 1984; *Paralastor* sp.: Smith and Alcock 1980). Unreceptivity can also be communicated with body movements if males mount, court, and attempt to copulate. Typically, unreceptive females curl their abdomen forward under their thorax, and males cannot gain intromission (e.g., *Euodynerus foraminatus*: Cowan 1986; *Euparagia richardsi*: Longair 1985). Female receptivity varies between two extremes. Females of species such as *Euodynerus foraminatus* quickly mate with the first male they encounter and then become permanently unreceptive (Cowan 1979, 1986). At the other extreme, females of *Abispa ephippium* mate repeatedly, without hesitation, with any male that finds them (Smith and Alcock 1980). *Ancistrocerus antilope* is intermediate: females mate with more than one male but do not accept all that court (Cowan 1984, 1986).

Females of some species may "prefer" to mate with a brother (Cowan 1979), as in *E. foraminatus* and *Paraleptomenes miniatus* (Jayakar and Spurway 1966a), where mating occurs at or near the natal nest. Inbreeding, combined with sex-ratio control (see Population Structure and Sex Ratio), would allow females to reduce the costs of sexual reproduction without becoming strictly asexual. A female that mates with her brother increases her relatedness to her offspring, and thus may increase her genetic representation in subsequent generations relative to outcrossing females. Once inbreeding has begun, females may be able to gain further by diverting resources to genetically more similar daughters and away from "excess" sons. While this results in a higher rate of increase for the inbreeding line, it incurs a cost in the form of reduced genetic variation among offspring. Occasional outcrossing occurs when females emerge from the nest and no brother is waiting, in which case they mate with unrelated males at flowers. Inbreeding depression due to the expression of deleterious recessive alleles in homozygous genotypes is probably not as severe a problem in the haplodiploid hymenopterans as in diplodiploid species because these alleles are regularly expressed and "weeded out" in the haploid males (Crozier 1975).

Multiple mating by females is of special interest with reference to the issue of female choice. If the amorous advances of males hinder females' nesting activities, but rejecting persistent males is more costly than simply mating and "getting it over with," then multiple mating presumably represents a complete loss of female choice. Females of *Abispa ephippium* appear to avoid detection by patrolling males at watering sites by drinking under the shelter of plants, but when discovered they readily mate (Smith and Alcock 1980). Females of *Pseudomasaris vespoides* mate only once and subsequently reject males, but sometimes only after grappling matches. Apparently as a result of this interference, females avoid flower patches where males are most active (Longair 1987).

## Courtship and Copulation

As with the other aspects of sexual behavior, courtship and copulation are strongly influenced by male-male competition and female choice. Among the solitary and presocial Vespidae sexual selection has resulted in considerable inter-specific diversity among species with regard to courtship behaviors, duration of courtship, manner and duration of copulation, the number of separate copulations engaged in by a single pair, and postcopulatory displays. In addition to striking differences among species, there is significant variation within species.

Males detect females visually, immediately fly onto their dorsum, assume a mounted position on the female facing in the same direction, initiate courtship, and attempt to copulate. When mounted, males of some species grasp the female's thorax with their front two pairs of legs but extend the hind legs out to the side (Smith and Alcock 1980); others grasp females with their front and hind legs, with the middle legs held out to the side (Cowan 1986). This grip leaves the females' wings free, and, in many species, once a pair has formed they fly from the place they met (water or flowers) to copulate in dense vegetation (Iwata 1953, Smith and Alcock 1980, Cowan and Waldbauer 1984, Longair 1984), probably to reduce interference by other males that are searching for mates in the same area. Males of *Odynerus reniformis* have a spine on the midcoxa, and males of *O. spinipes* have notches in the midfemur for grasping the females' wings during copulation (Miotk 1979).

As part of courtship, males tap or stroke the female with their antennae (Cowan 1986). Antennal movements vary from regular or synchronous (e.g., *Euodynerus foraminatus* and *Parancistrocerus pensylvanicus*) to irregular fluttering (e.g., *Ancistrocerus antilope*). During antennation, males of *E. foraminatus*, *Ancistrocerus adiabatus*, and *A. catskill* probe with their genitalia directly at the female's genital aperture, whereas males of *P. pensylvanicus* and *A. antilope* use their genitalia also to stroke or tap females between attempts to link genitalia.

Receptive females generally lift the end of their abdomens slightly, and males insert their genitalia. After intromission, males of some species usually remain mounted on the female, but males of other species immediately release their hold on a female's thorax and dangle behind her, attached only by the genitalia (Smith and Alcock 1980, Cowan 1986). Males of *E. foraminatus* remain dorsal throughout copulation except when a second male interferes. At these times, males that have linked genitalia fall back into the hanging position and apparently complete mating while the intruder mounts the female in the dorsal position and engages in courtship (Fig. 2.10). Intruders fail to copulate (Cowan 1986).

Courtship and copulation are brief (less than one or two minutes) in some species (Jorgensen 1942, Smith and Alcock 1980, Longair 1984, Cowan 1986); pairings of other species last 2–15 minutes. Copulation persists for up to 30 minutes in *Monobia quadridens* (Rau 1935), and pairs of *Ancistrocerus antilope* and *Paraleptomenes miniatus* remain together for about 1.5 hours and two hours, respectively (Jayakar 1966, Cowan 1986). These latter two species do not remain in copula throughout the entire pairing. Rather, they engage in repeated copulations with interspersed periods of inactivity or courtship. Smith and Alcock (1980) observed similar repeated copulations by single pairs of the species *Abispa ephippium*.

Separation terminating mating often seems to be precipitated by female behaviors that include wriggling and thrashing the abdomen and curling the abdomen forward to bite at the male's genitalia, as with *E. foraminatus*, *Ancistrocerus adiabatus*, and *A. catskill* (Cowan 1986). After unlinking genitalia, males of *Abispa ephippium* (Smith and Alcock 1980) and *Parancistrocerus pensylvanicus* (Cowan 1986) perform postcopulatory displays of biting and pulling or rhythmically bouncing on the female.

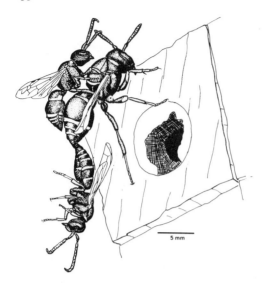

**Fig. 2.10.** A virgin female of *Euodynerus foraminatus* mates with her brother (suspended upside down by the genitalia) immediately after emerging from her natal nest, while a second male mounts to court. (From Cowan 1986, courtesy of Entomological Society of Washington.)

It may be that postcopulatory displays and multiple copulations with the same female are behaviors that help induce unreceptivity in females or influence sperm precedence in a male's favor (Smith 1979).

Sexual dimorphism among solitary vespids characteristically involves males being smaller than females, and eumenine males differ from females in having hooked antennae that are used during courtship. However, *Synagris cornuta*, *Parachilus insignis*, and *Delta alluaudi* are exceptional in that males are larger than females (Bequaert 1918, Gess and Gess 1976, Brooke 1981). Males of *S. cornuta* exhibit intrasexual polymorphism as well; some large males have huge horns on their mandibles (Fig. 2.9). These unusual morphological features hint that intrasexual combat is more intense in this species than in other wasps. Males of *Allorhynchium vollenhoveni* are polymorphic for a pair of spines on the second gastral sternite (Vecht 1963); the spines are found only on males above a certain size. Their role, if any, in sexual behavior is unknown. Many eumenine females have pits (cephalic foveae) on the top of the head that are secretory in nature (Cumming and Leggett 1985). Because the mouthparts of mounted males often touch females at this place, the secretions from these pits may play a role in sexual behavior.

## POPULATION STRUCTURE AND SEX RATIO

Solitary wasps often do not exist in large populations of randomly interbreeding individuals. Some populations are divided into many semi-isolated nesting aggregations that may persist for years. Because females may return to their natal area to nest (Evans and Matthews 1974, Jayasingh and Taffe 1982) and because males may concentrate their mate-searching efforts around their natal area, the extent of inbreeding within local populations may be significant. Even species such as twig nesters that do not live in persistent nesting aggregations may have high inbreeding coefficients because of frequent mating between siblings (Cowan 1979, 1981).

Variation in population structure may be reflected in variation in sex ratios. The null hypothesis for examining population structure via sex ratios was established with Fisher's Principle (Fisher 1958), which states that in large randomly breeding populations selection on parents will result in populationwide equal investment in sons and daughters. Because individual sons usually receive less investment (provisions) than individual daughters, most wasp species exhibit a numerically male-biased sex ratio (assuming that there is no sex-biased mortality). In a significant advance, Hamilton (1967) showed that if populations are fragmented so that there is inbreeding or local competition for mates among brothers, then mothers will be selected to reduce their investment in sons and produce female-biased sex investment ratios.

Because random samples of entire wasp populations are difficult to obtain, and because other factors may influence sex ratios (see below), apparent female biases in sex ratios do not necessarily imply local competition for mates or inbreeding. Even so, in the few species of Vespidae with well-documented inbreeding there is an effect on sex ratio. Females of the progressive provisioner *Paraleptomenes miniatus* always provision a female cell and then a male cell adjacent to it. The siblings emerge simultaneously and mate before dispersing. Even though this nesting pattern tends to produce an equal numerical sex ratio, mothers spend about 2.8 times as much time provisioning female larvae as they do provisioning male larvae (Jayakar and Spurway 1966a), so that the sex ratio of parental investment is considerably female-biased. Males of the twig-nesting *Euodynerus foraminatus* wait at their natal nest and mate with their sisters. When a female has nested in isolation, most of the competition for females emerging from the nest is between her sons (the emerging females' brothers). However, if adjacent nest sites are occupied by several females, the males emerging and competing for access to emerging virgins will not all be closely related. Consequently, nesting females tend to produce more female-biased broods when they nest in isolation than when they nest near other females (Cowan 1981).

Seasonal variation in sex ratio has been documented for a number of trap-nesting species in temperate North America (Medler 1964, Fye 1965, Longair 1981, Seger 1983, Cowan and Waldbauer 1984). These wasps overwinter as diapausing prepupae within the nest; in the spring they metamorphose to adults, mate, and produce a summer generation that develops directly into adults. The summer generation and any survivors of the overwintering generation produce the next overwintering generation in the late summer and fall. Typically, overwintering generations are numerically male-biased and summer generations are female-biased (Table 2.3).

Several hypotheses have been advanced to account for these seasonal swings in sex ratio. Longair (1981) proposed that overwintering generations are male-biased to compensate for possible increased male mortality in the more exposed outer cells of nests during winter. This hypothesis conflicts with Fisher's statement that regular immature mortality following the termination of parental care will not influence the primary sex ratio (Fisher 1958).

Seasonal swings in sex ratio might also be explained by seasonal changes in local competition for mates and inbreeding. If overwintering generations tend to mate randomly but summer generations engage in sibling mating, the observed changes in sex ratio would be expected. Heavy mortality during the winter might reduce the number of nests producing both sexes and thus reduce sibling mating. Alternatively, the costs and benefits of sibling mating may shift seasonally (somewhat like the seasonal alternation between sexual and asexual generations in some insects) so that individuals engage in behaviors that lead to outcrossing at some times and engage in behaviors that lead to inbreeding at others (Cowan 1979, Longair 1981). Overwintering males of the inbreeding wasp *E. foraminatus* are just as likely to remain at their natal nest and attempt to mate with sisters as are summer generation males (Cowan 1979, unpubl.), which is consistent with the available evidence suggesting that *E. foraminatus* does not have seasonal shifts in sex ratio (Table 2.3). However, the universality of this hypothesis is mitigated against by the fact that some species that have pronounced seasonal shifts in sex ratios, such as *Ancistrocerus antilope*, do not have local competition for mates (Cowan and Waldbauer 1984).

**Table 2.3.**   The proportion of males in overwintering and summer generations of four species of Eumeninae

| Species | Overwintering | Summer | Reference |
|---|---|---|---|
| *Ancistrocerus antilope* | 0.67 (107) | 0.09 (76) | Longair 1981 |
| | 0.84 (31) | 0.33 (12) | Fye 1965 |
| | 0.80 (20) | 0.39 (57) | Cowan and Waldbauer 1984 |
| *A. catskill albophaleratus* | 0.85 (40) | 0.30 (254) | Fye 1965 |
| *Euodynerus foraminatus* | 0.28 (75) | 0.43 (305) | Medler 1964 |
| | 0.59 (56) | 0.50 (133) | Longair 1981 |
| | 0.52 (318) | 0.47 (552) | Seger 1983 |
| *E. leucomelas* | 0.54 (44) | 0.32 (28) | Fye 1965 |

*Note*: Numbers in parentheses are sample sizes—i.e., the number of wasps.

Seasonal shifts in the sex ratio are also predicted by models based on partial overlap of generations in the bivoltine life history of many wasps (Werren and Charnov 1978, Seger 1983). If adults, especially males, of the overwintering generation live long enough into the summer to contribute genetically to the next overwintering generation, then investment will be male-biased in winter broods and female-biased in summer broods. Differences among species in the extent of seasonal sex ratio variation may result from differences in male longevity and the probability that males from the overwintering brood will live long enough to inseminate summer brood females. In tropical climates, breeding activity may be independent of season (Brooke 1981), but the population sizes of some of these species undergo seasonal fluctuations (Iwata 1964a), while wasps of other tropical species enter diapause during the dry season (Jayakar and Spurway 1965b). Female-biased investment ratios in nondiapausing species hint at possible local competition for mates (Brooke 1981), but numerically male-biased sex ratios in the diapausing generations of other species (Jayakar and Spurway 1966b) hint at partial bivoltinism.

Among renting species, brood sex ratios are also influenced by the physical nature of nesting cavities. In xylophilous species occupying trap nests, nests with small-diameter holes produce broods with a higher proportion of males than do nests with larger holes (Krombein 1967, Cowan 1981). To the extent that larger diameter holes are scarce, daughters (which require big brood cells) will predominate in large holes and sons (which need less space) in small holes (Charnov et al. 1981). *Pachodynerus nasidens* tends to rear more daughters when using empty mud pots of other wasps (Freeman and Jayasingh 1975b) than when using trap nests (Jayasingh and Taffe 1982). Jayasingh and Taffe (1982) also argue that *P. nasidens* has a more female-biased numerical sex ratio when nesting in shallower tubes than in deep tubes, but it is not clear whether their study controlled for the diameter of the nest hole.

## CONCLUDING REMARKS

A number of authors have written at length on the evolutionary stages of nesting behavior in the aculeate Hymenoptera (Ducke 1914; Roubaud 1916; Wheeler 1923, 1928; Evans 1958; Malyshev 1968; Evans and West-Eberhard 1970; Iwata 1976; West-Eberhard 1978a; Hansell 1987a). Each evolutionary stage or basic behavior pattern is associated with particular advantages and costs. Also, traits functioning in the context of one nesting pattern often are preadaptive for additional functions as nesting behavior evolves. By preadaptive I do not mean that evolution anticipates future contingencies, but rather that a particular behavioral trait can take on other adaptive functions as environments change. The progression of the hypothetical stages of nesting behavior as they apply to the solitary and presocial vespids is as follows:

1. *Mass provisioning by a solitary female*, with oviposition into an empty cell and then rapid storing of multiple prey items, is the predominant life history among noneusocial vespids and may be most advantageous in environments relatively rich in prey but largely lacking parasites and predators. Cells can be provisioned quickly and sealed before discovery by an enemy, and once sealed they may be overlooked or impenetrable. Egg deposition early in the cell-building cycle makes possible the evolution of progressive provisioning, and use of the mandibles to carry provisions can lead to the detection and destruction of parasites within the prey by kneading.

2. *Progressive provisioning by a solitary female* (subsocial behavior) occurs when a female brings in food at a slow rate influenced by the developmental state of her larva, so that a cell is open and the mother is in contact with her offspring for a significant portion of its development. The presumed function of this extended parental care is that the presence of the mother in the nest for long periods thwarts parasites and predators. As females spend longer periods sitting in the protected nest, different selective factors become important. Mortality due to accident and predation is likely to be reduced, there is selection to reduce rates of senescence (Williams 1957), life span increases, and mothers are more likely to be alive and nearby when some of their offspring reach adulthood (overlap of generations). Extended intervals between oviposition acts reduce the rate of egg maturation in the ovary so that females that oviposit and then lose their egg may be temporarily nonreproductive (West-Eberhard 1987c). Such nonreproductives are preadapted to help nearby relatives. Because provisions are brought to a larva shortly before they are eaten, prey need not be carefully paralyzed (preserved); they can be severely chewed to kill internal parasites, or even dismembered and ground to a paste to be distributed to several larvae. A sting that is no longer needed to carefully paralyze prey may be free to evolve into a more effective defensive weapon (Evans and West-Eberhard 1970).

3. *Overlap of generations* results from extended life spans, and presents wasps with opportunities for gains in fitness by associating with relatives of different generations.

4. *Communal behavior* results when young females, after emerging, remain associated with their natal (mother's) nest. Several females sharing a nest may more effectively repel parasites and predators, but there may also be increased reproductive competition from nestmates.

However, there are some particular adaptive opportunities available to communally nesting female relatives because temporarily nonreproductive females can help their relatives rear additional offspring at relatively little cost to their personal fitness. Newly emerged females can begin caring for larvae without having to wait for their own eggs to mature, and they may be able to occupy a vacant nest cell without the costs of building. Similarly, older females without a mature egg or larva of their own to care for can help rear the brood of other females. Communal societies are thus preadapted for the evolution of reproductive and worker castes because some individuals may coerce (dominate) others into extended nonreproductive periods. In this view, the initial evolution of castes and eusociality is primarily the outcome of competitive interactions and not an event involving increasingly greater cooperation (see also Michener 1985).

Even though all vespids with presocial and eusocial behavior practice progressive provisioning, this form of tending offspring is not necessary for the evolution of eusociality, as is evidenced by bees (Michener 1974, 1985) and sphecid wasps (Matthews, this volume). However, progressive provisioning seems to have led to selective pressures and opportunities for social interactions that significantly increase the chances that more complex social interactions will evolve. Because of its significance for vespid evolution, progressive provisioning must be studied in more detail. What environmental factors caused relatively few species to evolve progressive provisioning, and what factors resulted in the vast majority of vespids continuing to be mass provisioners? Because subsocial vespids occur only in the tropics (Krombein 1978), it seems likely that the constant environments, high rates of competition, or high rates of predation (primarily by ants?) often associated with tropical habitats are important.

The change from females nesting alone to communal nesting is critical because it must occur at some delicate balance between the dangers posed by heterospecific parasites and predators and the risks presented by associating with conspecifics. The relative costs and benefits of nesting communally will be influenced by the likelihood that an individual's cell or prey will be usurped, the likelihood that a nestmate will replace an egg, and by the relatedness of interacting nestmates. If relatedness is high, then the costs of feeding a nestmate's offspring will be reduced (West-Eberhard 1978a). For this reason, knowing the effects of the breeding system on relatedness and the population structure of solitary and presocial wasps is important for understanding the evolution of social behavior (see also Ross and Carpenter, this volume).

The haplodiploid genetic system often has been suggested as a significant factor in the evolution of sterility among the Hymenoptera, because females may gain by rearing sisters that are related to them by three-quarters rather than producing daughters that are related to them by only one-half. However, in order to reap these genetic benefits, sterile workers would need to avoid raising brothers that are related to them by only one-quarter. Thus it has been argued that the evolution of sterility will be most likely in haplodiploid species where sex invest- ment ratios are female-biased (Seger 1983) or where workers can con- trol the sex ratio (Hamilton 1972, Trivers and Hare 1976). These hypoth- eses based on three-quarters relatedness are sensitive to the assump- tion of single insemination of reproductive females (Alexander and Sherman 1977, Andersson 1984) and to the assumption of the subsocial route to sociality, which hypothesizes that protoworkers associate with their mother on their natal nest. Although we have some indications of the number of times individual females mate in a few solitary species, this information is virtually nonexistent for the socially more derived species. What is known about the natural history of communal wasps suggests that interactions within strictly subsocial groups are unlikely; interactions between more-distant relatives probably predominate. Thus, the most appropriate comparisons are not based on a choice be- tween sisters with a relatedness of three-quarters or daughters with a relatedness of one-half, but are based on a choice between aiding indi- viduals with a relatedness of one-eighth, one-sixteenth, or less or hav- ing a fitness of zero (West-Eberhard 1978a, 1987c).

Seger (1983) argues that for partially bivoltine species of Hymenop- tera in which only inseminated females overwinter (such as some bees), the summer generations will raise female-biased broods. Daugh- ters should then be more likely to remain at their natal nest and help raise their sisters without suffering the cost of raising brothers. Thus, eusocial behavior may be more likely to evolve in temperate-zone pop- ulations with seasonally fluctuating sex ratios than in tropical popula- tions that breed continuously year-round and have a constant sex ratio (Seger 1983). Temperate-zone wasps tend not to conform to this model because both sexes overwinter and mating occurs in the spring, but the temperate-zone eusocial vespids in the genera *Polistes* and *Vespula* do overwinter as inseminated females. The impact of seasonal tropical habitats on the life histories of wasps is little studied. Some wasps weather adverse seasonal conditions in the tropics with a prepupal dia- pause similar to that of temperate-zone species, but individuals of other species simply seem to disappear during such unfavorable times (Jayakar and Spurway 1965b). Maybe they migrate to more favorable habitats or seek shelter as adults and remain inactive. Perhaps these latter species have life histories that conform to Seger's assumptions.

A small number of eumenines (*Synagris*: Roubaud 1911, Bequaert 1918; *Calligaster*: Williams 1919; *Zethus*: Ducke 1914) are regularly cited in the literature as exemplifying the critical evolutionary stages of sub-social and communal behavior that connect solitary and eusocial wasps. Unfortunately, much of the information about these insects consists of the barest anecdotes, and the data often were not collected in the light of current ideas. Thorough reinvestigations of these insects, including careful attention to their life histories, nesting and mating behaviors, and population structures, are badly needed.

## Acknowledgments

I am grateful to R. Longair, K. Ross, and an anonymous reviewer for critical reading of the manuscript. The University of Michigan Biological Station provided a stimulating atmosphere during much of the writing.

# 3

# The Stenogastrinae

STEFANO TURILLAZZI

Until recently, the biology of wasps of the subfamily Stenogastrinae was almost completely unknown. Although regarded as very important for understanding the phases characterizing the early evolution of eusociality in wasps (West-Eberhard 1978a), studies of the Stenogastrinae were largely limited to the description of collection specimens or to occasional field observations. The dearth of studies on stenogastrine biology is primarily due to their distribution, which is restricted to the rainforests of the Indo-Pacific tropics from southern India to New Guinea. The uniqueness of these wasps, however, has been appreciated from the very beginning. Authors such as Williams, Jacobson, and Pagden described with wonder their hovering flight (for this they are commonly called hover wasps), their shy habits, and their strange, camouflaged nests hidden in the wet and dark parts of the jungle, hanging from roots and threadlike fungi along streams and near the spray of waterfalls. Iwata (1967) was the first Japanese researcher to study these wasps. His paper, which described the peculiar morphology of the larvae and pupae of many oriental species, was followed two years later by the first observations on stenogastrine social organization (of two *Parischnogaster* species) by Yoshikawa et al. (1969). Between 1969 and 1980, only two important papers, one by Spradbery (1975) on *Stenogaster concinna* from New Guinea and the other by Krombein (1976) on *Eustenogaster eximia* of Sri Lanka (formerly Ceylon), were published, along with a note in which Hansell announced his advancing research on the biology and social behavior of *Parischnogaster mellyi* (Hansell 1977). Since 1981, especially influenced by the increased inter-

est in sociobiology, entomologists from various countries have considerably increased our knowledge of these wasps.

## SYSTEMATICS

Stenogastrinae, a subfamily of the family Vespidae, contains about 50
described species divided into six genera (see Carpenter 1988a, this volume). Vecht revised the genera *Anischnogaster* (five species) (Vecht
1972) and *Stenogaster* (eleven species) (Vecht 1975) from New Guinea,
and the genus *Metischnogaster* (two species) (Vecht 1977a) from the Oriental region. However, the genera *Liostenogaster*, *Parischnogaster*, and
*Eustenogaster* (all from the Oriental region), which include most of the
species in the subfamily (34 described to date: J. Carpenter, unpubl., J.
van der Vecht, unpubl.), await revision.

## HABITAT, ADULT AND LARVAL MORPHOLOGY

Stenogastrine wasps live in the rainforests of the Indo-Pacific (Fig. 3.1),
from sea level to the oak forests of Mount Kinabalu on Borneo at a
height of 1,700 m. Usual nesting sites are protected places such as cliffs
overhanging forest roads and streams; caves; rocks near waterfalls;
vaults of small bridges, tunnels, or water pipelines; and other dark,
wet places with relatively constant temperature and humidity. Some
species can be found in more exposed situations, however. *Parischno-*

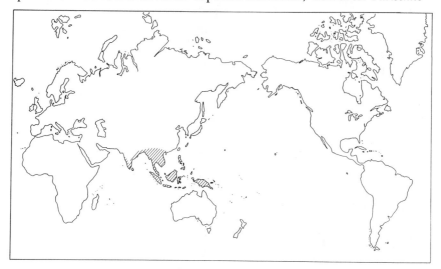

**Fig. 3.1.** Distribution of the Stenogastrinae. (Modified from Vecht 1967.)

*gaster nigricans serrei*, for example, often locates its nests on the branches of trees along avenues in Bogor, Java. *Parischnogaster mellyi* nests on threads of bamboo hanging from the roofs or ceilings of village huts, while the inverted-flask-shaped nests of various species of *Eustenogaster* can be found both in caves and on the walls of buildings in Malaysia.

Adult size varies from a total length of 1 cm (or less) in small species of *Parischnogaster* to about 2 cm in *Eustenogaster*. The body is slender and elegant with a very long petiole and short legs (Fig. 3.2). Adults are usually black or dark brown with variable-sized yellow areas (especially extended in *Liostenogaster*). Males are usually difficult to distinguish from females, but some *Parischnogaster* and *Metischnogaster* males have very evident white or cream-colored stripes on their abdomen.

The larvae of various genera and species look very similar; they possess conspicuous pleural blisters and dorsal protuberances, and, unlike other social vespid larvae, lie curled around the long axis inside their cells (Fig. 3.3f). Turillazzi (1985a), Hansell (1986a), and Samuel (1987) distinguished only four instars in the larvae of *P. nigricans serrei*, *Par-*

**Fig. 3.2.** Adults and nest of *Parischnogaster alternata*.

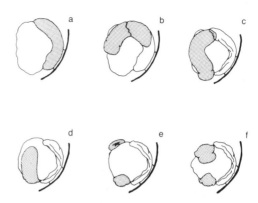

**Fig. 3.3.** Hatching sequence of the larva of *Parischnogaster mellyi*. The newly hatched larva (stippled) curls itself around the ball of abdominal secretion (a–e) and stays curled around the long axis of the cell for the remainder of its development (f). The cell bottom is to the right. (From Turillazzi 1985c, courtesy of *Monitore Zoologico Italiano*.)

*ischnogaster* (= *Holischnogaster*) *gracilipes*, and *Liostenogaster flavolineata*, whereas Hansell (1982a) reported three or four instars in *P. mellyi*. Carpenter (1988a) considered these underestimates of the true number. The pupae assume a bent position, with the tip of the abdomen almost touching the head, so that the pupal cells are shorter than the length of the entire animal inside them (Williams 1919, Iwata 1967, Spradbery 1975).

## THE ABDOMINAL SUBSTANCE AND REARING OF THE BROOD

If the morphology of the larvae and pupae of Stenogastrinae appears quite peculiar, the way in which the adults rear the brood is even more fascinating. A whitish, gelatinous substance found on the eggs and larvae of all species so far examined was interpreted as larval food by early authors and, at first, even by more-recent specialists (including myself). Williams (1919) and Pagden (1958a) thought this substance was of "external origin," while Iwata (1967) and Spradbery (1975) supposed that it was secreted by the adults. Hansell (1977, 1982a), studying *P. mellyi*, and Turillazzi and Pardi (1982), studying *P. nigricans serrei*, confirmed Jacobson's (1935) observations that the substance is in fact secreted by the adult females from a quite large Dufour's gland, and pointed out that the secretion is added to the egg at the moment of deposition, thus casting doubt on its purported nourishing function.

The secretion actually has many other functions in the social lives of these wasps: (1) it is used as a tool in oviposition; (2) it functions as a resting substrate for the small larvae; (3) it acts as a storage substrate for liquid and solid food for both larvae and adults; and (4) it constitutes the main part of the nest ant guards (defensive barriers against ants). These various functions are discussed in detail below.

## Oviposition

Among all the Vespidae oviposition behavior is perhaps the most peculiar in the Stenogastrinae. For instance, the female of *Parischnogaster*, just before oviposition, bends her abdomen and secretes the abdominal substance, which she then wads into a ball in her mouth with the aid of her anterior legs. She next extends her abdomen and bends it again; after a few seconds she lays an egg on the patch of secretion she holds in her mouth (Fig. 3.4). The egg is next attached to the bottom of the cell by means of a sticky secretion it bears on its convex surface, while the whitish abdominal substance remains on its concave surface (Fig. 3.4). The wasp is then free to collect another ball of secretion, which she deposits on the egg (Turillazzi 1985b). *Liostenogaster* and *Eu-*

**Fig. 3.4.** Egg laying by *Parischnogaster striatula*. Inset: an egg inside a cell with abdominal secretion (white mass). The egg is about 1.1 mm long.

*stenogaster* oviposit in an almost identical manner. Thus it appears that the substance serves as a tool to collect and place the egg in the cell (Turillazzi 1985c).

## Development of the Larvae and Larva-Adult Interactions

Adult females of Stenogastrinae attend to the brood from the egg stage to the emergence of the imago, and interactions between adults and larvae are quite complex. It seems strange that adults should provide "food" (in the form of the abdominal substance) five to seven days before the eggs hatch if they have no difficulty in furnishing the larvae with fresh food. We cannot completely exclude the possibility that the secretion has some trophic function, but if it is used as food it represents only a small part of the larval diet. The substance is refused if offered experimentally to ants, which, on the contrary, will readily eat eggs and small larvae. Adult stenogastrines do not eat the secretion, but sometimes they will chew bits of it for a considerable time before discarding them. The substance contains few proteins (Turillazzi 1985c, Delfino et al. 1988), raising further doubts as to its nutritive role. On the other hand, the secretion from the Dufour's gland of *Anthophora* bees, which is known to have a trophic function for the larvae, is composed solely of triglycerides (Norden et al. 1980).

The developmental period of the egg is almost a week. The newly hatched larva abandons the egg shell completely and slips over the ball of abdominal secretion, curling itself around it (Fig. 3.3). From the very first days after hatching the larvae are supplied with proteinaceous food, such as drops of liquid regurgitated by the adults and small pieces of chewed insects (primarily collected from spiders' webs by *Parischnogaster* and *Eustenogaster*: Williams 1919, Turillazzi 1983). Adults place food on the abdominal substance, where the larvae consume it little by little. This type of progressive provisioning resembles the brood rearing found in many subsocial wasps in that food is not placed directly on the larval mouthparts. The abdominal substance, therefore, appears to represent a suitable microhabitat where the small larva finds a moist supporting substrate and food provisions. Abdominal substance is given to the larvae for about two weeks following hatching, by which time the larvae are able to stay in the cells by themselves by pressing their dorsal surface against the walls. When these larger larvae are physically stimulated along their pleural lobes by the adults, the larvae open up like sphincters to receive food and fluid. This response can be elicited in some species by touching the larvae with an object such as a pin. According to Hansell (1987a:178), the larval mandibles are "extremely well developed" and are particularly important in the preparation of food to be ingested, in contrast to the "poorly developed" mandibles of the adults.

The average time of development from oviposition to emergence varies from 103 days (*L. flavolineata*: Samuel 1987) to 44 days (*P. nigricans serrei*: Turillazzi 1985a). The developmental period of *P. mellyi* is about 53 days (Hansell 1982a) and of *Eustenogaster calyptodoma*, 64 days (Hansell 1987b).

The adults in some species close the cells when the larvae are ready to pupate, then reopen the opercula after a few days to extract the larval meconium (fecal pellet). This unique behavior is favored by the characteristic position of the pupae in their cells (Turillazzi and Pardi 1982) and contrasts with the meconium-extraction behavior of some polistine species, in which the larval meconium is removed through the rear of the cell (Jeanne 1980a, Gadagkar, this volume). The cells of other stenogastrine species are simply narrowed at their openings at the time of pupation. Females of *Liostenogaster flavolineata* sometimes lay eggs on the opercula of pupal cells (Ohgushi et al. 1983a), but it is not known if these can develop or are used for some other purposes (for example as food for the adults or brood).

## Larva-Adult Trophallaxis

Spradbery (1975:317) notes that the "labial glands of *S. concinna* larvae are at least as well developed as in vespine larvae," but in this species and in *P. mellyi* (Turillazzi, unpubl.) larvae do not emit salivary fluid if stimulated by touching their labium, head, or pleural lobes with a pin, as do polistine and vespine larvae. Spradbery concludes from this that the labial gland functions primarily as a digestive organ, in addition to secreting an incomplete cocoon. However, this is not decisive evidence for stating that larva-adult trophallaxis is completely absent in this subfamily. Recently I observed that adults of *P. mellyi* frequently lick the fluid that collects in the middle of the coiled larvae (Turillazzi 1987a). The source of the liquid cannot be determined with certainty. It could be composed of larval secretions, but it could also include drops of fluid previously regurgitated by the adults. It is also true that food pellets that have not been completely consumed by the larvae always appear wet, as if subjected to extracorporeal digestion. Whatever the case, the adults often collect food or fluid from larvae to transfer to others or to keep for themselves.

## Storage of Liquid and Solid Food

Solid food and drops of liquid may be stored on the abdominal substance placed on eggs. Thus this secretion also functions as a food-storage medium for the entire colony, since the food is consumed also by the adults.

## Ant Guards

The abdominal secretion also seems to be the main component of the ant guards constructed by *P. nigricans serrei* and other species of the same and other genera (Turillazzi and Pardi 1981) (Fig. 3.5). Preliminary chemical analysis (Turillazzi, unpubl.) supports this opinion, as do the facts that the two substances look perfectly similar and that only one gland (the Dufour's gland) is large enough to produce such copious amounts of the substance as are observed. The ant guard substance of *Parischnogaster* is collected by the posterior legs in tiny drops

**Fig. 3.5.** Ant guard of *Parischnogaster jacobsoni*.

coming from the tip of the abdomen, passed through the middle legs to the mouth, and then applied to the substrate. This method of collection permits it to be applied continuously, while a different procedure ensures that the substance is supplied to the larvae in a single large application (see Oviposition). Perhaps secretions from the mouth are added to obtain different properties of the substance for different uses.

## Other Functions of the Abdominal Substance

The possibility that the abdominal substance has an antibiotic function against bacteria and fungi has not been confirmed experimentally (Turillazzi 1985c). We shall probably understand more about its roles in the colony when we know the exact chemical composition of the substance.

## COLONY DEVELOPMENT AND SOCIAL ORGANIZATION

Early authors considered some species of Stenogastrinae as social and regarded others as solitary mostly on the basis of the number of individuals found on the nests at the moment of collection or else on very scanty behavioral observation. At present we do not have sufficient information to state definitively that any species is truly solitary or even subsocial; moreover, all of the species studied closely so far have been shown to be eusocial in the sense of Wilson (1971), with generational overlap, cooperative care of the brood, and (at least temporary) reproductive division of labor (see also Cowan, this volume: Table 2.1; Carpenter 1988a, this volume). These species include *P. mellyi* (Hansell 1982a, 1983; Sakagami and Yamane 1983; Yamane et al. 1983a), *P. nigricans serrei* (Turillazzi and Pardi 1982, Turillazzi 1985d), *P. jacobsoni* (Turillazzi 1988a), *P. alternata* (Turillazzi 1985e, 1986a), *P. gracilipes* (Hansell 1986a), *L. flavolineata* (Hansell et al. 1982, Samuel 1987, Samuel and Hansell 1987), and *E. calyptodoma* (Hansell 1987b). Although Spradbery (1975) considered *S. concinna* subsocial, since the daughters of the solitary foundresses remain on the nests for only a short time, he did not furnish any detailed behavioral data. It is extremely important to study the colony cycle of species like this in order to assess the possibility that they include both subsocial and eusocial colonies.

## Nest Foundation

There is evidence that nest foundation by most stenogastrine species is haplometrotic (independent founding by a single female). This has

**Fig. 3.6.** Associative foundation by
*Parischnogaster alternata.*

been found to be the rule in *P. mellyi* (Hansell 1982a), *P. nigricans serrei* (Turillazzi and Pardi 1982), *P. jacobsoni* (Turillazzi 1988a), *P. timida* (C. Starr, unpubl.), *S. concinna* (Spradbery 1975), and *E. calyptodoma* (Hansell 1987b). However, two females of *P. mellyi* and of *P. nigricans serrei* have been reported to share a nest for brief periods (Turillazzi and Pardi 1982, Hansell 1983, Turillazzi 1985d), and I observed a case of cooperation and division of labor between two foundresses of *P. jacobsoni* that I followed for several days (Turillazzi 1988a).

Associative foundation (pleometrosis) is regularly present in *P. alternata* (Fig. 3.6), a species that forms dense clusters of colonies in wet, dark places (Turillazzi 1985e). Females of *L. flavolineata* also have been observed joining other single, and presumably unrelated, foundresses early in the colony's life (Hansell et al. 1982, Samuel 1987). *Liostenogaster flavolineata* also nests preferentially in dark and sheltered places, where it often forms large clusters of nests. Thus associative foundation may be a density-dependent phenomenon.

## Colony Size

Information regarding the colony cycle is available for only three species (*P. mellyi, P. nigricans serrei*, and *L. flavolineata*). Foundresses of *P. nigricans serrei* build an average of four to five cells before sealing the cell of the first larva to pupate (Turillazzi 1985d). Three to five females emerge before the first male, which usually develops from an egg laid by the foundress at about the same time as the first female emerges. In the early phases of the postemergence period (the period following the appearance of the first adult daughters) the number of nest inhabitants coincides approximately with the number of individuals that have emerged, but after some time young females begin to leave the colony. The number of resident individuals subsequently remains relatively constant, and the old foundress may be superseded by another female. External or internal disturbances, such as heavy predation or the residency of a great number of males on the nest, may disturb this equilibrium and many individuals may abandon the colony. Hypothetically, however, there is nothing to preclude the development of a perennial cycle in *P. nigricans serrei*, with continuous supersedure of the queen and fluctuation of colony population size.

Queen supersedure has been observed during the colony cycle of both *L. flavolineata* (Hansell et al. 1982) and *P. mellyi* (Hansell 1983, Yamane et al. 1983a). On the other hand, Samuel and Hansell (1987) observed only rare mother-daughter overlap for brief periods in the early colony life of *L. flavolineata*. The maximum total female and male population density of *L. flavolineata* coincides with the time of minimum rainfall, and the minimum population density with maximum rainfall, but there is no significant difference between the frequency of colony foundation in the wet and the dry seasons (Samuel 1987).

## Female Reproductive Status

As they grow older, females of *P. nigricans serrei* undergo ovarian development. Interestingly, reproductive potential differs among females of the same age in that some females may develop their ovaries in a few days while others require a longer period for eggs to mature. Moreover, almost all females mate 20–50 days after they emerge; in fact, among 18 mated and 19 unmated females of known age, none of the former was younger than 24 days and none of the latter was older than 46 days (Turillazzi 1985d). After mating, females may remain in the parental colonies for at least a short time. As the colony grows, not only does the ovarian development of the dominant female increase, but so does the number of potential egg layers present on the nest.

In mature colonies of *P. nigricans serrei* mated females represent, on average, about 69% of the adult female population ($N = 230$ adult females, 55 colonies), and mated females of *P. jacobsoni* represent about 63% ($N = 33$ adult females, 10 colonies) (Turillazzi 1987b). These two species also exhibit similar nest architecture and social organization (Turillazzi and Pardi 1982, Turillazzi 1988a). According to data reported by Yoshikawa et al. (1969), the percentage of mated females of *Parischnogaster striatula* is 81.4% ($N = 43$ adult females, 16 colonies), which is very similar to that found in the related species *P. alternata* (87%, $N = 24$, 8 colonies: Yoshikawa et al. 1969; and 85%, $N = 144$, 50 colonies: Turillazzi 1986a, 1987b). Of course, unmated females represented in these collections may have eventually mated at a later age.

*Parischnogaster nigricans serrei* and *P. jacobsoni* have quite equivalent percentages of females with mature oocytes in their ovaries (potential egg layers) (38 and 36%, respectively), while about 51% of *P. alternata* females are capable of laying eggs (Turillazzi 1986a). *Parischnogaster gracilipes* colonies seem to have about 39% of females with mature oocytes in their ovaries (Hansell 1986a), and thus this species is most comparable to *P. nigricans serrei* and *P. jacobsoni* in this regard. The number of potential egg layers of *P. nigricans serrei* increases with the number of females present on the nest. At present the real contribution such ovarian-developed females make toward egg and adult offspring production is still unknown for any species.

Unmated females of *P. nigricans serrei* usually have smaller ovaries than mated females (Turillazzi 1985d), a difference in reproductive development that is even more pronounced in *Liostenogaster vechti* (Turillazzi 1987b). This difference could mean either that females with larger ovaries are fertilized in preference to those with smaller ones or that mating stimulates ovarian development (Turillazzi 1985d). In contrast, some unmated females of *P. alternata* possess well-developed ovaries (Turillazzi 1986a). It should also be noted that, in captivity, unmated females of *P. nigricans serrei* (and probably also of *P. mellyi*) can found nests and produce broods of males.

Recent studies (Samuel 1987, Samuel and Hansell 1987) have shown that some *L. flavolineata* females exhibit greatly accelerated ovarian development and become foundresses or dominant females, while other females show retarded ovarian development and usually act as true workers.

## Social Interactions, Dominance Hierarchies, and Division of Labor

Williams (1928) reported that food sharing was the only social interaction between nestmate females of *Liostenogaster* (= *Stenogaster*) *varipicta*. Pagden (1962) reported a similar situation for *Metischnogaster cilipennis*. Yoshikawa et al. (1969) described a dominance hierarchy among females of an undetermined species belonging to the *P. jacobsoni* group, but they found no such hierarchy in *P. striatula*.

As suggested by the above, the social organization of stenogastrine colonies appears to differ from species to species. The females of both *P. nigricans serrei* and *P. jacobsoni* form linear dominance hierachies, identifiable at least in the higher ranks, and the hierarchy is well-correlated with the ovarian development of individual females (Turillazzi and Pardi 1982, Turillazzi 1988a, see also Yoshikawa et al. 1969). The dominant individuals walk unharrassed along the threadlike nests; subordinates give way to them or, if solicited, offer drops of regurgitated fluid. In these colonies a sharp division of labor exists: the dominant female almost always remains on the nest, patrolling and defending it, while the subordinates are often away foraging for food and collecting nest construction material (Fig. 3.7).

The dominance hierarchy of *P. mellyi* is hardly recognizable because there are so few interactions among females (Hansell 1983). However, Hansell (1983) describes a "senior female" as an individual that does not forage and always rests physically above the other females on the nest, and Sakagami and Yamane (1983) remark that only α-females (those that are behaviorally dominant) were observed to oviposit. It is also difficult to observe social interactions among females of *P. alternata* of the same colony (Turillazzi 1986a), but a division of labor can be detected after long periods of observation. Old and very young females remain on the nest, while middle-aged ones most often forage. This species forms great clusters of nests and exhibits associative foundation. Behavioral interactions among females during associative nest foundation appear to be rare and mild, and there is no observable divi-

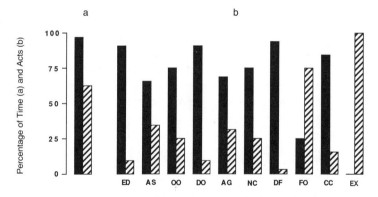

**Fig. 3.7.** Division of labor in colonies of *Parischnogaster nigricans serrei*. (a) Average percentage of time spent on the nest by dominant ($N = 6$) and subordinate ($N = 16$) females from seven colonies. (b) Average percentage of behavioral acts performed by dominant ($N = 6$) and subordinate ($N = 17$) females. Black bars, dominant females; striped bars, subordinate females. ED, egg deposition; AS, abdominal secretion supply; OO, oophagy; DO, dominance displays; AG, construction of ant guard; NC, nest construction; DF, defensive flights; FO, foraging; CC, cell closing; EX, extraction of the larval meconia. (Data from Turillazzi and Pardi 1982.)

sion of labor at this time. The two females of seven bigynous foundress associations exhibited equivalently developed ovaries upon dissection (Turillazzi 1985e).

According to Hansell et al. (1982) and Samuel (1987), *L. flavolineata* exhibits a "partially developed dominance system" where three (and occasionally up to six!) levels of females can be distinguished. The dominant female is the primary egg layer and plays the major role in nest defense; she has larger ovaries than the second-level females who, in turn, have larger ovaries than the third-level ones. In this species, co-foundresses seem to form dominance hierarchies of the *Polistes* type, with the dominant female remaining on the nest and subordinate (and younger) individuals foraging (Samuel 1987).

Hansell (1986a) reported a distinct correlation between ovarian development and time spent on the nest in females belonging to colonies of *P. gracilipes*. As in *P. alternata*, middle-aged females are the main foragers, and more than one potential egg layer can be found in some colonies.

The genus *Eustenogaster* is poorly known taxonomically (J. van der Vecht, unpubl.) but includes several species characterized by colonies that contain only a very small number of individuals. These species were considered to be solitary or subsocial by early authors, but Krombein (1976:305) observed that at least the first female of *E. eximia* to emerge "assists her mother in hunting for food and in caring for the developing brood." Hansell (1987b) found elements of eusociality in *E. calyptodoma*; he observed a division of labor between old and newly emerged females in colonies typically composed of only two females. Interactions between nestmates were not antagonistic.

Females of *P. mellyi* (Yamane et al. 1983a), *P. nigricans serrei* (Turillazzi 1985d), *E. calyptodoma* (Hansell 1987b), and *L. flavolineata* (Samuel 1987) may adopt different reproductive strategies depending on the particular situation in the colony in which they emerge, their reproductive potential, the availability of suitable nesting sites, and other factors. Remaining on the parental colony, founding a new nest, usurping an existing nest, joining an alien colony, and rejoining the parental colony are the principal (but not mutually exclusive) behavioral options. Usurpation or adoption is fairly common in many stenogastrine species, but detailed studies are wanting. An experiment conducted with *P. nigricans serrei* showed that a large pool of vagrant females is available to occupy both abandoned nests and active nests deprived of their adults (Turillazzi 1985d).

Samuel (1987) observed that single foundresses of *L. flavolineata* often move to other nests that have just been initiated but are vacant. Such drifting females thus care for alien brood.

## MALE BEHAVIOR AND MATING

The males of various species (*P. mellyi*: Turillazzi 1982; *S. concinna*: Spradbery 1975; *P. gracilipes*: Hansell 1986a; *P. striatula*: C. Starr, unpubl.) may form unisexual clusters on pendent substrates near nesting sites or, as in the case of *P. nigricans serrei*, they may remain on their parental nests or join other established or even incipient colonies (Turillazzi and Pardi 1982, Turillazzi 1985d). In contrast to these apparent social behaviors, the males of all species of *Parischnogaster* that I have observed (*P. nigricans serrei, P. mellyi, P. jacobsoni, P. alternata, P. striatula*) perform aerial patrolling away from nests during well-defined hours of the day. This patrolling behavior appears quite similar to that reported for *Metischnogaster* by Pagden (1962). It consists of a hovering flight near forest landmarks and the frequent display of whitish bands on the abdominal terga (Fig. 3.8), these bands usually being hidden from view when the abdomen is not fully distended. The display is particularly conspicuous if other males fly in the vicinity, in which case

**Fig. 3.8.** Male patrolling behavior of *Parischnogaster alternata*.

ritualized aerial combats may occur (Turillazzi 1982, 1983). In recent observations, captive males of *Parischnogaster* seemed to search particularly for well-illuminated spots, probably so that their displays would be more visible (Turillazzi and Francescato 1989). Aerial patrolling is most common in clearings in the forest. This patrolling behavior appears comparable to that shown by male orchid bees in the forest of Barro Colorado Island, Panama (Kimsey 1980a) and fits the definition of a lek given by Bradbury (1981). I have often observed swarms of males in captivity patrolling near spotlights, where competitive behavior occurs (Turillazzi and Francescato 1989). This suggests that territorial behavior may be influenced by density of males (see Thornhill and Alcock 1983 for a review of the matter).

Male integumental glands under the anterior part of the third gastral tergum of *P. mellyi* and *P. nigricans serrei* (Turillazzi and Calloni 1983) and under the fovea of the second gastral tergum of *P. alternata* (Turillazzi and Francescato 1989) are well developed and may play some chemical role during the visual display. Integumental glands also have been found under the gastral terga of males of species of *Metischnogaster*, *Liostenogaster*, and *Eustenogaster* (Turillazzi and Francescato 1990). Males of *L. flavolineata* have been observed to drag their abdominal terga on various prominent objects, probably along patrolling routes (Turillazzi and Francescato 1990). This action is suggestive of a chemical marking behavior. Males of *Eustenogaster micans* perform a kind of patrolling by "alighting again and again on the same leaf" (Williams 1919:175), but males of an undescribed species in this genus do not display as males of *Parischnogaster* do and often land on leaves with their abdomen in a raised position (Turillazzi and Francescato 1990).

That patrolling flights are an important component of the mating system of these wasps seems likely, because the only copulations or copulatory attempts observed to date occurred during patrolling (a species of the *P. striatula* group: Pagden 1962; *P. mellyi*: Turillazzi 1983, unpubl.). The fact that the daily patrolling time of two sympatric species in Java, *P. mellyi* and *P. nigricans serrei*, is completely distinct (Turillazzi 1982, 1983) is of special significance in this regard. Unfortunately, observed copulations are so rare that one can only speculate on the relation between reproductive success of males and patrolling behavior.

Hansell's (1982a) suggestion that males in colonies of *P. mellyi* contribute greatly to egg disappearance (by eating them) has been confirmed by recent observations (Turillazzi and Francescato 1989). Males also appear to interfere with the rearing of brood in colonies of *P. nigricans serrei* by robbing food loads from foragers and even from the larvae (Turillazzi 1985d). On the other hand, males of *P. mellyi* may indirectly contribute to a colony's success, since foundresses searching for a place to initiate a nest seem to use male clusters as indicators of suitable nesting substrates (Turillazzi and Francescato 1989).

## NEST ARCHITECTURE

Stenogastrine nest architecture differs from that of other social wasps and has such an incredible variety of shapes that it has in some instances been used as a systematic character. Pagden (1958a) divided these wasps into four groups according to the shape of the nest, and Sakagami and Yoshikawa (1968) distinguished the new species E. calyptodoma from E. micans on the basis of important differences in the nest architecture. Parischnogaster alternata was distinguished from the very similar P. striatula by Sakagami (in Yoshikawa et al. 1969) mainly by differences in the shape of the nest.

### Evolutionary Trends in Nest Architecture

None of the nests of any stenogastrine species has a petiole, which is one of the more striking differences with respect to the nests of other eusocial wasps. Rather, cells are built directly on various kinds of flat or threadlike substrates. The material used for the construction of the nest typically consists of mud or vegetable matter (see Wenzel, this volume). It is quite difficult at present to reconstruct the probable evolutionary stages of nest architecture for the entire group, but some speculation on individual genera can be hazarded and new entries included in the evolutionary scenario previously proposed by Ohgushi et al. (1983b).

Carpenter (1988a, this volume) suggests that the common ancestor of the social Vespidae may have already possessed the capability of using vegetable fragments. At the moment there are not many examples of convincingly primitive stenogastrine nests; a nest described by Ohgushi et al. (1983a) with five mud cells with no walls in common (Fig. 3.9a) cannot be ascribed with certainty to the Stenogastrinae as no resident was found on it. The same can be said for a nest with the cells arranged like organ pipes that I recently found in Malaysia (Fig. 3.9c). The most primitive architecture definitely associated with a stenogastrine species may be the mud nest of Liostenogaster varipicta described by Williams (1919), which consists of a row of cells placed side by side on a tree trunk (Fig. 3.9b).

Liostenogaster exhibits the greatest variety of architecture in the subfamily, with two main evolutionary features: nests are constructed either solely with heavy mud or mainly with vegetable matter. In the species that use mud we can see a progressive spatial concentration

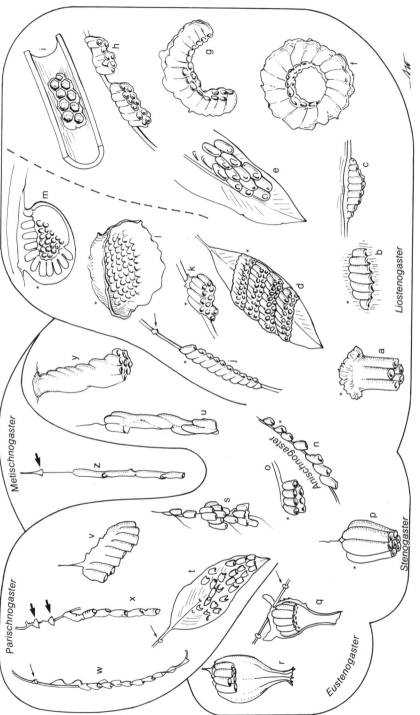

**Fig. 3.9.** Examples of nest architecture and probable intergeneric relations in Stenogastrinae. Asterisks indicate nests built with mud only. Thin arrows point to ant guards, bold arrows to structures that deflect raindrops. (a) *Liostenogaster* sp. ? (Ohgushi et al. 1983a); (b) *L. varipicta* (Williams 1919); (c) *Liostenogaster* sp. ?; (d) *L. nitidipennis* (Yoshikawa et al. 1969; (e) *L. nitidipennis*; (f) and (g) *L. vechti*; (h) *Liostenogaster* sp.; (i) *Liostenogaster* sp.; (j) *Liostenogaster* sp.; (k) *Liostenogaster* sp. (Yoshikawa et al. 1969); (l) *L. flavolineata*; (m) *Liostenogaster* sp.; (n) *Anischnogaster* sp. (Vecht 1972); (o) *A. iridipennis* (Vecht 1972); (p) *Stenogaster concinna* (Spradbery 1975); (q) *Eustenogaster fraterna*; (r) *E. calyptodoma*; (s) *Parischnogaster mellyi*; (t) *P. jacobsoni* group; (u) *P. striatula*; (v) *P. nigricans serrei*; (w) *P. gracilipes*; (x) *P. timida* (Williams 1928); (y) *P. alternata*; (z) *Metischnogaster drewseni*.

91

of cells, which culminates in comb nests provided with incomplete envelopes (*L. flavolineata*) (Fig. 3.9l) and globular nests protected by a
complete mud envelope (*Liostenogaster* sp.) (Pagden 1958a, Turillazzi,
unpubl.) (Fig. 3.9m). According to Hansell (1984a), these solid, heavy
mud nests could have evolved in response to predation by hornets.
One undescribed species of *Liostenogaster* builds a mud nest of two
rows of cells arranged on a thin twig, which it defends with an ant
guard of abdominal substance (Turillazzi, unpubl.) (Fig. 3.9j).

In the species that use vegetable fibers we find simple nests with
cells scattered or arranged in rows on flat substrates (*L. nitidipennis*:
Yoshikawa et al. 1969) (Fig 3.9e), and more advanced ringlike combs
with the cell openings facing the center (*L. vechti*: Pagden 1958a;
Yoshikawa et al. 1969, Ohgushi et al. 1983a, Turillazzi 1988b) (Fig.
3.9f,g). Comb nests made of light carton are built on small branches by
an undescribed species of *Liostenogaster* (Fig. 3.9h), while another undescribed species builds carton comb nests inside hollow tree branches
excavated in part by the wasps (Turillazzi, unpubl.) (Fig. 3.9i).

Only two types of nests belonging to species of *Anischnogaster* are
known at present (Vecht 1972, Spradbery 1989), and thus it is rather
difficult to compare them with nests of the other genera. We know
only the mud nest of *S. concinna* described by Spradbery (1975) for the
genus *Stenogaster*. Its bell-shaped architecture (Carpenter 1988a) is quite
similar to that of the nests of all species of *Eustenogaster*, especially in
the early stages of construction. The paper nests of this latter genus
include a comb of cells (usually fewer than 30) protected by an inverted-flask-shaped envelope that, depending on the species, is gar-
nished with various kinds of external protuberances as camouflage.
The envelope of nests of most *Eustenogaster* is simply an extension of
the external cell walls (pseudenvelope), but *E. calyptodoma* has a true
envelope that is completely separate from the comb of cells (Fig. 3.9r,
see also cover illustration). Some species in this genus also equip their
nests with viscous ant guards (Hansell 1984b, Turillazzi, unpubl.).

*Parischnogaster* builds its nests mainly with vegetable matter. The architecture of the nest of *P. mellyi* seems primitive, with cells arranged
in irregular combs placed on threadlike substrates such as roots (Fig.
3.9s). Hansell (1981) found that this species in Thailand has a tendency
to build linear series of cells along hanging substrates. More regular
combs are exhibited by *P. gracilipes* (Fig. 3.9v). This species, which lives
in mountain forests where predation from hornets is almost absent,
builds unusually large nests for the subfamily, containing up to 54 cells
(Hansell 1986b). Species belonging to the *P. jacobsoni* group (Fig. 3.9t)
and *P. nigricans serrei* (Fig. 3.9w) tend to build linear nests with the
cells arranged along hanging substrates (or on leaves) and defended by

sticky ant guards (Yoshikawa et al. 1969, Turillazzi and Pardi 1982, Ohgushi et al. 1983a, Turillazzi 1988a, unpubl.). As noted by Williams (1928), C. Starr (unpubl.), and Wenzel (this volume), the nest of *P. timida* (Fig. 3.9x) is extremely similar to nests of the two described species of *Metischnogaster* (Fig. 3.9z): cells are attached one under the other and cone-shaped structures, which deflect raindrops, are built from the same vegetable matter as the nest (Pagden 1962). The nests of *P. striatula* (Fig. 3.9u) and *P. depressigaster* have a characteristic spiral shape, whereas the cells of nests of species related to *P. alternata* (Fig. 3.9y) open into a tube formed by the elongations of the first cell walls (Ohgushi and Yamane 1983, Ohgushi et al. 1985, Turillazzi 1986a). Pseudenvelopes that originate as extensions of the cell walls are common in some *Parischnogaster* nests.

In general, the inferred trends in the evolution of stenogastrine nest architecture correspond quite well with the phylogenetic relationships of the genera proposed by Carpenter (1988a, this volume). It is interesting how in various genera different (but functionally equivalent) architectural solutions to common ecological problems have been reached—for example, in the independent evolution of envelopes and pseudenvelopes to protect the nest. On the other hand, identical architectural features also have arisen independently in response to the same ecological pressure, such as the viscous ant guards constructed of abdominal substance. New data and more-detailed studies on the systematics and nest architecture of these wasps should resolve the problem raised by Wenzel (this volume) regarding the great variation in stenogastrine nests relative to the other social vespid taxa, as well as address the validity of the tradition of separating species on the basis of the nest architecture.

## PREDATORS, PARASITES, AND NEST DEFENSE

Insectivorous vertebrates are potential predators of stenogastrine nests; for instance, an undetermined species of *Gekko* probably destroyed some colonies of *P. alternata* (Turillazzi, unpubl.). Detailed information on vertebrate predators is presently not available, however. Early authors observed that these wasps are extremely shy and are not aggressive toward human observers, even though the females have effective stings that they can use for personal defense. The sting, however, is a much less formidable weapon than that of other eusocial wasps. Color patterns of *Eustenogaster* and *Liostenogaster* are more striking than the patterns of other genera and presumably serve some aposematic function. The males of these genera are very similar to the females in their markings and exhibit pseudostinging behavior if disturbed (C. Starr, unpubl.).

Ants and hornets are the principal predators of the brood. I have observed many undetermined species of ants preying on stenogastrine nests, and Samuel (1987) reports the ant *Technomyrmex* as an occasional predator of *L. flavolineata*. *Vespa tropica* and other hornets have been observed preying on various species of Stenogastrinae (*P. mellyi*: Hansell 1982a, Sakagami and Yamane 1983; *P. nigricans serrei*: Turillazzi and Pardi 1982, Turillazzi 1985d; *L. vechti*: Turillazzi 1990). The main adaptations of stenogastrines to predation are choosing nesting sites where ants and wasps are rare or absent, camouflaging the nests, and constructing the viscous ant guards. The ant guards also protect the nest from other terrestrial predators such as nematodes (Turillazzi and Pardi 1982). *Parischnogaster nigricans serrei* (Turillazzi and Pardi 1981) and *P. jacobsoni* (Turillazzi 1988a) work on the ant guard particularly during the early evening hours, presumably to reinforce its effectiveness at night when active defense is more difficult.

One active defensive behavior stenogastrines use against ants looks very similar to that practiced by some polistine wasps. When intruders are discovered on the nest, the adults begin to patrol while simultaneously buzzing their wings. When faced with an ant, the female bends her abdomen laterally toward the intruder and then hits it with the dorsal tip of the abdomen, with the sting extruded, in a sort of abdominal slap. If the ant is large and does not abandon the nest, the females of various genera (including *Parischnogaster*, *Liostenogaster*, and *Eustenogaster*: Turillazzi, unpubl.) grasp the ant and fall with it from the nest. I twice observed a curious defensive behavior of *Metischnogaster* against ants. A single female on a three-cell nest strung along a thin aerial fungus, having tried unsuccessfully to get rid of a large ant that I had placed there, flew off the nest and then landed on it again so heavily that the nest shook violently. On one occasion this behavior managed to make the ant fall off the nest (Turillazzi 1987a).

The response to attacks by predators larger than ants is usually to flee. The adults of *Parischnogaster* drop passively from the nest if disturbed, and take flight a few centimeters below the nest. This behavior may make it more difficult for a predator to detect the exact position of the nest (Turillazzi 1987a).

On the other hand, a solitary foundress will defend her nest vigorously against attempts by conspecifics to usurp it. Females of various species of *Parischnogaster*, *Metischnogaster*, and *Liostenogaster* will attack in flight anything approaching the nest that is not too much bigger than they are (Sakagami and Yamane 1983, Turillazzi 1987a). (I never observed this behavior in *P. alternata* or in *Eustenogaster*.) In a successful attack, the wasps hit the intruder violently at full speed in midair, forcing it to flee. This behavior can easily be elicited by waving dead conspecifics or cardboard models stuck on the end of a stick in front of the nest (Turillazzi, unpubl.). Females of *P. mellyi* perform these attacks against any conspecifics that approach the nest with a hesitant flight, but not against their nestmates, which fly to the nest directly (Sakagami and Yamane 1983). *Polistes* wasps are, similarly, very aggressive toward individuals that approach their nest with a wavering flight, including heavily loaded forager nestmates (West-Eberhard 1969). Resident females of *E. calyptodoma* warn alien ones landing on their nests by use of an acoustic signal produced by beating their abdomens on the nest envelope (Hansell 1987b, Turillazzi, unpubl.)

Only a few parasites are reported from stenogastrine nests, and no particular defensive responses to them have yet been observed. Spradbery (1975) found tachinid flies (hyperparasitized by eulophid wasps) in a nest of *S. concinna* and a single adult wasp of this species with dozens of phoretic mites. Iwata (1967) reported a eulophid wasp of the genus *Syntomosphyrum* as parasitizing *P. striatula*. Williams (1919, 1928) found an ichneumonid fly in a nest of *P. timida*, and I found tachinid flies in a nest of *P. mellyi*. I also discovered quite strange undetermined flies, resembling in their morphology and flight *Parischnogaster* males, near a nest of *P. jacobsoni* (Turillazzi, unpubl.). Samuel (1987) found tachinid and bombyliid flies parasitizing brood of *L. flavolineata*. The resting of dominant females of *P. nigricans serrei* on closed cells and a unique wing-flipping behavior of the adults may be defensive mechanisms against parasitoids (Turillazzi 1987c).

## CONCLUDING REMARKS

Spradbery (1975) and Vecht (1977a) consider the morphological, biological, and ethological differences between Stenogastrinae and Polistinae + Vespinae to be extremely important. Vecht (1977a) maintains that the three subfamilies cannot be considered to be monophyletic, and Spradbery (1975) suggests that the Stenogastrinae originated from an ancient vespoid ancestor. Hansell (1987a), although uncertain as to the monophyletic or polyphyletic origin of the eusocial Vespidae, suggests that the weak jaw anatomy of the adults of Stenogastrinae could be more simply understood if these wasps were considered as directly derived from a mud-building ancestor. Carpenter, using cladistic methods to examine the phylogenetic relationships of the Vespidae (Carpenter 1982) and of the Stenogastrinae in particular (Carpenter 1988a), has concluded that the three eusocial vespid subfamilies actually are monophyletic. Carpenter's hypothesized phylogeny suggests that group living and temporary reproductive division of labor (with some individuals acting as temporary nonreproductive workers) were already characteristic of the common ancestor of the Stenogastrinae and the other eusocial vespids. This implies that the distinctive brood-rearing characteristics of stenogastrines (especially the indirect feeding involving the abdominal secretion), as well as their other peculiar ethological traits, evolved after their split from Polistinae + Vespinae and after the origin of a reproductive division of labor (Carpenter 1988a, this volume). This conclusion rests on the belief that if future studies of this as yet poorly known group reveal exceptions to the characterization of the stenogastrines as uniformly eusocial, these exceptions must represent secondary "reversions" to a more primitive level of social organization.

Presuming that a monophyletic origin of eusociality in the Vespidae is confirmed, we must still raise the question: did the distinctive mode of brood rearing by the Stenogastrinae contribute to the maintenance of group life and/or affect the particular kinds of social organization that evolved in this subfamily? The storage of liquid, highly nutritive food (Turillazzi 1985c) on the abdominal secretion (a substance that may originally have had only protective functions for the eggs and small larvae) and the presence of curled larvae may have been important factors favoring the maintenance of group living, in a sense replacing the direct provisioning, larva-adult trophallaxis, and storage of honey believed to have been important for social evolution in Polistinae and Vespinae. The nest, which was already a focal point for young females searching for ready-made cells in which to oviposit, became even more attractive as a cache of food for adults (Turillazzi 1989a). The use of liquid food for rearing the brood was, perhaps, also linked to the evolution of stimulatory mechanisms for inducing nestmates to regurgitate liquid, and thus may represent a step toward the evolution of communication among adults (Turillazzi 1989a). Once group living became established, the acquisition of complex social characteristics, as exemplified by a division of labor and dominance hierarchies, may have been relatively simple (Barnard 1983, Michener 1985). Once group living and a primitive reproductive division of labor originated, they were shaped independently in the Stenogastrinae and the Polistinae + Vespinae.

Hansell (1985, 1987a) has demonstrated how the construction material of stenogastrine nests is considerably poorer in quality than that of other eusocial wasps and suggested that this was the main factor preventing these wasps from evolving large colonies. Unsuitable material may have been an impediment to the construction of physically large nests, but it is less clear how it could have posed an insurmountable obstacle to the establishment of populous colonies. Species of *Lio-*

*stenogaster* and *Parischnogaster* construct large, dense clusters of nests with populations of hundreds of individuals per cluster, while some species of *Polistes* (Jeanne 1979a) and *Ropalidia* (Itô 1986b) appear to distribute their colonies over several contiguous nests.

Certainly, the rudimentary method of prey mastication by the adults also played a role in retarding social evolution in the Stenogastrinae (Hansell 1987a), but I suspect that the main limiting factor may be identical to the factor presumed to have promoted group living and the origin and/or persistence of eusociality in these wasps, that is, the abdominal secretion. The special form of brood rearing required with the use of this substance may strongly constrain the rate at which larvae can be reared (M. J. West-Eberhard, unpubl.), since the amount of abdominal secretion used is high and its production is limited (see Turillazzi 1985a). The result is that a large number of adults cannot build up on single nests, thus precluding the evolution of complex behavioral interactions. This also could have prevented the evolution of behavioral and morphophysiological mechanisms for active defense against large predators, so that the only defense available was to rely on the limited size of the colonies, nest camouflage, and the choice of particular environments for nesting.

On their own these factors probably limited the complexity of sociality in species that possess nests with specialized architecture and nests that are too small to permit more than a very limited number of adults to reside there at any one time (such as the two species of *Metischnogaster*: Pagden 1962, Turillazzi, unpubl.; *P. timida*: Williams 1919, C. Starr, unpubl.; and various species of *Eustenogaster*: Sakagami and Yoshikawa 1968, Krombein 1976, Hansell 1987b, Turillazzi, unpubl.). On the other hand, in the potential "supercolonies" formed by the cluster-nesting species, social regulatory mechanisms apparently are not sufficiently well developed to permit the emergence of a social organization including more than a single nest (Hansell et al. 1982, Turillazzi 1986a, Samuel 1987). Thus social organization in colonies of stenogastrine species studied so far can be characterized as being at the rudimentary-caste-containing stage of West-Eberhard (1978a) or the primitively eusocial level. Some species exhibit well-defined dominance hierarchies and reproductive division of labor, while others do not. The numerous behavioral options open to females suggest that in some species the transition from permanent subsociality to eusociality could be facultatively expressed in different colonies of the same species (C. Starr, unpubl.). On the other hand, the existence of associative nest foundation and the presence of several egg layers in mature colonies in some species raise questions concerning the possibility of permanent parasociality (Turillazzi 1986b, C. Starr, unpubl.; for definitions of types of social organization, see Cowan, this volume: Table 2.1).

The external and internal determinants likely to influence and limit the behavioral options of females for maximizing their fitness have only been tentatively identified in past studies. These include the probability of subordinates replacing the dominant female, the difficulty of finding good nesting sites, the potential for females to usurp or join colonies where they could become egg layers, and the existence of variable reproductive potential among females. A period of residence on the nest by young females until they reach a suitable stage of reproductive development could have been the prerequisite for the formation of a temporary worker caste, the existence of which could have been further favored by parental manipulation by dominant females (Yamane et al. 1983a; Turillazzi 1985d, 1989a). The differences in ovarian development in females of *P. nigricans serrei* of the same age (Turillazzi 1985d) and the greatly retarded ovarian development of some females of *L. flavolineata* reported by Samuel and Hansell (1987) could signal the onset of a caste of permanent workers. Such hypotheses must be tested in future research.

The stenogastrine wasps, hovering inhabitants of the Oriental rainforests, have evolved a unique kind of group living that represents a gold mine for future evolutionary investigations.

## Acknowledgments

L. Pardi's long discussions with me on the problems in the biology of Stenogastrinae stimulated much of my research. Professor Pardi also read and commented on the first draft of this chapter. J. M. Carpenter, M. H. Hansell, R. W. Matthews, K. G. Ross, and M. J. West-Eberhard devoted a great deal of their time to revising the manuscript and offered many useful suggestions for its improvement. I thank them all for their help and ideas. I also thank Christina Coster Longman for her revision of the English text.

# 4

# *Polistes*

## HUDSON K. REEVE

Since the 1940s, the eusocial paper wasps of the genus *Polistes* have served as model organisms for studies of animal social behavior. Studies of *Polistes* have appeared on the front lines of major socio-biological research efforts, including attempts to understand the nature of animal dominance hierarchies and the importance of ecological factors and kin selection in shaping cooperative interactions among conspecifics.

*Polistes* has been an attractive research target for several reasons. Evans (1958) regarded it as a "key genus" for understanding the evolution of insect eusociality because it exhibits only slight caste differentiation. In mature *Polistes* colonies, as in colonies of other eusocial insects, a reproductive division of labor exists between the queen and her workers. However, compared with workers of the advanced eusocial insects, *Polistes* workers differ physically from the queen only slightly (Eickwort 1969) and have greater opportunities for personal reproduction. Because foundress associations also are characterized by a rudimentary caste differentiation, these associations may offer clues about the context in which caste differences evolved (West-Eberhard 1978a, Carpenter, this volume). Thus, the *Polistes* social system may offer a glimpse of a social stage through which more-advanced eusocial forms have passed. Even if this system does not resemble a phylogenetic intermediate in the evolution of eusociality, and simply represents one in a continuum of evolutionarily stable endpoints, studies of it promise to elucidate how ecological and social pressures determine a species' position on the eusociality spectrum. In addition, because the relatively slight caste differentiation results in a high potential for reproductive conflict among colony members, *Polistes* offers an excellent opportunity to study the forms and consequences of conflict in insect societies (Pardi 1948, West-Eberhard 1969).

The study of *Polistes* societies also has been motivated by many practical advantages. *Polistes* colonies are found throughout the world and often are locally abundant, in both tropical and temperate regions. Because the nests are usually accessible and the combs are unenveloped and relatively small, researchers can study the intricacies of the social dynamics and reproductive consequences of specific social behaviors with relative ease. In addition, *Polistes* colonies are easy to rear in the laboratory, requiring in most cases only a simple container such as a cardboard box, a source of protein (e.g., mealworms) for the larvae, water, and honey. A final key advantage is the plasticity of *Polistes* social behavior. Different individuals in the same population often pursue alternative social strategies (e.g., joining a nest as a subordinate vs. nesting alone). This plasticity within a population allows researchers to analyze empirically the selective forces that operate on alternative social strategies.

The appeal of *Polistes* as a research organism has resulted in a large literature on its social biology; thus, I cannot hope to achieve a comprehensive review. My more limited aim is to summarize and synthesize, within a modern evolutionary framework, information from studies conducted since 1970. During this period, general natural history descriptions have branched into numerous detailed tests of specific evolutionary and mechanistic hypotheses, spanning such diverse topics as the evolution of foundress associations, adaptive significance of sex investment ratios, ecology of mating systems, ontogeny of nestmate recognition, mechanisms of intracolonial communication and activity regulation, and the physiological basis of dominance. A synthesis of this voluminous, and partly disconnected, recent literature is in order, especially given the emerging recognition of potentially important inter- and intraspecific variation in *Polistes* social structures. I will show that although *Polistes* stands as one of the best known eusocial taxa, we are just beginning to appreciate the factors underlying variability in key features of its social system. Introductions to the general biology of *Polistes* (and to the older literature) can be found in West-Eberhard (1969), Evans and West-Eberhard (1970), Wilson (1971), and Akre (1982).

My choice of citations is necessarily selective. I have tried to emphasize studies that address central issues in social evolution and employ rigorous analyses of data. Where possible, I have organized data around selectionist hypotheses about the evolutionary significance of social phenomena. Although I do not wish to deny that nonadaptive hypotheses (e.g., developmental constraint / phylogenetic inertia or neutralist hypotheses) may have to be invoked to explain the forms of particular social behaviors, I believe that selectionist hypotheses are the most illuminating for the majority of the social phenomena that I discuss, especially since my emphasis is on flexible social behaviors whose variants probably have markedly different fitness consequences in different ecological and social settings.

After briefly summarizing the systematics, distribution, nesting biology, and colony cycle of *Polistes*, I discuss recent data bearing on some of the more important issues in *Polistes* sociobiology. The major phases of the colony cycle provide a framework for these discussions; I examine in turn the adaptive significance of foundress associations, the nature of interactions and distribution of reproduction within these associations, worker reproduction and colony dynamics in post-worker-emergence colonies, sex investment ratios and the timing of production of reproductives, mating behavior, and social interactions in winter diapause associations.

## SYSTEMATICS AND DISTRIBUTION

*Polistes* wasps are medium-sized to large, rather stout, generalized vespids. The genus *Polistes* is one of 29 genera in the vespid subfamily Polistinae. Carpenter (this volume) has recently clarified phylogenetic relationships among the genera with a cladistic analysis of morphological and behavioral characters. Akre (1982) states that there are 206 recognized species of *Polistes* (including the three socially parasitic species of "*Sulcopolistes*"). The species-level taxonomy of *Polistes* is not yet well worked out; many of Bequaert's (1940) "color varieties" of North American species have since been recognized as distinct species. Studies of genetic differentiation among putative *Polistes* species have barely begun (Metcalf et al. 1984). Discussions of the taxonomic literature can be found in Wilson (1971), Akre (1982), and Carpenter (this volume).

*Polistes* is distributed throughout the world (Fig. 4.1), although it is most diverse in the Old and New World tropics (Richards 1978a). On the basis of its high paleotropical diversity, many entomologists believe that *Polistes* originated in Southeast Asia, later invaded the New World via Beringia and secondarily radiated into the Neotropics (Vecht 1965). There is an interesting association between *Polistes* chromosome number and geographic location. Eastern Asian species have markedly fewer chromosomes than either European or New World species (Fig. 4.2), although this trend is based on relatively few species. This pattern, in conjunction with a Southeast Asian origin, suggests that invasion of novel geographic regions was accompanied by an increase in chromosome number. This finding may bear on theories that relate the

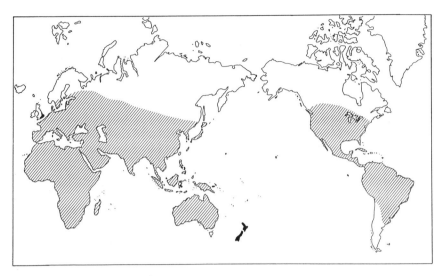

**Fig. 4.1.** Distribution of *Polistes* (Vecht 1967, Richards 1971, Akre 1982). Introduced populations (i.e., in England and New Zealand) are indicated by black shading.

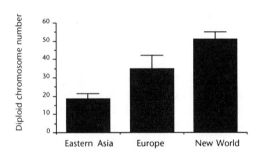

**Fig. 4.2.** Association between *Polistes* chromosome number and geographic location. The mean chromosome numbers (shown with standard errors) of European, eastern Asian, and New World species differ significantly ($P < 0.01$, Kruskal Wallis test). The mean chromosome number of New World species is significantly greater than that of eastern Asian species ($P < 0.05$, multiple comparisons test for Kruskal Wallis). New World species include *P. apachus* ($2n = 50$: Hung et al. 1981), *P. canadensis* ($2n = 32$: Kerr 1952), *P. carolina* ($2n = 38$: Hung et al. 1981), *P. exclamans* ($2n = 66$: Hung et al. 1981), *P. fuscatus* ($2n = 52$: Goodpasture 1974 [in Hung et al. 1981]), *P. metricus* ($2n = 52$: Hung et al. 1981), *P. simillimus* ($2n = 56$: Pompolo and Takahashi 1986), and *P. versicolor* ($2n = 62$: Pompolo and Takahashi 1986). European species include *P. dominulus* ($2n = 42$: Pardi 1942b) and *P. gallicus* (as *P. omissus*) ($2n = 28$: Pardi 1942b). Eastern Asian species include *P. jadwigae* ($2n = 18$: Machida 1934), *P. olivaceus* ($2n = 18$: Misra and Srivastava 1971), *P. rothneyi iwatai* ($2n = 12$: Machida 1934), and *P. snelleni* ($2n = 26$: Machida 1934).

102

evolution of genetic recombination rates to environmental novelty (Maynard Smith 1988).

## NESTING BIOLOGY

*Polistes* nests usually consist of a single petiolate (stelocyttarus), unenveloped (gymnodomous) comb made of chewed plant fibers from weathered wood and other sources (Jeanne 1975a, Richards 1978a), sometimes even including paper from old nests (e.g., tropical *P. erythrocephalus*: Young 1986). The nests vary in architecture from simple serial arrangements of cells suspended from a single petiole (*P. goeldii*) to large horizontal combs supported by multiple petioles (*P. fuscatus*: Downing and Jeanne 1986; see also Wenzel, this volume). Jeanne (1975a) suggested that the function of the narrow petiole is to restrict entry by ants and that it thus represents a good site for efficient deposition of a glandular ant repellent. Subsequent work has shown that females apply to the petiole an ant repellent secretion from a gland on the sixth gastral sternum (Post and Jeanne 1981); one active component of this secretion has been identified as methyl palmitate (Post et al. 1984a). Suzuki (1983) has suggested that foundresses tend to rest near the petiole at night to facilitate detection of ants, although this may also protect foundresses from vertebrate predation or perhaps even serve in egg thermoregulation.

Nest construction behavior of *P. fuscatus* females has recently been studied by Downing and Jeanne (1988, 1990). They found that wasps integrate information from a variety of cues during nest construction. These cues include, but are not restricted to, stimuli emanating from the last completed construction element. For example, construction of the first cell is guided jointly by location of the cell's edges, petiole location, and distance to the substrate. Thus, *Polistes* wasps appear to display greater information-processing abilities than those predicted from classical stigmergy theory (Downing and Jeanne 1988, 1990).

The petiole is almost always angled 90° from the substrate, but other features of the nest, such as the shape of the comb, the position of the petiole on the back of the comb, the frequency of multiple petioles, and the angle between the petiole and first cell, vary among species (Downing and Jeanne 1986). Multiple combs occur occasionally in the temperate-zone species *P. metricus*, *P. exclamans*, and *P. fuscatus* and commonly in the tropical species *P. canadensis*, where they are thought to serve as an antiparasite adaptation (Jeanne 1979a, Downing and Jeanne 1986). Parasite and pathogen pressures, along with physical deterioration of the nest, also presumably prevent frequent reuse of nests between seasons (West-Eberhard 1969, Starr 1978). However, up to 7% of nests of *P. annularis* are built the previous year, perhaps indicating a severe limitation on suitable nest sites (Queller and Strassmann 1988).

Nests occur naturally in or on vegetation, such as under leaves or on tree trunks, and also commonly on man-made structures, such as the undersides of bridges, roofs, or eaves of buildings. There appears to be some interspecific variation in preferred nesting sites (West-Eberhard 1969). For example, Reed and Vinson (1979) report that, in Texas, *P. carolina* favors highly concealed locations such as cavities in trees, whereas *P. exclamans* commonly nests in well-lit locations.

Mean sizes of mature nests (as measured by total cell number) exhibit at least an eightfold variation, ranging from 56 (*P. metricus* in North Carolina: Rabb 1960) to 492 cells (*P. annularis* in North Carolina: Downing and Jeanne 1986). This variation comprises a substantial intraspecific component (Downing and Jeanne 1986), as well as an interspecific component. However, mean nest sizes of most species are closely clustered around the overall mean for the genus, about 135 cells (Fig. 4.3). The largest reported nest of *Polistes* had 1,886 cells (*P. annularis*: Nelson 1968). As has been suggested previously (Evans and West-Eberhard 1970) but not documented, there is no significant correlation between the mean size of mature nests and latitude (Fig. 4.4); thus, despite experiencing longer periods of warmth, subtropical and tropical colonies do not tend to grow larger than temperate-zone colonies.

The similarity in mature nest sizes for temperate-zone and tropical colonies reflects similarity in durations of the colony cycles. The colony cycle of the tropical *P. erythrocephalus* (six to seven months) is only one to three months longer than the colony cycles of temperate-zone species (Evans and West-Eberhard 1970). The longevities of tropical colonies may be entirely determined by chance mortality events. For example, the frequency distribution of sizes of abandoned colonies of the tropical *P. erythrocephalus* (West-Eberhard 1969, fig. 20) is nicely fitted by the negative exponential expected if colony longevities were determined by random predation events. It is also possible that selection has programmed colony dissolution at six to seven months as an optimal trade-off between the chances of catastrophic predation before reproduction and number of reproductives produced if the colony survives. To test these hypotheses, more data are needed on colony growth patterns, colony cycle durations, predation pressures, and schedules of reproduction in a variety of tropical species (Jeanne 1982a).

## COLONY CYCLE

For convenience, I divide the colony cycle into four phases. (1) In the *founding phase* (also termed *preemergence phase*), young reproductive fe-

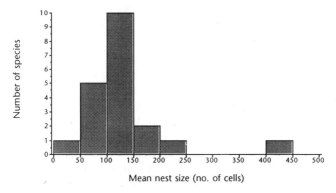

**Fig. 4.3.** Histogram of mean mature nest sizes (total cells) of *Polistes* species. Mean nest sizes from different studies of the same species were averaged. Mean sample size per species is 57 colonies. Ranked from smallest to largest mean nest size, the 20 species include *P. canadensis canadensis* (32 cells: Downing and Jeanne 1986), *P. gigas* (60 cells: Matsuura 1970a), *P. metricus* (74 cells: Rabb 1960; Nelson 1966 [in Akre 1982], 1968), *P. rothneyi iwatai* (74 cells: Yamane 1972), *P. erythrocephalus* (95 cells: West-Eberhard 1969), *P. dorsalis hunteri* (99 cells: Downing and Jeanne 1986), *P. riparius* (= *biglumis*) (100 cells: Yamane and Kawamichi 1975), *P. perplexus* (103 cells: Downing and Jeanne 1986), *P. carolina* (108 cells: Reed 1978 [in Akre 1982], Downing and Jeanne 1986), *P. peruvianus* (112 cells: García 1974), *P. lanio* (115 cells: Downing and Jeanne 1986), *P. instabilis* (128 cells: Downing and Jeanne 1986), *P. spilophorus* (138 cells: Richards 1969), *P. humilis* (140 cells: Cumber 1951), *P. fuscatus* (145 cells: Rabb 1960, Downing and Jeanne 1986), *P. dominulus* (149 cells: Turillazzi 1980); *P. exclamans* (162 cells: Rabb 1960; Nelson 1966 [in Akre 1982], 1968), *P. tepidus* (164 cells: Yamane and Okazawa 1977), *P. chinensis antennalis* (236 cells: Suzuki 1981a, Miyano 1983), and *P. annularis* (443 cells: Nelson 1966 [in Akre 1982],1968; Downing and Jeanne 1986).

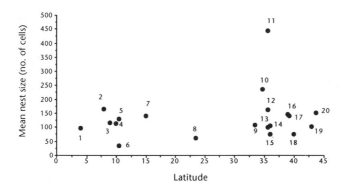

**Fig. 4.4.** Absence of association between mean mature nest size (total cells) of *Polistes* species and latitude (degrees from equator) (Pearson's $r = 0.21$; $P > 0.10$). Nest sizes and latitudes from separate studies of the same species were averaged (see Fig. 4.3 legend for nest size references). (1) *P. erythrocephalus*, (2) *P. tepidus*, (3) *P. lanio*, (4) *P. peruvianus*, (5) *P. instabilis*, (6) *P. canadensis canadensis*, (7) *P. spilophorus*, (8) *P. gigas*, (9) *P. carolina*, (10) *P. chinensis antennalis*, (11) *P. annularis*, (12) *P. exclamans*, (13) *P. dorsalis hunteri*, (14) *P. perplexus*, (15) *P. metricus*, (16) *P. fuscatus*, (17) *P. humilis*, (18) *P. rothneyi iwatai*, (19) *P. riparius* (= *biglumis*), (20) *P. dominulus*.

105

males (foundresses) initiate new nests, either singly (haplometrosis) or jointly with auxiliary foundresses (pleometrosis). In temperate-zone regions, this phase usually begins in early to late spring, shortly after the foundresses emerge from winter diapause. In tropical regions, founding begins shortly after the young foundresses disperse from their natal nests and can occur at any time of the year (West-Eberhard 1969, Yamane and Okazawa 1977, Young 1986). Mating may occur early in this phase in tropical colonies (West-Eberhard 1969). (2) The *worker phase* begins when the first adult workers eclose and lasts until the eclosion of the first reproductives (future foundresses, also known as gynes, and males) late in the colony cycle. (Production of "early" males along with the first workers is included in the worker phase rather than in the reproductive phase, because early males apparently mate primarily with workers that have replaced queens, not with future foundresses.) In temperate-zone areas, this period usually extends from late spring or early summer to early or mid summer. (3) The *reproductive phase* begins with emergence of the first reproductives and lasts until the period of colony decline (early or mid fall in temperate-zone colonies), at which time the reproductives begin to disperse from their natal nests.* (4) The *intermediate phase* encompasses the period between colony decline and the founding of new colonies. For temperate-zone colonies, this phase usually includes the period during which mating between males and gynes as well as winter diapause of gynes occur. The intermediate phase for tropical colonies may be quite short (West-Eberhard, 1969).

Temperate-zone colonies pass through these phases at roughly the same time, whereas tropical colonies display considerable asynchrony in development (West-Eberhard 1969, Young 1986, see also Gadagkar, this volume). As discussed below, the degree of synchrony may have important consequences for *Polistes* social behavior.

## FOUNDING PHASE

### Foundress Strategies: Founding, Joining, and Usurpation

In the founding phase, solitarily nesting females typically construct and oviposit in combs of 20–30 cells (Owen 1962, West-Eberhard 1969). When the first larvae hatch, the foundress temporarily suspends oviposition and progressively provisions the larvae with malaxated cater-

*The worker and reproductive phases are sometimes jointly referred to as the *postemergence phase*, which includes the period between first worker eclosion and colony dissolution.

pillar flesh. When several of the larvae (future first-generation workers) have pupated, the foundress resumes oviposition in empty cells.

A *Polistes* foundress has at least two reproductive options besides solitary nest founding. She can join conspecific females on another nest or attempt to take over a nest initiated by a conspecific female. The selective factors promoting joining have been the subject of intense study, especially since the co-occurrence of joining and solitary nesting in the same population presents an excellent opportunity to test or apply Hamilton's inclusive fitness theory (Hamilton 1964a,b, 1972; West-Eberhard 1975). An essential first step in inclusive fitness analyses of joining has been the estimation of genetic relatednesses among co-foundresses.

Substantial behavioral evidence indicates that co-founding females tend to be closely related. Future foundresses (gynes) marked on their natal nests in autumn are subsequently found to associate on spring nests in temperate-zone species (West-Eberhard 1969; Klahn 1979; Strassmann 1979, 1983; Noonan 1981; Gamboa 1988). Two behavioral mechanisms facilitate the association of former nestmates. First, foundresses tend to be philopatric, i.e., they return to the vicinities of their natal nests to found new nests in both temperate-zone and tropical species (West-Eberhard 1969; Klahn 1979; Krispyn 1979; Strassmann 1979, 1981b, 1983; Metcalf 1980; Hirose and Yamasaki 1984a; Young 1986; Makino et al. 1987; Gamboa 1988; Willer and Hermann 1989). However, some marked foundresses have been observed to disperse over 300 m from their natal nests (Makino et al. 1987). Philopatry may reflect selection for return to nesting sites of proven quality; for instance, Metcalf and Whitt (1977b) reported higher foundress nest failure at sites away from natal nest sites. A second mechanism promoting association of relatives is nestmate recognition. Foundresses are aggressive toward nonnestmates and, given a choice, preferentially cooperate with natal nestmates in building spring nests (Noonan 1981, Ross and Gamboa 1981, Post and Jeanne 1982a, Bornais et al. 1983, Strassmann 1983). Extensive laboratory studies (e.g., Shellman and Gamboa 1982, Pfennig et al. 1983a, Gamboa et al. 1986a) have shown that the nestmate recognition mechanism involves phenotype matching (Sherman and Holmes 1985). In the first few hours after eclosion, foundresses learn both environmental and heritable recognition odors from their natal nests, then later use a template based on memories of these odors to identify former nestmates (see Gamboa et al. 1986b for a review of nestmate recognition studies of *Polistes*).

Foundress associations are not always composed only of former nestmates. Hirose and Yamasaki (1984b) report that in one multiple-foundress association in the Japanese species *P. jadwigae*, five of six co-foundresses were marked on three different nests the previous fall.

Associations of probable nonrelatives have been documented at low frequencies in *P. fuscatus* (Klahn 1979, Noonan 1981) and *P. exclamans* (MacCormack 1982). Willer and Hermann (1989) found that former nestmates initiated nests together infrequently in a viscous population of *P. exclamans*; however, subsequent nest switching may have brought close relatives together. Associations of foundresses of low relatedness may result from recognition errors (Reeve 1989) or unavailability of highly related individuals for joining (joining nonrelatives could be advantageous provided joiners have significant prospects for personal reproduction in such associations: Klahn 1979).

Allozyme-based estimates of mean relatedness among co-founding females are now available for 14 *Polistes* species (see Ross and Carpenter, this volume: Table 13.2). In almost all cases, relatedness among females on mature nests was measured. Thus, these values reflect actual co-foundress relatedness only if it is assumed that foundress associations are random samples from the pool of gynes emerging on the same nest, that is, that foundresses do not preferentially associate with more closely related former nestmates. Queller et al. (1990) reported that such preferential association does not occur in *P. annularis*. With the assumption of no preferential association, the mean co-foundress relatedness across species is about 0.51, ranging from 0.31 (*P. annularis*) to 0.80 (*P. gallicus*).

The estimation of the coefficient of genetic relatedness ($r$) among co-foundresses is only the first step in understanding the selective value of foundress joining. Let $p$ be the proportion of the direct reproduction attributable to the joiner in an association; $B_j$ and $B_u$, the expected productivities of a surviving nest that had been joined or not joined, respectively, by the potential joiner; $B$, the expected productivity of a surviving nest founded solitarily by the potential joiner; $s_j$, $s_u$, and $s$, respectively, the survival probabilities of joined, not joined, and solitary (haplometrotic) nests; and $e$, the probability of successful establishment of a solitarily founded nest. According to Hamilton's rule (Hamilton 1964a, Grafen 1984), joining will be favored over nesting when

$$(ps_jB_j - esB) + r[(1 - p)s_jB_j - s_uB_u] > 0.$$

Thus, the benefits of joining will depend on a number of social and ecological factors underlying these variables (West-Eberhard 1975). Several studies have been directed at determining which of these factors are most crucial in favoring joining. I next consider the evidence for the seven (not mutually exclusive) leading candidates: joiner subfertility; joint defense against predators, parasites, or conspecifics; ergonomic synergism; survivorship insurance; and nest site limitation.

1. *Subfertility (low B, high B<sub>j</sub>)*. West-Eberhard (1967, 1969, 1975) advanced the hypothesis that joiners have, on average, lower reproductive potential than do dominant foundresses (perhaps because of inadequate larval feeding) and thus are selected to aid more fecund (dominant) relatives rather than to nest on their own. Alexander (1974) and Gibo (1978) proposed that queens (and, presumably, workers) even may have been selected to inadequately feed some future foundresses to ensure that these females would join and aid their sisters, thereby maximizing the total number of the queen's grand-offspring (or the workers' nieces and nephews).

As has been pointed out by Sullivan and Strassmann (1984) and Queller and Strassmann (1988, 1989), there is presently little support for these hypotheses. Subordinate foundresses indeed tend to be smaller than dominant foundresses, but they are significantly smaller than solitary foundresses in only some species (Dropkin and Gamboa 1981, Noonan 1981). In any case, smaller sizes of subordinates may only reflect size-correlated asymmetries in fighting ability, and do not necessarily indicate inherent differences in fecundity. Röseler (1985) found no significant differences in fecundity among former $\alpha$, $\beta$, and $\gamma$ foundresses of *P. dominulus* ( = *gallicus* of authors)* that were separated shortly after formation of the association and allowed to start new nests with subordinate helpers in the laboratory. In a recent field study (Queller and Strassmann 1989), both large and small *P. bellicosus* subordinate foundresses that were forced to maintain nests alone did not differ from similarly treated dominant foundresses in numbers of brood reared or in longevities of their nests. However, since this experiment was conducted during the period when oviposition is reduced as larvae are being reared, it is possible that fecundity differences between dominant and subordinate foundresses were not yet detectable.

The well-documented differences in ovarian development between the dominant foundress and her subordinates (Pardi 1948, Gervet 1962, Turillazzi and Pardi 1977, Dropkin and Gambo 1981, Turillazzi et al. 1982, Queller and Strassmann 1989) become elaborated after the associations are formed (Röseler et al. 1980, Röseler, this volume). This pattern may reflect declining investment in direct reproduction by losers in social competition and/or increasing investment in direct reproduction by winners (see West-Eberhard 1981). The tendency for dominant wasps to have slightly larger ovaries than subordinate wasps even be-

---

*A misidentification earlier in this century led to a three-way nomenclatural confusion about some European *Polistes* (Day 1979). Until very recently, the name *P. gallicus* was applied to a species that should have been called *P. dominulus*; and to complicate matters further, *P. gallicus* is the valid name for the species that has been called *P. foederatus*. Virtually all papers on "*P. gallicus*" published before 1988 in fact deal with *P. dominulus*.

fore the hierarchies are established (Röseler et al. 1984) may simply reflect a wasp's prior, partial assessment of her own competitive ability, not necessarily a causal effect of ovarian condition on reproductive strategy. For example, gynes might indirectly assess their competitive abilities through interactions with nestmates at their natal nests or with conspecifics in diapause aggregations (just before or after overwintering), then allocate energy reserves to reproductive development based on these assessments. Gibo (1974) provided evidence that a certain percentage of captive females fail to initiate nests (see also Gadagkar, this volume), but this result implies only a decision not to nest under the unnatural stimuli of the laboratory, not reduced fecundity. Noonan (1981) found that size of solitary-foundress nests in *P. fuscatus* was positively correlated with foundress body size in one year, but not in another. Haggard and Gamboa (1980) found no such correlation in *P. metricus*. Experimental deprivation of honey stores used by *P. annularis* foundresses reduced their survivorship and fecundity, but did not cause them to form larger foundress associations than controls (Strassmann 1979), contrary to the prediction of the subfertility hypothesis.

Finally, there is no evidence of differential feeding of gynes to produce a class of small joiners. Sullivan and Strassmann (1984) showed that the variance in size among *P. annularis* foundresses on spring nests was not significantly greater than that of all females from their natal nests (indicating that associations form randomly with respect to size) and that most of the variance in size of fall reproductive females was between nests rather than within them (which fails to support the hypothesis of strong differential feeding within colonies). In addition, Haggard and Gamboa (1980) found that fall gynes of *P. metricus* were not more variable in size than were workers. In sum, the subfertility hypothesis has received no decisive support, but should be further tested given that there is evidence of potentially important variation among diapausing gynes in physiological attributes such as protein levels in the hemolymph (Strambi et al. 1982).

2. *Joint defense against predators (high $s_j$; low s, $s_u$)*. *Polistes* colonies in many study populations are commonly preyed upon by vertebrate predators, both avian (Gibo 1978, Gibo and Metcalf 1978, Strassmann 1981b,c, Cervo and Turrillazzi 1985, Strassmann et al. 1988) and mammalian (raccoons, opossums, and foxes: Yamane and Kawamichi 1975, Strassmann et al. 1988). Vertebrate predators typically remove the entire nest carton and consume the brood (Strassmann et al. 1988). Ant predation has been recorded in some species (Richards and Richards 1951, West-Eberhard 1969, Miyano 1980, Strassmann 1981b, Strassmann et al. 1988), and, in eastern and southern Asia, colonies are also subject to predation by the hornet *Vespa tropica* (Miyano 1980). Post et al. (1984b) found that colony members are stimulated to attack objects

coated with *Polistes* venom, suggesting that wasps may be recruited to repel intruding predators. (Starr [1990] discusses the variety of anti-predator threat displays in *Polistes*.)

Despite these suggestive data, no study has yet documented that multiple-foundress colonies are more resistant to nest predation than are solitary-foundress colonies. Strassmann (1981b), Gibo (1978), Gamboa (1978), and Strassman et al. (1988) found no differences in rates of nest predation between solitary- and multiple-foundress colonies or between small and large worker-phase colonies.

Gibo (1978) is sometimes cited as providing evidence that the higher postpredation renesting success of multiple-foundress *P. fuscatus* colonies may be a major advantage of co-founding. However, this finding may simply reflect a larger association's greater probability of having at least one foundress survive a predator attack or the period before renesting; this alone does not provide a selective advantage for joining (see Appendix 4.1). In addition, wasps in Gibo's study were not marked, so it is not known whether solitary foundresses whose nests were lost to bird predation might have recouped their losses by joining or usurping other colonies. Strassmann et al. (1988) found that larger colonies (foundresses plus some workers) of *P. bellicosus* were more likely to renest after nest removal, and that marked foundresses rarely joined or usurped other nests in the study area after nest removal. Again, however, the higher renesting probability of larger associations does not necessarily favor joining if the probabilities of individual survival between predation and renesting are independent of original colony size, as they appear to be (Strassmann et al. 1988: fig. 2A; see also Appendix 4.1). Quantitative studies of the effect of foundress association size on ant predation are needed for tropical species, since indirect evidence suggests that ant predation on social wasp colonies may be especially heavy in the tropics (Jeanne 1979b).

3. *Joint defense against brood parasites (high $s_j$; low s, $s_u$)*. *Polistes* colonies are attacked by a variety of brood parasites, especially pyralid and cosmopterigid moths, sarcophagid flies, and ichneumonid and eulophid wasps (Nelson 1968, Gamboa 1978, Miyano 1980, Strassmann 1981c, Cervo and Turrillazzi 1985). Wasps exhibit alarm behavior (darting movements and "wing flipping") when a brood parasite is detected (West-Eberhard 1969). However, Gamboa (1978) found no evidence that larger foundress associations are more resistant to parasitism. Strassmann (1981c) found that larger worker-phase colonies of *P. exclamans* were *more* heavily attacked (per capita) by the parasitic wasp *Elasmus polistis* and the pyralid moth *Chalcoela iphitalis*.

4. *Joint defense against conspecifics (high $s_j$; low s, $s_u$)*. Gamboa (1978) and Klahn (1988) have shown that multiple-foundress colonies of *P. metricus* and *P. fuscatus*, respectively, are better able to resist usurpation

by conspecific females than are solitary-foundress colonies, evidently because (1) multiple-foundress nests are unattended for shorter periods of time than are solitary-foundress nests and (2) two or more females can simultaneously repel usurpers in multiple- but not solitary-foundress colonies (Gamboa et al. 1978). However, Queller and Strassmann (1988) found no differences in usurpation rates among different-sized foundress associations of *P. annularis*.

Factors 2, 3, and 4 are similar in their focus on the enhanced ability of larger associations to detect and repel threats to the colony. Although these factors have been emphasized in discussions of the adaptive significance of group nesting in the Hymenoptera (e.g., Lin and Michener 1972, Evans 1977a), in *Polistes* foundress associations only usurpation resistance has been shown to be important, and even then only in certain populations.

5. *Ergonomic synergism (high $B_j$).* Joining might be favored if multiple-foundress colonies are more ergonomically efficient than solitary-foundress colonies, perhaps as a consequence of division of labor. However, several studies have found that the per capita colony productivity of surviving colonies remains the same or declines as association size increases (Owen 1962, West-Eberhard 1969, Gamboa 1978, Gibo 1978, Noonan 1981, Hirose and Yamasaki 1984b, Queller and Strassmann 1988). Strassmann et al.'s (1988) conclusion that larger *P. bellicosus* colonies have a per capita advantage in nest success was based on an analysis that included colony survival probability in the calculation of per capita reproduction; among surviving colonies there is no consistent increase in per capita colony productivity with increasing colony size (Strassmann et al. 1988: fig. 2C). Gamboa (1980) found no evidence that *P. metricus* foundresses act synergistically to accelerate brood development. Lowered per capita reproductive efficiency with increasing colony size is a general phenomenon in social insects (Michener 1964).

6. *Survivorship insurance (high $s_j$; low $s_u$).* Several authors have noted that a major cause of colony failure in the founding period is the loss of all adult colony members (probably through foraging-related mortality: Strassmann 1981b) and that larger foundress associations are less likely than smaller ones to fail for this reason (Metcalf and Whitt 1977b, Gibo 1978, Noonan 1981, Strassman 1981b,c, Hughes and Strassmann 1988a, Queller and Strassmann 1988, Page et al. 1989a, Queller 1989, Strassmann and Queller 1989). Joining can be strongly advantageous if solitary-foundress survivorship is sufficiently low and if the foundresses who are joined experience increased survivorship (e.g., by foraging less: Gamboa et al. 1978; see also Appendix 4.2). Indeed, available data indicate that queens with subordinates have higher survivorship than queens without subordinates (Gamboa 1978, Pfennig and Klahn 1985).

This "survivorship insurance" advantage differs from a predator defense advantage in that it involves a reduction not in the risks of whole-nest predation, but in the risks that the colony work force will be reduced to zero because of predation on individuals.

7. *Nest site limitation (low e).* Joining is favored if there is a severe limitation in suitable nest sites. Queller and Strassmann (1988) suggested that the relatively high levels of nest reuse by *P. annularis* nesting on cliff faces may reflect nest site limitation, which could partly account for the unusually large foundress associations in their study population.

In summary, the principal selective factors maintaining a joining strategy are still incompletely understood (especially in tropical species), but, at present, defense against usurpation, survivorship insurance, and, perhaps, nest site limitation appear to be the best candidates for at least some species. Of these, the survivorship insurance advantage appears to provide the most general adaptive explanation for co-founding (a conclusion reached also by Strassmann and Queller [1989]). Of course, it is essential to know whether Hamilton's rule is satisfied when these and other factors (such as $p$, the joiner's share of direct reproduction) are simultaneously taken into account.

To date, four studies of *Polistes* have attempted to test Hamilton's rule with respect to foundress joining decisions. These studies either have assumed that the ecological constraint parameter $e$ (the probability of successful establishment of a solitarily founded nest) equals one (Metcalf and Whitt 1977b, Noonan 1981, Grafen 1984) or have used Hamilton's rule to predict what $e$ must be in order to explain joining (Queller and Strassmann 1988). Metcalf and Whitt's (1977b) inclusive fitness analysis showed that joining a single foundress was at least as favorable as haplometrotic founding in a *P. metricus* population. Noonan (1981) applied Hamilton's rule by pooling decisions to join different-sized associations into a single "joining" decision, and found that Hamilton's rule was satisfied for *P. fuscatus* in two out of three years as long as $r$, the relatedness between foundresses, was greater than 0.375. Grafen (1984) reanalyzed Noonan's data by considering separately decisions to join different-sized associations and found that joining was favored only when a solitary-foundress colony is joined (provided $r > 0.48$), yet associations of up to nine foundresses were observed. Queller and Strassmann (1988) found a similar pattern for *P. annularis*: Hamilton's rule (with $e = 1$) was satisfied only for associations of two and three foundresses in one year, which together accounted for less than 15% of all foundress associations in both study years. (Strassmann [1981a] also conducted an inclusive fitness analysis of *P. annularis* joining, but the inclusive fitness calculations assumed that each foundress contributes a constant amount and thus overestimated the benefits of joining.) Queller and Strassmann suggested that their results might be explained by nest site limitation (low $e$) in their study population, but they acknowledge that decisive support for this hypothesis is lacking. It is also possible that some foundress associations of *P. fuscatus* and *P. annularis* are composed of joiners that have lost incipient nests (as described in Strassmann 1983, Itô 1984a, Strassmann et al. 1987, Klahn 1988). These later joiners thus may face low expected productivities if they renest (low $B$), either because of depleted energy reserves or loss of time for colony growth. Time for colony growth may be an important variable, because colonies that are forced to renest after bird predation suffer marked losses in final productivity (Gibo 1978, Klahn 1988, Strassmann et al. 1988). This may result in an increase in the relative benefits of joining versus solitary founding as the founding period progresses (see Appendix 4.2). Even if productivity and energy losses were not significant for late-season foundresses, they might still gain by joining because their brood care efforts would less likely be wasted in a high-survivorship, multiple-foundress colony than in a low-survivorship, solitarily founded nest (Queller 1989). Accumulation of joiners throughout the founding period has been documented for a number of species, both tropical and temperate (West-Eberhard 1969, Gamboa and Dropkin 1979, Noonan 1981, Lorenzi and Turrillazi 1986, Klahn 1988). It is imperative that the expected productivities of solitarily founded and joined nests ($B$ and $B_j$) be estimated empirically for these later joiners.

An analysis of interpopulation variation in the sizes of foundress groups may also shed light on the selective factors favoring joining. Beginning with Ihering (1896a), numerous authors have asserted that the relative frequency of multiple-foundress associations in *Polistes* follows a latitudinal gradient, with co-founding occurring most frequently in the tropics. A correlational analysis of co-founding frequency and latitude, based on studies with adequate samples, supports this trend (Fig. 4.5). However, this result should be intrepreted with caution because it might arise from a sampling bias, particularly if tropical species with higher frequencies of associative founding are more likely to be found or reported by investigators (a potential problem, since data exist for a much greater fraction of temperate-zone than of tropical species). West-Eberhard (1969) suggested that such a correlation might arise from a greater variance in reproductive condition, and consequent larger fraction of potential joiners, among females in the tropics. However, the evidence against the subfertility hypothesis makes this explanation unlikely. Hamilton (1964b) proposed that greater nesting asynchrony in the tropics results in higher levels of inbreeding, which promotes joining by increasing genetic relatedness between joiners and their kin. No genetic data on tropical species exist to test this hypoth-

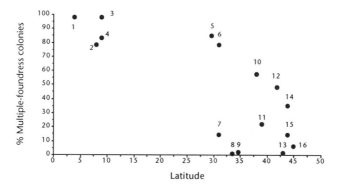

**Fig. 4.5.** Association between percentage of multiple-foundress colonies of *Polistes* and latitude (degrees from equator) (Spearman's $r = -0.76$; $P < 0.01$). (1) *P. erythrocephalus* (West-Eberhard 1969), (2) *P. tepidus* (Yamane and Okazawa 1977), (3) *P. canadensis* (Itô 1986a), (4) *P. versicolor* (Itô 1986a), (5) *P. bellicosus* (Strassmann et al. 1987), (6) *P. annularis* (Queller and Strassmann 1988), (7) *P. exclamans* (Strassmann 1981b), (8) *P. jadwigae* (Hirose and Yamasaki 1984b), (9) *P. chinensis antennalis* (Hoshikawa 1979), (10) *P. apachus* (Gibo and Metcalf 1978), (11) *P. metricus* (Bohm and Stockhammer 1977, Gamboa 1978), (12) *P. fuscatus* (West-Eberhard 1969, Gibo 1978, Metcalf 1980, Klahn 1981, Noonan 1981), (13) *P. riparius* (= *biglumis*) (Makino and Aoki 1982), (14) *P. dominulus* (Pardi 1942a, Turillazzi et al. 1982), (15) *P. nimpha* (Cervo and Turillazzi 1985), (16) *P. biglumis bimaculatus* (Lorenzi and Turillazzi 1986).

esis. Jeanne (1979b) showed that ant predation on social wasps is potentially greater in tropical than in temperate regions. Although it is tempting to believe that heavy ant predation favors co-founding in the tropics (Itô 1986a), as mentioned previously there is no evidence that multiple-foundress associations are more resistant to ant predation than are solitary-foundress colonies.

All the leading ecological hypotheses for the selective basis of joining predict higher frequencies of co-founding in habitats where survival of solitary-foundress nests is relatively low. Indeed, among all species, the relative frequency of multiple-foundress associations is significantly positively correlated with the probability of solitary-foundress nest failure (Fig. 4.6). This relationship may be partly attributable to the joining of nests by former solitary foundresses that have lost their nests (Strassmann et al. 1987, Klahn 1988), but this would still indicate a central role for ecological factors in promoting the evolution of multiple-foundress associations. To further pinpoint critical ecological factors, correlational studies such as those described above should be coupled with an examination of components of colony reproduction after *experimental manipulation* of foundress number. In such experiments, one could control for the possibility that naturally occurring larger associations have enhanced overall per capita reproduction simply because of the attraction of more foundresses to higher quality nesting sites.

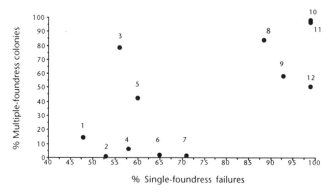

**Fig. 4.6.** Association between percentage of multiple-foundress colonies of *Polistes* and percentage of single-foundress failures in the founding (preemergence) period (Pearson's $r = 0.69$; $P < 0.05$). Data include only results of studies in which both the percentage of single-foundress failures and the percentage of multiple-foundress colonies were estimated for colonies in the same general area. These values were averaged across multiple studies of the same species. (1) *P. exclamans* (Strassmann 1981b, data in 1981c), (2) *P. jadwigae* (Miyano 1980, Kasuya 1981a), (3) *P. annularis* (Queller and Strassmann 1988; percentage of single-foundress nest failures in founding period calculated from fig. 6.2, under the assumption that 66% of all single-foundress failures up to the period of reproduction occurred before worker emergence [see Queller and Strassmann 1988:84]), (4) *P. biglumis bimaculatus* (Lorenzi and Turillazzi 1986), (5) *P. fuscatus* (Gibo 1978, Klahn 1981, Noonan 1981), (6) *P. chinensis antennalis* (Hoshikawa 1979, Miyano 1980), (7) *P. riparius* (= *biglumis*) (Yamane and Kawamichi 1975, Makino and Aoki 1982), (8) *P. bellicosus* (Strassmann et al. 1987; single-foundress failure probability in founding period calculated from failure probability for a nine-day period, with constant failure probability and 40-day founding period assumed), (9) *P. apachus* (Gibo and Metcalf 1978), (10) *P. canadensis* (Itô 1986a), (11) *P. versicolor* (Itô 1986a), (12) *P. metricus* (site 1; Gamboa 1978).

A foundress has another option besides nesting alone or joining relatives. She can also attempt to take over an existing nest, either by acquiring a newly abandoned nest (Kasuya 1982) or by actively usurping a nest from another foundress. This latter strategy has been reported throughout the genus (Yoshikawa 1955; Gamboa 1978; Klahn 1981, 1988; Strassmann 1981a,b; Makino 1985; Lorenzi and Turillazzi 1986; Queller and Strassmann 1988). Risk of usurpation seems to be highest in the latter part of the founding phase, just before or (less often) after the first workers emerge (*P. metricus*: Gamboa 1978; *P. exclamans*: Strassmann 1981b; *P. riparius*: Makino 1985; *P. fuscatus*: Klahn 1988). Evidence is accumulating that at least some usurpers are former solitary foundresses that have lost their own nests (Gamboa 1978, Strassmann 1981a,b, Makino 1985, Klahn 1988). Some usurpers displace females from neighboring nests, but the majority evidently leave the vicinity before attempting takeovers (Makino 1985, Klahn 1988). Since nesting foundresses tend to be strongly philopatric, this dispersal probably minimizes the probability that close relatives will be usurped (Klahn 1988).

Usurpation usually involves aggression directed by the usurper toward the resident foundress, ranging from biting to the most intense form of aggression, "falling fights" (Fig. 4.7), in which the owner and usurper grapple with, bite, and sting each other, often fatally (West-Eberhard 1969; Gamboa 1978; Klahn 1981, 1988). Surprisingly, available evidence suggests that successful usurpers do not tend to be larger than displaced females (Klahn 1981). In *P. fuscatus*, usurpers appear especially likely to displace residents of smaller nests (Klahn 1981), perhaps because residents of these nests have less to gain by engaging in potentially injurious combat with usurpers. Once usurpation has occurred, the usurper destroys the younger brood (eggs and small larvae), but retains the more advanced larvae and pupae (Klahn 1981, 1988; Klahn and Gamboa 1983; Makino 1985). Klahn and Gamboa

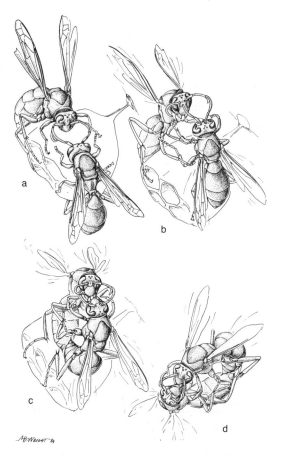

**Fig. 4.7.** Development of a "falling fight" between two *Polistes* foundresses during attempted usurpation. (a) The females antennate, (b) raise up on their hind legs, (c) grapple with front legs, with mutual stinging attempts, and (d) fall from the nest together.

(1983) have suggested that this brood destruction pattern selectively eliminates the brood most likely to develop into reproductives and spares the brood most likely to develop into workers.

Solitary-foundress colonies are significantly more subject to usurpation than are multiple-foundress colonies in some species (*P. metricus*: Gamboa 1978; *P. fuscatus*: Klahn 1988), but apparently not in others (*P. annularis*: Queller and Strassmann 1988). Consequently, there may be variation in the extent to which multiple-foundress associations offer resistance to usurpation. The overall incidence of usurpation varies both inter- and intraspecifically; the mean number of usurpations per colony ranges from fewer than 0.10 (*P. annularis*: Queller and Strassmann 1988; *P. biglumis bimaculatus*: Lorenzi and Turrillazzi 1986; *P. carolina* and *P. bellicosus*: Hughes and Strassmann 1988a; low-density populations of *P. metricus*: Gamboa 1978) to near 0.20 (*P. fuscatus*: 'Klahn 1981) to greater than 1 (high-density populations of *P. metricus*: Gamboa 1978).

## Foundress Strategies: A Synthesis

The relative frequencies with which foundresses nest alone, join related foundresses, or usurp nests of unrelated foundresses vary markedly among and within species. Can we make sense of this variation? A comprehensive theory of foundress strategies should predict how the relative fitness payoffs of solitary founding, joining, and usurpation will vary in different ecological circumstances, both between populations and over the nesting season within a population (e.g., when solitary foundresses lose their nests at different times in the pre-emergence period).

The necessary components for such a theory are already present. The fitness payoffs for solitary founding and joining can be derived from inclusive fitness theory (Hamilton 1964a,b, West-Eberhard 1975) coupled with recent models of the degree to which a dominant reproductive should skew the group's direct reproduction in its own favor (Emlen 1982a,b, Vehrencamp 1983a,b). The latter "optimal skew" models, originally developed to explain the distribution of reproduction within cooperatively breeding vertebrate groups, are useful in the present context because the payoffs for joining will depend in part on the share of the direct reproduction allotted to a subordinate foundress by the dominant foundress. Vehrencamp and Emlen showed that ecological factors may influence the extent of this sharing. For example, if constraints on solitary founding are weak, a subordinate might not join unless the dominant yields to it a significant fraction of the total direct reproduction (see also Itô 1986a). These models thus offer a way to relate the personal payoffs for joining to ecological conditions. The payoff for attempted usurpation of a nonrelative can be represented simply as the product of the probability that the given female will find and displace an unrelated nest owner and the expected reproductive output in the event of successful usurpation.

I have integrated these approaches into a simple ecological constraints model of the optimal nesting strategy when all three options are available to a potential subordinate foundress (Appendix 4.3). The model is completely described by three parameters: $K$, the ratio of expected productivity of a joined nest to the expected productivity of a previously established solitary-foundress nest; $x$, the probability of successful establishment of a new solitary-foundress nest times its proportional expected productivity relative to that of a previously established solitary-foundress nest; and $u$, the probability of successful usurpation. Ecological constraints on solitary founding are reflected by low $x$ (high nest site limitation, low solitary-foundress survivorship, small expected productivity due to lateness in the season, etc.). The model predicts that when ecological constraints are sufficiently weak ($x > K - 1$), only solitary nesting and usurpation are favored, the latter when the probability of successful usurpation ($u$) exceeds $x$. Under strong ecological constraints ($x < K - 1$), only joining and usurpation attempts are favored, the latter when $u$ exceeds $K - 1$ (Appendix 4.3). Since relatedness drops out of the three-strategy model, this model predicts either no association between mean co-foundress relatedness and frequency of joining or a negative association if strong ecological constraints are associated with low mean relatedness. Interestingly, the correlation between mean co-foundress relatedness (from Ross and Carpenter, this volume: Table 13.2) and percentage of multiple-foundress associations (from Fig. 4.6) is $r = -0.86$, which is significantly *less* than zero ($P < 0.05$; eight species).

This simple framework appears to explain much of the variation in *Polistes* foundress behavior. For example, solitary-foundress success rates of the Japanese species indicate relatively weak ecological constraints on solitary nesting (e.g., Fig. 4.6). Multiple-foundress associations are rare in these species, with solitary founding (Figs. 4.5, 4.6) and usurpation (Makino 1985) occurring as the principal foundress strategies. In agreement with the model, joining occurs more frequently in species with lower solitary-foundress success probabilities (Fig. 4.6). Intraspecific variation in the relative frequencies of foundress strategies can also be interpreted with the model. In *P. metricus* populations, the relative frequencies of usurpation and joining increase, and the frequency of solitary nesting decreases, with increasing density of preemergence colonies (Gamboa 1978, see also Cervo and Turillazzi

1985). This is consistent with the foundress strategy model if this higher density reflects greater nest site limitation (lower $x$) and/or greater opportunities for usurpation (larger $u$).

By predicting that usurpation becomes more easily favored over joining as the contribution of a joiner to colony reproduction, and thus $K$, decreases, the model also successfully explains shifts in foundress behavior over the course of the nesting season. Joining appears to be most likely in the earlier portions of the founding period ($K$ relatively high), whereas usurpation tends to be most likely in the latter part of the founding period ($K$ relatively low, approaching one) (Klahn 1988).

While this ecological model provides a useful framework for interpreting variation in foundress behavior, it will probably have to be extended in many ways. (1) As associations become larger, the dominant foundress may lose her ability to control subordinate reproduction, and the reproduction actually achieved by subordinates may exceed that predicted by simple optimal skew models. (2) When usurpation pressure itself provides a significant selective advantage to joining (Gamboa 1978), the parameters $u$ and $k$ will be negatively correlated, raising the interesting possibility that in some habitats joining and usurpation may be jointly maintained by frequency-dependent selection. (3) The model does not include the possibilities that foundresses join unrelated foundresses or that usurpers retain usurped unrelated foundresses as subordinates, both of which apparently occur (Gamboa 1978, Gamboa and Dropkin 1979, Klahn 1979, MacCormack 1982). (4) The frequency of solitary founding may be quite high, even if joining is highly advantageous, when individual colonies typically produce few surviving reproductives because of climatic factors or frequent predation.

## Foundress Interactions

Considerable attention has focused on interactions among foundresses within an association. Just after they have emerged from winter diapause, foundresses of temperate-zone species most commonly return to their natal nest sites and enter into frequent social interactions with conspecifics, including former nestmates (West-Eberhard 1969, Dropkin and Gamboa 1981, Strassmann 1981a). These interactions can be highly aggressive, often involving falling fights (West-Eberhard 1969, Gamboa and Dropkin 1979, Strassmann 1981a). The tropical *P. erythrocephalus* can engage in these contests for weeks after colony initiation (West-Eberhard 1969). Foundresses may be assessing relative fighting abilities during this period. Stable hierarchies can form among individuals with sufficiently disparate abilities, an example perhaps of settlement of animal contests by differences in resource-holding power (Maynard Smith 1982).

Foundresses of some species switch nests frequently early in the founding period (*P. fuscatus*: West-Eberhard 1969; *P. exclamans*: Strassmann 1983, Willer and Hermann 1989; *P. metricus*: Gamboa and Dropkin 1979; *P. versicolor*: Itô 1984a), but foundresses in certain populations of other species do not (*P. annularis*: Strassmann 1983; but see Hermann and Dirks 1975). It is tempting to hypothesize that shifting females are attempting to "trade up" in dominance rank by associating with less competitive nestmates. Gamboa and Dropkin (1979) found that in five of seven cases in which solitary foundresses of *P. metricus* were subsequently joined, the resident foundress became subordinate to the joiner. Since most associations of *P. annularis* are large at the beginning of the founding period and only highly ranked females achieve measurable direct reproduction (Strassmann 1981a), the benefits of nest switching may be reduced in this species.

A nest-switching foundress may be assessing more than just the relative competitive abilities of colonymates. For example, such a foundress may also be discriminating in favor of nests that (1) contain the most closely related foundresses, (2) will most benefit from the addition of a foundress, or (3) are most likely to survive so that the foundress's reproductive efforts will not be wasted. As a possible example of the third possibility, Gamboa (1980) found evidence that *P. metricus* foundresses preferentially join earlier established nests, presumably because these nests will experience greater survivorship as the result of the earlier appearance of workers.

Body size appears to be one correlate of competitive ability in foundress associations. Numerous studies have found that higher ranked foundresses tend to be larger (Pardi 1948, Richards and Richards 1951, West-Eberhard 1969, Turrillazzi and Pardi 1977, Dropkin and Gamboa 1981, Noonan 1981, Sullivan and Strassmann 1984, Queller and Strassmann 1989, but see Strassmann et al. 1988 for *P. bellicosus*). Röseler et al. (1984) claimed that size "did not contribute" to dominance in their laboratory studies of the hormonal basis of dominance, since size exerted no statistically independent effect on dominance when variation in endocrine state (corpora allata volume and ovarian development) was controlled (see also Röseler, this volume). Of course, this finding might simply have resulted from the fact that endocrine state is *part* of the proximate mechanism of behavioral dominance and thus obviously would be more strongly connected to the expression of dominance than would the *input* to the mechanism (assessed relative body size).

Size may not be the only factor that determines dominance ranking within foundress associations. The order in which foundresses join the nest may also play a role. West-Eberhard (1969) found that nest-initiating foundresses of *P. fuscatus* tended to be dominant to subsequent joiners, and Strassmann et al. (1987) showed that late joiners of *P. bel-*

*licosus* (often foundresses whose nests had failed) did not differ in size from, and were probably subordinate to, foundresses that had remained with their original nests. Sometimes joiners may rise in rank above nest initiators, as shown by Gamboa and Dropkin (1979), so it appears that dominance is a function of both order of joining and intrinsic competitive abilities such as those correlated with body size. Thus, in dominance interactions among *Polistes* foundresses, asymmetries in both ownership and resource-holding power (size) exist and can affect the outcomes of disputes. One possibility is that nest ownership determines dominance rank except when the joining foundress is sufficiently larger than the resident foundress, as would be predicted by game theory models (Hammerstein 1981). Hughes and Strassmann (1988b) have suggested that the dominance of earlier over later emerging workers (see Worker Phase) and the dominance of nest initiators over later joining foundresses are both examples of conventional settlement of dominance disputes by position in an arrival "queue."

There is considerable variability among and within species in the frequency and intensity of presumed dominance-related co-foundress interactions (e.g., West-Eberhard 1969, 1982a,b, 1986; Itô 1985a). In some multiple-foundress colonies, such interactions can be frequent and intense, involving, for example, chasing, lunging, biting, and "sting threats," and including submissive crouching by subordinates in response to aggressive mounting and antennation by dominants (Pardi 1942a, 1948; West-Eberhard 1969, 1982a,b, 1986; Gamboa and Dropkin 1979; Itô 1985a; see also Röseler, this volume) West-Eberhard (1986) has drawn a distinction between foundress associations with predominantly "confrontational" oviposition-inhibiting interactions (direct aggression, as in *P. canadensis* and *P. erythrocephalus*) and associations with predominantly "nonconfrontational" interactions (differential oophagy, as in *P. dominulus, P. metricus,* and *P. fuscatus*). Species with moderate or high rates of agonistic behavior usually exhibit a linear dominance hierarchy (at least among a subset of the foundresses), with dominant individuals preferentially directing their aggression toward those just beneath them in the hierarchy (Pardi 1948; West-Eberhard 1967, 1969, 1986; Downing and Jeanne 1985). This may function in dominance advertisement (West-Eberhard 1969) and/or activity stimulation (as in worker-phase colonies: Reeve and Gamboa 1983). Experimental evidence suggests that recognition of dominance rank is mediated by chemical cues from cephalic glands and the ovaries (Downing and Jeanne 1985; see also Downing, this volume). Ovariectomized *P. dominulus* foundresses maintain their behavioral dominance rank (although they lose their ability to inhibit reproduction by subordinates), implying that the presence of ovaries is not necessary for dominance recognition (Röseler and Röseler 1989).

Higher ranked subordinate foundresses are most likely to replace the dominant foundress when the latter disappears (Pardi 1948, Turrillazzi and Pardi 1977, Gamboa et al. 1978, Strassmann 1981a, Hughes et al. 1987). The probability of supersedure by a subordinate foundress varies among populations and may be related to colony size. In the large colonies of *P. annularis*, 35% of dominant foundresses are superseded by subordinates (Strassmann 1981a), compared with much lower values for the smaller colonies of *P. metricus* (6%: Gamboa et al. 1978) and *P. fuscatus* (3%: Noonan 1981). This may reflect increased competition for a bigger reproductive prize in a larger association. In support of this hypothesis, Strassmann (1983) found a positive relationship between colony size and percentage of foundresses with developed ovaries, even though the mean proportion of direct reproduction per subordinate in queenright colonies did not vary significantly with colony size (Queller and Strassmann 1988).

In species with frequent dominance-related aggression, there is usually also a well-defined division of labor. Dominant foundresses specialize in oviposition, nest petiole maintenance, cell initiation, cell inspection, and (occasionally) wood pulp foraging, whereas subordinate foundresses specialize in cell elongation and risky tasks such as nest defense and foraging for prey, nectar, and water (West-Eberhard 1969, Gamboa et al. 1978, Strassmann 1981a). Dominant foundresses also tend to receive a greater share of the food (prey and nectar) from incoming foragers than do subordinate foundresses (West-Eberhard 1969).

Well-developed hierarchies, frequent aggression, and sharply defined division of labor are not universal attributes of *Polistes* foundress associations. In colonies of some species, dominance hierarchies tend to be weakly developed, aggression infrequent, and/or division of labor ill defined ( *P. chinensis*: Yamane 1973, Hoshikawa 1979; *P. jadwigae*: Kasuya 1981a, Hirose and Yamasaki 1984b; *P. riparius* (= *biglumis*): Makino and Aoki 1982; *P. versicolor* and *P. bernardii*: Itô 1985a, 1986a).

The variation in frequency and intensity of intracolony aggression appears to be matched by variation in the distribution of direct reproduction between dominant and subordinate co-foundresses. In species with moderate or frequent aggression, the dominant foundress appears to have a near monopoly on direct reproduction (67–92% of the total reproductives produced, as inferred from allozyme data [*P. metricus*: Metcalf and Whitt 1977a] or observations of oviposition corrected for differential oophagy [*P. annularis*: Strassmann 1981a; *P. fuscatus*: Noonan 1981; *P. exclamans*: Hughes and Strassmann 1988a]). In species with moderate or frequent aggression, the share of reproduction by a subordinate foundress falls off rapidly with declining dominance status (Strassmann 1981a). In species with relatively infrequent aggression, co-foundresses appear to share in direct reproduction more equitably (Kasuya 1981a, Makino and Aoki 1982, Hirose and Yamasaki 1984b). Although oophagy is sometimes reported in these species, there is no evidence that it is unilateral (*P. riparius*: Yamane 1969, Makino and Aoki 1982; *P. jadwigae*: Kasuya 1981a).

Is it possible to make sense of this variation in dominance interactions and skew in direct reproduction? West-Eberhard (1982a, 1983) has advanced the hypothesis that variation in frequencies and kinds of dominance interactions results from the diverse outcomes of "runaway" competitive social coevolution, which is not necessarily related to local ecological conditions. However, as discussed previously, Vehrencamp (1983a,b) has developed a quantitative theory relating the optimal skew in reproduction from a dominant's perspective to ecological and social conditions (see also Itô 1986a). Her model predicts that the optimal reproductive skew increases as (1) the ecological constraints on solitary founding increase, (2) the relatedness between dominant and subordinate increases, and (3) the benefit of group living increases. I suggest that, as the optimal reproductive skew increases, there exists greater competition over reproductive rights within an association and, consequently, greater payoffs for agonistic advertisement and testing of competitive abilities among co-foundresses. Thus, one might predict that both the frequency and intensity of dominance interactions would increase with, for example, increasing ecological constraints on solitary founding. The occurrence of weak dominance and apparently equitable reproduction in the infrequent foundress associations of the Japanese *Polistes* (*P. jadwigae*: Kasuya 1981a, Makino and Aoki 1982), for which solitary founding seems only weakly constrained (Fig. 4.6), is in accord with this hypothesis. Relatively even distribution of reproduction in these species may also result from lower relatedness among co-foundresses (Hirose and Yamasaki 1984b), as also predicted by the optimal skew model.

Testing of these hypotheses requires rigorous inter- and intraspecific comparisons of social interactions and distribution of reproduction in foundress associations at similar stages of development. Several studies have documented that the frequency and intensity of intracolony aggression change as the founding period progresses. Aggression is initially high (*P. annularis*: Strassmann 1981a), subsides somewhat, and then may increase as the time approaches at which eggs giving rise to reproductives are laid (*P. metricus*: Gamboa et al. 1978, Gamboa and Dropkin 1979; *P. fuscatus*: Noonan 1981). Noonan (1981) interprets the latter half of this pattern as reflecting increased emphasis on the personal component of inclusive fitness by subordinates late in the founding period (see also West-Eberhard 1981). In accordance with this idea,

Noonan also found that foraging by subordinates declined as the founding period progressed. Thus, future interspecific comparisons of nestmate aggression should control for stage in colony development. In addition, more genetic data on the actual partitioning of direct reproduction among co-foundresses are required. Some of the above estimates of the distribution of reproduction were derived from oviposition data collected before the reproductive-destined eggs were laid.

Several studies have focused on the communicatory meanings of the distinctive "oscillatory" behaviors exhibited by foundresses (as well as by queens and some workers in the worker phase). Three distinct oscillatory behaviors have been described: abdominal wagging, lateral vibrations, and longitudinal vibrations (Gamboa and Dew 1981). Abdominal wagging has been reported in a wide range of *Polistes* species (Pardi 1942a; Morimoto 1961a,b; Esch 1971; West-Eberhard 1969, 1982b; Hermann and Dirks 1975; Maher 1976; Gamboa et al. 1978; Strassmann 1981a; Itô 1986a). The other two oscillatory behaviors may also occur widely, but may not always have been clearly distinguished.

Abdominal wagging, which involves repeated lateral movement of the gaster over brood-containing cells, is performed most frequently by the more dominant foundresses (West-Eberhard 1969, Gamboa et al. 1978, Gamboa and Dew 1981, Strassmann 1981a, Hughes et al. 1987). The function of abdominal wagging remains unclear and may vary among species. Gamboa and Dew (1981) found that abdominal-wagging rates in *P. metricus* were positively associated with the number of third-instar larvae and rates of prey foraging, but not with the number of foundresses on the nest. In addition, abdominal wagging occurred on solitary-foundress nests, i.e., in the absence of co-foundresses. These results suggest that, in *P. metricus*, abdominal wagging is involved in communication between foundresses and larvae (Gamboa and Dew 1981). In contrast, Hughes et al. (1987) reported that abdominal wagging occurred in early *P. annularis* colonies, before the appearance of larvae, and they suggest that abdominal wagging may serve as a dominance signal or may function in deposition of a dominance pheromone (as also hypothesized by Morimoto 1961a,b, West-Eberhard 1969, Hermann and Dirks 1975, Downing and Jeanne 1985). However, the reproductive domination hypothesis for the function of abdominal wagging is weakened by the finding that ovariectomized, dominant *P. dominulus* foundresses continue to wag but lose their ability to restrict reproduction by subordinates (Röseler and Röseler 1989).

Lateral vibrations, which tend to be performed mostly by dominant foundresses in some species (Downing and Jeanne 1985, West-Eberhard 1986) but evidently not in others (Gamboa and Dew 1981), are rapid, vigorous, often audible, lateral movements of the whole body, usually performed from a stationary position on top of the nest (Gamboa and Dew 1981). Lateral vibrations apparently are not distinct from abdominal wagging in *P. instabilis* and *P. annularis* (Strassmann 1981a, Downing and Jeanne 1985). Gamboa and Dew (1981) found that in *P. metricus* the rate of performance of lateral vibrations is positively associated with the presence of foundresses on the nest, and West-Eberhard (1986) reported that lateral vibrations sometimes preceded aggression by the performer or even seemed to provoke attacks on the performer. Although these observations suggest that lateral vibrations are dominance-related signals, Downing and Jeanne (1985) found that in *P. fuscatus*, rates of lateral vibration are higher on *single-foundress* colonies than on multiple-foundress colonies and occur in close temporal association with foundress-brood interactions.

Thus, lateral vibrations, like abdominal wagging, may be involved both in adult-adult and adult-brood communication, perhaps depending on the species or the context. One possibility consistent with the available evidence is that one or both of these oscillations simultaneously signal the larvae to reveal their nutritional states to the foundresses (including the queen in solitary-foundress colonies ) and stimulate subordinate foundresses to check the larvae. Aggression might be directed toward subordinates who do not so respond or who themselves oscillate rather than perform colony tasks, i.e., there may be intracolony conflict over task performance (Reeve and Gamboa 1983). When larvae are not yet present, one or both of these oscillations may stimulate subordinate foundresses to assess other colony needs. In support of these hypotheses, increases in queen activity, including abdominal wagging, in worker-phase colonies of *P. fuscatus* cause increases in worker nest activity and foraging for prey and pulp (Reeve and Gamboa 1983, 1987).

Longitudinal vibrations, which are rapid body oscillations along a longitudinal axis accompanied by antennal drumming on the rim of a cell that contains a larva, probably function to alert larvae that they are about to receive the crop contents of a foundress that has recently malaxated prey (Pratte and Jeanne 1984).

In summary, evidence is accumulating that the body oscillations exhibited by foundresses probably function in both adult-brood and adult-adult communication. The adult-adult communication may involve either dominance signaling or activity stimulation. These oscillations tend to be performed more frequently by more dominant foundresses (at least in some species), but this is not sufficient evidence that these behaviors function in dominance advertisement, since dominant wasps also tend to be the principal regulators of colony activity (Reeve and Gamboa 1983).

## WORKER PHASE

The modal developmental time for the first brood of *Polistes* is 45–50 days, but can range from 39 to 70 days, depending on the species and on environmental factors such as temperature (Strassmann and Orgren 1983). At the end of the first brood developmental period, the first adult workers emerge and assume colony tasks such as foraging, brood care, and nest maintenance. Worker-phase colonies grow as workers accumulate, but the relatively short adult life spans of workers (ranging from a mean of 14 days in *P. exclamans* [Strassmann 1985b] to 38 days in *P. chinensis* [Miyano 1980] ) usually prevent colonies from attaining populations of more than a hundred or so workers at any one time.

Initially, investigators focused less on social dynamics in worker-phase colonies than in founding-phase colonies, but recent studies have greatly increased our understanding of social interactions during the worker phase.

### Fates of Subordinate Foundresses

In many species, foundress associations tend to break up at or near emergence of the first workers, with subordinates disappearing at relatively high rates (Pardi 1946, 1948; Hermann and Dirks 1975; Gamboa et al. 1978; Krispyn 1979; Turillazzi 1980; Noonan 1981; Pfennig and Klahn 1985; Hughes and Strassmann 1988a). However, this pattern is not universal among species (Hughes and Strassmann 1988a) and even varies within species (*P. jadwigae*: Yoshikawa 1957, Kasuya 1981a; *P. chinensis*: Hoshikawa 1979).

The breakup of foundress associations at worker emergence in at least some cases appears to reflect intracolony reproductive conflict. Pfennig and Klahn (1985) showed that *P. fuscatus* foundresses tend to disappear in order of their dominance rank (highest first), suggesting that those posing the greatest reproductive threat to the queen were ejected first, either by queens or by workers, at a time when subordinates are less critical to colony success. However, Hughes and Strassmann (1988a) found no such correlation for *P. exclamans* subordinate foundresses, which also exhibit increased rates of disappearance after worker emergence. Noonan (1981) noted that aggression between *P. fuscatus* queens and subordinate foundresses increased when reproductive eggs were being laid (near the time of first worker emergence), and Gamboa et al. (1978) observed heightened aggression between *P. metricus* queens and subordinates shortly after initial worker emergence. Hughes and Strassmann (1988a) noted that rates of aggression toward foundresses do not appear to increase after worker emergence in *P.*

*annularis* and suggest that senescence may account for heightened worker-phase mortality of subordinate foundresses in this species. However, aggressive eviction of subordinates may have been missed if it was a rare interaction different in nature from the more frequent kinds of aggression involving co-foundresses.

Variability in subordinate nest tenures among and within species may result from variation in the costs and benefits of retaining subordinates to dominant foundresses or their workers. For example, Hughes and Strassmann (1988a) found that species that exhibited relatively low subordinate foundress disappearance rates were those characterized by high rates of colony failure due to loss of all workers, which suggests that evictions of subordinates are less likely when the survivorship insurance benefits of their retention are sufficiently great.

## Worker Dominance and Reproduction

Dominance hierarchies commonly occur among workers (Pardi 1948; Yoshikawa 1956, 1963; Morimoto 1961a,b; Strassmann and Meyer 1983; Miyano 1986; Reeve and Gamboa 1987), even in species in which foundress associations exhibit weakly developed hierarchies and equitable oviposition (e.g., *P. chinensis*: Morimoto 1961a,b; *P. jadwigae*: Yoshikawa 1963). Interestingly, the latter fact supports the prediction derived from Vehrencamp's optimal skew model that the queen's skew in reproduction, and thus the degree of intracolony reproductive competition, should increase as the benefits of solitary founding by her subordinates (relative to the benefits of continued cooperation) become smaller later in the colony cycle.

When both co-foundresses and workers are present, subordinate co-foundresses, which tend to be larger than workers, outrank workers and are more likely to replace the queen (Yoshikawa 1963, Klahn 1981, Hughes et al. 1987). Dominant workers are more likely than subordinate workers to replace the queen, and older workers tend to be more dominant (Pardi 1948, Yoshikawa 1956, Klahn 1981, Strassmann and Meyer 1983, Miyano 1986) and to have larger ovaries (Haggard and Gamboa 1980, Hughes and Strassmann 1988b, Strassmann et al. 1988).

Among workers, age is much more strongly positively correlated with dominance than is size, although the latter may have a small effect (Klahn 1981, Strassmann and Meyer 1983, Hughes and Strassmann 1989b). Since the first emerging (older) workers tend to be smaller than later emerging workers (see Seasonal Changes in Characteristics of Emerging Adults), it is not unusual for a smaller worker to dominate a larger one. The age-dominance correlation may result from (1) greater fighting abilities of older workers (which may have gained experience in social interaction) or (2) order of emergence serving as a cue for conventional settlement of potentially costly dominance contests (Maynard Smith 1982, Hughes and Strassmann 1988b). Hughes and Strassmann (1988b) suggest that size is less important in determining dominance among workers than among foundresses because the costs of escalated conflict in worker-phase colonies, in which closely related, needy brood are present, are greater than in early founding-phase colonies.

The frequency of queen supersedure by workers varies among species (19%, *P. fuscatus*: Klahn 1981; 52–63%, *P. exclamans*: Strassmann 1981b; 26%, *P. metricus*: Metcalf and Whitt 1977b; > 10%, *P. snelleni*: data in Suzuki 1985; 25%, *P. chinensis*: Miyano 1986). These values tend to be higher than the corresponding frequencies of supersedure by subordinates in multiple-foundress associations. For instance, the frequencies of queen supersedure by workers for *P. fuscatus* and *P. metricus* are, respectively, six and seven times the frequencies of queen supersedure by subordinant foundresses for the same species (see Founding Phase). This pattern is in accordance with the prediction from the extended optimal skew model that reproductive competition should be stronger in older colonies (assuming queen-worker conflict; see below). Since solitary founding by late-season workers is likely to be less successful than solitary founding by early-season subordinate foundresses, queens in worker-phase colonies should seek a higher reproductive skew than queens in foundress associations.

Indeed, available evidence indicates a much greater reproductive skew (hence potential for competition over reproductive rights) in worker-phase colonies than in foundress associations. The share of worker oviposition in queenright colonies tends to be very low (0%, *P. exclamans*: Strassmann and Meyer 1983; 0%, *P. fuscatus*: Metcalf 1980; 0.2%, *P. metricus*: Metcalf and Whitt 1977a), although it is relatively high in queenright colonies of *P. chinensis* (13–84%: Miyano 1980). The greater worker reproduction of *P. chinensis* may result from the queen's lower optimal reproductive skew, in turn a result of the apparently weaker constraints on solitary nest founding (Fig. 4.6) (Kasuya 1981a), or it may result from the lesser ability of the queen to limit worker reproduction (regardless of her optimal reproductive skew) in the relatively large colonies of this species (Suzuki 1986). Many *Polistes* species tend to lay a greater proportion of reproductive-destined eggs in the worker phase than in the foundress phase (Miyano 1980, Pfennig and Klahn 1985), further heightening potential reproductive competition in the worker phase. Klahn (1981) found that most supersedures occur near the time that workers first emerge, which is close to the time that reproductive-destined eggs are first laid (when queen death from senescence or foraging-related predation is especially unlikely). A conflict model of queen supersedure predicts that queen supersedure rates should increase as the relatedness between the queen and workers decreases (providing that queens do not compensate for lesser relatedness by forfeiting more reproduction to workers). In support of this hypothesis, Klahn (1988) showed that supersedure rates were higher in colonies that had usurper queens, whose workers were not their offspring, than on colonies that had original queens.

Why would a worker ever be in conflict with its queen over reproductive rights, given that workers are, on average, as related to a singly mated queen's offspring as to their own offspring (for an even sex ratio)? There are at least four ways in which such a conflict between the worker and the queen might be generated, all of which might be important in *Polistes*. (1) The queen mates multiply (Metcalf and Whitt 1977a) or (2) the worker is an offspring of a subordinate foundress. Both situations lead to a higher average relatedness of the worker to its own offspring than to the queen's offspring. (3) Queen fecundity is reduced relative to the worker's own fecundity. (4) The worker's assessed relatedness to the queen's offspring is devalued because of uncertainty over whether the queen is an unrelated female that usurped the original queen in the founding phase. These possibilities might be tested by examining interpopulational correlations between worker supersedure rates and frequencies of multiple mating, associative founding, and usurpation, or by conducting field experiments such as artificially reducing queen fecundity to elicit supersedure attempts.

The distribution of reproduction in colonies with supersedure by workers appears to be similar to, though perhaps slightly more equitable, than that in conspecific queenright colonies. When a worker supersedes the queen in *P. metricus*, this dominant worker can account for 95% of the direct reproduction (Metcalf and Whitt 1977a; see also Strassmann 1981d for *P. exclamans*), which is similar to the corresponding value for queenright colonies. In a single colony of *P. chinensis*, the dominant worker accounted for only 24% of the total oviposition, with at least seven others also ovipositing (Miyano 1986); this low skew is like that in queenright colonies of the same species (Miyano 1980). Workers that have replaced queens are often uninseminated (Metcalf 1980), although in some species they are inseminated by early-emerging males and thus can produce female offspring (Klahn 1981, Strassmann 1981b, Suzuki 1985). The mean productivity of colonies headed by laying workers has been estimated as half that of queenright colonies in *P. fuscatus* (data in Klahn 1981) and three-quarters that of queenright colonies in *P. chinensis* (Miyano 1986). The reduced productivities may result from greater worker conflict, and hence lowered ergonomic efficiency, in colonies headed by workers (Miyano 1986). Queen removal causes at least temporary increases in aggression among high-ranking workers (e. g., *P. exclamans*: Strassmann and Meyer 1983; *P. fuscatus*: Reeve and Gamboa 1983).

Strassmann (1981d) has described other contexts for worker reproduction (see also Page et al. 1989a). Sixteen to thirty-nine percent of worker-phase colonies of *P. exclamans* (particularly the larger colonies) produce nearby "satellite" colonies, evidently as an antiparasite adaptation made possible by the lengthy colony cycle of this species (Strassmann 1981b). Satellite colonies are usually initiated by the original queen, but workers also can successfully establish satellite nests, especially when the original queen is dead (see also Kasuya 1981b). Thus a worker can reproduce on (1) main nests left behind by satellite-initiating original queens, (2) worker-initiated satellite nests when the original queen is dead, (3) main nests when the original queen is dead.

When a main nest of *P. exclamans* is knocked down, workers from that nest are more likely to join a queen-initiated than a worker-initiated satellite. Thus, nonreproductive workers' decisions to help may depend on their relatedness to the queens they aid. Kasuya (1981c) found that workers of *P. chinensis* drift among colonies, raising the possibility that workers distribute their aid among different colonies based on relatedness or need. Although workers can discriminate among colonies of different mean relatedness (as above), and between adult females from their own nest and females from other nests (Pfennig et al. 1983b), no evidence currently exists that workers discriminate among differently related brood or adults *within their own nests*. (In fact, Gamboa [1988] recently found that females treat nonnestmate adult aunts and nieces just as tolerantly as nestmate sisters.) Coexistence of a worker's full siblings and her half siblings, nieces, and cousins in the same colony can occur as a result of multiple mating (Metcalf and Whitt 1977a), queen supersedure by workers (Strassmann 1985a), and reproduction by sister foundresses (Noonan 1981, Strassmann 1981a).

As in many multiple-foundress associations, in worker-phase colonies there is a division of labor that parallels the linear dominance hierarchy. Queens tend to forage much less than do their workers (West-Eberhard 1969, Reeve and Gamboa 1983, Strassmann and Meyer 1983), although in species with short colony cycles living at high altitudes, queens continue to forage at high rates after workers have emerged (Lorenzi and Turillazzi 1986). Among workers, more-dominant individuals forage significantly less than other workers in some species (*P. erythrocephalus*: West-Eberhard 1969; *P. fuscatus*: Reeve and Gamboa 1987; *P. chinensis*: Miyano 1986), but not in others (*P. exclamans*: Strassmann and Meyer 1983; *P. instabilis*: Hughes and Strassmann 1988b). Intermediate-aged workers of *P. metricus* (Dew and Michener 1981) and *P. fuscatus* (Post et al. 1988) forage at the highest rate, while relatively young and relatively old workers tend to spend a greater proportion of time on the nest. The tendency of more-dominant workers of *P. exclamans* to forage more is somewhat paradoxical, given the known risks of foraging (Strassmann et al. 1984a), but it may be related to the relatively high rates of queen supersedure in this species. In other words, foraging by dominants may represent selfish investment in a colony that they have a good chance of soon inheriting.

As in many foundress associations, dominant workers specialize in abdominal wagging and lateral vibrations, smearing of the pedicel, and cell inspection (Kasuya 1983a, Strassmann and Meyer 1983, Reeve and Gamboa 1987, Hughes and Strassmann 1988b), and they also are more aggressive, directing their aggression primarily at workers ranked just beneath them (Strassmann and Meyer 1983, Downing and Jeanne 1985, Miyano 1986, Reeve and Gamboa 1987, Hughes and Strassmann 1988b). Dominant workers of *P. chinensis* sometimes attempt to physically interfere when other high-ranking workers attempt to oviposit (Miyano 1986).

## Social Integration and Colony Growth

Recent studies have elucidated the ways in which colony activity is socially regulated in worker-phase *Polistes* colonies. The queen tends to be the most active individual on the nest and appears to be the central regulator of colony activity (Dew 1983; Reeve and Gamboa 1983, 1987). Both observational and experimental data suggest that the queen lies at the center of a social feedback control system (Fig. 4.8). When a forager returns to the nest, the queen becomes more active (i.e., walks over the nest and inspects brood cells). The queen's activity stimulates workers to become more active and, ultimately, to forage when colony needs are detected. When foragers leave the colony, queen activity, worker activity, and individual worker departure rates all decline (Reeve and Gamboa 1987). The queen thus both stimulates and synchronizes worker activity.

The mechanism of activity stimulation by the queen involves both directed and undirected signals (Reeve and Gamboa 1983). Most workers appear to respond to increased queen activity communicated by undirected cues such as nest vibrations, but direct queen aggression stimulates inactive workers to become more active and also elicits foraging departures by dominant workers, which tend to forage less than do subordinate workers. Such queen aggression possibly indicates conflict between the queen and some of her workers over how active the workers should be (Reeve and Gamboa 1983, 1987).

Much remains to be learned about the social regulation of activity and proximate determinants of labor specialization in *Polistes* colonies.

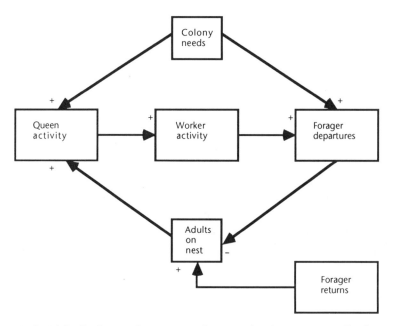

**Fig. 4.8.** Social feedback control system regulating worker foraging in small colonies of *Polistes*. +, stimulation/addition; −, inhibition/subtraction. (Modified from Reeve and Gamboa 1987, courtesy of *Behaviour*.)

The social stimuli that control nest construction, thermoregulatory fanning (Evans and West-Eberhard 1970), and foraging decisions (i.e., when to forage for prey, pulp, nectar, or water) have not yet been studied in rigorous fashion. Post et al. (1988) found quantitative differences among *P. fuscatus* workers in task specialization and suggested that workers can be divided into three groups: prey foragers, pulp and prey handlers on the nest, and nonforagers. Their data also indicated a crude age polyethism in *P. fuscatus* colonies. Workers less than seven to eight days old are either inactive or specialize in intranidal tasks such as receiving pulp and prey from foragers, feeding the larvae, and nest construction, while workers more than 13–14 days old specialize in foraging (workers 40 days or older appear to reduce their foraging activity).

Colony growth in *Polistes* appears to be limited by the availability of proteinaceous food (usually lepidopteran or, more rarely, dipteran larvae: Rabb 1960, García 1974). Klahn (1981) found that experimental protein supplementation in founding-phase colonies increased the mean number of reproductives reared by as much as 16 times. Foraging for prey takes longer than foraging for water, nectar, or pulp (Gamboa et al. 1978, Kasuya 1980, Strassmann 1981a) and seems to have a lower probability of success. Indeed, Kasuya (1983b) estimated that only 12.2% of attempted prey collections were successful in a population of *P. chinensis*. Competition for prey may occur among colonies nesting in the same area since foraging areas of neighboring colonies can overlap extensively (Kasuya 1980), and low prey abundance apparently causes some females to attempt to steal larvae from neighboring colonies rather than hunt for caterpillars (Kasuya 1983b). The need for defense against such robbers itself constrains the prey-foraging activity of foundresses. The occurrence of competition for prey is also consistent with evidence of resource partitioning among sympatric *Polistes*; Rabb (1960) found that *P. annularis* workers tended to forage for large caterpillars in wooded areas, while *P. fuscatus* and *P. exclamans* workers foraged for smaller (younger) caterpillars in the herbaceous vegetation of open fields. Similarly, Dew and Michener (1978) found that *P. metricus* workers tended to forage more often in trees than did workers of *P. fuscatus* nesting in the same area.

Suzuki (1980, 1981a) has traced the flow of nitrogen (prey flesh) through colonies of *P. chinensis* over the colony cycle. (For a discussion of short-term flow and processing of nutrients, see Hunt, this volume.) Suzuki estimated that a total of 9.35 g dry weight of prey flesh (10.52% nitrogen content) was obtained by the average colony over its lifetime (mean mature colony size = 258.1 cells). Approximately 51% of the total prey nitrogen was invested in workers, and 49% was allocated to male and female reproductives. The total dry biomass of reproductives was positively related to the total number of workers produced, but the biomass of reproductives produced per worker declined with an increase in the number of workers.

The latter trend is similar to the negative relationship seen between productivity per foundress and size of foundress association (see Foundress Strategies: Founding, Joining, and Usurpation), and may result from the tendency of workers in larger colonies (or of later emerging workers) to engage less in risky foraging (Strassmann et al. 1984a, Strassmann 1985b, Post et al. 1988). In support of this idea, Dew and Michener (1981) found evidence of foraging inhibition in later-emerged workers of laboratory *P. metricus* colonies, whereby removal of the principal foragers (earlier emerged workers) increased the foraging activity of the remaining workers. Reduced individual foraging in larger colonies may reflect increased emphasis on the personal versus the kin component of inclusive fitness when individual contributions are less critical to colony success (Strassmann et al. 1984a, Strassmann 1985b), or it may indicate increased ability of workers to achieve their personal reproductive options when the ability of the queen to force workers off the nest is reduced. Nonworking individuals that exercise such personal reproductive options in large colonies may be either potential replacement queens or, in mid- to late-summer colonies of temperate-zone species, potential future foundresses (gynes). An alternative to these "lazy worker" hypotheses is that larger colonies may require disproportionately greater protection from predators (such as ants) or robbers, resulting in an inverse relationship between colony size and proportion of workers that forage.

## Early Males

The production of males early in the colony cycle (i.e., along with the first workers) has now been reported for a number of species (Yoshikawa 1963, Klahn 1981, Strassmann 1981b, Suzuki 1981b, Kasuya 1983a, Pratte et al. 1984, Page et al. 1989a, Willer and Hermann 1989). The percentage of colonies that produce such males varies inter-specifically (2.3–4%, *P. jadwigae*: Yoshikawa 1963, Kasuya 1983a; 7–14%, *P. fuscatus*: Klahn 1981, Page et al. 1989a; 4.2–16.7%, *P. chinensis*: Suzuki 1981b, Kasuya 1983a; 20–38%, *P. exclamans*: Strassmann 1981b). Although the data are still quite limited, there appears to be a positive association between the frequency of early male production and the frequency of queen supersedure in worker-phase colonies. For example, the greatest frequency of early male production occurs in *P. exclamans*, which also exhibits the highest frequency of queen supersedure by workers. This pattern suggests that the advantage of early male production by foundresses lies in the ability of these males to inseminate virgin workers that have replaced queens in other colonies (Strassmann 1981b, Kasuya 1983a, Page et al. 1989a).

Early male production may incur a cost, however, because allocation of resources to early males reduces allocation to worker production (Strassmann 1981b, Page et al. 1989a). Although males may occasionally help fan the nest, feed the larvae, or engage in threat displays (West-Eberhard 1969, Yamane 1969, Hunt and Noonan 1979, Makino 1983, Cameron 1986, Starr 1990), their contribution to colony productivity is minimal compared with that of workers. A significant cost in producing early males would account for the fact that their production occurs relatively infrequently in smaller, presumably more vulnerable, colonies (Strassmann 1981b).

## Seasonal Changes in Characteristics of Emerging Adults

Numerous studies have examined changes in the physical characteristics of emerging *Polistes* adults over the course of the nesting season. One common pattern is that the first workers to emerge tend to be smaller than workers that emerge later (West-Eberhard 1969, Bohm 1972a, Haggard and Gamboa 1980, Turillazzi 1980, Klahn 1981, Turillazzi and Conte 1981, Miyano 1983, Strassmann and Orgren 1983) and to have shorter larval developmental times (Klahn 1981, Miyano 1983, Strassmann and Orgren 1983). It seems possible that (at an ultimate level) their relatively short developmental times reflect selection for early worker emergence on vulnerable preworker nests. The larger size of females that emerge later is proximately the result of both longer larval development times and increased rates of feeding, the latter mediated by increases in adult-larva ratios (West-Eberhard 1969, Haggard and Gamboa 1980). In *P. metricus*, workers that emerge in the middle of the worker phase are as large as the queen (Haggard and Gamboa 1980). In some species, females that emerge late are smaller than those that emerge early (Yamane 1969, Haggard and Gamboa 1980, Miyano 1983), perhaps in part because the worker-larva ratio declines as workers become less numerous during the period of gyne and male production (Bohm 1972a, Turillazzi and Conte 1981, Miyano 1983).

Gynes and males of temperate-zone species tend to be produced in a burst in the latter part of the colony cycle, as predicted by optimality models of annual colony reproductive strategies (Macevicz and Oster 1976). (Adequate data on the schedule of production of reproductives in tropical colonies are lacking.) Unlike late-season workers, gynes have significant stores of fat for overwintering (Eickwort 1969, Haggard and Gamboa 1980) and elevated levels of cryoprotectants such as fructose in the hemolymph (Strassmann et al. 1984b).

The proximate trigger for gyne production is still unclear. Increased larval nutrition alone is apparently not a sufficient cause for gyne differentiation (Deleurance 1948a, West-Eberhard 1969). Turillazzi and Conte (1981) were unable to confirm Deleurance's (1952a) result that gyne determination can occur as an effect of increased nocturnal temperature on nurse-workers. Instead, they found that low nocturnal temperature acted directly on brood to lengthen developmental time, resulting in relatively large "nonworking" adults with enhanced fat reserves. Turillazzi and Conte claimed that such females were not especially likely to become foundresses, but their data indicate that these females were more than three times as likely to survive overwintering and to terminate ovarian diapause after overwintering than were females reared at relatively high nocturnal temperatures. (Turillazzi and Conte's statistical analysis revealed no significant difference between the two treatments, but if one compares the proportions of colonies that produced *any* such surviving females, a significant difference is obtained; $P < 0.05$, G test: data in Turillazzi and Conte 1981: table 3.) Thus it seems possible that relatively low nocturnal temperature is at least one stimulus for differentiation of workers and future foundresses. Bohm (1972b) showed that the ovarian diapause characteristic of gyne differentiation may be induced by relatively short photoperiod and low temperature. The decision by late-season females of *P. annularis* to become gynes or workers may be influenced by both their relatedness to and their ability to feed the brood, because colonies experiencing early queen turnover and low prey abundance cease rearing brood relatively early (Strassmann 1989). In summary, it appears possible that temperature, photoperiod, larval nutrition, and social factors all affect the production of future foundresses, but the critical experiments teasing these and other factors apart have yet to be performed.

### REPRODUCTIVE PHASE

### Sex Investment Ratios and Timing of Male and Gyne Production

Trivers and Hare's (1976) seminal paper on queen-worker conflict over relative investment in reproductive females and males has stimulated research on sex investment ratios in a variety of *Polistes* species (Noonan 1978, Metcalf 1980, Strassmann 1984, Strassmann and Hughes 1986, Suzuki 1986; for critical discussion see Starr 1984; see also Cowan, Matthews, this volume, for solitary and presocial wasps). Suzuki (1986) found an association between sex investment ratio and relative timing of male and gyne production. In species with simultaneous (or at least considerable temporal overlap in) production of gynes and males, the sex investment ratio tended to be nearly equal. On the other hand, in species with protandrous (male first) or protogynous (gyne first) production, sex investment ratios tended to be female-biased (Table 4.1).

**Table 4.1.** Sex investment ratio and timing of male and gyne production in a variety of *Polistes* species

| Species | Sex investment ratio (female:male) | Reference |
|---|---|---|
| Simultaneous production | | |
| P. fuscatus | 0.99:1 | Noonan 1978 |
| P. jadwigae | 0.96:1 | Suzuki 1986 |
| P. metricus | 0.99:1 | Metcalf 1980 |
| Protandry | | |
| P. japonicus | 4.61:1 | Suzuki 1986 |
| P. mandarinus | 1.89:1 | Suzuki 1986 |
| P. riparius (= biglumis) | 4.55:1 | Yamane 1980, cited in Suzuki 1986 |
| P. snelleni | 5.62:1 | Suzuki 1986 |
| Protogyny | | |
| P. chinensis | 1.57:1 | Suzuki 1986 |
| P. exclamans | 1.94:1 | Strassmann 1984 |

*Source:* Modified from Suzuki 1986, *American Naturalist* 128: 366–378, © 1986, courtesy of The University of Chicago.

Suzuki (1986), drawing on the models of Bulmer (1981, 1983), proposed the following preliminary interpretations for these trends. Queens of species with simultaneous production control the sex ratio to produce their optimum (1:1 when workers do not produce males and there is no local competition for mates; see also Noonan 1978). Protandry presumably is associated with female-biased sex ratios because the laying of male-destined before gyne-destined eggs by the queen can give workers increased control over the sex investment ratio (Bulmer 1981). Workers in protandrous colonies consequently bias the overall sex investment ratio in favor of females, to which they are presumably more closely related. Queens lay male eggs first either to discourage male production by workers (Bulmer 1981) and/or because males that emerge earlier have a mating advantage when mating occurs soon after gynes emerge (Bulmer 1983).

Suzuki (1986) suggests that protogynous production in *P. chinensis* may result from the queen laying an excess of female eggs as an evolved response to later production of males by workers who escape queen control in the relatively large, late-season colonies of this species. Suzuki considered investment ratios only of colonies with foundress queens, thus he suggests that the female-biased investment ratio in his sample may reflect compensation for the relatively male-biased investment ratios of orphaned colonies. In accordance with the latter hypothesis, Metcalf (1980) provided evidence that *P. metricus* colonies with foundress queens exhibited significantly more female-biased investment ratios than colonies with workers acting as replacement queens (resulting in an overall even investment ratio). Strassmann (1984) suggests that protogynous, female-biased production in *P. exclamans* reflects selection for the initial production of females in a variable environment. If the warm season is longer than expected, the first gynes can serve both as potential future foundresses and as workers in the rearing of subsequent reproductives, whereas if the season is shorter than expected, the first gynes can immediately overwinter as future foundresses.

Many more data, including information on other factors predicted to affect sex investment ratios (e.g., the number of matings by a queen, local mate competition: Alexander and Sherman 1977), are required to test the above hypotheses and to elucidate why certain species have evolved simultaneous and others protandrous or protogynous production schedules (assuming that species are usefully categorized in this way). Large samples are needed for precise estimates of population-wide sex investment ratios, given the high heterogeneity in investment ratios of individual colonies (Metcalf 1980) and potential biases in estimates derived from small samples (Boomsma 1989). (The mean sample size in Suzuki's [1986] study was only nine colonies.)

At a proximate level, the relative timing of laying of male-destined and gyne-destined eggs appears to be influenced by photoperiod (Suzuki 1981c, 1982) and temperature (Turillazzi and Conte 1981). Behavioral mechanisms for sex investment ratio modification by workers of species with presumed worker sex ratio control are still undocumented.

## INTERMEDIATE PHASE

In both temperate and tropical regions, mature colonies eventually enter a state of decline that is often associated with the disappearance of the queen or cessation of her egg laying (usually in late summer or early fall in temperate colonies). During this period, idle gynes and males accumulate on the nest, brood care and colony growth decline as the number of foragers decreases, and cannibalization of larvae may occur (West-Eberhard 1969). This situation signals the beginning of the period during which gynes and males disperse from their natal colonies.

## Mating

Interspecific variation exists in the seasonal timing of mating, even among temperate-zone species. In some species mating occurs during, or just after, the reproductive phase (*P. fuscatus*: West-Eberhard 1969; *P. metricus*: Bohm and Stockhammer 1977; *P. snelleni* and *P. mandarinus*: Kojima and Suzuki 1986). In other temperate-zone species, mating tends to occur later in the fall (*P. chinensis* and *P. jadwigae*: Kojima and Suzuki 1986) or even in the following spring (*P. annularis*: Hermann et al. 1974, Krispyn 1979). Females of the tropical species *P. erythrocephalus* appear to mate just before or after they initiate new nests (West-Eberhard 1969). The factors underlying the variation in timing of mating are unclear and should be investigated further.

Males usually disperse sooner after emergence than gynes, sometimes after being aggressively ejected from their nests by resident females (West-Eberhard 1969, Kasuya 1983a). After a variable period, they pursue one of several mating strategies. (1) Males may defend mating territories near cavities (hibernacula) used by females for overwintering in temperate-zone species (West-Eberhard 1969, Lin 1972, Noonan 1978, Kasuya 1981d, Post and Jeanne 1983a) or near newly founded nests in tropical species (West-Eberhard 1969). Such a mating system has been categorized as "resource-defense polygyny" since males maximize their encounter rates with receptive females by defending areas at or near resources used by females (Emlen and Oring

1977, Thornhill and Alcock 1983). (This classification and those below assume polygyny [that is, multiple mating by males], but more data are needed on the actual distributions of mating frequencies for the two sexes.) (2) Males may seek matings at sites where females forage (Post and Jeanne 1983a), possibly an example of resource-defense polygyny (if territories are defended) or "scramble competition polygyny" (if territories are not defended). (3) Males may seek matings at non-resource-based sites such as bushes, large trees, or sites on hilltops (Alcock 1978, Turillazzi and Cervo 1982, Turillazzi and Beani 1985, Wenzel 1987a, Beani and Turillazzi 1988a,b). This "lek polygyny" (if territories are defended) or scramble competition polygyny (if territories are not defended) may reflect attempts by males to intercept females at prominent landmarks that are used as orientation cues by females. Presumably, selection will favor this mating tactic when the dispersion of resources used by females makes defense or patrolling of resource-based territories unprofitable or when competitively inferior males cannot compete with other males at those resources (Emlen and Oring 1977, Thornhill and Alcock 1983). Different strategies can be pursued simultaneously by different classes of males in the same population. For example, Post and Jeanne (1983a) found that larger males of *P. fuscatus* defended hibernaculum territories while smaller males patrolled foraging sites. Turillazzi and Beani (1985) and Beani and Turillazzi (1988b) reported that males of *P. dominulus* that owned non-resource-based territories were larger and copulated more frequently than males that drifted between territories. *Polistes dominulus* drifters sometimes later became territory defenders (Beani and Turillazzi 1988b).

Males may sometimes chemically mark their territorial perches or emit sex attractant pheromones. Males of *P. nimpha* drag their gasters over the substrate at territorial perches (Turillazzi and Cervo 1982, Turillazzi and Beani 1985, Beani and Turillazzi 1988b), as do *P. fuscatus* males (Post and Jeanne 1983a). *Polistes jadwigae* males raise their gasters at perches (Kasuya 1981d), and *P. major* males emit volatile pheromones from unusually large ectal mandibular glands (Wenzel 1987a). The precise functions of these pheromones have not been determined, but they may be involved in intermale competition, mate attraction, or both (Wenzel 1987a).

Males first orient toward, then pursue and grasp conspecifics that enter their mating territories; they grapple with male intruders and attempt to copulate with female intruders (Post and Jeanne 1984a). Territory owners usually win male-male contests (Kasuya 1981d, Post and Jeanne 1983a). The black and yellow striped color pattern of *P. fuscatus* elicits the pursuit response; cuticular and venom-borne pheromones mediate the recognition of species and sex when contact occurs (Post and Jeanne 1984a). Copulations last only about 20 seconds (Wenzel

1987a). Males attempt to copulate with any encountered conspecific female, regardless of her age, caste, or prior mating experience (Post and Jeanne 1985). Females of *P. metricus* usually mate at least twice, apparently using the sperm from different mates in an unequal ratio (Metcalf and Whitt 1977a). Unfortunately, the frequencies of multiple mating in other species remain unknown (see Ross and Carpenter, this volume), although Wenzel (1987a) found that *P. major* females never copulated more than once in the laboratory.

Current data suggest that preferential inbreeding does not occur widely. In laboratory arena tests, *P. fuscatus* males and females that have had prior exposure to colonymates of both sexes fail to discriminate among nestmates and nonnestmates in mating contexts (Larch and Gamboa 1981, Post and Jeanne 1983b), even though they make within-sex discriminations under similar conditions (Shellman and Gamboa 1982, Shellman-Reeve and Gamboa 1985). Ryan and Gamboa (1986) found that males preferentially copulated with nonnestmate females, but only if the latter had not been exposed to male nestmates. Metcalf and Whitt (1977a) and Metcalf (1980) found no evidence from population allozyme data of widespread inbreeding in *P. metricus*, although Davis et al. (1990) did report such evidence for this and another species. As mentioned earlier, data on possible inbreeding in tropical species are desirable, since asynchrony in colony initiation may increase the potential for inbreeding (Hamilton 1964b).

There is considerable interspecific variation in the degree of sexual dimorphism (Richards and Richards 1951, West-Eberhard 1969, Evans and West-Eberhard 1970, Starr 1990). For example, males of *P. fuscatus* tend to be smaller and exhibit more yellow coloration than females, but the sexes of *P. erythrocephalus* and *P. annularis* are virtually indistinguishable (West-Eberhard 1969, Krispyn 1979) (Fig. 4.9). In addition, males of sexually dimorphic species, unlike males of sexually monomorphic species, tend to have distally curved antennae, which are used to lift and stroke a female's antennae during courtship (West-Eberhard 1969).

West-Eberhard (1969) suggested that sexual dimorphism results from selection for females to more easily recognize and preferentially eject males from nests, forcing them to disperse to sites (e.g., hibernacula) where they are more likely to encounter future foundresses. However, this hypothesis assumes an unspecified conflict between males and females and requires that males would be unable to win this conflict by evolving female-like morphologies. Although males and females are expected to differ in their tolerance of inbreeding (Waser et al. 1986), it is difficult to see why males should be forced to develop discriminable phenotypes as the result of such conflict. One testable alternative is that relatively small size, with resultant increased maneuverability in

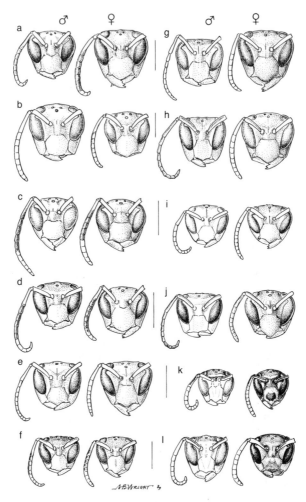

**Fig. 4.9.** Sexual dimorphism in facial characteristics and antennae among *Polistes* species. (a) *P. erythrocephalus*, (b) *P. major*, (c) *P. annularis*, (d) *P. kaibabensis*, (e) *P. apachus*, (f) *P. exclamans*, (g) *P. carolina*, (h) *P. metricus*, (i) *P. dorsalis*, (j) *P. fuscatus*, (k) *P. biglumis*, (l) *P. semenowi* (social parasite).

flight, is sexually selected when there is scramble competition among males for access to large numbers of receptive females. The synchronous production of gynes in the late-summer colonies of temperate-zone species would seem especially likely to induce scramble competition and may account for the apparent relatively high frequency of sexually dimorphic species among temperate-zone species (Evans and West-Eberhard 1970). Sexual dichromism might also benefit a male engaged in scramble competition if it prevents gender misidentifications by, and hence time-consuming interactions with, other males. The functional significance of curved antennae in sexually dimorphic males also remains an intriguing puzzle. In any case, our understanding of the selective basis of sexual dimorphism in these wasps may be deepened by studies of the link (if any) between the degree of dimorphism and mating system, as has been examined for vertebrates (Alexander et al. 1979). An alternative hypothesis that phylogenetic inertia determines the taxonomic distribution of sexual dimorphism might be tested by conjoining such a comparative study with a cladistic analysis.

## Winter Diapause

Females (and sometimes males) of temperate-zone species, gather into diapause aggregations in protected cavities (hibernacula) before the onset of winter (West-Eberhard 1969). These aggregations may usually be composed of nestmates, since nestmates preferentially hibernate with each other in arena choice tests (Allen et al. 1982, Shellman and Gamboa 1982). The kin-selective benefits of such preferential aggregation are unclear (Gamboa et al. 1986b). On warm winter days, *P. annularis* gynes return to their natal nests to feed on stored honey; nestmates defend their honey caches against nonnestmates (Strassmann 1979). Females undergo some ovarian development throughout the winter, but maximum development occurs in the early spring, at the beginning of the founding period (*P. dominulus*: Turillazzi 1980).

## CONCLUDING REMARKS

Perhaps the most important general conclusion emerging from a survey of studies since 1970 is that the *Polistes* social system is quite flexible, with its major features depending in large part on local ecological conditions. The relative frequencies of joint founding, solitary founding, and usurpation, frequency and intensity of dominance interactions, extent of division of labor, distribution of direct reproduction among colony members, rates of queen supersedure, fates of subordinate co-foundresses, frequency of early-male production, male and gyne production schedules, sex investment ratio, timing of mating, male mating strategy, and degree of sexual dimorphism all have been found to vary among species, among populations within a species, or even within populations (e.g., in different phases of the colony cycle).

Many of these features co-vary in ways that appear explicable within modern evolutionary frameworks. For example, I have suggested that the co-variance in frequencies of subordinate strategies, intensity of dominance interactions, division of labor, distribution of reproduction, and queen supersedure rates is interpretable through an ecologically based model of female strategies that incorporates the optimal skew in reproduction from the viewpoint of a queen (Vehrencamp 1983a,b). Increasing ecological constraints on solitary founding appear to lead to greater frequencies of joining, higher reproductive skew, more frequent or intense dominance interactions, sharper division of labor, and higher rates of queen supersedure (the latter resulting from increased intracolony competition). However, the intercorrelations among the social variables and ecological factors are imperfect and still only suggestive. In conjunction with detailed analyses of the genetic structure of colonies, comparative studies are needed to test the model of optimal foundress strategies and also models of sex investment ratios, mating strategies, and sexual dimorphism. Only comparative studies with large sample sizes, standardized methodologies, and rigorous, quantitative data analyses are likely to elucidate the factors underlying variation in the *Polistes* social system. Detailed studies of tropical species, in particular, are sorely needed.

Much is known about the forms of social interactions within colonies, such as the kinds of presumed aggressive interactions and the variety of oscillatory behaviors. Although it is often assumed that these behaviors serve to advertise dominance, the precise functions of many of these behaviors are still unclear. For example, some of the "aggressive" oscillatory behaviors may serve to stimulate colony activity, and some or all of the oscillatory behaviors may be involved in adult-brood communication. The current emphasis on dominance interactions may reflect a tendency to underestimate the importance of cooperation in *Polistes*. Low-ranked workers (losers in social competition) are likely to have genetic interests that overlap extensively with those of the queen. Thus, these workers may have been selected to enter into subtle cooperative interactions with the queen (Reeve and Gamboa 1987). Well-controlled studies with independent samples (preferably separate colonies) and comprehensive behavioral sampling techniques are needed to determine the communicatory meanings of the various intracolony interactions. Such studies should permit a more precise characterization of the balance of cooperation and conflict in *Polistes* societies.

*Acknowledgments*

I thank B. Alexander, J. Boomsma, G. Eickwort, G. Gamboa, N. Jacobson, U. Mueller, F. Ratnieks, K. Ross, P. Sherman, T. Seeley, C. Starr, J. Strassmann, and K. Visscher for valuable comments on an earlier version of this chapter. I also thank K. Murphy, D. Pfennig, J. Shellman-Reeve, F. Ratnieks, and M. Webster for insightful discussions on a variety of issues relating to the chapter and J. Shellman-Reeve for technical assistance in its preparation.

**APPENDIX 4.1** Renesting Model

Assume that colonies are subject to a single predator attack. Let $n$ be the probability of nest predation on either a single- or a two-foundress colony, and let $s$ be the probability that an individual foundress will not survive the predation event and the period until renesting. Values of $s$ are assumed to be the same for a solitary foundress and for each foundress of a two-foundress colony, thus any synergistic effects of foundress association on the probability of surviving nest predation are eliminated. The expected reproduction, $S$, of a single-foundress colony is thus

$$S = (1-n)B_1 + n(1-s)B_1',$$

where $B_1$ is the expected productivity of a surviving solitary-foundress colony and $B_1'$ is the expected productivity of a renesting colony. The expected direct reproduction of a subordinate ($M_s$) in a two-foundress colony is

$$M_s = 2pB_1(1-n) + ns(1-s)B_1' + 2pB_1'n(1-s)^2,$$

where $p$ is the joiner's proportion of the total direct reproduction. It is assumed that there are no synergistic effects of foundress association on productivity. The expected direct reproduction of the dominant foundress ($M_d$) is obtained by substituting $1-p$ for $p$ in the above expression. In this model, a solitary foundress is less likely to renest after predation (probability $= 1 - s$) than is at least one foundress of a two-foundress colony (probability $= 1 - s^2$), in accordance with Gibo's (1978) data. (Strassmann et al. [1988: fig. 2A] plot the probability of renesting vs. original colony size $N$ in *P. bellicosus* for colonies of 1–12 individuals; the observed relationship is closely approximated by the curve $1 - s^N$, where $s = 0.74$.) By Hamilton's rule, joining will be favored over founding when

$$M_s - S + r(M_d - S) > 0,$$

where $r$ is the relatedness of the dominant's offspring to the joiner divided by the relatedness of the joiner's own offspring to herself. It can be shown that Hamilton's rule cannot be satisfied in this model when $p < 0.5$. (If $p > 0.5$, it would pay the joined foundress to leave and found singly.) Hence, the finding that multiple-foundress colonies have greater renesting success than solitary-foundress colonies does not by itself indicate a selective advantage to joining.

The situation changes, however, when predation occurs separately on foundresses, not on the whole nest, and joiners increase the dominant foundress' survivorship. In this case, joining can be favored because the nest, including the portion attributable to the subordinate's efforts, is *less likely to fail* through loss of all foundresses when there are multiple foundresses (the survivorship insurance model: Appendix 4.2).

**APPENDIX 4.2** Survivorship Insurance Model

Assume that the founding period is of duration $T$ and that deaths of foraging foundresses are Poisson-distributed with mean rate $m$. If all foundresses in a colony die in the founding period, the colony fails. Assume for simplicity that joiners have no direct reproduction. When a foundress is joined by another individual, she stops foraging and has negligible mortality. If a subordinate joiner subsequently dies (as the result of foraging), the joined foundress resumes foraging. Let $B_1$ and $B_2$ be the productivities of surviving solitary-foundress and two-foundress nests, respectively. The expected reproduction of a solitary-foundress colony is $S = B_1(e^{-mT})$. The expected total direct reproduction of a two-foundress colony is

$$M = B_1 \int_0^T me^{-mt}e^{-m(T-t)}dt + B_2 \int_T^\infty me^{-mt}dt = e^{-mT}[B_2 - B_1 ln(z)],$$

where $z$ is the solitary foundress's survival probability in the founding period. By Hamilton's rule, joining will be favored when $-S + r(M - S) > 0$, where $r$ is the relatedness of the joined foundress to the joiner. This condition can be rewritten as

$$r > 1/[k - 1 - ln(z)],$$

where $k = B_2/B_1$. If $r = 0.65$ (as in Metcalf and Whitt 1977a) and if $k = 1.6$ (as estimated from Klahn 1981, Noonan 1981, Queller and Strassmann 1988), it follows that survivorship insurance favors joining if the solitary-foundress survival probability ($z$) is less than 0.39. This implies a "critical" solitary-foundress failure frequency of about 60%. Interestingly, most species with high frequencies of multiple-queen founding have solitary-foundress failure rates greater than 60% (Fig. 4.6). In sum, if solitary foundresses have sufficiently high mortality rates, kin selection favors joining as colony survivorship insurance even if joiners have no direct reproduction.

Suppose further that a foundress loses her nest at time $t$ in the founding period. If the foundress renests, her productivity is lessened by an amount linearly related to $t$. Klahn (1988) and Strassmann et al. (1988)

found that later nesters have lower reproductive success. The productivity of a solitary-foundress colony started at time $t$ relative to that of a solitary-foundress colony started at time 0 can be expressed as $1 - t/R$, where $R$ is the extent of the period of the production of reproductives (i.e., later renesters have shorter periods of reproduction before colony termination). Let $k = 2 - t/T$, meaning that a joiner's contribution to the joined colony also decreases as the founding period progresses (i.e., $k$ starts at 2 and decreases to 1 at time $T$). By Hamilton's rule, joining will be favored over renesting when

$$r\{[(T-t)me^{-m(T-t)} + ke^{-m(T-t)}] - e^{-m(T-t)}\} - (1 - t/R)e^{-mT} > 0.$$

This can be simplified to

$$r > e^{-mt}(1 - t/R)/[1 + mT - t(m + 1/T)].$$

It can be shown that the critical value of $r$ must decrease as $t$ increases if $T > R$. $T$ is approximately equal to the brood developmental time (about 45 days or longer: Rabb 1960, West-Eberhard 1969, Yamane and Kawamichi 1975, Metcalf 1980, Miyano 1980, Strassmann and Orgren 1983, Page et al. 1989a). For temperate-zone species, $R$ is approximately 30 days or less (Miyano 1980, Pfennig and Klahn 1985). Thus, joining becomes favored over solitary founding and renesting as foundresses lose their nests later in the founding period.

**APPENDIX 4.3** Foundress Three-strategy Model

Assume that a potential subordinate foundress can found a nest, join a dominant relative on an established nest, or usurp a nonrelative on an established nest. Let $p$ equal the proportion of direct reproduction by a subordinate foundress; $x$ equal the probability of successful nest foundation by a potential subordinate foundress times the proportional expected productivity of such a nest relative to that of an already established solitary foundress nest; $B_1$ and $B_2$ equal the expected productivities of a joined nest and an established solitary-foundress nest, respectively; and $u$ equal the probability of successful usurpation by a potential subordinate foundress. Ecological constraints on solitary founding are indicated by low $x$ (e.g., nest site limitation or relatively low expected productivity of a newly established solitary-foundress nest). Usurpation probability $u$ can be viewed as the probability that an unrelated usurpation target will be found times the probability that the usurper will acquire the targeted colony. (Usurped colonies may have lower productivities than solitary-foundress colonies [Klahn 1988].) An additional coefficient representing this loss of productivity can be absorbed into $u$.) By Hamilton's rule, strategy $i$ will be favored over strategy $j$ if

$$(P_i - P_j) + r(K_i - K_j) > 0,$$

where $r$ is the coefficient of relatedness, $P_i$ (or $P_j$) is the personal reproduction associated with strategy $i$ (or $j$), and $K_i$ (or $K_j$) is the dominant relative's reproduction if strategy $i$ (or $j$) is performed. The $P$'s and $K$'s for each of the three foundress strategies are as follows:

| Strategy | $P$ | $K$ |
|---|---|---|
| Join dominant kin | $pB_2$ | $(1 - p)B_2$ |
| Usurp nonkin | $uB_1$ | $B_1$ |
| Found nest | $xB_1$ | $B_1$ |

By Hamilton's rule, founding a nest is favored over usurping if $x > u$. The conditions favoring joining over the other two strategies depend on $p$, which in turn will depend on how much reproduction a dominant yields to the subordinate. Let $K = B_2/B_1$. Using Vehrencamp's (1983a,b) approach and Hamilton's rule from the viewpoint of a dominant, we can see that the dominant kin will yield non-zero reproduction to the subordinate so as to make joining by the subordinate favorable, provided $r(K - 1) < x < K - 1$ (if founding is favored over usurpation) or $r(K - 1) < u < K - 1$ (if usurpation is favored over joining). A subordinate will join even if the dominant allows no subordinate reproduction ($p = 0$), provided $x < r(K - 1)$ (if founding is favored over usurpation) or $u < r(K - 1)$ (if usurpation is favored over founding). Founding will be the best strategy if $x > u$ and $x > K - 1$; usurpation will be the best strategy if $u > x$ and $x > K - 1$ or if $u > K - 1$ and $x < K - 1$. These predictions are summarized in the text.

# 5

# *Belonogaster, Mischocyttarus, Parapolybia*, and Independent-founding *Ropalidia*

## RAGHAVENDRA GADAGKAR

Animals that live in colonies of individuals of more than one generation, cooperate in brood care, and relegate reproduction to one or a small number of colony members are said to represent that pinnacle of social evolution, eusociality (Michener 1969, Wilson 1971). Except for the naked mole-rat, which lives in underground tunnels in Africa (Jarvis 1981), eusociality has been achieved only by ants and termites and by some bees and wasps. With the exception of one or a few species of sphecids (Matthews 1968a, this volume), all eusocial wasps belong to the family Vespidae. This family has traditionally been divided into three subfamilies, namely Stenogastrinae, Polistinae, and Vespinae (Richards 1962, 1978a,b). Carpenter's (1982) recent classification recognizes three additional subfamilies within the family Vespidae—namely, Eumeninae, Masarinae, and Euparagiinae—but eusociality is restricted to the three previously mentioned subfamilies.

Stenogastrines have sometimes been thought to be rather different from the polistines and vespines (Spradbery 1975), and their phylogenetic position within the Vespidae has been debated (Vecht 1977a), although Carpenter's (1982) cladogram indicates that the Stenogastrinae form the sister group to Polistinae + Vespinae (see Carpenter, this volume). Stenogastrines are regarded as primitively eusocial (see Cowan, this volume: Table 2.1, for definitions of levels of sociality) or, perhaps in some instances, as presocial (Turillazzi, this volume). Vespines represent a uniformly highly eusocial group (Matsuura, Greene, this volume).

That leaves the Polistinae, a rather large group consisting of 29 genera and about 800 species. Richards (1962) subdivides the Polistinae into three tribes: Ropalidiini (consisting only of the genus *Ropalidia*), Polistini (consisting of *Polistes* and "*Sulcopolistes*"), and Polybiini (consisting of about 26 genera). However, there are problems with this classification (see Distribution and Systematics). In any case, further subdivision of the Polistinae based on phylogenetic considerations may be less rewarding to students of social evolution than subdivision based on behavior and nest architecture. An elegant example of the latter has been provided by Jeanne (1980a) who distinguishes two subgroups, the *independent-founding* and the *swarm-founding* Polistinae (see also Jeanne, this volume: Chap. 6).

Independent-founding Polistinae live in relatively small colonies (rarely more than 100 adult wasps) and construct small, simple, unenveloped combs that are normally suspended by a narrow pedicel. Queens initiate new colonies either singly or in small groups, but without the aid of workers (that is, independently). Swarm-founding Polistinae live in more populous colonies (often with 1,000 or more adults) and have correspondingly large nests that sometimes have several tiers of combs covered by an envelope. New colonies are always initiated by queens in the company of workers (that is, by swarms). This classification on the basis of behavioral and architectural features is supported by concomitant morphological specializations.

Queens of the independent-founding Polistinae, with their small colonies, do not normally use a pheromonal means of control over their nestmates; overt physical dominance is the rule. Queens in swarm-founding groups, however, must find it quite impossible to subjugate large numbers of nestmates by physical dominance and have, perhaps for this reason, evolved a pheromonal means of control. Independent-founding species have a special problem with ants. Direct physical resistance to marauding ants is difficult enough with small numbers of adult wasps, but when the nest and its brood have to be left completely unguarded while a single foundress is away foraging it is impossible. Not surprisingly, independent-founding species have evolved a chemical defense against ants. Independent-founding species have a well-developed gland (named van der Vecht's gland) on the sixth (terminal) gastral sternum from which an ant-repellent substance is secreted. The gland is associated with a tuft of hairs that presumably serves as an applicator brush (Jeanne et al. 1983). Swarm-founding species never leave their nests unattended and usually have a substantial adult population to physically rid the nest of scouting ants. Also, they typically build enveloped nests with only one or a small number of entrance holes. Most swarm-founding species studied appear not to have the specialized ant repellent–producing van der Vecht's gland. Instead, at least some of them have a well-developed Richard's gland on the fifth gastral sternum that is often used to lay an odor trail to guide the members of a swarm to their new nesting site (Jeanne 1981a, Jeanne et al. 1983; see also Downing, this volume).

Swarm founding is characteristic of 24 polistine genera, as well as some species of *Ropalidia.* Independent founding is seen in five genera: *Belonogaster, Mischocyttarus, Parapolybia, Polistes,* and some *Ropalidia.* It is easy to argue that the independent-founding polistines provide perfect model systems for studies of social behavior. Small colonies make it possible to mark and study every adult wasp, and the open combs hide nothing from the observer. The nests of many species are quite abundant and often are built in remarkably accessible places. A relatively primitive level of eusociality, characterized by a lack of morphological caste differentiation, an essentially behavioral mechanism of queen control, and considerable flexibility in social roles of the adult wasps make the independent-founding polistines especially attractive subjects. Indeed, they have received increasing attention since the 1970s and have contributed more than any other group of social wasps to the formulation and testing of ideas concerning the forces that mold the evolution of group living and sterile worker castes.

The genus *Polistes,* unrivaled in its wide distribution and in the attention it has received, is the subject of the previous chapter (Reeve, this volume). The present chapter deals with the remaining four genera of independent-founding Polistinae. Some earlier information regarding these wasps may be obtained from several general reviews (Richards 1971, Spradbery 1973a, Iwata 1976, Akre 1982).

## DISTRIBUTION AND SYSTEMATICS

*Belonogaster, Parapolybia,* and *Ropalidia* are restricted to the Old World, whereas *Mischocyttarus* is found exclusively in the New World (Vecht 1965, 1967) (Fig. 5.1). All four genera are best represented in the tropics, although at least a few species of each extend into neighboring temperate latitudes. *Belonogaster,* comprising 79 species, is widely distributed in Africa south of the Sahara, with scattered populations in northern Africa, Arabia, and India. The latest comprehensive taxonomic revision of the genus is that of Richards (1982; see also Hensen and Blommers 1987). Since Roubaud's (1916) classic study of *B. juncea,* the biology of only one species, *B. grisea,* has been studied in detail (Marino Piccioli and Pardi 1970, 1978; Pardi 1977; Pardi and Marino Piccioli 1970, 1981). More recently, several modern studies have been initiated on *B. juncea* and *B. petiolata* (Richards 1969, Keeping and Crewe 1983, Kojima and Keeping 1985, Keeping et al. 1986).

*Mischocyttarus* is one of the largest genera of social vespids. In spite

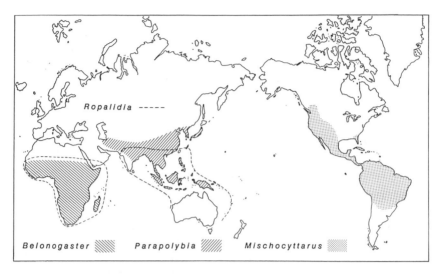

**Fig. 5.1.** Distributions of *Belonogaster, Mischocyttarus, Parapolybia,* and *Ropalidia.* (Redrawn from Vecht 1967, courtesy of the author and Koninklijke Nederlandse Akademie van Wetenschappen.)

of its taxonomic diversity (206 species), it is essentially restricted to tropical South America, although two species extend into the southern and western portions of North America. The most authoritative account of the systematics of this genus is that of Richards (1978a). A few species have been studied in considerable detail. Notable among these are *M. drewseni* in Brazil (Jeanne 1972), *M. labiatus* in Colombia (Litte 1981), *M. flavitarsis* in Arizona (Litte 1979), and *M. mexicanus* in Florida (Litte 1977) and in Georgia (Hermann and Chao 1984a). Preliminary information has been published on *M. angulatus* and *M. basimacula* in Panama (Itô 1984b).

*Parapolybia* is a refreshingly small genus with only five species distributed from Iran in the west, to Japan, the Philippines, and New Guinea in the east (Vecht 1966). Only three of the five species are known to any extent. Yamane (1980, 1984, 1985) has provided a detailed account of *P. varia* in Taiwan along with fragmentary information on *P. nodosa*, while *P. indica* in Japan has been studied by Sugiura et al. (1983a,b) and Sekijima et al. (1980).

*Ropalidia*, another large genus, with about 136 known species, occurs in tropical Africa, southern Asia, Australia, and Okinawa. The Indo-Australian species have been revised by Vecht (1941, 1962), the Australian and New Guinea species by Richards (1978b), the Philippine species by Kojima (1984a), the Nepalese species by Yamane and Yamane (1979), and the Indian species by Das and Gupta (1989). Although Vecht (1962) recognized three subgenera, it is now common to recognize six (Richards 1978b). Of these, the subgenera *Anthreneida, Icariola,*

*Polistratus,* and *Ropalidia* can be characterized as independent-found-ing, although at least a few species in the subgenus *Icariola* appear to be swarm founders (R. Jeanne, unpubl.). *Ropalidia marginata* (Gadgil and Mahabal 1974; Gadagkar 1980; Belavadi and Govindan 1981; Gad-agkar et al. 1982a,b; Gadagkar and Joshi 1982a, 1983), *R. cyathiformis* (Gadagkar and Joshi 1982a,b, 1984, 1985), *R. variegata* (Yamane 1986), *R. fasciata* (Itô 1983, 1985b, 1986b; Itô et al. 1985; Kojima 1983a,b,c, 1984b,c; Suzuki and Murai 1980), and *R. cincta* (Darchen 1976a) are the only independent-founding species studied in any detail, and all of these belong to the subgenus *Icariola.*

The phylogeny of social wasps is now being vigorously investigated (Carpenter 1982, 1987a,b, 1988a, this volume; Rasnitsyn 1988), partic-ularly with regard to the relationships within the diverse subfamily Polistinae. The main problem with Richards's (1978a) subdivision of Polistinae into tribes is that the four Old World polistine genera (in-cluding the swarm-founding *Polybioides*) appear to have affinities that unite them as a natural group, so that Richards's inclusion of the Old World genera *Parapolybia, Polybioides,* and *Belonogaster* with the New World swarm-founding genera in the tribe "Polybiini" is inappropriate. A fascinating behavioral trait linking the Old World polistine genera is their meconium extraction behavior. Usually in social wasps, the larval meconium (fecal pellet) is left at the bottom of the brood cell, but *Be-lonogaster, Polybioides, Parapolybia,* and *Ropalidia* have evolved an elabo-rate behavior of chewing a small hole at the bottom of the cells and removing the meconium (Marino Piccioli 1968; Marino Piccioli and Pardi 1970, Kojima 1983c, Gadagkar, unpubl.). Although differences in the details of this behavior and the means by which the holes are closed are evident, all four Old World genera have windows at the bottoms of cells from which adults have emerged, while such features have not been seen in any of the remaining 25 genera of polistines. This fact alone strongly suggests a monophyletic origin for the four Old World genera (Jeanne 1980a), as Carpenter's cladistic analysis (this vol-ume) confirms.

## NEST ARCHITECTURE

Social wasps show considerable variation in nest architecture (Jeanne 1975a, Vecht 1977b, Wenzel, this volume), but the independent-found-ing polistines all build simple stelocyttarus (suspended by a pedicel), gymnodomous (unenveloped) combs, so that there is relatively little variation in nest architecture within the group (Fig. 5.2). A single comb is the rule, and this may be either oriented in a vertical plane (held by a horizontal pedicel) or oriented in a horizontal plane with the cells and brood facing downward (held by a vertical pedicel). The comb is usu-

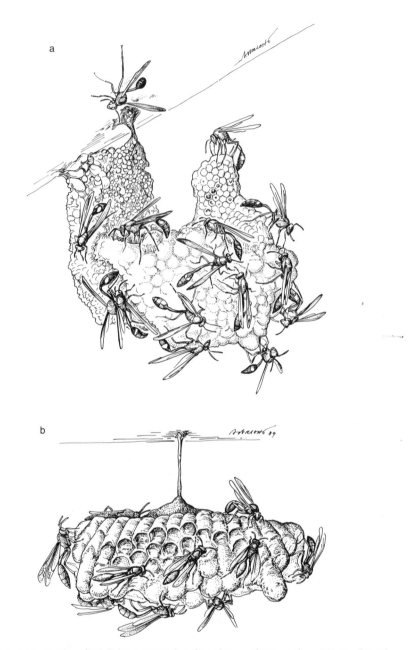

**Fig. 5.2.** Mature nest of (a) *Belonogaster grisea* (based on a photograph in Marino Piccioli and Pardi 1970, courtesy of L. Pardi and *Monitore Zoologico Italiano*), (b) *Mischocyttarus drewseni* (based on a photograph in Jeanne 1972, courtesy of the author and Museum of Comparative Zoology, Harvard University), (c) *Parapolybia varia* (based on a photograph in Yamane 1984, courtesy of the author and VEB Gustav Fischer Verlag), and (d) *Ropalidia marginata* (based on an original photograph). All views are from the side except (d), in which the comb face is viewed from the front. This comb is perpendicular to the ground and attached to the substrate by a horizontal pedicel that is not visible.

c

d

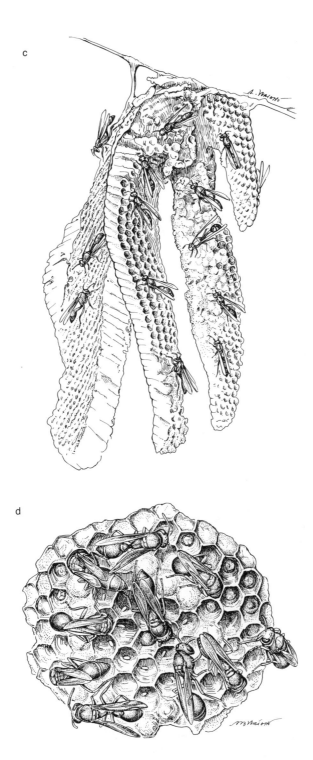

ally circular or oval, although it may be irregular or may have sharp corners to suit the substrate on which it is constructed. Some species invariably build long combs with two columns of cells (e.g., *R. variegata*: Davis 1966a,b, Yamane 1986, Gadagkar, unpubl.; *R. revolutionalis*: Hook and Evans 1982, Itô 1987a; *M. punctatus*: Richards 1978a) (Fig. 5.3).

**Fig. 5.3.** A typical long nest of *Ropalidia variegata* with two columns of cells.

The suspending of the comb by a long and narrow pedicel coated with an ant repellent is clearly a major defense of these wasps against predation by ants (Jeanne 1970a), although this form of defense appears to be compromised in many nests of *Ropalidia* where the pedicel is short and many secondary pedicels are constructed (see Enemies and Colony Defense). Furthermore, such a means of chemical defense remains to be demonstrated in *Parapolybia*.

Sometimes several combs are built close to one another and function as a single colony (Gadagkar and Joshi 1982b, Kojima 1984b, Herre et al. 1986, Itô 1986b) (Fig. 5.4). The advantages of this are not entirely

**Fig. 5.4.** Multiple combs of a *Ropalidia* nest from India.

clear, although this strategy might conceivably minimize damage from certain predators or parasites (see Jeanne 1979a). At least some instances of multiple-comb construction are for a rather different reason. When renesting after the original nest is destroyed, females working on a new nest start construction of several combs, thereby providing opportunities for many females to simultaneously perform the task of nest construction (S. Chandran and Gadagkar, unpubl.; see also Kojima 1984b).

*Belonogaster grisea* and *P. varia* show the greatest degree of deviation from the nest architecture typical of the independent-founding polistines. Nests of *B. grisea* are single-combed and are always suspended by means of a single eccentric pedicel attached to the first cell (Marino Piccioli and Pardi 1978). As the nest grows, it takes on a unique hammocklike appearance, with a highly convex ventral surface and a concave dorsal surface (see Fig. 5.2a). Another atypical characteristic of *B. grisea* nests is that cells are not reused but are often wholly or partly destroyed after the first use. In one population of *P. varia* in Taiwan (Yamane 1984), nests are always multicombed (Fig. 5.2c). The first comb is petiolate, and its cells may be reused. As more brood-rearing space is required, two or more lateral lobes are added, and later up to ten additional combs are constructed close to the original one. These subsequent combs are all without pedicels, and their cells are not reused.

Description of nest architecture is by necessity based on studies of a few isolated species, so we still have little appreciation of the relative extent of variation within and among species (see, for example, Carpenter and Wenzel 1988). *Ropalidia* colonies in India show considerable variation in nest architecture, and the wasps clearly vary the shapes and sizes of their nests to suit the available nesting site (Davis 1966a; Gadagkar, unpubl.). *Ropalidia marginata* is capable of at least sometimes constructing multicombed nests similar to those of *P. varia* in Taiwan (see Figs. 5.2c, 5.5).

## NESTING CYCLE

New nests are initiated independently (without workers) by one inseminated female (haplometrosis) or a small group of them (pleometrosis) (Table 5.1; see also Spradbery, this volume). In single-foundress colonies, the lone female forages for building material and for food for her larvae, builds the nest, lays eggs, and defends the nest until her first progeny become adults. Foundresses in multiple-foundress colonies, who are often sisters or other close relatives (e.g., Jeanne 1972, Litte 1977), engage in aggressive interactions that lead to the establish-

**Fig. 5.5.** An unusual multilobed nest of *Ropalidia marginata* from Bangalore, India. At the time of collection only one of the combs was occupied.

ment of a dominance hierarchy. One foundress at the top of this hierarchy often monopolizes all egg laying and does little else except occasionally forage for building material. Foraging for food and other duties of nest and brood maintenance are performed by one or more of the subordinate foundresses.

Immediately after nest initiation there is usually a rapid increase in the number of cells, all of which are filled with eggs. Foraging is mostly for building material during this so-called *egg substage*. When the eggs begin to hatch, the foundress(es) start feeding the larvae. Mature larvae spin a cap on their cells and undergo metamorphosis. The period from the hatching of the first egg to the spinning of the first cocoon is referred to as the *larval substage,* and the period from the spinning of the first cocoon to the emergence of the first adult is called the *pupal substage.* The entire period from nest initiation to the emergence of the first adult offspring is called the *preemergence phase* (*founding phase* of Reeve, this volume).

Subordinate foundresses usually begin to die or to disappear from the nest at about the time of the emergence of adult progeny. This first batch of offspring, nearly always females, become workers and take on the duties of foraging, nest building, feeding larvae, and other tasks involved in nest maintenance and defense. This is called the *worker production substage* (*worker phase* of Reeve, this volume) or, sometimes, the *ergonomic substage* (Oster and Wilson 1978). At this point, new cells are usually added to the nest. Later, males and nonworker females

**Table 5.1.** Single- and multiple-foundress colonies of *Belonogaster*, *Mischocyttarus*, *Parapolybia*, and independent-founding *Ropalidia*

| Species | Single-foundress colonies No.(%) | Multiple-foundress colonies No.(%) | Max. no. of foundresses | Time of nest founding | Locality | Reference |
|---|---|---|---|---|---|---|
| *Belonogaster petiolata* | 38(47) | 43(53) | 16 | ? | Transvaal, South Africa | Keeping and Crewe 1987 |
| *Mischocyttarus drewseni* | 20(69) | 9(31) | 8 | Throughout the year | Lower Amazon, Brazil | Jeanne 1972 |
| *M. flavitarsis* | 131(97.8) | 3(2.2) | 2 | March–June | Arizona, USA | Litte 1979 |
| *M. labiatus* | 14(77.8) | 4(22.2) | 9 | Throughout the year | Colombia | Litte 1979, 1981 |
| *M. mexicanus* | 114(66.7) | 57(33.3) | 20 | Throughout the year | Southern Florida, USA | Litte 1977, 1979 |
| *Parapolybia indica* | 49(100) | 0 | 1 | May | Southwestern Japan | Sugiura et al. 1983b |
| *P. indica* | 108(100) | 0 | 1 | May | Southern Japan | Sekijima et al. 1980 |
| *P. varia* | 2(100)[a] | 0[a] | 1 | March–May | Central Taiwan | Yamane 1980, 1985 |
| *P. varia* | 1(7) | 13(93) | 22 | December–February | Southern Taiwan | Yamane 1985 |
| *Ropalidia fasciata* | 331(75) | 109(25) | 13 | February–April | Okinawa, Japan | Suzuki and Murai 1980 |
| *R. fasciata* | 27(46.6) | 31(53.4) | 22 | February–April | Okinawa, Japan | Itô 1985b |
| *R. marginata* | 8(29.6) | 19(70.4) | 20 | Throughout the year | Pune and Bangalore, India | Gadagkar et al. 1982a |
| *R. variegata jacobsoni* | 37(63.8) | 21(36.2) | 4 | Throughout the year? | Sumatra | Yamane 1986 |

[a] Although only two nests were studied in detail, anecdotal evidence suggests that single-foundress colonies may be the rule (Yamane 1980, 1985).

(reproductives or gynes) are produced. In some species, this *male and nonworker production substage* (*reproductive phase* of Reeve, this volume) is well separated from the worker production substage. After the reproductives emerge, there is a considerable amount of brood destruction during the so-called *declining substage*, after which the colony usually is abandoned. The time from the emergence of the first worker to nest abandonment is called the *postemergence phase*.

The preemergence phase and the postemergence phase together constitute what might be called a colony cycle. If the nest is abandoned at the end of one such cycle, the colony cycle and the nesting cycle become equivalent; species that show this pattern may be called *determinate nesting cycle species* (after Jeanne, this volume: Chap. 6). The nesting cycles in a population of such a species are sometimes markedly seasonal and fairly synchronous (see Table 5.1). In other cases, nests are initiated asynchronously at all times of the year, but the nesting cycle is still determinate because every nest goes though the six substages of a colony cycle, after which the nest is abandoned. Finally, several colony cycles may be repeated on the same nest because the nest is not necessarily abandoned at the end of each declining substage. Species that show this pattern may be called *indeterminate nesting cycle species* (after Jeanne, this volume: Chap. 6).

## Seasonal Nesting Cycle

Most species occurring at temperate and subtropical latitudes show a markedly seasonal nesting cycle. Examples include *B. juncea* and *B. petiolata* in southern Africa (Keeping and Crewe 1983), *M. flavitarsis* in Arizona (Litte 1979), *M. mexicanus* in Georgia (Hermann and Chao 1984b), *P. indica* in Japan (Sugiura et al. 1983b), *P. varia* in Taiwan (Yamane 1980), and *R. fasciata* in Okinawa (Suzuki and Murai 1980, Itô 1983, Itô et al. 1985). Nests are typically initiated by inseminated, overwintered females early in the spring (see Fig. 5.6). Females that emerge early in the colony cycle become workers, at least partly because there are no males at this time of the year with whom they can mate. However, males of *R. fasciata* are produced at all stages of the colony cycle, and females of this species emerging from the first brood can mate and found their own nests (Itô and Yamane 1985). Colonies that show a seasonal cycle begin to decline in the fall; wasps that emerge in late summer remain on the nest for a while as nonworking individuals but later leave the nest and mate. The males die and the females hibernate, either on their natal nest (e.g., *R. fasciata*: Itô et al. 1985) or away from the nest in crevices of rocks or under the bark of trees.

The proximate causes of colony decline during the fall are not clear. The accumulation of nonworking males and females at this time may

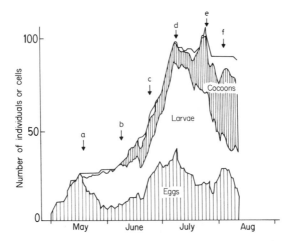

**Fig. 5.6.** Features of the colony cycle of *Parapolybia indica* in Japan illustrating the seasonal nesting cycle. a, first hatching of larva; b, first appearance of capped cell; c, first emergence of adult; d, death of critical number of workers; e, first male emergence; f, death of foundress. The uppermost line indicates total number of cells. (Redrawn from Sugiura et. al 1983b, courtesy of the authors and *Bulletin of the Faculty of Agriculture, Mie University.*)

be as important as declines in the food supply. However, seasonal changes in temperature, daylength, and humidity are likely to have a profound effect on the form and duration of the colony cycle, as evidenced by the relationship between duration of the nesting period and local climate. At the extremes, *P. indica* in Japan nests for only about three-and-a-half months, whereas *P. varia* in southern Taiwan nests throughout the year and does not hibernate (Fig. 5.7). Indeed, within species two examples are known in which populations follow seasonal cycles in one environment and aseasonal cycles in another (*M. drewseni drewseni*: Dantas de Araújo 1982; *M. mexicanus cubicola*: Hermann and Chao 1984b).

## Aseasonal Determinate Nesting Cycle

An excellent example of an aseasonal determinate cycle is provided by *M. drewseni* in Brazil (Fig. 5.8). Nests are initiated throughout the year, go through the typical cycle including the declining substage, and are always abandoned after about six months (Jeanne 1972). *Mischocyttarus mexicanus* in Florida seems to follow an identical pattern (Litte 1977). Because nests are asynchronously initiated, the population is expected to contain males throughout the year. It is thus quite common in these species for inseminated daughters to replace their mothers as queens or to usurp other colonies. Nevertheless, there seems to be a

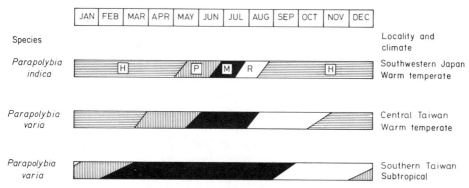

**Fig. 5.7.** Comparison of nesting cycles of *Parapolybia indica* in warm-temperate south-western Japan, *P. varia* in warm temperate central Taiwan, and *P. varia* in subtropical southern Taiwan. H, hibernating phase; P, preemergence phase; M, matrifilial phase (worker production substage); R, reproductive phase (male and nonworker production substage). (Redrawn from Sugiura et al. 1983b and Yamane 1980, courtesy of the authors and *Bulletin of the Faculty of Agriculture, Mie University*.)

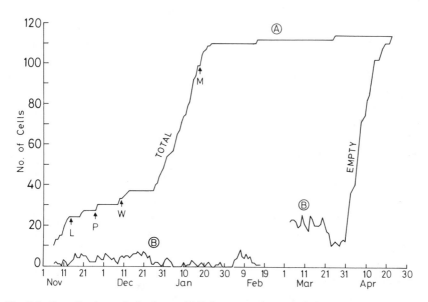

**Fig. 5.8.** Growth of a typical colony of *Mischocyttarus drewseni*. Colonies may be initiated at any time of the year. A, total cells; B, empty cells; L, first larva eclosed; P, first larva spun its cocoon to pupate; W, first adult worker emerged; M, first adult male emerged. (Redrawn from Jeanne 1972, courtesy of the author and Museum of Comparative Zoology, Harvard University.)

163

pronounced tendency to produce typical workers in early broods and nonworking females (potential reproductives) in later broods. What is the cause of this tendency, and why does the colony decline so regularly after six months of existence, independently of the season? Jeanne (1972) has argued rather convincingly that it is not environmental conditions but intrinsic factors that regulate the nesting cycle. Jeanne's hypothesis is that physical domination by the queen is necessary for a female to assume the role of worker, and as the colony grows larger more females escape such domination and become nonworkers. With a sharp increase in the ratio of nonworkers and males to workers, all go hungry, leading to brood abortion and adult dispersal.

## Indeterminate Nesting Cycle

*Ropalidia marginata* in peninsular India provides the only clearly documented case of an indeterminate nesting cycle among independent-founding polistines (Gadgil and Mahabal 1974, Gadagkar et al. 1982a,b, Chandrashekara et al. 1990). Nests are initiated throughout the year either haplometrotically or pleometrotically. Rates of colony failure are very high in young colonies, but successful colonies appear to go through all the stages of the typical colony cycle, including a declining substage involving brood abortion (Fig. 5.9). The main difference between the determinate nesting cycle of the *M. drewseni* type and the indeterminate nesting cycle of the *R. marginata* type is that dispersal of adults during the declining substage is not complete in the latter, so that a small number of females remain to start another colony cycle in the same nest. A given nest may be used for a series of such cycles.

Why are colony decline and abandonment not complete? One possibility is that these colonies never reach the stage where the queen is unable to dominate her daughters, either because many of the daughters leave their parental nests throughout the colony cycle or because some mechanism of queen control other than physical domination is important. If the queen, in fact, does not become incapable of dominating her daughters, then why is there a declining substage in the first place? It is conceivable that what appears to be a declining substage is really a period of intense reproductive competition and that the individual who wins in this competition may start the colony cycle anew. Alternatively, the indeterminate nesting cycle, which appears to encompass several repeats of the determinate nesting cycle, may have evolved in response to predation by *Vespa tropica*, which is especially relentless on large colonies. Colonies of *R. marginata* might find it adaptive to issue "swarms" of dispersing wasps to found new nests periodically before all is lost to *Vespa tropica*.

Future studies of such indeterminate nesting cycles are bound to

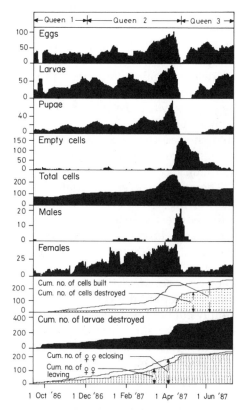

**Fig. 5.9.** Development of a *Ropalidia marginata* colony transplanted into a cage from which free foraging was allowed. This colony reached a maximum standing crop of 265 cells, 103 eggs, 96 larvae, 76 pupae, 21 adult males, and 47 adult female wasps. After considerable brood and cell destruction, the colony declined in late April to a standing crop of 164 cells, 4 eggs, 0 larvae, 0 pupae, 0 adult males, and 17 adult female wasps, only to continue a second colony cycle on the same comb.

provide many interesting new facts regarding the regulation of colony cycles in these wasps. Because daughters may leave their natal nests to initiate new nests more-or-less continuously, indeterminate nesting cycles may make the composition of a founding group quite heterogeneous in terms of age and social status. We know of at least one clear case in *R. cyathiformis* where a group of individuals who already had well-defined social roles on their natal nest left and initiated a new nest, largely retaining their previous roles (Gadagkar and Joshi 1984, 1985). This is not very different in principle from swarm founding. It therefore seems likely that more detailed studies of indeterminate cycles in independent-founding *Ropalidia* species will offer many surprises and may even obliterate the distinction between independent founding and swarm founding.

## ENEMIES AND COLONY DEFENSE

Adult wasps in the genera considered here seem to have few serious enemies, but a variety of parasites and predators occasionally maraud their nests (Appendixes 5.1 and 5.2). Birds, lizards, and bats some-times plunder the nests, and tachinid and phorid flies, ichneumonid and torymid wasps, and pyralid moths may seriously affect produc-tivity by parasitizing the brood. The wasps appear to have no particu-lar defense against vertebrate predators, with the exception of *Mischo-cyttarus immarginatus*, which nests in association with the much more aggressive wasp *Polybia occidentalis* and thereby presumably derives some protection (Windsor 1972, Gorton 1978). It has been suggested that choice of nesting site as well as coloration of nests may help in avoiding predators (Fitzgerald 1950, Kojima 1982a, Hermann and Chao 1984a,c).

There is also no specific defense against parasites, although the alert-ness of the adults and their efforts to chase away parasites undoubt-edly reduce the extent of parasitism. The most common response to parasitized brood is simply to ignore those cells until the parasite emerges, and then to reuse the cells. Somewhat surprisingly, the wasps seldom make any attempt to remove or destroy the affected lar-vae or pupae. *Mischocyttarus labiatus* is a clear exception to this rule. An attack by a phorid moth can be so devastating to a nest that the queen simply cuts the pedicel and lets the whole nest fall to the ground to be consumed by ants—brood, parasites, and all—and begins construction of a new nest in roughly the original spot (Litte 1981).

Perhaps the most serious natural enemy of *Ropalidia* in southern In-dia (Gadagkar, unpubl.) and *Parapolybia* in Taiwan (Yamane 1980) is the hornet *Vespa tropica*, whose workers systematically search for, locate, and consume almost the entire brood of large and conspicuous prey colonies (see Matsuura, this volume). Adults of the prey species are untouched, and they usually sit around the nest completely helpless until the predator departs. They then return to the nest, inspect the cells with great agitation, and either cannibalize any remaining brood and abandon the nest, or sometimes continue to produce more brood on the same nest. Often the queen and some of the workers stay while others leave.

In the Neotropics, army ants must pose some threat to wasp colo-nies, although their overall impact on wasp populations is difficult to estimate. Several species of *Mischocyttarus* have presumably been under sufficient selection pressure from army ants to evolve nesting associa-tions with "army ant-resistant" ant species. Herre et al. (1986) found that 29 out of 31 active social wasp nests, including those of ten species of *Mischocyttarus*, were built on plants occupied by *Allomerus* and *Phei-dole* ants. It is clear that by such association the wasps derive significant protection from army ants, which consistently avoid such ant-bearing plants.

What is the most important enemy of these social wasps? Should this distinction be given to the most devastating of the enemies seen in action today or to those against whom the wasps have found it necessary to evolve effective defenses? This conundrum is best illustrated by an examination of Appendixes 5.1 and 5.2, which may tell us little about those enemies that have decided the course of evolution of these wasps. The reference here is to a variety of species of ants where foraging is carried out by individual scouts who can quickly recruit additional workers upon finding a valuable source of food. The reason that the importance of such predators is not obvious is that most wasp species have evolved fairly effective defenses against them. To appreciate the overriding importance of ant predation one has only to break down such defenses: remove the adults or break the pedicel and drop the nest to the ground; it will be discovered and devoured by ants within minutes.

The first line of defense appears to be the single narrow pedicel suspending the comb, which restricts routes of access by ants. As a second line of defense, the pedicel is usually coated with an ant repellent produced from the van der Vecht's gland. *Mischocyttarus drewseni* is the prototype for these two forms of defense (Fig. 5.10). The nest pedicel is very narrow and long in this species, not only physically restricting access by ants but also making it easy to keep coated with ant repellent. The gland and the brush are well developed in this genus, and the efficiency of the ant repellent has been demonstrated (Jeanne 1970a).

Females of *B. grisea* (Marino Piccioli and Pardi 1970) and those of *B. petiolata* (Keeping 1990) have also been seen to rub secretions from their van der Vecht's glands onto their nest pedicels. Several recent observations (Keeping 1990) support the role of these secretions in chemical defense in this genus.

*Ropalidia* females rub their abdomens on the pedicels, the gland and brush are well developed, and the efficacy of the ant repellent has been demonstrated (Kojima 1982b, 1983a; Gadagkar, unpubl.). On the other hand, the pedicels are often so short that most ants can probably get to the nest without crossing the pedicel. Besides, secondary pedicels are often constructed at several points to strengthen the attachment of the nest to the substrate (contrast Fig. 5.11 with Fig. 5.2b), and the wasps do not appear to rub their abdomens on the secondary pedicels. All this probably diminishes the importance of the first two lines of defense in *Ropalidia*.

The third line of defense is the constant guarding of the nest by the adults, who become very agitated when ants are moving nearby. This is, of course, unavailable to a single foundress away foraging, but it appears that larger nests of *Ropalidia* depend primarily on the third line of defense. It is in recognition of the importance of the third line of

**Fig. 5.10.** Female *Mischocyttarus drew-seni* applying ant-repellent secretion by rubbing the tuft of hair on the terminal gastral sternum against the nest pedicel. (From Jeanne 1972, courtesy of the author and Museum of Comparative Zoology, Harvard University.)

defense that multiple-foundress associations are often considered an antipredator adaptation (see Origin of Social Life: A Perspective from Studying Independent-Founding Polistines).

In a 20-month study involving 113 nests, Yamane (1980) never once saw *P. varia* females rubbing the nest pedicel with their abdomens, although Kojima (1983b) reports that he saw one wasp doing it once! More important, these wasps enlarge their nest by adding several additional combs that have no pedicels. Clearly the third line of defense, namely, active defense by adults, is most important in this species too. It is probably true for all species that predation pressure from ants can be quite serious in the absence of defenses but that existing defenses are quite effective. It has therefore been rightly emphasized that predation pressure from ants has been a very important factor in the course of evolution of eusocial wasps (Vecht 1967, Jeanne 1975a), especially in the tropics (Jeanne 1979b).

A similar conundrum over the relative importance of historical versus current predation pressures concerns the function of the sting and venom. The most striking feature of wasps to anyone who comes

**Fig. 5.11.** A large comb of *Ropalidia marginata* viewed from the side. Notice the large number of pedicels and the narrow distance between the nest and the substrate.

into casual contact with them is their often formidable sting. But detailed studies show no profound role for the sting in the defensive biology of most independent-founding species. Deliberate attention to the role of the sting brings the verdict that "the venom apparatus appears to fall short as a defensive mechanism against their chief vertebrate predators" (Hermann and Chao 1984c:339; see also Starr 1985b, 1989a; Kukuk et al. 1989). Here again we must imagine that the concentrated resource that large nests with brood represent was exploited by a large number of vertebrate and invertebrate predators during the evolutionary history of the wasps, and that the sting evolved as a means of deterring these former enemies. Only a few specialized predators have been able to overcome this defense to become important present-day enemies.

## FOOD AND FEEDING HABITS

This section is intended to highlight our remarkable ignorance of the food and feeding habits of the independent-founding polistines. Barring Jeanne's (1972) study of *M. drewseni*, we have little to go by. Adult wasps forage for two classes of food: arthropods and nectar. It is a safe bet that there is no great specificity about which arthropod species are taken or where the nectar is collected from; easy availability is probably the deciding factor. These wasps are known to hunt lepidopteran larvae (Belvadi and Govindan 1981); spiders (K. Chandrashekara and Gadagkar, unpubl.); and moths, ants, spider eggs, hemipteran nymphs, and tettigoniid grasshoppers (Jeanne 1972). *Mischocyttarus* wasps not only seek out insects caught in spider webs but are actually able to land and walk along the strands of the web, presumably on account of a special adaptation of the tarsal lobes of the mid- and hind-legs (Jeanne 1972). Foragers returning to the nest with masticated solid food routinely share it with their nestmates, including males. The food is further masticated by the adult wasps, during which juices are extracted and imbibed. Larvae are fed with the solid lumps as well as by regurgitation after the lumps have been masticated. Males may also occasionally offer solid food to larvae (e.g., Pardi 1977) but have never been observed to regurgitate.

Nectar collected from a variety of flowers, and sometimes honeydew collected from mealybugs or other homopterans, is also brought to the nest and shared with nestmates and larvae. Nectar is occasionally stored in empty cells or in cells containing eggs or young larvae. This is not for direct consumption by the larvae but for later removal, distribution, and consumption by the adults. Such nectar storage is usually more frequent during times of food scarcity.

Larvae regularly produce droplets of a liquid imbibed by the adults, who often enter larval cells and solicit the secretion by mouthing the larvae. Such larval-adult trophallaxis is a rather striking feature of wasp societies but remains very poorly studied. It is clear from the studies of Hunt et al. (1982) that wasp larval secretion has enormous nutritive value for the adults, and it almost certainly is a glandular secretion rather than simply a regurgitation of food given to the larva. A recent attempt to better understand the significance of this form of trophallaxis is that of Hunt (1988, this volume), who seems to have shown that *Mischocyttarus* larvae in preemergence colonies are more likely to surrender saliva than are those in postemergence colonies. His hypothesis to explain this differential behavior is that larvae in preemergence colonies face a greater risk of being cannibalized by the adults, who are themselves likely to be undernourished during this phase of the colony cycle. On the other hand, the better nourished adults in post-emergence colonies are less likely to cannibalize larvae. In other words, surrendering of saliva is hypothesized to be a kind of appeasement behavior by larvae, meant to forestall their own deaths by cannibalism.


## MATING BEHAVIOR

Most studies of presocial and primitively eusocial insects aim either to unravel the mechanism of social organization and integration or to understand the origins of cooperative behavior. Quite understandably, males, which appear peripheral to these questions, have largely been neglected. Typically, colonies first produce several worker females and later produce males and nonworking (reproductive) females. Some species, such as *M. drewseni* (Jeanne 1972) and *P. indica* (Sugiura et al. 1983b), produce males only during a relatively short period toward the end of the colony cycle. Consequently, males of species with a seasonal nesting cycle, such as *P. indica*, are present during only a small part of the year and opportunities for mating are restricted. The aseasonal and asynchronous nesting cycle of *M. drewseni*, however, ensures the presence of males in the population throughout the year.

*Ropalidia fasciata* (Itô and Yamane 1985) and *R. cyathiformis* (Gadagkar and Joshi 1984) males are produced over a fairly long period, so that even if the nesting cycle is seasonal opportunities to mate are more widespread. This has important consequences for social organization, as females reared from the first brood can potentially mate and establish their own colonies or usurp the egg-laying position of their mothers (for examples in *Polistes*, see Reeve, this volume).

The behavior of adult males is quite diverse. The males of some species, such as *P. varia* (Yamane 1980), *R. marginata*, *M. drewseni*, and *M. flavitarsis*, remain on their natal nests for a few days after they emerge but then leave to spend the remaining several days of their lives attempting to mate. Males of other species, such as *M. labiatus* (Litte 1981) and *R. cyathiformis* (Gadagkar and Joshi 1984), appear to spend their entire lives on their natal nests, leaving only for several hours every day (presumably to attempt to mate). Males have never been observed to mate on their natal nests.

*Mischocyttarus drewseni* males in Brazil patrol areas where females are likely to forage, pouncing on conspecific females as well as conspecific males and the similar-looking females of *Polybia sericea* (Jeanne and Castellón Bermúdez 1980). These males make no attempt to mark or defend territories. Males of *M. labiatus* in Colombia similarly patrol areas that are likely to be frequented by females, but drag their abdomens and appear to apply glandular secretions to their perch sites. The function of this application seems not to be as much to mark territories and exclude other males as to attract females (Litte 1981). Males of *M. flavitarsis* in Arizona exhibit lekking behavior. They mark their perch sites and defend them against other males, sometimes using abandoned or active conspecific nests as perch sites (Litte 1979). Males of this species possess well-developed exocrine glands on their sterna, which open onto a dense brush of hairs, presumably aiding in pheromone application (see Downing, this volume). Males of other species in the genus that do not conspicuously drag their abdomens generally do not have such well-developed glands and lack the applicator brush (Post and Jeanne 1982b).

Patrolling males appear to orient visually to anything crudely resembling a conspecific female. Subsequent recognition and release of copulatory behavior is clearly mediated by chemical cues from the female's venom gland and possibly also head and thorax (Litte 1979, Keeping et al. 1986).

## SOCIAL ORGANIZATION

Any society, be it insect, avian, or mammalian, should have a set of rules that govern division of labor and access to resources if it is to function efficiently. Without doubt, insect societies deserve to be studied with a view to discover these rules, and, indeed, honey bee and ant societies have been used as model systems with great profit (Oster and Wilson 1978, Seeley 1985, Winston 1987). The study of social organization in insect societies such as those considered in this chapter has even more to offer. Because these primitively eusocial wasps are not obligately social, at least to the extent that single-foundress colonies are still possible, an appreciation of social organization is crucial for an understanding of the evolution of eusociality itself. The study of social organization is thus certain to shed light on the forces that mold the evolution of group living.

All independent-founding polistines, including *Polistes* (Reeve, this volume), have a fundamentally similar social organization. When colonies are founded singly there is, of course, no society until the first progeny emerge, but in multiple-foundress colonies social organization and division of labor are crucial issues in both the preemergence and postemergence phases. In species with seasonal nesting cycles, especially in temperate regions where hibernation is required, foundresses are always of the same generation (even-age cohorts). But with the emergence of the first workers foundresses other than the queen begin to die or disappear, so that the postemergence phase consists primarily of a mother queen and her daughters (see Yamane 1985). Social organization during the pre- and postemergence phases may thus be expected to be different, as has been emphasized in some studies of temperate-zone species (e.g., Yamane 1985). In less-seasonal environments, where asynchronous nesting cycles occur, foundresses may not necessarily belong to the same generation; thus differences in social organization between pre- and postemergence colonies are less pronounced. For instance, no consistent differences have been noticed between pre- and postemergence phases of the indeterminate nesting cycle in *R. marginata* (Gadagkar, unpubl.). I shall therefore consider social organization in general and mention pre- and postemergence phases only when striking differences are evident.

## Dominance Hierarchies

Independent-founding polistine wasps are characterized by a lack of morphological caste differentiation and consequent flexibility in the social roles of adults. When a group of wasps nest together it is of interest to know who will become the queen and who will take on the worker role. As Pardi (1948) showed in a classic study, role differentiation is determined largely by means of aggressive interactions leading to the establishment of a dominance hierarchy (see also Gadagkar 1980, Gadagkar and Joshi 1982b, Röseler, this volume). Dominance interactions are seen in all species, but the intensity and frequency of such interactions vary widely (e.g., Marino Piccioli and Pardi 1970, Gadagkar 1980, Kojima 1984c, Itô 1985a). Severe fights may involve grappling, biting, and stinging, occasionally leading to injury and even death. More frequently, however, one sees highly ritualized mock fights.

The most common such ritualized interaction in *R. marginata* and *R. cyathiformis* consists of one animal, dominant by definition, climbing on top of another and attempting to reach out and bite its mouthparts. The animal being so treated, subordinate by definition, becomes motionless, keeping its body as compact as possible and its mouthparts inaccessible (Gadagkar 1980, Gadagkar and Joshi 1982b, 1983). *Ropalidia cincta* (Darchen 1976a) and *B. grisea* (Marino Piccioli and Pardi 1970) appear to have a very similar form of dominance interaction. There are also other less frequent forms of dominance interaction, especially in *R. cyathiformis* (Gadagkar and Joshi 1982b), such as a dominant wasp nibbling, chasing, sitting on, or holding a wing or leg of a subordinate. Dominance interactions described by Yamane (1985) for *P. varia* and by Jeanne (1972) for *M. drewseni* appear not to be radically different from these. It seems probable that there are species-specific differences in dominance-related behaviors, but this has not yet been demonstrated.

Frequencies of dominance interactions vary greatly. Many colonies of *R. marginata* may be observed for days without evidence of aggressive interactions, while other colonies of the same species may show dozens of aggressive interactions per hour of observation. Itô (1983, 1984b, 1985a, 1986a,b) has attempted to discern some pattern in this variation, suggesting that increased levels of dominance interactions in postemergence as opposed to preemergence colonies constitute evidence of maternal manipulation, and that the inherently low levels of dominance interactions characteristic of some *Ropalidia* species may have facilitated the evolution of group living. Considering that so few quantitative data are available, the observed variation in dominance interactions of species such as *R. marginata* and *R. cyathiformis* leads one to suspect that intraspecific variation may often equal or exceed interspecific variation.

But what makes some females dominant and others subordinate? Although crucial, this question is almost entirely unanswered in this group of wasps. Body size, age, and hormone levels suggest themselves as factors that might influence an animal's probability of becoming dominant over another (Röseler, this volume), but only future work can tell us more.

## Role of the Egg Layer

Without exception, the most dominant female (queen) is the principal if not sole egg layer. Subordinates may lay no eggs (e.g., *R. marginata*: Gadagkar et al. 1990a; preemergence colonies of *P. varia*: Yamane 1985; *P. indica*: Sugiura et al. 1983a) or may lay some eggs, most of which are eaten by the queen (e.g., *B. grisea*: Marino Piccioli and Pardi 1970). In at least some cases, however, subordinates lay a substantial proportion of the eggs, which are not necessarily eaten (e.g., *M. mexicanus* fall nests: Litte 1977; *R. cyathiformis*: Gadagkar and Joshi 1982b).

The behavioral repertoire of the queen varies considerably from species to species. At one extreme is the queen of *R. marginata*, who does almost nothing other than lay eggs. In most colonies she rarely indulges in overt dominance interactions with her nestmates, let alone performs any other tasks, although her superior status is obvious— nestmates simply withdraw from her presence (Gadagkar and Joshi 1983). The queen of *R. cyathiformis*, on the other hand, is often the most aggressive and active individual in her colony, running about, physically attacking and challenging her nestmates, yet never leaving the nest (Gadagkar and Joshi 1984). Even more active are *M. drewseni* queens, who chew the silk caps off vacated cells and prepare them to receive fresh eggs; initiate new cells in postemergence colonies; elongate cell walls; solicit food, building material, and nectar from returning foragers; distribute food to larvae and nestmates; and even leave the nest to forage (Jeanne 1972). *Mischocyttarus labiatus* queens have been described as "the primary but not the sole egg-layers", "the only egg-eaters", and "frequent biters and solicitors of nestmates" (Litte 1981:11). *Parapolybia varia* queens are said to have "rarely left the nest and monopolized most ovipositions by physically disturbing the subordinates' attempts to oviposit" (Yamane 1985:27). In all species studied, subordinates (workers) have the primary responsibilities for tasks other than oviposition, such as foraging, brood care, nest building, and other nest-maintenance activities.

## Behavioral Caste Differentiation

The foregoing discussion is based on an emphasis of a queen-worker dichotomy modeled after the highly eusocial insects, but this is arguably an insufficient if not misleading analogy for primitively eusocial species. Variability among workers of the latter is often large, and the range of behaviors or other traits of workers may sometimes overlap that of the queen. More significantly, in primitively eusocial insects, caste is at least partly determined in the adult stage, and workers therefore have not completely lost their capacity for direct reproduction. Challenging and replacing the queen, laying eggs in the presence of the queen, or leaving the nest to found their own nest are various ways by which workers may realize direct reproductive success. One might therefore expect workers to adopt a variety of strategies to maximize their chances of direct reproduction. Consequently, variability in worker behavior must hold important clues to the nature of reproductive competition, a fact that has rarely received the attention it deserves.

As an example, Litte (1981) distinguished three types of females in preemergence colonies of *M. labiatus* in Colombia; queens, dominant co-foundresses, and foraging co-foundresses. Dominant co-foundresses performed fewer tasks than even queens and sometimes laid eggs, while the foraging co-foundresses had poor ovarian development and did most of the foraging. Most daughters became foragers on their natal nests but were nevertheless capable of founding new nests and were therefore considered potential reproductives.

Investigations of *R. marginata* and *R. cyathiformis* in India have been explicitly concerned with the elucidation of interindividual variability and its potential for understanding social organization, reproductive competition, and, ultimately, the evolution of eusocial behavior (Gadagkar and Joshi 1982a, 1983, 1984, 1985). Departing somewhat from typical approaches, time activity budgets have been constructed for individually identified wasps using all common activities. The behaviors documented include seemingly trivial ones, such as "sitting" or "sitting with raised antennae," behaviors that traditionally have not been the focus of much attention. Multivariate statistical techniques used to examine patterns of interindividual differences show that each colony has three distinct kinds of individuals or behavioral castes. Superimposing this cluster analysis on data for frequencies of traditionally studied behaviors has permitted the designation of these behavioral "castes" as "Sitters," "Fighters," and "Foragers" (Fig. 5.12). Sitters are those wasps that spend a relatively large proportion of their time sitting and self-grooming. Fighters spend a great deal of time sitting with raised antennae and also show high frequencies of dominance behaviors. Foragers are absent from the nest a great deal of the time and show high frequencies of bringing food and building material back to the nest.

The position of the queen among these clusters is of particular inter-

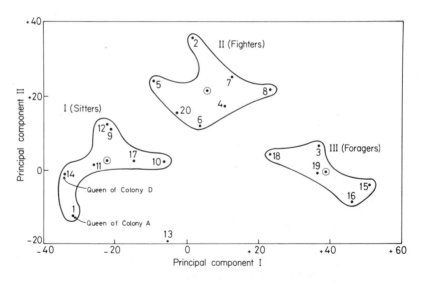

**Fig. 5.12.** Behavioral profiles of 20 individually identified females (numbered) from two colonies of *Ropalidia marginata* analyzed by principal components analysis. Each point represents one wasp plotted in the coordinate space of the first two principal components. The points fall into three clusters (referred to as behavioral castes) by the nearest-centroid (circled dot) criterion. (From Gadagkar and Joshi 1983, courtesy of *Animal Behaviour*.)

est in view of the fact that data on egg laying, the traditional criterion defining the queen-worker dichotomy, were not used in the cluster analysis. In 13 out of 14 *R. marginata* colonies studied, the queens were Sitters (Gadagkar and Joshi 1983, Chandrashekara and Gadagkar 1990). An obvious interpretation of this is that the queens are programmed to spend little time and energy doing anything other than laying eggs. Since there is normally only one queen per colony, the remaining Sitters who are not queens may be thought of as hopeful queens attempting to maximize their chances for future reproduction. Fighters are probably performing the function of keeping the colony active and guarding it against parasites. That Fighters may also be hopeful queens is suggested by the fact that they show the highest frequency of fighting, not with Foragers as might be expected, but among themselves. It seems likely that Foragers, who leave the nest to perform the risky tasks of gathering food and building material, have the least chance of becoming queens. Both Sitters and Fighters have significantly better developed ovaries than Foragers (Chandrashekara and Gadagkar 1990).

These hypotheses regarding the evolutionary significance of behavioral caste differentiation can be evaluated using both comparative and experimental methods. A comparative study of *R. marginata* and *R. cyathiformis*, both of which are common in peninsular India, has revealed that social organization in these two species is very similar, with an important difference being that in all *R. cyathiformis* colonies studied, queens belong to the Fighter caste. This means that the queen of *R. cyathiformis* does more than just lay eggs. In fact, in contrast to *R. marginata* queens, she is one of the most active individuals in the colony, routinely taking part in aggressive dominance interactions with her nestmates and spending much of her time sitting with raised antennae.

Why do queens of *R. marginata* and *R. cyathiformis* differ in this way? One possibility is that a queen of *R. marginata* is a Sitter because she faces relatively little reproductive competition from her nestmates, while a queen of *R. cyathiformis* is a Fighter because she faces relatively high levels of such competition. This idea is supported by several facts. First, *R. cyathiformis* colonies sometimes have multiple egg layers (Sitters and Foragers also occasionally lay eggs in this species), whereas *R. marginata* colonies usually have only a single egg layer. Second, in single-foundress colonies of *R. cyathiformis*, the queen or egg layer belongs to the Sitter caste rather than the Fighter caste. In other words, a queen of *R. cyathiformis* is also a Sitter when she faces no reproductive competition. (Notice that even solitary foundresses can be classified as Fighters because the delineation of a Fighter is done on the basis of time activity budgets of behaviors other than fighting.)

Evolutionary hypotheses concerning behavioral castes are also amenable to experimental testing (Gadagkar 1987). The approach here has been to examine behavioral caste differentiation in a colony, experimentally remove the queen, and study the consequent changes in social organization. Every *R. cyathiformis* colony seems to include one or two wasps who behave much like the queen (Fig. 5.13). These potential queens, who take over the role of the queen when she is removed, always belong to the Fighter caste, as does the queen in this species. Upon removal of the queen, not only does one of the other Fighters become the queen, but one of the Sitters may change her behavioral profile rather drastically to become a Fighter and, perhaps, the next potential queen. In short, every colony has a potential queen already differentiated even when the original queen is present. An interesting fact is that these potential queens are often young and aggressive individuals who seldom forage for food or do other work. This is in complete contrast to the North American paper wasp *Polistes exclamans*, where older foragers tend to become replacement queens (Strassmann and Meyer 1983).

Yet another line of work has provided additional insights into the social organization of *R. cyathiformis*. During a long-term study, a colony divided so that about half its members left to form a new colony

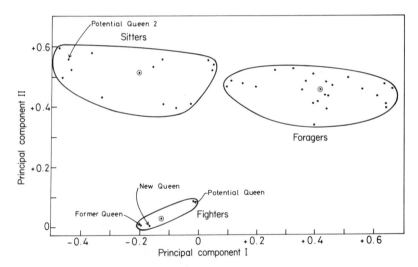

**Fig. 5.13.** Behavioral profiles of individually identified females in a *Ropalidia cyathiformis* colony analyzed by principal components analysis. Each point represents one wasp (either before or after the queen was removed) plotted in the coordinate space of the first two principal components. When the queen labeled *Former Queen* was removed, the wasp labeled *Potential Queen* became the replacement queen (her modified behavioral profile is labeled *New Queen*). When this new queen was in turn removed, the wasp labeled *Potential Queen 2* became the next replacement queen. (From Gadagkar 1987, courtesy of Verlag J. Peperny.)

just a few feet away from the parent colony. A behavioral analysis before and after colony fission revealed that inclusive fitnesses of the "Rebels" who left the colony, as well as of the "Loyalists" who stayed in the old colony, increased as a result of the fission (Gadagkar and Joshi 1985), probably because of the prevailing low efficiency of brood rearing before colony fission associated with high levels of aggression. The second most dominant individual before colony fission became the queen on the new colony. Here, therefore, is an instance where a wasp managed to establish herself as a queen, yet avoided the cost of challenging the original queen as well as the risk of failure that lone foundresses face. Interestingly, Loyalists and Rebels both behaved as internally coordinated groups well before the actual fission, synchronizing among themselves their times of being on or away from the nest and avoiding members of the other group (Gadagkar and Joshi 1985).

In summary, when viewed with an emphasis on queen-worker dichotomy, social organization appears to achieve efficient division of labor and colony harmony, but when viewed with an emphasis on interindividual variability a rich mosaic of complex behavioral strategies becomes evident, suggesting ways by which individual selection might mold worker behavior (see Origin of Social Life: A Perspective from Studying Independent-founding Polistines). It must be stressed, however, that the foregoing discussion of social organization is based on only a few studies of a few species. Future studies of these and the large number of as-yet unstudied species (see, for example, Itô and Higashi 1987, Itô et al. 1988) are bound to revise our current ideas.

## CASTE

Primitively eusocial wasps of the kind we are dealing with provide rich model systems for studying the role of interactions between the adults in determining social organization. But are interactions between adults sufficient to tell us all there is to know about social organization? In other words, are all females at emergence potentially capable of assuming any role in the colony? This has by and large been assumed to be so (e.g., Queller and Strassmann 1989, Reeve, this volume), but since there has never been a direct test there is no firm evidence one way or the other. An important exception is the study by Richards and Richards (1951), who clearly recognized the significance of this question and demonstrated slight preimaginal caste differentiation in a number of polistine species. However, most of their material, barring a few species of *Polistes* and *Mischocyttarus*, consisted of highly eusocial swarm-founding Neotropical polistines.

As far as the independent-founding genera are concerned, there appears to have been only one direct test of the null hypothesis that "all eclosing females are potentially capable of laying eggs" (Gadagkar et al. 1988:176; see also Gadagkar et al. 1990b). This hypothesis was tested by collecting a number of nests of R. *marginata* and isolating emerging females in individual cages. Of 299 virgin female wasps from 39 nests so tested, only 150 laid eggs. The remaining 149 died without doing so in spite of living, on average, longer than the time taken by the egg layers to lay their first eggs. This result clearly suggests that there is some preimaginal biasing of caste. But what factors determine which females will become egg layers and which will become non–egg layers? Among a large number of variables studied only two were significantly correlated with the probability of a wasp becoming an egg layer. One of these variables, the number of empty cells in the parent nest, is likely to be a strong indicator of the queen's declining influence on a colony. The second correlate, the rate of food consumption by wasps in the experiment, varied considerably in spite of all animals being housed in individual cages and being provided unlimited access to food.

On the basis of these results Gadagkar et al. (1988) proposed a model for preimaginal biasing of caste in primitively eusocial insects. With a decline in the queen's influence resulting from poor health, old age, or a temporary programmed shift in her physiology and/or behavior, a set of processes would be initiated that has two consequences: (1) the accumulation of empty cells and (2) the production of a class of female offspring programmed to feed more and so to have a high probability of becoming reproductively competent. In contrast, when the queen's influence is high there are two rather different consequences: (1) the absence of empty cells and (2) the production of a class of female offspring programmed to feed relatively less. Preimaginal caste bias is undoubtedly partial, leaving considerable potential in the adult stage for environmental and other social factors to influence caste.

Indirect evidence suggests the existence of a similar caste-biasing system in B. *grisea*. Pardi and Marino Piccioli (1970, 1981) found two distinct classes of females: (1) larger "queenlike" females who were fertilized more often, were more oophagous, and were mainly nest foundresses and (2) "workerlike" females who had the opposite traits and foraged more often. These authors postulated that preimaginal trophic factors determine the two "morphophysiological" conditions.

If the queen or other adults bias the future caste of their colony's brood, one might expect the brood in a preemergence colony to be channeled toward worker development and the brood in a mature postemergence colony to be channeled toward queenlike development. If such channeling is based on differential larval nourishment, one would expect larvae in preemergence colonies to be poorly nourished compared with larvae in mature postemergence colonies. Although no quantitative data on larval nourishment in pre- and postemergence colonies are available, there is suggestive evidence that *R. marginata* nests with relatively well-nourished larvae produce more egg layers, whereas those with relatively poorly nourished larvae produce more non–egg layers (Gadagkar et al. 1990c).

Hunt (1988) has pointed out a factor that can potentially exaggerate such differences in larval nourishment. This is the tendency of *Mischocyttarus* larvae in preemergence colonies to give up saliva to the soliciting adults more readily than larvae in postemergence colonies (see Food and Feeding Habits). The result is an exaggeration of the difference in nourishment between larvae from pre- and postemergence phases of the colony cycle, thus providing a more powerful mechanism of channeling emerging females into their respective worker and reproductive roles.

The above discussion suggests that some form of preimaginal caste biasing already occurs at the level of social evolution represented by the independent-founding polistines. While postemergence events undoubtedly influence caste, preemergence factors should be studied with greater vigor and better techniques. Whether all adult females are equally capable of assuming any role in the colony in the absence of postemergence social interactions is still an open question for most species. If the answer is negative, as it seems to be for some species, then we should be viewing these primitively eusocial insect societies from a rather different perspective and suitably modify our theories to explain altruism.

## ORIGIN OF SOCIAL LIFE: A PERSPECTIVE FROM STUDYING INDEPENDENT-FOUNDING POLISTINES

Why do social wasps live in groups? Why do some individuals accept the role of sterile worker? During colony founding, why do some individuals nest with others even if it means few or no opportunities to lay their own eggs? During the early postemergence phase of the colony cycle, why do many females stay to help their mothers produce more offspring? Primitively eusocial polistine wasps are attractive model systems in insect sociobiology because they seem to provide the opportunity to answer such questions.

In most highly eusocial insects such as honey bees, termites, and most ants, the simple answer to those questions could be that individuals who accept sterile worker roles simply have no other choice; over evolutionary time they have lost the ability to reproduce on their own. Workers in many species are capable of laying eggs, but with rare exception (e.g., Anderson 1963) they lay only haploid, male-producing eggs, and even then such worker reproduction is usually suppressed in the presence of the queen (Hamilton 1964b, 1972; Wilson 1971, Trivers and Hare 1976; Fletcher and Ross 1985; Ross 1985; Page and Erickson 1988; Ratnieks 1988; Ratnieks and Visscher 1989; Visscher 1989).

In primitively eusocial species there are several situations in which the animals almost certainly have a choice regarding the option of direct reproduction. Single- and multiple-foundress colonies coexist in *Mischocyttarus, Ropalidia,* and *Parapolybia* (Table 5.1); subordinate females adopt the role of the queen if the most dominant female is lost or removed (Jeanne 1972; Litte 1979, 1981; Gadagkar 1987); daughters sometimes challenge, drive away, and replace mother queens (Jeanne 1972, Yamane 1986, Gadagkar, unpubl.); and females leave their natal nests to found new colonies (Gadagkar and Joshi 1984, 1985). All these facts strongly suggest that individuals accept subordinate roles not because they are incapable of doing anything else but because social life, even if it means partial or full sterility, must sometimes be more advantageous than solitary life. Thus, the study of primitively eusocial species such as those considered in this chapter, which have real choices concerning reproductive roles, allows us to focus on the origin of social life rather than simply its maintenance.

How can social life be more advantageous than solitary life if the former means sterility? Workers in social groups may have opportunities to gain inclusive fitness by caring for their relatives' offspring if the group consists of close kin (for discussion of inclusive fitness and kin selection, see Ross and Carpenter, this volume). We may broadly generalize Hamilton's (1964 a,b) concept of inclusive fitness and say that group life is favored over solitary life if

$$\sum_{i=1}^{n} r_i > 1/2 \ m, \qquad (5.1)$$

where $n$ is the number of individuals (offspring or other relatives) reared in the group mode, $r_i$ is the coefficient of genetic relatedness between these individuals and the sterile workers, and $m$ is the number of offspring reared in the solitary mode. One way in which this inequality may be obtained is for the average $r_i$ ($\bar{r}$) in the group mode to be greater than 0.5, that is, for the genetic relatedness of relatives reared to be greater than that of offspring. This may be achieved with a male haploid (haplodiploid) genetic system, such as is found in Hymenoptera, in which genetic relatedness between a female and her full (= super) sisters is 0.75. Realization of this high $\bar{r}$ requires that colonies consist of a single egg layer mated to a single male, so that workers need not rear any half-sisters or more distantly related individuals. Furthermore, workers either must successfully skew investment in favor of their sisters or must be able to rear their own sons instead of brothers. (Brothers are related to workers by only 0.25, while sons are related to their mothers by 0.5.)

Many features of the biology of the genera considered in this chapter suggest that conditions necessary for high $\bar{r}$ between workers and the brood they rear may not be met (see also Ross and Carpenter, this volume). In *B. petiolata* only 15% of foundress associations consist exclusively of former nestmates (Keeping and Crewe 1987), so that cofoundresses probably are not often closely related. In *M. drewseni, R. variegata jacobsoni,* and *R. marginata,* queen supersedure is common (Jeanne 1972, Yamane 1986, Gadagkar et al. 1990a). Polygyny, the simultaneous presence of more than one egg layer, has been reported in *M. mexicanus* fall nests (Litte 1977), *R. cyathiformis* (Gadagkar and Joshi 1982b, 1984), *R. variegata jacobsoni* (Yamane 1986), and *R. fasciata* (Itô 1986b), and nests frequently are usurped by foreign conspecifics in *M. flavitarsis* (Litte 1979). Multiple mating by the egg layer is known in at least one species (*R. marginata*: Muralidharan et al. 1986). Among these genera, average relatedness between female nestmates has been measured in *Mischocyttarus basimacula* ($r = 0.44$), *M. immarginatus* ($r = 0.77$) (Strassmann et al. 1989), and *R. marginata* ($r = 0.53$; calculated from data in Muralidharan et al. 1986).

Polygyny or multiple mating by the queen should pose no great difficulty for attaining high levels of inclusive fitness if workers discriminate between full sisters and less-related individuals, giving preferential aid to the former (Gadagkar 1985b). However, studies of nestmate discrimination in *R. marginata* suggest that the labels and templates used in discrimination are not produced individually, but rather are acquired from a common external source, namely the natal nest or nestmates, making it unlikely that different levels of genetic relatedness can be effectively recognized among members of the same colony (Venkataraman et al. 1988). Thus, it seems likely that workers in these primitively eusocial wasps often rear complex mixtures of full sisters, half-sisters, nieces, daughters, brothers, nephews, sons, and cousins without the ability to discriminate on the basis of genetic relatedness. Because $\bar{r}$ in such societies rarely exceeds 0.5, it has been increasingly suspected that haplodiploidy is not as important a factor in the origin of insect sociality as was earlier thought (Evans 1977a, West-Eberhard 1978a, Andersson 1984, Stubblefield and Charnov 1986, Venkataraman et al. 1988, Gadagkar 1990a,b).

Even if $\bar{r}$ is rarely much greater than 0.5 (or even less than 0.5), the inequality in Eq. 5.1 can nonetheless be achieved if $n$ is sufficiently greater than $m$. The ecology of independent-founding polistine wasps suggests that there must be substantial benefits to group living, that is, $n$ must often be greater than $m$ (see also Reeve, this volume). For instance, the probability of survival of single-foundress colonies is certainly small, as the nest and its brood are extremely vulnerable to predators and parasites (e.g., Suzuki and Murai 1980). Yet the effectiveness of protection from such enemies by even one or two supernumerary adults is obvious to anyone who has watched these wasp colonies. Furthermore, multiple foundresses are more likely to be able to rebuild a damaged nest than are solitary foundresses. This latter factor favors multiple-foundress nests even when the destructive agent is itself indifferent to group size, as in the case of predation by birds or destruction by typhoons (Litte 1977, 1979, 1981; Itô 1983, 1984b, 1985a,b,c, 1986a, 1987b; Kojima 1989). A combination of high adult mortality and slow brood development in R. marginata has been shown to enhance the inclusive fitness of workers relative to that of solitary foundresses (Gadagkar 1990c).

Another way by which the inequality in Eq. 5.1 may be attained is if ecological conditions exist such that a parent who manipulates a fraction of her offspring into being sterile and helping to rear her remaining (fertile) offspring leaves behind more grandchildren than her wild-type counterpart (parental manipulation hypothesis: Alexander 1974). A significant problem with this is whether counterselection on the offspring would be successful in making them overcome parental manipulation. A related idea, which circumvents this problem, is that subfertile females produced by whatever cause (even by accidental variation in the quantity of food obtained as larvae) will find it "easier" to give up reproduction and accept a worker role (subfertility hypothesis: West-Eberhard 1975). That is, $m$ will be so small that the inequality in Eq. 5.1 may be satisfied rather easily. The general ideas embodied in the parental manipulation and subfertility hypotheses have found support in theoretical (Stubblefield and Charnov 1986), modeling (Craig 1979, 1983), and empirical (Michener and Brothers 1974) studies. The evidence for preimaginal caste bias in R. marginata (see Caste) suggests that at the level of primitive eusociality represented by the genera discussed here, subfertility may contribute to achieving the inequality in Eq. 5.1.

The presence of permanently sterile worker castes is the most prominent and seemingly paradoxical feature of the highly eusocial insects. Attempts to explain the evolution of eusociality have sometimes obscured the rather obvious fact that many social insect species do not possess permanently sterile workers but could nevertheless be forerunners of the highly eusocial state. The evolutionary forces that promoted the origin of such primitive levels of sociality may thus have been quite different from those discussed above. The statements that "social behavior in insects is in part mutualistic" and that "social colonies without altruism are therefore considered a probability" (Lin and Michener 1972:131) may well prove to have been prophetic. The theory of reciprocal altruism proposed by Trivers (1971), an individual-level selection theory that suggests that aid may be given today with the hope of getting a return tomorrow, has for some mysterious reason seldom been applied to social insect colonies. A related individual selection model has been discussed by West-Eberhard (1978a) and Gadagkar (1985a). Consider two wasps that come together and nest jointly. If their joint productivity (say, 21 offspring) is even slightly greater than the sum of their individual productivities (say, $10 + 10 = 20$ offspring) in the solitary mode, and if the roles of queen and sterile worker are assigned randomly, then wasps who take the risk of joint nesting will, on average, produce 10.5 offspring and thereby do better than those who shy away from the risks of joint nesting. In this "gambling" model, benefits of group living can be infinitesimally small, and abilities of parents to manipulate their offspring or of colony members to discriminate on the basis of differential genetic relatedness are not necessary.

## A POSSIBLE ROUTE TO EUSOCIALITY

Many features of the genera discussed in this chapter lend credibility to the idea that the initial incentive for group living comes from mutualism, reciprocal altruism, and the benefits of gambling (*the gambling stage*: Fig. 5.14). This is possible without any preadaptation for parental manipulation or for recognition on the basis of genetic relatedness, although group living among kin will evolve more easily (West-Eberhard 1978a, Schwarz 1988). The only prerequisite for the evolution of incipient societies by mutualism is a sufficiently complex behavioral repertoire to permit the necessary interactions; solitary wasps seem to possess an appropriately diverse array of behaviors (Tinbergen 1932, 1935; Tinbergen and Kruyt 1938; Brockmann and Dawkins 1979; Brockmann et al. 1979).

Once group living is established, the stage is set for manipulation (*the manipulation stage*). Accidental variations in food supply leading to subfertility can be exploited, and the ability to manipulate offspring can be selected for. As manipulation becomes increasingly effective, benefits of group living become increasingly unavailable to some individuals, who begin to lose reproductive options and to get trapped into worker roles. It is precisely at this stage that the ability to recognize

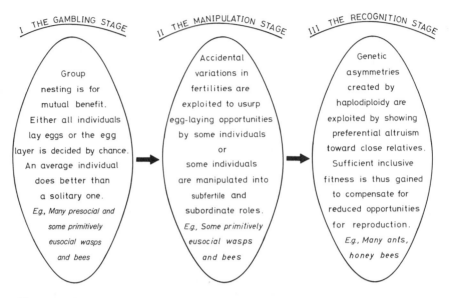

**Fig. 5.14.** The route to eusociality, a hypothesis concerning the evolution of the highly eusocial state from the solitary state through the gambling, manipulation, and recognition stages. The examples given for each stage are tentative, as our knowledge of the causes and consequences of group living in most social insect groups is rather sketchy.

and give preferential aid to closer relatives will begin to have selective value (*the recognition stage*). In other words, the benefits of haplodiploidy for social evolution become operative at this final stage.

This route to eusociality (Gadagkar 1990d) explains one otherwise curious fact. The ability to discriminate among the members of a colony on the basis of relatedness seems to be absent in all primitively eusocial species studied but present in the highly eusocial ants and honey bees (reviewed in Venkataraman et al. 1988). If haplodiploidy were important for the origins of insect eusociality one would expect workers in primitively eusocial species to exploit the genetic asymmetries thus created by discriminating between close and distant relatives. But if eusociality originated because of mutualistic benefits, as assumed here, and its subsequent maintenance in highly eusocial forms is due largely to haplodiploidy, the observed distribution of kinship discrimination abilities is no longer a paradox.

## CONCLUDING REMARKS

Any attempt to collate the rapidly accumulating information on polistine biology may appear aimless unless there is a well-defined structure around which this information is organized. Therefore, recognition of the existence of two behaviorally distinct groups by Jeanne (1980a), namely the "independent-founding Polistinae" and the "swarm-founding Polistinae," was perhaps the most significant recent conceptual advance in our understanding of polistine wasps. Much work since then has aimed to further consolidate such a classification. We may now have reached the stage, however, of beginning to question this classification and to start looking at exceptions to the characteristics of each group. To what extent do species of *Ropalidia* and *Parapolybia* depend on chemical defense against ants? To what extent is nest founding by several females of different ages, arriving simultaneously and sometimes having well-defined social roles in their previous colonies, really different from swarm founding? What means of communication, if any, do the independent-founding species use in choosing and reaching a new nesting site? This is not to imply that we have reached the end of the utility of Jeanne's (1980a) classification. Indeed, it is a tribute to its continuing utility that we expect significant new advances in polistine biology to be triggered by standing this classification on its head and looking for exceptions!

## Acknowledgments

It is a pleasure to thank Kenneth G. Ross, Robert W. Matthews, Robert L. Jeanne, Madhav Gadgil, and Seetha Bhagavan for many helpful comments on an earlier draft of this chapter and G. Jayashree, S. Vedha, and U. N. Shyamala for much assistance in its preparation. Robert L. Jeanne suggested the terms *determinate nesting cycle species, indeterminate nesting cycle species,* and *preimaginal caste bias.* My own work was supported by grants from the Department of Science and Technology, Government of India, and the Indian National Science Academy.

**Appendix 5.1.** Common Parasites of *Belonogaster*, *Mischocyttarus*, *Parapolybia*, and Independent-founding *Ropalidia*

| Host species | Parasite | Locality | Stage affected | Remarks | Reference |
|---|---|---|---|---|---|
| *Belonogaster juncea* | *Pediobius ropalidia* (Eulophidae: Hymenoptera) | Ghana | Pupa | High proportion of the pupae destroyed | Richards 1969 |
| *B. petiolata* and *B. juncea colonialis* | *Anacamptomyia* sp. (Tachinidae: Diptera) and *Camptotypus apicalis* (Ichneumonidae: Hymenoptera) | Transvaal, South Africa | Prepupa and pupa | | Keeping and Crewe 1983 |
| *Mischocyttarus drewseni* | Strepsiptera | Brazil | Adult | Only 2 out of 760 adults were affected | Jeanne 1972 |
| *M. flavitarsis* | *Chalcoela iphitalis* (Pyralidae: Lepidoptera) | Arizona, USA | Larva and pupa | Common cause of brood mortality | Litte 1979 |
| *M. flavitarsis* | *Monodontomerus* sp. (Torymidae: Hymenoptera) | Arizona, USA | Larva and pupa | Rare | Litte 1979 |
| *M. flavitarsis* | Strepsiptera | Arizona, USA | Pupa and adult | | Litte 1979 |
| *M. labiatus* | *Megaselia* sp. (Phoridae: Diptera) | Colombia | Brood | Responsible for a larger proportion of nest failures than any other single factor | Litte 1981 |
| *Parapolybia varia* | *Bakeronymus typicus* (Trigonalidae: Hymenoptera) | Taiwan | Larva | | Yamane and Terayama 1983 |

| Species | Parasite | Location | Stage | Remarks | Reference |
|---|---|---|---|---|---|
| *Ropalidia cyathiformis* | Unidentified species of Ichneumonidae (Hymenoptera) | Bangalore, India | Larva | | Gadagkar, unpubl. |
| *R. fasciata* | *Arthula formosana* (Ichneumonidae: Hymenoptera) | Okinawa, Japan | Larva | Very high percentage of cells parasitized | Itô 1983 |
| *R. flavobrunnea lapiniga* | *Pseudonomadina biceps* (Trigonalidae: Hymenoptera) | Philippines | Brood | In addition to larvae and pupae, live adults of the parasite were found in active wasp nests | Yamane and Kojima 1982 |
| *R. formosa* | *Hemipimpla pulchripennis* (Ichneumonidae: Hymenoptera) | Madagascar | Larva | | Brooks and Wahl 1987 |
| *R. marginata* | *Koralliomyia portentosa* (Tachinidae: Diptera) | Bangalore, India | Larva | | Belavadi and Govindan 1981 |
| *R. marginata* | Strepsiptera | Bangalore, India | Adult | | Belavadi and Govindan 1981, Gadagkar, unpubl. |
| *R. marginata* | Unidentified species of Tachinidae (Diptera), Torymidae (Hymenoptera), Ichneumonidae (Hymenoptera) | Bangalore, India | Larva | | K. Chandrashekara and Gadagkar, unpubl. |

**Appendix 5.2.** Common Predators of *Belonogaster*, *Mischocyttarus*, *Parapolybia*, and Independent-founding *Ropalidia*

| Prey species | Predator | Locality | Stage affected | Remarks | Reference |
|---|---|---|---|---|---|
| *Belonogaster petiolata* | *Hoplostomus fulgineus* (Scarabaeidae: Coleoptera) | Transvaal, South Africa | Pupa | | Keeping 1984 |
| *Mischocyttarus drewseni* | *Monomorium pharaonis*, *Camponotus abdominalis* (Formicidae: Hymenoptera) | Brazil | Brood | | Jeanne 1972 |
| *M. drewseni* | Spiders | Brazil | Adults away from the nest | | Jeanne 1972 |
| *M. flavitarsis* | Birds | Arizona, USA | Whole nest or pieces of the nest | Inferred | Litte 1977 |
| *M. flavitarsis* | Spiders, preying mantids | Arizona, USA | Adults away from the nest | | Litte 1977 |
| *M. labiatus* | Army ants | Colombia | Brood | Seen only once | Litte 1977 |
| *M. mexicanus* | Ants, birds | Mexico | Whole nest or pieces of the nest | Predation by blue jays was observed, and the involvement of other birds was inferred | Litte 1977 |
| *Mischocyttarus* spp. | Bats | Brazil | Whole nest | | Jeanne 1970b |
| *Parapolybia indica* | *Vespa tropica* (Vespidae: Hymenoptera) | Japan | Pupa and larva | Most important enemy | Sekijima et al. 1980 |
| *P. varia* | Ants | Taiwan | Brood | | Yamane 1980 |
| *P. varia* | *Vespa tropica* (Vespidae: Hymenoptera) | Taiwan | Pupa and larva | Most important factor regulating population levels | Yamane 1980 |
| *Ropalidia marginata* and *R. cyathiformis* | *Vespa tropica* (Vespidae: Hymenoptera) | Southern India | Pupa and larva | Most important factor regulating population levels | Gadagkar, unpubl. |

# 6

# The Swarm-founding Polistinae

ROBERT L. JEANNE

In temperate regions, the vespine genera *Vespula* and *Dolichovespula* are the dominant social wasps. Their colonies can become large, and at peak seasons *Vespula* species especially can become troublesome pests by dint of sheer numbers and their predilection for picnic foods. In most tropical regions of the world, however, the vespines are replaced by the swarm-founding Polistinae as the dominant wasps. This is especially true in the Neotropics, where one group of polistines has undergone a spectacular evolutionary radiation and achieved remarkable ecological success because of a mode of social organization unique in the social wasps—swarm founding. In this chapter, these wasps are examined, with particular attention being paid to the characteristics that have led to their success.

## WHAT ARE THE SWARM-FOUNDING WASPS?

The social vespids fall into three groups on the basis of their mode of colony founding and mechanism of reproductive dominance (Hölldobler and Wilson 1977, Jeanne 1980a). In the subfamily Vespinae, queens of most species found colonies alone, and reproductive dominance involves pheromones. Exceptions to the rule of independent

This chapter is dedicated to the memory of Martin G. Naumann (1939–1984), field biologist par excellence, good friend, and gentle person, who knew these wasps better than any of us and loved them more.

191

founding in the subfamily occur in the tropics of Southeast Asia (see Matsuura, this volume). There, colonies of *Vespa affinis indosinensis* and *V. tropica leefmansi* are frequently founded by several queens accompanied by one (or more?) workers (Matsuura 1983b, Matsuura and Yamane 1984), and the nocturnal *Provespa* founds nests by swarms of a single queen and several dozen workers (Matsuura and Yamane 1984).

The remaining social vespids—the subfamilies Stenogastrinae and Polistinae—can be divided into two groups, the *independent founders* and the *swarm founders*. Colonies of independent-founding species are initiated by one or several inseminated queens, independently of any workers. Soon after colony founding, one of the founding queens typically becomes the sole egg layer. This mode of founding occurs in five genera of Polistinae (*Polistes, Mischocyttarus, Belonogaster, Parapolybia*, and most species of *Ropalidia*) (Reeve, Gadagkar, this volume) as well as in the subfamily Stenogastrinae (Turillazzi, this volume). Reproductive dominance is based on direct physical attacks by the queens (Pardi 1948).

In the swarm-founding species, a colony is founded by a swarm consisting of a large number of workers accompanied by a smaller number of queens. Reproductive dominance appears to involve pheromones (West-Eberhard 1977a). Swarm founding is the exclusive reproductive mode in the majority of genera in the subfamily Polistinae.

Although in many species of independent-founding Polistinae colonies are founded by a small group of foundresses, these groups lack several characteristics of a true swarm. The queens of independent founders enter a solitary phase after insemination. This phase may last no more than a day or two in the equatorial tropics, where nesting occurs throughout the year (Jeanne 1972). But in more-seasonal tropical regions, the solitary phase lasts for a month or more, and in temperate regions it extends through the winter and into the spring months. Queens of swarm founders, on the other hand, are never without the company of workers (except, perhaps, when they leave the nest to mate). Thus, swarm-founding wasps are unique in having socialized all stages of the colony cycle, including dispersal and founding.

A second difference is that independent founders typically colonize a new nest site gradually. Co-founding females arrive one by one over a period of days, and co-foundresses often shift among two or more newly founded nests in an area during the early weeks of the nesting season (West-Eberhard 1969). In contrast, swarm founders move to a new nest site in a coordinated synchronous fashion with the aid of trail pheromones (Jeanne 1981a, West-Eberhard 1982c).

Finally, in independent-founding species, the inseminated founding queen(s) selects the nest site and initiates the nest, even though her co-foundresses and worker offspring later take over most of the labor of enlarging the nest. In the swarm founders, in contrast, the workers choose the nest site and build the nest, while the queens wait idly by until enough cells are completed for them to begin ovipositing (Forsyth 1978, West-Eberhard 1982c).

Swarm founding and independent founding appear to be distinct enough for us usefully to treat them as discrete groups. Species for which we know anything about founding can be assigned to one or the other group. Although swarming has not yet been directly observed in a few of the lesser known genera, on the basis of colony size and/or nest architecture we assume that all New World polistines except *Polistes* and *Mischocyttarus* are swarm founders (Table 6.1), as is the small Old World genus *Polybioides*. Greater uncertainty surrounds *Ropalidia*, the largest Old World genus, for it contains both independent-founding (the majority) and swarm-founding species. As we learn more about some of the lesser known New World genera and *Ropalidia*, we may discover that the differences between the two modes are of degree only, and that some species—or populations—are intermediate between the two forms (e.g., *Belonogaster*).

About one-fourth of the 800-odd polistine species, or slightly less than that proportion of all eusocial wasps, are believed to be swarm founders. The swarm-founding polistines are largely tropical in distribution, although the ranges of several species considerably exceed the limits of the tropics in the New World and in eastern Australia (Fig. 6.1). Although they are a relatively small group in terms of numbers of species, the swarm-founding wasps exceed all other eusocial wasp groups in terms of taxonomic diversity at the generic level, diversity of nest architecture, and range of colony size.

Richards has revised the New World (1978a) and the Australian (1978b) Polistinae, whereas Carpenter (this volume) has studied the phylogenetic relationships of the polistine genera. Vecht has revised the Indo-Australian *Ropalidia* (1962) and the eastern Asian and Indo-Australian *Polybioides* (Vecht 1966). For African *Polybioides*, unfortunately, there has been no revision since Bequaert's 1918 paper.

## THE SUCCESS OF THE SWARM-FOUNDING POLISTINAE

How successful are the swarm-founding genera relative to the independent-founding polistines and to one another? Perhaps the best criterion of success of a group is the proportion of energy flowing through a habitat that it manages to divert to its own use. Unfortunately, no such data exist for social wasps. Instead, as a first approximation, we can use the relative numbers of individuals in each genus

**Table 6.1.** Number of species and distribution of the swarm-founding and independent-founding polistine genera

| Genus | No. of species | Distribution | Reference |
|---|---|---|---|
| **Swarm founding** | | | |
| *Agelaia* | 21 | Neotropics: Mexico to N Argentina | Richards 1978a |
| *Angiopolybia* | 4 | Neotropics: Panama to S Brazil | Richards 1978a |
| *Apoica* | 8 | Neotropics: Mexico to N Argentina | Richards 1978a |
| *Asteloeca*[a] | 1 | Neotropics: Amazonia to Bolivia | Richards 1978a, Raw 1985 |
| *Brachygastra* | 16 | Neotropics: SW USA to Argentina | Richards 1978a |
| *Chartergellus* | 7 | Neotropics: Costa Rica to SE Brazil | Richards 1978a |
| *Charterginus* | 6 | Neotropics: Honduras to Amazonia | Richards 1978a |
| *Chartergus* | 3 | Neotropics: Guianas to Paraguay | Richards 1978a |
| *Clypearia* | 7 | Neotropics: Panama to Bolivia | Richards 1978a |
| *Epipona* | 3 | Neotropics: Mexico to Mato Grosso, Brazil | Richards 1978a |
| *Leipomeles* | 2 | Neotropics: Panama to Bolivia | Richards 1978a |
| *Marimbonda*[a] | 2 | Neotropics: Amazonia to Mato Grosso, Brazil | Richards 1978a |
| *Metapolybia* | 11 | Neotropics: Mexico to Paraguay | Richards 1978a |
| *Nectarinella* | 1 | Neotropics: Costa Rica to Colombia | Schremmer 1977, Richards 1978a |
| *Occipitalia* | 1 | Neotropics: Venezuela to Amazonia | Richards 1978a |
| *Parachartergus* | 17 | Neotropics: Mexico to Argentina | Richards 1978a |
| *Polybia* | 54 | Neotropics: SW USA to Argentina | Richards 1978a |
| *Polybioides*[a] | 6 | Old World: equatorial Africa, SE Asia | Bequaert 1922, Vecht 1966, Darchen 1976b |
| *Protonectarina* | 1 | Neotropics: S Brazil to Argentina | Richards 1978a |
| *Protopolybia* | 23 | Neotropics: Honduras to Argentina | Richards 1978a |
| *Pseudochartergus*[a] | 5 | Neotropics: Guatemala to S Brazil | Richards 1978a |
| *Pseudopolybia* | 4 | Neotropics: Nicaragua to Bolivia | Richards 1978a |
| *Synoeca* | 5 | Neotropics: Mexico to Argentina | Richards 1978a |
| *Synoecoides* | 1 | Neotropics: Amazonia, NE Brazil | Richards 1978a |
| **Independent founding** | | | |
| *Belonogaster* | 79 | Old World: Africa | Richards 1982, Hensen and Blommers 1987 |
| *Mischocyttarus* | 206 | Nearctic and Neotropics: SW Canada to N Argentina | Richards 1978a, Snelling 1983, Raw 1985, Carpenter and Wenzel 1988 |
| *Parapolybia* | 5 | Old World: Iran to Japan, Philippines, and New Guinea | Meade-Waldo 1911, Vecht 1966 |
| *Polistes* | 206 | Cosmopolitan | Akre 1982 |
| *Ropalidia*[b] | 136(?) | Old World: Africa, India, SE Asia to Australia | Vecht 1962, Richards 1978b |

[a]Swarm founding is assumed on the basis of colony size and/or nest architecture but has not been observed.
[b]Includes some swarm-founding species.

194

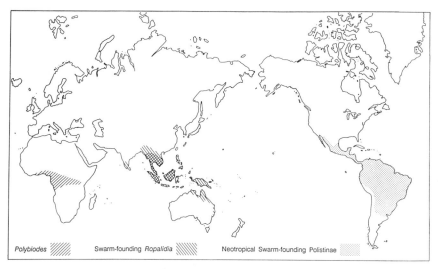

**Fig. 6.1.** Distribution of swarm-founding Polistinae (Bequaert 1918, 1922; Vecht 1962, 1966, 1967; Richards 1978a,b; Das and Gupta 1989). The Neotropical group comprises 23 genera (but see Carpenter and Wenzel 1989).

in an area. Richards and Richards (1951) reported such data for Guyana and Richards (1978a) for Mato Grosso, Brazil. The Richardses' sampling techniques, which were not designed to census the wasp species populations, may be biased (for example, they may have avoided collecting common species or species whose nests were high or otherwise difficult to get), but their data are the best we have. On the basis of numbers of individuals, *Polybia* is by far the most abundant polistine genus in both localities (Table 6.2). *Polistes* and *Mischocyttarus*, the two independent-founding genera, are close to last in the ranking. The swarm-founding genera are so overwhelmingly represented in Table 6.2 that it is probably safe to say that more carefully controlled sampling would not qualitatively alter the conclusion that swarm founders dominate the eusocial wasp faunas at these locations in terms of numerical abundance.

This conclusion is supported by Heithaus's (1979) finding that 86% of the eusocial wasps visiting flowers throughout the year in Guanacaste, Costa Rica, belonged to swarm-founding genera. Moreover, the swarm-founding visitors as a group tended to be less seasonal in their flower visitation than the independent-founding species.

Similarly, T. Erwin's data for rainforest canopy insects collected by insecticide fogging show swarm-founding wasps to be by far the dominant group. In Peruvian rainforest, 49 of 68 eusocial wasp species (72%) and 194 of 215 individuals (90%) were swarm founders, 75 of the individuals being in the genus *Polybia*. In Brazil (near Manaus, Amazonas), 40 of 55 species (73%) and 133 of 159 individuals were swarm

**Table 6.2.**  Individual abundance of polistine wasps at two
Neotropical locations

| Genus[a] | Guyana | Mato Grosso | Total adults | Individual abundance index[b] |
|---|---|---|---|---|
| Polybia | 8,441 | 59,690 | 68,131 | 34,065 |
| Protopolybia | 1,271 | 11,948 | 13,219 | 6,610 |
| Agelaia | 1,709 | 6,642 | 8,351 | 4,176 |
| Brachygastra | 7,379 | 592 | 7,971 | 3,986 |
| Chartergus | 0 | 3,301 | 3,301 | 1,650 |
| Parachartergus | 150 | 2,893 | 3,043 | 1,522 |
| Epipona | 0 | 2,061 | 2,061 | 1,030 |
| Chartergellus |  | 732 | 732 | 732 |
| Apoica | 58 | 1,035 | 1,093 | 546 |
| Angiopolybia | 841 | 0 | 841 | 420 |
| Synoeca | 616 | 0 | 616 | 308 |
| Pseudopolybia | 0 | 400 | 400 | 200 |
| Metapolybia | 262 | 0 | 262 | 131 |
| Mischocyttarus | 127 | 8 | 135 | 68 |
| Marimbonda |  | 56 | 56 | 56 |
| Polistes | 93 | 4 | 97 | 48 |

Source: Guyana: Richards and Richards 1951; Mato Grosso: Richards 1978a.
Note: Data are numbers of adults from all colonies censused at each location.
A zero indicates that the genus occurs at the location but that no colonies were
found. A blank indicates that the genus does not occur at that location.
[a]Ranked by individual abundance index.
[b]Total adults divided by the number of locations at which the genus occurs.

founders (84%); 54 of the individuals belonged to *Polybia* species (T. Erwin and C. Starr, unpubl.). These numbers may even be underestimates of the actual individual abundance of swarm founders; many swarm-founding species occupy enclosed nests, and therefore most of the adults that were killed by the insecticide may have been prevented from dropping onto the collecting sheets.

An alternative measure of relative dominance is colony abundance. This is an even less direct measure than individual abundance because it does not control for species differences in colony size. Despite this shortcoming, I include it here (Table 6.3) because it allows us to sample eight Neotropical sites instead of just two. Again, there are undoubtedly sampling biases that would be avoided by thorough quadrat sampling, but the data give us a crude idea of colony abundance. On the basis of colony abundance, *Polybia* is still the top-ranked genus, but now the two independent-founding genera, *Mischocyttarus* and *Polistes*, because they have relatively small colonies, rise to second and third place. *Protopolybia*, *Agelaia*, and *Brachygastra* are among the top eight swarm-founding genera in both individual and colony abundance.

Wilson (1976) ranked the three most species-rich ant genera in the world fauna in terms of "prevalence," a broader index of success than either individual or colony abundance, based on four biological proper-

**Table 6.3.**  Colony abundance of polistine wasps at eight Neotropical localities

| Genus[a] | Mex | CR:G | CR:H | Pan | G | B:Pa | B:M | E | Total colonies | Mean colony abundance index[b] |
|---|---|---|---|---|---|---|---|---|---|---|
| Polybia | 26 | 154 | 19 | 63 | 43 | 217 | 85 | 114 | 721 | 90.1 |
| Mischocyttarus | 5 | 65 | 1 | 7 | 75 | 216 | 24 | 44 | 437 | 54.6 |
| Polistes | 40 | 26 | 7 | 2 | 7 | 237 | 3 | 4 | 326 | 40.6 |
| Protopolybia |  | 0 | 1 | 78 | 9 | 8 | 2 | 2 | 100 | 14.3 |
| Angiopolybia |  |  |  |  | 5 | 5 | 5 | 30 | 45 | 11.2 |
| Agelaia | 9 | 0 | 1 | 6 | 2 | 5 | 3 | 28 | 54 | 6.8 |
| Metapolybia | 1 | 1 | 27 | 6 | 6 | 0 | 0 | 1 | 42 | 5.2 |
| Synoeca | 5 | 0 | 1 | 4 | 0 | 10 | 1 | 20 | 41 | 5.1 |
| Occipitalia |  |  |  |  | 0 | 8 |  | 3 | 11 | 3.7 |
| Brachygastra | 5 | 1 | 0 | 0 | 11 | 5 | 2 | 4 | 28 | 3.5 |
| Parachartergus | 0 | 3 | 1 | 0 | 2 | 5 | 8 | 7 | 26 | 3.2 |
| Chartergus |  |  |  |  | 0 | 4 | 2 | 7 | 13 | 3.2 |
| Apoica | 0 | 0 | 0 | 2 | 1 | 17 | 5 | 0 | 25 | 3.1 |
| Leipomeles |  |  |  | 9 | 0 | 1 | 0 | 5 | 15 | 3.0 |
| Clypearia |  |  |  | 0 |  | 5 | 0 | 6 | 11 | 2.8 |
| Charterginus |  | 0 | 0 | 0 | 0 | 0 | 0 | 17 | 17 | 2.4 |
| Pseudochartergus |  | 0 | 0 | 0 | 0 | 0 | 0 | 12 | 12 | 1.7 |
| Pseudopolybia |  | 0 | 0 | 0 | 0 | 3 | 3 | 3 | 9 | 1.3 |
| Epipona | 0 | 0 | 0 | 0 | 0 | 6 | 3 | 0 | 9 | 1.1 |
| Chartergellus |  |  |  | 0 |  | 0 | 3 | 1 | 4 | 1.0 |
| Marimbonda |  |  |  |  |  | 0 | 1 | 0 | 1 | 0.3 |
| Nectarinella |  |  |  | 0 |  |  |  |  | 0 | 0 |
| Protonectarina |  |  |  |  |  |  |  | 0 | 0 | 0 |

*Note*: Data are numbers of colonies found in extensive searching for all species of social wasps over varying periods of time from 2 to 28 months in each locality. Localities, habitats, and references are as follows: Mex: Mexico, Veracruz, Los Tuxtlas (rainforest and clearings) (Jeanne, unpubl.); CR:G: Costa Rica, Guanacaste, Cañas (seasonal deciduous forest, clearings) (Jeanne, unpubl.); CR:H: Costa Rica, Heredia, La Selva (rainforest, clearings) (Jeanne, unpubl.); Pan: Panama, Barro Colorado Island (rainforest, clearings) (Chadab 1979a); G: Guyana (rainforest, savannah) (Richards and Richards 1951); B:Pa: Brazil, Pará, Santarém (rainforest, varzea, clearings) (Jeanne, unpubl.); B:M: Brazil, Mato Grosso (gallery forest, cerrado, cerradão, camp, dry forest) (Richards 1978a); E: Ecuador, Napo, Limoncocha (rainforest, clearings) (Chadab 1979a). A zero indicates that the genus occurs in that locality but that no colonies were censused. A blank indicates that the genus does not occur in that locality.
[a]Ranked by mean colony abundance index.
[b]Total colonies divided by the number of sites at which the genus occurs.

ties: local colony abundance, species richness, endemic range, and ecological diversity. Table 6.4 shows a similar ranking for Neotropical polistine wasps. The genera are ranked according to individual and colony abundance, plus the two additional criteria of species diversity and endemic range. The geographic ranges of *Polistes* and *Mischocyttarus* extend from well into Canada to temperate South America, far exceeding the primarily tropical distributions of the swarm-founding genera. Of the latter, however, *Brachygastra*, which has succeeded in barely reaching into Texas and Arizona, south into 18 Argentine states, and to 2,600 m in the Bolivian Andes (Naumann 1968), and *Polybia*, with an almost equal range, have been most successful in exceeding the limits of the tropics. In terms of species richness, on the other hand, none can compare with the 206 species of *Mischocyttarus* and the 89 New World species of *Polistes*.

When we weight all four criteria equally and combine the rankings to yield an overall estimate of relative prevalence (Table 6.4), *Polybia* still retains its first-place position, followed by *Mischocyttarus* in a distant second place. *Protopolybia, Agelaia,* and *Brachygastra* are still ranked high and have to be considered successful genera by these criteria. But in terms of colony abundance and especially individual abundance, the two crude estimates of relative biomass, the swarm founders are clearly the dominant eusocial wasps in the Neotropics.

The Old World tropics appears to be another matter. Although the dearth of studies on social wasps in this region makes it difficult to document, it is my impression, on the basis of the literature and personal experience in the field in tropical Australia, that swarm-founding wasps are much less abundant in the Old World tropics than in Central and South America. O. W. Richards, in three weeks of collecting in the coastal hills west of Cairns in northern Queensland, Australia, found only three colonies of swarm-founding species (Richards 1978b). In intensive searches scattered over eleven months in the region between Townsville and Daintree, Queensland, I managed to find only 39 colonies, 30 of them of *Ropalidia romandi.* Although I have not had personal

**Table 6.4.** "Success" of New World polistine genera based on four measures of prevalence

| Genus[a] | No. of New World species | Abundance[b] Individual | Abundance[b] Colony | Range[c] | Diversity[d] | Total[e] | Overall rank[f] |
|---|---|---|---|---|---|---|---|
| Polybia | 54 | 1 | 1 | 4 | 3 | 9 | 1 |
| Mischocyttarus | 206 | 14 | 2 | 2 | 1 | 19 | 2 |
| Protopolybia | 23 | 2 | 4 | 10 | 4 | 20 | 3 |
| Agelaia | 21 | 3 | 6 | 7 | 5 | 21 | 4 |
| Polistes | 89 | 16 | 3 | 1 | 2 | 22 | 5 |
| Brachygastra | 16 | 4 | 9 | 3 | 7 | 23 | 6 |
| Parachartergus | 17 | 6 | 10 | 6 | 6 | 28 | 7 |
| Synoeca | 5 | 11 | 8 | 5 | 11 | 35 | 8 |
| Metapolybia | 11 | 13 | 7 | 9 | 8 | 37 | 9 |
| Apoica | 8 | 9 | 12 | 8 | 9 | 38 | 10 |
| Angiopolybia | 4 | 10 | 5 | 13 | 12.5 | 40.5 | 11 |
| Chartergus | 3 | 5 | 11 | 14 | 14.5 | 44.5 | 12 |
| Epipona | 3 | 7 | 14 | 11 | 14.5 | 46.5 | 13 |
| Chartergellus | 7 | 8 | 15 | 15 | 10 | 48 | 14 |
| Pseudopolybia | 4 | 12 | 13 | 12 | 12.5 | 49.5 | 15 |
| Marimbonda | 2 | 15 | 16 | 16 | 16 | 63 | 16 |

[a]Only the genera listed in Table 6.2 are included.
[b]The rankings of individual and colony abundance are taken from Tables 6.2 and 6.3, respectively.
[c]Ranked by decreasing geographic range (distribution data from Richards 1978a).
[d]Ranked by number of New World species (Richards 1978a; see also Snelling 1974, 1983; MacLean et al. 1978).
[e]Sum of the four rankings shown.
[f]Based on totals.

experience in other parts of the Old World tropics, published reports by those who have suggest that *Polybioides* and swarm-founding *Ropalidia* are not common anywhere within the areas studied (Carl 1934; Vesey-Fitzgerald 1950; Vecht 1962, 1966). Assuming these impressions reflect reality, why should swarm-founding wasps be less diverse and successful in the Old World than in the New World? The answer may ultimately reveal a good deal about the behavioral ecology of swarm-founding wasps in both the Old and New World tropics.

## BIOLOGY

We are much indebted to the late O. W. Richards and Maud J. Richards for their extensive studies of colony populations in Guyana (Richards and Richards 1951), Mato Grosso, Brazil (Richards 1978a), and Australia (Richards 1978b). Their results have provided the groundwork and inspiration for the more-detailed field studies of particular species that have been completed by numerous workers since 1970. Most of this work has been done on a few of the more common Neotropical species. In comparison, very little is yet known about the biology of the Old World genus *Ropalidia* and virtually nothing is known about *Polybioides*. For this reason, most of this chapter is based on the better known New World genera.

### Colony Founding

The colony cycle has been best studied in *Polybia occidentalis* (Forsyth 1978), *Metapolybia* species (West-Eberhard 1973, 1978b, 1981, 1982c; Forsyth 1978), and *Protopolybia acutiscutis* (Naumann 1970). The following account is compiled from those studies, as well as from my own observations on *Polybia sericea* and *P. occidentalis*.

The swarm is guided from the old nest to the site of the new nest by means of a scent trail (see Communication). In some species (*Polybia velutina*: Chadab 1979a; *P. sericea*: Jeanne 1981a; *Angiopolybia pallens*: Naumann 1975) by the time the main body of the swarm reaches the new nest site, the first workers to have arrived are already busy constructing the foundation of the nest. In others (*Polybia scutellaris*, *Synoeca septentrionalis*, *Agelaia* sp.), construction does not begin until after the swarm has settled (Bruch 1936; West-Eberhard 1982c, unpubl.). The nest is built of wood fiber, although some, particularly the smaller species, use plant hairs exclusively (Moebius 1856, Naumann 1970, Schremmer 1972, Jeanne 1973, Machado 1982), and the five species in the subgenera *Pedotheca* and *Furnaria* of *Polybia* use mud (Richards 1978a). The particles of material are cemented together with an orally secreted "glue," shown recently for *Pseudochartergus chartergoides* to contain chitin (Schremmer et al. 1985). (See Wenzel, this volume, for a discussion of nest architecture.) As a few of the workers continue construction at a rapid pace, the queens and the rest of the workers sit idly by in one or more tight clusters adjacent to the nest. Within hours after the first cells have been constructed the queens begin to oviposit in them; after several days the nest is complete and the swarm has moved inside. By this time many of the cells have received eggs.

Since egg developmental time is approximately six days (*Polybia occidentalis*: Machado 1977a), the first larvae appear about a week after nest initiation. As the larvae increase in number and size over the next two to three weeks, the rate at which workers forage for prey increases to keep pace with the rising demand of the larvae for food. Meanwhile, with no adults eclosing yet, the swarm population undergoes gradual attrition as workers die of senescence, accident, or predation while foraging (West-Eberhard 1981).

About 30 days after the first egg was laid, the first adult offspring begin to eclose (*Polybia occidentalis*: Schwarz 1931, Jeanne, unpubl.). Workers emerge in a pulse that mirrors the pulse of oviposition, although it is more diffuse because individual developmental times vary. The cells that are emptied as adults eclose are resupplied with eggs for the second round of brood rearing. This perpetuates the pulse in brood production, although it may be increasingly damped over several brood generations.

## Polygyny and the Fluctuation of Queen Number

Another distinguishing character of the swarm-founding polistines is that they are polygynous, that is, a swarm or colony has more than one egg-laying queen. Until recently it was thought that polygyny was permanent, because colonies that were collected at different stages of development usually contained several, often many, inseminated females with well-developed ovaries (see Table 6.5 and references therein). The occasional colony with only one such female was not seen as a serious challenge to this generalization.

Recent work, particularly that of M. J. West-Eberhard, has shown that the number of queens in a colony is not static, but rises and falls. Colonies newly established by reproductive swarms contain relatively large numbers of queens, but soon after founding these numbers begin to decrease. In a colony of *Metapolybia aztecoides* observed closely in Colombia, workers directed a distinctive shaking dance at queens (West-Eberhard 1978b). The dance was especially vigorous during periods of queen elimination and has been interpreted by West-Eberhard as a test of queen dominance. If a queen responded submissively to a dance (crouched or offered regurgitated liquid), she was attacked. If she responded dominantly (ignored or avoided the worker or solicited fluid), she was not attacked. In these attacks, which began two to three days after nest initiation, the workers bit the queens, pulled on their legs and wings, sometimes chased them off the comb, and vigorously solicited them for liquid. Sometimes dominant queens pushed or chased attacked queens off the comb (Forsyth 1975, West-Eberhard 1982c). Some of the attacked queens eventually began to behave as workers, while others disappeared. In West-Eberhard's small observation colony, queens were eliminated in this way until the colony contained only one queen, a condition that lasted for the next six months.

Reduction to one queen in *Metapolybia* may occur early in the development of the new colony (West-Eberhard 1973) or later (West-Eberhard 1978b). In any case, it precedes the laying of the eggs destined to produce virgin queens; thus, these offspring are sisters, and high genetic relatedness within the colony is presumably maintained (West-Eberhard 1978b).

Once a condition of monogyny is reached, how is polygyny restored? Whether a young *Metapolybia* female develops into a worker or a queen depends on whether a viable queen is present in the colony (West-Eberhard 1978b). All female wasps evidently go through a phase of slight ovarian development, peaking at one to two weeks of age in independent-founding species (Pardi 1948, Jeanne 1972). If a queen or queens are present, the ovaries of young individuals develop slightly, reach a peak, then regress as these individuals become functional workers. If the single queen dies, disappears, or becomes reproductively senescent, the young females continue to undergo ovarian development, mate, and become replacement queens, restoring the colony to polygyny. In colonies of *Metapolybia* and other swarm-founding genera, females with slightly developed ovaries but empty spermathecae, called intermediates (Richards and Richards 1951, Richards 1978a), appear to be young queens that have not yet been inseminated (Forsyth 1978).

Thus, queen number in some *Metapolybia* colonies cycles between polygyny and monogyny, a condition West-Eberhard (1973, 1978b) refers to as "periodic monogyny." On the basis of the presence of only one queen in at least one collected colony, periodic monogyny appears to occur in a number of other species, all of which form relatively small colonies (Table 6.5). There is no evidence of periodic monogyny in species that form large colonies, although there are data for some suggesting that queen numbers do oscillate. Four colonies of *Agelaia areata* in Mexico, for example, had an average of 8.9% queens (257–716 queens in populations of 4,227–7,950 adults) at the end of winter, when brood rearing was just beginning. In contrast, two colonies collected in the

**Table 6.5.** Colony population size and polygyny in swarm-founding Polistinae

| Species | No. of colonies | Population (no. of adults) | | % queens[a] | | Monogynous colonies known? | Reference |
|---|---|---|---|---|---|---|---|
| | | Mean | Range | Mean | Range[b] | | |
| *Agelaia areata* | | | | | | | |
| winter | 4 | 5,693 | 4,227–7,950 | 8.9 | 5.5–12.2 | No | 5 |
| summer | 2 | 13,570 | 5,339–21,800 | 0.3 | 0.07–0.53 | No | 7, 13 |
| *A. cajennensis* | 1 | 93 | | 13.0 | | No | 12 |
| *A. fulvofasciata* | 2 | 1,020 | 631–1,409 | 8.8 | 4.3–13.4 | No | 2, 9 |
| *A. lobipleura* | 1 | 400 | | 5.5 | | No | 9 |
| *Angiopolybia pallens* | 6 | 144 | 25–390 | 20.5 | 2.5–44.0 | No | 2, 9 |
| *Apoica gelida* | 3 | 217 | 172–272 | 4.7 | 2.6–7.6 | No | 9 |
| *A. pallens* | 3 | 170 | 56–329 | 6.1 | 0.9–12.8 | No | 9, 12 |
| *Brachygastra augusti* | 5 | 1,000 | 307–2,861 | 35.4 | 26.0–50.0 | No | 9 |
| *B. bilineolata* | 3 | 60 | 32–100 | 22.8 | 8.8–50.0 | No | 9 |
| *B. scutellaris* | 5 | 489 | 89–876 | 19.5 | 6.5–35.7 | No | 2, 9, 11 |
| *Chartergellus atectus* | 1 | 90 | | 8.3 | | No | 9 |
| *C. communis* | 3 | 244 | 74–343 | 10.7 | 1.0–23.0 | No | 9 |
| *Chartergus chartarius* | 1 | 3,226 | | 1.4 | | No | 9 |
| *Clypearia apicipennis* | 1 | 69 | | 1.4 | | No | 13 |
| *Leipomeles dorsata* | 1 | 149 | | 2.0 | | No | 9 |
| *Marimbonda albogrisea* | 1 | 56 | | 4.3 | | No | 9 |
| *Metapolybia azteca* | 24 | 200 | 17–1,106 | 12.9 | 1.3–34.6 | Yes | 8 |
| *M. cingulata* | 5 | 58 | 37–106 | 6.8 | 0–22.2 | Yes | 2, 11 |
| *Occipitalia sulcata* | 3 | 86 | 31–130 | 4.0 | 1.5–6.5 | Yes | 13 |
| *Parachartergus fraternus* | 8 | 280 | 166–534 | 11.8 | 0–24.0 | No | 9, 13 |
| *Polybia bicyttarella* | 10 | 77 | 25–143 | 6.4 | 1.8–21.9 | Yes | 2 |
| *P. bistriata* | 17 | 64 | 26–129 | 12.4 | 1.0–53.3 | Yes | 2, 9, 11 |
| *P. catillifex* | 5 | 28 | 6–51 | 13.4 | 4.2–23.5 | Yes | 2, 11 |
| *P. chrysothorax* | 2 | 34 | 32–36 | 3.0 | 2.9–3.2 | Yes | 9 |
| *P. dimidiata* | 3 | 3,528 | 447–6,389 | 4.7 | 0–7.0 | No | 6, 9 |
| *P. emaciata* | 5 | 87 | 16–165 | 18.1 | 4.0–54.0 | Yes | 9, 12 |
| *P. erythrothorax* | 8 | 383 | 23–1,327 | 8.9 | 1.5–48 | No | 9 |
| *P. gorytoides* | 1 | 145 | | 6.2 | | No | 13 |

| Species | n | Mean | Range | Mean | Range | Sperm[a] | Sources |
|---|---|---|---|---|---|---|---|
| P. jurinei | 6 | 501 | 28–1,037 | 7.6 | 0.4–32.0 | No | 9, 12 |
| P. micans | 2 | 114 | 30–197 | 2.2 | 0.5–4.0 | Yes | 9 |
| P. occidentalis (Mato Grosso, Brazil) | | | | | | | |
| dry season | 12 | 77 | 5–382 | 20.5 | — | No | 2, 9 |
| wet season | 12 | 37 | 0–152c | 19.5 | — | No | 2, 9 |
| P. occidentalis (other localities) | 47 | 235 | 23–1,650 | 10.5 | 0–48.4 | Yes | 2, 9, 13 |
| P. parvulina | 2 | 500 | 142–858 | 17.8 | 16.0–19.7 | No | 2 |
| P. platycephala | 4 | 381 | 58–642 | 12.7 | 1.4–43.1 | No | 9 |
| P. quadricincta | 2 | 943 | 69–1,817 | 1.4 | 0.2–2.5 | No | 9 |
| P. rejecta | 6 | 1,877 | 22–4,877 | 12.2 | 0.2–40.9 | No | 9, 13 |
| P. ruficeps | 17 | 1,212 | 100–4,609 | 5.0 | 1.0–15 | No | 9 |
| P. scrobalis | 2 | 54 | 46–61 | 13.9 | 6.5–21.3 | No | 9 |
| P. scutellaris | 2 | 4,687 | 3,074–6,300 | 6.3 | 1.1–11.5 | No | 1, 9 |
| P. sericea | 7 | 552 | 52–900 | 13.7 | 3.2–28.8 | No | 9, 13 |
| P. simillima | 1 | 716 | | 1.3 | | No | 13 |
| P. singularis | 3 | 1,157 | 1,000–1,269 | 12.2 | 6.0–21.0 | No | 9 |
| P. striata | 3 | 355 | 39–872 | 17.6 | 0–34 | No | 9 |
| P. velutina | 1 | 61 | | 1.6 | | Yes | 12 |
| Protopolybia acutiscutis | 16 | 6,654 | 696–16,570 | 3.3 | 0.3–11.7 | No | 2, 4 |
| P. minutissima | 4 | 270 | 51–693 | 27.1 | 3.0–73.3 | No | 2 |
| P. scutellaris | 1 | 389 | | 13.0 | | Yes | 9 |
| P. sedula | 2 | 5,974 | 556–11,392 | 11.2 | 0.2–22.1 | No | 9 |
| Pseudopolybia compressa | 1 | 335 | | 18.0 | | No | 13 |
| P. difficilis | 1 | 40 | | 20.0 | | No | 13 |
| P. vespiceps | 2 | 49 | 25–73 | 40.2 | 22.0–58.3 | No | 13 |
| Ropalidia kurandae | 3 | 173 | 95–237 | 6.1 | 5.1–7.4 | No | 13 |
| R. mackayensis | 1 | 590 | | 20.0 | | No | 13 |
| R. romandi | 10 | 4,098 | 459–17,182 | 9.0 | 1.4–33.2 | No | 13, 14 |
| R. trichophthalma | 3 | 1,361 | 280–3,288 | 17.2 | 7.5–22.3 | No | 13 |
| Synoeca septentrionalis | 1 | 275 | | 37.0 | | No | 13 |
| S. surinama | 28 | 257 | 12–861 | 6.8 | 0–25.1 | Yes | 9, 10, 13 |

Sources: 1, Ihering 1903; 2, Richards and Richards 1951; 3, Rodrigues 1968; 4, Naumann 1970; 5, Jeanne 1973; 6, Rodrigues and dos Santos 1974; 7, Jeanne 1975b; 8, Forsyth 1978; 9, Richards 1978a; 10, Castellón 1980a; 11, Carpenter and Ross 1984; 12, W. Hamilton, unpubl.; 13, Jeanne, unpubl.; 14, J. Kojima, unpubl.

[a]Includes females with at least one fully developed egg and/or with sperm in the spermatheca.

[b]A dash indicates that data are not available.

[c]Excludes one colony of 1,177 adults.

summer, when their populations were high and males were being produced, had only 0.07% queens (4 out of 5,339 adults) and 0.53%
queens (115 out of 21,800 adults) (Jeanne 1975b; see Table 6.5). Much of
the large variability in the percentage of queens among collected colonies (Table 6.5) therefore probably reflects these oscillations. Even
within *Metapolybia*, Forsyth (1978) found that three colonies that were
monogynous in the reproductive stage had a mean ($\pm$ SD) worker population of 45 $\pm$ 27 workers, while ten polygynous colonies had a mean
($\pm$ SD) of 170 $\pm$ 115 workers. Thus the level to which queen number is
reduced during the sexual-producing stage appears to be correlated
with number of workers (Forsyth 1978), and colonies with moderate to
large worker populations may never be reduced to monogyny. Nevertheless, the reduction to a low number, even if not to a single queen,
would have the effect of increasing the relatedness between workers
and the wasps they rear.

Colonywide oviposition rates appear to be independent of queen
number. When there is a high percentage of queens, each has relatively few eggs per ovariole, but when queens are relatively few, each
has a large number of eggs per ovariole (Richards and Richards 1951;
Jeanne 1973, 1975b). Richards and Richards (1951) show this nicely for
colonies of *Polybia bistriata* (populations of 30–130 adults). There were
means of 21.0 ovarial eggs per queen in monogynous colonies ($N = 5$
colonies), 11.3 in colonies with 3–6 queens ($N = 6$ colonies), and only
3.4 in colonies with 16–33 queens ($N = 3$ colonies). This is reflected in
an inverse relationship between queen number and per capita oviposition rate in *Metapolybia* (Forsyth 1978). The queens, whether many or
few, appear to oviposit at a rate that is determined by the number of
available cells. Since the number of cells in the nest is a function of the
number of workers (Forsyth 1978), oviposition rate is ultimately controlled by workers.

## Colony Growth and Reproduction

As do other eusocial wasps, swarm founders progressively feed their
larvae malaxated (chewed) prey, while the adults feed primarily on floral and extrafloral nectar and honeydew. Recent evidence, however,
indicates that in *Polybia occidentalis* these diets are not exclusive; not
only do adults imbibe hemolymph from prey, but a certain amount of
nectar finds its way into larval midguts (Hunt et al. 1987). Larva-adult
trophallaxis also occurs; the adults imbibe the larval salivary secretion,
which they apparently treat as food (Hunt et al. 1987). Naumann (1970)
found cuticular material in the guts of *Protopolybia acutiscutis* workers,
indicating that they eat prey. Unlike temperate-zone polistine and vespine wasps, and probably most swarm-founding wasps (including *P.*

*acutiscutis*: Naumann 1970), which malaxate their prey before bringing it to the nest, foragers of *Polybia occidentalis*, *P. paulista*, *P. ignobilis*, and *Agelaia pallipes* bring in some of their prey more-or-less intact. Thus, prey items can be identified to order or family. These four species rely most heavily on lepidopterous larvae (41–80% of identifiable prey); nonlepidopterous prey items include adults and immatures of eleven other insect orders and arachnids (Gobbi et al. 1984; Gobbi and Machado 1985, 1986; Machado et al. 1987; Jeanne, unpubl.).

Just as brood production occurs in pulses initiated by the rapid filling of the cells with eggs after nest initiation, nest expansion also is episodic. After the nest is built by the founding swarm, there may be no addition of cells to the nest for weeks or even months. When nest expansion does occur, it is typically a discrete event, lasting only several days (but see Wenzel, this volume). In this regard, the swarm-founding wasps differ from the independent-founding polistines and the vespines, in which nest expansion is much more continuous throughout the growth phase.

Because of the pulsing of brood production, the worker-larva ratio also fluctuates periodically. As a cohort of eggs in the nest begins to hatch into larvae, the worker-larva ratio drops dramatically, and the workers busy themselves with foraging for food for the immatures. But as the brood begins to pupate and emerge as adult workers, the worker-larva ratio rises. Nest expansion often occurs during these periods, when the workers have relatively few larvae to feed (Forsyth 1978). If worker mortality is low during the weeks following founding, a high worker-larva ratio followed by nest expansion may occur when the first cohort of brood begins to pupate. If, on the other hand, worker attrition is high, nest expansion may not occur until a subsequent episode of worker emergence (Forsyth 1978).

There is another factor in this formula, however: queen aggression. In one study, if aggression toward and among queens of *Metapolybia aztecoides* was low, indicating low queen competition (as when there was a single queen), then pulses of adult eclosion were followed by nest expansion. If, however, queen aggression was high after peaks of adult emergence, the colony produced swarms (West-Eberhard 1982c).

Male production is little understood in swarm-founding polistines. Most colonies collected for analysis lack males. In fact, males of a number of species remain undescribed (Richards 1978a). Forsyth (1978) claims that male production of *P. occidentalis* and *Metapolybia* species is not directly correlated with polygyny, monogyny, large colony size, or stage of development of the colony, but occurs in colonies that have old queens. Production of males of at least some species in some localities appears to be seasonally synchronized (see Latitudinal Patterns in Seasonal Synchrony).

Virtually nothing is known about the mating systems of swarm-founding wasps. Richards (1971) states without evidence his opinion that, in most genera, mating occurs at the natal nest before the founding swarms leave. However, despite many hours of observation of behavior at the nest by numerous ethologists, including West-Eberhard's extensive observations on a colony of *Metapolybia aztecoides*, mating has never been observed at the nest. On the other hand, I have seen large numbers of *Polybia occidentalis* males patrolling within restricted areas in shrubby vegetation away from nests in Guanacaste, Costa Rica, and West-Eberhard has observed this in *P. sericea* and species of *Parachartergus, Agelaia, Synoeca,* and *Epipona* (M. J. West-Eberhard, unpubl.). Presumably gynes come to such areas to mate. These observations suggest that copulation takes place away from the nest. Males apparently do not accompany swarms (except possibly in *Apoica*: Ducke 1905, 1910; Richards 1978a).

## Types of Swarms

It has been difficult to reach an understanding of the colony cycle of swarm-founding wasps in part because swarming can occur in several contexts, and when a swarm is collected in the field it is impossible to determine the context that gave rise to it. *Reproductive swarming* (also called "budding" or "fissioning") can be defined as division of the colony into two or more groups; all adults may disperse from the parent nest, or one group may remain behind. Reproductive swarms are probably normally stimulated by such intrinsic conditions as large population size, a new cohort of young females beginning ovarian development, and aggression among queens (Forsyth 1978, West-Eberhard 1982c).

*Absconding swarms* are triggered instead by an externally imposed stress on the colony. The entire adult population evacuates the nest and emigrates (usually without subdividing) to a more-or-less distant site where it constructs a new nest. Precisely when the colony absconds in response to a stress may be dictated by events within the colony. If the stress is severe, involving complete destruction of the brood, the adult population will abscond immediately. If the stress is not severe—the nest site becomes exposed to direct sunlight, say, resulting in frequent overheating—the colony may wait until the current cohort of brood has eclosed before absconding. The resistance of a colony to absconding under stress and how aggressively it will defend its nest if disturbed appear to be related directly to the level of the colony's investment in the brood currently in the nest. If a new pulse of brood production has just begun, with most cells containing eggs or young larvae, the colony will be relatively nonaggressive in its defense of the nest and will readily abscond (Forsyth 1978). The same point of brood rearing can be reached fairly quickly in a new nest, so the cost of absconding is low. On the other hand, if the nest contains a large population of older larvae and pupae, the workers will defend it aggressively, and the colony will be reluctant to abscond until the current brood have eclosed to adults, presumably because reaching the same point in brood production in a new nest involves a much greater investment of energy, not to mention the additional weeks of attrition without replacement suffered by the adult population. This reluctance to abscond was amply demonstrated to me by a colony of *Agelaia testacea* near Santarém, Pará, Brazil, whose brood-filled nest had fallen to the ground when the palm spathe to which it was attached tore loose. Despite the fact that the combs had suffered considerable damage in the fall, the adults (or at least 6,466 of them) had remained with the nest (Jeanne 1970c).

In some species, there appears to be a third type of swarming, which may be called *emigration swarming*. *Polybia occidentalis* colonies in Guanacaste appear to abandon their dry-season nests in July after raising one cohort of adult offspring early in the wet season (Jeanne, unpubl.). These swarms are not reproductive swarms because there is no fissioning of the adult population. Although two colonies that swarmed in this way had been subjected to minor repeated envelope removal and may have abandoned under stress, ten others that had not been manipulated moved at about the same time. Thus, emigration swarming may be a normal event in the colony cycle of this population (Jeanne et al. 1988). Forsyth's (1978) data, showing a bimodal annual distribution of new nests of *P. occidentalis* and *Metapolybia azteca* in Guanacaste, one mode from October through January and a larger one from May through July, corroborate this conclusion. Naumann (1970) found that *Protopolybia acutiscutis* in Panama moved to new nest sites twice a year. The move into the forest at the end of the dry season is made by reproductive swarms, but the move back to more-open habitat at the end of the wet season may not be, since male production has not yet begun. To determine whether these emigration swarms truly represent a third kind of swarm and, if so, what their function is will require further work.

## Independence of the Nesting and Colony Cycles

As is seen from this account, the colony cycle in swarm-founding wasps is essentially the same as in the more familiar temperate-zone *Polistes* and Vespinae: colony founding is followed by a growth (ergonomic) phase, which culminates in a reproductive episode. Colony cycles of temperate-zone wasps are clear-cut annual affairs, displaying a synchrony of events that is enforced by the extreme seasonality of the climate. At the end of the nesting season, the colony terminates as a social unit, the nest is abandoned, and the females mate and disperse, entering a solitary phase. Nests are not normally reused the following year. Colony and nesting cycles have a congruency that is enforced by the short favorable season (Fig. 6.2a).

In tropical climates, the seasonal constraints are not so rigorous, and the components of the cycle of many populations of swarm-founding wasps become dissociated, leading to considerable confusion over what we mean by the colony cycle; indeed, even the colony itself loses the temporal and spatial clarity it has in temperate-zone wasps. It seems necessary, therefore, to define more precisely some terms that have been used rather loosely in the literature.

We start with the simplest. The *nest* is simply the structure the colony builds to house itself. The *colony* is the social unit that occupies the nest, consisting of brood as well as adults. A *swarm* can be thought of as a colony that is temporarily without a nest (i.e., between nests) and therefore also without brood. *Colony reproduction* is the production of reproductives—males and/or gynes—often followed by the emission of *reproductive swarms*. The *colony cycle* is the period of development lasting from the end of one reproductive episode to the end of the next. A *brood developmental period* is a subunit of the colony cycle, namely, the period of development of the immature, from egg to adult. Several successive cycles of brood are typically produced in each comb. The *nesting cycle* is the period the colony spends in a particular nest, beginning with founding and ending with abandonment (except when absconding is induced by predation or accident).

So far so good. The confusion comes when we face the fact that in swarm-founding species the neat congruence between nesting cycle and colony cycle is often lost. In other words, colony cycles and nesting cycles are semi-independent, which is why it is useful to distinguish the two. As Fig. 6.2b shows, a colony may occupy a nest for several years, during which time it probably goes through annual episodes of reproduction and swarm emission (Richards and Richards 1951); that is, several colony cycles are included in one nesting cycle. Alternatively, colonies of other species—*Polybia occidentalis* in Guanacaste, for example—appear to routinely occupy two nests successively during a single colony cycle. In other words, there are two nesting cycles within a single colony cycle (Fig. 6.2c). What, then, is the colony? I propose to modify the definition of the *colony* given above with the following addition: the *colony* lasts for the duration of its occupancy of a single nest or for one colony cycle, whichever is longer.

The phenomenon of noncongruence of the colony and nesting cycles led Richards and Richards (1951) to distinguish two main types of be-

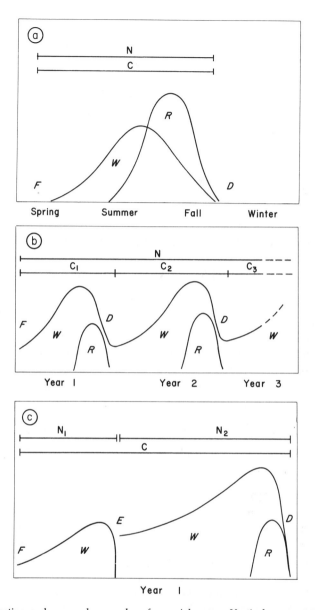

**Fig. 6.2.** Nesting cycles vs. colony cycles of eusocial wasps. Vertical axes represent adult populations. (a) Annual colony and nesting cycles of temperate-zone polistines and vespines. Colony and nesting cycles are temporally congruent. (b) Indeterminate nesting cycle. One nesting cycle may last for two or more colony cycles (e.g., *Polybia scutellaris* in northern Argentina). The colony produces several cycles of brood in the nest ($C_1$–$C_3$), each followed by a bout of reproduction, with budding off of reproductive swarms. A fraction of the population remains in the nest to repeat the colony cycle. (c) Determinate nesting cycle in which one colony cycle lasts for two nesting cycles (e.g., *Polybia occidentalis* in Guanacaste, Costa Rica). The colony produces one cohort of workers in nest 1 ($N_1$), emigrates to found nest 2 ($N_2$), produces several cycles of worker brood, then reproduces. Reproductive swarms disperse, and the nest is abandoned. F, nest initiation; W, worker production; R, reproductive production; D, dispersal of reproductives (as solitary gynes or gynes in swarms); E, emigration swarm from old nest to new; C, extent of colony cycle (between vertical ticks); N, extent of nesting cycle (between vertical ticks). Note the differences in time scale.

havior with respect to colony duration in swarm-founding polistines. "Short-cycle" species produce several generations of workers, then produce males and queens and abandon the nest. "Long-cycle" species occupy the nest for more than a year, often substantially longer. A colony of *Epipona tatua* on Trinidad, for example, was observed by Vesey-Fitzgerald (1938) to persist for more than two years, and a colony of *Brachygastra mellifica* in Guanacaste first seen in 1982 was still active in 1984 (Jeanne, unpubl.). Bruch (1936) observed a colony of *Polybia scutellaris* in Argentina that lasted for more than three years, and according to Lucas (1867, 1885) nests of this species may be occupied for 25 years. Richards (1978a) cites records of colonies of *P. scutellaris* lasting 10, 20, and 30 years in Piracicaba, near São Paulo, Brazil.

In fact, there may be no qualitative difference between the two groups of Richards and Richards (1951) with respect to their colony cycles; both probably give off reproductive swarms at regular intervals, at least once annually. The difference seems to be one of how many colony cycles are included in a nesting cycle. In view of the distinction made above between nesting and colony cycles, I propose to replace Richards and Richards's short- and long-cycle species with the following terms. *Determinate nesting cycle species* are those in which the nesting cycle does not exceed a single colony cycle, and *indeterminate nesting cycle species* are those that use the same nest for more than one colony cycle.

We will return for another look at the indeterminate nesters in the section Indeterminate Nesting Cycles.

## CAUSES OF SUCCESS

Why are the swarm-founding polistines able to so dominate the eusocial tropical wasp fauna? With the evolution of swarming came two important advantages over the independent-founding wasps: (1) the socialization of dispersal and founding and (2) a social infrastructure that enabled larger colonies to evolve. I shall argue that these two traits are in large part fundamentally responsible for the ecological success of the swarm-founding wasps.

### The Socialization of Dispersal and Founding

The polygynous swarm seems well adapted to social life. in the tropics. The presence of a large contingent of workers in the swarm reduces the risk of mortality to the propagule during its dispersal from the natal nest and founding of the new nest, for several reasons. First, the workers assume the risks of scouting out a new nest site, a process that can take a day or two of intensive flight activity with its consequent exposure to the risk of predation (Jeanne 1981a). The queens, who wait in the security of the natal nest until the new nest site has been selected, are exposed only during the short flight to the new site. Second, unlike independent-founding queens, swarm-founding queens avoid the additional risks of having to forage during the preemergence stage of colony development, that is, before the first-reared workers emerge as adults. Finally, the large worker force can provide a formidable defense against ants and other predators that might find the brood in the new nest a tasty meal (Fig. 6.3). This is especially important in wet tropical regions, where ant predation pressure is high (Jeanne 1979b). Forsyth (1981) reported, for example, that only about 20% of the population of *Metapolybia* swarms foraged during nest initiation, while most of the remainder stationed themselves around the periphery of the comb, effectively acting as guards against patrolling ants.

The polygyny of the swarm further reduces the risk of loss of the female reproductives by dividing the swarm's eggs up among many "baskets." Not only can each queen thereby remain relatively mobile

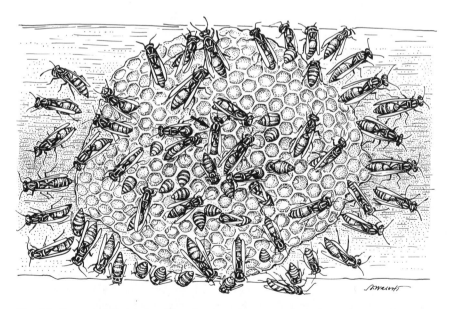

**Fig. 6.3.** Swarm of *Occipitalia sulcata* that has nearly completed the comb of cells but has not yet begun the envelope. A queen (arrow) is shown ovipositing in one of the cells. Many of the workers are arrayed around the comb, where they act as guards to prevent ants from reaching the nest.

by not having to be physogastric (Richards and Richards 1951), but the loss of one or two queens during dispersal or founding will not hinder the reproductivity of the young colony. In sum, the polygynous swarm, although a costly mode of reproduction because it requires a large number of workers to found each new colony and presumably rewards each worker's efforts with a relatively low genetic payoff (because of the multiple queens and the sometimes lengthy intervals between reproductive episodes), contains built-in cushions against its own extinction.

The heavy investment in adult numbers making up the swarm results in a dramatic reduction in colony mortality rate during the disperal, founding, and preemergence stages. These are precisely the stages in the independent-founding polistines and vespines that suffer the highest mortality rates (e.g., Archer 1980a, Gadagkar, Green, this volume). Low mortality rates (and concomitant heavy investment in workers) are reflected in the correspondingly low productivity of colonies of swarm-founding wasps. Defining productivity as the number of queen offspring per queen to which the colony gives rise, Richards and Richards (1951) estimated the productivity of swarm-founding polistines to be between one and four. Forsyth (1978) estimated it at 6.65 for *Metapolybia azteca* and, using Naumann's (1970) data, on the order of ten for *Protopolybia acutiscutis*. These values compare with 244–1,261 queens per founding queen for six species of Vespinae (Archer 1980a). Although the values of queen productivity are computed indirectly for the swarm founders and are confounded by the fact that a swarm containing dozens of queens may yield a colony whose reproductive offspring are the product of only one, the difference is clear: each colony of swarm-founding polistines produces only a few propagules per year (in the form of reproductive swarms), while a typical vespine colony in the temperate zone produces hundreds (in the form of individual queens). The lower productivity of swarm founders compared with independent-founders suggests that a relatively greater proportion of the energy budget of the former is allocated to tasks not directly involved in reproduction, such as defense and nest construction.

### Colony Size

A second effect of the evolution of the swarm is that it enabled much larger colonies to evolve. Colony size in swarm-founding polistines varies tremendously among species and genera (Fig. 6.4). Notable for especially wide ranges are the genera *Brachygastra*, *Polybia*, and *Protopolybia*, whose colony sizes span at least 2.5 orders of magnitude, and *Agelaia*, which spans nearly four. A colony of *A. vicina* nesting in an abandoned shack near Ribeirão Preto in the state of São Paulo, Brazil,

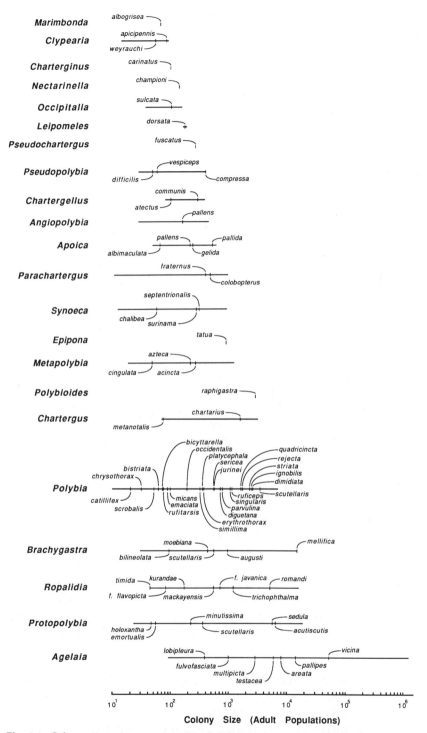

**Fig. 6.4.** Colony sizes of swarm-founding polistine genera, arranged roughly by increasing colony size. Horizontal lines represent the range of colony size of each genus. Vertical ticks give the mean colony size of the indicated species. A tick with no range indicates a sample size of one colony. (Ihering 1903; Schwarz 1929; Rau 1933; Hase 1936; Araujo 1940; Richards and Richards 1951; Pagden 1958b; Vecht 1962, 1966; Rodrigues 1968; Jeanne 1970c, unpubl.; Naumann 1970; Machado and Hebling 1972; Rodrigues and dos Santos 1974; Machado 1977a,b, 1985; Schremmer 1977, 1978a,b, 1983; Simões 1977; Forsyth 1978; Richards 1978a,b; Castellón 1980a; Höfling and Machado 1985; W. Hamilton, unpubl.; J. Kojima, unpubl.)

is by far the largest colony of social wasps ever recorded. When collected by D. Simões, R. Zucchi, and N. Gobbi, it contained an estimated 1.3 million adults, including 108,000 queens. The total comb area was estimated at 33 m² (D. Simões, unpubl.).

How do the swarm-founding wasps achieve such large colonies compared with the independent-founding polistines? Colonies of independent-founding wasps are initiated by solitary queens or by associations of several inseminated queens, one of which typically becomes the sole egg layer. The low fecundity of the egg-laying females keeps populations small. In addition, the relatively inefficient mechanism of queen control makes it necessary for the queen to expend energy on dominating her associates into subordinate roles as well as on reproduction.

The swarm-founding species have broken free of those constraints by virtue of the polygynous swarm. The size of the initial cohort of brood, and hence in large part the ultimate size of the colony, depend on two parameters: the number of workers and the rate of oviposition by the queens. Since it is the number of workers in the swarm that determines the number of cells in the initial nest, and it is the queens who fill the cells with eggs, either can constrain initial colony size. Because females of all polistines have just three ovarioles per ovary, a single queen clearly would not be able to service a very large swarm. The presence of multiple, at least temporarily mutually tolerant, queens ameliorates the limiting effect of oviposition rate on initial colony size. Thus, because the swarm is a highly organized social group of workers and queens, and physical dominance is not employed to maintain social structure, there are no known social limits to swarm size and hence to colony size of the swarm-founding wasps.

## CONSEQUENCES OF LARGE COLONY SIZE

What, then, sets colony size in these wasps? Optimal colony size for a species must represent the point at which the benefits minus the costs of colony size are maximized for a given habitat. The benefits of large colony size include the following: (1) At least within a limited range, large colonies are able to organize certain kinds of labor more efficiently (i.e., to specialize). For instance, large colonies (> 350 adults) of *Polybia occidentalis* in Costa Rica performed nest construction nearly twice as efficiently as did small colonies (< 50 adults) of the same species (Jeanne 1986a). (2) Large colonies have sophisticated forms of communication, which integrate the activities of the colony members. (3) Aggressivity increases with worker population, making for a more effective defense against predators. (4) Larger colonies are able to exert more homeostatic control over physical conditions within the nest.

There are, however, several costs to larger colony size. (1) Michener (1964) summarized evidence that increasing colony size is accompanied by a decrease in productivity per female. (2) Within a species, colony size has probably been limited during evolution by the nest structure, so that above a certain size, which varies with nest architecture, it may become more efficient for the colony to break up into smaller groups and start new nests than to continue enlarging the old one. In general, within a given architectural type, larger nests have lower "reproductive efficiencies," since an increasing proportion of nest material must go into supporting structures (Jeanne 1977a). (3) Foraging efficiency is depressed because the average foraging distance from the nest must increase if the workers are to gather food from a large enough area to support a larger colony. (4) Large nests are more difficult to hide from visually or olfactorily orienting predators and parasites.

In the following sections some of the benefits of large colonies are investigated.

## Specialization

Along with larger colony size there have evolved both greater specialization among colony members and more sophisticated integration of their activities. In this section I briefly examine polymorphism and polyethism among swarm-founding species. Polyethism is covered in detail in Chapter 11.

QUEEN-WORKER SPECIALIZATION   The degree to which castes are specialized behaviorally, physiologically, and morphologically is one measure of the level of sociality achieved by a taxon. In the independent-founding polistines, queens tend to be among the larger individuals, but there is no evidence of morphological caste differences. Among the swarm-founding wasps, however, there is a continuum, from species in which there is no size or morphological difference between queens and workers to species in which queens show a discontinuous size difference with respect to workers. In an extreme case, queens of *Agelaia areata* average 14% larger (thorax length) than workers (Jeanne and Fagen 1974). In general, species with larger colonies show the greatest dimorphism (Jeanne 1980a). Among species with queens larger than workers, the queens tend to show allometric growth with respect to workers. Surprisingly, queens of a few species—most notably *Polybia dimidiata*, *Apoica gelida*, and *A. pallens*—are smaller than workers (Rodrigues and dos Santos 1974, Richards 1978a). These species have small- to moderate-sized colonies.

At least in the species that lack queen-worker dimorphism, there is a considerable degree of caste flexibility. West-Eberhard (1978b, 1981) has shown, for example, that all newly emerged females of *Metapolybia* and *Synoeca* are potential queens and that caste is determined by contests among adults. If the colony is queenright, young females are dominated by the queen(s), and by the end of their first week they begin to behave as workers. If the colony becomes queenless, females in their first week become replacement queens (West-Eberhard 1981). This caste flexibility during the early days of adult life allows individuals to respond to changing reproductive opportunities when the colony cycle is disrupted by predation or accident (Forsyth 1978, West-Eberhard 1981).

To the extent that they can depend on always being surrounded by workers, swarm-founding queens can be more exclusive reproductive specialists than can independent-founding queens. Consequently, reproductive division of labor is carried further in swarm-founding polistines than in any other group of wasps. Activities of the queens are limited to oviposition, self-feeding, and behavior patterns that appear to be displays, while workers perform all other tasks (West-Eberhard 1978b). More of the activities related to control of colony development—selection of the nest site and determination of the size of the initial nest, for example—are in the hands of workers than is true for the independent-founding polistines or vespines (see also Jeanne, this volume: Chap. 11).

SPECIALIZATION AMONG WORKERS  As is true for all other social wasps, there is no evidence of morphologically specialized subcastes of workers in swarm-founding wasps. There is, however, age polyethism among workers, and it follows the pattern found in other social insects: young workers engage in activities on and in the nest, while older workers become foragers (Naumann 1970, Simões 1977, Forsyth 1978, Jeanne et al. 1988). In addition, individual workers differ with respect to level of activity and degree of specialization. Some individuals become "fixated" on certain tasks and perform them quickly. Fixated workers are more efficient but less responsive to changes in supply and demand than are nonfixated workers (Forsyth 1978). Workers show sophisticated and efficient task partitioning with respect to complex tasks such as nest construction.

## Communication

Swarm-founding species with large colonies have in some cases evolved sophisticated communication to coordinate the activities of the many colony members. In recent years, several forms of chemical communication have been demonstrated in these wasps, but there is still much to learn.

Alarm recruitment was recently demonstrated to exist in *Polybia occidentalis* (Jeanne 1981b). The venom of workers contains a pheromone that rapidly recruits large numbers of adults to the outer surface of the nest envelope (Fig. 6.5). Outside the nest, the odor of venom greatly reduces the threshold for the release of attack behavior. Attack itself— flying at the intruder and stinging—is released visually; dark moving objects are the most effective stimuli. Although it has not been demonstrated experimentally, a similar communication of alarm may exist in *P. sericea* (Jeanne, unpubl.), *P. rejecta* (Overal et al. 1981), *Protopolybia acutiscutis* (Naumann 1970), and *Apoica pallida* (Schremmer 1972). Chemical communication of alarm is no doubt widespread, if not universal, in the swarm-founding wasps, although it would be worth investigating whether it exists in species that form very small colonies.

Another chemical communication mechanism is use of a trail pheromone in the context of swarm emigration. Naumann (1975), working with *Protopolybia acutiscutis*, was the first to report evidence suggesting the presence of such a pheromone. During the movement of a swarm from the old nest to a new site, some of the workers alit on leaves and

**Fig. 6.5.** Alarm recruitment in *Polybia occidentalis*. The palm rachis bearing the nest was tapped several times, causing many of the adults to emerge rapidly from the nest and spread out over the nest envelope, wings spread and alert. The same response can be elicited by the odor of venom wafted over the nest. The alert defenders will fly to attack any moving nearby object. After 30–60 seconds, the wasps will begin to reenter the nest. (From Jeanne 1981b, courtesy of Springer-Verlag.)

other substrates along the route and walked over the surface while dragging the gaster. This behavior suggested to Naumann that a pheromone was being applied. Working with *Polybia sericea* in Brazil, I subsequently demonstrated experimentally that scouts use a secretion produced by an exocrine gland opening on the penultimate (fifth gastral) sternum to scent-mark the route between the two points (Jeanne 1981a). The same scent is used to mark the site on which the swarm temporarily assembles if the old nest has been destroyed by a predator. When the scouts have decided on a single site and have reinforced the trail leading to it, they somehow communicate this to the rest of the swarm. Individuals in the resting cluster lift off and begin to fly in broad looping arcs in search of scent marks (Fig. 6.6). They appear to be visually attracted to prominent objects away from the swarm, the same kinds of objects that are attractive to scent-marking scouts. By hovering just downwind of such points, or by landing on them and inspecting them, the swarm members determine whether the scent is present. If it is, they proceed farther from the cluster site in the same direction and repeat the search (Fig. 6.6). This behavior eventually brings them to the site selected by the scouts. The swarm in these species is diffuse, not at all like the dense swarms of *Apis mellifera*. It may be 30–60 minutes from the first lift-off until the last stragglers arrive at the new site. Even more extreme, swarms of a species of *Agelaia* (near *hamiltoni*) may recruit nestmates from the natal nest for many days before establishment of a new colony is complete (M. J. West-Eberhard, unpubl.). Swarms of *Agelaia areata* and *Brachygastra lecheguana* may move up to 100 m in a single step, and the former species may move on several days in succession (Jeanne 1975b, Forsyth 1981). Similar behavior has been observed in other species in several genera (Jeanne 1975b, Naumann 1975, West-Eberhard 1982c), and the presence of the sternal gland in most other swarm-founding genera suggests that the use of the emigration trail pheromone is widespread in this group (Jeanne et al. 1983).

This reliance of emigrating swarms on scent trails no doubt has limited the dispersal of swarm-founding wasps across water barriers. This could restrict gene flow in Amazonia, where broad rivers may only rarely be crossed by swarms. Dependence on chemical trails could likewise be responsible for the absence of these wasps on all Caribbean islands except Trinidad, Tobago, and Grenada (Jeanne 1981a).

Finally, behavioral evidence suggests that *Metapolybia* queens are recognized as a class by workers on the basis of odor alone, and that the source of the odor is the head (West-Eberhard 1977a). It seems likely that the odor substance plays some role in ovary suppression in workers, but how the effect is mediated is not known. There is also behavioral evidence that queens use pheromones to control worker

**Fig. 6.6.** Swarm emigration by *Polybia sericea*. The swarm is just beginning to leave the natal nest (left foreground). Swarm members are taking off from the cluster and flying in broad arcs in search of the chemical trail to the new nest site. Scout wasps are shown "dragging" the gaster on leaves (middle inset) to deposit a pheromone produced in the penultimate (fifth gastral) sternum. Naive swarm members are seen hovering downwind of potential scent-marked sites or landing on them and inspecting them in search of scent marks (lower inset). In the distance can be seen the early arrivals at the new nest (upper inset), which at this point consists of just a few cells.

ovarian development (or as signals of dominance) in *Polybia occidentalis*, (Forsyth 1978), *Protopolybia sedula, P. exigua* (Naumann 1970), and *Agelaia pallipes* (Simões 1977).

The social wasps remain the only one of the four groups of eusocial insects for which communication of distance and direction information in recruiting nestmates to a food source has not been demonstrated. Lindauer (1971) found evidence for a weak recruitment of nestmates to a food dish 150 m from the colony in *Polybia scutellaris*, but there was evidently no communication of distance or direction information. In preliminary experiments with *Polybia, Metapolybia, Synoeca, Brachygastra,* and *Agelaia,* Forsyth (1978) found little evidence for recruitment. If social wasps indeed do lack the ability to communicate distance and/or direction of food, then the interesting question becomes why. On the surface, the ability would seem to be as advantageous to species such as *Brachygastra mellifica, B. lecheguana, Polybia scutellaris, Protonectarina sylveirae,* and others that collect and store large amounts of honey as it is to *Apis* and the meliponine bees. Recruitment could conceivably also be advantageous to species that utilize concentrated, point-source food resources such as carrion, as does *Agelaia pallipes.* Lack of adequate genetic variability seems an unlikely reason for the absence of such communication about food sources: there has been enough variability for natural selection to have led to the evolution of the trail pheromone described above, a mechanism similar to that used by certain meliponine bees to recruit nestmates to a food source (Lindauer 1971).

## Defense

As one works in the field in the American tropics, it is impossible not to be impressed with the importance of ants to the ecology of wasps. I suspect that social wasps pay much more attention to ants in their environment than we think. Because army ant raids are spectacular and conspicuous, it is commonly assumed that these ants are the wasps' primary formicid foes. As overpowering as they are, however, often months go by between raids (Chadab 1979a), and many colonies probably escape army ant predation simply by chance. Meanwhile, myriad species of scouting and recruiting ants are omnipresent in the wasps' immediate environment, especially in the wetter tropical regions (Richards and Richards 1951, Jeanne 1975a). In these habitats, a defenseless wasp larva will fall prey to ants within hours, sometimes even in minutes (Jeanne 1979b). Thus these ants keep up a steady, if low-intensity pressure on wasp colonies, and they must be dealt with on a daily basis.

While independent-founding polistines depend largely on chemical repellents to keep scouting and recruiting ants from their brood

(Jeanne 1970a, Turillazzi and Ugolini 1979, Post and Jeanne 1981, Kojima 1983a, Gadagkar, this volume), swarm founders rely primarily on the nest envelope coupled with active guarding of the entrance by the workers (Jeanne 1975a). When ants approach a nest of *Protopolybia exigua*, worker wasps respond by rearing their bodies up and backward and buzzing the wings in short, semisynchronous 0.9-second bursts, 1.3 seconds apart (Chadab 1979a). Chadab demonstrated that the visual or olfactory stimulus of species of *Eciton* (army ants) and of *Camponotus sericeiventris*, as well as the odor of formic acid, elicited this "group fanning" behavior. She interpreted the behavior as an alarm signal that alerts nestmates to the approach of army ants so that the colony can evacuate quickly if the ants reach the nest.

I have seen *Polybia occidentalis* respond in the same way to fire ants (*Solenopsis*), but my interpretation of the function of the behavior is different. When fire ant workers descended the supporting twig onto the nest of a colony of *P. occidentalis*, several wasp workers ringed the upper part of the nest and performed the wing buzzing (Fig. 6.7). If an ant approached a worker closely, the wasp would dart at it, pick it up in its mandibles, and immediately fly off to drop it away from the nest. Each burst of buzzing, which was exactly as described by Chadab, directed a pulse of air forward at the ants, whose antennae were visibly deflected with each pulse. These little blasts of air were effective in causing the ants to stop momentarily. Persistent fanning by the wasps sooner or later succeeded in causing the foraging ants to turn around and retreat up the twig. I interpret the response as primarily defensive in function, rather than communicative.

After incursions by ants onto the substrate immediately adjacent to

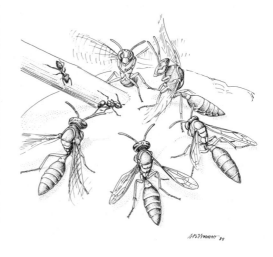

**Fig. 6.7.** Ant defense by *Polybia occidentalis*. Two fire ant (*Solenopsis* [probably *saevissima*]) foragers have found their way down the twig and onto the nest, eliciting defensive behavior in the wasps standing nearby. The wasp workers rear back and buzz their wings in short bursts, directing intermittent blasts of air at the ants. The antennae of the ant that has just stepped onto the nest are being blown rearward by the draft. Faced with this rhythmic mini-windstorm, the ants eventually retreat back up the twig. Any ant that succeeds in advancing to or past the phalanx of defenders will be picked up by a wasp in her mandibles, carried away from the nest, and dropped.

the nest or onto the nest itself, workers of *Occipitalia sulcata* and *Meta-polybia aztecoides* scrape the violated surfaces with their mandibles (Chadab 1979a, West-Eberhard 1989). I have seen *Polybia occidentalis* behave in the same way. Observations and manipulations by West-Eberhard (1989) suggest that the effect of this behavior is to remove the trail pheromone of the ants. By thus diverting the odor trail of the ants around the nest, the wasps reduce the likelihood that ants will approach the nest in the future.

Although the rule among swarm founders is active defense of the nest by workers, there are two known exceptions. *Nectarinella championi* (Skutch 1971, Schremmer 1977) and *Leipomeles dorsata* (Ducke 1910, Williams 1928, Jeanne, unpubl.) defend access to the nest with arrays of sticky-tipped stalks, which are apparently produced from an oral secretion (see Wenzel, this volume: Figs. 14.41 and 14.42). Both are extremely small wasps, for which actively defending against many ant species may not be feasible.

While active defense and sticky traps are adequate lines of defense against the smaller scouting and recruiting ants, army ants are another matter. Most wasp species appear to have no effective defense against army ants and make little effort to repulse a raid. Instead, the typical response to the arrival of these wolves of the insect world is for the entire adult population to evacuate the nest posthaste. Laggards are taken as prey by the raiding ants, which go on to relieve the now-defenseless nest of its larvae and pupae. Some polistine species, however, appear to have evolved special means of escaping army ant predation. *Epipona tatua*, *Chartergus chartergoides*, and others nest very high in the canopy, beyond the reach of most army ant raids. *Synoeca septentrionalis* workers can prevent army ants from entering by blockading the nest entrance with their bodies (Skutch 1971, Chadab 1979a). *Synoeca chalibea*, *Polybia emaciata*, *P. jurinei*, and others with tough nest carton may do the same (Chadab 1979a). The passageways between combs in nests of *Chartergus chartarius* are reduced to precisely the diameter of a brood cell, which could be an adaptation to enable a single wasp to block access by ants to the brood-bearing combs. *Polybia emaciata* workers retreat into their mud nest when disturbed (Jeanne, unpubl.) and may also physically block the reinforced, spoutlike entrance against army ants.

Still other species, including *Synoeca virginea*, *Agelaia myrmecophila*, and *Polybia rejecta*, nest in close association with *Azteca* ants, which are able to defend their nesting trees against raiding *Eciton* army ants (see also Indeterminate Nesting Cycles, below).

The nest envelope characteristic of most swarm founders may also be an effective line of defense against most parasites. Indeed, it is not unusual to find a mature colony completely free of brood parasites (Jeanne, unpubl.). When parasites are found, their numbers are low compared with the large infestations that can occur in the open nests of the independent founders (Nelson 1968, Starr 1976). Forsyth (1978), for example, found that only 7% of 141 colonies of *Metapolybia azteca* and *Polybia occidentalis* in the seasonally dry forest of Costa Rica had parasites, and only in the dry season was more than 1% of the brood affected. By comparison, Nelson found that nearly 60% of the nests of *Polistes metricus* in southern Illinois were infested with the parasitic moth *Chalcoela iphitalis*; in some nests, 90–95% of the brood were parasitized (Nelson 1968).

Vertebrate predators present a different sort of problem. The dogma is that aggressive group stinging is the main line of defense, and no doubt it is effective in deterring many potential vertebrate predators. Nevertheless, colonies of many species of swarm-founding wasps are successfully attacked by a variety of birds, bats, and primates (Jeanne 1975a). Windsor (1976), for example, described the heavy predation on dry-season colonies of *Polybia occidentalis* by gray-headed kites (*Leptodon cayanensis*) in Guanacaste. Again, certain species appear to have evolved means to minimize the risk of vertebrate predation. Species with small colonies appear in some cases to have evolved cryptic nests that may reduce detection by visually hunting vertebrates and parasitoids. *Leipomeles dorsata*, for example, chooses the underside of a large leaf on which to build its combs, which it covers with a flat envelope that is sometimes camouflaged green and/or provided with a pattern that mimics the venation of the leaf (Williams 1928, Richards 1978a, Jeanne, unpubl.). Others build large, tough nests that may be impregnable to most vertebrates (see Indeterminate Nesting Cycles; Wenzel, this volume), and the nasty-tempered *Polybia rejecta* of the varzea habitat of the lower Amazon valley sometimes nests in aggregations that probably daunt all but the most foolhardy would-be predators (Jeanne 1978).

## Homeostasis

Large colonies are better able than small ones to buffer themselves against environmental stress and to bring conditions inside the nest close to optimal for brood survival and development (Wilson 1971). Swarm-founding polistines, like honey bees, are able to exercise some homeostatic control over temperature in the nest. When a nest overheats, as when it is struck by full sun, workers on the nest envelope and at the entrance begin fanning their wings steadily to ventilate the nest. If this is inadequate, foragers begin to bring water to the nest, where it is spread on the surface of the combs and envelope, bringing about cooling by evaporation (Naumann 1970, Jeanne, unpubl.). Wasps cannot control the nest temperature to as narrow tolerances as honey bees, but they can prevent temperatures from rising to lethal extremes. Naumann (1970), for example, found that internal nest temperatures in the large dry-season nests of *Protopolybia acutiscutis* in Panama never exceeded 32°C and varied between day and night no more than 3.2°C.

After a rain, workers of *P. acutiscutis* (Naumann 1970) and *Polybia occidentalis* (Schwarz 1931, Jeanne, unpubl.) suck water from the wet nest surface and regurgitate it over the side of the nest. A similar bailing of excess moisture from within the nest occurs after rains and in the early morning after cool, humid nights.

Some species of wasps are also able to store food reserves in the nest, thus exerting a degree of homeostatic control over food supply. Many, perhaps most, species store small amounts of nectar as droplets on the walls of empty or egg-bearing cells, and *Brachygastra lecheguana* stores droplets in pockets in the envelope (W. Hamilton, unpubl. cited in Richards 1971). These stores presumably are used when foraging conditions are poor. Near the margins of the tropics, species that build large nests (*Brachygastra mellifica, B. lecheguana, Polybia scutellaris, P. paulista, P. diguetana, Protonectarina sylveirae*) typically store such large amounts of honey that it is sometimes collected for human consumption (Schwarz 1929). On the other hand, *Agelaia* and *Apoica* have long been reported not to store nectar in the nest (Richards 1971, Schremmer 1972, Jeanne, unpubl.), although M. J. West-Eberhard (unpubl.) reports that one of ten nests of a species of *Agelaia* (tentatively *A. multipicta*) she observed contained some stored nectar.

Wasps of the species related to *P. occidentalis* are also known to store certain prey when opportunity permits. When ants or termites have mass mating flights, workers collect the alates in large numbers and store them—minus wings, legs, and heads—in empty cells in the nest (Forsyth 1978, Richards 1978a, Jeanne, unpubl.).

Taken together, the biological traits of the swarm-founding polistines discussed here have enabled the group to dominate the Neotropical social wasp fauna. As I have tried to show, many of the traits are correlated with the large colony size uniquely achieved by these wasps. Indeed, the genera that produce the largest colonies—particularly *Agelaia, Protopolybia, Brachygastra*, and *Polybia*—are the most successful by most measures.

## ECOLOGICAL INFLUENCES ON COLONY SIZE AND CYCLE

Now let us examine some specific patterns of colony size and cycle exhibited by swarm-founding polistines and attempt to understand them in terms of how they adapt the species to local conditions, particularly seasonality and natural enemies. Although much remains to be learned about how the interaction between social biology and local environmental conditions affects colony size and cycle, we can gain some insights by studying the patterns that have begun to emerge. I think it will be clear from what follows just how important such environmental factors as seasonality and predation are in influencing colony size and cycle. Future work will no doubt reveal that they also influence the details of social behavior.

## Latitudinal Patterns in Colony Size

It has long been noted that in some species or groups of closely related species colony size increases with latitude (Ducke 1910, Schwarz 1931). The best data exist for *Polybia occidentalis* and its close relatives, as these are among the most common swarm founders at many sites in the American tropics. The data are summarized in Fig 6.8, in which three measures of colony size are given. The pattern is quite clear: for the group as a whole, colonies are largest at the edges of the tropics. Interestingly, however, *P. occidentalis*, which has larger colonies in Costa Rica and Mexico than closer to the equator, does not appear to increase colony size with increasing latitude south of the equator. Instead, the largest colonies in Mato Grosso, Rio Claro, and points south belong to *P. ruficeps*, *P. paulista*, *P. scutellaris*, and *P. erythrothorax*. It is intriguing to speculate that in the south there is competition among species in the group, of which one manifestation is character displacement with respect to colony size.

The pattern seems to be repeated in other genera. In general, the genera with a range of colony sizes at the small end of the scale (Fig. 6.4, *Marimbonda-Angiopolybia*) tend to have distributions that are well within the Tropics of Cancer and Capricorn (Table 6.1). Among the genera with very large colony sizes, typically the species with the largest colonies are at the northerly and southerly limits of the range of the genus. Thus, in the American tropics *Polybia scutellaris* extends farther south into Argentina than its congeners. *Brachygastra mellifica* is the one member of its genus that succeeds in crossing the U.S.-Mexican border. *Agelaia pallipes* and *A. vicina* are the only members of their genus that occur as far south as northern Argentina. In Australia, the same pattern holds for swarm-founding *Ropalidia*: *R. kurandae* and *R. mackayensis* are limited to the northern half of Queensland, well north of the Tropic of Capricorn. *Ropalidia trichophthalma*, with somewhat larger colonies, makes it to just south of the Tropic of Capricorn, while *R. romandi*, with the largest colonies, has been taken as far south as Sydney (34° S) (Richards 1978b).

The local mix of predator-parasite pressure probably has a major ef-

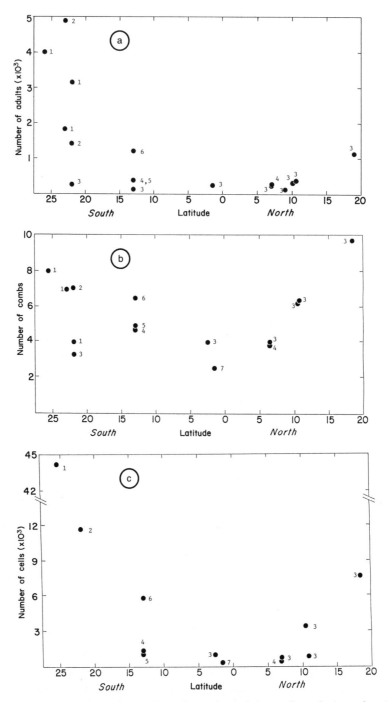

**Fig. 6.8.** Colony size (a, adult population; b, comb number; c, cell number) as a function of latitude for species related to *Polybia occidentalis*. Colonies sampled at the following locations: Brazil: Curitiba (Paraná) (25°26′ S); São Paulo (São Paulo) (23°30′ S); Rio Claro (São Paulo) (22°25′ S); 1968 Royal Society, Royal Geographical Society Xavantina-Cachimbo Expedition (Mato Grosso) (12°50′ S); Taperinha (Pará) (2°27′ S); Belém (Pará) (1°28′ S). Guyana: Mazaruni Settlement (6°24′ N). Panama: Barro Colorado Island (9°11′ N). Costa Rica: Cañas (Guanacaste) (10°25′ N); La Selva (Heredia) (11°26′ N). Mexico: San Andrés Tuxtla (Veracruz) (18°26′ N). (1) *Polybia scutellaris*, (2) *P. paulista*, (3) *P. occidentalis*, (4) *P. platycephala*, (5) *P. erythrothorax*, (6) *P. ruficeps*, (7) *P. scrobalis*. (Ihering 1896b, 1903; Schwarz 1931; Richards and Richards 1951; Rodrigues 1968; Machado 1977a, 1985; Forsyth 1978; Richards 1978a; R. Jeanne, unpubl.)

fect on colony size over much of the range of these species. In the wet habitats of the equatorial tropics, army ant predation is certainly a major force (Chadab 1979a). For many species, army ant raids appear to be virtually unstoppable, regardless of colony size. Heavy army ant pressure probably has the effect of keeping colony size small, since the frequent interruptions of the colony cycle imposed by this kind of predation ought to favor small colonies with higher intrinsic rates of growth (Richards and Richards 1951, Michener 1964, Forsyth 1978).

On the other hand, there is evidence that in dry tropical areas at higher latitudes, such as Guanacaste, Costa Rica, predation by army ants is low, while bird predation is relatively more important. Certain kinds of vertebrate predation can probably be resisted more effectively by large colonies, with their larger numbers of defending workers, greater aggressivity, and the construction of tough nests that are resistant to penetration.

The large colonies at the margins of the tropics may be adapted not only to high predation pressure by vertebrates and low pressure by army ants, but to the relatively long and harsh unfavorable season as well. Large colonies can store large amounts of honey and build sturdy nests in which it is easier to maintain homeostasis.

There are, of course, exceptions to these patterns. Colonies of *Agelaia myrmecophila* in Amazonia, for example, become moderately large. But these have another explanation, to be taken up below (Indeterminate Nesting Cycles).

## Latitudinal Patterns in Seasonal Synchrony

The pace and timing of the colony cycle vary with climate and probably with biotic factors. The best way to get at this is by means of long-term studies lasting a year or more, but there are very few such studies. In lieu of this, the presence of males in collected colonies, as well as the appearance of swarms, may be used as indicators that a colony is in the reproductive stage. Colonies of *Polybia sericea* collected along the Amazon River near Santarém, Pará, Brazil (2°27′ S), for example, were producing males in the months of April, June, July, and November (Jeanne, unpubl.). In Mato Grosso, Brazil (12°50′ S), *P. occidentalis, P. ruficeps,* and *P. emaciata* colonies commonly had males in September-October, but not in February-April (Richards 1978a). *Polybia singularis, P. rejecta,* and *Chartergus chartarius,* on the other hand, had males in February-April.

Farther south, in the Brazilian state of São Paulo (22° S), there appears to be substantial seasonal synchrony in the appearance of males in some species, but little or none in others. *Polybia paulista* is said to reproduce once a year, around January, at the height of the summer wet season (Machado 1985). *Polybia ignobilis* in the same region, however, produces reproductives twice a year, once in September-October and again in January-March (Höfling and Machado 1985). *Polybia occidentalis* swarms three times a year: May-June, September-October, and January-February (Machado 1977a). *Protopolybia sedula* goes through four reproductive cycles a year (Machado 1977b), but it is not clear how synchronous local colonies are. Colonies of *Epipona tatua* are evidently seasonal at Cayenne, French Guiana (5° N), according to Lacordaire (1838, cited in Richards and Richards 1951). New colonies were initiated beginning in mid-June, at the end of the wet season, and colony populations grew into September. Brood populations were high during this period (dry season) and low during the first half of the year (wet season). At Cali, Colombia (4° N), *Metapolybia aztecoides* swarming shows no obvious seasonality (West-Eberhard 1982c).

The most thorough longitudinal study of a swarm-founding wasp is that of Naumann (1970) on *Protopolybia acutiscutis* on Barro Colorado Island in Panama (9°11′ N), where the colony cycle is clearly seasonal. During the dry season (January-April), colonies nesting on palm fronds in semi-open habitat grow rapidly and become large, producing males and eventually reproductive swarms at the beginning of the wet season. These swarms move into the forest, where they construct relatively small nests. There they pass the wet season, apparently without producing much brood. At the end of the wet season (in December-January) swarms leave these nests and move to the dry-season habitat to found dry-season nests. As does *Epipona tatua* at Cayenne, *Protopolybia acutiscutis* on Barro Colorado treats the dry season as the favorable season during which rapid growth is achieved. Both areas are very wet, with more than 2,700 mm of precipitation per year (Wernstedt 1972).

In Guanacaste, Costa Rica (10°25′ N), reproduction in colonies of *Polybia occidentalis* is seasonal. The area is much drier than either Cayenne or Barro Colorado, and has a pronounced dry season that is treated by *P. occidentalis* as the unfavorable season. By the end of the dry season, colonies cease rearing brood, probably for lack of food to feed the larvae (Forsyth 1978). Cells in the nests at this time of year contain only eggs. As the eggs hatch, the larvae are eaten and are replaced with fresh eggs (Jeanne, unpubl.). When the rains begin in May, the larvae are allowed to grow and are fed with the flush of prey items now available. One cohort of worker offspring is reared in these nests during June and early July. When the last of them eclose, the colony abandons the nest, emigrates to a new site nearby, and constructs a new nest in which brood rearing resumes. In the new nest, workers are produced first, but by September-October males appear. In October, reproductive swarms are given off and these found new nests.

In southern Veracruz, Mexico (19°26′ N), *Polybia occidentalis* is seasonal. There the unfavorable season is the relatively cool "winter." Colo-

nies begin rearing worker brood in January and February and reproductives in July (Jeanne, unpubl.). Preliminary observations suggest
that colonies move twice annually: they nest low in forest understory
during winter, then move into high and exposed sites in summer.
*Agelaia areata* follows a similar colony cycle but continuously occupies
its nests throughout the year.

These observations suggest that in regions where there is an unfavorable season (a cool winter, or a harsh wet or dry season), colonies
are constrained to one synchronous episode of reproduction per year
(see also West-Eberhard 1982c). This is clearly the case in Veracruz,
Guanacaste, and Panama for the species discussed above. It is interesting that some species in São Paulo appear to lack seasonality in reproduction, even though winter is cooler there than at the Mexico site.
Closer to the equator, there is no seasonality.

The data on colony cycles, such as we have, suggest the following
pattern. Among determinate nesters, the frequency and timing of reproduction are probably influenced by both local seasonality and the
nature of predation pressure. If predation pressure (e.g., by army ants)
is high and seasonal constraints are low (at least some brood rearing is
possible year-round), as in the wet equatorial tropics, reproductive episodes appear to be timed intrinsically. For example, West-Eberhard
(1982c), summarizing data on a colony of *Metapolybia aztecoides* in Cali,
Colombia, suggested that colony fission is likely to occur when peaks
in the worker-larva ratio coincide with peaks in reproductive competition among females. These peaks seem to bear little relation to season.
In drier or more seasonal habitats, where seasonal factors become
stronger and predation by ants is reduced, determinate cycle colonies
tend to reproduce once per year at a specific season, although not synchronously.

## Indeterminate Nesting Cycles

Indeterminate-cycle species are those for which the benefits of continuing to occupy an old nest outweigh the costs. Benefits of staying in
the old nest include (1) the cost of constructing a new nest is avoided;
(2) the adult population is not exposed to the increased predation it
would face while searching for a new nest site and during emigration;
(3) the delay in brood rearing imposed by giving up the old nest is
avoided. This delay extends from the beginning of eclosion of the last
batch of brood in the old nest to the beginning of oviposition in the
new nest, a period of two weeks or more for *Polybia occidentalis*. The
extended swarming observed to occur in some species of *Agelaia* helps
to reduce this hiatus (M. J. West-Eberhard, unpubl.). On the other
hand, there are several costs associated with continued occupation of
nests. (1) The filling of the cells with the meconia (fecal material) of

successive cycles of brood may attract parasites and diseases. (2) The buildup of meconia reduces the depth of the cells. Eventually, as the cells are elongated slightly to provide enough length for each new cohort of brood, inadequate space remains between the face of one comb and the back of the next to accommodate the workers' access to the brood. (3) As the nest ages and increases in size and weight, effort must be put into maintaining and strengthening its supporting structures.

There appear to be at least two conditions in which the benefits of staying in an old nest outweigh the costs. The first is an elaborate nest structure, which makes the cost of rebuilding high. The nests of *Polybia scutellaris, Chartergus chartarius, Epipona tatua*, and *Polybia liliacea*, for example, are made of an extremely thick, tough carton, possibly as a defense against bat and bird predation. These nests are occupied for several years before they are abandoned (Lucas 1885, Vesey-Fitzgerald 1938, Richards 1978a). *Chartergus chartarius* does not desert its nest even when the combs become so full of meconia that there is not enough room to rear additional brood. Instead, the full combs are abandoned for brood rearing and new ones are added at the bottom of the nest (Jeanne, unpubl.).

The second condition that can lead to indeterminate nesting occurs when nest sites are rare or difficult to secure. In the wetter equatorial regions, where army ant predation is a major threat to social wasps, several species of wasps nest in close association with ants, particularly with certain species of the dolichoderine genus *Azteca*. Colonies of these ants build large carton nests and succeed in dominating the surrounding vegetation to the exclusion of other ants, including army ants (Chadab 1979a). Wasps that nest in association with *Azteca* apparently thereby gain immunity from army ant raids. *Polybia rejecta* and *Synoeca virginea* commonly place their nests close to or in contact with the nests of these ants, while *Agelaia myrmecophila* actually nests in cavities inside the *Azteca* carton nest. In the latter two species, the association may be obligate, as they have never been found nesting alone (Richards 1978a, Jeanne, unpubl.). My observations suggest that all of these wasps need to keep a constant vigil to keep *Azteca* out of their nests, so the alliance seems to be an uneasy one.

In an extensive study of army ant-wasp interactions in the Amazonian lowlands of Ecuador, Chadab (1979a) has shown that the only social wasp colonies that last a year or longer are those that nest with *Azteca* (although the converse is not true, i.e., not all wasp species that nest with *Azteca* occupy their nests for more than a year). According to Chadab (1979a), a swarm of *P. rejecta* took four days to find a nest site. It is likely that either suitable *Azteca* colonies are rare, or securing a beachhead in the midst of one is difficult enough to make the cost of moving very high, thus favoring indeterminate nesting in these species.

There is no potential nesting site in the tropics that is entirely free of ants, many of which readily accept wasp brood as food. It seems likely that the *Azteca*-wasp nesting associations are only the most conspicuous examples of ant-wasp interactions, and that further study will reveal that swarm-founding wasps have as many "words" for ants as Inuit have for snow.

## CONCLUDING REMARKS

The swarm-founding Polistinae occupy a major adaptive zone that sets them apart from their independent-founding relatives. By socializing the dispersal and colony-founding stages by means of the polygynous swarm, these wasps have been able to counter in unique ways the heavy predation pressures that are characteristic of the tropics. The resulting reduction in colony mortality rate has been coupled with the diversion of substantial proportions of the energy budgets of these wasps away from reproduction and into maintenance and defense. As a consequence, they have been able to take much greater control over their physical and biotic environment than have the independent founders. One could say that the swarm founders are *K*-strategists and the independent founders are *r*-strategists.

We stand to learn a good deal about the behavioral ecology and social biology of this group by studying the conditions that give rise to the varieties of cycles that are found among its members. The dissolution of the congruency between the nesting and the colony cycles, characterized by indeterminate nesting cycles of some species, is particularly interesting. It is already quite clear that the combination of predation pressures faced by a population has a great deal to do with the social and cyclic tactics it has evolved. Future work will probably show that factors such as parasitism, food availability, competition for food and nesting sites, and even body size also influence the details of social life. Especially worthwhile will be careful comparative studies within species or species groups in localities with different climates and mixes of natural enemies.

*Acknowledgments*

I thank Ken Ross and Mary Jane West-Eberhard for helpful comments on an early draft of this chapter. My research has been supported by the College of Agricultural and Life Sciences, University of Wisconsin, Madison, by grants from the National Science Foundation, and by the John Simon Guggenheim Memorial Foundation.

# 7

# *Vespa* and *Provespa*

## MAKOTO MATSUURA

Of the four genera that constitute the subfamily Vespinae—*Vespa, Provespa, Dolichovespula,* and *Vespula*—the former two contrast markedly with the latter two in that they exhibit more specialized habits and are principally inhabitants of the Oriental region. *Provespa* contains only three species, which are endemic to the tropics of eastern Asia (Fig. 7.1) (Vecht 1957). Almost nothing was known about the ecology of this genus until recently; new information has started to accumulate following investigations in Sumatra (Matsuura 1984) and Malaysia (Maschwitz and Hänel 1988). The genus *Vespa* (true hornets) contains 23 species, most of which are also distributed throughout eastern Asia (Fig. 7.1). Some species' ranges extend through the islands of the South Pacific to New Guinea, and the distributions of two species (*Vespa crabro* and *V. orientalis*) extend to the western regions of Eurasia.

Among *Vespa* species, the habits of *V. crabro* in Europe have been summarized in general accounts on eusocial vespids (Kemper and Döhring 1967, Spradbery 1973a, Edwards 1980), in addition to the classic achievements of Janet (1903). Recently, nesting data and some behavioral observations of this species have also been collected by Archer (1980b, 1984a, 1985a) and Nixon (1981, 1983a, 1984, 1985a). In North America, where this species has been introduced, its biology has been described by Akre et al. (1980) and Batra (1980). *Vespa orientalis* has been studied intensively by Ishay and his co-workers in Israel, mainly from behavioral and physiological viewpoints (e.g., Ishay 1964, 1965, 1973a, 1977a, 1981; Ishay et al. 1968, 1983a, 1986; Ishay and Schwartz 1973; Ishay and Perna 1979), and there are some reports on the general biology and nesting habits of this species (Rivnay and Bytinski-Salz

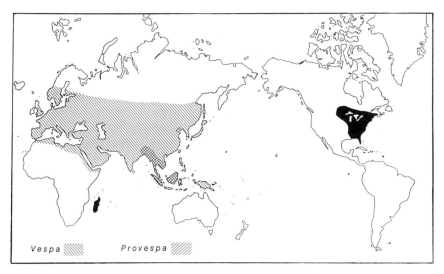

**Fig. 7.1.** Distributions of *Provespa* and *Vespa* (Vecht 1957, Akre et al. 1980, Matsuura and Yamane 1984). Black shading indicates introduced *Vespa crabro*.

1949, Bodenheimer 1951, Wafa et al. 1968, Wafa and Sharkawi 1972). Among the hornets of eastern Asia, a great amount of information has become available on the ecology of species in Japan and Taiwan (Matsuura 1969, 1971a,b, 1973a,b, 1984; Iwata 1971; Matsuura and Sakagami 1973; Yamane and Makino 1977; Makino et al. 1981; Matsuura and Yamane 1984; Ono and Sasaki 1987; Martin 1988), but reports on the *Vespa* species in the interior of the continent and forest-inhabiting species in tropical Asia are still rare (Vecht 1957, 1959; Seeley and Seeley 1980; Starr 1987). Recently, behavioral observations have been made on the species in tropical lowlands, resulting in much new information, such as descriptions of polygyny in *Vespa affinis* (Matsuura 1983b, Spradbery 1986) and *V. tropica* (Matsuura, unpubl.).

In this chapter, published and unpublished studies conducted to date on the biology of *Vespa* and *Provespa* in tropical lowlands are integrated with available information on the temperate-zone species of *Vespa* to produce an overview of the social biology of these remarkable vespine genera.

## LIFE HISTORY OF *VESPA*

### Life History Strategies

*Vespa* species in both temperate and tropical regions display annual nesting cycles. In temperate regions, the new colony is founded between spring and early summer by a single queen that copulated in the preceding autumn and has completed hibernation. The colony develops rapidly after the first workers emerge, and as the number of workers reaches its peak between late summer and late autumn, males and new queens are produced as reproductive individuals for the next generation, though some rearing of workers continues after production of reproductives (Greene 1984). Soon after the emergence of sexuals, the founding queen dies, worker emergence ceases, and consequently the colony starts to decline rapidly. If a queen dies in the midst of nesting activity, some workers in the colony undergo ovarian development and lay eggs that produce males. However, oviposition by workers in the presence of a founding queen has so far not been observed. Males and new queens leave their nests and copulate in the field or near the nest entrance, after which each of the new queens constructs its own hibernaculum underground or inside rotten wood to spend a long period of dormancy until spring.

The nesting period (from foundation to disintegration of the nest) varies significantly among the various species of *Vespa* in temperate regions, as it does among species in the other two temperate-zone genera in the subfamily Vespinae, *Vespula* and *Dolichovespula* (see Greene, this volume). Among the seven hornet species that occur in Japan, *V. tropica pulchra* has the shortest nesting period (around four months) and *V. simillima* has the longest (about seven months). The other five species are positioned between the above two extremes in the following ascending order: *crabro, dybowskii, analis, mandarinia, affinis* (Matsuura 1984).

While the wide variation in the nesting period of the three temperate-zone vespine genera is influenced by environmental differences associated with latitude and elevation, it is also related to the species-specific nest size scale. Generally, colony size (number of cells and number of workers) is large in the nests of species that have long nesting periods. In contrast, in the independent-founding Polistinae, which are taxonomically close to the Vespinae, the nesting periods of the component species in given climatic regions are almost identical in spite of the remarkable interspecific differences in colony size. For example, the seven *Polistes* species and two *Parapolybia* species that occur in the warm southwestern districts in Japan invariably found their nests from late April to early May and stop their nesting activities in mid to late August (Matsuura 1984). Such a difference between the Vespinae and independent-founding Polistinae is considered to be related partly to differences in their nest structures. That is, the combs in a vespine nest are covered with a thick envelope that provides a high degree of thermal insulation compared with a *Polistes* or *Parapolybia* nest, which has no envelope. This confers a *Vespa* colony with the ability to maintain a temperature of around 30°C during the cool autumn season (Himmer 1932, Ishay et al. 1968, Ishay 1973a,b, Matsuura 1984, Martin 1988), which is necessary for the rearing of new queens. It seems that such structural and functional characteristics of vespine nests have permitted the evolution of a large degree of flexibility in the nesting periods of Vespinae in comparison with the independent-founding Polistinae. As a result, vespines exhibit a differentiation in their life histories from short-period to long-period nesting types (see also Greene, this volume).

The general features of the life history strategies characteristic of these two types are as follows (Matsuura and Yamane 1984). In the long-period nesting type species, (1) the nesting period starts early and ends late; (2) nest size and the number of males and new queens produced are large; (3) the threshold value for number of cells and number of individual workers that must be reached before onset of production of males and new queens is high; and (4) the feeding habit is that of a generalist.

In the short-period species, every characteristic is opposite to the above: (1) the nesting period starts relatively late and ends early; (2) nest size and the number of males and new queens produced are small; (3) the threshold value for cells and number of individual workers before onset of production of males and new queens is low; and (4) the feeding habit is that of a specialist.

In *Vespula* and *Dolichovespula* these two life history syndromes can be recognized as generic or species group characteristics; the *V. rufa* group and the genus *Dolichovespula* generally exhibit those traits associated with the short-period nesting type, while the *V. vulgaris* group exhibits the syndrome characteristic of the long-period type (Spradbery 1973a, Akre et al. 1980, Edwards 1980, Archer 1981a, Matsuura 1984, Greene, this volume). In the case of *Vespa* in Japan, on the other hand, these life history syndromes do not characterize taxa above the level of the species; for instance, *V. simillima* is typical of the long- and *V. tropica* of the short-period nesting types. Indeed, it is often difficult to decisively sort *Vespa* species according to the duration of their nesting period or other characteristics because in many cases each species has its own idiosyncrasies. For example, *V. analis* has a long nesting period and can be characterized as a feeding "generalist," yet it builds relatively small nests (Matsuura 1984).

## Usurpation of Nests

No *Vespa* species is known to exhibit the most advanced form of social parasitism (interspecific, obligatory permanent parasitism: see Taylor 1939, Röseler, this volume), such as is known to occur in *Vespula* and *Dolichovespula*. However, a queen nest or small nest with workers is sometimes visited by a queen of the same species and may be taken over by the intruder after a violent battle (Makino et al. 1981, Matsuura and Yamane 1984). For instance, queens of *V. crabro* have frequently been observed to visit nests of conspecifics after emergence of the first workers (Janet 1903); the resident queen and workers cooperate to fight against foreign queens (Fig. 7.2), which are occasionally killed (Matsuura 1970b, Nixon 1983b). Furthermore, the queen of *V. dybowskii* not only founds her own nest but sometimes attacks and takes possession of worker-containing nests of *V. simillima* or *V. crabro* (Sakagami and Fukushima 1957). In addition, nests of *V. velutina* taken over by *V. basalis* queens have been observed in Taiwan (Matsuura, unpubl.), although details of the usurpation process are not yet clear.

## NESTING HABITS

### Nest Materials and Nest Sites

Hornets gather nest materials mostly from the xylem of rotten wood or from the dead parts of live trees, but sometimes they gnaw the bark of live trees (Spradbery 1973a, Edwards 1980, Matsuura 1984). *Vespa crabro*, *V. simillima*, and *V. analis* visit various logs and growing trees including Japanese cedar, pine, white cedar, bamboo, oak, and peach. *Vespa mandarinia* uses the bark of live white cedars as its main nesting material. An unusual example, *V. analis* has been observed to use newspaper.

The nest sites preferred by *Vespa*, be they above the ground in the open, above the ground in an enclosed space, below the ground in a cavity, or some combination of these features, differ characteristically

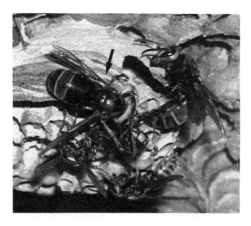

**Fig. 7.2.** Invasion of a *Vespa crabro flavofasciata* nest by a conspecific foreign queen (arrow). The resident queen and workers attack the intruder vigorously.

among the species (Matsuura 1971b, 1984). *Vespula* and *Dolichovespula*, in contrast, exhibit genus-specific preferences. The former mainly nest in cavities, both above- and underground, whereas the latter nest in open situations above the ground (Duncan 1939, Spradbery 1973a, Akre et al. 1980, Edwards 1980).

*Vespa luctuosa* (Kojima and Yamane 1980), *V. affinis*, and *V. analis* prefer aboveground, open sites. They construct their nests on the boughs of trees, under the eaves of houses, on rock walls, and in similar situations. *Vespa crabro*, *V. dybowskii*, and *V. tropica* prefer enclosed situations both above- and underground (Matsuura 1971b). In Europe and North America, *V. crabro* typically builds nests in hollow trees, but thatched roofs, abandoned beehives, holes in walls, bird nest boxes, and cavities in all types of buildings are also used (Duncan 1939; Akre et al. 1980; Archer 1980b; Nixon 1981, 1983a,b, 1984, 1985a). *Vespa orientalis* builds nests in similar situations (Bodenheimer 1951).

*Vespa mandarinia* is the only species of hornet known to nest exclusively in underground cavities. Nests of this species are generally built in subterranean cavities formed either around rotten tree roots or made by small vertebrates, such as moles and snakes. This species occasionally builds its nest inside a mud wall or in a tree hollow close to the surface of the ground (Matsuura 1984).

*Vespa simillima* is the only *Vespa* species known that shows adaptability to a wide range of nest sites in both open and enclosed situations aboveground and underground. Population densities of this species have recently been increasing in residential sections on the outskirts of cities in Japan. In this artificial environment, its nests are constructed under eaves, in attics, inside walls, and in ventilation holes. In a natural environment, however, it builds its nests in tree hollows, in underground cavities, in bushes, and on rock walls (Makino et al. 1981, Matsuura 1984). In addition, this species shows further marked adaptability in its nesting behavior through its ability to relocate colonies in the middle of the nesting period (Matsuura 1984) (see Relocation of the Nest).

## Structure of the Nest

The nest built by a foundress queen of *Vespa* consists of a pedicel, a comb, and an envelope. About three cells are first constructed on the tip of the pedicel, and then construction of the envelope starts. The pedicel is shaped like a club and possesses neither a hanging sheet nor a disc, structures that are common in the early queen nests of *Vespula* and *Dolichovespula* (Spradbery 1973a, Edwards 1980, Yamane et al. 1981, Matsuura and Yamane 1984). The comb consists of 40–60 cells by the time the first workers emerge (Janet 1903, Archer 1980b, Matsuura

1984) and shows no notable structural difference among various species, whereas the envelope has remarkable species-specific characteristics (see also Wenzel, this volume). The shape of the queen nest envelope is closely related to the nest site preference and its form can be classified into the following three types (Matsuura 1971a):

1. *Bowl-shaped envelope.* This type has an inverted bowl-like envelope made of a single sheet, which covers only the upper part of the comb. It is always found in the cavity-nesting species, including *V. tropica* (Fig. 7.3) and *V. mandarinia.*

2. *Ball-shaped envelope.* This type has a spherical envelope that completely encloses the comb except for an entrance left at the bottom. The cavity-nesting species *V. crabro* always builds such an envelope composed of a single sheet, while the ball-shaped envelope of *V. simillima*, which nests both in the open and in enclosed spaces, consists of three to five sheets (Fig. 7.4).

3. *Flask-shaped envelope.* The main body of the envelope is composed of a single sheet and is spherical, but a tubular vestibule is present at the bottom as an entrance. This type is always found in the open-space nesting species, including *V. affinis* and *V. analis* (Fig. 7.5).

After the first workers have emerged, the queen nest is transformed into its species-specific mature shape by the new architects (Matsuura 1984). Workers of *V. analis* and *V. affinis* completely remove the tubular entrance characteristic of their flask-shaped queen nests. The envelope is enlarged so that it covers the comb entirely, and a new entrance, usually consisting of one hole, is provided on one side of the nest. In the cavity-nesting species, on the other hand, the structure of the ma-

**Fig. 7.3.** Queen nest of *Vespa tropica pulchra* in a roof space.

**Fig. 7.4.** Complete queen nest of *Vespa simillima xanthoptera* in a citrus bush.

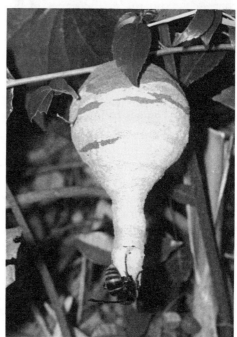

**Fig. 7.5.** Complete queen nest of *Vespa analis insularis* in a bush.

ture envelope is simpler; the bottom part of the nest is opened widely and the lowest comb is always fully exposed. The workers of all of the cavity-nesting species gnaw off the envelope of the queen nest little by little and reuse the material for further cell construction. More than two-thirds of the envelope of the queen nest of *V. crabro flavofasciata* is removed during a period of about two months after the first workers have emerged, during which time the entire comb is exposed (Matsuura 1973a). The envelope of *V. simillima* completely covers the comb whether nesting occurs in an open or an enclosed situation. On the basis of this characteristic it seems probable that this species evolved from ancestors that nested exclusively in the open.

## Relocation of the Nest

Colonies of *V. simillima* and *V. crabro flavofasciata* in temperate regions relocate their nests when they become large and more space for nest expansion is required (Edwards 1980). Most queen nests of *V. simillima* are built in narrow covered places above- or underground, but the relocated nests are constructed in more open situations such as on rock walls and under the eaves of buildings (Matsuura 1984). Queen nests of *V. crabro* are found in cavities similar to those used by *V. simillima*; however, sites for the relocation of *V. crabro* colonies are restricted to more expansive cavities such as tree hollows or attics (Archer 1984a, 1985a; Matsuura 1984; Nixon 1985a). Among nests observed during and after August in Japan, 88.6% of nests of *V. simillima* and 55.6% of nests of *V. crabro* were relocated nests. In England, Archer (1985a) showed that 25.0% (5/20) of *V. crabro* nests collected during August and September were such "secondary nests."

Both species relocate from July to August, after the number of workers in a colony reaches from several dozen to over one hundred but before the period of rearing males and new queens begins. Nest relocation is initiated when some of the workers begin to search for a new nest site by scouting within 20–200 m of the parental nest. When one of these scout workers has found a suitable site, it spends most of the day resting at the new site and only occasionally returns to the parental nest. Several such sites are often settled by different groups of scouts from a single colony. Other workers visit these candidate sites in succession, and the most suitable place gradually becomes filled with workers (Fig. 7.6), a process that seems quite similar to nest site selection by honey bees (Lindauer 1971). After three to four days of such activity the queen leaves the old nest to search on her own for the new nest site using cues that researchers have yet to define. Workers that remain in their natal nest do not follow the queen but start to

**Fig. 7.6.** Nest relocation by *Vespa simillima xanthoptera*. The workers have gathered at the new nest site.

move to the new site independently after the queen has left. As soon as the queen arrives at the new site, construction of a comb begins.

The new comb of *V. simillima* is covered with an envelope within two to three days, whereas *V. crabro* does not start construction of the envelope until two to three combs have been completed. At this point there are two nests relatively close to each other, one new and one old. Workers of both species fly between the two nests frequently, while the queen stays at the new nest and begins to lay eggs. In the old nest, building activities are terminated completely, but the larvae that were left behind are reared by workers who visit frequently from the new nest. Newly emerged workers leave the old nest in search of the new nest, again using unknown cues to find it. In this way, interchange between the two nests continues for about one month, until all eggs, larvae, and pupae that were left behind have been reared and the workers derived from them have completed their resettlement to the new nest. Nest relocation thus results in only a small loss to a colony's previous investment, since it does not involve sacrifice of the brood left behind in the old nest. Furthermore, by continuing to rear larvae in the old nest workers can gather larval secretion, which is an important food source for adults in the relocated nest (see also Hunt, this volume).

Relocation of *Vespa* colonies in other than the above two species has been observed only for *V. velutina* in a mountainous district in Taiwan (Matsuura, unpubl.). Rather than practicing nest relocation, *V. mandarinia*, which nests underground, has a well-developed habit of using the mandibles to remove soil surrounding the nest and depositing this at some distance from the nest when expansion of the cavity is required (Matsuura 1984), a habit typical also of yellowjackets. *Vespa simillima* and *V. crabro* hardly ever perform this type of soil excavation, even when their nests are built underground. Instead, they direct their labor toward the construction of new nests by relocating to more spacious situations.

The reason queens of *V. simillima* and *V. crabro* select such constrained spaces for nest sites in the first place seems to be related to their earlier nest-founding period compared with other sympatric *Vespa* species. The microclimate in such small cavities during the cool weather of spring appears to be well suited for brood development. Moreover, such places cannot be found easily by natural enemies.

## FEEDING HABITS

### Food Sources

The food of *Vespa* gathered from the field can be classified into two types, liquids and solids. These serve primarily as nutriment for the larvae, the former mostly as carbohydrate food and the latter as protein food. Liquids are distributed also to adult nestmates.

The main carbohydrate sources are tree sap, honeydew, flower nectar, ripe fruit, mushrooms, and discarded manufactured sweets (Duncan 1939, Vecht 1957, Kemper and Döhring 1967, Iwata 1976, Matsuura 1984). The most important source of carbohydrate for *Vespa* species in Japan is the tree sap exuding from insectmade injuries on living tree trunks, especially *Quercus* species (oak). In Europe and North America, however, it has been reported that *V. crabro* feeds on tree sap by gnawing the bark from young trees such as ash (Cory 1931, Santamour and Greene 1986). In addition, fruits with soft pericarps and high sugar levels, such as figs, grapes, peaches, apples, and pears, are sometimes fed upon in large quantity by many *Vespa* species, including *V. crabro*, *V. mandarinia*, and *V. orientalis* (Wray 1954, Vecht 1957, Butani 1979, Matsuura and Yamane 1984).

Insects and spiders are the main protein sources for hornets. All species are solitary predators, except for *V. mandarinia*, which also attacks its prey in groups (Matsuura and Sakagami 1973, Matsuura 1984). Because each species displays characteristic prey preferences and strategies for obtaining those prey, hornets can be distinguished as generalists, semi-specialists, or specialists (Matsuura 1984). The seven *Vespa* species in Japan are characterized as follows:

1. Generalists: *V. simillima, V. analis, V. affinis, V. dybowskii*
2. Semi-specialists: *V. crabro, V. mandarinia*
3. Specialists: *V. tropica*

Generalists prey on a wide variety of insects and spiders. For instance, more than 40 species from eight orders (mainly dipterans) and more than 50 species from eight orders (mainly dipterans and hymenopterans) have been recorded as the prey of *V. simillima* and *V. analis,* respectively (Matsuura 1984).

Semispecialists hunt a wide range of prey but usually prefer a specific group of insects. For example, more than 95% of the observed prey of *V. crabro* in Japan are various cicada species (Matsuura 1984). (The subspecies *V. crabro germana* in North America is also known to prey heavily on cicadas [Davis 1925].) *Vespa mandarinia* preys mainly on coleopterans, including scarabs and longhorn beetles, which are not used by the other *Vespa* species. However, this species also preys to a lesser extent on other large, slow-moving insects such as hornworms and mantids. The most remarkable foraging trait of *V. mandarinia* workers is that, in addition to hunting alone, they often mount group attacks on their prey, in this case other eusocial wasps and eusocial bees (see Group Predation).

*Vespa tropica* is a specialist predator; indeed, it is perhaps the most specialized predator to be found among the Vespinae. This species depends solely on the larvae and pupae of Polistinae and Stenogastrinae for protein food for its own larvae. Its polistinae prey include *Polistes* (Sakagami and Fukushima 1957, Matsuura 1984), *Parapolybia* (Sekijima et al. 1980, Yamane 1984), and *Ropalidia* (Ruiter 1916, Kojima 1982a). In the tropics of Southeast Asia, various species of Stenogastrinae are also important food sources (Matsuura, unpubl.).

## Group Predation

*Vespa* species generally hunt by individual initiative. An exception is *V. mandarinia,* which, in addition to foraging alone, has a specially developed foraging strategy in which workers attack nests of other eusocial hymenopterans en masse, exterminate workers in the prey nests, and carry back larvae, pupae, and surviving adults to their own nest as food for their larvae (Matsuura and Sakagami 1973; Matsuura 1984, 1988). This species preys on all sympatric species of *Vespa* and *Vespula,* in addition to *Apis mellifera* and *A. cerana.*

Group predation in *V. mandarinia* consists of three steps: the hunting phase, the slaughter phase, and the occupation phase (Matsuura and Sakagami 1973). In the hunting phase, a worker of *V. mandarinia* captures a worker at the prey's nest. The prey workers attempt an aerial counterattack, but are eventually killed by the lone predator and carried back to its nest. The slaughter phase begins as other *V. mandarinia* workers from the same nest (sometimes several dozen) gather around the prey's nest, then suddenly start a group attack around the nest entrance. The call-up mechanism for summoning fellow *V. mandarinia* workers into battle may involve a pheromone-like substance emitted from the battlefield, but details of this mechanism are not yet clear. In the case of such an attack on *V. simillima*, a life-or-death struggle continues for one to three days and the predators sometimes kill more than 1,000 workers in the prey nest. Often, however, more than half of the attacking *V. mandarinia* may also be killed in the battle. When counterattacks by prey workers cease, the raiders rush into and occupy the nest. They then carry away all larvae and pupae in the nest over a period of one to three weeks (Fig. 7.7). During this period, the entrance of the occupied nest is guarded day and night by a few of the raiding workers.

Group attacks by *V. mandarinia* take place from mid-August to early November. In some years, more than half of the colonies of *V. simil-*

**Fig. 7.7.** *Vespa mandarinia japonica* workers pulling out pupae in an occupied nest of *V. simillima*.

*lima*, *V. analis*, or *Vespula flaviceps* in a particular area may be exterminated by such attacks. Thus *Vespa mandarinia* is the most serious natural enemy of other sympatric vespine species (at the colony level) in the later stages of the nesting period (Matsuura 1984).

## SOCIAL STRUCTURE

### Caste Differentiation and Division of Labor

Female caste dimorphism is well developed in hornets, as queens and workers typically differ to a considerable extent in their morphological, behavioral, and physiological attributes. However, morphological differences are not always obvious; for instance, it is difficult to distinguish between queens and workers of *V. analis* and *V. tropica* strictly on the basis of body size (Vecht 1957). In fact, workers emerging late in the nesting period are sometimes larger than their queen in these species (Matsuura and Yamane 1984).

The number of ovarioles is similar between queens and workers in some species, such as *V. orientalis* (7/7 [left/right] for both castes: Kugler et al. 1976) and *V. analis insularis* (7/7 for both castes: Matsuura 1984). In other species, although ovariole number in queens varies specifically from 8/8 (in *V. simillima* and *V. crabro flavofasciata*) to 10/12 (in *V. mandarinia japonica*), workers consistently possess fewer ovarioles than conspecific queens (Iwata 1971, Yamane 1974a). Before and during hibernation, *Vespa* queens possess a large amount of fat body in their gasters, which functions to sustain them through the cold season. For example, the mean fresh weight of new queens of *V. mandarinia* just after emergence is 2.88 g (SD = 0.25), but reaches 3.46 g (SD = 0.21) when they are about to leave their nest. This increase (about 20%) is due, mostly, to the growth of the fat body (Matsuura 1966). Worker hornets display relatively little development of the fat body, even in the fall.

A basic division of labor is formed between the queen and the workers once the latter have emerged. Oviposition is performed by the queen, and all other tasks are performed primarily by the workers. However, since all *Vespa* colonies studied so far from temperate regions are founded independently by a single foundress (haplometrosis), the queen generally performs all tasks for 35–45 days until the first worker emerges (preemergence phase), as in the case of solitary wasps. As the workers appear (initiation of postemergence phase), the labor contributions of the queen change gradually in both their nature and the extent to which each task is conducted until, finally, her activities are largely restricted to oviposition and larval trophallaxis.

The behavioral changes that occur in the queen during the post-emergence phase can be summarized as follows (Matsuura 1984). Extranidal activities (those occurring outside the nest) that change include collection of food and nest materials. Food collection activity by the queen typically decreases rapidly after the emergence of workers, although there are exceptional records of such activity continuing by queens of *V. crabro, V. simillima, V. mandarinia,* and *V. analis* for 10–25 days after the appearance of workers. Collection of nest materials generally ceases within one to three days after worker emergence, although one example is known where such activity by a *V. tropica* queen continued for an additional 30 days. Intranidal activities (those occurring within the nest) such as food provisioning and nest building may also change, but less dramatically. The queen typically continues to provide food collected by foragers to larvae long after the emergence of workers. Food provisioning continues for over 50 days after worker emergence in *V. crabro* (Matsuura 1974), 40–70 days in *V. analis,* and 20–40 days in *V. mandarinia* and *V. simillima* (Matsuura, unpubl.). In almost all nests of *V. tropica,* food provisioning by queens continues throughout their lives. With regard to nest-building behavior, a queen of *V. crabro* was observed in this activity for about 40 days after the emergence of workers, with specific construction-related tasks being abandoned in a characteristic sequence (Matsuura 1974). Queens of *V. analis* terminate nest-building activity 30–50 days after the first workers emerge, and queens of *V. mandarinia* and *V. simillima,* 20–30 days after. Initiation of cells by queens of *V. tropica* ceases 20–40 days after worker emergence, but enlargement of cell walls generally continues throughout the queen's life (Matsuura, unpubl.).

The activities of queens of the species described above decrease gradually in frequency as the number of workers in a colony increases. The most remarkable change in the queen's behavior as workers emerge is a sharp increase in her resting time on the combs. About 70% of the time budget of a queen of *V. crabro* during the daytime is occupied by extranidal activities before the first workers emerge, 25% by various intranidal activities, and the rest (5%) by oviposition and resting. In contrast, her resting time on the comb increases to 98% of her daytime budget after termination of her extranidal activities (Matsuura 1974). Similarly, resting on the combs begins to occupy most of the daily time budget of queens of *V. analis, V. mandarinia,* and *V. simillima* within 30–60 days after the first workers emerge (Matsuura, unpubl.).

As described above, the extent of participation in various activities other than oviposition by the queen in a postemergence colony varies with species, and such variation is closely related to the typical size of a mature colony. That is, division of labor between the queen and worker castes has a tendency to develop more quickly after worker emergence in species that form large nests. Oviposition by queens of *V. simillima*, the Japanese species with the largest nests, becomes an exclusive activity earlier in the postemergence period than is the case for other species. Queens of *V. tropica*, a species that forms small nests, frequently take part in various types of work even long after the first workers have emerged (Matsuura 1984).

Division of labor among workers on the basis of age (temporal polyethism) is well known in eusocial insects (e.g., Jeanne, this volume: Chap. 11). In five Japanese *Vespa* species (*V. tropica*, *V. simillima*, *V. analis*, *V. mandarinia*, *V. crabro*: Matsuura 1984) a correlation between age and task specificity, such as is known in honey bees (Rösch 1925, Lindauer 1952), is scarcely recognizable, but the following tendencies have been observed. Workers spend most of the first day of their adult lives resting in their nest. On the second day, they begin to engage in most types of intranidal activity, such as feeding the brood, nest construction, nest thermoregulation, and guarding. Orientation flights begin as soon as their cuticle is sufficiently sclerotized for them to fly (most frequently on their third or fourth day).

The first extranidal task, a foraging trip, is performed after the second or third orientation flight. The young worker may collect various types of loads, such as nest materials, food, and water, but appears to exhibit no individual-specific tendency in this regard (Matsuura 1973b). Once a worker begins extranidal activities it may take on a variety of tasks within a single day, so that no specific correlation between age and task can be seen. However, older workers of *V. orientalis* (35–38 days of age) reportedly do not carry out extranidal activities, but instead are involved only in intranidal activities (Ishay et al. 1968). Any relationship between task specificity and development of particular glands, which is well known in honey bees (Rösch 1925), is unknown in hornets, since any food for larvae or nest material can be used essentially in the state in which it was collected from the field. The ability to conduct any form of labor at any time, which is possessed by all but the youngest *Vespa* workers, seems to be related to the fact that each task necessary for the colony to function and to be maintained is likely to be encountered by individuals of any given age.

A total of 1,033 foraging trips by many workers were observed from June to July in a single colony of *V. crabro fasciata* (Matsuura 1973b). During these trips, 57, 13, 17, and 7% of workers collected fluid, prey, pulp, and water, respectively, and 5% returned with no observable load. The mean foraging times spent in the collection of the above four types of loads were 13.3, 10.8, 6.3, and 2.9 minutes, respectively. A worker that collects a specific type of load can sometimes be seen to stick to this same type during most of a particular day or even for several days. This tendency is especially strong when an individual has

discovered an abundant food source in the field. Such a forager continues to collect the item until it is used up and then abandons the
foraging site in search of other items.

When prey or fluid collected in the field is carried back to the nest,
all or a part of each load is received by nestmates and supplied to
larvae. Nest materials are not delivered to nestmates, however, but are
treated and used for envelope and cell construction by the individual
that collected them. Water also is not delivered to nestmates; the collector disgorges it, mainly, on the surface of cocoons in a thin filmy layer
to cool the inside of the nest.

## Queen Control and Communication

In a populous colony of *Vespa* in the later stages of the nesting period, several workers form a "royal court" by clustering around the
queen on the comb and orienting their heads toward her (Fig. 7.8).
They lick her body and, sometimes, bite her violently. Such behavior
can be explained on the assumption that the queen secretes a pheromone that the workers ingest (Ishay 1965, Matsuura 1968). In *V. orientalis*, this pheromone has been identified as δ-*n*-hexadecalactone
(Ikan et al. 1969) and has been synthesized artificially (Coke and
Richon 1976). The queen pheromone of *V. orientalis* induces royal court
behavior and, in conjunction with appropriate environmental stimuli,
stimulates the building of large (queen) cells (Ishay et al. 1965; Ikan et
al. 1969; Ishay 1973a, 1975a; see also Green, Downing, this volume).

Among Japanese *Vespa* species, the presence of a royal court has
been confirmed in *V. crabro*, *V. analis*, *V. mandarinia*, and *V. simillima*,

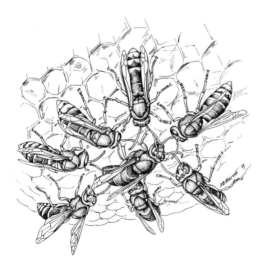

**Fig. 7.8.** Royal court of *Vespa crabro
flavofasciata*.

but it seems to be absent in *V. tropica pulchra* (Matsuura 1984). It seems that the queens of *V. tropica* may control their colonies by some other form of behavior (such as "tapping"; see below). The absence of a royal court in this species may be related to the fact that it has the smallest colonies among the hornets, with only 10–40 workers in a mature colony.

When a vespine queen is artificially removed from the colony or dies, or when she begins to senesce toward the end of the season, the ovaries of some workers develop, and they begin to lay eggs some days later (Ishay 1964, Matsuura 1984). The substitute "queen" is characterized by a loss of hairs on the gena and mandibular base and a polished head like that of the foundress queen (Yamane 1974a, Matsuura 1984). These facts suggest that a pheromone produced by the queen plays an important part in suppressing ovarian development in workers and in regulating worker behavior (Spradbery 1973a, Edwards 1980, Matsuura and Yamane 1984, Greene, this volume).

Since the queen pheromone was reported in *V. orientalis*, some additional pheromones have been proposed for several *Vespa* species. Ishay (1972, 1973b) suggested the presence of a pupal pheromone that stimulates the workers of *V. orientalis* to warm the pupae when exposed to low temperatures. This active compound in *V. crabro* proved to be *cis*-9-pentacosene (Veith and Koeniger 1978). Ishay and Perna (1979) have also proposed other pheromones, including a wasp assembly pheromone, a male-wasp assembly pheromone, a building-depressor pheromone released by the queen, and a building-initiating pheromone released by the workers (see also Ishay 1981). Batra (1980) suggested four probable pheromones in *V. crabro*: an aggregation pheromone from female feces, a defensive (alarm) pheromone in venom, a male cephalic territory-marking pheromone, and a contact pheromone on thoraxes of attractive queens that elicits copulatory behavior in males. The presence of sex pheromones that release male sexual behavior has been demonstrated in the queens of six sympatric Japanese *Vespa* species (*V. analis*, *V. mandarinia*, *V. tropica*, *V. simillima*, *V. dybowskii*, and *V. crabro*) by Ono and Sasaki (1987). Interestingly, clear interspecific cross-activities of the pheromones were found for all pairwise combinations between the species, a finding that suggests that the sex pheromones are not responsible for the maintenance of reproductive isolation between the species. Such isolation may instead result from a species-specific aggregation pheromone that attracts flying males to the nest entrance, where mating occurs (Ono et al. 1987). *Vespa* pheromones, though poorly known at present, would appear to play an important part in colony social life and reproduction (see also Downing, this volume).

It is well known that fifth-instar hornet larvae make an audible noise by scraping their mandibles against the cell wall (Ishay and Landau

1972, Ishay and Schwartz 1973, Yamane 1976, Matsuura and Yamane 1984). This noise appears to be a signal of a larva's hunger, and it occurs in all the *Vespa* species in Japan (Matsuura 1984) and in *V. orientalis* (Ishay 1975b). Although this behavior is especially conspicious in *Vespa* (Matsuura and Yamane 1984), among the other Vespinae it has also been noted in some species of *Dolichovespula* (Weyrauch 1935, Jeanne 1980a), *Vespula* (Akre 1980), and *Provespa* (Matsuura, unpubl.). Ishay (1975b, 1977a) compared the frequency of scraping sounds produced by the larvae of three vespine species—*Vespa crabro*, *V. orientalis*, and *D. sylvestris*—and showed that the sounds produced by *V. crabro* were bimodal in frequency, in contrast to the unimodal pattern exhibited by the others. Though the reason for this difference is unknown, it is possible that the type of nest materials or the construction of the cells has some effect on the frequency characteristics of the sounds produced (Edwards 1980).

Vibratory communication by *Vespa* adults is also known. Workers of *V. orientalis* tap their abdomens on the comb as morning activities begin; this is known as the larva-awakening tap or dance (Ishay et al. 1974, Ishay 1977a). The queen and workers of *V. tropica* also make a vibrating sound when they feed their larvae by tapping the comb with the mid- and hindleg tarsi for several seconds. Such behavior has not yet been observed in other *Vespa* species (Matsuura 1984).

## POPULATION DYNAMICS

### Survivorship Curves

The mean total developmental time, from egg to adult, of workers of five *Vespa* species in Japan varies among the species: *V. simillima*, 30.8 days (SD = 2.5); *V. crabro*, 32.4 days (SD = 4.7); *V. analis*, 32.4 days (SD = 4.1); *V. tropica*, 35.6 days (SD = 4.7); *V. mandarinia*, 40.1 days (SD = 4.6). There is a tendency for the developmental period to increase as adult body size increases in these five species (Matsuura 1984).

Survivorship curves for workers of five of the Japanese species are similar. They are all convex curves (Matsuura 1984); that is, mortality during the period from the egg stage to a few days after emergence is low, but the number of adults that die in the field increases at a constant rate after the commencement of foraging activities. Mean length of survival for adult workers of *V. simillima* is shortest (12.9 days, SD = 6.2) and increases in the order *V. mandarinia*, *V. crabro*, *V. analis*, and *V. tropica*. The longevity of *V. tropica* adults is remarkably long (34.9 days, SD = 11.5) in comparison with the other four species (Matsuura 1984). These interspecific differences in longevity reveal a general tendency for shorter mean adult longevity with increased colony size. Adult longevity differences may also reflect various ecological differences among the species, such as differences in their foraging habits.

## Colony Size

Mature colony size (number of cells) in *Vespa* varies remarkably among the species. The largest nest of *V. tropica* observed so far in Japan had only 313 cells (with three combs), whereas a nest of *V. simillima xanthoptera* had 14,272 cells (with ten combs) (Iwata 1971). I have proposed that the colony size of hornets in Japan be classified into the following three types on the basis of the final number of cells typically observed (Matsuura 1984): (1) small-scale nesting type: 1,000 or fewer cells with 3–4 combs (*V. tropica*, *V. analis*); (2) intermediate-scale nesting type: 2,000–4,000 cells with 4–12 combs (*V. crabro*, *V. dybowskii*, *V. mandarinia*, *V. simillima simillima*, *V. affinis*); (3) large-scale nesting type: 4,000–10,000 cells with 6–12 combs (*V. simillima xanthoptera*). Some mature nests of *V. orientalis* observed in Israel had 600–900 cells with 3–6 combs (Bodenheimer 1951), whereas nests of the European *V. crabro germana* typically contain 2,000–3,000 cells (Janet 1903, Kemper and Döhring 1967, Archer 1980b), a size similar to that of nests of the Japanese *V. crabro flavofasciata*.

The small-scale species, *V. tropica* and *V. analis*, are distributed over an extensive range from the tropical lowlands to the temperate zone. Two subspecies in the subtropics, *V. tropica pseudosoror* and *V. tropica loochooensis*, exhibit a colony size similar to that of *V. tropica pulchra* in temperate Japan (Matsuura, unpubl.). Since, as will be discussed later, the colony size of *V. tropica leefmansi* in the tropics is ten times larger than that of the above-described forms, it seems that *V. tropica* in temperate regions and the subtropics have a relatively smaller colony size as a result of adaptation to colder climates and the more limited available food resources.

The numbers of males and new queens produced by a colony are closely related to colony size. Among Japanese *Vespa* species, only several dozen reproductive adults per colony are produced in a small-scale species, *V. tropica*, but greater than 1,000 per colony are produced in a large-scale nesting species, *V. simillima xanthoptera*. Furthermore, the number of new queens produced varies greatly even within a species, again depending on colony size. For example, nests of *V. simillima xanthoptera* collected in October, when the emergence of new queens is at its peak, produced more than 1,000 new queens (total number of cocoons and adults) when the nests had over 8,000 cells (Matsuura 1984). In poorly developed colonies of this subspecies, however, only two and six queens were produced in nests with 746 and 949 cells, respectively (Matsuura 1984). Therefore, it appears that a minimum of 1,000 cells per nest is required for the production of a significant number of new queens in this hornet. For the other Japanese *Vespa* species, approximate numbers of cells necessary for the production of new queens are 800 in *V. mandarinia*, 300 in *V. crabro*, 200 in *V. analis*, and 100 in *V. tropica*, indicating that the smaller the typical colony size, the lower the threshold value among these species.

## Nest Survival

It is difficult to find *Vespa* nests in the early nesting period and to examine them continuously since the density of nests in the field is much lower than for most other eusocial wasps. There is, however, one example of periodic observation of 59 queen nests of *V. analis insularis*, whose nests can be found relatively easily because it nests in trees (Matsuura 1984). Among these nests, 28 (47.5%) lost their queens on foraging trips before worker emergence and the nests expired. Another eight (13.6%) lost their queens in similar fashion during the month after worker emergence and became orphaned. Subsequently, three of the nests with their original queens (5.1%) ceased nesting activities and failed to produce new queens, whereas two such nests (3.3%) were destroyed by *V. mandarinia* while rearing new queens. In the end, only 30.5% of the nests survived to produce new queens. As for the other hornet species in Japan, the likelihood of nest mortality is also highest from the time of solitary founding by queens to the time just after worker emergence (see also Greene, this volume).

Factors that contribute to colony mortality include loss of the queen, natural enemies, human interference, climatic factors such as storms, and unknown factors. Loss of the queen, which occurs most frequently at the early stage of nesting, is thought to be caused by predation during foraging, extreme weather, lack of food, human interference, or physiological death. Unfortunately, it is almost impossible to confirm the cause of death of a queen while she is on a foraging trip.

Natural enemies of *Vespa* that cause damage at the colony level include the congeneric species, *V. mandarinia*, which mounts group attacks, as well as the honey buzzard, *Pernis apivorus*. The former is the most significant predator of other sympatric *Vespa* species in eastern Asia.

## Sympatry and Mechanisms of Coexistence of
*Vespa* Species

Eastern Asia is the center of diversity of *Vespa*, and in this zone the number of species decreases toward the north and west, in marked contrast to *Vespula* and *Dolichovespula*. In Japan, *Vespula* and *Dol-*

*ichovespula* mainly inhabit high-latitude regions, and where they occur in warm temperate regions they prefer mountainous districts more than 1,000 m above sea level. On the other hand, *Vespa* species only rarely inhabit cold districts or highlands. Five species (*V. simillima*, *V. tropica*, *V. crabro*, *V. mandarinia*, and *V. analis*) are distributed sympatrically in areas extending principally from low mountainous regions to open fields at sea level. In most cases, these species occupy the highest rank in the arthropod food web, because they are large predators with well-developed mandibles and effective hunting strategies. It would be interesting to understand from the viewpoint of population ecology how they have achieved stable coexistence in sympatry.

Comparison of these five *Vespa* species reveals that each has developed species-specific nesting habits, foraging habits, and life histories (Matsuura 1984). First, it is apparent that each species has a characteristic nest site preference, which forms the basis of its life style: *V. analis* always nests in open situations above the ground, *V. mandarinia* nests underground, and *V. simillima* has wide adaptability to both above- and underground situations, and can relocate its colonies. Such diversity in nest site preferences not only reduces interspecific competition for limited nest sites, but also acts as a driving force for the development of species-specific nest structures and modifications of nest sites.

Second, interspecific differentiation of feeding habits, particularly with regard to protein food sources, occurs among these *Vespa* species, although they tend to use common carbohydrate sources. They range from extreme specialists to generalists as predators, thus effectively partitioning the available food sources. In addition to solitary predation, typical of vespines, *V. mandarinia* obtains high-quality protein food on a large scale by attacking colonies of other eusocial wasps and eusocial bees in large worker groups. Therefore, *V. mandarinia* is not only a competitor with the other species for common food sources, but it is also their most important natural enemy at the colony level.

Finally, each *Vespa* species has a characteristic life history that is closely related to the appearance and disappearance of its preferred prey. *Vespa tropica*, which is a specialist predator and exhibits a short-period nesting cycle, synchronizes its life history with that of its food source, various species of Polistinae, and has adapted to the limited availability of its food source in temperate regions by evolving a small-scale nesting strategy. In contrast, *V. simillima*, a generalist predator of the large-scale nesting type, is a long-period species whose food sources must be abundant for a considerable time; the nesting period of *V. simillima* is twice that of *V. tropica*. Although these two species exhibit the extremes in contrasting life histories, three other Japanese species also occupy unique individual niches, as evidenced by their rather slight overlap with regard to the main life history governing factors. As a result, these five *Vespa* species exhibit different peak periods

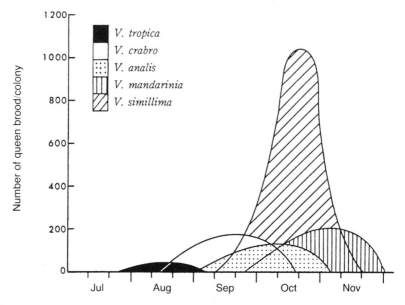

**Fig. 7.9.** Rearing time of new queens in five sympatric *Vespa* species in Japan. (After Matsuura 1984.)

in the production of reproductive adults (Fig. 7.9), a period during which demands for food by the larvae become maximal. These differences in timing in the production of sexuals also reduce interspecific competition for tree sap and mushrooms, which are essential sources of carbohydrates for newly emerged males and queens of all species.

It may thus be concluded that five sympatric species of *Vespa* in Japan have achieved coexistence by developing diversified species-specific life history strategies, thus securing individually unique ecological niches while simultaneously maintaining their dominant positions in the food web. Sympatric distributions of *Vespa* species can be seen in many regions in Southeast Asia. It seems that differences in life history strategies may function as an important factor in the maintenance of suitable conditions for coexistence of these species.

## LIFE CYCLE CHARACTERISTICS IN THE TROPICS

The tropical and subtropical areas of Southeast Asia are inhabited by two genera of vespine wasps—*Vespa*, which is in many ways the most primitive genus of the Vespinae (see Carpenter, this volume), and *Provespa*, which has some of the most specialized habits. In Southeast Asia, 16 *Vespa* species and 3 *Provespa* species occur, the former genus being especially species-rich in mountainous regions. In the tropical lowlands, three *Vespa* species, *V. affinis*, *V. analis*, and *V. tropica*, have differentiated into many subspecies (Vecht 1957). These three species are also common in subtropical or temperate regions, including Japan, but compared with the other *Vespa* species they show evidence of specialization to a tropical climate. The discussion below summarizes the characteristics of the life cycles of those three species in the tropics, mainly on the basis of data I obtained in Sumatra, Indonesia.

## Asynchronous Colony Cycles

In the tropics, the environment changes periodically (as in the dry and wet seasons), but there are no extremely restrictive periods for colony growth such as winter in temperate regions. Consequently, nest initiation by *Vespa* can be observed throughout the year. For example, I observed a total of 35 nests of *V. affinis* in the central region of equatorial Sumatra during the rainy season (mid-November); 11 of these were at the stage before emergence of workers (preemergence phase), 15 were at a stage between the first emergence of workers and emergence of males and new queens (worker phase), and 9 were at the stage of emergence of males and new queens (reproductive phase). I observed 15 *V. analis* nests in the same area in the dry season (August); five nests were in the preemergence phase, seven were in the worker phase, and three were in the reproductive phase (Matsuura, unpubl.). Similar examples have also been reported for *V. analis* and *V. velutina* in western Java (Vecht 1957).

## Polygyny and Nest Size

*Vespa* has typically been regarded as monogynous (having a single functional queen in each colony), irrespective of whether it inhabits temperate or tropical regions. An early record of polygyny in the tropics (Rothney 1877) was disputed by Vecht (1957), who conducted investigations at various locations in Indonesia. In Sumatra, however, I observed not only monogynous nests of *V. affinis indosinensis* but also polygynous nests containing two to eight mated queens; the latter type generally was more common than the former (Matsuura 1983a,b). There were always 30–50 cells in queen nests in this population, with no apparent difference between polygynous and monogynous nests. Polygyny was also subsequently observed for *V. affinis picea* in New Guinea (Spradbery 1986) and for *V. tropica leefmansi* in Sumatra (Matsuura, unpubl.) and Malaysia (N. Huda, unpubl.).

A polygynous nest is generally founded initially by a single queen, and then, during a short period before the emergence of workers, other queens arrive and form a polygynous association. Usually, there is no evident behavioral dominance hierarchy formed among these queens; however, before the emergence of workers only one or two of them oviposit. Resident queens do not leave the nest after worker emergence but maintain a state of coexistence, and they all begin to oviposit as the number of workers increases. As a consequence, a polygynous nest develops rapidly and produces new queens and males relatively early.

*Vespa* species in the tropics achieve a notable increase in nest size relative to congeners in temperate regions because of the climate of everlasting summer and the abundant food sources. For example, subspecies of the three primarily tropical *Vespa* species (*V. affinis*, *V. analis*, and *V. tropica*) are found also in temperate or subtropical Japan, but mature nests of every subspecies in the tropics have several times more total cells, new queens, and workers than their Japanese counterparts (Vecht 1957, Matsuura, unpubl.). Despite its specialized food dependency on the larvae and pupae of lesser social vespids, *V. tropica* colonies in the tropics reach a level of 5,000–6,000 total cells constructed and 500–1,000 new queens produced (Matsuura, unpubl.). Enormous numbers of prey nests of polistines and stenogastrines must be required to maintain such a large nest. Because I found four large nests of *V. tropica leefmansi* at their peak of activity within a radius of 1 km at a single site in Sumatra, this species would seem to enjoy remarkably abundant sources of food in some areas. *Vespa tropica pulchra*, which inhabits the northern periphery of the species' range, has been observed to attack about 150–200 polistine nests during its short (four-month) nesting period and to construct only about 200 cells on average and produce only a few dozen new queens (Matsuura 1984).

### Nesting Period

The process of colony development in the three *Vespa* species occurring in tropical lowlands is basically the same as that for hornets in temperate regions. That is, a newly founded colony undergoes a stage of slow development for about one month before the emergence of workers and for about 1–1.5 months thereafter, and then enters a stage of rapid growth as the number of workers increases. Worker emergence continues for four to six months until the colony enters the reproductive phase, when the males and new queens are produced. They leave their nest when they become sexually mature, and the nest in its final stage declines rapidly and disintegrates, in the same manner as observed for nests in temperate regions.

Hardly any differences can be recognized between tropical and temperate-zone *Vespa* populations in terms of brood developmental periods or the number of times that one cell is reused (up to three).

Therefore, the life cycle of a tropical colony is completed within one year, and it does not become perennial, although the entire nesting period is two to three months longer than in temperate regions.

## Structural Characteristics of Tropical Nests

Queen nests of the tree-nesting *V. affinis* and *V. analis* in the tropics are covered with a single-layer flask-shaped envelope showing hardly any structural differences from the nests of the same species in subtropical and temperate regions. However, a conical structure develops on the top of tropical nests as they are enlarged in the postemergence phase. The interior of this conical structure is filled with densely packed layers of cellular envelope but entirely lacks brood cells. Its outer surface is strengthened by a firm coating of nest materials forming a "peaked roof." Vecht (1957) has suggested that this conical structure represents an adaptation to climatic conditions in the tropics by protecting the nest from driving rain, which is ever-present in tropical rainforests.

The envelope surrounding the comb of tropical *Vespa* nests is only about 1 cm thick, which is several times thinner than any envelope constructed by hornets inhabiting temperate regions. Such a difference can be explained by the need for nests in temperate regions to retain heat through thickening of the envelope, whereas retention of heat in the nest is not important in the tropics.

## LIFE HISTORY OF *PROVESPA*

The genus *Provespa*, represented by only three species, occurs in the Oriental tropics, including the eastern Himalayas, Sumatra, and Borneo (Fig. 7.1). These wasps have a yellowish brown body and are endowed with features associated with a nocturnal habit (Maschwitz and Hänel 1988), such as well-developed ocelli and long antennae (Vecht 1936, 1957). The biological information reported for this genus is quite scanty; indeed, not a single reliable record on nesting has previously been available. This section describes some aspects of the social biology of two *Provespa* species, *P. anomala* and *P. nocturna*, that I have observed in central Sumatra (Matsuura, unpubl., see also Matsuura and Yamane 1984).

## Nocturnal Habits

Specialized nocturnal habits are known or suspected for very few eusocial vespids besides *Provespa* (one example is the polistine genus *Apoica*: Richards 1978a). Both *P. anomala* and *P. nocturna* are typical nocturnal wasps. They start extranidal activities at about sunset and continue these activities through the night; the peak occurs one to three hours after sunset. Most workers return to their nests before sunrise, and all of the wasps rest in the nest during the day. The entrance of a nest is guarded by several workers facing the outside. Many workers come out of their nest if it is disturbed and cover the entire surface of the envelope, even in the daytime; some of them fly about the nest to find and attack the enemy. Because a swarming colony may sometimes trespass into a house through an open window at night, attracted by lamplight, they are regarded as a nuisance in some parts of their range.

## Nest Foundation

Both species typically nest in shrubs or trees. Their nests are established at heights ranging from near the ground to more than 10 m aboveground and are constructed in the foliage to protect them from direct exposure to sunlight. Nests of *P. anomala* may occasionally be found in an enclosed cavity.

It is now known that both species found their nests by swarming, which is an unexpected means of nest foundation in the Vespinae. Nest foundation behavior of *P. nocturna* has not been observed. The nest foundation process in *P. anomala* is as follows: (1) Production of males starts (using empty cells that have previously been used for rearing workers) when a colony reaches 800–900 cells and about 150 workers. Shortly thereafter, 10–30 large cells for rearing new queens are constructed on the outer edge of the lowest or next-to-lowest comb, and oviposition in these begins. (2) Simultaneous rearing of workers, males, and new queens proceeds until there are 2,500–3,000 cells. During this period, new queens emerge intermittently, thus resulting in the temporary coexistence of the founding queen and new queens. Each cell that is vacated when a queen emerges is immediately reused for rearing another queen. (3) A single new queen and several dozen workers leave the parental nest to initiate a new nest by swarming. Subsequently, repeated intermittent swarming from the old nest continues. The swarming behavior of *Provespa*, in which a single queen accompanies the swarm, is in marked contrast to swarming by Neotropical polistines, in which many queens are in the swarm (Jeanne, this volume: Chap. 6).

## Nest Structure and Colony Cycle

The structure of the nests of *P. anomala* and *P. nocturna* is basically the same as for other Vespinae; that is, several combs are covered completely with an envelope (Fig. 7.10), with only a single entrance at the

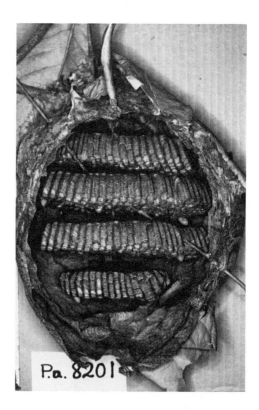

**Fig. 7.10.** Mature nest of *Provespa anomala* with the envelope partially removed.

bottom. Dead banana leaves and similar materials that are rich in fibrous tissues are used as construction materials. The nest is evenly brown or grayish brown, showing little evidence of the striped pattern common to the nests of *Vespa* and *Vespula*. The newly founded nest contains a club-shaped pedicel and a circular comb and is about 4 cm in diameter. By the time the first new workers emerge, the nest has developed to about 8 cm in diameter and generally contains three combs. A mature nest of *P. anomala* is 15–20 cm in diameter and 17–22 cm high; that of *P. nocturna* is approximately 30 cm in diameter and 40 cm high. Mature nests of both species contain four to six combs.

The upper half of the envelope consists of several dense layers of special air cells, with the surface strengthened by a thickly plastered mixture of nest materials and saliva. In the lower half of the envelope, however, the air cells are larger, and toward the bottom the envelope is formed from only a few sheets, as is also typical of *Dolichovespula* nests. Hardly any plastering of the surface can be seen in this area. The whole envelope is much more watertight and tough than envelopes known for other vespines. In addition, since the cells are reused several times, the entire comb becomes more solid with increasing frequency of cell reuse. Solidification of meconia (fecal pellets) from larvae that have completed the spinning of their cocoons further contributes to the toughness of the comb.

The nest-founding period of *P. anomala* seems to be unrestricted, because all growth stages of nests can be seen throughout the year. The colony cycle of *P. anomala* is estimated to last from seven to ten months; the colonies apparently do not become perennial. Production of new queens and males occurs in the latter half of the nesting period. Rearing of workers continues after the production of the new sexuals, which may be an adaptation for swarming. At the final stage of nesting, however, the ovipositional ability of the founding queen decreases, the number of empty cells increases, and disintegration of the nest begins as the queen reaches the end of her life.

The largest nests of *P. anomala* contain 2,000–2,500 cells. Adults inhabiting a colony always consist of one functional queen, several dozen to 700 workers, several dozen to 100 males, and sometimes, depending on the nest, one or a few new (virgin) queens. *Provespa nocturna* is the largest wasp in the genus *Provespa*, and its colonies are considerably more populous than those of *P. anomala*. For example, one of the largest nests found contained 4,946 cells, and the resident adults included 1 queen, 1,559 workers, and 135 males. In addition, this nest contained four cocoons of new queens, indicating that it was in its most active stage.

## Interactions among Individuals in a Nest

Larva-adult trophallaxis can be seen frequently in *Provespa* and appears similar to that in other Vespinae (Greene, Hunt, this volume). That is, fifth-instar larvae disgorge a large quantity of secretion from their mouthparts, and adults preferentially use this as a food source. In addition, fifth-instar larvae show a form of behavior in which they solicit workers for food by rubbing their mandibles against the interior walls of their cells to make audible sounds, in much the same way as has been observed in other Vespinae (Weyrauch 1935, Ishay and Schwartz 1973, Yamane 1976).

Dominance behavior, which is common in the Polistinae, has not been observed between a queen and her workers in *Provespa*. However, I have observed that when a queen is resting on a comb, 10–20 or more workers will cluster around her, orienting their heads toward the queen and thus forming a royal court (Fig. 7.11). Such behavior can be seen only in the later stages of nesting in *Vespa* (Matsuura 1984) and is closely related to the construction of new cells for new queens (Ishay et al. 1968). More workers of *Provespa* attend the royal court than is true

**Fig. 7.11.** Royal court of *Provespa anomala*.

for *Vespa* (Matsuura 1984), but the significance of this difference in behavior is still not clear.

## CONCLUDING REMARKS

*Vespa* and *Provespa*, along with the other two genera in the subfamily Vespinae (*Vespula* and *Dolichovespula*), exhibit the most diversified forms of social life found among the Vespidae. Because the former two genera exhibit greater diversity with regard to their distribution, life history, morphology, and ecology than do the latter two, and because they appear to be relatively primitive in the subfamily, records on the biology of *Vespa* and *Provespa* have great significance in studying the evolutionary history of the Vespinae. On the basis of observations of seven *Vespa* species occurring in temperate Japan, this review demonstrates for the first time that the habits of the genus *Vespa*, which seem outwardly to be uniform in appearance, actually have diverged to a considerable extent among the various species. This divergence provides the basis of a mechanism of sympatric coexistence of multiple species. Characteristics of the life histories of *Vespa* subspecies in the tropical lowlands of Southeast Asia, including the occurrence of polygyny, differ substantially from those of subspecies in temperate regions. The genus *Provespa*, the ecology of which has been little studied previously, is a group with well-developed ecological specializations to a tropical climate not found in other vespine genera, such as completely nocturnal habits and nest foundation by swarming.

*Acknowledgments*

I sincerely thank the following persons for their helpful comments and suggestions during the preparation of this manuscript: R. W. Matthews, K. G. Ross, A. Greene, and Sk. Yamane.

# 8

# *Dolichovespula* and *Vespula*

ALBERT GREENE

Inasmuch as the study of vespine wasps has been greatly aided by several major reviews, it is appropriate to begin this chapter with a listing of the principal surveys of *Dolichovespula* and *Vespula* biology published during the 1970s and 80s. By far the most extensive are the works of Spradbery (1973a) and Edwards (1980). Those with a relatively broad scope, in which vespines are discussed in conjunction with other social wasps, are by Richards (1971), Jeanne (1980a), and Akre (1982). Those with more of a regional emphasis are by Kemper and Döhring (1967), Akre et al. (1980), Matsuura and Yamane (1984), and Akre and MacDonald (1986).

Numerous other, more specific, reviews will be cited throughout this chapter. These references together constitute an impressive resource, and perhaps an intimidating one for those unacquainted with the field. My purpose here is to provide a compendium of information on the social biology of *Dolichovespula* and *Vespula*, with an emphasis on the more recent literature, on drawing conclusions as broadly as possible, and on identifying some of the most conspicuous areas where information is lacking. Space constraints necessitate a fairly conservative definition of "social biology": most aspects of foraging behavior, which is generally a solitary endeavor, are not included.

Any prolonged discussion of *Dolichovespula* and *Vespula* must immediately confront the problem of how to concisely refer to these wasps.

This chapter is dedicated to Roger Akre, whose energy and ingenuity began a new volume of yellowjacket study.

Since *Dolichovespula* is now almost universally recognized at the generic level, simply employing the older name *"Vespula"* as an umbrella term is inappropriate. Turning to common names, only the American word *yellowjacket* is available as a brief, unambiguous label for members of both genera (Greene and Caron 1980), and will therefore be used in this chapter. In fact, a good argument could be made that the switch to splitting the yellowjackets into two genera by North American researchers during the 1970s was based more on excessive enthusiasm than on a judicious appraisal of what constitutes generic-level variation in other groups of social wasps. Be that as it may, the increasing habit of also using *Paravespula* as a genus is totally unjustified by modern taxonomic criteria (see Carpenter, this volume).

The recent past has been a period of exceptional productivity for the study of yellowjacket social biology. Some of the greatest advances in our knowledge of the group during the 1970s and 80s have involved observations of the nest interior by means of various artificial containment systems. These have revealed a surprising interspecific diversity in behavior, particularly queen-worker interactions and other manifestations of reproductive competition and control. Maintenance of captive colonies also has provided a considerable amount of detailed information on the queen nest period and on social parasitism. Other significant developments have been the recognition of nest usurpation, particularly resulting from intraspecific competition, as a dominant component of the colony cycle; the collection and analysis of hundreds of nests belonging to a wide range of species, producing an abundance of demographic data and a renewed interest in the underlying causes of population cycles; the use of electrophoretic techniques to provide answers to long-standing questions of sperm utilization by queens and the origin of males in queenright colonies; and the realization that enormously productive perennial colonies are normal wherever yellowjackets occur in warm climates, challenging the traditional "north temperate bias" in our conception of their capabilities.

There have been disappointments as well. It is unfortunate that the recent well-publicized but little-studied spread of *Vespula germanica* throughout many urban areas of the United States and Canada represents a missed opportunity to examine the vanguard of one of the swiftest, most extensive insect colonization events in recent history. Also frustrating has been the lack of progress in learning the chemical bases of queen control, although the testing and rejection of the once-favored airborne pheromone hypothesis in *Vespula* species has narrowed the field for future investigators.

Because of the difficulties of working with these insects, much of our newly acquired stock of behavioral data represents results from only a limited number of replicates. Furthermore, it must be emphasized that, as with the other social wasps, virtually all of our knowledge of yellow-jackets has come from research on a minority of the known species. Since new information on even this small group is accumulating so rapidly, many of the following conclusions are clearly provisional.

## GENERAL LIFE HISTORY PATTERNS

Yellowjackets are the premier group for those wishing to explore the possibilities and limits of insect societies that have an annual reproductive cycle. Paragons of behavioral plasticity and ecological success, these wasps are widely distributed in the Northern Hemisphere (Fig. 8.1) and exhibit far more diversity in their social biology than any other group of annual eusocial insects. At one extreme, species that inhabit subarctic forests and polar tundra are represented for most of the year only by the solitary, diapausing queens. Mature colonies have several dozen workers, and the nest can be cupped in one hand. In contrast, new queens of some species inhabiting warm temperate to tropical climates may not even enter diapause, but instead return to their own or another colony after mating and begin to oviposit. At its peak during the next season, the resulting "composite" colony may have an es-

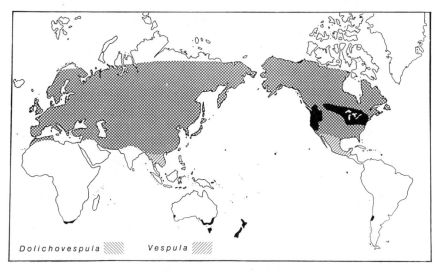

**Fig. 8.1.** Generalized distribution of *Dolichovespula* and *Vespula*. The geographical limits of the genera are approximately equivalent, although several *Vespula* species naturally occur slightly farther south, and at least two have recently become established in some regions of the Southern Hemisphere (black shading). Introduced populations of *Vespula* also occur in Hawaii and North America (black shading). Introduced *V. germanica* in the United States tends to be concentrated in urban areas (e.g., MacDonald and Akre 1984, Akre et al. 1989).

timated 300,000 workers and a nest weighing up to half a ton (Tissot and Robinson 1954, Thomas 1960, Vuillaume et al. 1969, Spradbery 1973a,b).

Such a pronounced behavioral divergence among close relatives is all the more remarkable in that it has evolved with very little divergence in readily observed features of the morphology. Most of the approximately 35–40 yellowjacket species are virtually indistinguishable from one another if color patterns and characters of the male genitalia are ignored.

With the exception of only two principal departures from the norm—perennial colonies with multiple queens and advanced social parasitism—the life history of all yellowjacket species is fundamentally the same. The nest is initiated in the spring by a single overwintered female that was inseminated during the previous season. Working alone, this foundress queen attempts to rear her initial brood to the point where several adult workers eclose and begin to forage. However, both intra- and interspecific competition for possession of young nests are intense, and a large majority of spring queens are probably involved. For the remainder of the season, colony growth depends on the labor of the workers and on the oviposition of the surviving queen, which spends the rest of her life inside the nest. In general, the onset of production of sexuals heralds a gradual deterioration of the colony, with decreasing social cohesion and increasing brood destruction. Most new queens and males leave the nest to mate, after which the queens enter protected locations to overwinter. All other colony members die, and the nest is not used again.

The major variations of yellowjacket life history within this basic framework are usually organized into a dichotomy. The most established division is based on nest site, which generally coincides with malar length, the principal morphological basis for the traditional taxonomic dichotomization of these wasps. *Dolichovespula* species are "long-cheeked" (long malar space) and typically build exposed nests (Fig. 8.2), although subterranean nesting frequently occurs in at least two species and others often build concealed or semiconcealed nests at ground level (Brian and Brian 1952, Greene et al. 1976, Archer 1976, Roush and Akre 1978, Yamane et al. 1980, Bunn 1982a, Akre and Bleicher 1985). *Vespula* species are "short-cheeked" (short malar space) and typically build concealed nests, either underground or in protected cavities above the ground such as in decayed logs, stumps, or the wall voids and attics of buildings. *Vespula germanica* is synanthropic in North America, and until recently colonies were found almost exclusively in human-made structures (Fig. 8.3). Some eastern and midwestern populations have shifted away from this restriction and now are also utilizing more typical subterranean sites (Morse et al. 1977,

**Fig. 8.2.** *Dolichovespula* nests. (a) *D. maculata* nest on traffic-light wire. A peaked dorsal extension, similar to that made by tropical *Vespa* species, is frequently constructed by exposed colonies of this species in eastern North America. (b) *D. arenaria* nest on building. A prominent, curved ventral extension characterizes large nests of this species.

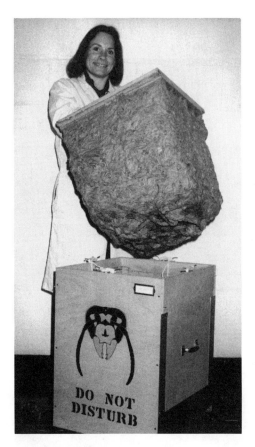

**Fig. 8.3.** Typical annual *Vespula germanica* nest initiated in a nest box (below), which was completely filled with envelope within four months. The combs occupy only about one-eighth of the volume of the paper mass.

MacDonald et al. 1980a, Keyel 1983, MacDonald and Akre 1984, Greene and N. Breisch, unpubl.). Nests of *Vespula* species in warm climates may be located high off the ground in vegetation (Tissot and Robinson 1954, Thomas 1960, Perrott 1975, Nakahara 1980).

A foundress queen's selection of nest site naturally has far-reaching consequences for her colony. Exposed or "aerial" nests face a much greater challenge from the weather and, particularly when small, greater difficulties with thermoregulation than protected nests in more buffered microenvironments. Accordingly, the nest envelope of those *Dolichovespula* species that often live high off the ground is thicker, tougher, and more flexible than that of *Vespula*, and is always made from fibers of sound, dead wood (Duncan 1939, Spradbery 1973a). Many *Vespula* species also use this type of pulp, but others construct nests with a delicate, brittle paper produced from well-rotted wood (Weyrauch 1936a, MacDonald and Matthews 1981). The adaptive value of using decayed pulp is unknown, and in these nests the cocoon silk and voided larval meconia (fecal pellets) are extremely important as binders that increase the combs' structural integrity. Yellowjacket nest architecture also varies in several other basic ways, including pattern and organization of the envelope paper, shape of the comb profile, distribution of cells among the combs, and type of comb suspensoria (Weyrauch 1935, 1936b; Duncan 1939; Spradbery 1973a; MacDonald et al. 1974, 1975a; Wenzel, this volume).

Another obvious consequence of aerial nesting is that a colony's energy budget need not include the laborious task of cavity expansion. Furthermore, birds may be an important mortality factor for young aerial colonies in some areas, whereas medium-sized and large mammals may present a risk for older nests that are subterranean or near the ground (Simmons 1973, Spradbery 1973a, MacDonald and Matthews 1981, Reed and Akre 1983a, Pallett et al. 1983, Pallett 1984).

Despite these and additional behavioral differences (to be discussed later) between *Dolichovespula* and *Vespula*, there is an even more meaningful way for the social biologist to divide the noninquiline yellowjackets into two fundamentally different groups. The basis is colony size, a criterion that represents social development and ecological niche without strictly reflecting phylogenetic groupings (see also Matsuura, this volume). More than two-thirds of the yellowjacket species can be classified in the small-colony group, with most mature nests containing fewer than 4,000 cells (often only several hundred) and maximum worker populations usually consisting of fewer than 800 individuals (often fewer than 100). Included in the small-colony category are all species of *Dolichovespula* and all species in the *Vespula rufa* species group of Carpenter (1987a, this volume) (MacDonald et al. 1974, 1975a; Greene et al. 1976; MacDonald and Matthews 1976; Akre et al. 1982; Makino 1982; Reed and Akre 1983a; Stein 1986).

Small-colony yellowjackets share several major characteristics. Most species rear workers on only a single comb, which is continuously expanded throughout the season from the nucleus of cells constructed by the foundress. All subsequent combs (usually one to three) consist of larger cells in which only reproductives are reared. One of the best known exceptions to this pattern is the North American *Dolichovespula arenaria*, which often builds two or three worker combs and is capable of some of the greatest productivity within this group (Greene et al. 1976). The little known *Vespula sulphurea* (*V. squamosa* species group), traditionally considered a small-colony species, also may construct at least three worker combs and amass worker forces of more than 1,000 individuals (Akre et al. 1980).

Another principal correlate of small colony size is a relatively short duration of colony life, irrespective of the length of the warm season. The period from colony initiation to dissolution ranges from 15 to 17 weeks for many *Dolichovespula* species in temperate areas, to 23 weeks for *D. maculata* in eastern North America (Archer 1980a, Greene 1984). Production of reproductives starts relatively early in the cycle of most species and, as is typical for smaller eusocial insect colonies, constitutes a larger percentage of total colony production (Wilson 1971, Archer 1980a).

Finally, it is not only their demographics that make most small-colony species ecologically inconspicuous relative to those in the other yellowjacket group. Nearly all proteinaceous larval nutrition in small-colony species is provided by the capture of live arthropod prey and is rarely, if ever, supplemented by necrophagy and other scavenging. Thus, although they are considered generalist predators and utilize natural carbohydrate sources as well, their foraging options are quite limited in comparison with large-colony species (MacDonald et al. 1974; Akre et al. 1976, 1982; Greene et al. 1976; Greene 1982; Reed and Akre 1983a; Heinrich 1984).

The large-colony yellowjackets consist of (as far as we know) all species in the *Vespula vulgaris* (and *V. koreensis*?) species group(s) and *V. squamosa*, an aberrant North American species related to the *V. rufa* group. Mature annual nests of large-colony species in north temperate regions often contain over 4,000 cells and may exceed 20,000 cells. Maximum worker populations are often greater than 1,000 and may exceed 5,000 individuals (Kemper 1961; Spradbery 1971; MacDonald et al. 1974, 1980b; Archer 1980a; MacDonald and Matthews 1981, 1984).

What accounts for this considerable increase in productivity beyond what most yellowjackets can attain? The first principal divergence in development between large-colony species and nearly all of the small-colony ones is the initiation of a second comb soon after the onset of adult worker eclosion in the former group. This is almost certainly a derived behavioral character among these wasps, a precocious manifestation of a switching point that, in typical small-colony species, is linked with the onset of large-cell construction much later in the season. It is difficult to imagine, from a purely architectural standpoint, how the greater fecundity of large-colony queens could have evolved without this simple prerequisite.

In all yellowjacket nests, once a second comb has been initiated, others continue to be added (if space permits) until colony dissolution. The number of worker combs in large-colony nests depends on species, colony vigor, and configuration of the nest cavity. Usually three to eight worker combs are followed by a smaller number of large-cell combs. Single-season colonies of introduced species in warm climates often achieve unusually rapid growth rates and may construct nests with 18 or more worker combs, far exceeding the largest north temperate colonies in total cell number (Thomas 1960, Nakahara and Lai 1981).

Obviously, the construction of multiple combs to accommodate additional worker brood is not the only mechanism by which large colony size is achieved. Queen longevity is increased, and duration of colony life is prolonged, ranging from 26 to 36 weeks in north temperate regions. The bulk of reproductive production takes place relatively later in the cycle of large-colony species and constitutes a smaller percentage of total colony production than in small-colony species (Archer 1980a). Queen production, which is particularly delayed in typical large colonies, lags behind male production by about a month. The delay of queen production contributes further to the greater percentage of fertilized eggs deposited in worker cells of large-colony species than of small-colony species (Greene 1984).

Despite all of these factors, it would be misleading to imply that the greater productivity of large-colony species is simply a result of having the means to continue some standard growth curve beyond the point where it peaks in small-colony species. In fact, large-colony yellowjackets initially operate with a much greater efficiency, constructing more cells and rearing more brood per adult worker early in the season than do typical small-colony species (Greene 1982). One likely reason for this substantially greater growth rate, as well as the ability to maintain their populations much later in the season, is the greatly expanded spectrum of food sources used by large-colony species. These wasps feed their brood one of the most eclectic diets known for insects. In addition to preying on virtually any invertebrate that they can overcome, they also take flesh and produce wounds to imbibe blood from living vertebrates (e.g., humans, livestock, nestling birds), scavenge on soft tissues of vertebrate and invertebrate carrion, feed on many different fruits at varying stages of maturity as well as other natural carbohydrates (Fig. 8.4), and consume a wide array of processed human food, both carbohydrate and proteinaceous (Duncan 1939, Spradbery 1973a, Edwards 1980). One of the most important qualitative differences between the two yellowjacket groups, the ability to collect from large, stationary food sources in addition to hunting for small, active ones, enables a worker force to forage with a far greater return for its time and energy investment.

The last major distinction between small- and large-colony species to be discussed in this section involves polygynous, perennial colonies, one of the two major deviations from the basic yellowjacket life history plan. Once considered a curiosity, it is now clear that this phenomenon is widespread. It has been recorded only for large-colony species: *V. germanica* (Thomas 1960, Vuillaume et al. 1969, Spradbery

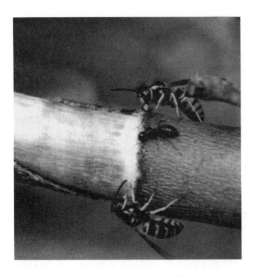

**Fig. 8.4.** Workers of *Vespula mac-ulifrons* imbibing sap on an ash branch stripped of its bark by the hornet *Vespa crabro*.

1973a,b, Perrott 1975); *V. pensylvanica* (Nakahara 1980, Nakahara and Lai 1981); *V. vulgaris* (Duncan 1939, Gambino 1986); *V. maculifrons* (Ross and Visscher 1983); and *V. squamosa* (Tissot and Robinson 1954, Ross and Matthews 1982) (see also Spradbery, this volume: Table 10.7). Since the basic environmental prerequisite is a winter season warm enough for continual foraging, perennial colonies occur either in the southern part of the wasps' natural ranges or where they have been introduced into regions with mild climates (Fig. 8.5).

The process is initiated during the reproductive phase, when numerous mated queens return to their own or another conspecific nest rather than entering their normal, solitary diapause. They begin to oviposit soon afterward, and the colony continues to produce large numbers of reproductives until the following warm season, when it shifts back again to primarily worker production. In exceptional cases, the cumulative effect of possibly hundreds of functional queens can result in nests 5 m long and containing several million cells (Thomas 1960, Spradbery 1973a,b). It is not known whether such colonies typically survive to produce reproductives past the second year, so perhaps their life history could be more precisely referred to as biennial.

Why this striking behavior occurs is puzzling, particularly since it is believed that the majority of conspecific colonies in many of the same areas display normal annual cycles and monogyny. Furthermore, the tendency for nests to become perennial seems to vary greatly among the different large-colony species. The simplest explanation is that induction of diapause in these yellowjackets is mediated at least in part by low temperatures, with the critical threshold varying somewhat among species and individuals. However, most published speculation

**Fig. 8.5.** Nest of perennial *Vespula germanica* colony in Tasmania, Australia, with envelope removed. Combs become increasingly irregular as nests of large-colony species expand. (Courtesy of CSIRO, Division of Entomology, Australia.)

has centered on the emergence of new queens during atypically short photoperiod regimes as a crucial factor. This was first suggested for introduced yellowjackets in the Southern Hemisphere (Spradbery 1973b), but there is also evidence that midsummer nest initiation occurs in Florida, offsetting the colony cycle and resulting in winter queen production (Tissot and Robinson 1954, Ross and Matthews 1982). Returning temporarily to the parent colony is not unusual for young yellowjacket queens (Thomas 1960, Spradbery 1973a, Makino 1982, Haeseler 1986), nor is bypassing the risky, solitary nest initiation period the next season (see Colony Initiation and Queen Competition). Whatever the stimulus that enables queens to bypass reproductive diapause as well, it is clearly advantageous to parasitize the resources of an established colony, particularly if the initial production of these joining queens consists primarily of viable reproductives.

How polygynous colonies operate is equally uncertain, considering that queens of haplometrotic species are notoriously intolerant of each other during nest initiation. However, annual yellowjacket colonies with more than one functional queen are occasionally discovered (Akre and Reed 1981a), indicating that resistance to coexistence wanes later in the colony cycle (cf. Ishay et al. 1983b).

In view of a recent trend to interpret yellowjacket polygyny as reflecting, or at least resembling, an ancestral condition of the group (Ross and Visscher 1983, Spradbery 1986), it is important to determine whether or not these colonies have anything like the organization that characterizes other polygynous wasp societies. How integrated are these enormous systems? Do the various queens range freely throughout the entire nest, or are they largely confined to their own "spheres of influence" within it? The high incidence of supernumerary brood (multiple eggs or larvae per cell) and worker ovarian development in polygynous colonies of *V. germanica* and *V. maculifrons* (Spradbery 1973a,b, Perrott 1975, Ross and Visscher 1983) suggests a noncohesive, ad hoc condition that may in fact bear little resemblance to the ancestral social organization (see also Carpenter, this volume). On the other hand, polygynous *V. squamosa* nests with no supernumerary brood have been reported (Ross and Matthews 1982).

## COLONY INITIATION AND QUEEN COMPETITION

The short, critical period in which an overwintered vespine queen establishes her nest and rears the first workers is perhaps the most distinctive part of the colony cycle. Paradoxically, however, very little of what the queen does during this period is intrinsically different from normal worker activity throughout the season. The traditional fascination among researchers with the diminutive "embryo nest" lies in the solitary execution of familiar stereotyped routines that are usually considered to be social behavior. Not only can the processes of nest construction and colony growth be studied in microcosm, but the implication that each of the queen's daughters also works from this same self-contained "blueprint" offers a useful conceptual theme for analyzing colony organization.

Yellowjacket queen nest architecture is remarkably similar among the dozen or so species for which it has been described (Freisling 1938; Duncan 1939; Brian and Brian 1948, 1952; Potter 1964; Greene et al. 1976; Yamane and Makino 1977; Yamane et al. 1981; Makino 1985a). The circular comb of relatively small cells is suspended by a central petiole. Surrounding the comb are several more-or-less spherical envelopes, varying in number among individuals, species, and nest environments. Usually a total of two to six envelopes have been constructed by the time the first workers emerge, although some may remain incomplete and the innermost ones are often progressively torn down to accommodate comb expansion. The outer completed envelope of some species may be prolonged into a short, spoutlike extension at the ventral entrance hole (see also Matsuura, this volume). This structure is incongruously exaggerated in many *D. maculata* queen nests into a tube of up to 11–12 cm, about five times as long as the queen's body and barely wide enough for her to crawl through (Davis 1919, Rau 1929, Greene et al. 1976, Akre et al. 1980). Both defensive and thermo-regulatory functions have been proposed for the tube (Yamane 1976, Yamane and Makino 1977, Greene 1979).

Queen nests are cryptic, difficult to adequately sample, and are seldom inspected sans envelope precisely at the time of first worker emergence. For these reasons there has been a long-standing misconception about a typical queen's total productivity during her period of solitary labor. Although about 40 records involving mature, unmanipulated queen nests of eleven species show a broad range for this parameter, most nests contain between 30–55 cells, with examples containing 60–75 cells having been observed for four species (Brian and Brian 1952, Greene et al. 1976, Jeanne 1977b, Archer 1981a, Matthews et al. 1982, Makino 1985a, Greene, unpubl.). This achievement is particularly impressive considering that the solitary stage lasts only 19–24 days under natural conditions (Greene 1984).

Throughout it all, the queen nest is characteristically a system in stress. Brood abortion, oviposition in occupied cells, and widely varying brood developmental times all attest to the foundress's difficulty in balancing expansion with consolidation, dealing with the simultaneous demands of larval feeding, ovipositional drive, and thermoregulation (Brian and Brian 1952, Potter 1964, Brian 1965, Gibo et al. 1977, Ross 1983a, Makino 1985a, Greene, unpubl.). The nest is dangerously lacking in security as well as stability, for superimposed on its internal economic problems is the high-risk nature of foraging, the often harsh springtime weather, and vulnerability to predation. Queen nests of *D. maculata* in some areas are susceptible to decimating levels of parasitism by two species of chalcidoids (Greene, unpubl.), and the ichneumonid *Sphecophaga vesparum* may heavily parasitize queen nests of other species (MacDonald et al. 1975b, Archer 1984b).

This stage is thus considered to be one of the two primary periods of yellowjacket queen mortality. Winter diapause is the other, although with no empirically determined death rates its significance is largely intuitive (Archer 1980a, 1984b). In contrast, at least six cohorts of queen nests have been followed, with mortality ranging from 73 to 100% (Brian and Brian 1952, MacDonald and Matthews 1981). The actual demise of the queen is rarely witnessed; she often simply does not return to her nest. Investigators have noted that individual queens differ greatly in several indicators of efficiency, and that a significant decrease in energy and attention to the nest frequently precedes her disappearance (Brian and Brian 1952, Bunn 1982a). Thus, physiologically based "queen quality" is a critical factor in some models of yellowjacket population regulation (Lord and Roth 1978; Lord 1979; Archer 1980a,c, 1985b; Roth and Lord 1987). Inclement weather is probably also of major importance in many habitats as a drain on time and energy budgets, in addition to the more direct effects of flooding on subterranean nests and wind on aerial ones. Poor weather thus plays the central role in several other models of population cycles (Beirne 1944, Fox-Wilson 1946, Döhring 1960, Long et al. 1979, Akre and Reed 1981b, Madden 1981, see also Pallett 1984).

Although, as described so far, the prospects facing a typical foundress seem dire enough, she has still more obstacles to overcome. Implicit in the discussion up to this point has been the traditional concept that the strongest of these individuals survive the winnowing-out process and become the mothers of the next generation. However, nest initiation may actually be exceptional behavior for yellowjacket queens, the least frequently exercised option in a classic example of mixed reproductive strategies.

It has been well known for many years that queens of annual, haplometrotic eusocial insects (i.e., those with nests started by lone foundresses) compete with others of their species for young nests (Röseler, this volume). The importance of this competition in yellowjacket biology cannot be overemphasized. Virtually every observer who has studied incipient colonies in the field has recorded successful or attempted usurpation by conspecific queens, and virtually every investigator who has excavated subterranean nests has found corpses of dead conspecific queens in the entrance tunnels (Brian and Brian 1952; Akre et al. 1976; Matthews and Matthews 1979; Archer 1980a, 1985b; MacDonald and Matthews 1981; Bunn 1982b, 1983; Makino 1985b; Nixon 1985b). Several accounts of 3–25 corpses at a single location, and a grand record of 56 queens attracted to a decoy nest over a $5^{1}/_{2}$ week period, lead to the suspicion that fighting for established nests involves the large majority of spring queens in many populations (Cottam 1948, Nixon 1955, Evans 1975, Matthews and Matthews 1979, Akre and Reed 1981b, MacDonald and Matthews 1981). In fact, usurpation attempts and the adoption of "orphan" colonies that have no current queen persist well past the incipient nest stage, and queens without colonies are on the wing for much of the summer in some areas (Akre et al. 1976, Greene et al. 1976).

Intraspecific competition for nests is undoubtedly density-dependent and must frequently be of crucial significance in regulating yellowjacket abundance (Lord and Roth 1978; Lord 1979; Archer 1980a,c, 1985b; Matthews 1982; Roth and Lord 1987). It is likely to be a major selective factor in their evolution, and is at the core of the two principal departures from their basic life history pattern (i.e., the multiple "invasions" of established nests that lead to polygyny, and the evolutionary specialization that has culminated in inquilinism). Unfortunately, the most intriguing questions about this behavior remain unanswered.

The first such question is whether attempted usurpation typically is an option of first or last resort. At one extreme is the possibility that "very few queens ever establish nests" (Akre et al. 1976:68) and that the remainder represent well-defined subpopulations that can be considered as incipient parasitic species. The opposite scenario is that most queens establish nests but then abandon them for a variety of reasons and have no choice but to fight for possession of another. There are a few records of usurpation behavior by queens that either were stressed by manipulation or had their own nest destroyed (Jeanne 1977b, Greene, unpubl.).

The second question concerns the identity of the winner of these contests. Although introductions of queens into laboratory nests most often result in victory for the resident individual (Nixon 1936, Akre et al. 1976, Matthews and Matthews 1979, Matthews 1982), usurpation in the field has been recorded often, with the most extensive data documenting successful takeovers in 13 of 36 monitored colonies (Brian and Brian 1952; MacDonald and Matthews 1981; Bunn 1982b, 1983; Greene, unpubl.). If nothing else, these reports demonstrate that the common practice of referring to a colony's queen as the foundress is unjustified unless that individual has been tracked since nest inception. Many field observations of unsuccessful usurpation attempts involve nests in which adult workers were present (Evans 1975, Jeanne 1977b, Bunn 1982b, Makino 1985b, Nixon 1985b).

The third and most important question concerns the ultimate fitness payoffs for usurpation (e.g., Reeve, this volume). Do good fighters produce vigorous colonies, or are they just one more mortality factor with which nonusurping foundresses, the prime contributors to the next generation, must contend? Available data that indicate substantial mortality of usurped nests may be irrelevant, since most nests at this stage fail anyway because of other factors (MacDonald and Matthews 1981; Bunn 1982b, 1983). The issue's significance very much depends on the answer to our first question. If successful invaders are mostly foundresses whose original nests have recently met with misfortune, then it would not be surprising if many are able to resume where they left off and productively manage their new acquisition. On the other hand, if yellowjacket populations are renewed in large part by legions of aggressive queens that routinely achieve reproduction through usurpation as a primary strategy, then the life history pattern of these wasps is indeed more complex than we had once believed.

Conflict among vespine queens is not limited to conspecifics, and the various categories can be arranged in what seems to be a logical evolutionary progression (Taylor 1939). The drive to usurp a conspecific colony is usually regarded as the basis for the much more infrequently observed cases of mixed-composition nests involving two free-living species that often are closely related (Nixon 1935, Akre et al. 1977, Roush and Akre 1978, Bunn 1982b, Akre and Reed 1984, O'Rourke and Kurczewski 1984). It is not known whether such mixed-composition nests typically result from actual usurpation or from the adoption of orphan colonies, or whether they merely represent mistakes by the invading individuals. However, studies in which wild queens are induced to initiate nests in laboratory window boxes suggest that interspecific struggle during this period is relatively common among free-living vespines and often involves competition over the nest sites themselves (Greene, unpubl.; see also Spradbery 1973a). Such competition is widespread among nesting animals in general (Ricklefs 1987, Alcock 1989), and, although the environment may appear to be replete with an almost infinite selection of suitable sites (Akre and Reed 1981b, Makino 1985b), the wasps may well perceive it differently. Even the aerial nesting *Dolichovespula* may be more constrained than is usually supposed, as particular areas on structures and vegetation are repeatedly occupied (Greene et al. 1976, Pallet 1984).

It is thus possible that much conspecific nest usurpation is actually only an epiphenomenon of nest site competition, that it is essentially territorial behavior rather than parasitic, and that the frequent initiation of vespine nests in unsuitable locations (Duncan 1939; Spradbery 1973a; Lord and Roth 1978, 1985; Lord 1979; Edwards 1980) is largely due to the exclusion of foundresses from preferred sites that have already been occupied.

The turning point in the evolution of interspecific usurpation is when this behavior approaches the rule rather than the exception and becomes more clearly parasitic in nature, with the invading species distinguished by some specialization for that purpose. The Nearctic *V. flavopilosa* is apparently at this threshold, and the large size of its queens is possibly an adaptation for combat with its close relatives (Jacobson et al. 1978, MacDonald et al. 1980b, Ross et al. 1981). *Vespula flavopilosa* queens become active several weeks after their hosts (MacDonald et al. 1980b, Keyel 1983), ensuring that nest establishment by these species has already commenced.

Rather than being an adaptation for social parasitism, however, delayed spring emergence may be one of the fundamental reasons for it, applying to both conspecific and heterospecific queen conflict. The most widely accepted mechanism for the origin of heterospecific queen conflict (originally proposed for bumble bees) is the relatively late release from diapause of a southern species spreading into the range of a more northern relative (Richards 1927, 1971; Taylor 1939). This scenario is probably best illustrated by the southern *V. squamosa*, which even exceeds the model's specifications in that its parasitism is not confined to close relatives.

This enigmatic species is perhaps the most distinctive yellowjacket. Its queens are among the largest of the group and have an atypical orange gaster that differs strikingly from that of the workers. The coloration of both castes appears to have originated as part of a more southerly mimicry complex than that involving most other Nearctic vespines. But the most fascinating aspect of the biology of *V. squamosa* is that its late-flying queens routinely usurp the colonies of virtually every cavity-nesting vespine species they encounter, including *V. vidua*, *V. maculifrons*, *V. flavopilosa*, *V. germanica*, and even the comparatively enormous hornet *Vespa crabro* (Taylor 1939; MacDonald and Matthews 1975, 1984; Matthews and Matthews 1979; Matthews 1982; Greene, unpubl.).

*Vespula squamosa* queens often—perhaps primarily—invade young colonies with adult workers, and rely on physical combat to subdue the occupants. As is probably the case with all vespines, including the more advanced social parasites, the usurping queen is in turn susceptible to conspecific usurpation (Matthews and Matthews 1979, Matthews 1982, MacDonald and Matthews 1984). It is unknown whether its success in both usurpation and eventual reproduction varies with its different hosts. Also a matter of conjecture is the extent of independent colony founding throughout its range. Although independent founding is almost certainly retained as an option in the more southern populations, social parasitism is probably obligatory or nearly so in the northern part of its distribution (Matthews 1982, MacDonald and Matthews 1984).

The most highly evolved form of social parasitism is inquilinism (e.g., Röseler, this volume), in which the worker caste has been dispensed with and the parasitic strategy thus presumably specialized to the point of no return. How yellowjacket inquilines manage to avoid production of workers is a critical question, since vespine caste determination depends almost totally on the size of the cell in which a fertilized egg is deposited. If large-cell construction by workers at least partly depends on a stimulus provided by the colony's queen (see Queen Control and Reproductive Competition: Queen Production), an inquiline female that has disposed of the latter must therefore be able to mimic or circumvent this aspect of the host queen's control. The problem illustrates well how inquilinism requires a far greater intimacy between parasite and host than does any other form of interspecific exploitation.

Inquilinism has so far been described for only four yellowjacket species: *Vespula austriaca*, *D. arctica*, *D. adulterina*, and *D. omissa*. All seemingly adhere to "Emery's rule," a long-standing generalization that very highly specialized social parasites are more closely related to their host than to any other species (Wilson 1971, Varvio-Aho et al. 1984, but see Carpenter 1987b). Studies of *V. austriaca* and *D. arctica* indicate that females are morphologically adapted for combat, have a brief but intense oviposition period, physically dominate the colony members with hyperactive, aggressive behavior (Fig. 8.6), but apparently have some allomonal control over them as well (Jeanne 1977b; Greene et al. 1978; Reed et al. 1979; Reed and Akre 1982, 1983b,c; Schmidt et al. 1984; see also Eck 1979). Neither species has achieved anywhere near the degree of integration with its hosts as is exhibited by the more advanced ant inquilines (Wilson 1971), and thus, for what seem like obvious logistic and demographic reasons, yellowjacket inquilinism is a phenomenon restricted to small-colony, short-cycle species.

Despite the basic similarity of their niches, the two better known inquilines display considerable differences in their behavior. *Vespula austriaca* starts its season in the same manner as the less specialized social parasites, becoming active well after its host, forcibly invading nests with workers present, and quickly dispatching the queen. Managing the colony afterwards is accomplished with a higher degree of sophistication, however, as the inquiline female can inhibit (presumably with allomones) both worker ovarian development and aggressive behavior directed toward her (Reed and Akre 1983b,c).

*Dolichovespula arctica*, on the other hand, has evolved a much more advanced stratagem for acquiring a colony, becoming active at about the same time as its host and "passively" invading a nest prior to worker eclosion by submissively enduring the resident queen's attacks. These attacks rapidly subside, again presumably because of allomonal pacification. The tactic is probably far less dangerous than fighting multiple defenders, but early invasion necessitates a coexistence period

**Fig. 8.6.** Aggressively soliciting for trophallaxis, a female of the inquiline *Dolichovespula arctica* pursues a foundress *D. arenaria* queen around the periphery of her young nest. The ventral envelope of this captive colony has been removed.

in which the host queen is allowed to oviposit and thereby increase the future labor force. Curiously, the inquiline female's control over the colony is relatively limited, and she cannot suppress worker ovarian development after eliminating the queen. This leads to frequent and escalating bouts of physical conflict, which ultimately result in the inquiline's death (Evans 1975, Jeanne 1977b, Greene et al. 1978).

All the categories of usurpation described up to this point involve the efforts of single reproductive females. There is circumstantial evidence that another variation exists that not only transcends this norm but also challenges a basic assumption about the limits of yellowjacket worker communication. The behavior might be termed "parasitic colony translocation," as distinguished from the nonaggressive translocation exhibited by at least two *Vespa* species (Matsuura 1984, this volume). It entails the apparent elimination of a young colony (*D. maculata*) and the occupation of its nest by an invading colony of a different species (the much smaller *D. arenaria*) (Pallett 1984). The actual invasion process remains to be observed, but in view of the enormous role that usurpation plays in yellowjacket affairs, its occurrence should not be considered as totally surprising.

## WORKER AND REPRODUCTIVE PRODUCTION

Despite overwhelming odds, some incipient yellowjacket nests survive and develop to maturity. Inside its envelope, the growth of the comb mass can be roughly likened to that of an inverted woody plant, with an actively expanding periphery and the core gradually going out of production. The principal meristematic zone is at its apex, the ventral margin where new combs arise and cell construction is most rapid. The lateral meristems are the edges of the older combs, which expand at progressively slower rates with age. The oldest areas toward the base of the nest become increasingly ignored and eventually are abandoned. Most of the older small cells are used to rear two or three, and sometimes four individuals in succession, but the majority of reproductive cells are used only once.

Hidden from view and dispersed throughout the moderately complex and well-defended nest, the brood is thus impossible to track under normal conditions, and colony dynamics must be deduced rather than directly observed. This has been done in three basic ways. The first is to monitor a colony maintained in an observation nest box with sufficient vertical space for only one comb. Although data are indeed gathered by direct observation, they must then be interpreted in light of the inevitable retardation of colony growth that such confined conditions produce. The most detailed use of this technique was a study of a *D. arenaria* colony parasitized by *D. arctica* (Greene et al. 1978).

The second method involves the dissection of wild colonies throughout the season. Such sampling yields precise analyses of nest populations at one moment in time, but the range of developmental sequences of individual colonies must still be inferred. Several recent studies along these lines have provided most of our knowledge of yellowjacket colony growth (Spradbery 1971; Archer 1972a, 1981b; MacDonald et al. 1974, 1980b; Greene et al. 1976; MacDonald and Matthews 1976, 1981, 1984; Roush and Akre 1978; Akre et al. 1982; Makino 1982; Reed and Akre 1983a; Stein 1986).

The third way to examine colony dynamics is to model them, producing idealized summaries that aid in the interpretation of field data and generate possible solutions to questions that have not yet been answered empirically. Such models range from relatively brief studies with simplified assumptions to an intricate computer program with 253 parameters and variables (Lövgren 1958, Long et al. 1979, Rowland and McLellan 1979, Archer 1981a).

A yellowjacket colony produces workers for most of its life, following a typical sigmoidal population growth curve (Brian 1965, Spradbery 1973a, Archer 1980a). As the colony becomes larger and more efficient, its worker-larva ratio increases, more food is available for the larvae, and the adult workers typically increase in size (Archer 1972b, Spradbery 1972). Most workers take about a month to complete their immature development and have adult lives averaging two to three weeks (Archer 1980a, Greene 1984).

Production of reproductives begins relatively early in colony life, overlaps substantially with the rearing of workers, and occupies an increasing proportion of colony effort as the season progresses. There are two basic ways in which it is accomplished. In small-colony nests with one worker comb reproductive production is abruptly launched with the initiation of a second, large-cell comb. The queen subsequently lays both fertilized and unfertilized eggs on this comb. Male and queen brood generally are intermixed, although either may account for a considerable majority of the reproductive production for varying periods of time. Numerous males, and sometimes a smaller number of queens, eventually are reared on the first comb as well (MacDonald et al. 1974, MacDonald and Matthews 1976, Akre et al. 1982, Makino 1982, Reed and Akre 1983a, Greene 1984, Stein 1986).

Reproductive production by the large-colony species begins in a more protracted manner, with male brood gradually appearing in small cells and beginning to mature in substantial numbers at about the time that large-cell construction is initiated. These large cells constitute the last comb(s) of the nest and/or are added to the periphery of the last small-cell comb(s). Large cells typically are used to rear queens, although in some cases many of the occupants are males. Nevertheless, most of the male production normally takes place in the small cells (Spradbery 1971; MacDonald et al. 1974, 1980b; Archer 1980a; Mac-Donald and Matthews 1981, 1984; Greene 1984).

The atypical small-colony species *D. arenaria* reproduces in a manner distinct from that of other yellowjackets. Reproductive production consists of discrete phases during which individuals of virtually only one sex are reared. Since a colony's reproductive effort is normally dominated by (or even exclusively composed of) only one of these unisexual phases, this species appears to have evolved a nearly dioecious system (Greene et al. 1976).

In fact, the heavily queen-biased production of some *D. arenaria* colonies is this species' most unusual reproductive trait in comparison with other species. In general, precise ratios of vespine queen-male output (i.e., actual number of adults leaving the nest) are almost impossible to obtain for the same reason that makes accurate worker-reproductive output ratios so elusive: the accumulated larval meconia and pupal cocoon layers that serve as a colony's gross production record do not reveal the sex of the occupants of the small cells (and sometimes of large cells). However, cell-utilization data in combination with counts of reproductive adults and pupae in sampled nests of many yellowjacket species indicate that total colony reproductive output is usually numerically male-biased, often by at least a factor of two (Spradbery 1971; MacDonald et al. 1974; MacDonald and Matthews 1976, 1981, 1984; Akre et al. 1982; Makino 1982; Reed and Akre 1983a; Stein 1986). At the population level, the predominance of males is even more evident. The unsurprising fact that larger colonies with more workers produce more queens has long been known. What has only recently been appreciated is that these are also the nests most likely to be sampled and that typical populations probably include a sizable cryptic element of less successful, smaller colonies, including those that have lost their queens, whose production of sexuals is largely or exclusively of males (Archer 1980a, Greene 1984).

What makes a colony less successful? In some cases, extrinsic factors can be identified, such as a restrictive nest site that hinders comb expansion or one that subjects the wasps to some other stress. Although yellowjackets are susceptible to a wide array of parasites, these usually are not serious mortality factors after the queen nest stage. A notable exception is the ichneumonid *Sphecophaga vesparum*, which may frequently retard colony development in some populations of *Vespula atropilosa* in North America and *V. vulgaris* in Europe (Spradbery 1973a, MacDonald et al. 1975b).

However, it is likely that most colonies that never approach the growth potential of their species are products of intrinsically weak queens with incomplete insemination, inadequate pheromone production, or some other disability. In many instances it appears as if queens simply run out of stored sperm and prematurely begin producing males. In other nests, widespread brood abortion by workers suggests an ineffective level of queen control (Archer 1981b), although, considering our almost total ignorance of the control mechanisms, this is hardly a satisfying explanation. It is also not unusual for queens to die relatively early in the season, perhaps doomed the previous year by malnourishment before diapause.

## COLONY MAINTENANCE

The life of adult worker yellowjackets consists of repeatedly performing several major activities whose functions are usually obvious but whose organization and control remain largely obscure. Much of the flow of information is invisible, and many of the immediate cues that stimulate specific behavioral responses are unknown. In general, workers appear to perform tasks in relative isolation from their nestmates, with direct interaction being limited mostly to food transfer. To a great extent, each individual resembles a miniature, self-contained "foundress," reacting primarily to her physical environment and the needs of the brood. It is thought that communication among workers and between workers and brood consists in large part of broadcasting vibrations or pheromones that serve to attract or to alert. Division of labor is relatively weak and somewhat correlated with age; younger workers tend to concentrate on larval feeding, pulp collection, and construction, and older workers on foraging (Brian and Brian 1952; Potter 1964; Montagner 1967; Akre et al. 1976, 1982; Reed and Akre 1983d).

The major components of colony maintenance (other than foraging) can be grouped into six general categories.

### Nest Site Modification

Most yellowjacket colonies must continually remove some amount of surrounding material as they expand the nest, be it soil, decayed wood, living vegetation, or building insulation. Many species are primarily subterranean nesters, with colonies initiated in cavities barely larger than the queen nest itself. Excavation may therefore be a principal worker activity throughout the season.

Soil is primarily scraped up and fashioned into a bolus with the mandibles, as is wood pulp, and there seems to be a general response to gnaw on any material immediately adjacent to the nest (Duncan 1939). However, fossorial and related behaviors have not been well studied, and no observation nest-box system has been employed in which colony expansion is impeded by soil or some other removable substance. It is unknown if work on the nest site is primarily accomplished by a specific age class, but the atypically large amounts of pulp gathered by nonexcavating, unconstrained colonies of normally subterranean species (Fig. 8.3) suggests it is mainly done by younger foragers.

## Sanitation

Like excavation, sanitation involves gathering and removing unwanted material, and is probably also undertaken principally by younger individuals. Perhaps the most distinctive aspect of this behavior is the tremendous variation in its degree. Until colony decline, *Dolichovespula* and *V. vulgaris* group species are relatively meticulous in gathering up refuse from beneath the combs, although in large *Dolichovespula* nests particulate detritus is often sequestered in areas remote from the brood rather than carried out of the nest (Greene et al. 1976, Stein 1986). In contrast, *V. rufa* group species characteristically allow much of their waste to accumulate on the nest floor throughout the season, an unusual condition for eusocial insects and especially surprising in that corpses of colony members frequently are included (Akre et al. 1976, 1982; Reed and Akre 1983d).

## Construction

Nest construction is probably the most complex category of yellowjacket behavior, made all the more remarkable by the simplicity of each individual application of building material (Duncan 1939, Brian and Brian 1948, Balduf 1954, Potter 1964). Never transferred or shared, loads of malaxated wood pulp (and occasionally other substances, such as fragments of cell cappings) are fashioned into three distinct structural elements: envelope, cells, and comb suspensoria. The pulp is either collected from external sources of fiber or taken from the nest's inner envelope, which must be continually removed in many places to accommodate comb expansion. Cells are built almost entirely from the recycled paper, while incoming pulp from the field is usually applied to the outer envelope (Duncan 1939; Akre et al. 1976, 1982; Reed and Akre 1983d; Makino 1985a).

Despite the gradual accumulation of observational data, little progress has been made in modeling the construction process, particularly the interplay of stimuli for pulp collection and the immediate cues that determine its placement. For example, pulp foraging and subsequent envelope construction are stimulated by light and cold, but one of the proximate reasons the envelope is still slowly extended in darkness and optimal temperatures may simply be chance encounters of pulp-laden workers with a freshly applied edge (Weyrauch 1936a, Potter 1964). This type of direct response is undoubtedly the reason for the "runaway" application of pulp over surfaces far from the main body of the nest displayed by large, unconstrained colonies. A further complication is that envelope paper may vary substantially in structure and configuration according to its location on the nest, as well as the colony's age, size, and environment (see also Wenzel, this volume).

Cell construction presents an additional suite of problems, including how uniformity is achieved in cell shape and what factors are responsible for cell initiation. Curved when they are first started, cell walls are reworked and straightened as they are lengthened and adjoined to others. It is assumed that combs are so precisely modular because the building wasps employ their antennae and possibly their front legs as measuring tools, and use adjacent dimensions as their standard. However, the mechanism by which worker cell size is gradually increased throughout the season and the basis for the quantum leap to reproductive cell construction are still unknown (Duncan 1939, West-Eberhard 1969, Ishay et al. 1982).

The fundamental criterion for cell initiation on the comb periphery seems to be a principle that can be termed the "rule of 3," a manifestation of the general response of building wasps to bridge gaps and to even out depressions (Weyrauch 1936a, Balduf 1954). Rather than beginning a new row by isolating a cell in the shallow notch between two walls, most workers prefer to extend an existing row if possible, thus spanning the deeper pocket formed by three walls (Duncan 1939, Ishay et al. 1982). If growth is unimpeded, evenly dispersed additions around the border result in the comb maintaining its symmetry as all sides simultaneously expand row by row.

One of the most distinctive aspects of nest construction is the start of a new comb, a major initiative that is essentially a repetition of the nest's origin. The incipient structure is often highly attractive to workers (especially in *D. maculata*), which may either revolve excitedly on it, antennating and nibbling the cells, or cling to it without moving for extended periods. Usually a new tier originates from only one centrally located pedicel, but reproductive combs in large North American *V. germanica* nests are typically formed from numerous separately initiated groups of cells. The same phenomenon can be artificially produced by maintaining colonies of other species in crowded conditions with restricted vertical clearance (Ishay et al. 1986, Greene, unpubl.).

Initiation of comb construction thus appears to be stimulated, at least in part, by the accumulation of an attractive chemical secreted by sternal or tarsal glands onto the substrate (see Downing, this volume). This may be one of several persistent pheromones involuntarily deposited by walking or resting workers, but it is also possible that only a single principal compound is produced that serves different functions in different concentrations or contexts: assembly, building initiation, and trail marker for guiding adults approaching the nest. A colony-specific substance retained on the comb also apparently mediates nest recognition by *D. maculata* (Butler et al. 1969, Ishay and Perna 1979, Landolt and Akre 1979a, Edwards 1980, Akre and MacDonald 1986, Ferguson et al. 1987).

## Thermoregulation

Yellowjackets endeavor to maintain a temperature of about 29–32°C within their nests (Himmer 1927, Sailer 1950, Gibo et al. 1974a,b) and display a number of different responses when this optimum is not met (Seeley and Heinrich 1981). Throughout much of their range, efforts to warm the brood are required more frequently than efforts to cool it. Incubation is particularly critical in the early queen nest; the foundress spends long periods with her body curled around the pedicel, transmitting heat through the comb roof to the eggs beneath her. The larvae themselves soon become the primary source of heat, and queen brooding behavior then decreases (Gibo et al. 1977, Makino and Yamane 1980). Direct warming of pupae by workers pressed against cell caps also occurs (Ishay 1972, 1973b; Reed and Akre 1983d), usually in the case of reproductive pupae in small colonies or, later in the season, when larval and adult populations are declining and the colony once again has difficulty heating the nest interior.

Aside from their body heat generated when inside the nest, the principal contribution of most workers to warming the brood is envelope construction. In general, subterranean colonies, which are better buffered thermally than ones situated elsewhere, construct the thinnest envelopes. However, rate of envelope construction as a function of nest temperature has not been well studied. In one intriguing series of tests, foraging for pulp was apparently stimulated by temperatures slightly lower than optimum but then was inhibited at 25°C, a puzzle that awaits further experimental clarification (Potter 1964).

Cooling an overheated nest in summer often requires a persistent effort by numerous adults. Fanning behavior, particularly at the nest entrance, is the most commonly observed response, supplemented in extreme conditions by spreading water and even larval salivary secretion over cell caps and other interior surfaces. Aerial colonies may substantially widen the nest entrance, and a large percentage of the workers may move to the outer envelope on hot nights (Weyrauch 1936c; Gaul 1941, 1952; Akre et al. 1982; Reed and Akre 1983d).

## Food Transfer

Most direct interactions in a yellowjacket nest involve the exchange of food, and, at least among adults and from adults to larvae, the nutritional pathways seem relatively unambiguous. The extent to which solid food is processed in the field depends on the item, the species, and the individual forager. Prey may be left virtually intact or reduced to an extensively masticated bolus (Archer 1977). It is ground up even further inside the nest, both by the forager and other workers, which chew pieces off the original mass. Larvae are then given portions roughly commensurate with their size, with large pieces sometimes taken back for further distribution. Transfer of regurgitated crop fluid takes place much more frequently than transfer of solid food and follows basically the same pattern, but involves a more precise repertoire of solicitation-donation behavior among the adults that is influenced strongly by the relative dominance of the solicitor (Montagner 1964, 1966a). Toward the end of the season, adult reproductives intercept much of the incoming liquid food (and solid as well in some cases, particularly in *Dolichovespula*), contributing to larval starvation and accelerating colony decline (Greene et al. 1976, 1978; Akre et al. 1982; Reed and Akre 1983d).

Two principal aspects of food transfer remain somewhat enigmatic. The first involves what might be termed classical trophallaxis, the nutritional pathway from larvae to adults. The significance of this behavior in social evolution has been vigorously debated for decades. All adult colony members frequently solicit the larvae for a clear salivary secretion from the labial glands by nibbling at their heads and mouthparts. This secretion contains sugar, protein, and free amino acids, and although it may serve several functions, it is now generally assumed to have a major role in colony nutrition (Hunt, this volume). The continual need for carbohydrates by adult yellowjackets is obvious, and although they are capable of digesting protein from fluids extracted during prey malaxation, the constant supply of nitrogenous compounds obtained from larvae is probably crucial for oogenesis by the queen (Brian and Brian 1952, Montagner 1963, Ikan and Ishay 1966, Maschwitz 1966a, Wilson 1971, Spradbery 1973a, Grogan and Hunt 1977, Hunt et al. 1982).

Even more uncertain is the role of acoustical communication in food transfer. Two of the most noticeable forms of sound production inside the nest are gastral vibration against the comb by adults and mandibular scraping of cell walls by larvae (Ishay and Brown 1975; Ishay and Nachshen 1975; Akre et al. 1976, 1982; Greene et al. 1976; Ishay 1977a; Greene 1979; Ross 1982a; Reed and Akre 1983d; Makino 1985a). Individual workers sometimes tap or vibrate their gasters between feeding visits to larvae, and this has been interpreted as a type of alerting signal. However, the behavior occurs only sporadically, is apparently more frequent in young colonies, and also occurs independently of larval feeding, often in conjunction with tapping by other workers and the queen. Scraping by larvae has been interpreted as a hunger signal, but here, too, there may be more than meets the eye, as large groups of larvae sometimes repeatedly contract and scrape in perfect synchrony.

## Defense

A wide array of phenomena can be thought of as "defensive," including cryptic nesting habits, various aspects of nest architecture, and even the presence of norepinephrine in the envelope, which might confer protection against bacteria or fungi (Bourdon et al. 1975, Lecomte et al. 1976). However, yellowjacket defense as usually conceived is virtually synonymous with the use of their sting, particularly against potential vertebrate predators of the colony.

The overt elements of this behavior have been well documented (Gaul 1953, Spradbery 1973a, Edwards 1980, Akre and Reed 1984). Vibration of the nest results in agitated workers running to the outer envelope or taking flight, with actual stinging typically carried out by only a small minority of the alarmed individuals. Spraying of venom in the direction of the disturbance has been reported for two *Vespula* species and is frequently done by *Dolichovespula* when stinging through mesh (Maschwitz 1964a,b, Greene et al. 1976). Yellowjacket stings often become fixed in leather clothing and are sometimes ripped out of the gaster if the wasp is brushed away, suggesting that the familiar, rapid sting withdrawal from human skin may not be typical for all nest predators. Indeed, the sting of at least *V. maculifrons* routinely holds fast in human skin. Although spontaneous autotomy of the embedded weapon does not occur, it is usually more easily torn from the wasp than from the victim (Greene, N. Breisch, D. Golden, and K. Schuberth, unpubl.).

There are several unresolved problems about stinging behavior. It is widely assumed to be crucial for colony survival (cf. Schmidt 1982, Starr 1985b), but in view of the numerous accounts of active yellowjacket nests being destroyed by common omnivores (Simmons 1973, Spradbery 1973a, MacDonald and Matthews 1981, Reed and Akre 1983a, Akre and Reed 1984), it would be enlightening to discover the usual circumstances (other than human interference) in which stinging is clearly an adaptive defense against an attacking vertebrate. Rodents are likely candidates for readily deterred potential enemies. Nevertheless, the effectiveness of the sting as a weapon of competing spring queens is presently a lot more evident than its effectiveness in colony protection (Evans 1975, Schmidt et al. 1984).

Probably the most puzzling issue in yellowjacket defensive behavior is the relative importance of vibrational versus pheromonal signals in communicating alarm to nestmates. Alarm pheromones are so widespread among eusocial insects (Maschwitz 1966b, Blum 1969, Wilson 1971) that their presence in most or all yellowjackets would not be surprising, especially since venom spraying and sting anchoring suggest pheromonal marking. So far, however, a compound in the venom that rapidly produces mass recruitment and a prolonged, frenzied attack has been isolated and chemically identified only in *V. squamosa* (Landolt and Heath 1987, Heath and Landolt 1988). A similar reaction to venom has been described for three European species, while three North American ones have tested negative (Maschwitz 1964a,b, 1984; Edwards 1980; Akre 1982). Darting and wing buzzing in response to disturbance may be the sole way these three species transmit alarm throughout the nest, but it is also possible that nest inhabitants react to pheromone concentrations that are reached only at a relatively high level of disturbance.

For example, in light of the dramatic effects of the *V. squamosa* alarm pheromone, it is puzzling that this species has been described as particularly docile (Tissot and Robinson 1954). Strikingly incongruent evaluations have also been made of the aggressiveness of *D. arenaria* (Evans 1975, Greene et al. 1976). A third problem therefore involves the identification of factors that regulate individual response thresholds for various disturbing stimuli, including alarm signals by other colony members. Worker age, colony size, previous disturbance, and meteorological conditions are some commonly discussed possibilities (Gaul 1953, Balduf 1954, Potter 1964).

## QUEEN CONTROL AND REPRODUCTIVE COMPETITION

Queen control is a pervasive, multidimensional influence in a yellowjacket colony, with its scope defined by the essential aspects of the society's function (e.g., Spradbery, this volume). Most important is that the colony is the queen's method of reproduction, an elaborate instrument whose raison d'etre is to rear her sexual offspring. She is able to accomplish this because the innate brood-rearing drives of most of the queen's daughters are harnessed first to develop an effective labor force and then to tend the individuals that constitute the end product. The society is therefore based on the reproductive imperative of all the adult females, a potentially double-edged sword for the queen.

Queen control is consequently often regarded as synonymous with its most conspicuous element, the suppression of the workers' ability to directly reproduce, principally by means of an assumed pheromonal inhibition of their ovarian development. Although the most obvious purpose of this attempted castration is to monopolize oviposition sites (Fig. 8.7), brood-rearing labor, and male production, it is also the foundation of another of the queen's major regulatory functions: maintenance of the social cohesion necessary for cooperative brood rearing. In addition, her control is believed to extend to the crucial initiation of the queen production process. The queen's role is therefore a complex one, and, at least superficially, the manner in which its interrelated demands are met varies considerably among species.

Understanding of the physiological bases of intraspecific diversity in "queen strength" currently appears to be far from our grasp. Much of the interspecific variation, however, reflects the well-established concept that eusocial insects with relatively small colonies are generally characterized by relatively weak caste differentiation, readily observed reproductive conflict, and queen control involving physical aggression. Species with large colonies exhibit strong caste differentiation, muted reproductive competition, and queen control achieved almost totally by chemical means (Wilson 1971, 1985; Fletcher and Ross 1985).

**Fig. 8.7.** An aging *Dolichovespula arenaria* queen forces her head into the interior of a cell. Older queens often repeatedly alternate head and gastral insertions before finally ovipositing.

Among the small-colony yellowjackets, the weakest caste differentiation is exhibited by *Dolichovespula* species. For example, *D. maculata* is
notable for the intensity of ovipositional conflict during the reproductive production phase, as well as its minimal caste dimorphism compared with other yellowjacket species (Fig. 8.8) (Greene 1979). Fall
queens of *D. media* display remarkably workerlike behavior, including
regular foraging trips for nectar over a period of days or weeks, extensive liquid exchange with other adult colony members, and even trans-
portation of dead larvae and adults out of the nest (Haeseler 1986).
New *D. arenaria* queens often feed larvae with the remains of prey they
have seized from workers and extensively malaxated (Greene et al.
1976). Caste divergence at such a rudimentary level has not yet been
observed in any *Vespula* species, although small-colony queens in this
genus rely on several physical control measures that are not present in
the more socially advanced large-colony group.

The following discussion organizes the elements of yellowjacket
queen control into three, somewhat overlapping, functional categories:
reproductive dominance, colony cohesion, and queen production.

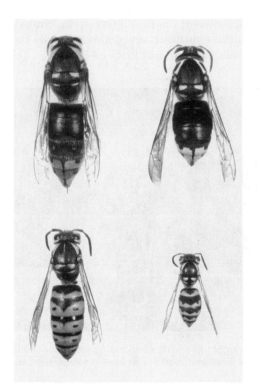

**Fig. 8.8.** Queen (left) and worker
(right) of *Dolichovespula maculata*
(top) and *Vespula flavopilosa* (bottom), illustrating the range of caste
dimorphism in yellowjackets.

## Reproductive Dominance

### Nutritional Aspects

All physical and pheromonal mechanisms of reproductive control rely to some extent on the fundamental nutritional basis of caste differentiation. Yellowjacket workers are stunted adults, the result of a quantitatively limited larval diet (see below). Precisely how colony biomass is divided may be analyzed as a problem in ergonomic efficiency, with a balance struck between the advantages of having a few large workers versus many small ones (e.g., Oster and Wilson 1978). However, smaller workers are undoubtedly easier to dominate physically and probably pheromonally as well. Because worker-cell size is essentially fixed by the foundress, selection for stronger queens may thus be the primary reason for the increase in yellowjacket caste dimorphism— or, more precisely, the decrease in worker size relative to that of the queen—that accompanies increased social complexity (Greene 1979).

Worker dwarfing does not directly result in what has been termed "alimentary castration," as natural loss or experimental elimination of the queen results in the immediate ovarian development of many workers, culminating in oviposition within eight to ten days (Marchal 1896a, 1897; Landolt et al. 1977). In some workers, the hypertrophied ovaries produce noticeable physogastry, and their production of mature, viable eggs is prodigious. However, such extreme ovarian development occurs only in workers that have ceased foraging and assumed a less strenuous routine within the nest (Landolt et al. 1977, Greene et al. 1978). This is consistent with Marchal's (1897) suggestion that "nutricial [in the sense of "nurturing"] castration" resulting from the physiological stresses inherent in brood care (thus sometimes called work castration) is also important in limiting worker ovarian growth (also Hunt, this volume).

### Physical Aspects

During the early part of the season, activities inside a *D. maculata* nest generally appear no different from those of any other yellowjacket species, and interactions between workers and queen are limited to brief episodes of food transfer. As the reproductive production phase approaches, however, the colony reaches a critical behavioral turning point. While the queen spends much of her time seeking new cells on the still rapidly expanding periphery of the single worker comb, an unknown (but probably small) percentage of the worker force begins ovipositing in recently vacated central cells, eating one another's and sometimes the queen's eggs (Balduf 1954), and having their eggs in turn eaten by the queen when she discovers them. In what has been interpreted as a preemptive effort to deny laying workers access to newly available oviposition sites, the queen monopolizes cap trimming following eclosions (a routine function normally performed by workers) and immediately oviposits upon completion of the task (Greene 1979).

The *D. maculata* queen also chases ovarian-developed workers, pins them to the comb, and bites them repeatedly while manipulating them with her legs. Remaining motionless throughout the ordeal, the workers are not noticeably damaged and typically resume normal activity after being released. This stereotyped, agonistic behavior frequently occurs among workers as well, and has been termed *mauling* (Akre et al. 1976, Greene et al. 1976).

A *D. maculata* queen uses overt methods of control to a greater extent than has been observed with any other nonparasitic yellowjacket species, and her reproductive dominance clearly entails ovipositional as well as ovarian suppression. Nevertheless, considering that she holds in check a worker force of several hundred individuals, it seems likely that oophagy, cap trimming, and worker mauling serve as supplementary measures in a process that is primarily pheromonal. Plausible explanations for the delayed onset of the conflict are that after a certain "critical mass" has been reached in colony size, the queen's relatively primitive chemical control loses effectiveness, or that her production of a particular compound begins to wane with time. However, the most important question involves the efficacy of this composite control system as a whole. What are the extent and degree of worker ovarian development? Does this typically result in a significant contribution to male parentage, or do the workers continue to be stymied by the queen's efforts (as well as by competition among themselves)?

As one progresses from *D. maculata* along the yellowjacket social continuum, most of the same physical elements of queen control can be found in varying degrees among the remaining small-colony species of *Dolichovespula* and the *V. rufa* group (Akre et al. 1982, Reed and Akre 1983d) but are absent in the large-colony species. Queen effectiveness seems to be consistently high in *Vespula*. Although a few workers invariably manage to elude castration and successfully oviposit without interference (Montagner 1966b, Akre et al. 1982, Reed and Akre 1983d, Greene 1984), dissections of extensive samples from nests belonging to both small- and large-colony groups, as well as electrophoretic analyses of two large-colony species, have indicated that total worker contribution to male parentage is typically insignificant in queenright colonies (Reed and Akre 1983b,d; Ross 1985, 1986). However, since there is equally abundant evidence for substantial male production by workers after the queen dies, the summed effect of worker reproduction on yellowjacket population genetics is probably considerable (Greene et al.

1978, Greene 1984, Ross 1986). It has also been suggested that social insect workers may subvert the queen's control and exert an indirect reproductive influence by selectively destroying brood that share fewer of their genes (Trivers and Hare 1976). Although worker removal of apparently healthy larvae and pupae of both sexes often occurs in yellowjacket colonies (Duncan 1939; Balduf 1954; Akre et al. 1976, 1982; Archer 1980a, 1981b), and kin recognition among vespines and other eusocial wasps has been demonstrated at least at the level of distinguishing siblings from nonsiblings (Ryan et al. 1985, Michener and Smith 1987), there is presently no empirical evidence that yellowjacket immatures are killed or spared on the basis of relatedness to their older sisters.

Fighting between adults that includes attempted stinging occasionally takes place in the nest; usually it involves workers with developed ovaries in either queenright or queenless colonies (Landolt et al. 1977, Greene et al. 1978, Akre et al. 1982). However, the most puzzling agonistic worker behavior is the normally noninjurious mauling. Frequent episodes of this unilateral aggression occur among workers of all yellowjacket species studied to date (Montagner 1966a; Akre et al. 1976, 1982; Greene et al. 1976, 1978; Reed and Akre 1983d). Mauling is particularly noteworthy in the large-colony species, where the absence of any other physical discord makes it all the more incongruous.

The affinities of mauling with the stylized harassment of general dominance behavior are obvious. Attacks are often associated with liquid food solicitation and exchange, interactions that normally include a strong component of dominance and submission. Indeed, a variation of mauling displayed by both small-colony *Vespula* queens and inquiline females is forcible trophallaxis with workers (Landolt et al. 1977, Greene et al. 1978, Akre et al. 1982, Reed and Akre 1983b,d). Actual or attempted forcible trophallaxis is also a feature of some maulings between workers (Montagner 1966a).

It is therefore not surprising that in queenless yellowjacket colonies much of the mauling is clearly an expression of reproductive dominance (Landolt et al. 1977, Greene et al. 1978), appearing no different in function than when practiced by small-colony queens or inquilines. However, the aggressive and subordinate roles in mauling interactions between workers in queenright colonies have so far not been correlated with size, ovarian development, or any other manifestation of reproductive status. Also, some of the most vigorous and prolonged maulings in small-colony species involve workers attacking males (as well as new queens in *Dolichovespula*), and there is often no trophallactic payoff at the conclusion of maulings by workers, even though the attacked individual may proffer a regurgitated droplet (Akre et al. 1976, 1982; Greene et al. 1976; Reed and Akre 1983d).

Because it is still unknown whether mauling in queenright colonies is primarily a response of typical workers directed toward marked (i.e., distinctive) individuals, a response of atypical workers expressed more or less at random, or a highly variable activity somewhere between those two extremes, interpretations of its role in the colony must remain highly speculative. It may well be correlated with degree of ovarian development after queen removal, and thus a manifestation of potential reproductive conflict. Perhaps, like the efforts to monopolize cap trimming displayed by *V. acadica* queens even in the absence of widespread worker ovarian development (Reed and Akre 1983d), mauling is merely a persistent behavioral relic with limited functional significance while the queen is alive. For that matter, the paradox of physical control measures in general is that for all their readily observed and dramatic qualities, the "sound and fury" is probably never the major component of the overall control process in queenright yellowjacket colonies.

## Pheromonal Aspects

The nature of queen pheromones remains one of the last major gaps in our basic understanding of yellowjacket social biology. Inasmuch as their importance is almost taken for granted, it is admittedly uncomfortable to invoke what at present seems more like an article of faith than concrete, chemical entities.

Once again, *D. maculata* provides a convenient starting point. As the queen moves over the comb after the onset of the reproductive production period, she regularly elicits several stereotyped responses from workers within 2–4 cm of her (Fig. 8.9): (1) prolonged antennation either on her body or very close to it, particularly near her head; (2) sudden, short rushes toward her, each immediately followed by an abrupt jump backwards (Fig. 8.10), with workers sometimes repeatedly darting forward and flinching back in extremely rapid succession; (3) erratic dashing away from her, with buzzing wings and vibrating body, by agitated workers that may career across the entire nest interior before returning to the queen's vicinity; (4) akinesis, with workers remaining almost totally immobile for up to six minutes; and (5) a remarkable "dance," in which a worker's gaster is arched away from the comb and vigorously shaken from side to side as the individual rotates in place. After the vibrations and turning have run their course, the individual remains akinetic with appendages tucked close to its body for several minutes before suddenly resuming normal activity (Greene 1979, unpubl.).

This assortment of localized responses with its conflicting elements of attraction and repulsion, excitation and immobility, constitutes some of the most distinctive behavior yet observed in yellowjacket nests,

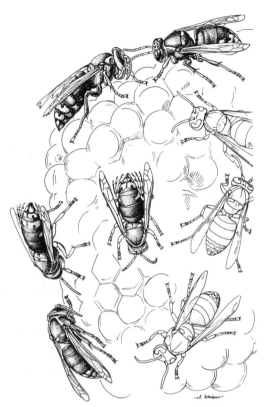

**Fig. 8.9.** Worker reaction to the queen in a mature *Dolichovespula maculata* nest. As the queen (upper left) searches for oviposition sites, two nearby individuals begin to "dance," turning in place while laterally vibrating their arched gasters and drumming their hind legs on the comb. A third worker has passed through the active phase of the dance and now remains motionless with appendages tucked close to its body (lower left). A fourth worker excitedly approaches the queen and touches her head with outstretched antennae. Other individuals continue normal activities with no apparent reaction to the queen.

**Fig. 8.10.** An excited *Dolichovespula maculata* worker lunges at the queen, barely making contact with the tip of her antennae before springing back.

although much of it bears a strong resemblance to queen-induced reactions of *Vespa crabro* workers (Greene, unpubl.). Frequency and intensity of the responses are usually greatest during the queen's preovipositional cell inspections and actual egg laying. It seems likely they are elicited by her release of a volatile pheromone at these times, although workers may also be reacting to brief antennal contact with a compound on her body. The resulting disruption of normal behavior in her immediate vicinity may ultimately serve to facilitate oviposition by the queen or it may be an incidental by-product of another function.

Whatever the various *D. maculata* worker responses signify, such conspicuous, periodic reactions to the queen do not occur in nests of *Vespula* species. Workers of *Vespula consobrina*, *V. acadica* (both small-colony species), and *V. vulgaris* (a large-colony species) often avoid approaching the queen closely (Akre 1982, Akre et al. 1982, Akre and Reed 1983, Reed and Akre 1983d), a possible manifestation of pheromonal release similar to the "reactive zone" around *D. maculata* queens. Nevertheless, there is still a strong possibility that the chemical agents of vespine worker ovarian suppression are totally surreptitious in both dissemination and effect. "Royal courts" of workers surrounding the stationary queen occur in some *Vespa* species (Matsuura, this volume) and apparently serve, as in honey bees, to facilitate pheromonal transfer from her body surface by worker tongue and antennal contact (Ishay et al. 1965, Seeley 1979, Ishay 1981, Matsuura 1984). No such courts have been observed in yellowjacket nests, and only in *D. maculata* does there seem to be sufficient queen-worker interaction for direct contact to be an effective means of chemical dispersal. However, a major aspect of these events both in *Vespa* species and in *D. maculata* is that the prominent worker responses occur mainly during later colony life, being strongly correlated with queen rearing but (despite the implication by some authors) not at all associated with worker ovarian suppression (Ikan et al. 1969, Matsuura 1984). Conversely, groups of *Vespula germanica* workers that were maintained with a *Vespa orientalis* queen were reproductively inhibited but did not participate in court behavior (Ishay et al. 1986), further suggesting that these two manifestations of queen control are not directly related.

Experiments in which a captive colony and its nest are divided by a double screen that prevents all physical contact between individuals on the queenright and queenless sides have demonstrated the absence of an airborne ovarian suppressant in two *Vespula* species (Landolt et al. 1977, Akre and Reed 1983). At present, a leading candidate for the mechanism of queen pheromone dispersal is deposition of a substance on the comb, perhaps principally during oviposition, which must be continually renewed to remain effective. Preliminary evidence for this hypothesis is that experimental switching of queenright and queenless worker groups to opposite sides of a divided nest partially inhibits ovarian development in the queenless individuals (Akre and Reed 1983), and that nonovipositing queens quickly fail to maintain control (Marchal 1896a, Spradbery 1973a). The controlling pheromone or pheromones might be deposited directly onto the comb from the Dufour's or sternal glands, or it might be secreted elsewhere and transferred over the queen's body by her frequent grooming (Akre et al. 1976, Greene et al. 1978, Landolt and Akre 1979a, also cf. Downing and Jeanne 1983).

## Colony Cohesion

Many authors have noted that removal of a vespine queen results in a striking decline in what used to be termed "colony morale." Social organization deteriorates markedly. Workers become more aggressive toward one another and begin to spend more time in the nest. Diminished foraging for pulp and food results in decreased construction and larval feeding, leading to retardation of brood development and increased brood mortality (Potter 1964, Ishay et al. 1965, Montagner 1967, Landolt et al. 1977, Akre and Reed 1983).

It is therefore often suggested that worker pacification and the stimulus to forage are direct effects of "queen pheromone" beyond its premier task of suppressing reproductive competition. On the contrary, considering the relation between oogenesis and dominance behavior in *Polistes* and other eusocial insects (Röseler 1985, Strambi 1985, Wheeler 1986), it is more likely that much of colony cohesion is determined by worker ovarian physiology and associated hormonal activity rather than by a separately evolved, stimulatory aspect of the queen's control. Undeveloped ovaries—or perhaps more fundamentally, less active endocrine glands such as the corpora allata—produce a set of interrelated behavioral traits in an individual that might be termed "suppressed ovarian syndrome" (or, "workerlike behavior"), characterized by a preoccupation with brood care and the foraging necessary for it. Release from reproductive control immediately results in the progressively intensifying expression of "developed ovarian syndrome" (or, "queenlike behavior"), characterized by a preoccupation with direct reproduction and the search for oviposition sites on the comb necessary to achieve it. Yellowjackets with suppressed ovaries are relatively oblivious to their nestmates, and the colony functions largely as the sum of their individual responses to the physical environment and the brood. Ovarian development results in struggles for reproductive dominance and a reduced capacity for colony maintenance in general.

This hypothesis of a primarily endogenous determinant of individual behavior under the extrinsic influence of the queen essentially (1) uncouples colony activity integration from the mechanism of queen control and (2) is the converse of Marchal's (1897) proposal that a compulsion to care for brood results in ovarian suppression, although the reduction and eventual cessation of flight activity must constitute an important anabolic contribution to ovarian development.

Since one of the most prominent characteristics of a queen is her inhibition of the reproductive ability of other colony members, it is not surprising that ovarian-developed yellowjacket workers that exhibit queenlike behavior also can do this to some extent. The stabilizing influence of such workers is readily observed in the close quarters of small, queenless colonies, where it is typical for only a few individuals to attain maximum ovarian development. Under their control, most of the colony remains at least partially ovarian-suppressed, and thus is capable of a greater degree of foraging, construction, and brood rearing than would otherwise be possible (Landolt et al. 1977, Greene et al. 1978, Yamane 1982, also cf. Matsuura 1984). The cohesive effect of queenlike workers is harder to evaluate in queenless nests of large-colony species, where it is typical for a much larger percentage of individuals to undergo ovarian development (Marchal 1896a, Ross 1985). However, small groups of large-colony workers can produce brood-filled nests that are quite normal in appearance (e.g., Perrott 1975).

Regulating the reproductive physiology of her daughters is undoubtedly not the only means by which a vespine queen maintains an effective level of colony activity. For example, her "ovipositional control" may have a somewhat mechanical component in addition to its hypothesized association with pheromonal release. As a rule, vespine workers will not construct new cells on even a rapidly growing comb periphery until the most recently built adjacent cells have been occupied with an egg (Ishay 1973a; Ishay and Perna 1979; Ishay et al. 1982, 1986). Thus, one reason for the almost immediate drop in construction and its attendant foraging after the queen has been removed is probably the sudden appearance of vacant cells acting as barriers to further comb expansion (Fig. 8.11). The underlying basis for this aspect of construction behavior, however, may well involve a pheromonal presence in newly filled cells.

## Queen Production

Vespine queens are virtually identical to workers in gross morphology except for their larger size and a slight allometric increase in gastral dimensions (Blackith 1958). They are produced by rearing females in large rather than small cells and feeding them a commensurately greater quantity of food so that the mature larvae completely fill the cells. Regardless of cell size, most larval growth takes place during the

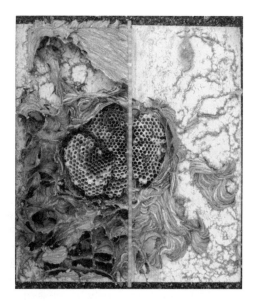

**Fig. 8.11.** Vacant *Dolichovespula maculata* nest in a single-tier nest box divided by a barrier that allowed free access by workers but confined the queen to one half. After the comb was divided, envelope and cell construction took place primarily on the queen's half, while the queenless half was gradually abandoned.

fourth and (especially) fifth instars, and it is during this time that the crucial extra food necessary to achieve queen size is consumed (Potter 1964, Montagner 1967, Spradbery 1973a, Ishay 1975a).

An enhanced feeding rate alone would naturally have no effect on larval size if the molting schedule remained unchanged, that is, if molting occurred at the same stages of larval development. It seems likely that an increased titer of juvenile hormone in the potential queen larvae during certain critical periods is responsible for delaying their fourth and fifth molts (cf. Nijhout and Wheeler 1982), but the exogenous stimuli that affect the hormone's secretion are unknown. Any hypothesis of the mechanism must reconcile the following observations. (1) In general, queen production depends on the prior construction of large cells. Particularly in small-colony yellowjacket nests, however, it is not uncommon for a few female larvae in small cells to attain queen status while neighboring larvae remain worker size. The cell walls around these isolated queens are continuously extended to keep pace with their growth (Greene et al. 1976). (2) Males developing in large cells often grow much bigger than those in small cells (Potter 1964), and even fully queen-sized males are occasionally produced. However, at least in *D. maculata*, many males in large cells grow no bigger than small-cell individuals (Greene, unpubl.). Isolated worker production in large cells also occurs in some species, especially when these females are surrounded by male brood (Greene et al. 1976).

Since increased larval growth and large cell size are thus not inextricably linked, the (presumably chemical) stimulus that induces the endogenous delay of the last two larval molts may be selectively provided by adult workers and/or the queen during food transfer with the larvae (Spradbery 1973a). There is also the possibility that a high rate of feeding may in itself affect corpus allatum activity (Nijhout and Wheeler 1982).

The supposition that large-cell construction is directly governed by the queen rests primarily on two as yet unreplicated experiments conducted in the 1960s, both of which were reported with very little detail. Potter (1964) found that exchanging queens between a large-cell-producing and a non-large-cell-producing colony quickly reversed this aspect of the two colonies' behavior. This suggests that queen production is initiated when the queen reaches a certain physiological state, perhaps as a feedback reaction after some threshold of colony development has been attained (Spradbery 1973a). Ikan et al. (1969) discovered that workers in queenless *Vespa orientalis* colonies were stimulated to produce large cells when given the same attractive, synthetic queen pheromone that produced artificial royal court behavior. This corresponds to the previously discussed association between prominent worker reaction to the queen and the reproductive production period that has been observed in other *Vespa* species (Matsuura 1984) and *D. maculata*. Some of the distinctive responses of *D. maculata* workers in close proximity to the queen are also displayed by individuals perched on newly initiated reproductive combs without the queen nearby (Greene, unpubl.).

As might be expected in a process that presumably depends on internal colony variables, conspecific yellowjacket colonies in the same location typically begin large-cell construction at widely varying times during the season (Potter 1964, Spradbery 1973a, Greene 1984). Although extrinsic factors therefore do not seem to be a primary determinant of vespine reproductive production schedules, a direct, causal relationship between photoperiod and large-cell construction that is essentially independent of queen control has been postulated for *Vespa orientalis* (Ishay et al. 1983a). This study apparently contradicts the influential earlier work by Ikan et al. (1969), raises many more questions than it answers, and will, one hopes, be repeated and expanded by other laboratories before another few decades have elapsed.


## MATING

Captive yellowjacket reproductives will mate readily, and the coupling process has been described in detail (Spradbery 1973a, MacDonald et al. 1974, Greene et al. 1978, Edwards 1980, Post 1980, Akre et al. 1982, Reed and Akre 1983d, Ross 1983b). However, most of the fundamental aspects of mating in natural populations are still unknown. Modern analyses have tended to examine this behavior from a genetic perspective as part of a broader effort to evaluate kin- and sexual-selection theory (e.g., Ross and Carpenter, this volume), and thus the subject is currently organized around two central problems: the normal degree of relatedness of mating pairs (the extent of inbreeding in a population), and the origin of a queen's total supply of utilized sperm (the extent of effective multiple matings).

Much of the answer to the first question hinges on where mating typically occurs. In contrast to the obvious presence of spring queens in most yellowjacket habitats, fall queens are surprisingly cryptic in spite of their far greater numbers. Nevertheless, it is well established that males of many species form loose, conspecific aggregations around prominent vegetation and other landmarks while flying back and forth throughout the area (Spradbery 1973a, Edwards 1980, Post 1980). Mating has been observed in these "swarms," suggesting that fall queens that have left the nest are inconspicuous because they spend most of their time in the canopy before seeking overwintering sites, and that outbreeding is probably common in the more abundant species. Outbreeding is undoubtedly further promoted by emission of numerous males considerably prior to queen output in nests of large-colony species, and especially by the imbalanced reproductive crop of many *D. arenaria* colonies.

Other observations of mating or courtship at, on, or even in the nest (the latter only under artificial conditions) suggest some degree of inbreeding is also widespread (Spradbery 1973a, MacDonald et al. 1974, Greene et al. 1976, Akre et al. 1982, Reed and Akre 1983d). Siblings were preferred as mates by *V. maculifrons* queens in laboratory tests (Ross 1983b). However, the same study also determined that copulation was inhibited by darkness, confirming field observations that vespine mating activity occurs principally on sunny days and casting doubt on whether it normally occurs in the nest interior.

As might be expected in a numerically male-biased reproductive production system, competition for queens is intense. Receptive queens presumably release a pheromone, which in some species results in mobbing by compact masses of excited males. Successful coupling appears to be attained only with difficulty in many cases, and mounted individuals are frequently dislodged before it can occur. There is a tendency for males to resist uncoupling, which has been interpreted as postcopulatory guarding against competitors (Post 1980, Ross 1983b). An electrophoretic study has determined that queens of two large-colony species may be inseminated by five to ten or more males whose sperm probably mixes freely within the spermatheca (Ross 1986, Ross and Carpenter, this volume), and multiple matings by queens are probably the rule for other species as well. Single males of *V. maculifrons* have been shown to be capable of inseminating several queens in the laboratory (Ross 1983b), but, considering the competitive odds, it seems unlikely that most males would have the opportunity to mate even once under natural conditions.

## CONCLUDING REMARKS

It is worthwhile to summarize the major conclusions of this chapter before suggesting how future progress can be best accomplished. At this point in our understanding, most of the fundamental variation in yellowjacket social biology is between two major groups of species. Those in the small-colony group typically have maximum nest populations of several hundred workers, which are all reared on a single comb; a relatively short duration of colony life, much of which is occupied with reproductive production; intermixed queen and male brood on the reproductive combs; a relatively limited foraging spectrum, which is restricted to live arthropod prey and mostly natural carbohydrate sources; and some degree of physical queen control behavior. Inquilinism, one of two principal departures from the general yellowjacket life history pattern, is associated only with small-colony species.

Large-colony species typically have maximum nest populations of up to several thousand workers, which are reared on multiple combs; a relatively long duration of colony life, somewhat less of which is occupied with reproductive production; mostly queen brood on the reproductive combs of vigorous colonies; an extremely broad foraging spectrum, which includes predation, hematophagy, necrophagy, frugivory, and a wide utilization of processed human food; and queen control with virtually no physical component. Perennial (or at least biennial) polygynous colonies, the second major deviation from the typical yellowjacket life history pattern, are restricted to large-colony species in warm climates.

The solitary nest initiation period is a time of severe stress for queens, which must contend with both external (inclement weather and predation) and internal (conflicting needs of the young colony) challenges. In addition, mortal combat for possession of young nests, particularly among conspecifics, is likely to involve the majority of spring queens in many habitats. Reproductive production of most colonies is numerically male-biased, and typical populations probably include a large, but difficult to sample, percentage of relatively small nests, which produce mainly or exclusively male reproductives. Most mating appears to follow the normal pattern for monogynous eusocial insects, with multiple insemination of queens by unrelated males.

Much of colony cohesion is probably a direct function of worker ovarian development and corpus allatum activity. Reproductively suppressed individuals respond principally to stimuli provided by the brood and the physical environment (including the nest itself), and overt interaction between them is limited mostly to food transfer. The stereotyped, usually noninjurious attacks by some suppressed workers on others (mauling) seem inconsistent with the rest of their behavior, and may be the first manifestation of a slightly enhanced gonadotropic hormonal secretion.

Queen inhibition of worker ovarian development is probably accomplished mainly by pheromones even in the small-colony group. Her control is very effective in the limited number of species that have been investigated, and despite the inevitable presence of some ovipositing workers, she is responsible for almost all male production during her life. However, colonies often remain functional long after queen death, allowing ovarian-developed workers to produce (and ovarian-suppressed workers to rear) a substantial crop of additional males. The physiological mechanisms that stimulate queen production are unknown, but a suite of distinctive, extremely conspicuous worker reactions to the queen in *D. maculata* is apparently associated, like the royal court in some *Vespa* species, with large-cell construction rather than worker ovarian suppression.

In view of the frustratingly tentative nature of this summary, how much is really known about yellowjackets? Although their recorded natural history has become fairly detailed, it should be obvious that what is visible represents only the surface of their world. Transparent nest boxes may reveal activity in the very heart of the colony, but it is activity determined by an invisible matrix of chemical information and control. If we wish to gain a substantially deeper understanding of social wasp life, we must attempt to perceive their world as they perceive it.

Only an intensive collaboration between chemists and behaviorists will be able to create the Rosetta stone for deciphering what we see. Our primary focus should now be on identifying glandular secretions and devising artificial delivery systems for them so that behavior can be experimentally manipulated. This approach has been advocated by social biologists for years, and it is long overdue for the study of yellowjackets.

# PART II

*Special Topics in the Social Biology of Wasps*

# 9

# Reproductive Competition during Colony Establishment

PETER-FRANK RÖSELER

According to natural selection theory, each individual strives to produce many descendants and thereby to transfer as many copies of its genome to the next generation as possible. However, this principle is very much restricted in the eusocial insects because only relatively few females, the queens, lay eggs. The non-egg-laying females (workers, auxiliaries) can contribute to reproduction only indirectly by helping the queen to rear her offspring, and they profit from such helping behavior only if they share with the queen genes identical by descent (Hamilton 1964a,b, West-Eberhard 1975).

The sterility of non-egg-laying females is not irreversible in most eusocial species. Such females remain potentially fertile, and, under certain conditions, they can lay eggs and compete for direct reproduction. Hence conflicts of interest exist between females of a colony—that is, between queens and their daughters as well as between females of the same generation. For most species, this conflict is not overt because the egg-laying female is largely able to inhibit direct reproduction by other females, and thereby to pacify the colony. Inhibition can be effected by subtle ritualized agonistic behavior or by pheromones. Since workers and auxiliaries must perform all the various labor tasks in the colony, the queen can invest a relatively greater proportion of her energy in egg production. Reproductive dominance enables a queen to fully utilize her reproductive capacity and to monopolize reproduction in her colony (Fletcher and Ross 1985).

Most highly eusocial species exhibit conspicuous morphological queen-worker dimorphism, which results in differential fertility (Brian

1979); the fertility of workers is reduced, the fertility of queens is increased. Since, moreover, the fertility of workers of most highly eusocial species is more or less perfectly controlled by pheromones of the queen (Bourke 1988), workers can lay eggs only when the queen is no longer able to inhibit egg formation, either because of her death or senescence or because the colony has become too large for her to control. Under these circumstances workers may also compete for direct reproduction.

The conflicts of interest among females appear to be greatest in species in which several females initiate a colony together. Multiple-foundress associations are widespread among species of eusocial Vespidae, especially in the tropics. The advantage is seen as better protection of the nest against predators and against intra- or interspecific usurpation, because the nest is only seldom, if ever, left unattended. Moreover, destroyed nests are more likely to be successfully reestablished when several females are present (e.g., Gibo 1978, Reeve, this volume).

Multiple-foundress associations of social wasps can be formed in several ways (Jeanne 1980a). (1) A foundress can join a nesting female as a passive companion (auxiliary), or she can try to usurp the nest to become the egg layer (West-Eberhard 1969, Gamboa 1978, Reeve, this volume). (2) Several foundresses can start a nest together (Strassmann 1981a, Reeve, Gadagkar, this volume). This mode of founding, together with passive joining, is often referred to as pleometrosis or associative foundation. (3) Multiple-foundress associations can be formed during the process of swarming, which occurs in many polistine species, particularly in the New World tropics, as a regular feature of the colony reproductive cycle (West-Eberhard 1982c, Jeanne, this volume: Chap. 6).

Each female of a multiple-foundress association attempts to become an egg layer and to prevent direct reproduction by other females. In small associations, such as those of *Polistes*, eventually only one foundress becomes the principal egg layer, and all other foundresses become more or less sterile. Even in species in which several females with developed ovaries coexist, some females may lay eggs at significantly higher rates than others. On the other hand, at least one species of Stenogastrinae that has very small colonies (*Parischnogaster alternata*) apparently exhibits no reproductive dominance among foundresses (Turillazzi 1985e, 1986a).

As a rule, competing for reproductive dominance is equivalent to competing for social dominance in that a foundress that has achieved the dominant social position becomes the principal egg layer (Pardi 1948). A subordinated foundress has several options: she may escape and try to start her own nest, she may become the dominant egg layer in another association, or she may surreptitiously lay eggs when the queen is absent. If she remains as a subordinate, she may become an auxiliary, in which case she assumes a workerlike role and helps to rear the offspring of the egg-laying foundress. Because the ovaries of a subordinate foundress regress more and more through time, this individual gains in inclusive fitness mainly by rearing relatives. Moreover, the subordinate has the prospect of waiting for the death of the dominant to assume her reproductive role.

Reproductive competition between females of species that found nests solitarily (haplometrosis) may also exist during the period of colony founding, when females attempt to usurp established nests. In contrast to species with associative nest foundation, competing foundresses of strictly haplometrotic species never form dominance hierarchies. Rather, the conflicts regularly end with the death of one or both females (Matthews and Matthews 1979, Akre and Reed 1984, Akre and MacDonald 1986).

## OCCURRENCE AND DISTRIBUTION

Competition for reproduction always means competition for direct reproduction (egg laying). Competition for indirect reproduction—that is, exclusion of nonrelatives as helpers—is not known. Even in species in which foundress associations are formed exclusively of former nestmates (e.g., Strassmann 1983), the expulsion of nonrelatives presumably represents competition for direct reproduction.

Reproductive competition during colony establishment is widespread among eusocial wasps. Foundresses of many species may compete only for suitable nest sites, but foundresses of species with multiple-foundress colonies compete directly for reproductive privileges. At present it is not possible to present a complete review of the distribution and frequency of usurpation (e.g., Akre and Reed 1984) or of reproductive dominance in multiple-foundress associations (e.g., West-Eberhard 1978a, Itô 1986a), because only a few species have been investigated rigorously. Detailed knowledge of the relationships among foundresses requires continuous observation from the establishment of the association onward. Such observation is difficult to accomplish in the field because associations are usually discovered only after the construction of the first cells. At that stage, however, reproductive dominance has already been established. Moreover, observations in the field are often casual observations performed for only a limited time, so that many studies are fragmentary. Since the composition of foundress associations can change after nest initiation, reproductive competition during the total preemergence period—that is, from initiation until the emergence of the first offspring—is considered in this review.

## Nest Usurpation and Social Parasitism

Nest usurpation may occur in most social wasp species because nests with brood are attractive to females searching for a nesting site. It may be most frequent in the temperate zone, however, because colony cycles are synchronized there. After hibernation all queens look for a suitable nest site during roughly the same period, leading to competition for nest sites. Temperate-zone vespines, in which haplometrosis is obligatory, compete for nest sites as the overwintered queens emerge from diapause and consequently large numbers are present. They usurp young nests of the same species (intraspecific usurpation) or of other species (interspecific usurpation) (Taylor 1939, Akre and Reed 1984, Akre and MacDonald 1986, Greene, this volume). Which queen wins the confrontation is probably largely a matter of chance, although *Dolichovespula sylvestris* and *Vespula vulgaris* nest owners are usually killed by the conspecific invader (Archer 1980a), whereas resident queens of *Vespula maculifrons* (Matthews and Matthews 1979) and *Vespa crabro flavofasciata* (Matsuura 1970b) usually kill the intruder. Interspecific usurpation has been observed in *Vespula vulgaris* nests, which are taken over by *Vespula germanica* queens (Nixon 1935) or by *Vespula pensylvanica* queens (Akre et al. 1977, Roush and Akre 1978), and also in *Vespula maculifrons* nests, which are usurped by *Vespula flavopilosa* queens (MacDonald et al. 1980b). Both intra- and interspecific usurpation can occur several times in the same nest (Matthews and Matthews 1979). Some observations show that most of the older brood of the former queen(s) is reared by the invader and used to rear her own progeny. Temperate-zone *Polistes* also usurp conspecific nests when many overwintered foundresses are present (Gamboa 1978). Unlike vespine queens, however, successful *Polistes* foundresses do not kill the others, but instead use them as subordinate helpers (see also Reeve, this volume).

Usurpation as a manifestation of reproductive competition ultimately evolves into interspecific social parasitism, in which parasite queens are specialized for taking over young nests of the host species (Taylor 1939, Spradbery 1973a, Akre 1982). Queens of *Vespula squamosa* invade colonies of *Vespula vidua* (Taylor 1939), *Vespula flavopilosa* (MacDonald and Matthews 1975), and especially *Vespula maculifrons* (Matthews and Matthews 1979) when workers are already present. The host workers rear the brood of the parasite queens and, with time, they are gradually replaced by *V. squamosa* workers. With this strategy, *V. squamosa* queens avoid the high risk of failure of young workerless nests. This behavior has also been described for queens of the hornet *Vespa dybowskii*, which sometimes usurps nests of *Vespa crabro* and *Vespa xanthoptera* (Sakagami and Fukushima 1957). Queens of both *Vespa dybowskii* and *Vespula squamosa* (in the southern parts of its range) are also capable of starting their own nests, so they exhibit facultative social parasitism.

Obligate social parasites have no worker caste, so only sexual progeny of the parasite are reared by host workers. Obligate parasitic vespine species include *Dolichovespula adulterina*, *D. omissa*, *D. arctica*, and *Vespula austriaca* (Edwards 1980, Akre 1982). Three obligate parasitic species have also evolved in *Polistes* in Europe (*Polistes atrimandibularis*, *P. semenowi*, *P. sulcifer*). Again, in contrast to the vespines, these interspecific parasites do not kill the *Polistes* host queens but subordinate them and use them as non-egg-laying helpers (Scheven 1958, Turillazzi et al. 1990). Sometimes, however, *P. semenowi* females leave the nests of *P. nimpha* after a period of egg laying to usurp another nest. The *P. nimpha* queens rear the *P. semenowi* brood and then start ovipositing again (G. Demolin, unpubl.). The obligate parasitism is thus only temporary for a particular nest, and it does not completely exclude the host queen from reproducing.

## Competition among Foundresses of Different Nests

Normally, only foundresses of the same nesting association compete for reproduction. Whether there also exists competition among foundresses of different nests, aside from usurpation and social parasitism, is not yet clear. Some observations indicate that this might occur, if only exceptionally.

Stenogastrine foundresses sometimes try to land on a foreign conspecific nest. As a rule, approaching females are rejected by the resident foundress. When foundresses succeed in landing on a nest they sometimes eat an egg and lay their own in that cell (Turillazzi 1985d). Interestingly, foundresses may also lay eggs in nests of other species, but it is not known whether these are reared (Turillazzi and Pardi 1982). At present, it is not possible to distinguish this behavior from nest usurpation. Are these vagrant females looking for a chance to usurp a nest, or are they foundresses that have successfully started a nest but are also trying to enhance their reproductive success by parasitizing another nest?

In Sumatra, Yamane (1986) observed an egg-laying *Ropalidia variegata* foundress of one nest also laying all the eggs in another nest, which she visited from time to time. Neither of the two resident foundresses of the visited nest was observed to oviposit, but both seemed to rear the brood of the foreign foundress. Probably, both nests belonged to the same foundress association, for Itô (1986c) has found that 70% of all colonies of *R. variegata* in Australia have multiple combs. The dominant foundresses in these unusual colonies tend to stay on the central combs with most of the larval cells, whereas low-ranking foundresses preferentially stay on peripheral combs. Subordinate foundresses of *R. cyathiformis* in India have been observed to start new nests (termed satellite nests) in the vicinity of their natal nests. It is thought that this kind of nest initiation results from reproductive competition (Gadagkar, this volume). On other nests of this species with closely situated combs, foundresses have been observed to move about and oviposit on all the combs, but it is not known whether these were satellite combs or nests started independently by foundresses that were not members of the original association (Gadagkar and Joshi 1982b, 1984, 1985).

Foundresses of some *Polistes* species start their nests in close proximity. During the early preemergence period, foundress groups can change composition daily (e.g., *Polistes erythrocephalus*: West-Eberhard 1969; *P. exclamans*: Strassmann 1983; *P. fuscatus*: West-Eberhard 1969, Noonan 1981; *P. dominulus* [= *gallicus* of authors]: Pratte 1980), with the associations becoming stable only after some days. Such nest-switching behavior may eventually result in foundress groups sharing several combs. Five to six foundresses of *Polistes metricus* were observed to share three combs, with individual foundresses preferring particular combs (Gamboa 1981). It is not known whether these preferences reflect different reproductive hierarchies on different combs. Itô (1984a) has described a *Polistes versicolor* association in Panama consisting of more than ten foundresses sharing two combs 50 cm apart. He observed that one foundress laid eggs in both combs. Subordinate foundresses of *P. dominulus* held in captivity occasionally start a second comb in the cage. The dominant foundress of the group controls both combs and eats the eggs laid by subordinates (Röseler, unpubl.). In contrast to these examples, other *Polistes* species show great nest fidelity, even if nests are started close together. For example, foundresses of *P. annularis* join other nests only when their own is destroyed and always become subordinates in the new nest (Strassmann 1983).

*Polistes canadensis canadensis* in Brazil starts constructing satellite nests near the original nest during the preemergence period. As many as four combs have been found at the end of this period, but mature colonies may have up to 38 combs. All of the associations have only one egg layer, however (Jeanne 1979a). Satellite nests are also constructed by *P. exclamans* (Strassmann 1981b) and by *Vespa* species (Kemper and Döhring 1967, Hamilton 1972, Edwards 1980, Kulike 1987), but in the postemergence period exclusively (e.g., following first worker emergence). Kulike (1987) reported an interesting case of competition between two *V. crabro* queens heading nests that were established 20 m apart and already contained some workers. One queen attempted to invade the other nest and was killed.

## Competition among Associates on
## Multiple-foundress Nests

Nests of most stenogastrine species are started by lone foundresses, but *Liostenogaster flavolineata* and *Parischnogaster alternata* commonly exhibit associative nest foundation (see Turillazzi, this volume). *Parischnogaster alternata* associations contain two to three foundresses, and *L. flavolineata* up to seven (Hansell et al. 1982; Turillazzi 1985e, 1986a,b). Associations of *Parischnogaster nigricans serrei* can also be formed when foundresses reuse abandoned nests (Turillazzi 1985d). In most such cases, two foundresses are present on a nest. The associations are not stable, however, as foundresses may disappear after a few days and other females may join the nest. Nothing is known of the fate of the eggs laid by the different foundresses.

Among stenogastrines, reproductive competition may be widespread, but linear dominance hierarchies (see The Dominance Hierarchy) have thus far been observed only in *L. flavolineata* and *P. nigricans serrei* (Hansell et al. 1982, Turillazzi and Pardi 1982, Turillazzi 1985d). The dominant foundress of both species is the only egg layer. Subordinate foundresses have developed ovaries, but oocytes are likely to be resorbed. When a nest is destroyed, foundresses initiate a new nest in the vicinity of the old one while maintaining the previous dominance hierarchy.

Nests of independent-founding polistines often are started by multiple-foundress associations (reviewed by Reeve, Gadagkar, Spradbery, this volume). In the case of independent-founding *Ropalidia*, nests are initiated by a foundress group or by a single foundress that later is joined, or has her nest usurped, by others. In Japan, *R. fasciata* foundresses sometimes overwinter in groups on the natal nest, and in the next spring the group reuses the old nest (Itô et al. 1985). Very large groups of up to 20 *R. marginata* foundresses have been observed in India (Gadagkar et al. 1982a). The dominant foundress in multiple-foundress *Ropalidia* associations is either the only egg layer or the principal egg layer (Gadagkar, this volume), but the sometimes considerable development of subordinates' ovaries indicates that they might also lay eggs. Several foundresses have been observed to oviposit in single *R. cyathiformis* associations, but the dominant one lays eggs most frequently. Since eggs, larvae, and pupae are eaten, there must be considerable reproductive competition among foundresses of this species (Gadagkar and Joshi 1982b).

Multiple-foundress nests of *Belonogaster* and *Mischocyttarus* are invariably started by lone foundresses, which are later joined by others. A dominance hierarchy has been observed in *Belonogaster grisea*, but all foundresses, even those that are uninseminated, may have developed eggs in the ovaries (Pardi and Marino Piccioli 1981). Multiple-foundress nests of *Mischocyttarus labiatus* in Colombia have a sole or principal egg layer (Litte 1981). *Mischocyttarus mexicanus* has single-foundress nests or small associations with only one egg layer during winter and spring; in fall, nests are initiated by larger groups of as many as 20 foundresses, and up to three of them may have developed ovaries (Litte 1977). The dominance hierarchy of *M. angulatus* and *M. basimacula* seems to be flexible, as the relationship between foundresses changes frequently (Itô 1984b). It is not known whether these foundresses oviposit at comparable rates or whether they compete for reproduction.

Many *Polistes* species, especially in the tropics, form multiple-foundress associations. The foundresses of some species are ranked in a hierarchy, which is sometimes established only after severe fighting. The dominant foundress is the principal egg layer, but occasionally subordinates also oviposit during early colony development. Since eggs laid by subordinates may or may not be eaten by the dominant foundress, direct offspring production by subordinates in these associations is a possibility. The degree of direct reproduction by subordinates probably varies according to species and may even differ among colonies of the same species (see Reeve, this volume). With time the hierarchy among foundresses becomes stabilized, so that subordinates only rarely oviposit by the end of the preemergence period.

Associations of the temperate-zone species *Polistes annularis* sometimes contain more than 20 foundresses (Strassmann 1981a, 1983). The dominant foundress starts constructing cells and becomes the principal egg layer, but the larger the group, the greater is the probability that one or more subordinates will also oviposit. Since not all of these eggs are eaten, a certain percentage of early-emerging adults is likely to be derived from subordinates. *Polistes canadensis canadensis* in Brazil also forms large associations, with up to 29 foundresses, but in nearly all colonies only one foundress with well-developed ovaries has been observed (Jeanne 1979a).

Most temperate-zone *Polistes* species form smaller associations, with fewer than five foundresses as a rule (*P. exclamans*: West-Eberhard 1968, Strassmann 1981b; *P. fuscatus*: West-Eberhard 1969, Gibo 1978, Noonan 1981; *P. dominulus*: Pardi 1941, Turillazzi et al. 1982; *P. metricus*: Gamboa et al. 1978). The percentage of nests with multiple foundresses, listed by Röseler (1985) and Spradbery (this volume: Table 10.1), seems to depend on the density of the foundress population. Eggs deposited by subordinates are often eaten by the dominant; for instance, all eggs laid by subordinates of *P. dominulus* in the early stages of nest development meet such a fate (Gervet 1964a). On the other hand, Noonan (1981) estimated that in the late preemergence and early postemergence periods of *P. fuscatus*, 33% of all brood reared were derived from eggs laid by subordinates. For all species, the dominant foundress is only seldom replaced by subordinates during the preemergence period. Very rarely, foundresses of different species cooperate; two nests containing foundresses of both *P. fuscatus* and *P. metricus* were observed by Hunt and Gamboa (1978).

New nests of swarm-founding polistines are started by swarms that contain workers and several to many egg-laying queens (see Jeanne, this volume: Chap. 6). During the colony cycle of *Metapolybia aztecoides*, polygyny and monogyny alternate (West-Eberhard 1978b). Early in the colony cycle, queens tolerate egg laying by others, but with the increasing development of each queen's ovaries competition among them increases. When queens attacked by other queens or by aggressive workers exhibit submissive behavior they subsequently begin to act as workers, or they are driven from the nest or leave with swarms. In this way, the colony becomes functionally monogynous. Supernumerary queens can also be induced to assume workerlike roles when workers are lost from the colony.

Other than West-Eberhard's (1978b) detailed study on *M. aztecoides*, little is known about reproduction in swarm-founding polistines. Observations of some species show that reproductive swarms frequently contain multiple queens, and studies of the composition of mature nests often reveal the presence of multiple queens with developed ovaries, thus confirming functional polygyny (Jeanne 1973, 1975b, 1980a, this volume: Chap. 6; Akre 1982; West-Eberhard 1982c; see also Queller et al. 1988). The reproductive contribution of different queens is unknown, however, and the possibility that colonies do not remain polygynous over the whole nesting cycle is virtually unstudied.

Nests of temperate-zone Vespinae are started by single queens, which vigorously defend their nests against others. However, several multiple-foundress nests of the Southeast Asian hornet *Vespa affinis* with up to 15 queens each have been discovered (Matsuura 1983b, this volume; Spradbery 1986). The data indicate that nests are founded by a single queen, which is later joined by others. Most of the queens are inseminated and possess developed ovaries, but the few available observations suggest that there may be differential rates of egg laying and undertaking of specific tasks among queens. Dominance behavior was not detected.

Colonies of some *Vespula* species may not perish at the end of the annual cycle, so that under favorable climatic conditions they are able to perennate (see Greene, this volume). Multiple young queens apparently are recruited to requeen these nests when the founding queen is still present, leading to secondary polygyny in the second season of development. These perennial colonies grow rapidly, sometimes to more than a million cells and more than 100,000 workers (Spradbery 1973b, 1986). It is unknown how the normally hostile fertile queens come to tolerate each other and whether they are separated in reproductive territories in the large nests.


## THE DOMINANCE HIERARCHY

Dominance means social dominance as well as reproductive dominance. As a rule these are associated—the behaviorally dominant foundress is also the principal egg layer—however, the two kinds of dominance can be separated experimentally. Ovariectomized foundresses of *Polistes dominulus* are able to achieve the socially dominant α-position in groups of normal foundresses and to maintain it during colony development. These females remain dominant in behavioral interactions and perform all tasks specific to dominants except egg laying. The β-foundresses in the social hierarchy become the principal egg layers in such experimental groups; that is, they achieve reproductive dominance (Röseler et al. 1985, Röseler and Röseler 1989).

In groups of only a few foundresses, the structure of the dominance hierarchy is linear. The α-foundress is dominant over all other foundresses, the β-foundress is dominant over foundresses other than the α-foundress, and so forth ($\alpha > \beta > \gamma > \ldots$). In large associations of *P. annularis* and of *P. canadensis canadensis*, the exact position of low-ranking foundresses is difficult to recognize because they do not show clear reactions to other low-ranking foundresses (Strassmann 1981a, West-Eberhard 1986). Therefore, the hierarchy in such large groups apparently more closely resembles a pyramid than a chain, although definitive conclusions about dominance structure can be drawn only if the relationships of each foundress to all others are accurately analyzed.


## Social Dominance

### Characteristics

A dominant foundress is characterized by several behaviors that were first observed in *Polistes dominulus* by Heldmann (1936a) and later analyzed in great detail in this species by Pardi (1942a, 1948) and Pratte (1989). The rank of a foundress can be inferred from the nature of specific interactive behaviors between pairs of foundresses and from the tasks an individual performs. A dominant foundress spends most of her time on the nest. Indeed, the dominant foundress of some species never leaves the nest when associates are present. Thus dominant foundresses typically do not undertake foraging tasks, although they may occasionally collect wood pulp for nest construction. Dominant individuals characteristically participate in feeding larvae and building cells; the dominant foundress of some species initiates most cells as well (West-Eberhard 1969). As a rule, dominant foundresses vigorously defend the nest against potential usurpers and, in the case of *P. dominulus* (Turillazzi et al. 1990) as well as *Ropalidia fasciata* (Itô 1985a), also against predators and parasites. Dominants of some species have preferred resting places on the comb. Dominant *P. dominulus* and *P. erythrocephalus* foundresses prefer to sit on the nest face (Pardi 1948, West-Eberhard 1969) and *R. variegata* dominants prefer the comb center (Itô 1986c). Queens of *Metapolybia aztecoides* cluster at the edge of the comb (West-Eberhard 1978b).

A characteristic behavior of *Polistes* species that may be related to dominance is abdominal wagging, a short-period vibration of the gaster from side to side (see Reeve, this volume). Foundresses perform this behavior as they inspect cells that have eggs or as they walk over cells with larvae, either before or after they have been inspected and fed (Gamboa and Dew 1981, Strassmann 1981a, Downing and Jeanne 1985, Röseler and Röseler 1989). This behavior is sometimes also performed by a dominant *P. dominulus* foundress during antennation of a subordinate (Röseler and Röseler 1989). Foundresses of *Mischocyttarus drewseni* also exhibit abdominal oscillations, but, in contrast to *Polistes*, these are dorsoventral vibrations of the gaster. This behavior is frequently displayed during feeding of larvae and sometimes during dominance interactions with a subordinate (Jeanne 1972).

The meaning of abdominal vibrations by these wasps is not yet clear. They are likely to be vibration signals for adult-larva and adult-adult communication, but it is also possible that a "dominance pheromone" is distributed by means of these behaviors (West-Eberhard 1969). In *P. canadensis canadensis* abdominal vibration is thought to be an aggressive signal (West-Eberhard 1986). The "pecking" behavior of *Mischocyttarus* and *Ropalidia*, where a foundress inspects a cell containing a larva and vibrates her head violently within the cell (Jeanne 1972, Itô 1983), seems to be analogous to abdominal wagging.

A further characteristic feature of dominance is the favorable status dominant individuals enjoy during food sharing or trophallaxis. A dominant foundress solicits food from a subordinate, which sometimes spontaneously offers it. Thus, the flow of food within an association seems always to be directed toward the dominant. The "kisses" (Itô 1983) observed in some species of *Ropalidia* (Darchen 1976a, Gadagkar 1980), *Belonogaster* (Marino Piccioli and Pardi 1970), and *Mischocyttarus* (Itô 1984b) seem to be a special kind of trophallaxis. Returning foundresses are immediately mounted by another foundress, usually the dominant one, their mouths touch, and liquid food is shared. It is probable that this represents a highly ritualized dominance behavior (Turillazzi and Marucelli Turillazzi 1985).

The dominance behaviors listed above are not exclusive characteristics of a dominant foundress, but they are exhibited more frequently by foundresses of higher social ranking. Thus, quantitative differences in performance of the behaviors exist among foundresses (Gadagkar and Joshi 1983, 1984), with the relative frequencies of the behaviors changing through the preemergence period as the dominance hierarchies are formed (Strassmann 1981a).

### Dominant-Subordinate Interactions

When two wasp foundresses encounter one another on the comb they interact with a characteristic dominant-subordinate behavior, which can be more or less ritualized depending on the species and on the stability of the hierarchy. Such interactions between *Polistes* foundresses have been extensively described by Pardi (1942a, 1946, 1948), among others. When two foundresses of *P. dominulus* meet for the first time, they approach each other with raised antennae and, upon contact, begin to fight, as illustrated in Figure 9.1. When the fights escalate, the foundresses bite each other and bend their abdomens while extruding the sting. In severe fights, foundresses can become so entangled that they lose their grasp and fall from the comb or other substrate (falling fights). In such cases, they separate, fly back to the nest site, and resume fighting. After a brief period of such agonistic interactions, one of the foundresses escapes or adopts a subordinate posture (Fig. 9.2), in which she remains motionless. In all future encounters, the subordinate foundress immediately adopts the subordinate posture, sometimes offering regurgitated fluid as well. When an association contains several foundresses, the α-foundress dominates the β-foundress most frequently, while dominating the other foundresses less frequently and in accordance with their rank (Pardi 1946, 1948).

Dominance interactions become more ritualized once the hierarchy has been established. In the later stages, dominance among foundresses is seen, for instance, in avoidance behavior when two wasps approach each other on the nest: the subordinate retreats while the dominant advances. Sometimes the dominant performs a rapid jerk toward the subordinate, but without contact. In a newly established nest of *P. canadensis canadensis* in Colombia, West-Eberhard (1986) observed frequent aggressive interactions, but these did not occur as frequently in older colonies dominated by a "despotic" queen. Weak dominance interactions in an established hierarchy, therefore, say nothing about the repertoire of behaviors used during hierarchy formation.

Generally, aggressive interactions occur in all multiple-foundress

**Fig. 9.1.** Dominance interactions between foundresses of *Polistes dominulus* at their first meeting one day after emergence from hibernation. The foundresses intensively antennate one another at first contact (a). In severe fights both foundresses straighten up and grapple with their forelegs (b) and attempt to bite one another (c).

associations of eusocial wasps, but the behaviors vary considerably among different taxa (West-Eberhard 1982b). In Stenogastrinae, the α-female dominates her subordinates by slapping them with her antennae or darting aggressively at them. During severe aggression wing buzzing, biting, and stinging may occur (Hansell et al. 1982, Turillazzi and Pardi 1982, Turillazzi 1986a). Dominant-subordinate interactions of *Ropalidia* seem to be generally less violent than those of *Polistes* while containing many of the same elements (Gadagkar and Joshi 1982b; Itô 1983, 1985a; Turillazzi and Marucelli Turillazzi 1985; Yamane 1986; see descriptions in Gadagkar, this volume). In *Mischocyttarus mexicanus* only

**Fig. 9.2.** Dominance interactions between foundresses of *Polistes dominulus* at their first meeting. After fighting, the subordinate remains motionless, pressing her body to the substrate and lowering her head and antennae (a), while the dominant climbs on her body and intensively antennates and mouths her (b).

biting has been observed (Litte 1977), whereas in *M. immarginatus* falling fights similar to those of *Polistes* have been reported (Gorton 1978). Attacked foundresses of these species behave passively, as do *Polistes* subordinates, by crouching, avoiding the dominant, or leaving the nest. A subordinated foundress of *M. drewseni* lowers her head but raises her gaster (Jeanne 1972). Interactions between queens of the swarm-founding polistine *Metapolybia aztecoides*, consist of bending displays and rapid jerks toward subordinates by dominants. Overt fights have not been observed (West-Eberhard 1977a, 1978b).

## Reproductive Dominance

### Characteristics

A reproductively dominant foundress is first of all the principal egg layer. Studies of foundress associations of several *Polistes* species have shown that the dominant foundress invariably has the best-developed ovaries (Pardi 1946, 1948; Dropkin and Gamboa 1981; Turillazzi et al. 1982; Strassmann 1983). Therefore, the gaster of the principal egg layer of some species is more swollen than that of subordinates. Egg formation in subordinates is not completely inhibited but occurs at different rates corresponding to the social rank. The higher the social position, the more eggs the ovaries contain (Pardi 1946). This scheme, based on findings in several *Polistes* species, might also be at least partly correct for Stenogastrinae (Hansell et al. 1982, Turillazzi and Pardi 1982) and for independent-founding Polistinae other than *Polistes* (Litte 1977, 1981; Turillazzi and Marucelli Turillazzi 1985; Yamane 1986), although few data are available for the preemergence period in these latter groups.

Since at least some subordinates usually are capable of laying eggs, one way the dominant foundress monopolizes direct reproduction is by eating their eggs. This behavior, first observed by Heldmann (1936a), was termed "differential oophagy" by Gervet (1964a,b). A reproductively dominant foundress is able to recognize eggs derived from other females, although the mechanisms involved have not been completely analyzed. Gervet (1964a,b) postulated that a *Polistes* foundress learns the unique smell of her own eggs during oviposition or subsequent antennation. West-Eberhard (1969) proposed that these wasps can only recognize the age of an egg. In either case, the recognition odor seems to be volatile and short-lived, since eggs of other foundresses can be recognized for only a relatively short time after they are laid. Eggs of subordinate *P. fuscatus* are eaten within 40 minutes of deposition (West-Eberhard 1969), whereas *P. dominulus* eggs may be eaten as long as one day after being laid (Gervet 1964b). Eggs of subordinate *Mischocyttarus flavitarsis* are eaten on the same day (Litte 1979), and eggs of *M. labiatus* within 90 minutes of oviposition by subordinates (Litte 1981). Egg eating is not observed in species in which the dominant foundress is the sole egg-layer (West-Eberhard 1969, Jeanne 1979a).

A dominant foundress normally detects a newly laid egg of a subordinate within a short time because she spends most of her time on the comb and regularly inspects the cells. It has been observed that not all eggs laid by subordinates of several *Polistes* species are eaten (Gervet 1964a, Noonan 1981, Strassmann 1983). It remains unclear, however, whether these eggs go undetected by the dominant or whether she simply tolerates them. As a rule, differential oophagy only occurs immediately before the dominant foundress lays an egg, so she may not eat eggs unless she is ready to oviposit. Foundresses of some species prevent their associates from eating their eggs by guarding them after oviposition (West-Eberhard 1981, 1986), a behavior that can last up to three hours in the polygynous wasp *Metapolybia aztecoides*.

Oophagy may also occur in a quite different context of reproductive competition, that is, when a foundress usurps a foreign nest. Usurping foundresses of the stenogastrine wasp *Parischnogaster mellyi* eat not only eggs but also small larvae (Turillazzi 1986b).

Differential oophagy must be distinguished from nutritional oophagy, in which the eggs serve to provide nutrition for the larvae. For instance, when the first larvae of *Polistes dominulus* hatch they are fed with eggs. Sometimes the dominant foundress lays an egg, turns to retrieve it with her mandibles, chews it, and then feeds young larvae. During this period, even subordinate foundresses of this species have been observed to eat eggs (Heldmann 1936a, Röseler and Röseler 1989), a behavior also observed in *P. annularis* (Strassmann 1983). Presumably, eggs are used as larval food when no arthropods are collected. Nutritional egg eating also has been observed in *Mischocyttarus drewseni* after eclosion of the first larvae (Jeanne 1972).

## Inhibition of Egg Formation

Although differential oophagy is an important means of monopolizing direct reproduction, a reproductively dominant foundress maintains her position primarily by inhibiting oogenesis in her associates. Thus, subordinates have less-developed ovaries than the dominant egg layer. But the inhibition is effective only in relatively small groups. The larger the associations become, the greater is the probability that there will be more than one foundress with developed ovaries (Litte 1977, 1981; Strassmann 1983).

Inhibition of oogenesis is accomplished by inhibition of gonadotropic juvenile hormone (JH) production. The titer in the hemolymph in adult hymenopterans is mainly regulated by synthesis and excretion, so that decreased synthesis of JH by the corpora allata (CA) results in a low hormone level in the hemolymph, and egg formation either is not induced or it is stimulated only at a reduced rate. Studies of *Polistes annularis* and *P. dominulus* have shown that the synthetic activity of the CA is lower in subordinate than in dominant females. The lower the position of a foundress in the hierarchy, the lower is the activity of her CA and the greater the inhibition of oogenesis (Röseler et al. 1980, Vawter and Strassmann 1982).

After a dominance hierarchy has become established a divergent development between foundresses takes place: endocrine activity and oogenesis in the dominant become elevated, while in subordinates these become increasingly diminished. Interestingly, the endocrine activity of the α-foundress is greater when more subordinates are present. The same is true for β- and γ-foundresses. Furthermore, an α-foundress of a multiple-foundress association has more-active CA than a lone foundress. The endocrine activity of an individual, therefore, is higher when more foundresses are arranged beneath her. Thus the existence of a dominance hierarchy not only inhibits endocrine activity and oogenesis in subordinates, but it also has a stimulatory effect on the reproductive physiology of more-dominant foundresses (Dropkin and Gamboa 1981, Turillazzi et al. 1982, Röseler et al. 1984).

The factors controlling endocrine activity and oogenesis are not yet clear. Dominance behavior alone seems to be insufficient for maintaining reproductive dominance. Ovariectomized foundresses of *P. dominulus* are able to achieve social dominance and maintain it. Although these foundresses dominate their associates in the same manner as do α-foundresses with ovaries, the β-foundresses are able to lay eggs and, indeed, their egg-laying capacity corresponds to that of unmutilated α-foundresses. The eggs are not eaten by the ovariectomized dominant, but are tolerated (Röseler and Röseler 1989).

Several factors related to task allocation are likely to contribute to the inhibition of oogenesis in subordinates and to the increased reproductive activity of dominants. The energy-expensive foraging flights are performed by subordinates, whereas the α-foundress seldom leaves the nest. Subordinates, therefore, rarely lay eggs during the period when larvae are present, as they must frequently collect food. Moreover, the dominant holds an important nutritional advantage because of inequities in trophallaxis and, to some extent, because of differential egg eating.

Empty cells stimulate females to oviposit (Deleurance 1950, Gervet 1964b, West-Eberhard 1986). By laying eggs in all cells immediately after their construction the dominant effectively prevents oviposition by subordinates. The dominant foundress of some species defends regions with empty cells against competing foundresses (West-Eberhard 1986). This lack of oviposition opportunities probably also leads to ovarian regression in subordinates.

In the highly eusocial swarm-founding polistines and vespines, chemical queen recognition and suppression of reproduction almost certainly occur (West-Eberhard 1977a, Jeanne 1980a, Fletcher and Ross 1985, Bourke 1988, Spradbery, this volume). However, there is no indication yet that chemical cues are involved in the control of reproduction in primitively eusocial wasps as well. The observation that inhibition of egg formation is relatively less effective in large foundress groups certainly does not support the widespread involvement of such pheromones. At present, it seems that several factors could affect the maintenance of reproductive dominance, but the exact contribution of each of these factors in various taxa is still poorly understood.

## ESTABLISHMENT OF A DOMINANCE HIERARCHY

In multiple-foundress associations formed by absconding swarms or during reconstruction of nests, a previously existing hierarchy may be maintained. But when a new association is formed at the time of nest initiation a hierarchy among the foundresses must be newly established. A lone foundress that has started a nest tries to defend the nest and to drive away other foundresses. The resident foundress has the advantage of nest ownership and of greater endocrine and ovarian development. Therefore, as a rule, a foreign foundress can join only as a subordinate. When a joiner does become the principal egg layer the process is considered to be usurpation.

Foundresses of some *Polistes* species return to the vicinity of their natal nest to start a new nest (philopatry); thus, the probability is high that former nestmates (i.e., relatives) group together (e.g., Reeve, this volume). Philopatry is thought to be the only cause of grouping of former nestmates of *P. dominulus* (Pratte 1982), whereas, because nonnestmates of *P. annularis* that try to join a group are severely attacked (Strassmann 1983), some form of recognition and discrimination is implicated in this species. Associations of former nestmates have also been observed in *P. fuscatus* (West-Eberhard 1969, Noonan 1981, Post and Jeanne 1982a) and *P. erythrocephalus* (West-Eberhard 1969), as well as in *Mischocyttarus drewseni* (Jeanne 1972), *M. mexicanus* (Litte 1977), and *Ropalidia fasciata* (Itô et al. 1985).

Experimental studies have shown that foundresses of several *Polistes* species are able to discriminate between nestmates and nonnestmates and to recognize kin by specific odors (Gamboa et al. 1986b). Such abilities enable these wasps to decide which foundress group to join or, alternatively, which females to exclude from a group. Surprisingly, nestmate recognition has also been reported among queens of *Dolichovespula maculata* (Ryan et al. 1985), a species in which it would seem to serve no purpose because nests are started by single queens and remain monogynous. Ryan et al. suggested that nestmate recognition abilities could help a usurping queen find nests of unrelated queens, which she might preferentially usurp.

Dominance hierarchies are established on new nests after a period of occasional severe fighting. During that time, the subordinated foundresses of some *Polistes* species switch among associations, so the composition of a group can change daily (West-Eberhard 1969, Pratte 1980, Noonan 1981, Strassmann 1983). After a hierarchy has been established, however, the associations become stable and overt fights are never observed. The foundresses have differentiated into egg-laying dominants (queens) and more-or-less sterile helpers (auxiliaries, workers).

Caste differentiation in primitively eusocial wasps may occur at different stages. At the brood stage, females can develop into either workers or potential foundresses (gynes), depending to some extent on the quantity and probably also the quality of food obtained as larvae (Strambi 1985, see also Gadagkar et al. 1988, 1990b and Hunt, this volume). In temperate-zone forms, only potential foundresses are able to overwinter and to start a new nest. At the adult stage, reproductive competition among associated overwintered foundresses leads to imaginal differentiation into egg-laying queens and workerlike helpers. The main question about imaginal caste differentiation is, What factors are responsible for determining which female will become the dominant egg layer?

Investigations of *Polistes* foundresses in which body size was measured and the sizes of the ovaries and corpora allata (CA) were used as measures of physiological activity have been carried out at different points during nest development. There is a tendency for the dominants of many *Polistes* species to be larger than their subordinates (Turillazzi and Pardi 1977, Dropkin and Gamboa 1981, Noonan 1981, Strassmann 1983, Sullivan and Strassmann 1984). Differences in body size, of course, favor a large female in aggressive interactions. But there also are associations in which the dominant foundress is not the largest wasp; about 50% of associations of *P. annularis* (Strassmann 1983) and 30% of *P. dominulus* (Turillazzi and Pardi 1977) are of this sort. Large body size, therefore, is not an absolute predictor of attainment of the dominant position. Moreover, large dominants do not necessarily have more subordinates than do small dominants (Sullivan and Strassmann 1984).

Dominant foundresses are characterized by having better-developed ovaries and more-active CA than their subordinates. However, these differences among foundresses have generally been observed only after the hierarchy has been established. Therefore, recent studies of *P. dominulus* have been directed toward elucidating whether these differences are the consequence of the hierarchy or whether they may be causally involved in its formation (Röseler et al. 1984, 1985). In these studies, dominance among foundresses was tested at their very first meeting after hibernation. Immediately after the dominance relationship between two foundresses was established, morphological and physiological measurements were taken. The data show that body size does not play a significant role in eventual foundress ranking, but that the attainment of dominance is positively associated with the activity of both the CA and ovaries. For instance, 91 and 83% of foundresses that became dominants possessed larger CA and oocytes, respectively, than did their subordinates. Also, when foundresses of a group were tested in pairs, a linear hierarchy reflecting the activity of their CA was established (Fig. 9.3). Aggressive interactions between two foundresses were invariably most severe when the combined activity of their CA was great, but foundresses with small, inactive glands immediately adopted submissive postures when paired with foundresses with more-active glands.

The CA synthesize juvenile hormone (JH). The ovaries produce ecdysteroids, a portion of which is released into the hemolymph, leading

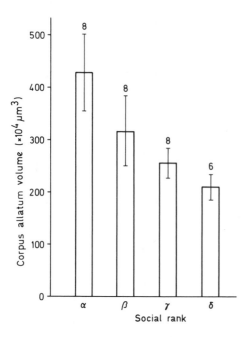

**Fig. 9.3.** Volume of the corpora allata in differently ranked foundresses at the time of dominance hierarchy establishment in foundress groups of *Polistes dominulus*. Numbers above each column indicate sample size; bars indicate standard deviations.

to increases in hemolymph ecdysteroid levels during oogenesis (Strambi et al. 1977). Thus, foundresses that achieve the dominant position have higher titers of JH and ecdysteroids in their hemolymph than subordinated foundresses. The influence of both hormones on dominance behavior can be demonstrated by hormone treatment (Röseler et al. 1984). More foundresses with relatively small CA and oocytes became dominant after they had been injected with JH I, 20-hydroxyecdysone, or both simultaneously, than did controls (see Fig. 9.4 for effect of JH I on dominance). The hormones seem not to act synergistically, since the influence of the two together is not greater than the influence of either alone. A relative deficiency of hormones can be compensated for only if the natural endocrine activity is not too low. Thus, foundresses with very inactive CA (e.g., parasitized foundresses) cannot achieve the dominant position even after hormone treatment.

At present, the specific roles of juvenile hormone and 20-hydroxyecdysone in governing dominance behavior remain unclear. JH treatment of *P. annularis* workers results in oocyte growth and in elevated levels of dominance interactions (Barth et al. 1975), as it does in *P. dominulus* queens. On the other hand, aggressive behavior of *P. canadensis canadensis* does not depend on ovarian development and is exhibited only by mated foundresses (West-Eberhard 1986), suggesting that factors released by mating (prostaglandins?) might be involved (Loher et al.

**Fig. 9.4.** Influence of juvenile hormone on dominance in *Polistes dominulus*. When two foundresses meet at the nest site after hibernation, the foundress with the higher corpus allatum activity usually becomes dominant. The greater the difference in hormonal activity (greater + values), the higher the probability that the foundress with more-active glands will achieve the dominant position (hatched bars). Foundresses with relatively lower endocrine activity (− values) only seldom become dominant (hatched bars). Injection of juvenile hormone I (white bars) increases the probability of achieving the dominant rank for foundresses with relatively small corpora allata, but only if the activity of this gland is not too much lower than that of the other foundress. Numbers next to the bars are sample sizes (pairs of wasps).

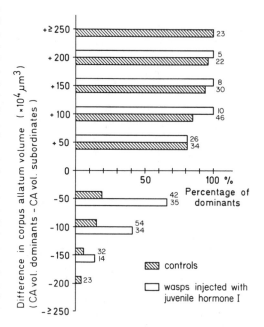

1981). In ovariectomized foundresses of *P. dominulus*, the ecdysteroid titer in hemolymph remains low, but these foundresses nonetheless can display clear dominance behavior (Röseler et al. 1985). However, dominance of ovariectomized foundresses is much more strongly associated with the activity of the CA than it is in unmutilated foundresses, an effect observed recently in bumble bee workers as well (Doorn 1989). It seems, then, that dominance behavior is influenced by several endocrine factors, and that a deficiency in one system can be compensated for by overproduction in another.

Endocrine activity rapidly increases after termination of hibernation in *P. dominulus*, and foundresses with the highest hormone levels at the time of group formation become dominant. It is, therefore, advantageous for a foundress to leave the hibernaculum as early as possible and to arrive first at the nesting site. In a recent study, *P. dominulus* foundresses were exposed to sunlight and elevated temperatures for one hour daily during the last ten days of overwintering. Foundresses that received this treatment had significantly larger CA than did control foundresses and a probability of achieving the dominant rank estimated at 86% (Röseler 1985).

Why do hormonal differences exist among foundresses at the end of hibernation? Studies of *P. dominulus* foundresses indicate that the activity of the CA may change periodically during overwintering (Strambi 1969), and hibernating foundresses differ in the levels of total protein in their hemolymph (Strambi et al. 1982). On the basis of these results we can postulate that foundresses with inherently elevated endocrine activity exhibit a more sensitive response to changes in the environment and leave hibernation earlier than foundresses with low endocrine activity. Such physiological differences among foundresses may be genetically controlled or they may be induced by larval feeding; they may also be influenced by date of adult emergence in the fall or by other events during the prehibernation period. However, it remains to be shown definitively that these factors ultimately have any influence on the establishment of dominance in the spring.

## MAINTENANCE OF DOMINANCE

Once a stable hierarchy has been formed, aggressive interactions among foundresses occur rarely. Subordinated foundresses immediately behave submissively when they encounter a dominant foundress. The foundresses appear to recognize each other by brief antennations, and it seems that they also recognize each individual's rank in the hierarchy relative to their own (Pardi 1947, Jeanne 1972, Downing and Jeanne 1985).

A dominant *Polistes* foundress spends most of her time on the comb, but in some cases permanent attendance on the comb is not necessary to maintain social dominance. For instance, when the α-foundress was removed for up to 24 hours from 14 four-foundress nests of *P. dominulus*, this individual invariably became dominant again upon her return (Röseler et al. 1986). The α-foundress of one colony even regained the top rank after she had been absent for seven days. The longer the foundress is absent, however, the more severe and the more frequent are her interactions with subordinates when she returns.

Other studies of *P. dominulus* have confirmed that frequent acts of domination are not required for retention of dominance. In one experiment, α-foundresses could remain dominant for seven days after their first domination act even if they were permitted to dominate the β-foundresses only once a day on days 1, 4, and 7 (Röseler et al. 1986). A change in dominance occurred in only 20% of the cases. It seems that the queens of other species, however, must continually contact associates to subordinate them (confrontational dominance), and aggressive interactions may persist until the postemergence period (Itô 1985a, West-Eberhard 1986).

Results of these studies of *P. dominulus* suggest that the experience of the first domination by the α-foundress probably is imprinted on a subordinated foundress so that she will immediately behave submissively in all future encounters with the dominant (nonconfrontational dominance). The cue for individual recognition might be the individual's odor, which the subordinates could learn while being intensively mouthed during the first encounter. Downing and Jeanne (1983) have found that dominant foundresses of *P. fuscatus* often have larger and better developed abdominal exocrine glands than subordinates, so that the products of these glands may be involved in communication of dominance (Downing, this volume). The same authors have performed experiments in which they tested the reaction of β-foundresses toward dead or cooled dominant or subordinate foundresses and toward single parts of their bodies. They conclude that chemical cues from the head and ovaries might communicate status in *P. fuscatus* (Downing and Jeanne 1985). In *P. dominulus* chemical cues related to the development of the ovaries seem not to be the only cues that contribute to maintaining social dominance, since ovariectomized foundresses are still able to maintain the α-position until the postemergence period (Röseler and Röseler 1989).

## REPRODUCTIVE SUCCESS

Foundresses can transmit their genes to the next generation by direct means (egg laying) or by indirect means (helping relatives). Given that workers are generally sterile, direct reproductive success must be considered in terms of the reproductive offspring an individual produces. In many species, eggs laid in the beginning of the colony cycle will not develop into reproductives, but rather into sterile workers, so these early eggs are probably of little consequence in terms of reproductive competition. The probability that the reproductives reared by a colony will survive until initiation of new nests is generally not known, nor is the degree of relatedness among foundresses in most cases (Ross and Carpenter, this volume). Thus, it is necessary to make many assumptions to estimate the relative genetic success (inclusive fitness) of foundresses in multiple-foundress groups. Calculations of gains and losses in inclusive fitness in *Polistes* have been presented by several authors and are summarized by Reeve (this volume). General reviews of inclusive fitness estimation, particularly with reference to social wasps, can be found in West-Eberhard (1975), Grafen (1984), Queller and Strassmann (1988, 1989), and Strassmann and Queller (1989). In the following discussion only direct reproduction during the period of colony establishment is considered, regardless of the sex and caste of the offspring.

The dominant foundresses of some species completely monopolize direct reproduction. Furthermore, their reproductive capacity may be enhanced by exploiting their sterile subordinates; they seldom undertake the energy-expensive foraging flights, they eat the eggs laid by subordinates, and they receive a disproportionate share of food during trophallaxis. Therefore, dominant foundresses can invest most of their energy in egg formation and reach a high level of egg production.

Subordinates of other species may reproduce directly to some extent, which may markedly increase their inclusive fitness if their eggs are reared as reproductives. Noonan (1981) has estimated that 33% of all surviving eggs in the late preemergence period of *P. fuscatus* are laid by subordinates. However, little is known about the variability in the contributions of individual foundresses, although the probability of successful direct reproduction clearly is diminished for low-ranking females in larger associations (e.g., Strassmann 1981a). Subordinates of several species may have some chance of taking over the dominant role and reproducing directly if the α-foundress disappears or loses dominance. In most *Polistes*, however, such a situation is rather hypothetical: the α-foundress, spending most of her time on the comb, is well protected from predation; a change in dominance has seldom been observed to occur; and, moreover, subordinates are sometimes driven from the nest at the time the first workers emerge (Pardi 1948, Gamboa et al. 1978, Pfennig and Klahn 1985, Röseler 1985).

Multiple-foundress nests of primitively eusocial wasps generally have a higher survival rate than single-foundress nests of the same species. Also, the total reproductive output of a colony is increased when several foundresses cooperate, probably in part because in associations with only one egg layer the upper limit of growth is determined by the reproductive capacity of that egg layer. For instance, a single female of *P. dominulus* can lay at most nine eggs per day in the postemergence period (Röseler 1985). Turillazzi et al. (1982) have shown that the number of cells constructed as well as the number of offspring reared by this species increase with the number of foundresses. Similarly, the nest growth rate of *P. canadensis canadensis* is greater when more foundresses are present (Jeanne 1979a), as is that of *Mischocyttarus* (Jeanne 1972, Litte 1977).

Apparently, there is an optimal number of foundresses for achieving the highest reproductive output. As a rule, the maximum reproductive output of colonies with a single egg layer is reached in associations of two to four foundresses (*P. fuscatus*: West-Eberhard 1969, Gibo 1978; *P. erythrocephalus*: West-Eberhard 1969; *P. dominulus*: Röseler 1985; *M. labiatus*: Litte 1981). Larger groups do not appear to have higher reproductive rates, although exceptions may exist in some unusually large associations of *P. fuscatus* (West-Eberhard 1969). This limit to the benefits of increasing numbers of foundresses has also been observed in *Ropalidia* species (*R. fasciata*: Itô 1983; *R. marginata*: Gadagkar et al. 1982a).

## CONCLUDING REMARKS

In wasp species whose nests are typically established by single foundresses (haplometrosis), reproductive competition occurs only when
numerous foundresses compete for nest sites. However, in the multiple-foundress (pleometrotic) nests of many species, reproductive selfishness regularly leads to conflict. Foundresses compete for direct reproduction by more or less intense agonistic interactions, the most
severe of which may result in injuries. Because the survival of the individuals and productivity of the colonies may be threatened by continued overt conflicts, aggressive interactions generally are limited to a
short period and are finally suppressed.

One way to avoid aggressive interactions is to tolerate egg laying by
others. Another way is to inhibit the reproductive efforts of others by
some means. In multiple-foundress groups of nonswarming species, a
reproductive dominance hierarchy usually develops in which one foundress becomes the principal egg layer. The egg-laying dominant of some
species (e.g., *Polistes canadensis*: Itô 1985a, West-Eberhard 1986) must
continue to suppress subordinates by direct physical contact, although
the frequency of interactions diminishes somewhat with time. In other
species (e.g., *Polistes dominulus*), overt competition is greatly reduced
through time so that reproductive dominance is maintained by a ritualized threat display. In any case, overt aggression is largely restricted
to the period of group establishment in most species, although it is
exhibited anew when the dominant dies or is lost. Aggressive interactions among foundresses of swarm-founding species probably do not
arise during nest foundation because the previous hierarchy is likely to
have been maintained.

Reproductive dominance is connected to social dominance. Dominance behavior is induced by the same element of the endocrine system that regulates oogenesis, so the foundress with the highest endocrine activity achieves both behavioral and reproductive dominance in
the social group. This association between reproduction and dominance behavior by means of endocrine activity apparently results in
selection for dominant foundresses with a high reproductive capacity,
as has previously been hypothesized (West-Eberhard 1967, 1981).

The evolution of dominance hierarchies is only possible once submissive behavior has evolved (Pratte and Gervet 1980), because dominance
behavior alone leads inexorably to elimination of all but one competitor, as is seen in usurping vespine queens. Thus a major unsolved
problem is why some foundresses in primitively eusocial forms exhibit
subordinate behavior and help others instead of starting their own
nests. Gervet (1964b) and West-Eberhard (1967) first postulated that
subordinates are foundresses with inferior reproductive potential that

would not be able to rear offspring as lone foundresses and would thus benefit most by helping a related foundress with a high reproductive capacity. To date, however, there is no evidence that foundresses differ inherently in reproductive potential (*Polistes fuscatus*: Bendegem et al. 1981; *P. dominulus*: Röseler 1985; *P. annularis*: Queller and Strassmann 1988, Strassmann and Queller 1989; *P. bellicosus*: Queller and Strassmann 1989). On the basis of these studies and experimental work on the reproductive physiology of *P. dominulus*, we can hypothesize that dominance is not conferred on foundresses with a high reproductive potential per se, but rather on individuals that happen to have the highest endocrine and reproductive capacity at the time of nest foundation.

A subordinate foundress gains inclusive fitness through the indirect component only if she helps a closely related foundress. Return to the natal nest site and the ability to recognize former nestmates (or even kin) facilitates the association of subordinates with their relatives (Gamboa et al. 1986b, 1987). The natal nest site might be preferred because of its proven suitability for successfully rearing offspring (protection of nest, favorable microclimate, availability of food). Foundresses, therefore, may be expected to reserve that site for related foundresses only and to reject nonrelatives.

Dominant foundresses benefit most from group nesting and are therefore generally willing to accept other foundresses as subordinates. Indeed, haplometrotic foundresses of *P. metricus* may even seek out subordinates by deserting their own successful nest, usurping another, and retaining the usurped foundress as a subordinate (Gamboa and Dropkin 1979). The benefits to nest owners of having helpers during the critical preemergence period are such that they may even tolerate some egg laying by subordinates. To avoid the risk that a nonrelated subordinate could reproduce in such circumstances or even supersede the dominant, foundresses of some species chase away such individuals or allow them to join only at a low position in the hierarchy (e.g., *P. annularis*: Strassmann 1983). Subordinates of other species may be driven away from the nest when the first offspring emerge, to prevent them from laying eggs that could be reared as reproductives (*P. fuscatus*: Pfennig and Klahn 1985; *P. dominulus*: Pardi 1948, Röseler 1985; *P. metricus*: Gamboa et al. 1978). Subordinated egg-laying queens of the swarm-founding *Metapolybia aztecoides* are tolerated in young nests only until sufficient worker brood is present to ensure the success of the colony (West-Eberhard 1978b).

It seems that the problem of reproductive competition is resolved in different ways among the various groups of eusocial wasps. In most Vespinae, competition results in the elimination of foundresses; in independent-founding Polistinae it results in the formation of multiple-foundress groups. In multiple-foundress groups, the surplus found-resses rear offspring so that the probability of successful nest develop-ment is markedly increased. Many gaps remain in our knowledge of wasp societies, and further research will be required to determine more precisely the mechanisms by which social interactions in multiple-found-ress associations are coordinated and regulated. These mechanisms hold the key to understanding how reproductive competition molds social organization during the early development of wasp colonies.

## Acknowledgments

I am very grateful to H. A. Downing, R. W. Matthews, K. G. Ross, C. K. Starr, and M. J. West-Eberhard for greatly improving the manuscript with their comments.

# 10

# Evolution of Queen Number and Queen Control

J. PHILIP SPRADBERY

"All social insects have evolved from solitary ancestors via a long history in which the development of reproductive dominance by one or a few individuals is a major theme" (West-Eberhard 1981:14). Some of the most basic questions relating to the social biology of wasps concern the mechanisms of queen control and the regulation of queen number, both key elements in the evolution of eusociality in vespid wasps.

Some 100–200 million years ago the stage was set for the extraordinary evolutionary events that resulted in the almost bewildering array of forms of social organization found in wasp communities today. The richness of this diversity, from solitary to highly eusocial, is such that theories of the genetic and environmental determinants of social evolution find compelling support in examples from the contemporary wasp fauna.

There has also been an evolution in our appreciation and understanding of wasp biology, reflected in the change of theories proposed to explain social behavior. In the mid-twentieth century, our knowledge of wasp biology was based largely on studies of the temperate-zone fauna of the Northern Hemisphere. Today, there is an accelerating accumulation of knowledge about the presocial and primitively eusocial wasp communities that occur in the tropical and subtropical regions of the world.

I begin this chapter with an account of the classical studies of Leo Pardi on social hierarchies of *Polistes* wasps in Italy. A survey of queen control and queen number in the primitively eusocial Stenogastrinae and the remaining Polistinae follows. I have followed an essentially

336

phylogenetic framework and, where possible, have tabulated the rapidly accumulating data on the subject. A review of queen number and queen control in the highly eusocial Vespinae concludes the descriptive component of the chapter. Finally, a synthesis of the evolutionary aspects of queen number and control is presented, together with a scenario for the evolution of eusociality in vespid wasps.

## QUEEN CONTROL IN *POLISTES*

### Dominance Hierarchies: A Case Study

In the 1940s, Leo Pardi published his classic account of dominance hierarchies in the European paper wasp *Polistes dominulus* ( = *gallicus* of authors) (Pardi 1942a, 1946, 1947, 1948), and it remains the definitive, pioneering study of queen control in primitively eusocial wasps. In Italy, nests of *P. dominulus* are initiated in the spring, frequently by several females that coexist until the first workers eclose. During this preemergence period, one female, the α-individual or queen, remains on the nest, lays eggs, does less building than her associates, and dominates them. The other females, the auxiliaries, forage more, construct the nest, lay fewer eggs, and are subordinate to the principal egg layer. Before long, the ovaries of the auxiliaries decrease in size. After the first workers emerge (postemergence period), the auxiliaries are finally driven from the nest, and the colony consists of the remaining α-foundress and her workers.

When two founding females (or two workers) make contact on the nest, one usually displays a dominant behavior, while the other assumes a submissive posture (Reeve, Röseler, this volume). It is rare for two females to be exactly equivalent, thus in every encounter there is invariably a dominant-subordinate reaction. If two individuals are roughly equivalent in status, the behavioral interaction between them generally is violent.

The net result of such interactions among groups of foundresses or workers on a nest is the development of a linear hierarchy in which each female occupies a specific rank in the dominance scale. The hierarchy is headed by the α-female, which is dominant over all others, the β-female, which is dominant over all except the α-female, and so on. The frequency of dominance behavior between pairs of wasps depends on the social status of the individuals; an individual most often dominates the one that occupies the place immediately below it. Social status can change if a female dies or if new individuals are added. For instance, disappearance of the queen or α-individual is quickly perceived by the β-female, which promptly displays highly aggressive behavior to stake her claim to the α-position. In this way, a succession of wasps can become α-females (Fig. 10.1).

The physiological status of the ovaries contributes to dominance, such that the ovaries of the dominant individual in any contact are almost always more developed than the ovaries of the subordinate. The ovaries of wasps have age-related developing and regressing phases, and, all things being equal, the female with her ovaries in a developing phase will normally dominate the one with regressing ovaries.

The establishment of the dominance hierarchy results in (1) a distinct trophic advantage for the dominants, which ingest more food during the frequent trophallactic exchanges with the less dominant females, and (2) dominants remaining on the nest for longer periods and doing less foraging than the subordinates. The overall consequence of these differences is that the better-fed, infrequently foraging dominants develop larger ovaries at the expense of the subordinates, which become increasingly sterile. Thus in a foundress association, the α-female always lays the majority of eggs.

Most species of *Polistes* studied so far exhibit the same or similar elements of dominance behavior as described by Pardi (see Table 10.1, Fig. 10.2a, also Reeve, this volume), although the exact structure of the hierarchy may vary. *Polistes canadensis* is an extreme case, in that the queen exercises such complete dominance over all the other females that dominance interactions among subordinates are rare (West-Eberhard 1986). Thus no linear hierarchy can normally develop. West-Eberhard (1986) showed that the queen of this species exerts her control by overt physical dominance and that her sphere of influence is limited by her ability to interact directly with other members of the colony.

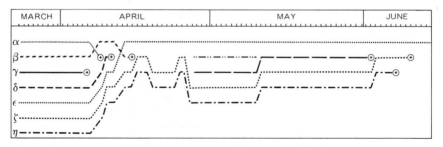

**Fig. 10.1.** Dominance relationships during a three-month period in a colony of *Polistes dominulus*. α–η, dominance ranks of individual foundresses; +, death/loss of female. Two females that were not members of the original association joined the colony in late April. (Redrawn from Pardi 1948, *Physiological Zoology* 21: 1–13, © 1948 by The University of Chicago, courtesy of the author and The University of Chicago.)

**Table 10.1.** Colony foundation, colony composition, and social organization in *Polistes* species

| Species | Locality | No. of foundresses | Percent of nests pleometrotic (N) | No. of laying females | Dominance behavior | Oophagy[a] | Reference |
|---|---|---|---|---|---|---|---|
| *P. annularis*[b] | Texas, USA | 1–28 | 92(25) | 1 + subordinates | — | — | Strassmann 1983, Sullivan and Strassmann 1984 |
| *P. apachus* | Oklahoma, USA | 2+ | — | — | — | — | Strassmann 1983 |
| *P. bernardii richardsi* | Australia | 5–8 | 100(2) | 3 of 5 | ?Absent | — | Itô 1986c |
| *P. biglumis bimaculatus* | Alpine Europe | 1(rarely 2) | 0 | 1 | Aggressive | — | Deleurance 1946, Lorenzi and Turillazzi 1986 |
| *P. canadensis* | Brazil | 1–29($\bar{X}$ = 9.1) | 68(12) | 1 | — | — | Jeanne 1979a |
| *P. carnifex* | Mexico | 3 | — | — | — | — | Rau 1940 |
| *P. chinensis* | Japan | ?1 | 0 | 1 | Linear | — | Morimoto 1961a,b |
| *P. dominulus* | Marseille, France; Pisa, Italy | 2–7 | 82(17) | 1 | Linear | + | Pardi 1942a, 1948; Gervet 1964a |
| *P. dominulus* | Torino, Italy | 1 | 35(29) | 1 | — | — | Turillazzi et al. 1982 |
| *P. erythrocephalus* | Colombia | 1–13($\bar{X}$ = 4.9) | 98(90) | Serial polygyny | Nonlinear | 0 | West-Eberhard 1969, 1986; Itô 1987b |
| *P. exclamans* | Oklahoma, USA | 1 | — | 1 | — | — | West-Eberhard 1968 |
| *P. exclamans* | Texas, USA | 1 or 2 + | — | — | — | — | In West-Eberhard 1968, Strassmann 1981d |
| *P. fuscatus* | Michigan, USA | 1–7 | 61(46) | 1 + subordinates | Cell guarding, linear | + | West-Eberhard 1969, Noonan 1981 |
| *P. fuscatus* | Wisconsin, USA | 3–4 | 43(113) | 1 | — | — | Gibo 1978, Downing and Jeanne 1985 |

**Table 10.1**—*continued*

| Species | Locality | No. of foundresses | Percent of nests pleometrotic (N) | No. of laying females | Dominance behavior | Oophagy[a] | Reference |
|---|---|---|---|---|---|---|---|
| P. humilis | Australia | 1 or 16–36 | 0 or 100[c] | 1 or 9–14 | Absent | — | Itô 1986d |
| P. instabilis | Mexico | 1 or 2–7 | 33(12) | — | — | — | Rau 1940 |
| P. jadwigae | Japan | 1 or 2–6 | Temporary pleometrosis | 1 | Absent | — | Yoshikawa 1957 |
| P. metricus[b] | Kansas, USA | 1 or 2–5 | 38(119) | 1 | Linear, aggressive | — | Gamboa 1978, Gamboa and Dropkin 1979, Dropkin and Gamboa 1981 |
| P. nimpha | Italy | 1 or 2+ | 15(211) | 1 | Linear | — | Cervo and Turillazzi 1985 |
| P. riparius (= biglumis) | Northern Japan | 1 | 0(10) | 1 | — | — | Yamane 1969 |
| P. snelleni | Northern Japan | 1 | 1(100) | 1 | — | — | Yamane 1969 |
| P. spilophorus[b] | Africa | 1–9 | 50(6) | — | — | — | Richards 1969 |
| P. tenebricosus hoplites | Sumatra | 1–5 | 25(12) | — | — | — | Abbas et al. 1983 |
| P. tepidus | New Guinea | 1–7 | 78(18) | 1? | — | — | Yamane and Okazawa 1977 |
| P. versicolor | Panama | 3–7 | 78(23) | — | Mild | — | Itô 1985a |

*Note:* A dash indicates no data available.
[a] +, present; 0, absent.
[b] Caste dimorphism recognized.
[c] New nests established by solitary queens (haplometrosis); old nests reused in spring by groups of overwintered queens (pleometrosis).

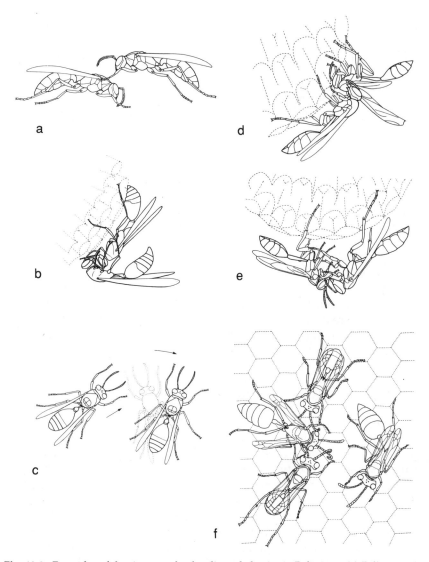

**Fig. 10.2.** Examples of dominant and subordinate behavior in Polistinae. (a) *Polistes versicolor*. Submissive posture of subordinate female (left) (redrawn from Itô 1985a, *Zeitschrift für Tierpsychologie* 68: 152–167, courtesy of the author and Verlag Paul Parey, Berlin and Hamburg). (b) *Ropalidia fasciata*. Food sharing between two foundresses—the "kiss" (redrawn from Itô 1983, courtesy of the author and *Journal of Ethology*). (c) *R. fasciata*. "Dart and stop short without direct contact," a mild dominance act (redrawn from Itô 1985a, *Zeitschrift für Tierpsychologie* 68: 152–167, courtesy of the author and Verlag Paul Parey, Berlin and Hamburg). (d) *Mischocyttarus drewseni*. Wasp at left chewing thorax of subordinate, which adopts a submissive posture (redrawn from Jeanne 1972, courtesy of the author and Museum of Comparative Zoology, Harvard University). (e) *M. drewseni*. Queen (right) soliciting nectar from forager (redrawn from Jeanne 1972, courtesy of the author and Museum of Comparative Zoology, Harvard University). (f) *Metapolybia aztecoides*. Queen (right) performing an aggressive "bending display" while two workers (left) "dance" at a second queen (between them) and touch her with their antennae (redrawn from West-Eberhard 1978b, *Science* 200: cover, 441–443, © 1978 by the AAAS, courtesy of the author and the American Association for the Advancement of Science).

341

## Differential Oophagy

During the early stages of multiple-foundress associations of *Polistes*, before ovaries of the auxiliaries regress, lower ranking females frequently lay eggs. These eggs are recognized by the dominant female, which invariably eats them (e.g., Table 10.2), a phenomenon termed "differential oophagy" (Pardi 1942a, Gervet 1964a). Recognition of eggs laid by rivals is closely related to the age of the egg, which is presumably marked with a specific odor (West-Eberhard 1969, Downing and Jeanne 1983). Dominant females are more efficient than subordinates in seeking out and destroying competitors' eggs. This is probably related to their greater supply of mature ova, for once an egg is eaten the dominant female invariably replaces it with one of her own (West-Eberhard 1969). It is relatively rare for eggs laid by the queen to be eaten by subordinate co-foundresses, while sterile females (workers) never practice oophagy. Gervet (1956) found that cooling the queen overnight reduced her oviposition rate, and the ovarian activity of the auxiliaries increased as the number of empty cells increased. However, there was no change in dominance status, as the queen practiced oophagy to maintain her dominant position in the hierarchy.

## Cell Building and Other Activities

An important role of the primary egg layer or queen in a *Polistes* colony is the control of cell-building rates. If the queen of *P. erythrocephalus* or *P. fuscatus* is lost or stops ovipositing, new cell construction stops completely (West-Eberhard 1969). Queens of these species act as the "primary initiator" of new cells, with 23 of 26 and 22 of 28 new cells begun by the queen of *P. erythrocephalus* and *P. fuscatus* colonies, respectively. Cell initiation behavior is probably stimulated by the numerous mature eggs in the ovaries of queens. Dew (1983) found that queens of *P. metricus* initiated more new cells (0.15 cells per queen per hour) than did the workers (0.01 cells per worker per hour).

*Polistes* queens are much more active than other members of the colony and initiate many more interactions than the average worker. They thereby regulate the level of colony activity by changes in their own activity; that is, regulation is accomplished by behavioral control rather than control by chemical pheromones (Reeve and Gamboa 1983, Reeve, this volume). If a queen is removed from the nest, activity is greatly diminished; returning a queen that has been cooled to the nest also causes a decrease in the synchronized activities of the workers. The queen apparently directs her aggressive actions more toward the inactive workers than toward the active members of the colony, thus stimulating their participation in the activities of the colony.

**Table 10.2.** Frequency of oviposition and oophagy acts by dominant (α) and subordinate (sub.) females of four polistine species

| Species | Stage | No. of nests | Oviposition | | | Oophagy | | | Reference |
|---|---|---|---|---|---|---|---|---|---|
| | | | By α | By sub.[a] | % by α | By α | By sub. | % by α | |
| *Polistes dominulus* | Preemergence | 2 | 55 | 31[b] | 64 | 31 | 7[b] | 82 | Pardi 1942a |
| | Preemergence | 5 | — | — | — | 41 | 6[b] | 87 | Gervet 1964a |
| | Postemergence | 2 | 131 | 69[b] | 66 | 164 | 36[b] | 82 | Pardi 1946 |
| *P. fuscatus* | Preemergence | 1 | 9 | 5 | 64 | 4 | 2 | 67 | West-Eberhard 1969 |
| *Ropalidia fasciata* | Preemergence | 4 | 3 | 4 | 43 | 0 | 1 | 0 | Itô 1987b |
| | Postemergence | 23 | 11 | 10 | 52 | 1 | 6 | 14 | Itô 1987b |
| | Total | 27 | 14 | 14 | 50 | 1 | 7 | 13 | Itô 1987b |
| *Belonogaster grisea* | Preemergence | 1 | 43[c] | 3 | 93 | 20 | 0 | 100 | Pardi and Marino Piccioli 1970 |

*Note:* A dash indicates no data available.
[a] At postemergence stage category includes foundresses and progeny females.
[b] By β-foundress only.
[c] Inseminated females.

## Physiological Basis

Because dominance rank is closely correlated with degree of ovarian development, factors that influence reproductive morphology and physiology can reasonably be assumed to play a role in queen control (see Röseler, this volume). Röseler et al. (1984) found that the most critical factor determining hierarchical position of *Polistes* females is the juvenile hormone (JH) titer in the hemolymph. In populations of overwintering *Polistes* females there appear to be slight differences in the size of the corpora allata (glands that produce JH) and the degrees of ovarian development, suggesting that differences in endocrine activity may exist before nest initiation (Röseler et al. 1984). Furthermore, because the endocrine system is activated during the first few days after overwintering, the females that emerge from hibernation earliest have an advantage in dominance contests when nests are initiated (Röseler 1985).

During the establishment of dominance hierarchies, endocrine activity in subordinates is inhibited by the dominant female, which then monopolizes oviposition. Thus, small differences in endocrine activity at the time of nest initiation become more and more pronounced during the course of early nest growth. Also, nutritional advantages accrue to the dominant during trophallactic exchanges and through differential foraging activities. The result is that the foundress with the highest level of endocrine activity, and thus reproductive capacity, becomes the sole or principal egg layer in the association (see Röseler, this volume).

Recognition of a dominant female by subordinates is likely to be mediated by an individual odor, learned by the subordinates during their early encounters. There is a suggestion that endocrine glands associated with the gastral sterna could be involved in chemical recognition and communication of dominance (Downing and Jeanne 1983); sternal gland secretions could be emitted during the characteristic lateral vibrations of the gaster by the queen that have been described in several *Polistes* species, as well as in *Ropalidia* and *Vespula* (see West-Eberhard 1982c). Dufour's gland secretion could possibly identify deposited eggs to circumvent oophagy (e.g., Downing, this volume).

## QUEEN NUMBER AND QUEEN CONTROL IN THE STENOGASTRINAE AND POLISTINAE

The following is a survey of the Stenogastrinae and Polistinae other than *Polistes* that will help set the stage for discussions of the evolution of queen number and queen control in eusocial wasps. Most of the

studies described here were published in the 1980s, and I will emphasize those groups, such as the Stenogastrinae and *Ropalidia*, that have,
until recently, received little attention. This survey highlights the close
interrelationship between number of queens and the manner in which
they control the activities of the colony.

## Stenogastrinae

The stenogastrine wasps range from species with temporary reproductive castes to those with more-advanced eusocial organization,
some with a hint of physical caste dimorphism. The group as a whole
is generally considered primitively eusocial (Carpenter 1988a, this volume). Most nests of the least-specialized species, such as *Stenogaster
micans* (Williams 1919), *Parischnogaster nigricans serrei* (Turillazzi 1985d),
and *S. concinna* (Spradbery 1975), have a single female, but occasionally
a second female, presumably a daughter, is found on the nest. It is
likely that these temporary associations prevent usurpation and help
reduce parasitism. For example, a returning female of *Anischnogaster
iridipennis* stimulates a violently aggressive response by the resident
female (Spradbery 1989). Similar protective behavior is exhibited by
*Parischnogaster striatula* (Hansell 1982b) and *P. mellyi* (Hansell 1982a), in
which nests are actively defended by colonies of fixed membership
against marauding conspecifics.

The great majority of stenogastrines found nests with a single female
(Table 10.3, also Turillazzi, this volume). Most mature colonies of Stenogastrinae have more than one resident female (Table 10.3), although
the individual contributions to oviposition are largely unknown. Ohgushi et al. (1983a) found that most females of *P. mellyi* had developed
ovaries, but generally one female had better developed ovaries than
the others and was probably the sole egg layer. Thus, while being ana-
tomically polygynous, *P. mellyi* is functionally monogynous (Yamane et
al. 1983a).

Turillazzi and Pardi (1982) found that the inseminated *P. nigricans
serrei* females they examined had slightly larger heads and better developed ovaries than the uninseminated females. Yoshikawa et al. (1969),
however, detected no morphological caste differences in any *Parischnogaster* species.

Mild dominance behavior unrelated to hierarchies or ovariole development has been described for *P. alternata* (Turrillazzi 1986a), while
classical linear dominance hierarchies are known in several *Parischnogaster* species (see Table 10.3, Fig. 10.3). The linear dominance hierarchy of these species is correlated with ovariole development, with a
dominant α-individual being the principal egg layer. She exhibits
strong dominance behavior and defends the nest, while the subordi-

**Table 10.3.** Colony foundation, colony composition, and social organization in the Stenogastrinae

| Species | Locality | No. of colonies sampled | No. of foundresses | No. of adults Females[a] | Males | Dominance behavior | Maximum no. of cells | Reference |
|---|---|---|---|---|---|---|---|---|
| Anischnogaster iridipennis | New Guinea | 16 | 1 | 1-2(1) | 1 | Absent[b] | 18 | Spradbery 1989 |
| Eustenogaster sp. 1 | Sumatra | 1 | 1 | 2(1) | 0 | — | 7 | Ohgushi et al. 1983a |
| E. calyptodoma | Sarawak, Malaysia | 1 | 1 | 1 | 0 | — | 7 | Sakagami and Yoshikawa 1968 |
| E. calyptodoma | Malaysia | 24 | 1 | 1-4(1-2) | 0-1 | Absent | 14 | Hansell 1987b |
| E. fraterna | Thailand | 6 | 1 | 1-4(1-3) | 0 | — | — | Yoshikawa et al. 1969 |
| Liostenogaster flavolineata | Sumatra and Malaysia | 54 | 1 | 1-7 | 2 | Partial dominance | 89 | Pagden 1958a, Hansell et al. 1982, Ohgushi et al. 1986 |
| Parischnogaster alternata | Sumatra | 9 | — | 1-4(1-4) | — | — | 18 | Yoshikawa et al. 1969 |
| P. alternata | Sumatra | 1 | — | 13 | 16 | — | 47 | Ohgushi et al. 1985 |
| P. alternata | Sumatra | 1 | — | 2 | 2 | — | 19 | Ohgushi et al. 1983a |

| Species | Locality | | | | | | Reference |
|---|---|---|---|---|---|---|---|
| *P. alternata* | Malaysia | 52 | 2–5 | 1–9(51%) | 0–6 | Very mild, no hierarchy | Turillazzi 1986a |
| *P. gracilipes* | Borneo | 17 | ?1 | 1–8(1–2) | 0–2 | — | Hansell 1986a,b |
| *P. jacobsoni* | Sumatra | 2 | 1 | 6(3) | 3 | Linear | Yoshikawa et al. 1969, Ohgushi et al. 1983a |
| *P. mellyi* | Thailand | 35 | 1 | 1–3(1) | 0–3 | ?Absent | Iwata 1967, Hansell 1982a |
| *P. mellyi* | Sumatra | 32 | 1 | 2(1) to 3(3) | — | Serial polygyny | Ohgushi et al. 1983a, Yamane et al. 1983a |
| *P. nigricans serrei* | Java | 54 | 1 + ♂ | 4–6(43%) | — | Linear | Turillazzi 1985d |
| *P. nigricans serrei* | Java | 10 | 1 + 4 ♂ | 2–5(1–5) | 0–5 | Oophagy | Turillazzi and Pardi 1982 |
| *P. serrei* | Thailand | 10 | 1 | 1–2(1) | — | — | Iwata 1967 |
| *P. striatula* | Thailand | 14 | 1 + ? ♂ | 1(1) | 1–2 | — | Iwata 1967 |
| *P. striatula* | Thailand | 25 | 1 | 1–5(1–3) | 0 | Mild | Yoshikawa et al. 1969 |
| *P. striatula* | Sumatra | 9 | 1 | 1–3 | 1–9 | — | Ohgushi et al. 1983a |
| *Parischnogaster* spp. 1–4 | Sumatra | 13 ♂ | 1 | 2–7 | 1–9 | — | Ohgushi et al. 1983a |
| *Stenogaster concinna* | New Guinea | 12 | 1 | 1–2(1) | 0 | Absent | Spradbery 1975 |

*Note:* A dash indicates no data available.
[a] Number of egg-laying females or percent potential egg layers shown in parentheses.
[b] But aggressive to intruders.

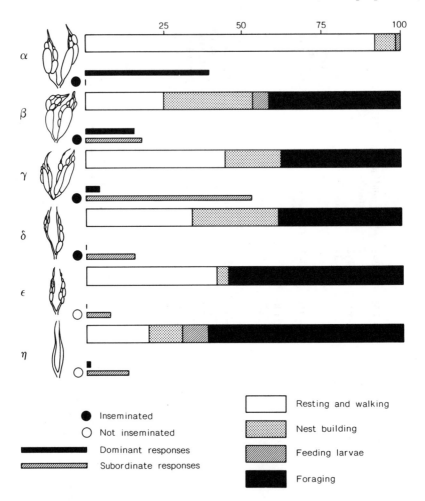

**Fig. 10.3.** Ovarian development, number of behavioral responses that are dominant or subordinate, and percent of time spent in different activities in a colony of a stenogastrine wasp in the genus *Parischnogaster*. (Redrawn from Yoshikawa et al. 1969, courtesy of Faculty of Science, Kanazawa University.)

nates do virtually all the foraging. If the α-female is lost, she is supplanted by the β-female. Dominance acts of stenogastrines tend to be relatively mild, with subordinates avoiding contact, although moreovert behaviors such as wing buzzing, biting, and even stinging do sometimes occur.

A form of sequential or serial polygyny has been described for *P. mellyi*, in which female nestmates usurp the α-individual position on a regular basis (Ohgushi et al. 1983a). Usurpation by nonnestmates has been described for *Eustenogaster calyptodoma* in Malaysia (Hansell

1987b). When young inseminated females of this species leave the parental nest they frequently usurp nearby nests or take over vacant ones, continuing to tend the adopted brood as well as their own. Similarly, single-female nests of *P. mellyi* often are abandoned but can be reoccupied by other females, resulting in stepmother-stepdaughter associations (Yamane et al. 1983a).

## Old World Polistinae

### Ropalidia

In the mid-1970s, there was virtually no biological information on *Ropalidia*. It has often been suggested that knowledge of this genus, which includes both swarming and nonswarming species, would throw considerable light on aspects of social evolution in the Vespidae (Spradbery 1973a, Jeanne 1980a). Recently, nearly 30 species have been subjected to varying intensities of study, and the results more than justify our expectations. Nests of *Ropalidia* species are initiated independently by single or multiple foundresses or by swarms, and they range in size from a dozen to hundreds of thousands of cells. Frequently these wasps exhibit rather unusual means of colony fission.

While many nonswarming (independent-founding) species can initiate a nest with one individual, the majority of such species are pleometrotic with up to 20 or more founding females (e.g., Table 10.4, also Gadagkar, this volume). As seems de rigueur for most pleometrotic wasps, nests are generally initiated by a single female, which is soon joined by co-foundresses to form the group, although many species (e.g., *R. fasciata*, *R. gregaria*, *R. plebeiana*, *R. revolutionalis*, *Ropalidia* near *variegata*: Table 10.4) found nests pleometrotically from the outset. Monogyny is typical of smaller colonies, although it is not always very durable, with usurpation and serial polygyny ocurring often. For instance, in the case of *R. variegata jacobsoni* a daughter replaces the founding female and, in turn, is replaced by a younger sister (Yamane 1986). Gadgil and Mahabal (1974) found that in *R. marginata* monogyny prevailed in multiple-foundress nests that had fewer than 80 cells, but, in general, long-term monogyny appears to be relatively rare in *Ropalidia* (Table 10.4).

There is still very little information on the swarming species of *Ropalidia*. Pagden (1976) observed a swarm-established nest in Malaysia, and Yamane et al. (1983b) described two colonies of the swarming species *R. montana* in India, in which 0.2 and 0.6% of females were queens. This is the only *Ropalidia* species so far described with structural caste dimorphism, as distinct from behavioral or physiological caste differences or the size differences seen between foundresses and

**Table 10.4.** Colony foundation, colony composition, and social organization in *Ropalidia* species

| Species | Locality | No. of foundresses | Percent of nests pleometrotic (N) | Dominance behavior in preemergence colonies | No. of laying females/total females in postemergence colonies | Maximum colony size | | Swarming (S), Nonswarming (NS) | Reference |
|---|---|---|---|---|---|---|---|---|---|
| | | | | | | No. of adult females | No. of cells | | |
| *R. bambusae* | New Guinea | 2+ | 100(5) | — | 13/43 | 194 | 705 | S | Spradbery and Kojima 1989 |
| *R. bensoni* | New Guinea | — | 100(2) | — | — | 472 | 12,727 | — | Spradbery and Kojima 1989 |
| *R. cincta* | Africa | 1 | 0(6) | — | 1/3–19 | 26 | 129 | NS | Darchen 1976b |
| *R. cristata* | New Guinea | 1 | 0(8) | — | 1/11 | 11 | 115 | NS | Spradbery and Kojima 1989 |
| *R. cyathiformis* | India | 1 or 2+ | — | Linear | 5/10 (1 dominant) | >30 | 74 | S ("fission") | Gadagkar and Joshi 1982b |
| *R. fasciata* | Okinawa | 1 or 2+ | 25(440) | — | — | — | — | NS | Suzuki and Murai 1980 |
| *R. fasciata* | Okinawa | 1–10 | 35(28) | Weak | — | — | — | NS[a] | Itô 1983 |
| *R. fasciata* | Okinawa | 1–22 | 53(58) | Mild | 2.6/10 | — | — | NS | Itô 1985b |
| *R. fasciata* (preemergence) | Okinawa | 2–8 | — | Linear | — | — | — | NS | Kojima 1984c |

| | | | | | | | | | |
|---|---|---|---|---|---|---|---|---|---|
| R. fasciata | Java | 1 or 2+ | 67(3) | Linear | Serial polygyny | 11 | 56 | NS | Turillazzi and Marucelli Turillazzi 1985 |
| R. fasciata gregaria R. g. gregaria | Okinawa Australia | $\bar{X}$ = 3.2 2–8 | 55(251) 100(3) | — Weak | 3/7 1–5/9 | — — | — — | NS NS | Itô 1987b Itô 1986c |
| R. g. spilocephala | New Guinea | 1 or 2+ | 90(21) | Linear | 1–12/2–35 | 35 | 331 | NS | Spradbery and Kojima 1989 |
| R. guttatipennis | Africa | 6–7 | — | — | 1/6 or 7 | — | — | NS | Richards 1969 |
| R. horni | Thailand | — | — | — | 10/20 | 40 | 85 | NS | Yoshikawa et al. 1969 |
| R. kurandae | New Guinea | — | — | — | 18/109 | 109 | 1,013 | ? | Spradbery and Kojima 1989 |
| R. leopoldi | New Guinea | — | 0(1) | — | 1/8+ | 8+ | 121 | — | Spradbery and Kojima 1989 |
| R. loriana | New Guinea | — | — | — | 18/289 | 289 | 733 | ? | Spradbery and Kojima 1989 |
| R. maculiventris | New Guinea | ?>2 | 100(5) | — | 1–45/52–477 | 477 | 1,935 | NS/?S | Spradbery and Kojima 1989 |

**Table 10.4** —*continued*

| Species | Locality | No. of foundresses | Percent of nests pleometrotic (N) | Dominance behavior in preemergence colonies | No. of laying females/total females in postemergence colonies | Maximum colony size | | Swarming (S), Nonswarming (NS) | Reference |
|---|---|---|---|---|---|---|---|---|---|
| | | | | | | No. of adult females | No. of cells | | |
| R. marginata | India | 1 or 2–20 | 70(27) | Linear | 1(< 80 cells) 3+ (> 80 cells) | >40 | 500 | NS/S "Mass exodus" | Gadgkar 1980 Gadgil and Mahabal 1974, Gadgkar et al. 1982a |
| R. m. jucunda | New Guinea | 2–6 | 100(3) | — | — | 100 | 1,718 | NS | Spradbery and Kojima 1989 |
| R. montana[b] | India | 2+ | 100(2) | — | 11/5,189;49/8,414 | 8,463 | 96,000 | S | Yamane et al. 1983b |
| R. ornaticeps | Malaysia | — | — | — | 1/20 | 100 | 356 | ? | Yoshikawa et al. 1969 |
| R. plebeiana[c] | Australia | N: 1–15 (X = 2.6) O: 1–32 (X = 8.0) | 69(317) 96(187) | Very weak | 3–9/3–12 | 32+ | 800+ | NS | Richards 1978b; Hook and Evans 1982; Itô 1985c; 1987a; Itô and Higashi 1987; Itô |

352

et al. 1988; Yamane et al. 1991

| Species | Location | | | | | | | | Reference |
|---|---|---|---|---|---|---|---|---|---|
| R. revolutionalis | Australia | 5–7 | 89(27) | Rare | 1–5/4–15 | 30 | 147 | NS | Hook and Evans 1982, Itô 1987a,b |
| R. romandi | Australia | — | — | — | 13–29/4,400 | — | — | — | S.-N. Shima, Sô. Yamane, Y. Itô, and B. Jenkins, unpubl. |
| R. stigma stigma | Malaysia | — | — | — | 1/8 | 10 | 96 | NS | Yoshikawa et al. 1969 |
| R. taiuana koshuensis | Formosa | 1 | 0(6) | — | 1/3 | 3 | 21 | NS | Iwata 1969 |
| R. timida | Malaysia | ?>2 | — | — | — | — | 250 | S | Pagden 1976 |
| R. variegata jacobsoni | Sumatra | 1 or 2–4 | 36(58) | Very weak, linear | Serial polygyny | 45 | 145 | NS | Yamane 1986 |
| Ropalidia near variegata variegata | Australia | 1–5 | 86(7) | Mild | 1–2/3–10 | 10 | 76 | NS | Itô 1986c |
| R. sp. (nigra?) | New Guinea | 2+ | 100(2) | — | >2 | — | — | S | Spradbery, unpubl. |

Note: A dash indicates no data available.
[a]Emergency swarms/rebuilding.
[b]Female caste dimorphism.
[c]N, newly founded nests; O, reused nests. Reused nests frequently are cut into two or more pieces, each of which becomes an independent nest (Yamane et al. 1991).

the first-generation females in most wasp colonies. I have examined swarms of an unnamed, black *Ropalidia* species in New Guinea, one with 30 females, of which 5 (16.7%) were queens, and another with 57 females, of which 6 (10.5%) were queens. Mature colonies of this species contain more than 500 adults with about 4% queens, confirming that most if not all swarm-founding *Ropalidia* species are likely to be permanently polygynous (see also Jeanne, this volume: Chap. 6).

Although not swarm-founding in the usual sense, several species can rebuild nests that have been destroyed by predators or by natural disasters such as fire or typhoons. These include *R. fasciata* in Okinawa (Itô 1985b, 1986b) and *R. maculiventris pratti* in New Guinea (Spradbery and Kojima 1989). Both species can have multiple foundresses, although *R. fasciata* generally becomes secondarily monogynous while *R. m. pratti* is usually polygynous. *Ropalidia marginata*, though generally regarded as an independent-founding species, tends to produce "swarms" when nests contain more than 40 adults—then a "mass exodus" takes place (Gadagkar et al. 1982a,b).

Dominance hierarchies among founding females or between a foundress and her offspring typically develop along classical linear lines, with the most dominant female being the principal or sole egg layer and rarely leaving the nest to forage, while subordinates rarely or never lay eggs. However, little overt dominance behavior is seen among foundresses of *R. plebeiana* and an unnamed *Ropalidia* species in Australia (Itô 1985c, 1986c; Itô and Higashi 1987). With words such as "pecking" and "kissing" used to describe dominance behavior in *Ropalidia*, the dominance repertoire appears to be among the mildest of the Polistinae so far studied (Fig. 10.2b,c). Cell guarding has never been recorded for *Ropalidia*, although oophagy has been for *R. cyathiformis* (Gadagkar and Joshi 1982b, 1984) and *R. fasciata* (Itô 1987b) (see Table 10.2).

### Belonogaster

*Belonogaster* is an Old World polistine genus in which only three species have been subjected to any observational study. Nests are usually initiated by a single female that is soon joined by others to form foundress associations of 2–16 females (Table 10.5). The exposed single-comb nests can become quite large, with up to 600 or more cells and 60 adults (Roubaud 1916, Iwata 1966, Richards 1969, Pardi and Marino Piccioli 1970).

It appears that in the more tropical areas such as Somalia and Ghana, colonies of *B. grisea* and *B. juncea* are polygynous, while in the more temperate regions of southern Africa, *B. grisea* and *B. petiolata* are monogynous after linear dominance hierarchies have been established (Keeping and Crewe 1983, 1987). For *B. petiolata* at least, the sole func-

**Table 10.5.** Colony foundation, colony composition, and social organization in *Belonogaster*, *Parapolybia*, and *Mischocyttarus* species

| Species | Locality | No. of colonies sampled | No. of foundresses | Percent of nests pleometrotic | No. of laying females/ total females | Dominance behavior in preemergence colonies | Oophagy[a] | Maximum no. of cells | Reference |
|---|---|---|---|---|---|---|---|---|---|
| *Belonogaster grisea*[b] | Central Africa | 2 | 1+? | — | 2–18/35 | Strong, linear | + | 665 | Pardi and Marino Piccioli 1970, Marino Piccioli and Pardi 1978 |
| *B. juncea*[b] | Central Africa | 13 | 1–2 | 38 | 1–2/1–7 | — | ? | 450 | Roubaud 1916, Richards 1969 |
| *B. petiolata* | South Africa | 81 | 1–16 | 53 | 1/2–16 | Aggressive, linear | + | 240 | Keeping and Crewe 1983, 1987 |
| *Mischocyttarus angulatus* | Panama | 11 | 1–5 | 82 | 2–6/3–10 | Absent | 0 | 72+ | Itô 1984b |
| *M. basimacula* | Panama | 18 | 1–6 | 61 | 2–3/3–5 | Mild[c] | — | 100+ | Itô 1984b |
| *M. cubensis mexicanus* | Mexico | 3 | 1–6 | 33 | — | — | — | 40 | Rau 1940 |

Table 10.5—continued

| Species | Locality | No. of colonies sampled | No. of foundresses | Percent of nests pleometrotic | No. of laying females/ total females | Dominance behavior in preemergence colonies | Oophagy[a] | Maximum no. of cells | Reference |
|---|---|---|---|---|---|---|---|---|---|
| M. drewseni | Brazil | 29 | 1–8 | 31 | Serial polygyny = 1 | Linear | Rare | 210 | Jeanne 1972 |
| M. flavitarsis | Arizona, USA | 134 | 1–2 | 2 | Serial polygyny = 1 | — | 0 | 305 | Litte 1979 |
| M. immarginatus | Mexico | 15 | 1–10 | 60 | — | — | — | 65 | Rau 1940 |
| M. labiatus | Colombia | 58 | 1–9 | 22 | 1 + ? | Parallel hierarchy | + | — | Litte 1981 |
| M. mexicanus | Florida, USA | 117 | 1–20 | 35 | 1–4 | | + | 46+ | Litte 1977, 1979 |
| Parapolybia indica | Southern Japan | 129 | 1 | 0 | 1 | — | 0 | 291 | Sekijima et al. 1980 |
| P. varia | Southern Taiwan | 14 | 1–22 | 93 | 4/11[d] | Linear | 0 | 6,096 (28 combs) | Yamane 1980, 1985 |
| P. varia | North-central Taiwan, Japan | — | 1 | 0 | 1 | — | — | — | Yamane 1980, 1985 |

Note: A dash indicates no data available.
[a] +, present; 0, absent.
[b] Castes dimorphic.
[c] In postemergence stage, aggressive dominance behavior occurs.
[d] Mean values.

tional queen can remain reproductively dominant until the end of the season, when the new gynes are being produced (M. Keeping, unpubl.). The interactions between foundresses that result in the most dominant female becoming the primary egg layer (queen) and the subordinate females becoming nonreproductive auxiliaries are relatively fierce and frequent (Keeping and Crewe 1987). All colonies of tropical *B. grisea* studied (22) contained more than one mated female, and these sometimes accounted for more than 50% of the female population. These queenlike females laid all or most of the eggs, practiced oophagy (Fig. 10.4, Table 10.2), foraged infrequently, and, as in *B. petiolata*, were larger than the workerlike uninseminated females (Pardi and Marino Piccioli 1970). That *Belonogaster* occupies some transitional position in the development of castes with all females mated and capable of laying eggs, as proposed by Roubaud (1916), no longer appears valid.

### Polybioides

*Polybioides* is a small Southeast Asian and central African genus that constructs large, complex nests. Nests with up to 18,800 cells and adult

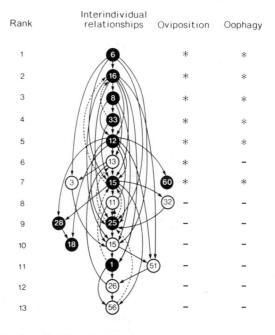

**Fig. 10.4.** Hierarchical dominance relationships in a colony of *Belonogaster grisea*. (Redrawn from Pardi and Marino Piccioli 1970, courtesy of L. Pardi and *Monitore Zoologico Italiano*.)

● fertilized females

○ unfertilized females

⟶ dominant /subordinate relationships

----▸ 7 cases of 52 in which a female of lower rank dominated a female of higher rank

∗ oviposition and/or oophagy observed

populations in excess of 2,500 (75% females) have been described for *P. raphigastra* (Pagden 1958b). Richards (1969) described colonies of *P. tabidus* with 3,000 females, of which 2–8% were queens that were morphometrically distinct from other females. In a sample of eight nests of this species, Darchen (1976b) found that two were monogynous, and that up to 88 queens and 817 workers resided in the polygynous colonies. Nests apparently are established by swarms consisting of one to several queens together with a retinue of workers.

### Parapolybia

*Parapolybia* is a small genus found in southern and eastern Asia and the Indo-Pacific. In the northern temperate parts of its range, *Parapolybia* is haplometrotic, and in the more tropical parts of its range it is generally pleometrotic (Table 10.5). In one instance of pleometrosis, four females of *P. varia* in southern Taiwan initiated a nest as a group (Yamane 1985), in contrast to many Polistinae in which foundress associations arise when single females are joined by one or more co-foundresses shortly after nest initiation.

Yamane (1985) has described the dominance interactions between foundresses as extremely violent; some encounters result in the death of rivals. There is also mild dominance behavior involving a nontactile sequence ("darting and stopping short") and an intermediate behavior in which a female uses the head to tap another female's head or thorax. The violence and frequency of dominance behavior decrease dramatically over a period of a few days after nest initiation. One female ultimately establishes dominance over the remaining co-foundresses and behaves "despotically" toward them. Nevertheless, this despotic behavior is inadequate to establish complete ovarian suppression in co-foundresses during the early stages of colony development, although after a month or two most colonies end up with a single egg-laying female. A further interesting observation by Yamane (1985) was that if a queen was lost and oviposition ceased, cell construction stopped abruptly—an example of queen control of building activity in this group.

## New World Polistinae: *Mischocyttarus*

*Mischocyttarus*, one of the largest genera of eusocial vespid wasps, is essentially a Neotropical group, although one species extends as far north as British Columbia, Canada. In the tropical and subtropical parts of their range, most nests appear to be initiated by a single female, which is sooner or later joined by several co-foundresses to form pleometrotic groups (Table 10.5). In temperate Arizona, however, 98% of nests were founded by a single female only (Litte 1979). In Florida,

*M. mexicanus* tends to found nests with single females in the spring but with multiple foundresses in the autumn (Litte 1977). Survival of single-female nests is generally low; none of the haplometrotic *M. angulatus* or *M. basimacula* nests in Itô's (1984b) study survived. There is no obvious caste dimorphism in *Mischocyttarus*, although queens are generally somewhat larger than workers (Jeanne 1972).

Dominance behavior among co-foundresses (Fig. 10.2d,e), which generally results in the presence of a single dominant egg layer, tends to be extremely mild or even absent in *Mischocyttarus*. Only 20–30% of females of *M. drewseni* engage in dominance acts (compared with 95% of *Polistes dominulus* females), with a low hourly rate of less than 0.1 act per female (compared with 1.9 in *P. dominulus*) (Pardi 1946, Jeanne 1972). In *M. angulatus* and *M. basimacula*, which are both polygynous species, dominance behavior is absent or very rare (Itô 1984b) and most females retain the capacity to become egg layers; indeed, although one female of *M. angulatus* tends to be queenlike, she is frequently replaced as the principal egg layer by different females on different days. More than 20% of nests of *M. flavitarsis*, which generally establishes nests with a single female, are usurped by nonnestmate conspecifics, up to six times on a nest (Litte 1979). The founding female of *M. drewseni* is generally superseded by a daughter, which in turn is superseded by a sister (Jeanne 1972). This serial polygyny can be interpreted as short-term monogyny (West-Eberhard 1978a), with the reproductive life span of the queen being barely long enough to allow an overlap of generations.

Once the first workers emerge, dominance activity increases as queens establish their reproductive superiority during the first 10–15 days of the young adults' lives, when the ovaries of the workers are in a developing phase. Once the emerging females have been "workerized," overt dominance behavior is no longer required to inhibit reproductive development (Jeanne 1972).

### New World Polistinae: Swarm-founding Genera

It is generally assumed that all polistine genera endemic to Central and South America except *Mischocyttarus* establish nests by swarming. Although female caste dimorphism is usually slight or absent, of 33 species listed by Richards (1978a), 22 (67%) exhibit some morphometric differences, with queens larger than workers in 18 species and queens smaller in four species. Intermediates—that is, females with slightly enlarged ovaries that may be suppressed queens or potential egg-laying workers—sometimes occur. Colonies of swarming species are frequently very large and long-lived; some nests have as many as 100,000 cells and more than 20,000 adults (see Jeanne, this volume: Chap. 6).

An *Agelaia vicina* colony contained 3,000–4,000 queens with more than one million morphologically distinct workers (Y. Itô, unpubl.).

The size and composition of swarms have been determined for few species. Some swarms are very large, such as one *Polybia occidentalis* swarm with 1,117 females (Rau 1940), but most confirmed swarms are smaller. Swarms may contain from 1 to 20% queens; however, a few *Polybia* species apparently emit swarms with a single queen (Richards 1978a).

Although most mature colonies appear to be polygynous, there is evidence that polygnous colonies, of some species at least, regularly become monogynous because of interqueen conflict (e.g., *Metapolybia cingulata*: Forsyth 1975), persecution of queens by workers (e.g., *Metapolybia aztecoides*: West-Eberhard 1978b), or losses through swarming. Colonies of *M. aztecoides* undergo cycles of alternating polygyny and monogyny, although in mature male-producing nests polygyny is normally permanent (West-Eberhard 1978b).

The possible occurrence of dominance behavior in swarming polistines has been studied in only a few species: the large, enveloped nests are not conducive to long-term observational studies. No evident agonistic behavior occurs between queens of *Polybia acutiscutis* (Naumann 1970). Workers of *M. aztecoides*, a species that has no morphological caste dimorphism, perform a "shaking dance" toward queens, which in turn can be distinguished by a characteristic aggressive "bending display" (Fig. 10.2f). The worker dance is performed most intensely during periods of queen elimination and may serve as a test of queen dominance (West-Eberhard 1978b).

## QUEEN NUMBER IN VESPINAE

Until quite recently, social organization in the highly eusocial Vespinae was assumed to be characterized exclusively by haplometrotic founding by independent queens and permanent monogyny throughout the annual colony cycle. While this generally remains true for the majority of Vespinae in temperate areas, recent studies of hornets in the tropics and invasive *Vespula* species in recently colonized countries have unearthed a far richer diversity of social organization, indicating a hitherto unappreciated behavioral plasticity in the subfamily.

### Pleometrosis

In 1974 I discovered an embryo nest of the hornet *Vespa affinis picea* in Port Moresby, Papua New Guinea, with six founding queens present. Although I saw only one queen oviposit, there was no apparent dominance hierarchy among the foundresses, and dissections of queens from 15 other polygynous colonies indicated that all queens were egg layers, even in mature colonies (Spradbery 1986). The discovery of pleometrosis and permanent polygyny was confirmed for *Vespa affinis indosinensis* in Sumatra (Matsuura 1983a,b) and *V. a. nigriventris* in the Philippines (C. Starr, unpubl.), and has also been recorded for *Vespa tropica leefmansi* (Matsuura and Yamane 1984). Thus, at least two *Vespa* species exhibit multiple-queen founding and permanent polygyny in the Southeast Asian tropics (Table 10.6). In two of Matsuura's eight pleometrotic embryo nests, a single worker was also present (Matsuura 1983b). Although the origin of these workers is unknown, there is the suggestion of a primitive form of swarming behavior.

During the same expedition to Sumatra, Matsuura (1983c) did in fact discover true swarming behavior in two species of *Provespa*. In these cases, however, single queens accompanied by 46–152 workers swarmed at night. Productivity of queens in such *Provespa* colonies is very low; a maximum of six new (virgin) queens was found in a nest at any one time, with the total production of new queens fewer than 20 per colony. Low queen productivity is presumably compensated for by the high survival rate of reproductive swarms in *Provespa*.

The advantages of pleometrotic founding by some tropical *Vespa* species are presumably similar to those accruing to pleometrotic polistines, that is, reduced parasitism and predation and the capacity for nest reconstruction. Even in temperate regions, hornets readily rebuild damaged nests or even reconstruct in a new site if the original nest site proves inadequate (see Matsuura 1984, this volume; Nixon 1985a). Despite the potential for continuous guarding of a pleometrotic nest

**Table 10.6.** Examples of pleometrosis and primary polygyny in the Vespinae

| Species | Locality | No. of pleometrotic colonies/ total no. of colonies sampled | No. of functional queens | Reference |
|---------|----------|------------------------------------------|------------------|-----------|
| *Vespa affinis affinis* | India | 1/1 | 7 | Rothney 1877 |
| *V. a. indosinensis* | Sumatra | 13/17 | 1–5 | Matsuura 1983a,b |
| *V. a. indosinensis* | Malaysia | 2/2 | 4 | Siew and Sudderuddin 1982 |
| *V. a. nigriventris* | Philippines | 1/1 | Several | C. Starr, unpubl. |
| *V. a. picea* | New Guinea | 15/19 | 1–11 | Spradbery 1986 |
| *V. tropica leefmansi* | Sumatra | — | Several | Matsuura and Yamane 1984 |
| *Vespula vulgaris* | USA | 1 only | 2 | Akre and Reed 1981a |

*Note*: A dash indicates no data available.

against natural enemies by one or more foundresses, in a five-queen *Vespa affinis picea* colony kept under observation for a complete day, there was no queen present on the nest for 17% of the time (Spradbery 1986).

There has been one recorded case of pleometrosis in a temperate area; a *Vespula vulgaris* colony with two founding queens was discovered in Washington, United States, by Akre and Reed (1981a). But, as a general rule, there is great antagonism between founding queens of *Vespula, Dolichovespula,* and *Vespa* species outside the tropics (e.g., Ishay et al. 1983b, Greene, this volume).

### Secondary Polygyny

The incidence of secondary polygyny—the recruitment of new laying queens—in *Vespula* species forming large colonies appears to be fairly widespread, accentuated by the accidental introduction of *Vespula germanica* into Australasia, Africa, and South America. The first record of secondary polygyny was that of Duncan (1939) for a *V. vulgaris* nest in California with 22 laying queens. Vuillaume et al. (1969) and Roland (1976) found several such nests of *V. germanica* in Morocco and Algeria, with up to 400 functional queens per nest. These overwintering or perennial colonies can attain a prodigious size, with millions of cells and hundreds of thousands of adult workers (Thomas 1960, Spradbery 1973a,b) (Table 10.7). Perennial multiqueen colonies are very common where *Vespula* has recently been introduced (such as Australia, New Zealand, and Chile). However, the northern African data, plus a recent report of perennial *V. germanica* colonies on the Greek island of Skopolos (Caïnadas 1987, unpubl.), newspaper accounts of 2-m long nests in southern France, and several case studies in North America (see Table 10.7) indicate that the propensity for polygyny is widespread in this group.

In perennial colonies of *V. germanica* in Australia, the number of laying queens can be very high, with several hundreds present per colony in the early overwintering stages. Rather surprisingly, Thomas (1960) reported only single queens in overwintering *V. germanica* colonies in New Zealand. The sequence of events leading to secondary polygyny can include recruitment before the loss of the foundress queen. Two nests of *V. germanica* were found with 20 and 23 newly recruited queens together with the original foundress (Spradbery 1973a), although the physiological and pheromone-producing status of the old queens was unknown. Ross and Visscher (1983) suggest that recruitment is made possible by the senescence or death of the original queen, with a concomitant reduction or loss of queen pheromone, which may aid in the acceptance of new queens by the adult popula-

**Table 10.7.** Examples of secondary polygyny in *Vespula* species

| Species | Locality | Nest size | | | No. of adults | | No. of laying queens[a] | Reference |
|---|---|---|---|---|---|---|---|---|
| | | No. of combs | No. of cells | No. of queen cells | Workers + males | Males | | |
| V. germanica | Algeria | 9 | — | — | 2,000 | 500 | 400 | Vuillaume et al. 1969 |
| V. germanica | Morocco | — | — | — | — | — | 117 | Roland 1976 |
| V. germanica | Morocco | — | — | — | — | — | 390 | Roland 1976 |
| V. germanica | Australia | 8 | 7,297 | 1,578 | 1,476 | 789 | 1 + 20 | Spradbery, unpubl. |
| V. germanica | Australia | 9 | 12,565 | 2,415 | 4,705 | 1,559 | 398(561) | Spradbery, unpubl. |
| V. germanica | Australia | 60 | 215,416 | 6,816 | 122,280 | 32,104 | 232(552) | Spradbery, unpubl. |
| V. maculifrons | Florida, USA | 15 | 100,120 | 744 | 11,793 | 3,044 | 6(23) | Ross and Visscher 1983 |
| V. pensylvanica | Washington, USA | 8 | 14,300 | — | 1,594 | 1,084 | 1 + 2(1,084) | Akre and Reed 1981a |
| V. squamosa | Georgia, USA | 11 | 65,000 | 0 | 300 | 20 | 7(7) | Ross and Matthews 1982 |
| V. squamosa | Florida, USA | 15 | 100,000 | 600 | 160 | 40 | 119(134) | Ross and Matthews 1982 |
| V. squamosa | Florida, USA | 12 | 30,000 | — | — | — | 12 | Tissot and Robinson 1954 |
| V. squamosa | Florida, USA | — | — | — | 1,385 | 1,013 | 1 + 13 | Tissot and Robinson 1954 |
| V. vulgaris | California, USA | 21 | — | — | — | — | 22 | Duncan 1939, Gambino 1986 |

*Note:* A dash indicates no data available.
[a] Total queens in parentheses; 1+ indicates original foundress present.

tion. There is evidence, however, that there is increased tolerance toward alien conspecifics in late-season vespine colonies (see Greene, this volume).

While I have previously assumed that newly recruited queens invariably originate from the parental nest (Spradbery 1973b), I have since found one example with five young laying queens in a nest that apparently had not yet produced gynes. Ross and Matthews (1982) found seven laying queens in a *Vespula squamosa* colony that had not even begun queen-cell construction. Until further evidence as to their genetic relatedness is forthcoming, it must be assumed that young laying queens can be recruited from the parental colony or from unrelated colonies—an assumption that poses some interesting questions about the genetics of perennial populations (see Ross and Carpenter, this volume).

## QUEEN CONTROL IN THE VESPINAE

Dominance hierarchies, which are so characteristic of the polistine wasps, play a far less significant role in the larger societies of vespines. Nevertheless, dominance behavior does occur in some circumstances, such as during trophallactic exchanges in *Vespula* species (Montagner 1966a) and among foraging queens or workers at highly attractive food sources, where intra- and interspecific dominance hierarchies have been observed in several Japanese *Vespa* species (Matsuura 1969, 1984).

### *Vespa* and *Provespa*

In *Vespa* colonies, workers typically crowd around the queen as she pauses between bouts of oviposition. Groups of 2–30 workers surround the queen of *Vespa orientalis* and *V. analis*; new arrivals orient toward the queen's head and later arrivals cluster around her thorax or gaster, where they collectively form a "resting circle" or "royal court" (Ishay 1964, Matsuura 1984, this volume). The workers lick the queen's body, especially her head and mouth parts, while some tap their gasters on the comb cells, suggesting that they are communicating the presence of the queen to the rest of the colony (Ishay and Schwarz 1965). Similar aggregations around queens have been recorded in several Japanese *Vespa* species (Matsuura 1968, 1984). Nixon (1985a:194) observed five or six workers in a *V. crabro* nest crowding around the heavily gravid queen, "nudging and jostling her with jerky and frenetic movements." Nixon noted that workers of *V. crabro* did not actually lick the queen and that, in the early part of the season, the queen seemed to attract little or no attention from her offspring. Queens of the swarm-founding *Provespa* attract a royal court similar to that of *Vespa* (Matsuura 1983c).

According to Ishay (1977b), an irreversible and permanent bond between workers and their queen is established early in the life of the callow hornet. Newly emerged workers from all over the nest are attracted to the queen, where they form the typical resting circles. This behavior occurs during the first 36 hours after eclosion and results in their attachment to the queen and to their siblings (imprinting), resulting in recognition of siblings both inside and outside the confines of the nest.

There is ample evidence that hornet queens influence building behavior. Workers initiate new cells, building the base and extending the walls to half their normal length until the queen lays an egg in the cell, after which the cell wall is completed. The queen thus regulates the rhythm of cell building (Ishay 1973a). If a queen is absent or if she fails to oviposit in new cells, empty cells accumulate and further new-cell construction is inhibited (Ishay et al. 1986). (A "building-depressor" pheromone released by the queen had previously been suggested as a possible means of regulating building activity [Ishay and Perna 1979].) The queen of *V. orientalis* also controls queen-cell construction, for queenless worker groups in the laboratory generally fail to construct queen cells irrespective of the season. However, if workers are removed from a queenright colony in which queen-cell construction is under way, they will continue to build queen cells in the queen's absence (Ishay 1973a). Ishay et al. (1983a) have suggested that the rate of change of daylength may also influence the onset of queen-cell construction, as even queenless worker groups build a few queen-sized cells at the appropriate season if exposed to natural daylength. This is contrary to the situation in *Vespula*, in which photoperiod appears to be an unlikely trigger for queen-cell initiation (Potter 1964, Spradbery 1971).

If the queen of *Vespa orientalis* is removed for more than 48 hours, she may be attacked by the workers on her return, suggesting that regular contact with the queen is necessary to reinforce the bond between queen and workers and to maintain the cues for queen recognition (Ishay et al. 1965). In the absence of the queen, several significant changes occur in the hornet colony. The workers become quarrelsome, a general state of unrest prevails in the colony, and normal brood development is disrupted. The ovaries of the workers begin to develop and they start to lay eggs some five days after the queen is lost (Motro et al. 1979). In the Japanese *V. analis*, queen loss is common, representing 39% of all colony failures. In these orphan nests a dominance battle among the workers ensues, resulting in one worker becoming the dominant or exclusive egg layer, which Matsuura (1984) labels a "substitution queen." Such orphan colonies may survive for some months, with several hundred unfertilized eggs being laid by the substitution queen (up to 24 eggs per day and over 400 in two months in one case). Once the substitution queen has established her dominant position in the colony there is very little aggressive behavior among the remaining workers, and the colony continues to function virtually normally.

In *Vespa* colonies, a physiologically competent queen suppresses ovarian development in workers, has a calming influence on the adults, and, directly or indirectly, controls the rate of cell building and initiation of queen cells. Since the queen displays almost no dominance behavior, clearly she must exert some form of pheromonal control. Ishay et al. (1965) prepared ethanolic extracts of functional queens of *V. orientalis* which, when presented to queenless worker groups on wads of cotton wool, attracted the adults and stimulated licking and antennating responses. The extract also had a tranquilizing effect on workers, in contrast to the restless behavior of hornets without access to the queen extract. Analysis of the extract revealed a strongly acidic component with these potent effects on worker behavior.

In further experiments with this species, Ikan et al. (1969) prepared ethanolic extracts of dismembered queen bodies and found that the head extract, but not extracts from the thorax or gaster, had an effect similar to whole-queen extract. They isolated a pheromone component that was subsequently identified as δ-*n*-hexadecalactone. The lactone had tranquilizing properties similar to the crude extract and also stimulated queen-cell construction. This pheromone component has since been synthesized (Coke and Richon 1976, Bacardit and Moreno-Manas 1983, Mori and Otsuka 1985).

## *Dolichovespula*

*Dolichovespula* species are characterized by relatively small nests and simple social organization among vespines (Greene 1979, this volume). Foundresses of *D. maculata* appear to be far less effective in controlling ovarian development and subsequent oviposition by workers than other vespine species, with workers regularly laying eggs about a month after they first begin to emerge. In contrast, worker oviposition in other groups is rare, generally occurs during colony decline, and then usually in the uppermost combs of large nests, well away from the queen. *Dolichovespula maculata* queens practice oophagy of eggs laid by workers and patrol the outer edge of the comb where new cells are built, where they monopolize oviposition. The queen also spends considerable time trimming pupal caps after adults have eclosed, ensuring that she is present to oviposit in newly vacated cells. Workers attempting to oviposit are chased away by the queen and may be mauled if they fail to escape.

In the close presence of the queen, *D. maculata* workers become "greatly excited" and occasionally antennate or "nuzzle" her with their heads (Greene 1979). Gastral tapping by the queen sometimes initiates simultaneous tapping by workers all over the comb. A spectacular reaction to the queen is a vigorous dance by small numbers of nearby workers involving extremely rapid gastral and leg vibration (see Greene, this volume: Fig. 8.9). These queen-stimulated responses by workers seem more typical of the primitively eusocial polistine manifestation of reproductive dominance, in which there is an overt, physical response to egg-laying females by nearby subordinates, than of the subtle, chemically mediated reproductive dominance of other vespines.

Queens of the social parasite *D. arctica*, which parasitize *D. arenaria* colonies, seem to be unable to suppress ovarian development in the host workers, for they may begin to oviposit within nine days after the death of their foundress despite the presence of the *D. arctica* queen (Jeanne 1977b, Greene et al. 1978). This phenomenon is echoed by the bumble bee parasite *Psithyrus ashtoni*, which is unable to inhibit workers of the host, *Bombus affinis* (Fisher 1983). Among *D. arenaria* host workers a rather typical hierarchy of ovarian development in a colony was discovered by Greene et al. (1978), which included an α-worker whose gaster was packed with developing oocytes and mature eggs, a β-individual with most of the gaster filled with oocytes, 5 workers with developing ovaries, and 31 with no ovarian development.

## *Vespula rufa* Species Group

The *Vespula rufa* species group has rather small nests with up to 400 worker adults and 2,500 cells in mature colonies. Investigations into the pheromonal basis of queen control of worker behavior and suppression of ovarian development have been made on three species: *Vespula atropilosa* (Landolt et al. 1977), *V. acadica* (Reed and Akre 1983d), and *V. consobrina* (Akre et al. 1982).

In the Landolt et al. study, two colonies of *V. atropilosa* were split in half and separated by a double-gauze screen that prevented any physical contact between the half that contained the foundress queen and the half that had no queen. The immediate reaction was a higher degree of agitation and increased level of aggressive behavior among workers on the queenless side. Longer-term effects included much reduced foraging and nest-building activities and fewer brood reared to pupation without the queen. Under the experimental conditions, queen suppression of ovarian development in workers on the queenless side was almost immediately lost, for these workers began to lay eggs eight days after the colony was split (at 30°C, workers of *V. germanica* begin to oviposit six days after emergence in the absence of a queen [Spradbery, unpubl.]). The ovipositing workers behaved very aggressively to nearby wasps, while competing laying workers frequently removed other eggs and laid their own immediately afterward—again a behavior typical of primitively eusocial polistines. The laying workers remained on the comb and never foraged. Although the laying workers functioned in a queenlike manner, they were unable to suppress ovarian development as effectively as the queen. A recovery in the foraging rate of the queenless group occurred after a week, which may have been due to the fact that laying workers became more queenlike as they became increasingly physogastric (Landolt et al. 1977).

Queens of *V. atropilosa* typically treat their workers roughly, pinning them against the comb while soliciting food; but the most aggressive behavior in this species is displayed by laying workers in a queenless situation. Workers of *V. consobrina* tend to avoid the queen if approached and, in queenless colonies, fighting among workers is more frequent and vigorous than in queenright colonies. There is little direct contact between the queen of *V. acadica* and her workers, although workers will occasionally antennate and nibble the queen's head, gaster, and appendages. The workers are not actively attracted to the queen, but neither do they actively avoid her. She rarely solicits food from her workers until the colony is in decline, although Reed and Akre (1983d) observed a few instances of queen aggression directed at workers, especially when they were in the vicinity of vacant cells or just before adults emerged from cells. Workers trimming the remains of the pupal cap or antennating an emerging adult were pushed aside by the queen. In two instances, workers competed with their queen to lay eggs in cells. Aggressive mauling of one worker prevented it from ovipositing, but on another occasion a worker eventually laid an egg after the queen left the area. The queen is apparently attracted to newly vacated cells or even cells with an emerging adult, ensuring that she oviposits in such cells promptly and thus prevents a worker from ovipositing there (Reed and Akre 1983d).

Queen control in the *Vespula rufa* group would appear to be a combination of physical intervention and pheromone action. That a double-gauze screen prevented a *V. atropilosa* queen from suppressing ovarian development in adjacent workers suggests that if a pheromone is responsible for ovarian suppression in workers it is probably not volatile (Landolt et al. 1977).

## *Vespula vulgaris* Species Group

Species of the *V. vulgaris* group produce the largest colonies among vespine wasps (with up to 5,000 workers and 15,000 cells per colony) and exhibit the greatest dimorphism between female castes. There is virtually no physical contact between the queen and her workers. There is no crowding or licking of the queen of *V. vulgaris* or *V. germanica* (Archer 1972c), and no manifestation of queen dominance behavior in *V. pensylvanica* or *V. vulgaris* (Akre and Reed 1983). Indeed, workers in this group tend to avoid the queen and retreat when they encounter her (Akre 1982), suggesting that a queen pheromone may act as a repellent or arrestant at close range, not as an attractant (Akre and Reed 1983).

All evidence to date suggests an exclusively pheromonal control of social organization in these populous colonies, with suppression of worker ovarian development the key component in queen control. The evidence for a queen pheromone is compelling. As early as 1896, Marchal experimented on the effects of queens on workers, having noted that laying workers sometimes occur in normal, queenright colonies but were much more numerous in queenless nests (Marchal 1896a). The principal conclusions drawn from Marchal's study are that (1) workers in the presence of a queen remain sterile, (2) in worker groups without a queen about one-third develop their ovaries, (3) young workers are more likely to develop their ovaries than older workers, and (4) within a laying worker group, a small proportion become physogastric and develop brown stains on the gaster which are typical of old, laying queens (Spradbery 1973a). An important observation made by Marchal was that if the queen failed to lay eggs, worker ovarian development would proceed. Montagner (1966a, 1967) confirmed Marchal's findings, but noted that the introduction of a fecund queen to a group of laying workers did not cause regression of their ovaries. Montagner concluded that laying workers spend most of their time on the upper combs well away from the queen, whose activities are largely confined to the more recently built, lower combs. Laying workers never engaged in foraging or nest construction duties and only rarely fed larvae, but there was no apparent worker hierarchy in relation to degree of ovarian development. A detailed study of several *Vespula* species by Ross (1985, 1986) failed to detect any significant contribution to male production by workers, with worker oviposition confined to large, late-season colonies. Queen control and her monopolization of reproduction appear to be highly effective throughout most of the colony cycle.

Using *V. vulgaris* and *V. pensylvanica* colonies, Akre and Reed (1983) separated groups of workers with or without a queen by single or double gauze screens in nest boxes. There was no inhibition of ovarian development on the queenless side, suggesting that any queen pheromone that inhibited ovarian development in the queenright group of workers could not be volatile. By switching combs from a queenright worker group to queenless workers every 24 or 48 hours, Akre and

Reed (1983) demonstrated a substantial decrease in ovarian development in queenless worker groups, especially with the 24-hour treatment. The implication is that the queen deposits some substance on
the comb (gland secretion or fecal material) that suppresses ovarian
development in workers.

Although there is little direct contact between queen and workers in
*V. vulgaris* group species, they do occasionally engage in trophallactic
exchanges (Fig. 10.5a). By feeding queens radioisotope-labeled honey
solution, I have determined that these infrequent exchanges of food
result in a rapid dissemination of labeled material among the worker
population (Fig. 10.5b) (Spradbery, unpubl.). If a pheromone is produced by head glands of the queen it could certainly be transmitted
throughout the colony effectively and rapidly.

What is the nature of queen control in large, perennial *V. germanica*
colonies? From examination of such nests, the laying queens appear to
be fairly evenly distributed throughout the combs, suggesting that each
maintains a sphere of influence within the nest. If several such queens

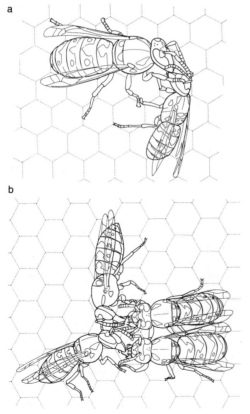

**Fig. 10.5.** Trophallactic exchange in
*Vespula germanica*. (a) Queen-worker
contact. (b) Worker-worker contact.

are maintained together in the laboratory, however, there is little indication of aggressive behavior among them. Competition for oviposition sites (empty cells) in a perennial colony with hundreds of functional queens must be intense—I have found up to 17 eggs per cell in such nests (with no evidence of worker oviposition). If pheromone emission is associated with oviposition activity (as seems likely for *V. germanica* at least) the concentrations produced by 200–300 queens could be very high in nests little bigger than a normal annual nest. Do queens compensate with reduced individual emission? Is there a threshold level above which aggressive behavior in workers could be released, as in some ants (Fletcher and Blum 1983)? Answers to these and the many other intriguing questions posed by perennating *Vespula* colonies must await further investigations.

## EVOLUTION OF QUEEN NUMBER

In setting the stage for a discussion of evolution of queen number in Vespidae, we must consider the embryonic social life among wasps that was unfolding 100–200 million years ago. What characteristics predisposed this group to sociality, and what advantages ultimately led to its genesis and diversification over the millenia?

### Predisposing Features

The presocial vespoids of the Mesozoic era would have displayed a variety of characteristics displayed by the presocial wasps of today (Carpenter and Hermann 1979, Cowan, Matthews, this volume). Among the most important were their manipulative skills in prey capture and nest construction and the skills required to maintain a nest. The ability to orient to a nest site after foraging for building materials and prey indicates a high degree of nervous system organization associated with the complex behaviors involved in locality learning, nest construction, and prey capture, manipulation, and carriage—indispensable attributes for future progress. Nests, and especially aggregations of nests at favorable sites, would also result in considerable conspecific contact and the possibility of competition among the wasps, while also acting as foci for potential predators and parasites.

Because all nonparasitic wasps feed their young by providing food in bulk (mass provisioning) or throughout larval life (progressive provisioning), the immatures are totally dependent on the tending adult, which can thus influence the size and fecundity of its offspring by the amount and quality of food it provides.

If primitive wasps had a sufficiently well-developed sensory apparatus and nervous system whereby they could recognize their own nests and even eggs, this capacity would have considerable significance for the future, leading to an ability to distinguish related conspecifics from nonrelatives (Brockmann 1984).

Modification of the ovipositor that resulted in its function changing from an egg-delivery system to a sting for paralyzing prey was undoubtedly a prominent feature of solitary wasps and would have important implications for future development of the sting for nest defense.

The haplodiploid method of sex determination that occurs in the order Hymenoptera results in males developing from unfertilized eggs, which are haploid. This alters the degrees of relatedness for some classes of relatives compared with diploid organisms, such that, for instance, a female is more closely related to a full (= super) sister (0.75) than to her own offspring (0.50) (Hamilton 1964a,b). This phenomenon has gained considerable significance in the development of genetic theories of social evolution (see, e.g., Ross and Carpenter, this volume).

## Benefits from Social Nesting

The costs and benefits from changes in structure or strategy would have been subjected to the competitive pressures of selection during the course of social evolution. Some of the more obvious advantages of group living will be described here, with examples drawn from recent studies of polistine wasps, especially species that exhibit both single- and multiple-queen founding.

Single-female nests would appear to be far more vulnerable to failure than those with several co-foundresses, because in the former instance loss of a single foundress results in total loss of the nest investment. Indeed, for many species there is no recorded survival of haplometrotic colonies, compared with variable but generally high rates of survival of conspecific pleometrotic colonies (Table 10.8). Overall, the survival rate of single-foundress nests in the studies listed in Table 10.8 is 17.6% compared with 65.3% for multiple-foundress colonies. Protection from putative predators and parasites is likely to be more effective if some foundresses remain on the nest while others forage, and this behavior would also be advantageous in situations where conspecifics attempt to usurp nests. When predation or bad weather such as heavy rains or high winds result in loss of a nest, the capacity to reconstruct a new nest is far greater in multiple-foundress groups. Compared with single-foundress nests of *Ropalidia fasciata*, in which successful nest construction occurred in only 6% of cases, 94% of multiple-foundress colonies rebuilt damaged nests (Sô. Yamane and Y. Itô, unpubl.). Similarly, 7%

**Table 10.8.** Pleometrotic founding and polygyny, dominance behavior, and colony survival in independent-founding Polistinae

| Species | Pleometrotic founding | | | Dominance behavior (no. of dominance acts per female per hour) | | Percent survival of preemergence colonies | | Reference |
|---|---|---|---|---|---|---|---|---|
| | Percent of colonies pleometrotic | No. of colonies | Percent of pleometrotic colonies polygynous | Preemergence | Postemergence | Haplometrotic | Pleometrotic | |
| *Belonogaster petiolata* | 53 | 81 | 0 | — | — | 0 | 42 | Keeping and Crewe 1983, 1987 |
| *Mischocyttarus angulatus* | 82 | 11 | 100 | 0 | 1.9 | 0 | 100 | Itô 1984b, 1987b |
| *M. basimacula* | 61 | 18 | 83 | 0.56 | 2.6 | 0 | 75 | Itô 1984b, 1987b |
| *M. drewseni* | 31 | 29 | 0 | 0.06 | 0.04 | — | — | Jeanne 1972 |
| *M. flavitarsus* | 2 | 134 | 60 | — | 0.76 | 52 | — | Litte 1979 |
| *M. labiatus* | 47 | 58 | 17 | 0.5 | — | 39 | 63 | Litte 1981 |
| *M. mexicanus* | 35 | 117 | 34 | 0.35 | — | 62 | 78 | Litte 1977 |
| *Parapolybia varia* | 93 | 14 | 100 | 0.3–2.4 | 0.21 | — | — | Yamane 1985 |
| *Polistes annularis* | 96 | 43 | 53 | 1.59 | — | — | — | Strassmann 1981a, 1983 |
| *P. canadensis* | 98 | 34 | — | 1.36[a] | 2.61 | 0 | 50 | Itô 1985a |
| *P. dominulus* | 82 | 17 | 0 | 1.95 | 4.43 | — | — | Pardi 1942a, 1946 |

**Table 10.8**—continued

| Species | Pleometrotic founding | | | Dominance behavior (no. of dominance acts per female per hour) | | Percent survival of preemergence colonies | | Reference |
|---|---|---|---|---|---|---|---|---|
| | Percent of colonies pleometrotic | No. of colonies | Percent of pleometrotic colonies polygynous | Preemergence | Postemergence | Haplometrotic | Pleometrotic | |
| P. fuscatus | 48 | 159 | 40 | — | — | 75 | 100 | West-Eberhard 1969, Gibo 1978 |
| P. humilis synoecus | 0 or 100[b] | 10 | 100 | 0 | — | — | — | Itô 1986d |
| P. versicolor | 78 | 23 | 50 | 0.09 | 0.88 | 0 | 96 | Itô 1985a, 1987b |
| Ropalidia fasciata | 35, 55 | 251 | 58, 60 | 0.32–1.06 | 0.72–0.89 | 34 | 77 | Itô 1987b |
| R. variegata jacobsoni | 90 | 10 | 0 | 0.59 | 1.23–2.05 | — | — | Yamane 1986 |
| Ropalidia near variegata | 86 | 7 | 100 | 0 | 1.0 | — | — | Itô 1986c, 1987b |
| R. revolutionalis | 88 | 25 | 0 | 0.04 | 2.95 | — | — | Itô 1987a,b |

Note: A dash indicates no data available.

[a] In presence of despotic queen, 0.17; in absence of despotic queen, 12.0 (West-Eberhard 1986).

[b] All reused nests are pleometrotic.

of single-female nests of *Polistes fuscatus* were rebuilt compared with 43% of multiple-female nests (Gibo 1978) (see also Reeve, this volume).

Although survival rates of nests tend to increase with increasing numbers of foundresses in both pre- and postemergence colonies, there appears to be a corresponding decrease in mean productivity per female. For example, Gibo (1978) found that single-female *P. fuscatus* nests produced 6.1 pupae per female, while multiple-female nests produced 5.9. These data were based on surviving nests. A similar situation also occurs in *R. fasciata* (Sô. Yamane and Y. Itô, unpubl.), but if data for all nests, including failed nests, are analyzed, productivity (number of cells per female) is found to increase with increasing number of females per nest throughout the colony cycle (Fig. 10.6).

The increased survival and productivity of multiple-foundress colonies are undoubtedly due to the protection afforded by several females and the partitioning of activities. With more than one foundress there is the opportunity for specialization of duties, particularly foraging activities, and greater opportunities to exploit food sources and to ensure regular supplies of materials for nest building and for feeding the larvae (see also Jeanne, this volume: Chap. 11).

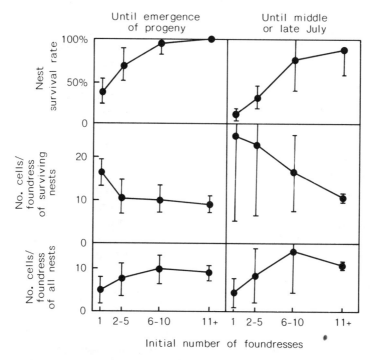

**Fig. 10.6.** Comparison of mean productivity and survival (±SD) in single- and multifoundress colonies of *Ropalidia fasciata*. (Modified from Sô. Yamane and Y. Itô, unpubl.)

Although cooperation among foundresses results in a greater likelihood of some individuals surviving to reproduce, such associations also are likely to exhibit varying degrees of social competition and genetic conflict.

## Possible Routes to Eusociality

On the basis of comparative studies of life histories of wasps, two different but not necessarily mutually exclusive theories for the genesis of sociality in the group have been postulated (Fig. 10.7). At a time when the most familiar social wasps were the temperate-zone vespines, in which single queens established nests unaided and produced morphologically distinct sterile workers, Wheeler (1922) proposed a sequence of events leading from casual and brief female-offspring contacts to longer-term associations between the two generations as adult longevity increased. This *subsocial* route to eusocial nesting behavior, in which mothers and daughters share the nest and cooperate in nest construction, defense, and provisioning of brood, was further refined by Evans (1958), who illustrated a multistep progression from solitary to eusocial wasps with each stage illustrated by extant species.

As knowledge of the natural history of wasps, especially the more primitive social wasps of the tropics and subtropics accumulated, an alternate route to eusociality, via *parasocial* nesting, was proposed by Michener (1958) and Lin and Michener (1972). Here, unrelated females of the same generation shared a nest because of mutualistic advantages to each, and progressed through a *semisocial* organization, with cooperative brood care and some reproductive dominance, to the *eusocial* condition, in which there is strong reproductive dominance and, ultimately, female caste dimorphism (see Cowan, this volume: Table 2.1 for definitions of levels of sociality). A variation on the parasocial theme is the *polygynous family* hypothesis of West-Eberhard (1975, 1978a), in which several related females form polygynous associations that may be maintained for one or several generations. Again, examples of the different steps in the parasocial route to sociality are accumulating as students of social organization delve deeper into the behavior of nest-sharing hymenopterans in the tropical and subtropical regions of the world.

## Genetic Basis for the Evolution of Eusociality

The evolution of sterile or semisterile castes plays the pivotal role in discussions of the evolution of eusociality in the Vespidae. Factors that regulate queen number and the mechanisms of queen control are synonymous with the evolution of sterility in eusocial wasps because

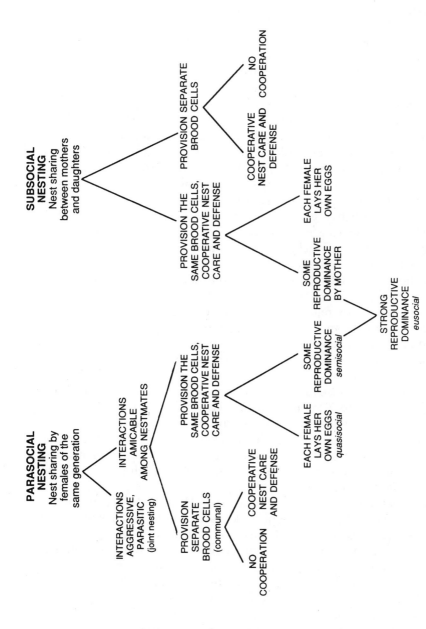

**Fig. 10.7.** The parasocial and subsocial routes to eusociality. (From Brockmann 1984, courtesy of the author and Blackwell Scientific Publications.)

pleometrotic founding often results in associating foundresses relinquishing their fertility in favor of a single dominant female, while queen control is the process whereby sterility is maintained in subordinate foundresses or offspring. Two major solutions to the apparent dilemma of inherited sterility have been proposed.

In the 1960s, Hamilton (1963, 1964a,b) published his theory of kin selection, suggesting that female hymenopterans can more successfully reproduce their genes in future generations by rearing full sisters that are related by 0.75 instead of their own daughters, which are related by only 0.50. Thus, genes that result in females rearing sisters or other close kin in their parent's colony, rather than daughters in their own nest, could be selected for despite the apparent loss of reproductivity in the individual. Where an individual influences the reproductive success of others in a group, the sum of personal reproduction together with the reproductive gains to others (and taking into account their relatedness) provides a measure of *inclusive fitness*. Thus, the number of offspring raised and inclusive fitness are two different but valid measures of reproductive success, with inclusive fitness accounting for the production of a relative's offspring if the nonreproductive (*altruist*) is responsible for their existence.

Hamilton's theory relies heavily on the coefficients of relatedness between individuals and the progeny of relatives, such that selection would favor social behavior that results in increased inclusive fitness to the participant. The theory is most robust when analyzing the subsocial (*matrifilial*) setting for the evolution of eusociality, with a single (haplometrotic) female inseminated by one male sharing the nest with her daughters. If there is more than one laying female in a nest or if a female has mated with several males, the degrees of relatedness can be substantially reduced. Nevertheless, genetic studies of relatedness in primitively eusocial and presocial wasps have shown that relatedness in such communities may often be comparatively high (Ross and Carpenter, this volume).

West-Eberhard (1975) has reformulated Hamilton's genetic theory and proposed that the argument for kin selection be less dependent on the three-quarters relatedness idea and instead consider factors that make altruism more profitable than selfishness, such as the potential reproductive capacity of altruistic females in relation to the parent's reproductivity. West-Eberhard has pointed out that small differences in physiology, size, or reproductive capacity among members of a group could well have far-reaching implications in the evolution of sterility in wasp societies. A female with little chance of reproduction in a nest would be expected to leave and found her own colony. However, if the subordinate were related to the dominant, she could gain inclusive fitness by helping to raise the dominant's offspring rather than attempting to establish a new nest, especially if solitary nesting has a low or zero chance of being successful. Also, if the dominant foundress should die, subordinates may assume the dominant role and lay their own eggs ("hopeful reproductives"; see Fig. 10.1). Small or otherwise inferior females probably have fewer reproductive options, and they may well derive greater inclusive fitness by staying on the maternal nest and helping than by attempting to establish nests alone.

Once group life is established, selection also could apply to the colony as a whole, assuming there is genetic similarity within the group. As spelled out by West-Eberhard (1981:13), "Colonies, like species in geological time, differentially multiply and differentially persist or become extinct, and this must affect the rate of selection and extinction of alleles."

A more recently developed theory to explain the evolution of eusociality is that of parental manipulation (Alexander 1974, Michener and Brothers 1974). As emphasized earlier, all vespid larvae are totally dependent on the tending adult for supplies of food. Thus, the size and fecundity of offspring (as well as their sex) are not just a reflection of genetic and external environmental factors, but are regulated by the parent. An extreme example is the construction of small worker-producing cells by vespine wasps; queen-cell production is withheld until late in the colony cycle, and their initiation is primarily under the pheromonal control of the queen (Spradbery 1973a). Parents can thus manipulate their progeny into being helpers rather than reproductives, offspring that are unable to maximize their inclusive fitness except by enhancing the reproductivity of parents. Again, the theory is most relevant in the matrifilial route to eusociality and seems less appropriate where several co-foundresses start a nest. It seems unlikely that a parent can manipulate what appear to be equivalent progeny that will form a pleometrotic group, even if they are very closely related to one another (sisters), when they are no longer in contact with that parent. But how equivalent are members of a founding group? Do they all have similar reproductive potential? All the evidence to date suggests that differences in attributes such as size, ovarian development, and other, more subtle traits do occur—at least in the nonswarming species. In *Belonogaster grisea*, Pardi and Marino Piccioli (1970) consider caste to be due to preimaginal inequality such that slightly larger females are preferentially inseminated by males and thus become the new queens. Gadagkar et al. (1988) have presented additional evidence for preimaginal biasing of caste (= parental manipulation) in *Ropalidia*, noting that only half of all potential foundresses lay eggs, even when isolated from the influence of other co-foundresses.

Although parental manipulation is generally discussed in terms of influencing growth and development of immatures, manipulation of adult offspring during the course of intracolony conflict is likely to be just as important. Itô (1984b, 1987b) has noted that, in several primitively eusocial polistine species, there is an escalation of dominance conflict when the first female progeny emerge (see Table 10.8). Because the increase in dominance behavior is apparently directed toward the newly emerged females, the daughters are forced to work off the nest and fail to develop their ovaries (Itô 1984b).

## The Polygynous Route to Eusociality

The predisposing conditions and advantages accruing from some form of group living in wasps have been amply demonstrated in both descriptive and theoretical studies. While the value of cooperative groups in terms of nest defense and reconstruction, exploitation of habitat resources, and partitioning of activities are readily appreciated, the debate continues over the genetic implications of diminished average reproductivity accompanying the evolution of increasing caste dimorphism. Kin selection and parental manipulation theories find their best expression when applied to the monogynous, subsocial (matrifilial) setting of eusocial evolution. While short-term monogyny characterizes the stenogastrines and eusocial wasps of the temperate regions, there is an overwhelming preponderance of nest founding by groups of associating females in eusocial vespids, especially within the tropics and subtropics.

The majority of polistine nests are founded by groups of potentially reproductive females. Eighteen of 22 *Polistes* species (82%) found nests pleometrotically at least occasionally (Table 10.1), and the majority of non-*Polistes* species of Polistinae, including all the swarming species, obligatorily or facultatively start colonies with multifemale groups (see, e.g., Tables 10.4, 10.5). Among *Polistes*, however, most species (15/18 = 83%) become secondarily monogynous during the preemergence period through dominance activities. Even when monogyny appears to be the rule, however, the longevity of many wasp foundresses may result in only a brief overlap of generations, leading to serial polygyny (= short-term monogyny) (Jeanne 1972, West-Eberhard 1978a, Yamane 1986).

A tropical venue for the origins of social life seems most likely, irrespective of the particular route to eusociality, because overlapping generations could occur more readily (e.g., Cowan, this volume). Of the 22 species of *Polistes* whose life histories have been described (Table 10.1), 12 occur in temperate regions, with 33% of these at least occasionally pleometrotic. Of the 10 tropical species, 80% are characterized by nests founded by associations of females. Colonies of three wide-ranging species of *Polistes* (*P. exclamans, P. dominulus, P. metricus*) and of *Para-*

*polybia varia* that are initiated in the more temperate areas of their distributions are generally haplometrotic, while in the warmer regions they tend to be pleometrotic (see Tables 10.1, 10.5). Among the Vespinae, which are characteristically haplometrotic, pleometrosis in tropical *Vespa* species apparently is not uncommon (Table 10.6), and secondary polygyny is rather frequent in some *Vespula* in more equable climates (Table 10.7).

Thus, if we discount the temperate-zone species, in which overwintering of inseminated females and haplometrosis have evolved as climatic adaptations, the vast majority of eusocial wasps display pleometrotic founding, with many retaining a polygynous society throughout the pre- and postemergence phases of the colony cycle.

In her review of social organization in presocial and eusocial Sphecidae and Vespidae, West-Eberhard (1978a) proposed that polygynous family groups were the likely setting for the evolution of eusociality in wasps. This "polygynous family" hypothesis has gained increasing support during the past decade (Fletcher and Ross 1985; Itô 1985a, 1986c, 1987b; and see Carpenter, this volume). From a *solitary stage*, in which the potential for future caste differences is manifested in variations in relative reproductivity, a *primitively social stage* is reached in West-Eberhard's model, characterized by nest sharing and increased reproductive competition among members of the group. To the extent that there is significant genetic relatedness among members of such nest-sharing groups it is most likely due to philopatry, a widespread though not invariant trait in extant primitively eusocial vespids (e.g., Klahn 1979, Strassmann 1983, Keeping and Crewe 1987). In such associations, reproductive competition among foundresses appears ubiquitous (e.g., Röseler, this volume), even though overt conflict among cofoundresses may be mild or even absent.

In West-Eberhard's model an intermediate stage between the casteless, primitively social and eusocial societies is a *rudimentary-caste-containing stage*. A rudimentary worker caste could be achieved by a temporal or partial division of labor, with some females behaving as workers during the earlier part of their lives or being more workerlike throughout their lives compared with more queenlike females. West-Eberhard (1978a) cites *Belonogaster* in central Africa (Pardi and Marino Piccioli 1970) and *Parischnogaster* (Yoshikawa et al. 1969) as taxa representing this stage, while Itô (1986a, 1987b) presents several examples of *Ropalidia*, *Polistes*, and *Mischocyttarus* species in which there is some partitioning of reproductive activities but without any obvious caste differences in the polygynous founding groups. This stage may also represent the groundplan of the Stenogastrinae (Carpenter 1988a).

In the final, *eusocial stage*, some individuals completely fail to reproduce and behave exclusively as workers. With increasing reproductive dominance within such groups, there is likely to be a continuing reduction in the number of reproductives (queens), leading ultimately to the evolution of monogyny. The characteristic polygyny of the swarming polistines should perhaps be considered as a stable alternative strategy to monogyny, selected for under conditions of frequent nest relocation or reconstruction. Hence, several queens and greater reproductive flexibility remain a considerable strategic advantage under some circumstances (West-Eberhard 1977a).

## EVOLUTION OF QUEEN CONTROL

A central theme in the evolution of eusocial wasps from solitary ancestors is the development and maintenance of reproductive dominance by one or a few females, the queens. Queen control is manifested within groups of founding females and also between foundress and offspring and thus involves mechanisms whereby queen number is regulated and a worker caste is maintained. As wasps evolved from simple societies composed of relatively few individuals to the highly populous colonies seen in contemporary swarming polistines and in vespines, mechanisms of queen control have inevitably increased in their complexity, with a trend toward a more indirect and pervasive influence that extends beyond exclusively behavioral dominance.

### Confrontation, Competition, and Compromise

When two females meet at a nest site or on a nest structure, they are faced with a variety of options—they may ignore and pass by one another, they may confront each other with the departure of one of them, they may coexist amicably and share the nest, or, through aggressive encounters, one may dominate the other and enhance its reproductive success at the other's expense. Some of these options represent milestones along the path to eusociality, and, with subtle variations in between, most of these stages in the evolution of queen control in wasps are represented by examples from the contemporary vespid fauna.

The peaceful coexistence of several foundress females that share a nest on apparently equal reproductive terms has been observed in *Polistes humilis* (Spradbery, unpubl.) in Australia, and, perhaps surprisingly, occurs in the hornet *Vespa affinis* in New Guinea, where founding queens apparently share reproductive responsibilities equally throughout the life of the colony (Spradbery 1986). Coexistence between parent and offspring with equality in reproductive contribution is much rarer. Several examples have been noted in the Stenogastrinae, although the relative contributions by mother and daughter to reproductive output remain unclear. Such cooperative societies represent a primitive stage in the development of social organization, portraying an early stage in social evolution that preceded a reproductive division of labor. The more familiar and frequently observed phenomena are, however, the conspicuous conflicts between associating females during the establishment of nests.

When competition occurs, overt physical dominance behavior characterizes the early encounters among foundresses. An array of behavioral components has been described, from the relatively mild confrontations seen in many tropical *Ropalidia* and *Polistes* to extremes of physical combat whose violence can lead to the death of competing queens. Such extreme dominance behavior can result in one female becoming a total despot over the remaining co-foundresses, as in *P. erythrocephalus* (West-Eberhard 1969). More often, such encounters between females result in a more-or-less linear hierarchy becoming established in which the dominant female is overtly aggressive toward one or a few subordinates immediately below it in the hierarchy, who in turn dominate those below them—economizing in both time and energy (Pardi 1942a, 1946, 1947, 1948). Some dominance hierarchies can be more circuitous than linear (e.g., *R. cyathiformis*: Gadagkar and Joshi 1982b), or intermediate (e.g., *Belonogaster*: Pardi and Marino Piccioli 1970; see Fig. 10.4). All such hierarchies imply some basic inequalities among the competing individuals, a point dealt with in more detail above. The development of more-ritualized physical dominance results in even more economizing of energy. Recognizable but subtle body movements by dominants result in the reinforcement of the social hierarchy as queens advertise their dominance status with threat postures. Examples include *Metapolybia aztecoides* (West-Eberhard 1973, 1977a) and *Protopolybia acutiscutis* (Naumann 1970).

While physical dominance behavior is the hallmark of primitively eusocial wasps, other behavioral characteristics have evolved whereby queens can maintain their status, and these are often accompanied by a decrease in overt fighting (e.g., Röseler, this volume). Although the principal manifestation of queen control is the inhibition of ovarian development in subordinates or workers, when subordinates do lay eggs dominant females in many species recognize and eat them ("differential oophagy"). Apart from destroying competitors' eggs, queens of some species of swarming Polistinae guard the cells in which they have recently oviposited.

Among other measures to control reproduction in the colony is the queen's influence over cell building. To ensure that few vacant cells ("oviposition resource") are available to potential competitors, queens may be the sole or predominant initiators of the cell base (e.g., *Polistes erythrocephalus*: West-Eberhard 1969; *P. metricus*: Dew 1983) and thus have first call on the oviposition resource. Alternatively, queens may ensure that there are few vacant cells by ovipositing in all newly constructed and recently vacated cells. Perhaps a paucity of empty cells stimulates continued new-cell construction by subordinates or workers in an apparent attempt to subvert the queen's reproductive dominance. However, if vacant cells do accumulate, construction rates tend to decrease—an apparent negative feedback phenomenon. Concurrent with their role as reproductive pacemaker, queens most likely control the activity of the nest by their interactions with the other members of the colony, raising activity levels generally and stimulating foraging and nest building in particular (e.g., Reeve and Gamboa 1983, Reeve, this volume).

In large nests with dozens of adults, the capacity to physically dominate so many females may well elude even the most active and dominant queen, and there is evidence that in such cases some degree of queen control is lost (e.g., Strassmann 1983). A reduction in the amount of direct dominance effort required by queens in large colonies can be achieved by the use of "worker control," such that workers help suppress reproduction in competing or newly emerged queens through worker dominance or oophagy (West-Eberhard 1977a). In large colonies, queens must increase their activity to maintain control or perhaps focus their attention on particular adults who in turn regulate the activity of subgroups of other adults. These "subsidiary regulators" may have a high dominance rank and even share in personal reproduction (Reeve and Gamboa 1983).

## From Physical to Chemical Dominance

While queen recognition through behavioral characteristics such as postures and dances may considerably reduce the effort required in queen control, close physical contact among colony members is still necessary. Recognition of queen status by means of chemical cues introduces a new dimension with enormous potential for increasing the sphere of influence of the queen. *Protopolybia acutiscutis* workers recognize queens and lick them before initiating a swarm (Naumann 1970), while *Metapolybia aztecoides* workers distinguish dismembered and squashed queens from other colony members, indicating that queen recognition by odor alone is likely in large polistine colonies (West-Eberhard 1977a). Chemical cues also may function to maintain dominance hierarchies of *Polistes*. In Downing and Jeanne's (1983) survey of glandular activity in *P. fuscatus*, the Dufour's gland and the fifth sternal gland were most active in aggressive dominant females, suggesting that secretions from such glands may be involved in the communication of dominance. However, there is as yet no conclusive experimental evidence that *Polistes* queens control ovarian development by chemical means (Reeve and Gamboa 1983).

Communication of status by ritualized postures can be effectively reinforced by a chemical odor cue, as in *M. aztecoides* (West-Eberhard 1978b). Recognition of the dominant female may then be sufficient to ensure the maintenance of hierarchical status. There is accumulating evidence that such indirect, chemically supplemented control also occurs in other swarming polistine species such as *Protopolybia acutiscutis*, *Agelaia pallipes*, and *Polybia occidentalis* (Naumann 1970, Forsyth 1978). *Protopolybia exigua* workers form courts around queens, there is little or no aggressiveness between castes or among queens, and oophagy is not practiced (Simões 1977).

With the advent of queen-recognition chemicals acting as warning signals to declare rank and status, the evolution of mode of action of such chemicals from a releaser (recognition) effect to a primer (inhibition) effect is but a short step. Indeed, these two effects could be interpreted as one and the same phenomenon if the end result is reproductive suppression (Fletcher and Ross 1985). As the chemical or pheromonal component of queen control increases, the capacity of the queen to extend her sphere of influence beyond direct physical contact becomes possible.

Queen control by a combination of pheromonal and physical elements, with the pheromonal element playing the lead role, seems to occur in the vespine genus *Dolichovespula*, although direct experimental evidence for a primer pheromone is lacking. *Dolichovespula* has relatively small colonies of some hundred or so workers. Queens display some of the behavioral elements of polistine queen control, and suppression of worker ovarian development is not fully effective even in the early part of the colony cycle when the queens are relatively young and vigorous (Greene, this volume). A further decrease in physical dominance with a presumed increase in pheromonal control occurs in the *Vespula rufa* group, which is also characterized by small colonies with up to 400 workers. Even in this group, however, physical acts of dominance have been observed in *V. atropilosa* (Landolt et al. 1977), and incomplete suppression of worker ovarian development occurs in *V. acadica* (Reed and Akre 1983d).

The source and method of dissemination of presumed pheromones are better known in the hornets (*Vespa* species) than in *Dolichovespula* and the *V. rufa* group. In *Vespa* (and to some degree in *D. maculata* [Greene 1979]), the queen stimulates nearby workers to antennate and lick her body and especially her head, with groups of workers forming royal courts. The significance of these courts has not been resolved. Although it would appear that such contacts could provide an ideal method for disseminating a queen pheromone, there is no experimental evidence that queen control of worker ovarian development is a direct result of such contacts. In some *Vespa* species, such as *V. crabro*, direct worker-queen contacts do not even occur in the early stages of nesting—a time when ovarian suppression in workers is wholly effective (Nixon 1985a). Furthermore, Ishay et al. (1986) found that ovarian development of *Vespula germanica* maintained with a *Vespa orientalis* queen was suppressed, although there was no direct contact between the alien workers and the hornet queen. There is no doubt that the pheromone isolated from *V. orientalis* queens controls some aspects of social organization such as queen-cell construction, but no evidence has so far been presented that the hexadecalactone suppresses worker ovarian development (Ishay et al. 1965, Ikan et al. 1969).

The *Vespula vulgaris* group contains species with the most populous of all vespine colonies. Although reciprocal feeding between the queen and the workers occurs sporadically, there is no consistent or widespread physical contact, yet the queen's influence in suppressing ovarian development in workers is completely effective, at least until the annual colony development cycle passes its peak. Experimental evidence to date suggests that the queen exerts control by means of a chemical or chemicals and that this pheromone is not very volatile (Akre and Reed 1983, Spradbery, unpubl.). If the pheromone is not airborne, how can the queen disseminate the substance so widely throughout the nest, which may have up to 14 or more tiers of combs with as many as 15,000 cells?

During her oviposition activities, the queen moves over most combs in a nest, laying up to 300 eggs per day in a mature colony. It is during these activities that the queen must deposit her pheromone. The pheromone could be transmitted through the eggs themselves, by reciprocal feeding contacts with the larvae, or by applying the pheromone directly to the comb structure. In recent unpublished experiments on *V. germanica*, I removed queen-laid eggs from the cells prior to presenting queen comb to queenless worker groups at twelve-hour intervals. The lack of eggs did not reduce the effectiveness of queen-contaminated comb in fully suppressing ovarian development in the queenless worker groups. Conversely, presenting the queen's eggs in uncontaminated comb had no effect in suppressing ovarian development. When comb with no larvae was used in such comb-swap experiments, the combs that had been exposed to queens were totally effective in suppressing worker ovarian development. The queen pheromone must therefore be applied directly to the carton of the nest and is not disseminated through eggs or larvae.

Many social wasps, including the swarming and independent-founding polistines, use sternal glands in the gaster to produce substances applied to the nest and to other structures (see Jeanne et al. 1983,

Downing, this volume). It is tempting to speculate that the paired van der Vecht's glands and associated brushlike structures of the sixth sternum are the source and applicator of the queen pheromone that suppresses ovarian development in *Vespula* workers. The glands are proportionally better developed in queens than in workers and the brush of hairs is remarkably similar to homologous structures in the Polistinae. There is little doubt that during the long evolutionary history of the Vespidae the sternal glands have played prominent roles in chemical communication—perhaps they have reached their peak of development in *Vespula* species with the production of the most potent of all social pheromones.

## CONCLUDING REMARKS

In 1963, when I was completing postgraduate studies on British vespine wasps at Rothamsted Experimental Station, a few miles away W. D. Hamilton was correcting the proofs of his seminal papers on the genetic theory of social behavior. Since then, Hamilton's concepts of inclusive fitness, altruism, and the importance of haplodiploidy have played pivotal roles in the development of social insect evolutionary theory. While competing, or perhaps more correctly, complementary theories have been developed, notably by Alexander (1974) and by West-Eberhard (1975, 1978a, 1981), Hamilton's (1963, 1964a,b) contributions remain the cornerstone of modern theory on social evolution in insects. In part, this chapter represents a celebration of the Hamilton era.

And what of the future? Naturalists will continue to build up that most important resource—observations of biology, ecology, and behavior in the field—despite the growing trends against such studies in favor of short-term laboratory projects. Perhaps the world's growing concern with the accelerating deterioration of the environment will help stimulate and finance such research before too many pristine habitats are lost forever. Modern science with its powerful computing capacity will continue to stimulate the development of increasingly sophisticated population and quantitative genetic models. Undoubtedly, novel genetic approaches using recombinant DNA methods will be applied to studies of population change and genetic differentiation, resulting in significant progress, especially in the areas of evolution and behavior. We are already on the brink of defining the chemical structure of pheromones that influence social behavior, especially the queen pheromones that suppress ovarian development.

I believe that the present imbalance between theoretical and empirical studies on social wasps will be corrected as modern technology stimulates more effective, faster, and more reproducible research from the field and laboratory. The next 25 years will certainly be remarkably productive, and I hope that I will be around to marvel at and enjoy the pleasures of wasp research well into the twenty-first century.

## Acknowledgments

I am grateful to the following for their comments on an earlier draft of this chapter: J. Carpenter, J. Ishay, Y. Itô, R. Jeanne, J. Kojima, and K. Ross. The artwork was done by A. Carter and C. Hunt. I thank I. Pumpurs for typing the manuscript.

# 11

# Polyethism

ROBERT L. JEANNE

One of the most striking features of a colony of eusocial insects is the existence of extreme behavioral differences among its members. Not only is there the fundamental divergence into reproductive and worker castes, but the workers themselves may be specialized into age or morphological subcastes. Even within a subcaste, individuals vary in frequency, rate, sequence, and ontogenetic timing of task performance. Variability among individuals may be so extreme that it is probably safe to say that no two individuals in a colony have the same lifetime behavioral repertoires. Yet despite—or because of—such varied behavior among its members, the colony manages to function in a seemingly coordinated way. In fact, the remarkable ecological success of the social insects (Jeanne and Davidson 1984) has been attributed largely to the colony's ability to feed, defend, and reproduce itself through the coordinated activities of its specialized members (Oster and Wilson 1978, Wilson 1985). There is at least a modicum of order to the variability, so that we can speak of a division of labor, or polyethism, among colony members.

Playing, as it does, such a central role in the organization of insect colonies, polyethism has not surprisingly attracted considerable attention from insect sociobiologists. Caste specialization has been described in numerous species in all four groups of eusocial insects, and much effort has been devoted to understanding the ontogenetic pathways leading to caste differentiation, particularly of morphological castes (for reviews see Wilson 1971; Schmidt 1974; Brian 1979, 1980, 1985; Wheeler 1986). In recent years, increasing attention has been given to the mechanisms by which the activities of various castes are integrated (Wilson 1971) and to the ecology and evolution of castes (Wilson 1971; Oster and Wilson 1978; West-Eberhard 1979, 1981). We are learning that how an individual colony member behaves is influenced by many factors, including its genetic makeup, trophic history, age, experience, social environment, and external environment.

Polyethism in social wasps has received relatively little attention compared with that in termites, ants, and bees. Most of the attention has been concentrated on the cosmopolitan and easily studied genus *Polistes*; the remaining genera are either tropical in distribution or difficult to observe because of their enclosed nests, or both. Yet it is well worth the effort to continue to extend the study of polyethism to the rest of the social wasps, for the group includes within it a broad range of polyethic specialization, from no specialization at all in solitary forms to morphological castes and well-developed age polyethism in the most advanced social forms. Not only are there numerous primitively eusocial species that can yield insights into how polyethism evolved, but higher levels of social behavior in wasps appear to have had several independent phylogenetic origins (see Carpenter, this volume), providing a rich field for comparative analysis.

This chapter is divided into two main sections. The first section surveys the distribution of polyethism in social wasps and identifies patterns. The focus is on the tasks performed by different castes in all taxa. The second section considers the evolution of polyethism. Throughout, the social wasps are divided into four behavioral groups: the Stenogastrinae, the independent-founding Polistinae, the swarm-founding Polistinae, and the Vespinae (see Jeanne 1980a).

## PATTERNS OF POLYETHISM

Reproductive division of labor, in which one or a few individuals specialize on reproduction while the remainder concentrate on non-reproductive tasks, is the most fundamental form of polyethism. As one of the three defining criteria of eusociality, along with cooperation in brood care and overlap of adult generations, it is universal in the eusocial insects (Wilson 1971). Reproductive division of labor in the wasps ranges from an incomplete functional separation among females that are morphologically alike to a division of reproductive and worker functions into two discrete castes that have evolved substantial physiological and morphological specializations (Jeanne 1980a), albeit not as extreme as in the most advanced bees, ants, or termites.

In addition to reproductive division of labor, there is worker polyethism, that is, the differential performance of tasks by the workers.

Again, the amount of specialization varies in degree among species of social insects, from no discernible worker specialization, through age polyethism, or temporal division of labor, in which workers change roles as they age, to the extreme levels of specialization correlated with the morphologically specialized worker subcastes found in some of the ants and in most termite species (Wilson 1971). There is no evidence for morphological worker subcastes among the social wasps, but age polyethism does occur.

This section is divided into four parts. The first deals with reproductive division of labor and examines the task divergence between the queen and her subordinates, the second reviews the evidence for age polyethism among workers, and the third discusses the concept of task partitioning. Finally, variability among individuals in performance of tasks is considered.

### Reproductive Division of Labor

Reproductive division of labor occurs in two contexts. First, there is polyethism within groups of colony-founding females. Such groups of associating foundresses occur facultatively in many populations of stenogastrines, independent-founding polistines, and in some of the tropical vespines, and obligately in the swarm-founding polistines (Richards and Richards 1951, Richards 1978a, West-Eberhard 1978a, Matsuura 1983b, Turillazzi 1986a, see also Spradbery, this volume). Their distinguishing feature is that the foundresses belong to the same generation. The other context is reproductive division of labor between the colony foundress(es) and the nonreproductive workers reared by the colony. Reproductive division of labor in this context normally separates along generational lines, such that the reproductives are of the parental generation, while the workers are the offspring.

Foundress groups of the independent-founding Polistinae are characterized by conflict over reproductive rights. Agonistic interactions lead to the establishment of a dominance hierarchy, in which rank is positively correlated with degree of direct reproductive success (Pardi 1946, 1948; West-Eberhard 1969; Marino Piccioli and Pardi 1970; Jeanne 1972; Litte 1977, 1979, 1981; Pardi and Marino Piccioli 1981; Kojima 1984c; Yamane 1985, 1986; but see West-Eberhard 1986). The top-ranked female (the $\alpha$-female) has the best-developed ovaries and lays the most eggs; the $\beta$-female lays the next most eggs, and so on down to the $\Omega$-female, who may lay no eggs. As the hierarchy becomes better established, the divergence between the $\alpha$-female and her subordinates in degree of ovarian development and rate of oviposition becomes increasingly pronounced, until eventually the $\alpha$-female may become the exclusive reproducer (Röseler, this volume). In at least some species of independent-founding polistines, the first offspring are also dominated physically by the α-female and her higher ranking associates, and after a day or two the offspring begin to perform nonreproductive tasks (Pardi 1946, 1948; Jeanne 1972).

A hierarchy of task performance accompanies the dominance hierarchy. The females on the nest diverge along this task hierarchy such that tasks less directly related to reproduction are abandoned by the more-dominant individuals and are taken up by the least dominant females. The tasks of the colony can be arranged roughly in the sequence in which dominant females perform them or give them up and subordinates or workers assume them (Table 11.1). First in this sequence is oviposition, the task most directly identified with reproduction. Following it are tasks progressively less related to direct reproduction or to inhibition of reproduction by nestmates. To what extent the queen retreats from performing worker tasks varies among taxa, as shown in Table 11.1.

A similar task displacement occurs as workers eclose, and is most clearly seen when a queen founds alone, as do most vespine queens. As workers appear, the queen first gives up foraging for prey and nectar, although the queen of many species of independent-founding polistines may continue to solicit loads from returning foragers, malaxate the loads, and feed them to larvae (Owen 1962, West-Eberhard 1969, Sugiura et al. 1983a, Turillazzi 1983, Yamane et al. 1983c, Kojima 1984c, Yamane 1986) (Table 11.1). Queens typically do not forage for nectar, although *Mischocyttarus drewseni* and *Polistes instabilis* queens do so rarely (Jeanne 1972, S. O'Donnell, unpubl.).

Queens in most taxa also abandon nest construction except for cell initiation. Many species of independent-founding Polistinae exhibit a close temporal association between cell initiation and oviposition, such that the queen constructs a new cell just before ovipositing in it (Spieth 1948, Deleurance 1950, West-Eberhard 1969). West-Eberhard (1969:21) emphasized that "cell initiation and pulp foraging by queens are characteristics directly associated with developed ovaries, as is dominance as shown by the experiments of Pardi" (1946, 1948). In some of these species the queen forages for the pulp she uses to initiate cells (Table 11.1); in others she takes pulp from returning foragers.

Queens of some species of independent-founding polistines continue to rub the petiole more frequently than subordinates or workers (Table 11.1). For several species (see Jeanne et al. 1983) this behavior has been shown to function in applying an ant repellent secretion to the petiole of the nest. However, the possibility that the secretion may have a communication function among nestmates as well has not been investigated (e.g., Downing, this volume).

Swarm-founding polistines are a special case in that inseminated

**Table 11.1.** Reproductive division of labor and task specialization in eusocial vespids

| Species | Social acts[a] | | | | | | | | | | | | Reference |
|---|---|---|---|---|---|---|---|---|---|---|---|---|---|
| | 1 | 2 | 3 | 4 | 5 | 6 | 7 | 8 | 9 | 10 | 11 | 12 | |
| **Stenogastrinae** | | | | | | | | | | | | | |
| *Parischnogaster jacobsoni* | QQ | ? | ? | ? | ? | ? | — | ? | ? | ? | ? | ? | Turillazzi 1988a |
| | ? | ? | ? | ? | ? | ? | WWW | ? | ? | ? | ? | ? | |
| *P. mellyi* | QQ | ? | QQ | NA | Q | QQQ | — | Q | ? | — | — | — | Yamane et al. 1983c |
| | — | — | S | NA | S | — | S | SS | ? | SS | SS | SS | |
| *P. nigricans* | QQ | — | ? | ? | ? | ? | — | ? | ? | ? | ? | ? | Turillazzi 1983 |
| | ? | | ? | ? | ? | ? | WWW | ? | ? | ? | ? | ? | |
| **Independent-founding Polistinae** | | | | | | | | | | | | | |
| *Belonogaster grisea*[b] | QQQ | QQ | QQ | QQ | QQ | QQ | ? | ? | ? | ? | ? | ? | Marino Piccioli and Pardi 1970, Pardi and Marino Piccioli 1981 |
| | W | — | W | W | ? | ? | ? | ? | ? | ? | ? | ? | |
| *B. petiolata* | QQ | ? | ? | Q | ? | ? | ? | ? | ? | ? | ? | ? | M. Keeping, unpubl. |
| | ? | ? | ? | SS | ? | ? | ? | ? | ? | ? | ? | ? | |
| | ? | ? | ? | W | ? | ? | ? | ? | ? | ? | ? | ? | |
| *Mischocyttarus atramentarius* | QQ | QQ | QQ | ? | ? | QQ | Q | ? | ? | ? | ? | ? | Silva 1981 |
| | S | S | S | ? | ? | S | S | ? | ? | ? | ? | ? | |
| | ? | ? | ? | ? | ? | ? | WW | ? | ? | ? | ? | ? | |
| *M. drewseni* | QQ | QQ | QQ | Q | ? | QQ | Q | Q | Q | Q | — | — | Jeanne 1972 |
| | S | S | — | S | ? | S | S | S | S | S | S | S | |
| | | | | WW | WW | | WW | WW | WW | WW | WW | WW | |
| *M. flavitarsis* | QQ | QQQ | QQ | ? | ? | QQ | Q | Q | Q | ? | — | — | Litte 1979 |
| | S | — | S | S | S | ? | S | S | S | S | S | S | |
| | | | W | ? | W | ? | WW | WW | WW | WW | WW | WW | |
| *M. labiatus* | QQ | QQQ | QQ | QQ | QQ | QQ | Q | Q | Q | Q | — | — | Litte 1981 |
| | SS | — | S | SS | SS | S | S | S | S | S | S | S | |
| | | | — | ? | W | — | WW | WW | WW | WW | WW | WW | |

**Table 11.1**—*continued*

| Species | Social acts[a] | | | | | | | | | | | | Reference |
|---|---|---|---|---|---|---|---|---|---|---|---|---|---|
| | 1 | 2 | 3 | 4 | 5 | 6 | 7 | 8 | 9 | 10 | 11 | 12 | |
| *Parapolybia indica* | QQ | ? | ? | ? | ? | QQ | Q | ? | ? | ? | ? | ? | Sugiura et al. 1983a |
| | ? | ? | ? | ? | ? | ? | WW | ? | ? | ? | ? | ? | |
| *P. varia* | QQ | — | QQ | ? | ? | Q | — | ? | ? | ? | — | — | Yamane 1980, 1985 |
| | S | — | S | ? | ? | S | S | ? | ? | S | S | S | |
| | — | — | W | ? | ? | WW | WW | ? | ? | WW | WW | WWW | |
| *Polistes annularis* | QQ | Q | QQ | QQ | ? | QQ | ? | ? | ? | ? | ? | — | Hermann and Dirks 1975, Strassmann 1981a |
| | S | S | S | S | ? | S | ? | ? | ? | ? | ? | SS | |
| *P. chinensis* | QQ | — | QQ | — | ? | — | Q | QQ | ? | Q | ? | Q | Kasuya 1983c, Miyano 1986 |
| | W | WW | W | WW | ? | WW | WW | W | ? | WW | ? | WW | |
| *P. dominulus* | QQ | ? | QQ | Q | ? | QQ | ? | ? | ? | ? | ? | ? | Pardi 1948, Deleurance 1950, Turillazzi and Ugolini 1979 |
| | S | ? | ? | S | ? | S | ? | ? | ? | ? | ? | ? | |
| *P. dorsalis* | QQ | ? | ? | ? | ? | QQQ | — | ? | ? | ? | ? | ? | Spieth 1948 |
| | — | ? | ? | ? | ? | — | WWW | ? | ? | ? | ? | ? | |
| *P. erythrocephalus* | QQ | ? | ? | ? | ? | QQ | Q | ? | ? | ? | ? | ? | West-Eberhard 1969 |
| | S | ? | ? | ? | ? | S | S | ? | ? | ? | ? | ? | |
| | ? | ? | ? | ? | ? | ? | WWW | ? | ? | ? | ? | ? | |
| *P. fuscatus* | QQ | ? | QQ | ? | ? | QQ | Q | ? | ? | ? | — | — | Owen 1962, West-Eberhard 1969 |
| | S | ? | S | ? | ? | S | S | ? | ? | ? | ? | ? | |
| | — | — | — | ? | ? | — | WW | WW | ? | WW | WW | WW | |

Note: Division of labor among queens, subordinates, and workers is shown for colonies that have produced offspring.

[a] 1, oviposit; 2, eat eggs laid by others; 3, dominate nestmates (degree); 4, rub petiole (apply ant repellent); 5, defend nest against nonnestmates; 6, initiate cells; 7, forage for pulp; 8, feed prey to larvae; 9, feed nectar to larvae; 10, nest construction other than initiate cells (includes build ant guard); 11, forage for nectar; 12, forage for prey. For each species there are up to three rows, each representing one of the three classes of adults: Q, top dominant or queen; S, subordinate co-foundress; W, worker offspring. The level of performance of each task by each class of adult is indicated by the number of symbols: e.g., QQQ, queen performs all, or virtually all, of the tasks; QQ, queen performs most such tasks; Q, queen performs the task only rarely; —, task never performed by that class of adult; ?, data unavailable; NA, task not performed by the species.

[b] Q, inseminated females; W, uninseminated females.

| Species | Class | 1 | 2 | 3 | 4 | 5 | 6 | 7 | 8 | 9 | 10 | 11 | 12 | Reference |
|---|---|---|---|---|---|---|---|---|---|---|---|---|---|---|
| P. metricus | Q | QQ | QQQ | QQ | QQ | Q | ? | ? | Q | ? | ? | ? | — | Gamboa et al. 1978, Dew 1983 |
| | S | S | — | S | S | S | ? | ? | S | ? | ? | ? | SS | |
| | W | ? | ? | ? | WW | ? | ? | ? | WW | ? | ? | ? | ? | |
| P. snelleni | Q | QQQ | ? | QQQ | Q | Q | ? | ? | ? | ? | ? | ? | ? | Suzuki 1987 |
| | S | ? | ? | — | WW | WW | ? | ? | ? | ? | ? | ? | ? | |
| | W | ? | ? | ? | ? | ? | ? | ? | ? | ? | ? | ? | ? | |
| P. versicolor | Q | QQ | ? | QQ | Q | Q | ? | ? | ? | ? | ? | ? | ? | Gobbi 1977 |
| | S | S | ? | S | S | S | ? | ? | ? | ? | ? | ? | ? | |
| | W | ? | ? | ? | WW | WW | ? | ? | ? | ? | ? | ? | ? | |
| Ropalidia fasciata | Q | QQ | ? | QQ | ? | QQ | ? | ? | QQ | ? | ? | ? | ? | Kojima 1984c |
| | S | ? | ? | S | SS | S | ? | ? | S | ? | ? | ? | ? | |
| | W | ? | ? | ? | ? | ? | ? | ? | ? | ? | ? | ? | ? | |
| R. variegata jacobsoni | Q | QQ | ? | QQ | QQ | Q | ? | ? | QQ | ? | ? | ? | ? | Yamane 1986 |
| | S | W | ? | W | WW | WW | ? | ? | W | ? | ? | ? | ? | |
| | W | ? | ? | ? | ? | ? | ? | ? | ? | ? | ? | ? | ? | |
| **Vespinae** | | | | | | | | | | | | | | |
| Vespa crabro flavofasciata | Q | QQQ | — | NA | — | — | Q | ? | Q | ? | Q | — | — | Matsuura 1974, 1984 |
| | W | — | — | NA | WWW | WWW | WW | ? | WW | ? | WW | WWW | WWW | |
| V. orientalis | Q | QQQ | — | NA | — | — | — | — | — | — | — | — | — | Darchen 1964 |
| | W | — | — | NA | WWW | WWW | WWW | WWW | WWW | WWW | WWW | WWW | WWW | |
| V. tropica | Q | QQQ | — | NA | — | — | Q | ? | Q | ? | Q | — | — | Matsuura 1974 |
| | W | — | — | NA | WWW | WWW | WW | ? | WW | ? | WW | WWW | WWW | |
| Vespula vulgaris | Q | QQQ | — | NA | — | — | — | — | — | — | — | — | — | Spradbery 1965 |
| | W | — | — | NA | WWW | WWW | WWW | WWW | WWW | WWW | WWW | WWW | WWW | |

queens are accompanied by workers even during colony founding (Jeanne, this volume: Chap. 6). Thus, functional queens never have to perform worker duties, so that reproductive division of labor is more complete than in other groups. As far as is known, workers do all construction and other tasks, except oviposition (Naumann 1970, Simões 1977, Forsyth 1978).

Vespine queens found colonies alone (except for certain tropical species: Matsuura and Yamane 1984; Matsuura 1985, this volume) and so must perform the full range of tasks until workers begin to eclose. The queen does not physically dominate her worker offspring, yet they soon relieve her of most of the nonreproductive tasks.

According to Matsuura (1984), no *Vespa* queen initiates cells after workers have emerged. Queens of *Vespa crabro flavofasciata* continue to forage for pulp for about three weeks after the first workers emerge, but they do not use this pulp to initiate cells (Matsuura 1974). After workers emerge, the queen ceases building activities in the following order: cell initiation, cell and envelope enlargement with self-foraged material, the same with material she collects from within the nest, re-working of wet cell walls without adding new material. The queen of the social parasite *Dolichovespula arctica* does construct new cells, using fiber chewed from inner envelope layers (Greene et al. 1978), but it has not been reported whether this behavior is closely followed by oviposition.

In summary, as queens acquire helpers in the form of either subordinate co-foundresses or worker offspring, they give up tasks roughly in the order: defending the nest against intruders, foraging for food, foraging for pulp (especially to heighten cells and build the envelope), feeding larvae, initiating cells. In other words, queens give up first those tasks that are risky and/or serve the role of food procurement, while retaining those most closely related to enhancing their own personal reproduction and inhibiting that of their nestmates. In general, the larger the species-typical colony size, the more quickly and completely the queen gives up nonreproductive tasks (Matsuura 1984, this volume).

## Age Polyethism among Workers

Several species of independent-founding polistines show evidence of a rudimentary age-related biasing of task performance. Pardi (1950) recognized three temporal stages in workers of *Polistes dominulus* ( = *gallicus* of authors): the juvenile phase, age 1–5 days, during which the young female works on the nest; the construction phase, age 6–20 days, during which construction activity peaks and foraging occurs; and the senile phase, from age 20 days, when construction activity declines while foraging for nectar and water continues. Post et al. (1988) found that young workers of *Polistes fuscatus* tend to remain on the nest, receiving pulp and prey. At age 7–14 days they begin to forage for pulp and prey as well as to build and to feed larvae. The first worker of *Polistes metricus* to eclose onto the nest begins to forage within just a few days and typically remains the most active forager. In comparison, later emerging workers are slower to start foraging and forage at lower rates (Dew and Michener 1981). Evidence of a similar tendency has been reported for *Polistes jadwigae* (Yoshikawa 1963) and *P. fuscatus* (West-Eberhard 1969). In one colony of *P. fuscatus*, all foraging observed on August 31 was performed by nine workers that were at least 44 days old plus an unmarked worker with frayed wings, evidently also an old individual. These females were much more active and aggressive than younger ones. In contrast to those species, *Belonogaster petiolata* appears to lack age polyethism or any other kind of task specialization among workers (M. Keeping, unpubl.).

In the Vespinae, the occurrence of worker age polyethism apparently also is variable. Matsuura (1984), who has done thorough studies on the Japanese species of *Vespa*, concludes that there is minimal worker age polyethism in the genus. Workers of five Japanese *Vespa* species begin foraging at age 2–4 days, and it is clear that as workers age they continue to carry out intranidal tasks along with foraging, rather than giving them up as younger workers emerge, as might be expected under age polyethism. Even though workers in these *Vespa* species begin to forage at age 2–4 days and continue to do so while performing nest duties the rest of their lives, there is a tendency in *Vespa simillima* and *V. analis* for foraging to increase in frequency with age, reaching a peak at age 7 days, while nest duties drop in frequency. *Vespa orientalis* shows a more clear-cut age polyethism. Workers begin to fly out of the nest at about six days of age, and prey foragers share their loads with nest workers that are generally less than six days old (Rivnay and Bitinsky-Salz 1949, cited in Darchen 1964). A. Greene (unpubl.) states that there is a weak but detectable tendency for pulp foraging to be done early in the foraging life of a *Vespa crabro* worker.

A similar picture of weakly developed age polyethism emerges for *Dolichovespula*. Gaul (1948) found that the first workers of *Dolichovespula arenaria* to emerge engage almost exclusively in nest construction and in foraging and brood nursing combined, implying that there is no specialization into nest workers and foragers. He went on to say, however, that about two weeks after the first workers have emerged distinct brood nursing specialists appear in the worker population. Gaul presumably meant that foraging and nursing are carried out by different individuals (see below). This description suggests that a change occurs; there is little or no age specialization among the first workers in the colony, but separation of on-nest duties from foraging becomes more pronounced among workers emerging later, when the colony population is larger. Brian and Brian (1952) found that workers of *Dolichovespula sylvestris* in Great Britain begin to forage on their first or second day and concluded that there was no age polyethism, although they did note that workers gradually change from pulp collection to food collection with age. For example, during a single day, pulp accounted for 59% of the loads brought back by young workers, while only 15% of the older workers' loads were pulp. Older workers more frequently foraged for fluid (Brian and Brian 1952).

Age polyethism in which workers start as nurses then progress to foraging does occur in at least some species in the genus *Vespula*. This pattern has been shown to occur in *Vespula flaviceps* (Shida 1959, cited in Matsuura 1984), *V. vulgaris* (Potter 1964), *V. atropilosa*, and *V. pensylvanica* (Akre et al. 1976) (Fig 11.1). *Vespula germanica* workers help in nest building during their first few days, then begin to feed larvae; finally some workers add foraging to their repertoire (Montagner 1966a). Shida (1959, cited in Matsuura 1984) reported that nest materials are collected primarily by older workers of *Vespula flaviceps*. In contrast, early workers of *V. atropilosa* and *V. pensylvanica* (Akre et al. 1976) begin foraging for pulp, and add prey and liquid only when they are older. As they continue to age they drop prey, then pulp, specializing in liquid, and finally they become inactive (Fig. 11.1). Similarly, Potter

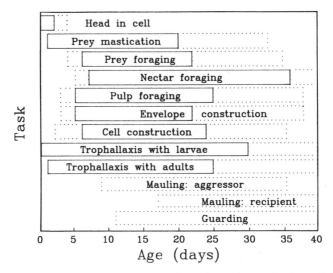

**Fig. 11.1.** Age polyethism among early workers of *Vespula pensylvanica* and *V. atropilosa*. Solid boxes represent age range during which most early workers perform the task. Dotted boxes represent the maximum age range observed. Prey mastication and trophallaxis with larvae are brood-nursing tasks. (Modified from Akre et al. 1976.)

(1964) showed that *V. vulgaris* workers spend their first week as nurses, then begin to forage, specializing in pulp, with a few trips for fluid. After a few days the rate of pulp foraging drops, while prey and fluid foraging increase. Pulp foraging usually ceases by age 18 days, while prey collection is heaviest for workers 14–22 days old. Fluid collection rates are low initially, but increase steadily as workers age. Beyond 24 days of age pulp and prey foraging cease, and workers forage for fluid only. After 30 days, foraging rates slow, and workers spend increasing amounts of time near the nest entrance acting as guards.

The conclusion of some authors that *Vespa* and *Dolichovespula* workers may generally lack age polyethism must be accepted cautiously. There is evidence for both *Polistes* and *Vespula* that the first-emerging workers begin to forage at several days of age, while those emerging when the colony is larger do not begin for a week or more (Potter 1964, Akre et al. 1976, Dew and Michener 1981, Strassmann et al. 1984a). Therefore, if a study focuses on the first cohort of workers it may yield the conclusion that the typical worker begins to forage when very young, when in fact later-emerging workers may show a clear-cut sequence of nest work followed only later by foraging. Brian and Brian's (1952) study of *Dolichovespula sylvestris*, for example, appears to be a case in point. Their data, collected between June 11 and 22, evidently came from the first 15 workers to emerge.

Similarly, the tendency, reported for several species, for foragers to concentrate on pulp when young and on prey when older, may in part be a reflection of changing colony needs. As these annual colonies reach peak size, it is likely that the need for pulp drops while the demand for prey to feed the large population of larvae is still growing. Studies that follow a single, narrowly defined cohort of workers as it ages are particularly susceptible to the potentially biasing influences of shifting resource needs as the colony ages. Nevertheless, the frequent reports of such an age sequence suggest that it may exist. Future studies should be designed to control for the effects on worker roles of shifting needs as the colony develops.

Matsuura (1984) suggests that the apparent lack of age polyethism among *Vespa* workers may be related to the lack of glandular activity associated with particular tasks in these wasps. This explanation does not seem probable. In the first place, it is likely that the development of particular exocrine glands is indeed associated with certain activities in wasps. Stenogastrine wasps secrete a milky abdominal substance that serves as a substrate for the eggs and larvae (Turillazzi and Pardi 1982; Turillazzi 1985c, this volume). The labial gland, which appears to secrete the chitin-containing saliva used as a binding matrix to form the nest carton of *Pseudochartergus* (Schremmer et al. 1985) and probably other genera as well, is more developed in workers than in queens of

*Protopolybia* (Machado and Rodrigues 1972) and may be active only in workers engaged in building. Finally, in the Polistinae and/or Vespinae, certain glands associated with the mouthparts may play a yet-undiscovered role in brood alimentation, as Simões (1977) has suggested. In the second place, in view of the observation that wax glands of older honey bee workers can redevelop and become active again if younger, wax-secreting bees are removed from the colony (Rösch 1930), it is more likely that gland activity responds to, rather than causes, the particular behaviors a worker is engaged in.

Accumulating evidence suggests that swarm-founding Polistinae have the most clear-cut worker age polyethism of all the social vespids. Schwarz (1931) reported evidence of a division of labor between builders, water foragers, and pulp foragers in *Polybia occidentalis*, but did not quantify it. In the first detailed study of swarm-founding Polistinae, Naumann (1970) showed that workers of *Protopolybia acutiscutis* tend to participate in nest construction during their first week, then patrol the nest surface and nearby leaves for about a week, fly from the nest without foraging, and finally forage, first for water and pulp, then for nectar and arthropod prey.

In a detailed study of *Protopolybia exigua* and *Agelaia pallipes* in southern Brazil, Simões (1977) demonstrated worker age polyethism in both species. He suggested that as workers age they move progressively to tasks less and less directly related to brood care. Thus, workers aged 6–10 days spend most of their time on the comb, workers aged 11–15 days move to the walls and floor of the nest cavity, and workers older than 16 days switch to foraging. Simões suggested that the hypopharyngeal gland may be developed in young workers and that it may play a role in brood alimentation (analogous to honey bees), but he provided no evidence to support this idea.

Similarly, the first few days of adult life of a *Metapolybia* worker is a relatively idle period, during which she is occasionally attacked and physically dominated by the queens. By the age of 6–10 days she begins nest construction and brood care activity, and by 10–15 days she begins to forage (Forsyth 1978; West-Eberhard 1978b, 1981).

Finally, good evidence has recently been obtained for age polyethism among workers of *Polybia occidentalis* (Forsyth 1978, Jeanne et al. 1988). In one study, workers switched from nest work to foraging at a mean age of 19.6 days in one colony and at 25.7 days in a second (Jeanne et al. 1988). Although the ages at which individual workers switched to foraging ranged from 6 to 40 days, the typical individual made the change relatively abruptly (Fig. 11.2). Thus there are at least two age castes in *Polybia occidentalis*—nest workers, which engage in building, nest maintenance, defense, and brood care, and foragers, which collect pulp, prey, water and/or nectar. There is some evidence for a third caste, an initial in-nest phase in which brood care is prominent, but

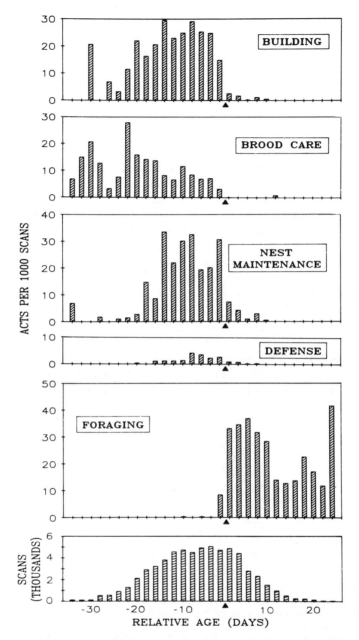

**Fig. 11.2.** Age polyethism among workers of *Polybia occidentalis*. Frequency distribution of on-nest tasks (building, brood care, nest maintenance, and defense) and foraging. The task distribution for each individual in the sample is plotted so that its age of switching from on-nest tasks to foraging is centered on the same point (arrow) on the *x* axis. Relative age is thus age before (negative values) and after (positive values) the switch to foraging for each individual. Switch age for individuals was defined as the two-day age interval nearest the point at which the number of on-nest acts recorded after it equaled the number of foraging acts recorded before it. Data are based on 38 individuals, scan-sampled at five-minute intervals (distribution of the scans by relative age shown in bottom panel). (From Jeanne et al. 1988, courtesy of Westview Press.)

401

because observations on workers inside the nest are limited by the nest envelope, existence of this caste has not been proven. The possibility that foragers of different ages specialize on different materials has not yet been investigated.

## Task Partitioning

So far we have examined how females specialize in certain tasks over relatively long periods of time—days or weeks. Specialization at this level distinguishes queens from workers and divides workers into age-related or temporal castes. But to describe fully how wasps organize work we must elucidate how materials-handling tasks are organized in ways that are independent of the division of labor among workers. Materials-handling tasks typically have two or more sequential components. Thus, collecting pulp in the field is distinct from using pulp at the nest, but the two tasks are sequentially linked. They can be performed sequentially by the same individual, or they can be uncoupled and each performed by a different individual, with the material handed off from one to the other. That such *task partitioning* (Jeanne 1986b) can be independent of division of labor is shown in Fig. 11.3. It is possible to have partitioning of materials-handling tasks in the absence of division of labor (upper right cell of Fig. 11.3), but division of labor into

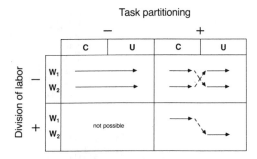

**Fig. 11.3.** Logically possible relationships between division of labor and task partitioning. Task partitioning addresses whether a resource (nest material, food, water) collected in the field is used at the nest by the individual that collected it ( − , no task partitioning) or by another worker ( + , task partitioning present). Division of labor occurs ( + ) when the two tasks are carried out by different castes, as defined by Oster and Wilson (1978), but is absent ( − ) when a given individual is equally likely to perform both tasks or there is no task partitioning. Short solid arrows indicate that the task in question is partitioned from its companion task (i.e., the collector and the user of a load are different individuals). Dashed arrows represent transfer of a load from the collector to the user. Long solid arrows mean that the collector uses her own load at the nest (i.e., collection and use are not partitioned). C, collection; U, use; $W_1$ and $W_2$, two different workers. (From Jeanne 1986b, courtesy of *Monitore Zoologico Italiano*.)

foragers and nest workers is not possible without task partitioning (lower right cell).

Table 11.2 summarizes data on task partitioning for each of the four materials used by social wasps: pulp, water, prey, and nectar. Three patterns are evident. The first is related to level of social organization. In the primitively eusocial species—Stenogastrinae and independent-founding Polistinae—task partitioning is incompletely developed. That is, returning foragers typically use all or part of their loads themselves (Keep and Share in Table 11.2). In the highly eusocial Vespinae, there is a greater tendency toward task partitioning, and in the highly euso-cial swarming species *Polybia occidentalis* task partitioning is virtually complete for all four materials.

The second pattern has to do with the type of material handled. There is a greater tendency to partition food-handling tasks (prey and nectar) than non-food-handling tasks (pulp and water). That is, even primitively eusocial genera typically share prey and nectar loads or give them up entirely to nestmates. In contrast, even the socially advanced Vespinae resist partitioning pulp- and water-handling tasks. Only *Polybia occidentalis* partitions materials-handling tasks in all four categories.

The third pattern is related to colony size. Even within a species, the size of the colony appears to influence the degree of task partitioning. For example, pulp handling is virtually completely partitioned in large colonies of *Polybia occidentalis*, but less so in small colonies (Fig. 11.4). In the genus *Vespula*, nectar handling is partitioned in old colonies but not in young ones, an apparent ontogenetic difference that may in fact be more directly a function of colony size.

### Individual Variability

Little studied to date is behavior at the level of the individual. Most studies of division of labor in wasps pool the behavior of a cohort of workers to come up with an average behavior for workers of a given age. This approach is valid for many purposes, but it overlooks the variability that may occur among individuals otherwise closely matched for age, size, or some other parameter. A few recent studies show that individuals vary tremendously with respect to (1) activity level, (2) degree of specialization, (3) rate of moving through the age polyethic sequence, and (4) steps in the sequence.

Recent studies of *Polybia occidentalis*, for example, indicate that the activity levels of water and pulp foragers vary considerably. In one study of 123 foragers, 94% of all water foraging was carried out by two

**Table 11.2.** Task partitioning in eusocial vespids

| Species | Act[a] Keep | Act[a] Share | Act[a] Give | N[b] | Reference |
|---|---|---|---|---|---|
| | *Pulp* | | | | |
| **Independent-founding Polistinae** | | | | | |
| *Mischocyttarus drewseni* | 0.80 | 0.20 | 0 | 195 | Jeanne 1972 |
| *M. mexicanus* | ** | | | | Litte 1977 |
| *Parapolybia varia* | ** | * | * | | Yamane 1985 |
| *Polistes fuscatus* | 0.49 | 0.45 | 0.06 | 203 | Post et al. 1988 |
| *P. instabilis* | 0.21 | 0.71 | 0.08 | 189 | S. O'Donnell, unpubl. |
| *Ropalidia cincta* | | *? | *? | | Darchen 1976a |
| *R. fasciata* | ** | | | | Kojima 1984c |
| *R. turneri* | ** | | | | Jeanne, unpubl. |
| **Swarm-founding Polistinae** | | | | | |
| *Polybia occidentalis* | | | | | |
| three large colonies | 0 | 0.002 | 0.998 | 920 | Jeanne 1986a |
| three small colonies | 0 | 0.18 | 0.82 | 485 | Jeanne 1986a |
| **Vespinae** | | | | | |
| *Vespa* spp. | ** | | | | Matsuura 1984 |
| *Vespula atropilosa* | ** | | | | Akre et al. 1976 |
| *V. germanica* | ** | | | | Weyrauch 1939 |
| *V. pensylvanica* | ** | | | | Akre et al. 1976 |
| *V. squamosa* | ** | | | | Gaul 1947 |
| *V. vulgaris* | ** | | | | Weyrauch 1939 |
| | *Water* | | | | |
| **Independent-founding Polistinae** | | | | | |
| *Polistes instabilis* | 0.93 | 0.07 | 0 | 377 | S. O'Donnell, unpubl. |
| *Ropalidia cincta* | | *? | | | Darchen 1976a |
| **Swarm-founding Polistinae** | | | | | |
| *Polybia occidentalis* | 0 | 0 | 1.00 | 103 | L. Phelps, unpubl. |
| **Vespinae** | | | | | |
| *Vespula atropilosa* | ** | | | | Akre et al. 1976 |
| *V. germanica* | ** | | | | Spradbery 1973a |
| *V. pensylvanica* | ** | | | | Akre et al. 1976 |
| | *Prey* | | | | |
| **Stenogastrinae** | | | | | |
| *Eustenogaster calyptodoma* | | ** | | | Hansell 1987b |
| *Liostenogaster varipicta* | | ** | | | Williams 1928 |
| **Independent-founding Polistinae** | | | | | |
| *Mischocyttarus mexicanus* | | ** | | | Litte 1977 |
| *Polistes fuscatus* | 0.21 | 0.35 | 0.44 | 240 | Post et al. 1988 |
| *P. instabilis* | 0.13 | 0.28 | 0.59 | 185 | S. O'Donnell, unpubl. |
| *Ropalidia cincta* | | ** | *? | | Darchen 1976a |
| *R. fasciata* | | ** | | | Kojima 1984c |
| *R. turneri* | | ** | | | Jeanne, unpubl. |
| **Swarm-founding Polistinae** | | | | | |
| *Polybia occidentalis* | | | ** | | Hunt et al. 1987, Jeanne, unpubl. |

**Table 11.2** —*continued*

| Species | Act[a] | | | N[b] | Reference |
| | Keep | Share | Give | | |
|---|---|---|---|---|---|
| Vespinae | | | | | |
| *Vespula atropilosa* | | * | ** | | Akre et al. 1976 |
| *V. germanica* | | ** | *? | | Weyrauch 1939 |
| *V. pensylvanica* | | * | ** | | Akre et al. 1976 |
| *V. squamosa* | | ** | | | Gaul 1947 |
| *V. vulgaris* | | ** | *? | | Weyrauch 1939 |
| | | Nectar | | | |
| Independent-founding Polistinae | | | | | |
| *Polistes instabilis* | 0.16 | 0.38 | 0.46 | 454 | S. O'Donnell, unpubl. |
| *Ropalidia cincta* | | *? | | | Darchen 1976a |
| Swarm-founding Polistinae | | | | | |
| *Polybia occidentalis* | | | ** | | Hunt et al. 1987, Jeanne, unpubl. |
| Vespinae | | | | | |
| *Vespula germanica* | | | ** | | Weyrauch 1939 |
| *V. pensylvanica* and *V. atropilosa* | | | | | |
| young colonies | ** | * | | | Akre et al. 1976 |
| old colonies | | | ** | | Akre et al. 1976 |
| *V. vulgaris* | | | ** | | Weyrauch 1939 |

[a]Keep, forager keeps the load and uses it herself at the nest. Share, forager transfers part of the load to nestmates, but retains a portion, which she uses herself. Give, forager transfers the entire load to nestmates. **, the usual fate of foraged loads returned to the nest, as determined from studies without quantified data. *, less-frequently observed. ?, seems to occur, but not entirely clear from the description. Values are proportions for each option, from studies with quantified data.
[b]Number of acts observed, when reported.

**Fig. 11.4.** Changes in degree of task partitioning of pulp handling with colony size in *Polybia occidentalis*. Line thicknesses are proportional to the magnitude of the conditional probability of performing the next task, except for very small values (horizontal dashed arrows). Solid arrows represent behavioral transitions for a given worker handling a given pulp load. Vertical dashed arrows represent the transfer of a pulp load from forager to nest worker. Upper figure is based on data compiled from three large colonies (> 350 adults) (N = 920 observations); lower figure is based on four small colonies (< 50 adults) (N = 485 observations). (From Jeanne 1986a, courtesy of Springer-Verlag.)

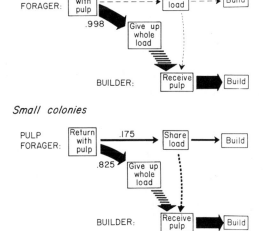

active foragers, and 75% of all pulp foraging was performed by three
others (S. O'Donnell, unpubl.) (Table 11.3).

My (1986a) study of *Polybia occidentalis* suggests that some individuals
engaged in nest construction are specialists, because in the course of a
day they performed only one of the three component tasks (water foraging, pulp foraging, and building), while others are generalists, because they performed two or even all three. Forsyth (1978) described
what he called "task fixation" in the same species and in *Metapolybia
azteca*, in which certain individuals may come to specialize on a given
task and continue at it for the duration of their lives. For example, one
worker of *M. azteca* became fixated on water foraging early in her life
and continued doing it for 41 days, which exceeds the mean forager life
span. In a similar vein, Simões (1977) stated that some foragers of *Protopolybia exigua* specialize on either nectar or pulp.

The rate at which individual workers progress through the temporal
task sequence also varies. Sixty-nine workers in two *Polybia occidentalis*
colonies switched from on-nest tasks to foraging at ages ranging from 6
to 40 days (Jeanne et al. 1988). Although most workers in these colo-

**Table 11.3.** Individual variability in rate of foraging for
water and pulp in a colony of *Polybia occidentalis*

| Water | | Pulp | |
|---|---|---|---|
| Individual[a] | No. of trips[b] | Individual[a] | No. of trips[b] |
| 207 | 581 | 4 | 176 |
| 36 | 291 | 198 | 144 |
| 198 | 32 | 52 | 99 |
| 184 | 5 | 184 | 52 |
| 52 | 3 | 170 | 37 |
| 173 | 3 | 32 | 12 |
| 296 | 2 | 164 | 8 |
| 188 | 2 | 296 | 7 |
| 159 | 2 | 173 | 5 |
| 19 | 2 | 196 | 5 |
| 170 | 1 | 28 | 5 |
| 32 | 1 | 271 | 3 |
| 35 | 1 | 183 | 2 |
| 44 | 1 | 207 | 1 |
| | | 188 | 1 |
| | | 200 | 1 |
| | | 102 | 1 |
| | | 131 | 1 |
| | | 84 | 1 |

*Source*: Data from S. O'Donnell, unpubl.
[a]Identified individuals, ranked in order of frequency of foraging
for water and for pulp.
[b]Number of trips by each individual during eight consecutive
days of observation.

nies progressed from working inside the nest, to working on the nest, to foraging as they aged, others omitted the on-nest phase, and still others did no foraging, despite living well beyond the mean age for beginning this role (Jeanne et al. 1988).

## POLYETHISM AND REPRODUCTIVE STRATEGIES

Because in eusocial species most members of the colony do not reproduce, reproductive division of labor forms the crux of the question of how natural selection has favored the evolution and maintenance of eusocial behavior. For Charles Darwin this was a dilemma that posed "by far the most serious special difficulty, which my theory has encountered" (Darwin 1964:242). Darwin resolved the difficulty by invoking the colony as the unit of selection. This view was influential in the application to the social insects of the "superorganism" concept, the notion that the colony can be likened to the body of an individual organism, whereby the reproductives represent the germ plasm and the workers represent the soma (Wheeler 1911, Emerson 1959, see also Wilson 1971). The superorganism concept encouraged the view of the colony as a smoothly functioning whole, whose members, especially its sterile workers, contribute altruistically to the common good.

But of course there is a fundamental difference between organisms and insect colonies. While the somatic cells of the body of an individual organism possess identical genomes, the members of a social insect colony do not. Lack of genetic identity among closely interacting individuals, whether parasite and host, mutualists, a courting male and female of the same species, or members of a social insect colony, is grounds for conflict of interest between them. Conflict is not conspicuous in the interactions of colony members in the more advanced eusocial insects—the honey bees, ants, and termites—with which Darwin, Wheeler, and Emerson were most familiar, and this may be why the superorganism concept gained such ground early on.

In recent decades, increasing attention has been paid to the primitively eusocial species of bees and wasps. The early work on wasps by Pardi (1946, 1948) and later work by Owen (1962), Montagner (1966a), West-Eberhard (1969), and myself (1972), among others, has called attention to the frequent aggressive interactions among colony members. On the small, open nests of *Polistes* and *Mischocyttarus*, for example, physical conflict among adults is common and easily observed. Colonies of these species are obviously not the "smoothly running machines" that the superorganism concept would lead us to expect. Since the landmark publications of Hamilton (1964a,b) and Williams (1966), insect sociobiologists have tried to interpret the characteristics of insect sociality in terms of selection acting on individual colony members (Trivers and Hare 1976; West-Eberhard 1975, 1979, 1981, 1982d). This approach remains a dominant paradigm.

West-Eberhard (1981) has made a strong argument that intragroup competition was a major force in the evolution of worker behavior as an alternative reproductive strategy. The four points of her argument can be paraphrased as follows:

1. When group life is advantageous, as it is for social insects, the reproductive success of individuals within the group depends increasingly on their ability to win in social competition.

2. Competition may lead to significant disparities in reproductive success among members of the colony.

3. The "losers" may salvage some reproductive success by facultatively adopting helping behavior.

4. Despite the availability of the helping option, the direct reproduction option is likely to remain the more productive one.

West-Eberhard's line of reasoning is largely adopted here and applied first to foundress groups and then to foundress-offspring relationships in interpreting the polyethic patterns described above.

## Options for Co-foundresses

Consider a *Polistes* founding group of related females at the start of the nesting season. In temperate-zone populations each female is inseminated (but see West-Eberhard 1986) and has moderately developed ovaries (Pardi 1948). Therefore, the personal reproductive potential of each female initially is moderately high. It is in the best interest of each to attempt to reproduce directly, and, if possible, exclusively, since each is more closely related to her own offspring of either sex than to the offspring of relatives. This is the source of the conflict that manifests itself as dominance interactions among founding females, the result of which is that the head of the hierarchy, the α-female, experiences increasing ovarian development and is able to assert increasingly exclusive reproductive rights over her associates.

It is clear why the α-female should favor such a situation. The probability of success of a colony increases with the number of foundresses in it (see Reeve, Spradbery, this volume), as may the eventual productivity of the colony. Furthermore, having co-foundresses to help her is to the α-female's advantage because it relieves her of much of the dangerous foraging required to produce adult offspring.

The perplexing question is, Why do the subordinate co-foundresses continue to stay on as nonreproducing helpers rather than leaving to found nests on their own? To understand why a subordinate should opt to stay and help, it is useful to consider the reproductive success that she can gain by helping instead of leaving and founding her own nest. The reproductive value ($RV$) associated with each option can be broken down into its components as follows.

If the subordinate stays, the indirect reproductive value ($RV_i$) of helping her relative rear offspring is $RV_i = s_i b_i r_n$, where $s_i$ is the probability that she survives to contribute to the rearing of her relative's offspring; $b_i$ is the amount, in adult equivalents, by which she augments her relative's reproductive output; and $r_n$ is the genetic relatedness of the offspring produced to the subordinate.

If the subordinate stays to help her relative she may get the opportunity to reproduce personally, if the $\alpha$-female dies, for example, or if she can partially escape the $\alpha$-female's ovarian-suppressing influence. Her direct reproductive value ($RV_d$) then is $RV_d = s_d b_d r_o$, where $s_d$ is her probability of reproducing directly, $b_d$ is her productivity in adult equivalents if she reproduces, and $r_o$ is the relatedness of her offspring to herself.

If she opts to quit her relative's nest and found her own, her reproductive value is $RV_f = s_f b_f r_o$, where $s_f$ is the probability that she will survive to produce adult offspring; $b_f$ is her productivity, given that she survives; and $r_o$ is the relatedness of her offspring to herself.

For a co-foundress to opt to stay on as a subordinate on the nest of a relative, Hamilton's inclusive fitness inequality, $RV_f < RV_d + RV_i$, must be satisfied (see also Reeve, this volume). The left-hand side of the formula represents the payoff in reproductive value of solitary founding, $RV_f$, and the right-hand side is the payoff for group founding. The two components of the right-hand side represent the inclusive reproductive success gained indirectly by helping ($RV_i$) and directly by any reproduction she may achieve on the relative's nest ($RV_d$) (West-Eberhard 1981). As the colony develops, the values for the various fitness components shift, as depicted in Fig. 11.5. Early on, when the

**Fig. 11.5.** Hypothetical curves of reproductive values for subordinate co-founding females. $x$ axis represents preemergence period, from colony founding to emergence of first offspring. $RV_d$, reproductive value via direct reproduction on nest with $\alpha$-female; $RV_i$, reproductive value via helping $\alpha$-female rear more offspring; $RV_f$, reproductive value via leaving $\alpha$-female's nest and founding a nest alone. Below is depicted the sequence of roles the subordinate takes on as its curve profile changes. (Modified from West-Eberhard 1981.)

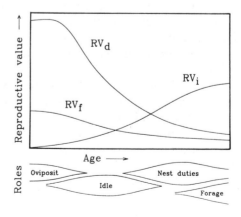

personal reproduction potential is high for all co-foundresses, it may be more advantageous for a given subordinate to stay because she still has a chance of winning the competition for reproductive rights and realizing direct reproductive success (high $RV_d$), while if she leaves to found a colony of her own it may have a low probability of succeeding because she may fail to attract helpers of her own (low $RV_f$). On the other hand, the helping option ($RV_i$) at this time is relatively unattractive as a route to inclusive fitness, simply because there is as yet little opportunity to provide help, since the nest is small and the brood population consists of just a few eggs.

During early colony development this picture gradually changes. As a subordinate is dominated by a stronger relative, her likelihood of realizing reproductive success by direct personal reproduction on the nest ($RV_d$) diminishes, although there is always the chance that if the α-female dies the subordinate may be able to accede to the dominant position. On the other hand, the feasibility of gaining at least some reproductive success indirectly, by helping on her relative's nest, increases with the growth in age and number of brood through which such behavior can be expressed.

Embedded in the survivorship terms, $s_f$, $s_i$, $s_d$, is a subtle difference between the two strategies that confers an advantage to helping, as pointed out by Queller (1989). If a solitary foundress dies before producing any adult offspring, her reproductive success will be zero, because her immature brood will almost certainly succumb in her absence (Metcalf and Whitt 1977a). As a helper, on the other hand, even though her own survivorship may be no higher than it would have been on her own nest (in either situation she will be heavily involved in the risky business of foraging), whatever contribution she makes on her relative's nest is more likely to be realized as inclusive reproductive success, since the colony, and therefore her investment, is likely to survive her own death. In other words, a solitary foundress surviving to less than the duration of egg-to-adult development will realize an inclusive fitness of zero. In contrast, a helper surviving that long will have a fitness greater than zero. Whatever contribution she makes before she dies will accrue to her inclusive fitness because her surviving nestmates greatly increase the probability that the colony will produce adults.

Fig. 11.5 emphasizes the continuity of changes in reproductive values through time. As West-Eberhard (1981) put it, individual females behave as if they were consulting a curve of reproductive value to determine the cost of queenlike, idle (waiting), or high- versus low-risk worker behavior. The behavior exhibited by a subordinate is thus, in effect, a reflection of her perception of the relative levels of her $RV_d$ and $RV_i$. As long as $RV_d$ exceeds $RV_i$, the subordinate should remain relatively idle. As $RV_i$ passes $RV_d$, she will engage in tasks on the nest that have a low cost and therefore do not seriously compromise her still-high $RV_d$. As $RV_i$ continues to ascend with the increase in numbers and size of larvae through which helping behavior can be expressed, the subordinate moves into riskier tasks such as foraging. Meanwhile, the α-female (not shown in Fig. 11.5) retreats from workerlike tasks to specialize increasingly in oviposition, domination, and cell initiation. This gives rise to the displacement of the females along the task hierarchy, as shown in Fig. 11.6. The tasks in the hierarchy are arranged from top to bottom roughly in order of increasing cost in energy and physical risk of mortality due to predation, parasitization, or accident. Tasks at the top are most directly associated with increasing the number of new offspring in the nest, while those at the bottom are involved with maintenance of the existing brood population.

West-Eberhard (1981) considers direct reproduction on a relative's nest and helping on that nest as alternative strategies. Strictly speaking, two strategies will not coexist if one is more successful than the other (Dawkins 1980), and it seems clear that subordinate "losers" achieve less inclusive reproductive success than do "winners" (Metcalf and Whitt 1977b). An alternative is to consider the options of group founding versus solitary founding to be alternative strategies, and the two together as a mixed evolutionarily stable strategy (ESS). According to this view, the group-founding option is but a single conditional strategy (Dawkins 1980). The conditions are the individual's changing $RV_d$:$RV_i$ ratios. As the dominant female closes down the subordinate's $RV_d$ value, the latter begins to salvage inclusive reproductive value by opting to help, a case of making the best of a bad situation (Dawkins 1980). $RV_d$:$RV_i$ ratios vary along a continuum and, consequently, so does the behavior the subordinate should adopt: If the ratio is very high, she should oviposit; if slightly greater than 1, wait for the chance to oviposit; if slightly less than 1, work at low-cost tasks; if very much lower than 1, forage.

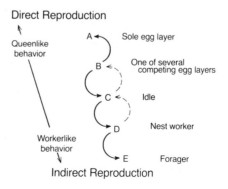

**Fig. 11.6.** Hierarchy of tasks performed by females, arranged along a gradient from direct reproduction (high $RV_d$) to indirect reproduction (high $RV_i$). Solid arrows indicate the more common transitions, dashed arrows less common transitions. (Modified from West-Eberhard 1981.)

## Options for Offspring

Should a daughter prefer to try to reproduce directly or to help rear her mother's offspring? Early kin-selection arguments predicted that greater inclusive fitness ought to be gained by helping because of the relatedness asymmetries produced by haplodiploidy (Hamilton 1964a, b, 1972; Wilson 1971; Starr 1979; Crozier 1982). These arguments, however, required rather restrictive assumptions in order for the full potential of haplodiploidy to elevate relatedness to be realized—such as single matings by queens, no supernumerary queens, and ability of workers to bias sex investment ratios—assumptions known to be invalid for many social wasp species (see Ross and Carpenter, this volume). Thus, haplodiploidy is not an exclusive explanation of why female offspring should behave as workers.

In fact, there is ample evidence to suggest that helping her mother is not a daughter's first choice, but that she will actively compete to reproduce directly. She stands to realize much greater fitness if she can do so. The production of even one son will yield her an offspring equivalent of 1.0, providing the egg survives to become an adult (West-Eberhard 1981). On the other hand, the inclusive fitness to be gained through helping often appears to be low. Reproductive value has been estimated at only 0.04 for workers of the Africanized honey bee (*Apis mellifera scutellata*) (Winston 1978, West-Eberhard 1981), and there is reason to suspect that indirect reproductive success in terms of offspring equivalents produced by aiding the queen(s) is low for wasps also. Under these conditions a worker gains a tremendous fitness advantage if she can escape the inhibiting effects of the queen and lay eggs.

Workers of all social wasp species have ovaries, are reproductively competent, and can develop and lay eggs during at least part of their adult lives (West-Eberhard 1978a, Bourke 1988, but see also Gadagkar et al. 1988). Within hours after the queen dies or is removed, many of the workers begin to display aggressive behavior toward one another and eventually begin to oviposit (Ishay 1964; Landolt et al. 1977; Ross 1985, 1986; Bourke 1988). In fact, there is some evidence that vespine workers lay eggs even in queenright colonies (Bourke 1988), indicating strong selection on workers to reproduce directly and a partial failure of queen-control mechanisms. These observations indicate that the presence of an active queen or queens is necessary to inhibit ovarian development and queenlike behavior among workers (e.g., Spradbery, this volume). In other words, the colony can be thought of as being under constant competitive tension, maintained from the top of the reproductive hierarchy.

The potential for workers to reproduce directly suggests that workers

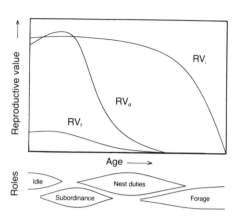

**Fig. 11.7.** Hypothetical curves of reproductive values for workers. $x$ axis represents period from worker eclosion to death. $RV_d$, reproductive value via direct reproduction on mother's nest; $RV_i$, reproductive value via helping (worker) behavior on mother's nest; $RV_f$, reproductive value via leaving mother's nest and founding a nest alone. Below is the sequence of roles the worker takes on as its curve profile changes.

play much the same conditional strategy that co-foundresses play. As depicted in Fig. 11.7, when a worker is just a few days old and has the physiological capacity for ovarian development and oviposition (i.e., her ovaries are in the "ascending phase"; see The Role of Dominance, below), her direct reproductive value is relatively high. However, the presence of a mature, reproductively active queen—her mother—inhibits the continued development of the worker's ovaries through dominance, eventually causing them to regress. As her ovaries regress, the offspring's $RV_d$ drops. At the same time, the $RV_i$ she can realize through helping behavior stays relatively constant and high, since the nest contains a large population of brood of all stages in which she can invest by helping. When her $RV_d{:}RV_i$ ratio is around 1, she remains idle. As the ratio falls below 1, she should begin to work at low-cost, on-nest tasks. That is, she enters the task hierarchy of Table 11.1 and Fig. 11.6 toward the top. As she ages, the capacity of her ovaries to recover from their undeveloped state diminishes, and her $RV_d$ continues to drop. Meanwhile, younger workers are eclosing and taking her place at tasks in the nest. As a consequence, she moves down the task hierarchy to costlier tasks, ultimately to foraging. In other words, age polyethism follows a centrifugal pattern in the sense that as they age, workers take up tasks progressively farther from the reproductive centers of the nest (Wilson 1985).

This analysis makes it clear that age polyethism is not programmed according to a rigid age schedule. Instead, the role a female performs is a function of curves of reproductive value attained by various alternative routes. The values of these curves in turn depend on dominance relations among adult females in the colony. Females with relatively low $RV_d$ values and high $RV_i$ values salvage a small amount of inclusive reproductive value by behaving as workers to augment the output of a reproductive relative in the colony. Thus, age polyethism is a misnomer, since role is influenced only indirectly by age. Nevertheless, since age is easier to measure than is relative reproductive value, the term will probably continue to be used.

By this reasoning, the roles females adopt in the colony—from queen to forager—constitute conditional strategies set by relative reproductive success attainable by different routes. Therefore, the notions of "reproductive division of labor" and "temporal castes" create artificial categories imposed on what is really a single, continuous variable.

If a worker's inclusive reproductive value is depressed to such low values through dominance, why doesn't she leave the nest, found her own colony, and raise her own offspring? In other words, why is worker behavior maintained under the stringent costs to reproductive success that appear to exist? There are probably several reasons. First, for a female to realize any reproductive success through founding a colony, she must survive at least until her first offspring reach adulthood. As a worker, on the other hand, the contribution she has made toward augmenting her relative's reproductive output is likely to be realized as inclusive reproductive success even if she dies short of the egg-to-adult developmental time (Queller 1989).

Second, the probability that a worker could successfully found a colony and carry it through to the production of adult offspring must often be extremely low. For instance, workers of many species have characteristics, conferred in the pre-imaginal stages, that bias them against successful founding (e.g., Gadagkar, Greene, this volume). They may be smaller than foundresses, lack adequate body fat, or be unable to mate (vespines).

Third, because many populations contain no males for workers to mate with during most of the colony cycle, laying workers could produce only males. If significant numbers of workers did this, it would decrease the value of males, selecting against their production (Fisher 1958). Yet the exercising of this strategy by "workers" is not completely unknown in eusocial wasps. Hamilton (1972) reported that small groups of *Vespa crabro* workers can found new nests in which they rear males. The "satellite nests" that are sometimes founded in midseason by *Polistes exclamans* in Texas (Strassmann 1981b) may also represent such a phenomenon. In fact, it is conceivable that the postponing of male production in early broods may have been selected early in the evolution of eusociality in part as a means by which the mother could manipulate her first daughters into opting to become workers.

## The Role of Dominance

A close look at the evidence for dominance behavior suggests that it is more widespread in eusocial wasps than previously thought (e.g.,

Röseler, Spradbery, this volume). In some independent-founding polistines in which overt dominance is a conspicuous component of reproductive competition among foundresses, female offspring also show
dominance behavior. Workers of *Ropalidia marginata*, for example, establish a dominance hierarchy that influences division of labor such
that subordinate individuals do most of the foraging (Gadagkar 1980,
Gadagkar and Joshi 1983). Although in this species dominance status is
not correlated with an individual's age, in *Polistes dominulus* it is primarily the younger offspring that are subjected to domination by the
queen and by any dominant co-foundresses that may still be present
(Pardi 1948). Such an age dependence is also seen clearly in *Mischocyttarus drewseni* (Jeanne 1972) (Fig. 11.8). By her second or third day as an
adult, each worker becomes involved in dominance interactions, usually as the object of domination by the queen. Sometimes, however,
the offspring also display dominance behavior toward each other. As
they reach the age of 10 days, the amount of domination they receive
drops dramatically.

Although dissections were not made in Jeanne's (1972) study, it is
reasonable to hypothesize that the amount of domination each worker
receives is correlated with the level of threat she constitutes to the reproductive dominance of the queen. Pardi (1946, 1948) has shown that
*Polistes dominulus* workers undergo an ascending phase of ovarian
development during the first few days of adulthood, followed by a
descending phase in which their ovaries atrophy as a result of domination received. By the time a worker ceases to be involved in dominance
interactions, her ovaries have regressed to the point where she no
longer challenges the queen's reproductive role. A convincing demonstration that the offspring are serious competitors for direct reproductive rights is seen in the *M. drewseni* colony that I observed (Jeanne
1972) (Fig. 11.8). The founding queen, number 1, was successfully chal-
lenged and superseded as queen by number 8, her own daughter.
Number 8 was herself subsequently superseded by number 36, her
younger sister.

Reeve and Gamboa (1987) have argued for the pervasive influence of
dominance in their "dominance inhibition" hypothesis of the queen's
effect on worker foraging. Working with *Polistes fuscatus*, they found
that the queen strongly stimulated foraging by workers, causing 50–
70% of all worker departures, and that the tendency to leave the nest
to forage was significantly negatively correlated with dominance rank.
In searching for a mechanism, Reeve and Gamboa found no evidence
that the queen's lateral vibration behavior, thought to communicate
dominance status (Gamboa and Dew 1981), provided the stimulus.
Similarly, there was no evidence that either her solicitations of returned foragers or her other interactions with them stimulated subsequent departures. They concluded that the stimulus to forage is an

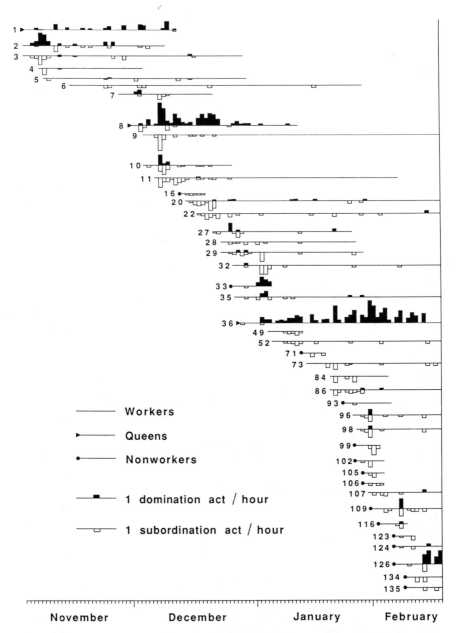

**Fig. 11.8.** Dominance and subordinance among females in a colony of *Mischocyttarus drewseni*. Each numbered female is represented by a horizontal line extending from the date it emerged as an adult to the date it disappeared from the nest. The daily rate at which each female dominated nestmates is represented by a solid bar extending above her line. The rate at which each female was dominated by nestmates (displayed subordinate behavior) is represented by an open bar extending below the line. Twenty-seven females that stayed on the nest ten days or less and were involved in dominance interactions on only one day are not included. The colony declined within one week after the end of observations. (Modified from Jeanne 1972, courtesy of Museum of Comparative Zoology, Harvard University.)

indirect result of the inhibition of dominance behavior in workers that perceive the queen's presence.

There is suggestive evidence from the vespines and from the swarm-founding polistines that interactions among workers involve dominance. Montagner (1966a) concluded that a dominance system existed among the workers of a *Vespula* colony and hypothesized that it is the mechanism by which workers divide labor. In a manner resembling the less-violent forms of dominance interaction described by Pardi (1946) for *Polistes*, the more-dominant workers beg for food in an erect posture, while subordinate ones assume a recumbent position. Materials usually pass from subordinate to dominant. Montagner suggested that this "contact dominance," the exchange of information through posture and antennal contact during trophallaxis, communicates relative dominance status. If a worker has repeated successful trophallactic contacts (i.e., is able to receive food from others in the nest by virtue of high dominance status), she confines herself to the nest, feeding larvae and building, using the material foraged by others. If she frequently yields food materials to others, she becomes a forager. Montagner pointed out that the level of dominance of a worker is not always correlated with its age or its degree of ovarian development.

A related phenomenon, mauling in *Vespula atropilosa* and *V. pensylvanica*, may be age-related, with younger (but not the very young) workers tending to be maulers and older ones mauling recipients (see Greene, this volume). Returning foragers are often mauled by nest workers. As the colony matures, the overall frequency of mauling increases, and some workers become heavily involved in mauling others. Akre et al. (1976) suggested that mauling may be a manifestation of dominance relations among the workers, but cautiously concluded that more data are needed. They did not look at ovarian development among workers, although no laying workers were evident.

Workers of *Protopolybia acutiscutis*, a swarm-founding polistine, show trophallactic interactions similar to those reported by Montagner for *Vespula* (Naumann 1970). A soliciting worker may take an elevated or lowered position, and antennal movements vary as described by Montagner. Naumann did not attempt to analyze these interactions in terms of dominance, nor did he correlate them with ovarian development. Simões (1977) reported that there was no dominance hierarchy among workers in the colonies of *Agelaia pallipes* and *Protopolybia exigua* that he studied, yet he did state that in the latter species a worker is occasionally persecuted by others. Simões never found a worker of *A. pallipes* with developed ovaries, while in *P. exigua* worker ovarian development was common, with some workers laying eggs. These presumably were the younger workers, because the ovaries of foragers are atrophied.

Finally, West-Eberhard (1977a, 1978b) observed that workers of *Meta-*

*polybia aztecoides* perform a dance toward queens that appears to function as a test of the queens' dominance. If a queen responds by regurgitating crop fluid, the workers attack; queens that are frequently attacked either disappear from the nest or become workers. West-Eberhard suggested that persecuted individuals are queens of limited reproductive capacity.

These observations suggest two things. The first is that our conception of "dominance" in social wasps has been confined to too narrow a range of behavior patterns in the past. When we think of dominance behavior we tend to call to mind the physical attacks first described by Pardi (1946, 1948) for *Polistes dominulus*. However, competition need not manifest itself in overt conflict, and dominance need not involve physical aggression (see Vehrencamp 1983a). Similarly, the determination of relative status need not require learned, pairwise relationships, but may be based on cues that are correlated with an opponent's competitive ability (Barnard and Burk 1979). The above observations suggest that if we expand the definition of dominance behavior to include not only the overt physical attacks common in *Polistes* and other independent-founding polistines, but also the more subtle, nonviolent interactions among workers of the more advanced eusocial wasps, we will find that dominance interactions involve virtually all individuals in a colony.

The second point is that if dominance interactions in such a broader sense involve the workers as well as the reproductives, it suggests that the competitive tension among them is not imposed exclusively from above by the queen(s), but is maintained by dominance interactions among all adults in the colony. This view is shared by West-Eberhard (1977a), who summarized evidence that the suppression of reproduction in both competing queens and newly emerged females in the Vespinae, as well as in the swarm-founding Polistinae, is the effect of dominance behavior on the part of workers. Thus dominance interactions may communicate and influence not only relative reproductive status, but may serve as the mechanism by which the structure of division of labor is maintained.

One way that such a dominance system can effect polyethism is through the mediation of spacing on the nest. The influence of an individual's rank on its position on the nest was first pointed out by Pardi (1946, 1948). *Polistes dominulus* females' positions in the hierarchy are positively correlated with the proportion of time spent on the face of the comb. Workers with regressed ovaries rest on the back of the comb. Similarly, in a founding-stage colony of *Polistes erythrocephalus* the dominant egg layer spends much of her time on the nest face, usually on the new cells, where egg laying takes place (West-Eberhard 1969, 1986). She chases subordinate females from this area, and, consequently, subordinates spend most of their time on older parts of the comb, near the pedicel, or entirely off the comb. Queens of swarm-founding polistines, too, spend most of their time on new combs, where new oviposition sites are being created (Simões 1977). Thus, a female's position in the larger dominance hierarchy determines where on the nest she spends her time, and this in turn plays a major role in determining the tasks that she performs. The importance of location on the nest in determining the roles of workers has been documented for honey bees (Seeley 1982) and ants (Wilson 1984) and is no doubt equally important in wasps.

These observations support the notion of a hierarchy of dominance mechanisms among eusocial wasps (see also Spradbery, this volume). A given species may use more than one form. The most direct and least sophisticated means of controlling reproduction by competitors may be postovipositional, that is, differential egg eating. The simplest preovipositional means is active exclusion of competitors from territories on the nest that contain empty cells. The monopolizing of behavior that creates new oviposition sites is yet another way the queen inhibits reproduction by subordinates. Thus, queens of many independent-founding polistines not only initiate cells but forage for the pulp they use for this purpose, even after workers are available to take over these tasks.

Some independent-founding species appear to have a more sophisticated means of queen control, in which the dominant female tolerates subordinates anywhere on the comb but inhibits their reproduction by milder attacks combined with differential egg eating. Even more subtle forms of queen control are found in some *Polistes* and *Ropalidia* in which conspicuous dominance interactions are virtually nonexistent (Itô 1985b, 1986d, 1987a). The possibility that dominant females transfer a primer pheromone to the body of subordinates in independent-founding polistines has not been investigated.

Vespine queens, as well as many of the swarm-founding polistines, evidently produce pheromones that bring about reproductive inhibition in workers without the need for direct physical attack (West-Eberhard 1977a, Jeanne 1980a, Akre and Reed 1983). It is interesting that in these species, despite the lack of any outward aggressivity on the part of the queens, physical dominance behavior often does manifest itself among at least a segment of the workers farther down the dominance hierarchy—in the mauling behavior of workers of *Vespula pensylvanica* and *V. atropilosa*, for example (Akre et al. 1976, West-Eberhard 1977a). Yet in other advanced eusocial wasps, workers show little sign of overt aggression. West-Eberhard (1981) argues that queens and workers have been involved in an "arms race" for control of reproduction, the queen evolving new means of ovarian suppression in workers as the workers evolve the means to escape from the older, less effective forms. Clearly, we need to pay closer attention to the more subtle forms of reproductive and social dominance that may exist in these socially advanced species.

## Releasers of Worker Behavior

By limiting a female's opportunity for direct reproduction on the nest, the dominance structure of the colony sets the conditions within which a female struggles to maximize her inclusive fitness. Through its effect on a female's location in the nest, domination leads to a divergence of roles between queens and workers and to age polyethism among the workers. But this dominance-imposed structure only broadly defines a female's role. There are other sources of stimuli within the colony that determine in the more immediate term what tasks a worker will do. These stimuli issue from the nest, the brood, and the behavior of other adults.

Physical conditions within the nest provide cues that release specific kinds of behavior. For example, if the nest of *Polybia occidentalis* overheats, certain workers engage in wing fanning on the nest surface. If the temperature continues to rise, foragers will start to collect water and transfer it to nest wasps, who spread it over the surface of the combs inside to speed cooling by evaporation (L. Phelps, unpubl.). Each of these behavior patterns is likely to be elicited in workers belonging to a specific age category (Jeanne, unpubl.). Similarly, reconstruction of the envelope of *Vespula vulgaris* after it has been damaged is elicited by both low temperature and penetration of light into the nest (Potter 1964).

The brood also play a role in releasing specific tasks in workers. There is some evidence that at least part of the stimulus for workers to forage for prey comes directly from the larvae (Potter 1964). Workers of *Polistes* are stimulated to lengthen cell walls by the larvae therein (Deleurance 1947, 1955a, 1957; Spieth 1948).

Finally, the activities of other adults in the colony modulate workers' responses. We have already seen how the queen of *Polistes* stimulates foraging by her workers. There is evidence that workers also respond to what other workers are doing. The presence of foragers, for example, appears to inhibit the onset of foraging by younger workers. The first worker to eclose onto a *Polistes* nest typically becomes the hardest working forager of the colony (Dew and Michener 1981, Strassmann et al. 1984a). If she is removed, however, some of the remaining (younger) foragers increase their rate of foraging precociously, in effect taking up the slack (Dew and Michener 1981). Whether the recruited foragers somehow directly perceive that the rate of foraging has dropped and increase their own rates accordingly, or whether they increase their rates in response to the queen's having redirected her stimulatory influence toward them is not known.

A similar phenomenon has been demonstrated for *Polybia occidentalis* during a round of nest construction. When the two most active pulp foragers were temporarily removed, replacements were recruited from among other foragers and/or previously idle workers, with the net effect that the overall rate of nest construction was maintained (Jeanne 1987). It seems likely that the replacements were either responding directly to the activities of other workers or were in some way monitoring the rate of inflow of material, rather than responding to some direct influence of the queens in deciding whether to become active or not. *Polybia* foragers are extreme specialists; they may make many trips for the same material, while coming into contact with the nest and its inhabitants only long enough to transfer each load to a nestmate—as little as seven seconds in the case of pulp loads (Jeanne 1986a). This suggests that the younger nest workers—the recipients of foraged material—provide the feedback that keeps foragers active. The nature of this feedback is not known, although it is possible that it may take the form of "eagerness" for the foraged material, as has been shown for honey bees (Lindauer 1971, Seeley 1989). This eagerness may be a manifestation of dominance.

Still unexplained is the source of the tremendous variability in rate of work and degree of specialization among individual workers. Differences in individual experience—such as reinforcement schedules at critical points in a worker's life—may play a role. However, virtually no investigations of such differences in social wasps have been made.

## The Regulation of Colony Cycle and Size

Any explanation of what keeps the colony together ought to include how colonies eventually dissolve. Colonies of vespines and independent-founding polistines in temperate climates have annual cycles divided into four phases: founding, growth, reproductive production, and decline. But surprisingly little work has been done on the factors that trigger each phase of the cycle. The assumption that external signals—daylength and temperature, for example—regulate the cycle may be too facile. The absence of any seasonal synchrony in many tropical populations of *Mischocyttarus* and *Polistes* (West-Eberhard 1969, Jeanne 1972) strongly argues for the colony cycle's being controlled by the feedback of cues intrinsic to the colony, rather than by environmental cues. Since temperate-zone populations of social wasps appear to be derived from tropical ancestors (Richards 1971), it is likely that the mechanism that regulates temperate-zone colony cycles is also intrinsically regulated, albeit entrained by local seasonal constraints.

If intrinsic mechanisms regulate the colony cycle in wasps, it is possible that these mechanisms are ultimately the function of changes in the dominance structure as the colony develops. West-Eberhard, for example, has suggested that a *Polistes* colony declines when the queen's reproductive and social dominance weaken (West-Eberhard 1969, 1986). Within the context of the argument that it is ultimately dominance that imposes and maintains the social structure in the colony, the queen would certainly seem to play a critical role in maintaining colony integrity. But there must be more to it than merely the weakening of the queen's dominance, because a weak queen early in the colony cycle will be superseded and the colony will not decline (see Fig. 11.8).

An alternative explanation brings the rest of the colony into the equation. Early in the colony cycle there are relatively few offspring, all female. The queen's dominance is effective in depressing the direct reproductive value of these females to below the level of their inclusive reproductive value as helpers, so they opt to become workers. If a queen weakens at this stage, a strong offspring whose direct reproductive value is not adequately depressed may realize greater reproductive success by superseding her mother and inheriting the nest and staff, because she has a better chance of realizing a higher productivity with the workers already present than she would have if she founded her own nest (Queller 1989). As the colony develops, however, this picture changes, because as the rate of eclosion of offspring increases, the effectiveness of the queen in suppressing their direct reproductive value is diluted by numbers. The dilution of the queen's effectiveness, plus the presence on the nest of older, active foragers, causes at least some of the later offspring to delay or avoid behaving as workers because their direct reproductive value remains high. These are the idle "nonworkers" that appear late in the season (West-Eberhard 1969, Jeanne 1972). At the same time, many of the offspring now eclosing are males. Thus, the population of nonworking mouths to feed increases faster than the population of workers to feed them. Nest growth slows and ultimately stops, and cannibalism of the brood begins, possibly to relieve nutrient stress experienced by the adults, although Deleurance (1952b) claimed that cannibalism is triggered by the inability of older workers to produce a trophic secretion essential to normal larval development. By this stage the colony is no longer a resource through which a young queen can realize much reproductive success, since she would have to exert a tremendous effort to impose enough domination on the large population to convert enough of them to worker behavior to turn the situation around. When the reproductive value to be gained by founding a new colony exceeds that available by reproducing on the old, females will depart and initiate new colonies.

Late-emerging females of most species are pre-imaginally biased toward enhancing their success as foundresses. Late-emerging *Polistes* females, for example, tend to be larger than early-eclosing workers, and in temperate-zone populations the fat body of gynes is more abundant and qualitatively different from that of workers (Eickwort 1969). These traits are most pronounced in vespine species, in which gynes are much larger than workers and differ from them in body proportions (Blackith 1958). In many swarm-founding polistines, especially those that form larger colonies, there is a similar, but less extreme, divergence in size and form between queens and workers (Jeanne 1980a).

Species-typical colony size, which varies tremendously among the social wasps (Jeanne 1980a), may be set by the reproductive values attainable by females. Theoretically, the limit to which the queen(s) can depress a female's inclusive reproductive value ($RV_d + RV_i$) before the female should opt to leave is the direct reproductive value the female would have if she were to found her own nest ($RV_f$) (see Vehrencamp 1983a). Therefore, assuming that inclusive reproductive value for workers decreases as colony size increases, a high value of $RV_f$ would set a low limit to colony size. Larger colony size could then evolve only if $RV_f$ were reduced in some way. This could come about through particular combinations of ecological factors, such as those that make founding a high-risk proposition at times other than the normal founding season, as well as through parental manipulation to reduce the likelihood of success of worker-founded colonies (e.g., by imposing small body size or a lack of males). When such ecological factors and parental manipulation are accompanied by the evolution of more efficient means of queen control (West-Eberhard 1977a), enabling the queen(s) to depress the inclusive reproductive value ($RV_d + RV_i$) of larger numbers of workers, the evolution of larger colony size would follow.

Queen-worker dimorphism can evolve only after colonies have taken control of their environment and colony cycle to the extent that disruptions in the colony cycle due to predation, accident, and so on, are infrequent or can be remedied. Such control reduces the need to retain queen-worker role flexibility and allows morphologically specialized castes to evolve.

## CONCLUDING REMARKS

Although the concept of the superorganism has lost favor, the idea that selection occurs at the colony level in social insects is still alive. Inclusive fitness theory is a way of expressing the operation of natural selection on both the interindividual and the intergroup levels (Hamilton 1975, Wilson and Sober 1989). Wilson and Sober (1989) have argued that maximizing inclusive fitness actually *requires* intercolony selection.

They illustrate this by considering an allele that causes workers to behave so as to increase the fecundity of the queen. Although the allele increases the inclusive fitness of the worker, its frequency can increase only by intercolony selection: colonies that carry the allele out-reproduce colonies that do not (Wilson and Sober 1989). Wilson and Sober argue that we should not think in terms of a contest between two competing theories of evolution—one that selection occurs at the individual level, the other that it occurs at the colony level—with one theory eventually to be proven wrong. Instead, we should view the two levels of selection as different conceptual frameworks, that is, alternative ways of analyzing evolution in structured populations that converge on the same conclusions regarding units of functional organization (see also Ross and Carpenter, this volume). Although the discussion in this chapter interprets the patterns of polyethism in eusocial wasps in terms of selection acting on individuals, it does not argue against the notion that selection operates between colonies. Selection operates in both ways, but the relative influence of each selective mode may be expected to vary with level of social organization. To the extent that workers have a chance to lay eggs (Bourke 1988) or to supersede a dead or weakened queen, they have a direct reproductive value that is greater than zero, and their behavior is shaped by selection acting on them. On the other hand, to the extent that they achieve fitness indirectly by augmenting their relatives' reproductive output, their behavior is shaped by intercolony-level selection. Insofar as these modes of selection conflict, selection acting on individuals may in many cases act to inhibit the refinement of social behavior by intercolony selection (Wilson 1985).

Nevertheless, as a worker's $RV_d$ is reduced through a combination of domination and aging and the possibly irreversible changes in endocrine and reproductive physiology that follow, her fitness increasingly comes to be realized indirectly, by helping boost her relatives' reproductive output. Assuming that the queen or queens have mated more than once, bringing the sex ratio favored by a worker into line with the 1:1 ratio favored by the queen (Trivers and Hare 1976), it is clear that as $RV_d$ is suppressed toward zero the worker's interests come to converge with the interests of the queen(s). This should lead to the evolution of worker behavior that gives every appearance of being performed for the "good of the colony." At this point the group emerges as the integrated unit of selection (Alexander and Borgia 1978; West-Eberhard 1979, 1981).

Inasmuch as individual-level selection appears to continue to exert an important influence in many, if not most, species of eusocial wasps, and given that intragroup competition plays an important role in polyethism, it becomes imperative to consider the behavior of individuals.

Many of the studies of division of labor completed to date lose the individual in the crowd by focusing only on the average behavior of a population of workers or on a sample of an age cohort. Such an approach can obscure the often extreme degrees of task specialization in time or in space that may be manifested by individuals. This is unfortunate, since a full understanding of the functioning of the colony will be achieved only when we can explain it in terms of the behavior of its individual members, however varied they may be.

## Acknowledgments

My research has been supported by the College of Agricultural and Life Sciences of the University of Wisconsin, Madison, by grants from the National Science Foundation, and by the John Simon Guggenheim Memorial Foundation. I thank C. K. Starr for comments on an earlier version of the manuscript.

# 12

# Nourishment and the Evolution of the Social Vespidae

JAMES H. HUNT

Scientific progress is often galvanized by the search for a Holy Grail: the structure of the atom, the molecular basis of inheritance, the rediscovery of *Nothomyrmecia macrops*. In the small, quiet corner of science explored by readers of this volume, no search has stimulated as much activity as has that for the origins of eusociality. The energies applied in that search to the four major taxa of social insects are not uniform, however. Termites and ants lack living nonsocial forms, thus the insight-generating process of comparison and contrast is inhibited. Bees present an array of eusocial lineages that can obscure both pattern and process, and "reversals" in social evolution cloud the picture further. Wasps appeal to contemporary crusaders both in the economy of their eusocial lineages and in the richness of presocial and primitively eusocial taxa. Surely, it seems, if a robust resolution to the problem of the origins of eusociality can be found, it will be found first among wasps.

Taxon-specific answers to questions of general interest are less than satisfying, and so attention has long been focused on a hypothesis that promised a general solution: the kin-selection hypothesis of the evolution of eusociality (Hamilton 1963, 1964a,b; reviewed in Ross and Carpenter, this volume). It is an unfortunate denouement of the high drama of the late 1970s and early 1980s that, although kin selection remains a strong general paradigm, the hypothesis that haplodiploidy offers an easy means of capitalizing, by inclusive fitness effects, on cooperative effort is inadequate as a singular explanation of eusocial evolution in the Hymenoptera (West-Eberhard 1975, Crozier 1982, Andersson 1984, Stubblefield and Charnov 1986). Although some crusaders still search among the rubble of haplodiploidy, others are moving in new directions or have abandoned the quest for want of a focal paradigm. In this new intellectual landscape, then, taxon-specific answers seem better than none at all.

There is no lack of alternative, or at least partially alternative, hypotheses. Nor is there a lack of reviews of the relevant literature. Those emphasizing social wasp natural history include Spradbery (1965, 1973a), Evans and West-Eberhard (1970), Richards (1971), Iwata (1976), Jeanne (1980a), and Akre (1982); those with a more theoretical bent include Alexander (1974), West-Eberhard (1975, 1978a), Starr (1979), Gadagkar (1985a,b), Itô (1986a), and Carpenter (1989). Andersson (1984) and Brockmann (1984) offer summaries of relevant literature on requisite preconditions for eusociality.

Considerable importance has been ascribed to nest making accompanied by transport of larval foods to the nest as a precondition for eusociality. Wilson (1971) and Crozier (1982) note the possession of mandibulate mouthparts (or, more generally, the ability to modify the insect's immediate environment) as significant; clearly this is an antecedent condition to the ability to make nests. Andersson (1984) and Brockmann (1984) cite references that address the importance of communal nest defense against predators and parasites, while Starr (1985b, 1989a) places special emphasis on the antipredator value of the stings of social aculeates (but see Kukuk et al. 1989). All of these are extrinsic factors (Evans 1977a) and so represent a focus different from the intrinsic factor of kinship.

In an earlier short article (Hunt 1982), I added another extrinsic factor to the list of requisites for eusocial evolution: a source of proteinaceous nourishment in liquid form adequate to sustain the longevity and oviposition of a social insect queen. The importance of this factor has failed to attract much support or even attention. I attribute this lack of interest to deficiencies in the original presentation of the idea rather than to shortcomings in the idea itself. The present forum offers an opportunity to test that hypothesis, for what follows is a full exposition of the foundations, operation, and ramifications of suggested nutritional components of eusocial evolution in the Vespidae.

My approach is descriptive and mechanistic, and it thus addresses Francis Crick's lament that evolutionists all too often try to identify the value of something before knowing how it is made (Gould 1987). I find a pocket calculator of value without knowing how it is made. I make no attempt, though, to use a calculator as a model system to understand the workings of my desktop computer. Social insects have aesthetic and ecological value that is independent of our knowledge of how their societies are established and maintained. However, when the value of social insects is as a model system to understand the operating principles of higher animal societies (Wilson 1971: chap. 22), it becomes critical that we understand the silicon chips and circuitry of how insect societies are made.

## FOUNDATIONS OF THE NUTRITIONAL PERSPECTIVE

Underpinnings of the importance that can be ascribed to nourishment lie in three areas: anatomy, evolutionary history, and life history. Independently, each can be dealt with in a straightforward way. Collectively, I believe, they define the inevitable involvement of nourishment in hymenopteran eusocial evolution.

In assessing features of hymenopteran anatomy that have been critical to this evolution, I take an intermediate position, literally, between the previously cited emphases on mandibles and sting: I stress the importance of the thread waist (petiole) that characterizes higher Hymenoptera. The thread waist, which functions admirably as a fulcrum of oviposition in parasitoid Hymenoptera, imposes a dietary constraint of considerable importance. Thread-waisted Hymenoptera, though mandibulate insects, are restricted as adults to obtain nourishment only in liquid form (although for bees and a few wasps this may contain pollen). No adult Aculeata are herbivores, detritivores, or even carnivores in the strict sense. Although relevant literature on food habits of adult aculeate wasps is fragmented and largely anecdotal (but see Spradbery 1973a), it is safe to generalize that nectar, nectarlike liquids, and body fluids of prey or carrion are the typical, and nearly exclusive, sources of adult nourishment. Facets of internal anatomy that reflect this diet are the distensible crop and restrictive proventriculus (Eisner 1957) (Fig. 12.1). These components of the alimentary tract serve, respectively, as reservoir and valve to facilitate rapid feeding while regulating flow into the midgut. That these features can be exploited to regurgitate ingested liquids is a secondary adaptation that contributes significantly to the characteristic nutritional interdependence among individuals of most hymenopteran societies (Fig. 12.2).

The juxtaposition of mandibulate mouthparts and restriction to a liquid diet reflect in part the facets of evolutionary history relevant to this argument. Aculeate Hymenoptera evolved and radiated synchronously with Angiospermae (Baker and Hurd 1968). In the initial stages of that radiation (and still in living solitary forms), aculeates continued the dichotomized life-style of their parasitoid ancestors: provisions for developing young constituted a class of resource independent of resources for adult nourishment. Requirements for adult nourishment of aculeates could be met by floral nectar. Nectars of all kinds have been

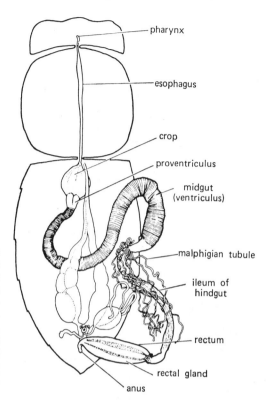

**Fig. 12.1.** Alimentary anatomy of a queen of the eusocial wasp *Vespula germanica*. Relevant features include the thread waist (petiole) between the thorax and gaster, the distensible crop, and the valvelike proventriculus. (From Spradbery 1973a, *Wasps: An Account of the Biology and Natural History of Solitary and Social Wasps*, © 1973 by J. Philip Spradbery. Courtesy of the author and University of Washington Press.)

shown to contain not only carbohydrates but also free amino acids, and patterns of abundance of those free amino acids strongly suggest adaptation of plants to supply the specific nutrient needs of nectar-gathering animals (Baker and Baker 1973a,b; Baker 1975, 1978). The subsequent interdependence and adaptive radiation of plants and nectar gatherers has of course made our world as beautiful as it is interesting. The principal historical consideration, then, is that nectar is the fundamental food of adult Aculeata.

The life history component of the tripartite foundation of the nutritional perspective depends wholly upon the intersection of the other two components: these insects, whose anatomy restricts them to a liquid diet and whose hereditary diet is nectar, may in the course of life history evolution only show an elaboration of dietary patterns within the bounds of two constraints. Novel food resources must be liquid, and they must be fundamentally similar to nectar in kind of nutrients. The variety of foods that meets these criteria undergirds the radiation and development of the various eusocial lineages of Hymenoptera. Five nourishment sources that meet the criteria have previously been identified (Hunt 1982): larval saliva, crop regurgitate from a conspecific

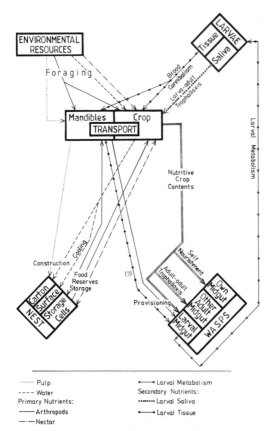

**Fig. 12.2.** Flow diagram of the movement of nutrients, water, and pulp (for construction) in nests of the swarm-founding polistine wasp *Polybia occidentalis*. Primary (environmental) and secondary (within-nest, e.g., larval) sources are shown at top; transport (by mandibles or crop) is in the center; utilization and storage are at the bottom. Similar patterns of materials flow are expected in primitively eusocial wasps such as *Polistes*. (Based on data in Hunt et al. 1987 and prepared in collaboration with R. L. Jeanne.)

adult, hemolymph of malaxated (chewed) larval provisions, eggs (by oophagy), and nectar fortified with pollen. Three other important sources should be added to the list: exudates of Homoptera and similar phytophagous insects, plant sap, and honey.

In one sense it is trivial to say that these foods have been involved in a fundamental way in social evolution in the Hymenoptera since, on the face of it, these merely are the foods that adult eusocial hymenopterans eat. A convincing argument must be presented that involvement of these foods contributes to the evolutionary process and is not just coincidentally correlated with it. In defense of the primary role of nourishment in eusocial evolution, a convincing argument must also be given that there are novel components to these ideas and that they are not merely reiterations of the nutritional hypotheses that have gone before.

## "NUTRITIONAL CASTRATION" AND THE TWO
## KINDS OF WORKERS

Two well-known nutritional hypotheses in the early literature of wasp social evolution are the "nutritional castration" hypothesis of Marchal (1896a, 1897) and the trophallaxis hypothesis of Roubaud (1916). Both are probably best known by their encapsulations in Wilson (1971).

Roubaud (1916) recognized the saliva of larval polistine and vespine wasps to be nutriment, and he proposed the term *ecotrophobiosis* to characterize the reciprocal exchange of foods between adults and larvae. (William Morton Wheeler's substitute term, *trophallaxis* [Wheeler 1918], survives while ecotrophobiosis does not.) Roubaud proposed that attractiveness of the larval saliva to adults was the raison d'etre of vespid sociality. He further proposed a theory of "alimentary castration" in which overexploitation of larval saliva could lead to undernourished, smaller adults with underdeveloped ovaries. Nutritional costs of working behavior as an adult, superimposed on deficient larval nourishment, then maintained worker sterility.

Trophallaxis was a cornerstone for expositions by Roubaud, Wheeler, and others on what became known as the *subsocial route* to eusociality in Hymenoptera. The generality that Wheeler presumed for the subsocial route began to unravel, though, when Michener (1958) effectively made the case that a much different *parasocial route* to eusociality provided a better fit to the data for most lineages of social bees (see Spradbery, this volume, for a discussion of these models). Unproductive jousting between subsocial and parasocial schools of thought continued until West-Eberhard (1978a) made an explicit plea that trophallaxis theory, and with it the subsocial hypothesis, be discarded for wasps as well as for bees. However, I and my coauthors (1982) have proposed that trophallaxis theory be retained for some, but not all, eusocial lineages. In addition, the theory, as further developed here, has clear subsocial components. The focus of West-Eberhard's discontent was Roubaud's proposition that attractiveness of larval saliva to adults was the main cause of social cohesion among them, and, indeed, that idea should be dismissed. However, I will explicitly argue here that the alimentary castration component of Roubaud's views is probably correct, though I will place it in a broader context of selection acting on larvae as well as on adults.

Throughout the earlier discourse on trophallaxis, Marchal's (1896a, 1897) trophic theories often garnered little more than passing interest. The usual term applied to his ideas is *nutritional castration*, and Wilson (1971) has suggested that Marchal's proposition that the worker caste in wasps is nutritionally determined, though simplistic, "may well be close to the truth in social wasps generally" (Wilson 1971:180). However, Wilson's ensuing discussion clouds the issue by mingling discussion of wasps that were undernourished during larval development with discussion of wasps that were undernourished during adulthood. Marchal intended to specify that both modes of worker determination (i.e., as larvae and as adults) occur, each under specific temporal and evolutionary scenarios. He wrote:

> Many authors, among them Herbert Spencer and Carlo Emery, . . . viewed the partial starvation of larvae as important in the ontogenetic production of workers (*trophic castration*), and indeed this appears to be the key factor. An additional important factor arises from the effects of *brood-care* on the adults which are raising the brood. . . . One of the prime causes of worker sterility in wasps . . . is the compulsion of . . . younger individuals to devote themselves to caring for a large group of larvae. Experiments have shown that this brings about the supression of ovarian development. We propose to call this mechanism in caste determination *nutricial castration*. Nutricial castration is much less evident as we go from primitive to more advanced social levels. We can consider it to have been a necessary condition in the earliest stages of caste differentiation, such as *Polistes* occupies today. In more socially advanced species, such as the honey bee, on the other hand, trophic castration is a sufficient mechanism, so that nutricial castration plays no part under normal conditions. (1897:556–557; translation by C. K. Starr)

In other words, Marchal (1897) specified two trophic modes of worker determination, neither of which he called nutritional castration. He proposed that caste determination through differential larval nourishment (trophic castration) is both self-evident and commonplace in advanced eusocial species (e.g., in the Vespinae). Independently, differential nourishment among adults of a primitively eusocial colony (e.g., independent-founding Polistinae) manifests itself by way of little or no ovarian development among the individuals that perform worker behaviors (nutricial castration). Essential elements of this dichotomy are reiterated in a recent review of caste determination in hymenopterans (Wheeler 1986).

The *trophic castration* of Marchal and *alimentary castration* of Roubaud are synonymous. Marchal viewed trophic castration (of larvae) and nutricial castration (of imagos) as largely separate phenomena, with only nutricial castration found in primitively eusocial polistines. Roubaud combined Marchal's ideas when he proposed that the nutritional costs of worker behavior reinforce the disadvantage of poor larval nourishment in primitively eusocial species. In my argument, I will echo Roubaud's position.

At the risk of generating more jargon, I propose the terms *ontogenetic worker* for workers determined by differential larval nourishment and *subordinate worker* for workers determined by differential adult nourishment. To do so, I feel, adds both clarity and perspective to the problems of eusocial evolution. Subordinate workers were the focus of West-Eberhard's (1978a) plea that the trophallaxis theory be discarded. I find no fault with her proposition that a parasocial model of eusocial evolution can be applied to polistine wasps as well as to most lineages of bees (but see Carpenter 1989, this volume). Differential nourishment among cooperating adults was, after all, an explicit component of the dominance hierarchy among *Polistes* co-foundresses first described by Pardi (1948). The subordination of reproduction in subdominant co-foundresses is both readily observable and evolutionarily significant, but it does not directly address the question of why offspring of a colony should also be workers rather than (potential) reproductives. Marchal, and perhaps a majority of recent researchers, would ascribe the production of such workers in primitively eusocial species to the compulsion of younger individuals to perform worker duties; that is, Marchal would have believed that offspring workers are behaviorally subjugated to their role by dominant reproductives just as are subdominant co-foundresses. In contrast, and in agreement with Roubaud, I will argue here that offspring workers in primitively eusocial species are initially cast into their role by virtue of nutritional deprivation during larval growth, with poor nourishment as a working adult contributing to the maintenance of the system. In other words, I believe that ontogenetic workers are found in primitively eusocial as well as advanced eusocial species, and it is in the development of ontogenetic workers in primitively eusocial species that I feel a role exists for trophallaxis as a major factor in vespid social evolution.

## ONTOGENETIC WORKERS IN THE EVOLUTION OF EUSOCIALITY

I give here a brief, uninterrupted overview of the evolution of primitive eusociality in vespid wasps as I envision it. Documentation and discussion of the various parts of the scenario follow.

### The Nutritional Scenario

The lineage that yielded eusocial vespids possessed adaptations to parasitoid and predator pressure such as compact nests, open cells, and progressive provisioning. Pleometrosis (cooperative nest foundation) may have been present, as may a subsequent division of reproductive labor in such pleometrotic associations.

Exploitation by foragers of provisions for larvae larger than could be transported in a single flight was competitively advantageous and was elaborated. Provision partitioning was accompanied by malaxation and ingluvial transport of the provision hemolymph. Some of the ingluvial hemolymph passed into the midgut of the transporting wasps (or others receiving it by interadult trophallaxis) and served as supplemental nourishment to the fundamental diet of nectar. The enhanced nourishment facilitated greater longevity and fecundity.

Larvae in a colony under food stress were at risk of being cannibalized. Larvae mandibulated by potentially cannibalistic adults expelled droplets of saliva as a potential appeasement to forestall cannibalism. Adults found this saliva similar to nectar and so drank it. Specific solicitation for saliva ensued. Through the selective agency of differential survival of cannibalism, saliva volume and concentration of nutrients increased.

In the early stages of a nesting cycle all nestmates were relatively poorly nourished, and so solicitation for saliva was relatively intense. Poor larval nourishment exacerbated by high levels of saliva solicitation exacted a developmental cost on larvae. Upon emergence, adults that had experienced such deficient nourishment as larvae faced poor prospects in any attempts to found nests independently. Worker behavior pursuant to a general model of inclusive fitness thus offered the prospect of rescuing some fitness in spite of a developmental setback. Worker behavior may have been preferentially directed to nestmates according to degree of relatedness.

Through the combined efforts of foundress(es) and workers, all nestmates present during the later portions of the nesting cycle enjoyed a higher level of nourishment than those present earlier. Better-nourished larvae could retain saliva at lowered risk of being cannibalized. Adults emerging late in the nesting cycle, which had enjoyed relatively rich larval nourishment, could pursue colony founding with a correspondingly higher probability of success.

Substantial discrepancies in success between colonies having workers and those not having workers drove selection in favor of behavioral repertoires that contributed to the production of ontogenetic workers.

## Elements of the Scenario

THE EUSOCIAL LINEAGES    Carpenter (1988a, this volume) argues that the three subfamilies of eusocial Vespidae are all derived from a common, primitively eusocial ancestor. The scenario to be given here, however, explicitly addresses the primitively eusocial Polistinae, as exemplified by *Polistes* and *Mischocyttarus*. While fundamental components of this scenario may have typified the common ancestor of all eusocial Vespidae, the full elaboration of sociality has very different elements in the Stenogastrinae (Turillazzi 1988a, this volume) from those in Polistinae and Vespinae. Thus, while application of early steps of the model to Stenogastrinae is assumed at present, the full model can be applied with confidence only to the common ancestor of the Polistinae + Vespinae. The scenario presented here is based on the ontogeny of a living colony of primitively eusocial polistine wasps, and it has been conceived as a vespid corollary of Haeckel's dictum that ontogeny recapitulates phylogeny.

PARASITIZATION AND PREDATION    The subsocial hypothesis is usually presented as a series of graded steps leading from solitary to eusocial nesting (Wheeler 1923, Evans 1958, Evans and West-Eberhard 1970). Wheeler's seven steps are characterized by their reference to adaptations that increase the protection afforded the brood. That is, the construction of a nest, the habit of remaining at that nest when not foraging, and progressive provisioning (with inspection of cells and young) are all adaptive with regard to parasitoids and brood predators and preadaptive with regard to the emergence of eusociality. Protection against vertebrate predators by use of the sting may have been rapidly selected for after eusociality was achieved, but at the threshold of eusociality the sting venom doubtless retained its provision-paralyzing toxins, that is, the sting was still best suited for securing larval provisions. Kukuk et al. (1989) have argued persuasively that the sting was of minimal importance as a preadaptation facilitating the evolution of wasp sociality. While stings may have enabled the elaboration of wasp sociality once it had originated (Starr 1985b, 1989a,b), it seems clear that if stings had not existed to be modified into a defense system then some other defense system would have been developed. One only need look at termites, stingless bees, and formicine and dolichoderine ants for examples of some possibilities (stings were lost in an ancestor that was already eusocial in the latter three cases).

PLEOMETROSIS AND DIVISION OF LABOR    The roles of pleometrosis and division of reproductive labor in wasp social evolution have been addressed most directly and forcefully by West-Eberhard, who emphasizes pleometrosis as a salient feature of mutual nest defense that is preadaptive for sociality. In her polygynous family hypothesis, West-Eberhard (1978a) argues that eusociality can be achieved in wasps in a semisocial manner by division of reproductive labor among related cofoundresses. Carpenter (1989, this volume) uses phylogenetic analysis to show that some components of the polygynous family hypothesis are likely to be robust while other components are not. Although these aspects of wasp social evolution are not unimportant, I agree with Evans (1958) that it is not necessary to determine whether haplometrosis (colony founding by a single female) or pleometrosis is the primitive condition in wasps in order to determine the important features in the origin of a worker caste. The critical question of worker evolution among offspring is probably independent of queen number and of any partitioning of reproductive effort among co-foundresses.

LARGE PROVISIONS  Provision carriage behavior in solitary wasps is the interesting subject of a large comparative literature (Iwata 1942, 1976; Evans 1962, 1966a,b). Provisioning in eusocial species, which usually subdue arthropods without stinging them and often transport only fragments to the nest, seems superficially less interesting. Nonetheless, I contend that the initiation of this mode of provision collection marks a watershed in the evolution of social wasps. Capture of large provisions that are fragmented and malaxated conferred two major advantages almost simultaneously: (1) the wasps gained access to a category of resource that was both abundant and not subject to direct competition from other aculeates; and (2) the phylogenetically established dichotomy between adult and larval nourishment sources was erased. These advantages were significant in that, in ecologists' parlance, mandibular capture and fragmenting of larval provisions marked the entry of social wasps into a new adaptive zone. The subsequent adaptive radiation of eusocial Vespidae can be viewed as a textbook example of ecological success and taxonomic diversification by a lineage that successfully passed through an evolutionary keyhole.

LARVAL PROVISIONS AND ADULT NOURISHMENT  I like to draw a clear distinction between *provision* taken for larvae and *prey* taken for self-nourishment. Almost all literature references to wasp prey and predation (including Spradbery 1973a) are actually records of provision collection. There are a few explicit references to prey items taken for adult nourishment, but they are scattered and largely anecdotal (e.g., Rau and Rau 1918:281, Lin 1978). The nature of this literature, together with my own observations, convinces me that taking of prey exclusively for self-nourishment is relatively rare in both solitary and social wasps (cf. Keyel 1983 and Ross 1983a, for vespine foundresses). A universal feature in eusocial Vespidae, however, is malaxation of larval provisions. It has been demonstrated that female *Polistes* and *Mischocyttarus* ingest (into the midgut) at least a portion of the hemolymph of malaxated provisions (Yoshikawa 1962, Jeanne 1972, Hunt 1984), and the ingested portion is digested (Kayes 1978) and rapidly assimilated (Hunt 1984). This is surprising only in that its importance has been vastly unappreciated. Its importance is perhaps best exemplified by a passage in Evans (1958:452): "The short life of the female in all Sphecidae seems to be one reason why none of them have become truly social. A second reason is that none of them macerate the prey before presenting it to the larva." I suggest that adult longevity and adult nourishment by means of provision malaxation in vespids are intimately and directly related. I further suggest that provision malaxation provides nourishment for sustained reproductivity during the course of the enhanced adult life span.

CANNIBALIZATION OF LARVAE    The weight of proof of the contention that larvae are at risk of being cannibalized lies in demonstrating that brood cannibalism occurs with more than trivial frequency and that it can be induced by conditions of low adult nourishment. Anyone who observes social wasp nests closely for extended periods will have observed occasional brood cannibalism both in Polistinae (e.g., Hunt and Noonan 1979, Hunt et al. 1987) and Vespinae (e.g., Akre et al. 1976, Greene, this volume). (Instances of mass destruction of brood and associated nest carton in midseason *Polistes* or *Vespula* colonies are not the same as occasional cannibalism and often seem explainable as a response to infestation by parasitoid or brood predator insects. Mass destruction of brood is sometimes characteristic of the close of a nesting cycle [Duncan 1939, Deleurance 1955b, Archer 1980a].) I have argued elsewhere that cannibalism is a major selective force in the origin and elaboration of larva-adult trophallaxis in social vespids (Hunt 1988). Hard data supporting this point are meager. Fig. 12.3, however, presents data that are suggestive. Shown are brood maps of eleven *Polistes metricus* and two *P. fuscatus* nests collected at a single site in Missouri, United States, in mid-May. Four of the *P. metricus* nests and one of the *P. fuscatus* nests contained eggs or early-instar larvae in cells nearest the nest petiole; later-instar larvae occupied nest cells farther from the petiole. No silk or meconium (fecal pellet) was in the petiole cells to indicate that imagos had emerged from them. Since eggs are laid earliest in cells at the petiole, I interpret the pattern in these five nests as evidence that the first brood had been cannibalized and replaced by eggs in the cells nearest the petiole. Three of eleven nests of *Mischocyttarus flavitarsus* collected in Colorado, United States in mid-June showed brood patterns similar to those of the *Polistes* nests, suggesting a similar cause. I encourage observers who frequently map social wasp nests to record these kinds of data bearing on the frequency of larval cannibalism.

LARVAL SALIVA    The copious production of larval saliva that adult eusocial wasps drink has stimulated the curiosity and speculation of many observers. Suggested roles for the saliva include regulation of nest temperature and humidity (Weyrauch 1936c), regulation of larval water balance (Brian 1983, M. J. West-Eberhard, unpubl.), larval excretion (Brian and Brian 1952), serving as an aid for the ingestion and digestion of provision masses (Roubaud 1916, Spradbery 1965, Hunt et al. 1982), serving as nutriment for adult wasps that imbibe it (Spradbery 1973a; Hunt et al. 1982), and serving as an appeasement to forestall cannibalism (Hunt 1988). While Brian and Brian's (1952) excretory hypothesis was based on incorrect experimental evidence (Maschwitz

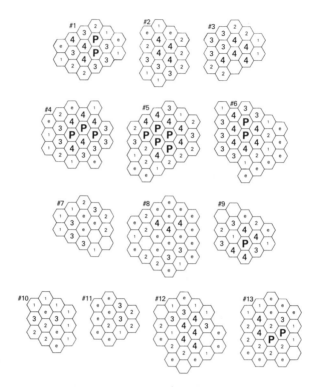

**Fig. 12.3.** Brood maps of eleven nests of *Polistes metricus* (#1–#11) and two nests of *P. fuscatus* (#12, #13) collected on May 19, 1987, at a single site in Adair County, Missouri, United States. The nest petiole and, hence, the first cells are centrally located on the combs of both species. Five nests show typical brood development patterns, with the oldest offspring in cells nearest the petiole (#1, #2, #4, #5, #6). Five nests have younger brood in cell(s) at the petiole than in more-peripheral cells (#7, #8, #9, #10, #13); this pattern is inferred to have been caused by cannibalism of larvae that were the first occupants of those cells. Three other nests (#3, #11, #12) also have cells from which larvae may have been cannibalized. e, egg; 1–4, increasing size categories of larvae; P, pupal cocoon.

1966a), none of the other hypotheses has been robustly disproven, and indeed they are not mutually exclusive. Some or all of the suggested roles may be met by larval saliva under varying circumstances of the colony and its environment. I believe, though, that the most robust combination of hypotheses is that larval saliva was first expressed as an appeasement to forestall cannibalism and that it was accepted by adults as nutriment resembling floral nectar.

SALIVA NUTRIENT CONCENTRATION Co-workers and I found that there is 50 times the mean concentration of free amino acids in larval saliva of seven wasp species than in the richest of floral nectars (Hunt et al. 1982). Similar data for an additional seven wasp species are given in Hunt et al. (1987). Attractiveness of the larval saliva to adult wasps is thus easily explained, but why should the saliva be so nourishing? We did not address this question (Hunt et al. 1982, 1987), nor did we propose any adaptive advantage to larvae for producing the trophallactic saliva. I have since suggested that intense solicitation by adults led to selection for nutritional richness in the saliva of larvae that yielded it as a successful appeasement to forestall being cannibalized (Hunt 1988).

NOURISHMENT AND SALIVA RETENTION Larval saliva can be collected easily by teasing the mouthparts of mid- to late-instar larvae with a capillary or Pasteur pipet. Some larvae in each of the 17 species from which I have collected saliva were "reluctant" to yield saliva, and some gave none at all. Last-instar larvae lose the ability to give saliva when the labial gland (the saliva-producing gland: Maschwitz 1966a) changes function to produce silk. However, I have described a remarkable behavior of larvae of *Mischocyttarus* (Hunt 1988) (Fig. 12.4), in which refusal to give saliva is accompanied by engorgement and erection of the abdominal lobe, a morphological structure characteristic of the genus (Reid 1942). Preliminary data suggest that the lobe engorgement behavior occurs primarily in well-nourished larvae and that it is not simply an indicator of incipient pupation. After seeing and reflecting on the *Mischocyttarus* behavior, I believe that many of the instances of "reluc-

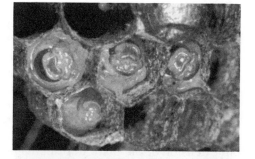
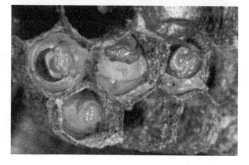

**Fig. 12.4.** Final-instar larvae of *Mischocyttarus immarginatus* in their nest cells. Final-instar larvae have a cuticular lobe on the ventral surface of the first abdominal segment that, although of varied morphology in different species, is characteristic of the genus. In the upper photograph, all larvae are in the normal resting posture. The larva in the central cell was artificially solicited (by a micropipet) for saliva; it refused to give saliva and instead withdrew its head into the cell. Head retraction (lower photograph) results in the abdominal lobe rotating anteriorly and centrally in the cell, where it becomes turgid and erect. This behavior has now been observed in all five *Mischocyttarus* species in which it has been sought.

tance" in taxa that do not have lobes also were expressions of saliva retention rather than of an inability to give it. These observations, together with the well-known cell-scratching solicitation signals of larval Vespinae (Matsuura, Greene, this volume), provide strong evidence that larvae of eusocial vespids are not simple inert sacs of protoplasm, but that they are as variably interactive with other colony members as leglessness and confinement to a cell will allow.

POOR NOURISHMENT   As Wilson (1971:180) notes, the life cycles of social wasp species generally are marked by a gradually increasing food supply for individual larvae as the colony grows older. Components of this pattern include improved worker-larva ratios with advancing colony development (Richards and Richards 1951, Spradbery 1965, West-Eberhard 1969) and a higher percentage of food loads among total forage loads (West-Eberhard 1969). That poor nourishment occurs naturally early in a *Polistes* nesting cycle is easy to demonstrate. Rossi and I (1988) placed small droplets of dilute *Apis* honey into the nest cells of preemergence colonies (those that have not yet reared adult workers) of *Polistes metricus*, in a manner analogous to the naturally occurring placement of honey droplets by the adult wasps. Although the first broods of female offspring from nests that were receiving supplemental honey did not differ in size from the first broods of control colonies, the offspring from supplemented colonies did have significantly higher body fat content—higher, in fact, than that of the foundresses. Body fat content is a correlate (and often an experimental discriminant: Eickwort 1969) of reproductive potential in offspring of *Polistes* colonies. It is logical to speculate that the behavior and reproductive activities of first-brood offspring that receive supplemental nourishment might well differ from those of typical workers. Rossi and I (1988) interpreted our data as a demonstration of food limitation of the wasps in their natural environment.

The developmental cost of poor nourishment was self-evident to Marchal (1897) and, in Wilson's (1971) view, was likely to be generally true (if simplistic). These authors were explicitly referring to Vespinae, but extension of the point to primitively eusocial Polistinae seems justified. To thus extend the argument can, in fact, be viewed as the crux of the entire scenario. Some recent relevant evidence on primitively eusocial Polistinae exists. Enhanced fat deposition in first-brood offspring of *Polistes metricus* colonies as a consequence of supplemental nourishment has been demonstrated (see above). Research in the Soviet Union (Grechka 1986) asserts that 80% of individuals of *Polistes dominulus* (= *gallicus* of authors) are predetermined as to reproductive caste in their preimaginal stages, and the amount and quality of food are reported as contributing to this preimaginal determination. Gadagkar et al. (1988) documented differences in colony-founding ability among isolated *Ropalidia marginata* females. A positive correlate of nest-founding success was the number of empty cells in the nest from which the experimental wasps emerged. Although Gadagkar et al. (1988) discuss this correlation in terms of declining queen influence, I suggest that a large number of empty cells also indicates a mature colony, and, as argued above, mature colonies have higher nourishment levels than young ones.

A component of Roubaud's alimentary castration hypothesis is that differential larval nourishment does not completely determine caste; nutritional costs associated with working were hypothesized to maintain adult worker sterility. Effects of differential adult nourishment on reproductive potential are indicated by the Gadagkar et al. (1988) study, in which the second and only other positive correlate of nest-founding ability was the amount of food consumed by a potential foundress during the period of isolation. Indeed, differential adult nourishment can engender continuously varying levels of reproductive potential among members of the morphological worker caste in Vespinae. Preliminary results from experimental feedings of isolated groups of *Vespula germanica* workers show a clear relationship between adult nourishment and ovarian development (J. Spradbery, unpubl.). Workers fed pureed blow flies had a mean ovarian index (proportional to the summed lengths of the three largest oocytes) more than three times as large as controls fed 30% honey solution. Worker oviposition is in fact a regular feature of the later stages of vespine colony cycles (see Ross 1985, Bourke 1988, Greene, Spradbery, this volume). Thus, level of nourishment as a larva may influence adult reproductive behavior in Polistinae and is a major determinant of the morphological worker caste in Vespinae, but rich nourishment while an adult can mitigate or even reverse the constraining effects of poor larval nourishment on subsequent reproductive ability.

EMERGENCE DATE   In all temperate areas and in much of the tropics, colony cycles of primitively eusocial polistines are synchronized locally by climate patterns. That is, colony founding is keyed by the onset of a favorable (e.g., warm or rainy) season, and colony decline is correlated with the waning of that season. This is generally true whether the ensuing unfavorable season is passed in physiological quiescence, as in the temperate zones, or by reduced activity, as in the seasonal tropics. Probability of successful founding may therefore depend less on capability to construct a nest, oviposit, and provision young than on the time constraint of doing so within the fragment of the favorable season that remains.

*Polistes metricus* is the most common *Polistes* in eastern Missouri, United States, where nest founding commonly occurs in April. In 13 years of fieldwork there I have occasionally seen preemergence nests in mid-July, which is well after the first workers have emerged from these nests founded in April. The number and circumstances in which the atypical nests occur strongly suggest that very few were refoundings after destruction of nests founded at the usual time; rather, they appear to have been nests founded by early-emerging offspring. Since colony dissolution regularly begins in August and is complete by early October, it is likely that most such nests founded in July produce few, if any, adult offspring. However, their occurrence compels me to believe that inability to found nests is not the reason that most early-emerging *Polistes* females do not do so. The less apparent but more likely reason is that early-emerging individuals lack the fat reserves required to sustain them through the unfavorable season until the start of the next spring. Away from the nest, foods would be largely nectar, which can certainly sustain life but may not contribute greatly to fat deposition. At the nest, larval saliva and provision hemolymph offer enriched nourishment, but it can be obtained only in the milieu of established social ranks. Early-emerging wasps, then, face limited options: to found immediately but with a shortened favorable season, to live independently pending a next season's founding but with a meager endowment of high-quality nourishment, or to remain on the home nest as a member of an established social unit. Most, apparently, pursue the latter course.

INCLUSIVE FITNESS AND KIN SELECTION   A wasp remaining on its natal nest still faces two options that are well documented in every study of colony development in primitively eusocial polistines: to participate as a worker at the nest or to forego working but remain as a consumer of colony resources. Wasps fitting the first description are, in the terminology introduced here, ontogenetic workers, and those fitting the second description are usually designated potential queens, although other behavioral descriptors (e.g., "nonworkers": West-Eberhard 1969; "fighters and sitters": Gadagkar and Joshi 1982a, 1984) have been applied. Ontogenetic workers that facilitate the growth and eventual reproductive success of kin may, of course, achieve some positive fitness through inclusive fitness effects.

Two points require emphasis here. The first is that kin selection comes into consideration only *after* the expression of worker behavior. West-Eberhard (1987a:370) asserts: "Kin selection may act primarily to maintain facultative aid by affecting the evolution of regulatory mechanisms assuring that worker behavior is expressed only when likely to be adaptive," but "kin selection is not necessary to explain the origin of sterile workers."

The second point requiring re-emphasis is that haplodiploidy is not required either (e.g., Crozier 1982, Stubblefield and Charnov 1986). On the other hand, the burgeoning literature on kin recognition in social species (see Fletcher and Michener 1987) suggests that mechanisms exist that can be exploited to preferentially direct worker effort to closely related individuals within a brood having varied relatedness to the worker. Therefore a possible role for haplodiploidy (after the expression of worker behavior) is that worker hymenopterans might more easily discriminate near kin than might worker termites, since haplodiploid species can have higher variance in the coefficient of relatedness (r in Hamilton's rule; see Ross and Carpenter, this volume) than can diplodiploid species. Evidence from empirical studies on kin recognition in eusocial hymenopterans is mixed (Gadagkar 1985b). For example, though preferential investment according to kinship has been shown in honey bees (e.g., Noonan 1986, Page et al. 1989), a rigorous examination of kin recognition in the primitively eusocial wasp *Ropalidia marginata* concluded that different genetic lines within a colony are unlikely to be discriminated by the wasps, so that preferential investment according to kinship is unlikely (Venkataraman et al. 1988, see also Queller et al. 1990).

SELECTION FOR BEHAVIORAL REPERTOIRES   The foregoing arguments, if taken at face value, would suggest that worker behavior is a function of options taken independently by every offspring that emerges from every primitively eusocial wasp colony, and that worker behavior is invented anew in every generation. Evidence in support of this speculation is not, in fact, lacking. That the reproductive future of the experimentally manipulated wasps in Rossi's and my (1988) study (see Poor Nourishment, above) probably would differ from that of unmanipulated wasps was suggested by observations of laboratory-reared paper wasps. When rearing wasps in cages with limited daily feedings, I have had little difficulty in producing large, vigorous colonies. However, when the same species (*Polistes metricus, Mischocyttarus flavitarsus, M. mexicanus cubicola*) are raised in flight rooms with food abundantly available, nests remain small with few resident offspring. Offspring in these circumstances tend not to stay on the natal nest but instead, after a few days, depart to another area of the room where they remain relatively sedentary (Hunt, unpubl.). Controlled experimentation is now needed to test the speculation that abundant nourishment early in the nesting cycle can reduce or even eliminate worker behavior in first-brood offspring of primitively eusocial polistines.

Wheeler (1986:17) approaches the heart of the problem when she says: "A central question in the evolution of caste differences is how incidental differences in larval nutrition are channeled efficiently into castes." That question actually can be fractioned into several parts: (1) How is caste determined among individuals? (2) How is caste determination partitioned during colony ontogeny? (3) How are the mechanisms and colony pattern of caste determination maintained by selec-

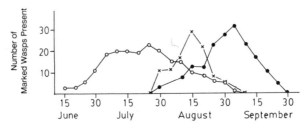

**Fig. 12.5.** Seasonal pattern of change in the resident adult population of a successful *Polistes fuscatus* colony from Michigan, United States. The transition from emergence of worker females (open circles) to emergence of nonworker females (gynes, closed circles) is characteristic of successful colonies. Appearance of males (x) in some (but not all) species of *Polistes* coincides with the transition to emergence of nonworker females. (Modified from West-Eberhard 1969, courtesy of *Miscellaneous Publications of the Museum of Zoology, University of Michigan.*)

tion? This chapter has until this point addressed the first two of the questions and has attempted to link them in a conceptual framework. Resolution of the third question lies, I believe, in the largely unexplored realm of social wasp population demographics.

## SOCIAL WASPS AS ANNUALS

The dynamics of successful temperate-zone *Polistes* colonies (Fig. 12.5) are often thought of as typical for primitively eusocial polistines in general. Few *Polistes* colonies founded at the onset of the favorable season are fully successful, however. The meager population data known to me (DeMarco 1982, D. Windsor, unpubl.) indicate that Type II survivorship (Deevey 1947) is characteristic of *Polistes* foundresses. Superimposing a survivorship curve for foundresses on a figure modeling the dynamics of a successful colony (Fig. 12.6) reveals that only a few successful *Polistes* foundresses in one season will produce the majority of foundresses for the following season. "Successful" and "typical" are not synonymous in primitively eusocial paper wasps.

Solitary wasps probably have low variance about the mean population fitness. That is, each produces few offspring, but the probability of each doing so is probably relatively high. Selection on life history traits associated with their demography thus may not be very strong in these wasps. In contrast, paper wasp queens have high variance about the mean population fitness. That is, a given *Polistes* foundress has a very low probability of producing any daughter queens, but a few foundresses will produce large numbers of them. Andersson (1984:166) was thus entirely correct in identifying "low success of young adults or solitary pairs that attempt to reproduce" as a condition crucial to the evo-

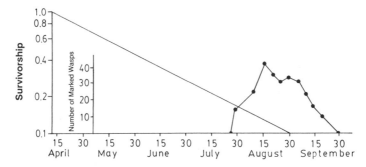

**Fig. 12.6.** A hypothetical Type II survivorship curve for queens (straight line) and production of potential reproductives (males plus nonworker females; closed circles) for the colony described in Fig. 12.5. The ordinate for wasp numbers is placed on the date of nest founding. The hypothetical queen survivorship curve is drawn from an ordinate at the approximate date of emergence from hibernacula to the latest date mentioned by West-Eberhard (1969) for seeing a resident queen on a nest.

lution of eusociality, although he did not directly address the matter from the present perspective.

Schaal and Leverich (1981) have presented an insightful model of seed production in annual plants that offers much to the resolution of problems in wasp eusocial evolution. The straw man for the Schaal-Leverich model can be characterized as a conventional wisdom that states that the most direct route to high fitness in $r$-selected organisms, such as annual plants, is to reproduce early (e.g., Pianka 1988: table 8.2). Under the Schaal-Leverich model, highest fitness accrues to those annuals that reproduce late. The model can be illustrated by a survivorship curve for the full life span of a single generation, from propagule formation to death (Fig. 12.7). Early life for an annual is spent as a seed in a "dormancy pool." In that dormancy pool, survivorship depends on the action of random mortality factors (especially seed predators) and is expressed by a Type II survivorship curve. Probability of surviving until germination, then, is inversely correlated with time spent in the dormancy pool. For seed crops having different dates of entry into the dormancy pool, the latest date of entry yields the highest survivorship to germination (Fig. 12.7).

Schaal and Leverich (1981) argued that delayed reproduction will be favored in species having two-stage (i.e., active and dormant) life cycles with the following characteristics: (1) initiation of the active phase is synchronous for the population; (2) mortality during dormancy is a function of time spent in dormancy; and (3) an age-structured dormancy pool exists (i.e., individuals in the pool differ in age). I contend that these criteria precisely describe primitively eusocial Polistinae in seasonal environments. Colony founding (initiation of the active

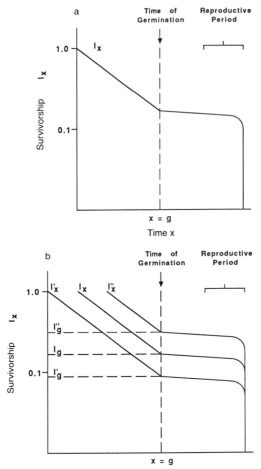

Fig. 12.7. (a) Idealized aggregate survivorship curve for an annual plant. Average survivorship of an even-age cohort is plotted on a logarithmic $y$ axis vs. age on the arithmetic $x$ axis. Mortality is at a constant rate prior to the time of germination at $x = g$. (b) The plot from a is disaggregated to show the effects of different times of entry into a seed pool on survivorship values at the time of germination ($x = g$). The plot $l_x$ is the average survivorship, as in Fig. 12.7a. Early seeds have curve $l'_x$, while late seeds have curve $l''_x$. Clearly, $l'_g < l_g < l''_g$. (Figure and legend from Schaal and Leverich 1981, *American Naturalist* 118: 135–138, © 1981 by the University of Chicago.)

phase) is keyed by onset of the favorable season; quiescent wasps awaiting the onset of the next favorable season experience mortality factors throughout their wait; and not all quiescent wasps are the same age.

According to the Schaal-Leverich model, traits that favor delayed production of potential reproductives could be under strong selection. The selective agency is increased survivorship of individuals in the dormancy pool from colonies with traits that foster delayed production of potential reproductives. Candidate traits for such selection, in accord with the model of ontogenetic worker production presented here, include high levels of larval saliva solicitation by foundresses, high levels of donorism of nutrient-rich saliva by first-brood larvae, retention of saliva by later-generation larvae, propensity for early-emerging offspring to remain at the natal nest, and proclivity of these individuals to participate in colony work.

## GENERALIZATION OF THE MODEL

The pieces are now at hand to outline the major features of an integrated view of vespid social evolution. Selective pressures from parasitoids and nest predators put into place the necessary preconditions of nest construction, progressive provisioning, and brood inspection. Exploitation of provisions that required partitioning for transport opened a new adaptive zone characterized by relaxed competition with other aculeates and by provision malaxation. Provision malaxation facilitated enhanced adult longevity and reproductive capacity. Features of adult anatomy and dietary preferences, coupled with the expelling of larval saliva as an appeasement to forestall cannibalism, led to larva-adult trophallaxis. Patterns of differential nourishment, which are correlated with colony development, contributed to an inequitable distribution of nourishment among larvae. Poor nourishment of larvae early in a nesting cycle was exacerbated by high levels of saliva donorism. Later-developing larvae were both better nourished and more predisposed to retain larval saliva. Female offspring emerging early in a nesting cycle, with low food reserves carried forward from larval development, faced limited options: to found nests in midseason, with little chance of success; to remain quiescent until the following season, again with little chance of success; or to remain as helpers at the natal nest. Most made the best of a bad situation by serving as helpers at the nest, which may have salvaged some fitness for the helpers through inclusive fitness effects. The pattern of delayed production of potential reproductives that resulted from worker behavior in early offspring was strongly selected for under a demographic model that articulates adaptiveness of delayed reproduction in organisms having two-stage life histories in seasonal environments. Linkage of the demographic selection model with the model of ontogenetic worker production presented here, I believe, brings us near a long-sought, though taxon-specific, resolution to the search for the origins of eusociality.

The model developed here can be only partially applied to the Stenogastrinae. These wasps apparently lack ontogenetic workers (Turillazzi, this volume), and so the trophallaxis model of ontogenetic worker production would seem to be limited to the common ancestor of Polistinae + Vespinae (see Carpenter 1982, 1989, this volume, for discussion of the phylogenetic relationships of the vespid subfamilies). However, early steps of the scenario, especially the very significant step of prey capture and malaxation that addresses adult reproductive longevity, may have been characteristic of the common ancestor of all three subfamilies of eusocial Vespidae. If stenogastrine larvae produce saliva as an aid to the ingestion and digestion of provisions (Spradbery 1975, Turillazzi, this volume), then the *origin* of expelling saliva as an appeasement to forestall cannibalism in the social vespids is questionable. However, the *elaboration* of trophallaxis as an appeasement behavior in the primitively eusocial ancestor of the Polistinae + Vespinae seems to be an idea that merits rigorous examination.

The evolution of a morphological worker caste in Vespinae goes a step beyond the model presented here. Components of vespine caste differentiation include larval nourishment, size of the cell in which a larva is reared, and queen control (reviewed in Spradbery 1973a, this volume; Brian 1980; Greene, this volume). The regularity of worker oviposition following queen death or at colony dissolution (e.g., Landolt et al. 1977, Matsuura 1984, Ross 1985, Bourke 1988) underscores the importance of queen control as a major factor in the elaboration of vespine eusociality after its initial establishment in the common ancestor of the eusocial vespids. Queen control is a subject not at all fully explored in primitively eusocial Polistinae, and close attention to the mechanisms and ramifications of queen control can add much to our understanding of wasp social evolution (Spradbery, this volume).

## CONCLUDING REMARKS

The closest fit between preexisting theory and the scenario for ontogenetic worker production in wasps outlined here is with the parental manipulation theory, best articulated by Alexander (1974). In a subtle but important distinction, the development of ontogenetic worker offspring as envisioned here is not so much a matter of foundresses manipulating (before the fact) as of exploiting (after the fact) their poorly nourished offspring, for whom life's opportunities are limited.

To reiterate an ancillary concern: it does not matter in this scenario whether colony founding is haplometrotic or pleometrotic, nor does it matter if there is a partitioning of reproduction among co-foundresses. What are critical are the mode and pattern of the production of workers among the colony's offspring.

West-Eberhard (1987a) stresses that worker behavior (be it ontogenetic workers or, in her argument, subordinate workers) does *not* represent the expression of an alternative set of alleles in wasps. It is instead a different developmental expression of alleles also present in reproductives. The model presented and developed here suggests that, at the threshold of eusociality for the common ancestor of eusocial vespids, the switch between worker and reproductive expression of those alleles was labile, sensitive to environmental cues, and probably reversible during both larval development and adulthood. Vespinae show a higher degree of control over the developmental switch, but even in vespines worker behavior is not genetically fixed.

To repeat another concern: it should be stressed that haplodiploidy is not necessary to explain the origin of sterile workers. This does not invalidate kin selection as a general paradigm, but even kin selection is applicable to social evolution only after the establishment of a colony (West-Eberhard 1987a). The emphasis in most studies of haplodiploidy has been on the opportunity for kin in haplodiploid species to capitalize on the asymmetries in their relatedness to siblings and so accrue high inclusive fitness. Evidence against the opportunity to exploit the theoretically possible high average relatedness to full sibling female brood (= super sisters) has been substantial from its first full consideration (Wilson 1971), and over a decade of investigations has confirmed that low average nestmate relatedness is as characteristic of hymenopteran societies as is high relatedness (Crozier 1980, Ross 1988a; but see also Ross and Carpenter, this volume). Continuing production of data that document low relatedness within colonies is therefore not surprising. Inclusive fitness effects under Hamilton's haplodiploidy hypothesis might be rescued if worker aid can be shown to be preferentially directed toward closer kin. Available data, though mixed, are not particularly encouraging on this point.

One role that might be salvaged for haplodiploidy is that suggested by Alexander (1974): this genetic system can be exploited to produce a sex-biased early brood in a taxon where only females provide useful brood care (Hunt and Noonan 1979 and Cameron 1985, 1986, notwithstanding). That is, delayed production of males could be an adaptive trait under selection as in the Schaal-Leverich model if queens that produce early males at the expense of an equal number of workers consequently place fewer potential reproductives into the dormancy pool at the end of the nesting season.

Another role that can be argued for haplodiploidy is that the sex-determining mechanism can be part of a strategy of queen control. For a queen to refrain from producing males early in the colony cycle potentially precludes the possibility of mating by early-emerging females that might otherwise be reproductively competent. Dramatic evidence of the role that mating can play in caste determination and subsequent behavior is available for the primitively eusocial bee *Halictus rubicundus* (Yanega 1988, 1989). First-brood *H. rubicundus* females that mate within a few days of their emergence enter early diapause and become foundresses at the start of the following season; unmated first-brood females or those that mate more than a few days after emergence remain at the natal nest and do not enter diapause. Production of early males in primitively eusocial polistines might thus be doubly selected against under the Schaal-Leverich model: not only would fewer potential workers be present in the first brood if males were produced, but any first-brood females that mated and subsequently did not work would further reduce the labor force and so limit the end-of-season production of potential reproductives that could enter the dormancy pool with the highest probability of survivorship. Relevant models, field studies, and manipulative experiments on the role of early males or (more to the point) the significance of their absence in primitively eusocial wasps are sorely needed.

For unmated female wasps, whose reproductive options are greatly constrained, producing male offspring or working at the natal nest are the only two routes to any realized fitness. The former represents the direct component and the latter the indirect component of inclusive fitness. The associated costs and payoffs of these alternative reproductive tactics merit continuing investigation. However, the model articulated here for delayed production of potential reproductives suggests that the gene pools of primitively eusocial polistines in seasonal environments may be locally swamped by the offspring of relatively few queens. If this is true, the fitness of unmated females, even in successful colonies (that may include the mixed brood of several mothers and fathers), may be insignificant relative to the fitness of successful mated reproductives. A driving motivation in many attempts to understand the evolution of eusocial insects has been to explain how workers can be sterile and yet have positive fitness. Maybe the most accurate answer for wasps is that most can not. After all, nobody ever said life was fair.

### Acknowledgments

In preparing this chapter I have profited greatly from the careful reading and critical evaluation of a preliminary draft by R. L. Jeanne, K. G. Ross, J. P. Spradbery, and C. K. Starr. J. M. Carpenter kindly provided relevant pre-publication manuscripts for reference.

# 13

# Population Genetic Structure, Relatedness, and Breeding Systems

KENNETH G. ROSS
JAMES M. CARPENTER

The theory of kin selection and its allied concept of inclusive fitness have played a preeminent role in discussions of the evolution of social behavior since the 1960s. This is particularly true for social evolution in the Hymenoptera, both because of the potentially favorable conditions for kin selection imparted by the male-haploid genetic system of the order and because of the difficulties that worker sterility in the social taxa poses for traditional genetic fitness models. Hamilton (1964a,b, 1972) expanded the traditional concept of fitness to include not only an individual's personal fitness, but also the fate of genes identical by descent found in others with which the individual interacts. Hamilton's rule shows that behavior will be under positive selection when $\Delta w_x + \Sigma r_{xy} \cdot \Delta w_y > 0$. The quantity on the left side, the *inclusive fitness effect* of a behavior, comprises the change in an actor's personal fitness ($\Delta w_x$, *direct fitness effect*) plus the sum of the changes in fitness of its interactants ($\Delta w_y$) weighted by the genetic relatedness ($r_{xy}$) between the actor and recipients (this sum being the *indirect fitness effect*) (West-Eberhard 1975, Crozier 1979, Starr 1979, Michod 1982, Queller and Goodnight 1989). Studies of social evolution in the Hymenoptera have usually employed Hamilton's rule in attempts to understand the evolution of *altruistic* behavior, defined as the situation in which $\Delta w_x < 0$. In this case $\Delta w_x$ represents the cost ($c$) of altruism and $\Sigma \Delta w_y$ the benefit ($b$), leading to the common alternative formulation of Hamilton's rule: $\bar{r} > c/b$.

As an adjunct to this inclusive fitness approach, kin selection can be studied from the rather different perspective of exact genetic models that examine gene frequency dynamics in structured populations (Crozier 1979, Abugov and Michod 1981, Michod 1982, Queller 1985). These population genetic models divide kin selection into two components, within-kin-group (individual-level) selection and between-kin-group selection. The former component is always nonpositive—that is, altruism is maladaptive or neutral within groups—because the recipients of altruism (i.e., queens) reproduce at the expense of altruists (i.e., workers). Nonetheless, between-group selection, mediated by differential contributions of kin groups to the population breeding pool, can override local selection against altruist genotypes, with the conditions under which this occurs specified by Hamilton's rule (Wade 1980, 1982, 1985; Wade and Breden 1981; Pollock 1983). Thus inclusive fitness and population genetic approaches to studying the genetic evolution of social behavior can be viewed as being generally equivalent and complementary (Abugov and Michod 1981, Queller 1985).

How can these genetic models serve to guide empirical studies of kin selection in social Hymenoptera? Taking the inclusive fitness approach, Hamilton's rule may be used to investigate the role of kin selection if the fitness parameters ($w$) for altruists and recipients, as well as their genetic relatedness ($r$), can be estimated (e.g., Metcalf and Whitt 1977b, Noonan 1981, Queller and Strassmann 1989, see also Queller and Strassmann 1988, Reeve, this volume). Alternatively, the rule simply may be applied to set a minimum value on the ratio of direct to indirect fitness changes necessary for the operation of kin selection once relatedness is known (e.g., Crozier et al. 1987, Queller and Strassmann 1989, Ross and Matthews 1989a). In either case, the importance of relatedness estimation in empirical studies of inclusive fitness is clear. This has led to what many authors argue is an overemphasis on relatedness in much of the literature on kin selection in social Hymenoptera, and a corresponding de-emphasis on extrinsic (ecological and social) factors influencing the fitness parameters in Hamilton's equation (West-Eberhard 1975, Evans 1977a, Andersson 1984, Strassmann and Queller 1989). This strong emphasis on relatedness is further attributable to the following factors: (1) unusually high levels of relatedness between female interactants may be generated in social Hymenoptera because of male haploidy (Fig. 13.1) (Hamilton 1964a,b, Trivers and Hare 1976, Crozier 1979, Andersson 1984); (2) powerful analytical tools exist for measuring relatedness (Pamilo 1984a, Ross 1988a, Queller and Goodnight 1989); (3) substantial obstacles remain to the measurement of fitness in the field (e.g., Gadagkar 1985a, Queller and Strassmann 1989).

Exact genetic models also tend to emphasize the primary importance of relatedness for empirical evaluations of kin selection by employing the concept of local population genetic structure, defined as the nonrandom spatial association of like genotypes (Michod 1982). Local ge-

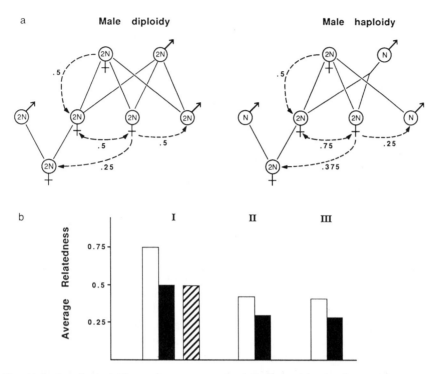

**Fig. 13.1.** A male-haploid genetic system can lead to elevated levels of relatedness between female interactants in social groups compared with a male-diploid system and can produce asymmetries of relatedness as well. (a) Similar pedigrees for male-diploid and male-haploid species illustrating how, in the latter case, the facts that the unreduced gametes of a male are identical with one another and that males are impaternate can lead to pedigree coefficients of relatedness quite different from those for the same classes of relatives in male-diploid species. Straight lines are pedigree links; curved dashed lines connect classes of relatives for which relatedness values are shown. For the latter, the arrow points to *y* where values are for the relatedness of *y* to *x* (e.g., Crozier and Pamilo 1980, Pamilo and Crozier 1982). (b) Average relatedness of 10 first-generation offspring for male-haploid (white columns) and male-diploid (black columns) species in colonies that contain only females, with three different types of social organization. I, a single mother produces all of the progeny; II, two full sisters (= super sisters in the male-haploid case) share equally in reproduction; III, a mother and her two daughters share equally in reproduction. All reproductive females are singly inseminated. The hatched column depicts the case in which half of the offspring produced by a single mother in a male-haploid species are sons, illustrating how the absence of female-biased sex allocation can compromise potentially high nestmate relatedness with male haploidy.

netic structure mediates kin selection through its influence on the relative strengths of within-group and between-group selection, such that any factors that increase genetic similarity within groups (relatedness) and concomitant genetic differentiation among groups favorably affect the probability of origin and rate of spread of altruism (Wade 1985, see West-Eberhard 1987a,c, for another perspective).

Factors that influence local genetic structure largely constitute the breeding and dispersal systems of organisms. Features of the breeding system known to have important effects on within-group relatedness include the frequency of matings by females, the apportionment of paternity for the progeny of single females, the number of reproducing females in social groups, the apportionment of maternity among those females, and the genetic similarity between mates and among laying females within groups (Table 13.1). Basically, an increase in female mating frequency or number of female reproductives within a group lowers relatedness by increasing genetic heterogeneity in the group (and decreasing differentiation among groups), an effect that becomes increasingly pronounced with more even apportionment of paternity and maternity. Increased genetic similarity between mates (e.g., inbreeding) or among queens in a colony reduces genetic heterogeneity within the group and boosts relatedness. More will be said about each of these features of breeding systems below as pertinent data are introduced. Complete discussions of their impact on genetic structure and relevance to kin selection can be found in Hamilton (1964a,b, 1972), Trivers and Hare (1976), Crozier (1979), Michod (1979, 1982), Starr (1979), Crozier and Brückner (1981), Wade and Breden (1981), Page and Metcalf (1982), Wade (1982, 1985), Uyenoyama (1984), Gadagkar (1985a), Page (1986), Nonacs (1988), and Strassmann et al. (1991).

The dispersal habits of an organism may lead to a hierarchically organized population genetic structure superimposed over the smaller scale structure arising from the presence of family groups (Crozier 1980, Pamilo 1983). Although such "population viscosity" has commonly been invoked as a facilitating factor for kin selection (e.g., Hamilton 1964b, Matthews 1968b, Kukuk 1989), it is now increasingly viewed as limiting the scope of competitive interactions among the progeny of various family groups and thereby hampering the crucial between-group component of kin selection (see Pollock 1983, Crozier et al. 1984, Pamilo 1984a, Ross and Matthews 1989a).

The importance of local genetic structure in mediating kin selection is influenced by the population sex investment ratio and the capacity for

**Table 13.1.** Components of the breeding system that influence relatedness in social groups

---

Factors acting through the number and nature of patrilines
  Number of matings by each reproducing female
  Apportionment of paternity for offspring of each female
  Relatedness of mates (e.g., inbreeding)
Factors acting through the number and nature of matrilines
  Number of reproducing females
  Apportionment of maternity among reproducing females
  Relatedness of reproducing females in group

---

*Source*: Modified from Wade 1985.

kinship discrimination. Because of asymmetries of relatedness created by a male-haploid genetic system (Fig. 13.1a), potential indirect fitness gains realized through high relatedness between female reproductives and the females that care for them can be more than offset by low relatedness between males and these brood care providers (Trivers and Hare 1976, Charnov 1978b, Craig 1980, Stubblefield and Charnov 1986). However, high average relatedness of reproductives to workers can be maintained if sex investment ratios are biased toward females (Fig. 13.1b), thus restoring more favorable conditions for social evolution through kin selection under some circumstances (e.g., Andersson 1984, Pamilo 1987, Frank and Crespi 1989). Population sex allocation data for social wasps, particularly presocial or primitively eusocial forms for which relatedness is known, are largely unavailable and are not considered further here (but see Metcalf 1980, Strassmann 1984, Ross and Matthews 1989a, Cowan, Matthews, this volume).

Whereas biased sex investment ratios seem to be necessary to preserve indirect fitness benefits associated with high female nestmate relatedness, kinship discrimination can act to enhance indirect fitness even when mean nestmate relatedness is rather low (Gadagkar 1985b, Page et al. 1989b). This is because a behaviorally mediated genetic "microstructure" involving nonrandom distribution of aid is superimposed on the existing colony genetic structure, such as when workers preferentially give food to their closest relatives in a colony composed of a mixture of patrilines (offspring of the same father) and/or matrilines (offspring of the same mother). Investigations of the kinship discrimination abilities of social wasps and their influence on social evolution are increasing in number (see Gamboa et al. 1986b, 1987; Michener and Smith 1987; Venkataraman et al. 1988 for reviews) but are largely beyond the scope of the present review.

Genetic models of kin selection provide a framework for empirical population studies concerned with hymenopteran social evolution by focusing attention on the need to understand local genetic structure and its determinants. This research program has thus far been dominated by behavioral, ecological, and physiological studies that infer the most obvious properties of colony reproductive structures and population mating systems. However, the increasing use of genetic markers from electrophoresis is now transforming this area of research (Ross 1988a, Queller and Strassmann 1989), largely because protein electrophoresis has become a readily accessible technology that can be undertaken at modest expense and with minimal special expertise, and because the procedures for estimating relatedness and other measures of genetic structure from the data generated by this technique are now well developed (see Pamilo and Crozier 1982, Crozier et al. 1984, Pamilo 1984a, Weir and Cockerham 1984, Wilkinson and McCracken 1985, Nei 1987, Queller and Goodnight 1989).

In this chapter we survey empirical studies contributing to our understanding of population genetic structure, relatedness, and breeding systems of social wasps. We hope that this review of what are largely natural history data will provide renewed impetus to the undertaking of additional direct genetic investigations, as a combination of observational and genetic approaches will yield the most useful new insights into kin selection and its role in wasp social evolution.

## PRIMITIVELY EUSOCIAL VESPIDAE

### Stenogastrinae

The Stenogastrinae (hover wasps) is a pivotal group for understanding vespid social evolution (see Carpenter, Turillazzi, this volume). Unfortunately, our knowledge of the social attributes that affect relatedness and local genetic structure remains fragmentary.

Despite early reports to the contrary (e.g., Williams 1919), it appears that stenogastrine colonies usually constitute family groups of varying composition and complexity. Colonies are founded by a single female (haplometrosis) in most circumstances (e.g., Williams 1919; Spradbery 1975; Hansell 1983, 1987b; Yamane et al. 1983a; Turillazzi 1985d), although multiple-female groups arising either by cooperative foundation (pleometrosis) or attempted usurpation have been reported during the period of colony establishment and early growth (Hansell et al. 1982, Turillazzi 1985e, C. Starr, unpubl.). Average genetic relatedness ($r$) of female nestmates may be relatively high in the matrifilial societies that develop from haplometrotic founding; if foundresses are effectively singly inseminated (monandrous) then the colonies they found will be simple family groups, with emerging females full sisters to one another (= super sisters of Page and Laidlaw 1988; $r = 0.75$).

Several events that may dilute the relatedness of female nestmates have been reported to occur during the development of haplometrotic stenogastrine colonies. These include internidal (between-nest) drift and usurpation (Yoshikawa et al. 1969, Yamane et al. 1983a, Turillazzi 1985d, Hansell 1987b), abandonment and reoccupation of nests with brood (Yamane et al. 1983a), oviposition by daughters of the foundress as nests mature, and eventual replacement of the foundress by one or more of these egg-laying daughters (Hansell 1983, 1986a, 1987b; Yamane et al. 1983a; Turillazzi 1985d; C. Starr, unpubl.). On the other hand, the occurrence of dominance interactions among nestmate females (Yoshikawa et al. 1969, Turillazzi and Pardi 1982, Yamane et al. 1983a, Turillazzi 1985d), as well as some ability to recognize and discriminate against nonnestmate females (Hansell 1982b), may act to preserve functional monogyny (reproduction by a single female) and high nestmate relatedness for some period of time.

The presence of two or more females during colony establishment is reported from species in which nests occur in dense local aggregations (Hansell et al. 1982, Turillazzi 1985e). Such associations usually arise when supernumerary females join a solitary foundress on her nest. If adult vagility is low in these dense local aggregations, then associated foundresses may be closely related; indeed, entire aggregations may represent extended family groups. This possibility attains particular significance in species such as *Parischnogaster alternata*, in which pleometrosis may lead directly to functional polygyny (multiple female reproductives), as suggested by the presence of several inseminated, ovarian-developed females on mature nests (Yoshikawa et al. 1969, Turillazzi 1985e). In such circumstances it is conceivable that nesting aggregations rather than individual nests constitute the most significant foci for the action of kin selection, presuming that there is some dispersal phase to effect the between-group competitive component (Pollock 1983).

As discussed fully below, parentage of males is an important component of the breeding system that may influence the way kin selection mediates social evolution. Little is known of the parentage of males in stenogastrine colonies, but males typically may be the sons of the foundress of *Parischnogaster nigricans serrei* (Turillazzi 1985d), as well as other haplometrotic small-colony species (Krombein 1976; Hansell 1983, 1987b).

Male stenogastrines appear to exhibit at least two quite distinct strategies for securing mates: (1) patrolling areas near landmarks, where they engage in ritualized combat with rivals and presumably intercept foraging females for mating (Pagden 1958a, 1962; Turillazzi 1982, 1983, 1986a,b), and (2) gathering at incipient nests, where they probably mate with foundress females (Turillazzi and Pardi 1982). It is also possible that males fly to established nests in order to find receptive mates. Several authors have noted the occurrence of aggregations of males ("male clubs") near nests (Williams 1919, Spradbery 1975, Turillazzi 1982, Hansell 1986a), but the significance of these is not clear. Although stenogastrine mating systems are not well known, it appears that competition among males for mates is intense (Turillazzi 1983). None of the mating systems described for these wasps would appear to promote a significant level of inbreeding.

## Independent-founding Polistinae

Independent-founding polistine species were described by Jeanne (1980a) as having no unmated "workers" in attendance during colony founding (even if founding is pleometrotic), minimal morphological differentiation of reproductive castes, and queen control by behavioral dominance. Recent studies have emphasized the heterogeneity of social structures and reproductive modes in the group and also the apparent continuity between the social characteristics of some of these wasps and species of swarm-founding Polistinae (Gadagkar, this volume). Nonetheless, Jeanne's classification remains a useful framework for ordering the increasing volume of data on the social biology of polistine wasps; we must bear in mind, however, that it does not faithfully reflect the phylogeny of the group (see Carpenter, this volume).

### *Ropalidia, Parapolybia, Belonogaster,* and *Mischocyttarus*

Three of the Old World polistine genera—*Ropalidia, Parapolybia,* and *Belonogaster*—as well as the New World genus *Mischocyttarus* are conveniently treated as a group because each has received only a modest amount of study compared with *Polistes,* the remaining polistine genus with independent founding. Modes of colony founding of *Ropalidia, Parapolybia,* and *Mischocyttarus* are diverse; indeed, colony founding within most species subjected to detailed study seems to occur by either haplometrosis or pleometrosis (Jeanne 1972, Darchen 1976a, Litte 1979, Suzuki and Murai 1980, Gadagkar and Joshi 1982a, Kojima 1984c, Yamane 1986). Although determinants of mode of colony founding are poorly understood, climatic factors may be of some importance for *Mischocyttarus* and *Parapolybia,* as indicated by the prevalence of haplometrosis in populations in more temperate areas (Litte 1977, 1979; Sugiura et al. 1983b; Yamane 1985; Gadagkar, this volume).

In species of *Belonogaster* studied to date, associative founding is the rule and occurs when single foundresses are joined by other females (Pardi and Marino Piccioli 1981; Keeping and Crewe 1983, 1987). The occurrence of unmated workerlike females in these associations of *B. grisea* suggests that this species may be intermediate in its founding habits between typical independent- and swarm-founding species (Pardi and Marino Piccioli 1981). A similar situation has been reported for *Ropalidia variegata* (Yamane 1986).

In the haplometrotic species *Mischocyttarus flavitarsis* and *Parapolybia indica,* future foundresses disperse from their natal nests and aggregate in large groups of presumably unrelated individuals to overwinter (Litte 1979, Yamane and Maeta 1985). Further dispersal takes place as foundresses search for appropriate nest sites in spring. The homogenizing effect of gene flow associated with this dispersal, in conjunction with the inferred prevalence of outbreeding (see below), would seem to indicate a lack of significant genetic structure (relatedness) above the level of the nest in populations of these and similar species.

When associative nest foundation predominates, the way in which foundress groups are formed and their resulting composition are important factors influencing genetic structure and the opportunity for kin selection to act. High relatedness of co-foundresses may favor adoption of subordinate roles by some as well as provide a means of maintaining high relatedness in colonies with multiple egg layers (Hamilton 1972, Nonacs 1988, but see also Pollock 1983). Significant relatedness between co-foundresses has been suggested for *Mischocyttarus drewseni* (Jeanne 1972) and *M. mexicanus* (Litte 1977), *Ropalidia fasciata* (Itô et al. 1985) and *R. marginata* (Gadagkar et al. 1982a), and *Belonogaster grisea* (Pardi and Marino Piccioli 1981) and *B. petiolata* (Keeping and Crewe 1983, 1987). On the other hand, there is no indication that co-foundresses are close kin in the well-studied *M. labiatus* (Litte 1981), and Itô (1985b) reports that co-foundresses of *R. fasciata* may in some instances be derived from different natal nests. Association of related co-foundresses, when it occurs, is invariably achieved by philopatry, that is, by reusing the natal nest or establishing a new nest close to it.

Philopatry may lead not only to high relatedness among co-foundresses on single nests but also to significant relatedness among individuals on neighboring nests. Such local genetic structure at a level above the nest does not depend on inherent philopatric tendencies of foundresses, however, but may simply result from a patchy distribution of appropriate nest sites coupled with low dispersal capabilities. The construction of multicombed or satellite nests (common in some species of *Ropalidia* and *Polistes*: Itô 1986b), as well as the existence of extensive nesting aggregations (common in some *Ropalidia* species: Richards 1978b, Itô and Higashi 1987, Wenzel 1987b, Itô et al. 1988), also suggest significant genetic structure at these higher levels. The complexity of genetic and social organization in these situations may blur the distinction between colony and noncolony, analogous to the difficulties encountered in studying polydomous ant populations (e.g., Crozier 1979, 1980; Pamilo 1983; but see Starr, this volume). Detailed microgeographic genetic studies are necessary to identify the most significant levels of genetic structure in such populations and thus to clarify both the nature of social organization and the probable focus of selection in mediating social evolution.

The genetic composition of colonies of independent-founding polistines can be dynamic in that it is affected by changes in the membership of foundress groups, the extent of which may be tied to nestmate recognition and discrimination abilities. Few relevant data are available for the genera discussed here. Itô (1985c) described the acceptance of an alien female into an established foundress group of *Ropalidia plebeiana*, whereas Litte (1977, 1981) considered the composition of pleometrotic groups of *M. mexicanus* and *M. labiatus* to be stable. Venkataraman et al. (1988) described nestmate recognition abilities in *R. marginata* workers. Significant genetic consequences of joining by unrelated females are of course manifested only if these individuals succeed in becoming egg layers (see Röseler, this volume).

As with modes of colony founding, the number of functional queens in mature nests of independent-founding polistines is often quite variable (Spradbery, this volume). *Parapolybia* seems to be characteristically monogynous in the postemergence phase (the period following first worker emergence) (Sugiura et al. 1983a,b, Yamane 1985); indeed, monogyny is established early on in pleometrotic *P. varia* associations by the behavioral dominance of a single queen (Yamane 1985). Subordinate foundresses in these colonies may occasionally oviposit, but because they disappear early in the postemergence phase nestmate relatedness is only temporarily diluted. The eggs laid by subordinate queens develop into workers, and thus subordinates do not appear to achieve any meaningful measure of direct reproductive success. If queens effectively mate with only one or two males, then *r* for female nestmates in mature *Parapolybia* colonies will be greater than 0.50.

The genera *Belonogaster*, *Mischocyttarus*, and *Ropalidia* include species with colonies that may be functionally monogynous or polygynous. Monogyny, when it occurs, is imposed by behavioral dominance and selective oophagy. On the basis of observational data monogyny appears to be invariant in *Belonogaster* populations found in temperate southern Africa (Keeping and Crewe 1987, Spradbery, this volume), in three *Mischocyttarus* species (Jeanne 1972; Litte 1979, 1981), and in two *Ropalidia* species (Darchen 1976a, Itô 1986c) among those studied thus far, whereas polygyny is facultative or invariant in *Belonogaster* occurring in more-tropical areas (Roubaud 1916, Pardi and Marino Piccioli 1981) as well as in at least three *Mischocyttarus* (Litte 1977, Itô 1984b) and five *Ropalidia* species (Gadagkar and Joshi 1982b, 1984; Gadagkar et al. 1982a; Itô 1985b, 1987a; Yamane 1986). Results of an electrophoretic study of *M. immarginatus* from Yucatán, Mexico (Strassmann et al. 1989), strongly implicate monogyny in the study population (Table 13.2). Among the polygynous forms, queen number varies seasonally in *M. mexicanus*, while in *R. revolutionalis*, *R. variegata*, and *R. marginata* it varies with the age (size) of the colony (polygynous colonies are older). For *R. marginata*, average nestmate relatedness must certainly decrease with the inception of polygyny, and it has been suggested that this stimulates mass emigrations of future foundresses from their natal nests (Gadagkar et al. 1982a). Queens of this species have been shown to mate multiply (Muralidharan et al. 1986), so that even within matrilines relatedness must be less than 0.75. Nonetheless, because relatively few females are recruited as new functional queens and because they may in any case be sisters, relatedness is not expected to

**Table 13.2.** Nestmate relatedness (r) of social wasps as determined by use of electrophoretic genetic markers

| Species | Predominant mode of colony founding[a] | r | Class(es) of nestmates[b] | Number of loci used in estimation | Method of relatedness estimation[c] | Reference |
|---|---|---|---|---|---|---|
| Polistinae: Primitively eusocial | | | | | | |
| Mischocyttarus basimacula | P[d] | 0.44 | Females[e] | 2 | IBD | Strassmann et al. 1989 |
| M. immarginatus | H,P[f] | 0.77 | Females[e] | 3 | IBD | Strassmann et al. 1989 |
| Polistes annularis | P[g] | 0.31 | Fall queens[h] | 2 | IBD | Strassmann et al. 1989 |
| P. annularis | P | 0.47 | Co-foundresses | 2 | IBD | Queller et al. 1990 |
| P. apachus/bellicosus | P | 0.43 | Workers → female brood | 5 | P | Lester and Selander 1981 |
| P. bellicosus | P[g] | 0.34 | Fall queens[h] | 1 | IBD | Strassmann et al. 1989 |
| P. canadensis | P[f] | 0.34 | Females[e] | 3 | IBD | Strassmann et al. 1989 |
| P. carolina | P[g] | 0.63 | Fall queens[h] | 6 | IBD | Strassmann et al. 1989 |
| P. dominulus | P[i] | 0.65 | Fall queens[h] | 4 | IBD | Strassmann et al. 1989 |
| P. dorsalis | P[g] | 0.61 | Fall queens[h] | 4 | IBD | Strassmann et al. 1989 |
| P. exclamans | H | 0.39 | Workers → female brood | 6 | P | Lester and Selander 1981 |
| P. exclamans | H[g] | 0.69 | Fall queens[h] | 3 | IBD | Strassmann et al. 1989 |
| P. exclamans | P[g] | 0.56 | Fall queens[h] | 1 | IBD | Strassmann et al. 1989 |
| P. fuscatus (= variatus) | H,P | 0.47–0.65[j] | Workers → fall queens | 4 | P | Metcalf 1980 |
| P. fuscatus (= variatus) | H,P | 0.20–0.25[j] | Workers → males | 4 | P | Metcalf 1980 |
| P. gallicus | H[i] | 0.80 | Fall queens[h] | 2 | IBD | Strassmann et al. 1989 |
| P. instabilis | P[g] | 0.53 | Fall queens[h] | 2 | IBD | Strassmann et al. 1989 |
| P. metricus | H | 0.66[j] | Fall queens[k] | 5 | P | Metcalf and Whitt 1977a |
| P. metricus | P | 0.49[j] | Fall queens[k] | 5 | P | Metcalf and Whitt 1977a |
| P. metricus | H[g] | 0.57 | Fall queens[h] | 5 | IBD | Strassmann et al. 1989 |
| P. metricus | | 0.25[j] | Workers → males | 5 | P | Metcalf 1980 |
| P. nimpha | H[f] | 0.54 | Fall queens[h] | 3 | IBD | Strassmann et al. 1989 |
| P. versicolor | P[f] | 0.37 | Females[e] | 4 | IBD | Strassmann et al. 1989 |

**Table 13.2** —*continued*

| Species | Predominant mode of colony founding[a] | r | Class(es) of nestmates[b] | Number of loci used in estimation | Method of relatedness estimation[c] | Reference |
|---|---|---|---|---|---|---|
| *Polistinae: Highly eusocial* | | | | | | |
| *Agelaia multipicta* | | 0.69 | Queens | 1 | IBD | West-Eberhard 1990 |
| *A. multipicta* | | 0.27 | Workers | 1 | IBD | West-Eberhard 1990 |
| *Parachartergus colobopterus* | | 0.11 | Females[e] | 2 | IBD | Queller et al. 1988 |
| *Polybia occidentalis* | | 0.34 | Females[e] | 4 | IBD | Queller et al. 1988 |
| *P. sericea* | | 0.28 | Females[e] | 4 | IBD | Queller et al. 1988 |
| *Sphecidae* | | | | | | |
| *Cerceris antipodes* | | 0.25–0.64[l] | Females[e] | 1 | C | McCorquodale 1988a |
| *Microstigmus comes* | | 0.63–0.70 | Females[e] | 2 | C | Ross and Matthews 1989a,b |
| *M. comes* | | 0.60–0.65[m] | Females[e] | 2 | C | Ross and Matthews 1989a,b |
| *M. comes* | | 0.42–0.51 | Co-foundresses[e] | 3 | IBD | Ross and Matthews 1989a |
| *M. comes* | | 0.20–0.21 | Females → males[e] | 2 | C | Ross and Matthews 1989a |
| *Vespinae* | | | | | | |
| *Vespula maculifrons* | | 0.32 | Workers → fall queens | 3 | C | Ross 1986 |
| *V. squamosa* | | 0.40 | Workers → fall queens | 2 | C | Ross 1986 |

[a]For primitively eusocial polistine populations only. H, haplometrosis; P, pleometrosis.

[b]For asymmetric cases, $r_{x→y}$ should be read as the relatedness of y to x (see Crozier 1970, Crozier and Pamilo 1980, Pamilo and Crozier 1982, and Queller and Goodnight 1989 for discussion).

[c]C, genotypic correlation procedure (Crozier et al. 1984, Pamilo 1984a); IBD, identity by descent procedure (Queller and Goodnight 1989); P, pedigree estimation procedure (Metcalf and Whitt 1977a, Lester and Selander 1981).

[d]Nests founded haplometrotically suffer early mortality (Spradbery, this volume).

[e]Reproductive and nonreproductive females were not distinguished.

[f]Spradbery, this volume.

[g]C. Hughes, D. Queller, and J. Strassmann, unpubl.

[h]Considered equivalent to the relatedness of co-foundresses in pleometrotic associations (C. Hughes, D. Queller, and J. Strassmann, unpubl.).

[i]S. Turillazzi, unpubl.

[j]Significant local genetic structure was also inferred (Metcalf 1980).

[k]Considered equivalent to the relatedness of fall queens to workers that rear them (Metcalf 1980).

[l]Values differed substantially among aggregations; local genetic structure was determined to have little effect on nestmate relatedness.

[m]Values adjusted for local genetic structure.

462

approach zero in secondarily polygynous *Ropalidia* colonies (see also Venkataraman et al. 1988, Gadagkar, this volume).

Colonies of *Belonogaster juncea* and *B. grisea* are apparently often highly polygynous (Roubaud 1916; Pardi and Marino Piccioli 1970, 1981). Fertilized reproductive females of *B. grisea* exhibit oophagy (egg eating) and are behaviorally dominant to unfertilized females, but the latter seem to lay a limited number of eggs that may give rise to males (Pardi and Marino Piccioli 1981). Relatedness among female nestmates of these polygynous wasps is expected to be fairly low, although it may be significantly greater than zero by virtue of co-foundresses being close kin. The tendency of *B. grisea* foundresses to initiate nests in the vicinity of their natal nest may lead to some degree of genetic structure (relatedness) at a level above the nest.

Queens of several species of *Mischocyttarus* (Jeanne 1972; Litte 1977, 1979, 1981) and *R. variegata* (Yamane 1986) in monogynous colonies are regularly replaced by a single co-foundress or daughter, a phenomenon known as serial polygyny. As is also true in the case of nest usurpation by alien females (e.g., Litte 1979, Turillazzi and Marucelli Turillazzi 1985), colony relatedness is diluted during the period of transition from one queen's offspring to another's, but the drop in relatedness is expected to be less for serial polygyny because the old and new queen are likely to be close kin. The magnitude of the effect on relatedness of such queen turnover depends on factors such as brood developmental periods, duration of reproductive dominance by single females, and extent of reproductive overlap between a queen and her replacement. These factors must receive a great deal more study before we can hope to understand the full genetic consequences of serial polygyny.

Mating systems of independent-founding polistines have received considerable attention and are becoming increasingly well understood (see also *Polistes*, below). Males of many species are only produced late in colony development (Jeanne 1972, Pardi and Marino Piccioli 1981, Gadagkar and Joshi 1983, Sugiura et al. 1983b, Keeping et al. 1986), but they are occasionally or regularly produced early in the colony cycle of *Mischocyttarus flavitarsis* (Litte 1979, T. Stiller, unpubl.), *M. mexicanus* (Litte 1977), *M. labiatus* (Litte 1981), *Ropalidia cyathiformis* (Gadagkar and Joshi 1984), and *R. fasciata* (Itô and Yamane 1985). The population-level production schedule of males is determined not only by this colony schedule but also by the degree of synchrony in colony cycles (e.g., Gadagkar, this volume). The nature of the population schedule is likely to be important in shaping social organization because mating will be precluded during particular phases of the colony cycle if male production is not continuous and colony cycles are synchronous. If females are sexually receptive only up to a certain age this could lead to the generation of a class of permanently subfertile (unmated) females.

Newly emerged males remain on their natal nests for varying periods before searching for receptive females. They may permanently abandon the natal nest while engaging in mating-related activities (e.g., Litte 1979, Jeanne and Castellón Bermúdez 1980, Gadagkar and Joshi 1983), or they may depart for only some period of the day and return to their nest to spend the night (e.g., Litte 1981, Gadagkar and Joshi 1984). Males of *Belonogaster petiolata* fly to alien nests, where they attempt to mate with resident females (Keeping et al. 1986). Mating occurs at unknown sites away from nests in *Parapolybia indica* (Sugiura et al. 1983b) and several species of *Ropalidia* (Darchen 1976a, Gadagkar et al. 1982a, Gadagkar and Joshi 1984).

Mating systems in the genus *Mischocyttarus* have been the subject of several studies. Males of *M. drewseni* patrol sites that contain flowering plants; these are used by females as nectar sources or as sites for finding prey (Jeanne and Castellón Bermúdez 1980). Males of *M. mexicanus* and *M. labiatus* similarly patrol sites rich with floral resources (Litte 1981, West-Eberhard 1982c), although patrol routes of the latter species contain one or more nonnatal nests as well. Male patrol routes of both *M. drewseni* and *M. labiatus* are established at a considerable distance (> 50–100 m) from the natal nests.

Males of *M. flavitarsis* use at least three strategies for locating mates. Males emerging early in the colony cycle (summer) patrol and mark territories, which they defend against conspecific males (Litte 1979, T. Stiller, unpubl.). Males attempt to mate with foragers coming to these areas to collect food or water—this is probably how replacement queens in serially polygynous nests are inseminated. Fall males congregate around potential overwintering sites to mate with females seeking suitable hibernacula. Early males that survive until fall do not switch to hibernation sites (T. Stiller, unpubl.), indicating that at least these two mate-searching strategies are distinct, rather than being mixed or conditional strategies employed by single males under different circumstances (see Thornhill and Alcock 1983). A final rare alternative mating strategy observed by Litte (1979) involved male occupation and defense of nonnatal nests, with the males attempting to mate with the resident females. There have been no observations of males of *M. flavitarsis* or any other *Mischocyttarus* species attempting to mate on their natal nests. This fact, in conjunction with the observed distance over which males disperse to their patrolling sites, suggests that *Mischocyttarus* populations are largely outbred.

The only data available regarding frequency of mating by foundresses of the four genera discussed in this section are those of Muralidharan et al. (1986), in which genetic markers were used to demonstrate the occurrence of multiple mating (polyandry) in *Ropalidia marginata*.

*Polistes*

As is true for the other independent-founding polistines, colony founding in *Polistes* takes diverse forms (Spradbery, this volume). Several species can be characterized as predominantly haplometrotic (e.g., Hoshikawa 1979, Makino and Aoki 1982, Hirose and Yamasaki 1984b, Strassmann 1985a), while others exhibit variable levels of pleometrosis or found colonies exclusively in this manner (e.g., Pardi 1948; West-Eberhard 1969, 1986; Hermann and Dirks 1975; Gamboa et al. 1978; Lester and Selander 1981; Noonan 1981; Cervo and Turillazzi 1985; Itô 1985a; Queller and Strassmann 1988; also Itô 1986d). That there is some association between climate and mode of colony founding is suggested by the fact that pleometrosis predominates among tropical members of the genus whereas both pleometrosis and haplometrosis occur among temperate-zone forms (Hughes 1987, Reeve, this volume).

Observations of marked foundresses have revealed a striking level of philopatry in most species that have been examined (Heldmann 1936b, West-Eberhard 1969, Hermann and Dirks 1975, Klahn 1979, Noonan 1981, Strassmann 1981a, Hirose and Yamasaki 1984b, see also Makino et al. 1987, Willer and Hermann 1989). The result of philopatry is that co-foundresses in pleometrotic groups are often close kin (demonstrated for *Polistes annularis* with genetic markers by Queller et al. [1990]; see Table 13.2), as are foundresses on neighboring nests. Metcalf (1980), using electrophoretic markers, reported significant genetic correlations between females on neighboring nests of *Polistes metricus* and *P. fuscatus* ( = *variatus*), a result that would be expected in species with this limited dispersal. Queller et al. (1990) used genetic markers to show that average relatedness between spring foundresses of *P. annularis* from neighboring nests (derived from the same parental nest) is 0.45.

Several electrophoretic and long-term behavioral studies to examine *Polistes* nestmate relatedness have been undertaken. These studies highlight the complex factors underlying colony genetic composition and the variability that may occur in such factors in natural populations. In primarily haplometrotic forms, relatedness among female nestmates is expected to be consistently high if foundresses fertilize eggs predominantly with the sperm of a single male and if usurpations by alien foundresses or supersedures by daughters are rare. However, polyandry has been reported, and several studies indicate that both usurpation and supersedure may be common in some populations (Gamboa 1978; Lester and Selander 1981; Strassmann 1981d, 1985a; Strassmann and Meyer 1983; Makino 1985c; also Klahn 1988). The expected result of these events is some decrease in average relatedness between workers and the female sexuals they rear from the value of

0.75 characteristic of simple families (i.e., a singly-mated queen and her offspring). Nonetheless, judging from the genetic data available (Table 13.2), at least some populations of *Polistes exclamans*, *P. fuscatus* (= *variatus*), *P. gallicus* (= *foederatus*), and *P. metricus* appear to match this model of colonies as simple families derived from haplometrotic founding. Female nestmate relatedness in haplometrotic forms rarely falls below 0.50.

In pleometrotic populations, quite low levels of nestmate relatedness may be expected to result from the presence of multiple potential egg layers during early colony growth, while supersedure and intercolony drift of foundresses (e.g., West-Eberhard 1969, Noonan 1981, Strassmann 1983, Itô 1984a) may lead to its further erosion. The results of Metcalf and Whitt (1977a), Strassmann (1981a), and Noonan (1981) suggest that subordinate foundresses may make a small but significant contribution to the parentage of sexuals, that is, that some level of functional polygyny and direct reproductive success of subordinates may occur often. Long-lived colonies in areas with favorable climates seem particularly likely to experience replacement of the primary egg layers (Lester and Selander 1981; Strassmann 1981a,d, 1983; Queller and Strassmann 1988) and even to become secondarily polygynous after a period of monogyny (Hoshikawa 1979; Kasuya 1981a; Strassmann 1981a; Itô 1984a, 1985a; West-Eberhard 1986; also Reed et al. 1988). In opposition to these factors diluting nestmate relatedness, reproductive competition among co-foundresses may create functional monogyny in pleometrotic colonies for at least the early period of colony development. Such competition is exemplified by dominance interactions (Pardi 1948; West-Eberhard 1969, 1986; Hermann and Dirks 1975; Gamboa et al. 1978; Strassmann 1981a; Itô 1985a; Hughes et al. 1987), differential oophagy (Gervet 1964a, West-Eberhard 1969, Gamboa et al. 1978), and the disappearance of subordinate foundresses early in the postemergence period (Hermann and Dirks 1975, Gamboa et al. 1978, Pfennig and Klahn 1985, Hughes and Strassmann 1988a). The sum result of these events is that average relatedness between nestmate females of various types from pleometrotic *Polistes* colonies may be as high as 0.65 but has not been found to fall below 0.30 (Metcalf and Whitt 1977a, Metcalf 1980, Lester and Selander 1981, Strassmann et al. 1989) (see Table 13.2).

The origin of males of *Polistes* and other eusocial Hymenoptera is currently viewed with great interest in the context of kin-selection theory. Because male haploidy leads to low relatedness of males to their sisters ($r = 0.25$: see Fig. 13.1), female reproductives may often be more closely related to workers providing brood care than are males; in such situations a bias in sex investment ratios toward females may promote the spread of altruism (Trivers and Hare 1976; Charnov 1978b;

Craig 1979; Wade 1979; Andersson 1984; Pamilo 1984b, 1987). Alternatively, workers may produce sons ($r = 0.50$) from their unfertilized eggs in the absence of female-biased sex allocation to compensate for the low relatedness of brothers, thus restoring conditions favorable to kin selection in some situations (Hamilton 1964b, Trivers and Hare 1976, Andersson 1984, Pamilo 1984b, but see Starr 1984, Woyciechowski and Łomnicki 1987, Ratnieks 1988). Because of the role worker production of males may play in social evolution by means of kin selection, increased effort has been given to determining the source of males in eusocial wasp colonies along with the patterns of female nestmate relatedness that presumably influence this (see Fletcher and Ross 1985, Bourke 1988, Ratnieks 1988).

Studies of queenright *Polistes* colonies have revealed considerable diversity in the maternity of males; in several species it appears that they are the sons of queens only (West-Eberhard 1969, Metcalf and Whitt 1977a, Metcalf 1980, Strassmann and Meyer 1983, Suzuki 1985), while in others all or a variable proportion may be derived from worker-laid eggs (Lester and Selander 1981; Miyano 1983, 1986; West-Eberhard 1986; Reed et al. 1988). If a queen is lost or dies and there is no subordinate foundress, one or more workers become functional reproductives. In queenless *P. metricus* colonies one of these substitute reproductives achieves the greatest apportionment of maternity of males (Metcalf and Whitt 1977a). Such reproductive workers of *P. exclamans* and *P. snelleni* mate with early-appearing males (Strassmann 1981b, Suzuki 1985, Willer and Hermann 1989) and so produce eggs that develop into females. In the case of *P. exclamans*, early males are derived exclusively from queen-laid eggs (Strassmann 1981b).

Males of tropical species with asynchronous colony cycles may be available year-round for mating (Richards 1978a), while males of temperate-zone forms almost always are produced near the end of the favorable season and do not overwinter (but see Hermann et al. 1974), leading to synchronization of mating activities in a population. Thus, with the exception of populations that produce early males (Strassmann 1981b, Kasuya 1983a, Gervet et al. 1986), workers of temperate-zone *Polistes* usually do not have the option of mating and establishing viable colonies.

Mating rarely occurs on mature nests of *Polistes* (West-Eberhard 1969, Noonan 1978, Strassmann 1985a, cf. Hook 1982); rather, males disperse to mating sites near resources of value to females or to symbolic or landmark-based mating sites. The most common pattern among temperate-zone species seems to be for males to establish territories near female overwintering sites, where they await receptive females and interact aggressively with competing males (West-Eberhard 1969, Lin 1972, Kasuya 1981d, Post and Jeanne 1983a, Strassmann 1985a). The advantages to larger males of *P. fuscatus* in competitive interactions at hibernation sites have apparently led to adoption of an alternative strategy by many of the smaller males, that is, patrolling patches of flowers where females forage (Post and Jeanne 1983a, see also Beani and Turillazzi 1988b for a similar example in *P. dominulus* [= *gallicus* of authors]). In the tropical species *P. erythrocephalus*, in which neither male production nor nest founding are synchronized among nests in a population, males occupy perches on vegetation near incipient nests, where they attempt to mate with virgin foundresses (West-Eberhard 1969). Males of several species defend perches that do not provide direct access to females on nests and are not associated with known resources of value to females (Alcock 1978, Turillazzi and Cervo 1982, Turillazzi and Beani 1985, Wenzel 1987a, Beani and Turillazzi 1988b); because these areas are visited by females for mating they are considered to represent leks. Finally, males of *P. gallicus* (= *foederatus*) form large groups at prominent landmarks, where they patrol common routes and aggregate at common perches (Beani and Turillazzi 1988a). Because males do not defend territories, yet they actively compete for access to females encountered in their patrol areas, this system has been described as "scramble-competition polygyny" (Beani and Turillazzi 1988a). Judging from the little we know of *Polistes* reproductive behavior, it is clear that an impressive diversity of mating systems occurs in this single genus.

The observed dispersal behaviors of males, the presence in mating aggregations of sexuals from various colonies, and the absence of significant deficiencies of heterozygotes at enzyme loci (as determined for *P. metricus* and *P. fuscatus* (= *variatus*): Metcalf and Whitt 1977a, Metcalf 1980) suggest that *Polistes* populations are often outbred (see also West-Eberhard 1969, Lin 1972, Noonan 1978), although significant local inbreeding has been detected in two species by use of genetic markers (Davis et al. 1990). The fact that neither sex has been observed to discriminate against nestmates in controlled mating trials (Larch and Gamboa 1981, Post and Jeanne 1983b, Turillazzi and Beani 1985) suggests that nestmate recognition does not serve a role in inbreeding avoidance.

Multiple mating of females has been observed by Hook (1982) in *P. tepidus* and demonstrated for *P. metricus* (Metcalf and Whitt 1977a) and for *P. fuscatus* (Metcalf 1980) from inspection of genotype arrays in natural colonies. The latter two studies reported the predominant use of sperm of a single male by polyandrous females, and furthermore, that this pattern of sperm usage was stable through time in *P. metricus* (Metcalf and Whitt 1977a). Values of nestmate relatedness determined for haplometrotic *Polistes* (Table 13.2) suggest that these queens effectively mate with no more than two males. It should be emphasized here that the robustness of estimates of the actual number of queen matings and of descriptions of patterns of sperm utilization depends on the number of marker loci studied and the nature of polymorphism at each marker. Also, to clearly disentangle the effects on colony genetic structure of polyandry from those of polygyny, either a large number of markers must be available for field studies or monogynous colonies must be established under controlled conditions to study progeny genotype arrays.

## HIGHLY EUSOCIAL VESPIDAE

### Swarm-founding Polistinae

The swarm-founding Polistinae form large, complex societies in which queen control seems to be pheromonally mediated (but see Röseler, this volume). These traits are regarded as hallmarks of advanced eusociality, yet caste dimorphism, a final characteristic of highly eusocial forms, is only poorly developed or absent in most swarm-founding polistines (see Richards 1978a; West-Eberhard 1978b; Jeanne 1980a, this volume: Chap. 6; Carpenter and Ross 1984). Colonies are founded by swarms, in which dozens to several hundred wasps, including a low but variable proportion of queens (1–15%) (Naumann 1970; Jeanne 1975b, 1980a; Forsyth 1978; Richards 1978a; West-Eberhard 1978b), establish a new colony at some distance from the parental colony. This distance has rarely been recorded, but seems to be quite restricted in some instances (Jeanne 1981a, West-Eberhard 1982c). Since it is necessarily restricted to the foraging range of worker scouts (Forsyth 1978), limited dispersal may result in significant local genetic structure in some populations. Indeed, Forsyth (1975) speculates that worker exchange may occur between neighboring related nests in such structured populations of *Metapolybia cingulata*.

Some species of *Ropalidia* are permanently polygynous and establish colonies by means of swarming rather than independent founding (Jeanne 1980a, Yamane et al. 1983b, Kojima and Jeanne 1986). Because permanent polygyny and swarming are thought to be derived from monogyny and independent founding in *Ropalidia* as well as in the Neotropical polistines (Jeanne et al. 1983, Kojima and Jeanne 1986, Carpenter, this volume), the occurrence of both social syndromes in *Ropalidia* has engendered the hope that comparative studies in the genus may shed light on how this fundamental restructuring of social organization occurs and on the nature of its genetic consequences. Unfortunately, no thorough studies of social organization or genetic structure are yet available for the swarming members of this genus.

Several Neotropical swarming species exhibit a rather rapid reduction in number of functional queens after nest establishment, and in some cases this reduction culminates in a prolonged period of functional monogyny (Forsyth 1975, West-Eberhard 1978b). Supernumerary queens of *Metapolybia aztecoides* that accompany the founding swarm are forced off the nest or assume workerlike roles when subjected to aggression or dominance displays by workers or other queens (West-Eberhard 1978b). It seems likely, however, that colonies of all swarm-founding polistines eventually become polygynous as they mature. Nonetheless, the "bottleneck" imposed by the presence of a single ma-triline of females during the monogynous phase means that newly reared queens of species such as *M. aztecoides* recruited for the poly-gynous phase are likely to be super sisters or half sisters (West-Eberhard 1978b; see also West-Eberhard 1990 for *Agelaia* [= *Stelopolybia*] *multi-picta* and Table 13.2). Such colonies may thus constitute more-or-less extended families, in which relatedness is variable but substantial at all times. Relatedness estimates for *Polybia occidentalis* and *P. sericea* (Quel-ler et al. 1988) (Table 13.2), which seem high in view of the number of mated nestmate queens commonly found in these species, could be explained on the basis of such periodic reductions in queen number. Studies contrasting relatedness at various phases of the colony cycle in a diversity of swarm-founding species are essential to clarify the impor-tance of such variation in queen number for elevating relatedness, es-pecially with reference to other factors with similar effects, such as con-sistently low queen number or large variance in reproductive output among queens.

In species that exhibit high levels of polygyny throughout the colony cycle, nestmate relatedness is expected to be quite low if outbreeding predominates and one or a few queens are not monopolizing reproduc-tion to a significant extent. Studies of variance in reproductive output among queens in polygynous wasp societies, akin to those undertaken in ants (Ross 1988b), have only begun (Strassmann et al. 1991). Rela-tively equivalent rates of observed oviposition and an absence of overt dominance interactions among queens (Naumann 1970, Jeanne 1980a) may be taken to suggest that numerous queens contribute to colony reproduction in at least some species of swarming polistines. The rela-tively low relatedness reported for some *Parachartergus colobopterus* colo-nies (Table 13.2) is that expected if eight to ten singly-mated queens produce daughters at roughly equivalent rates (Queller et al. 1988).

The mode of requeening in polygynous societies also will influence the genetic composition of a colony. Available data are consistent with the interpretation that functional queens of swarm-founding polistines generally are recruited from among individuals that were reared in the colony (West-Eberhard 1978b, Jeanne, this volume: Chap. 6), a habit likely to maintain within-nest relatedness at a level higher than would a strategy of random recruitment of new egg layers. It will be interest-ing to learn if relatedness in this group of wasps ever drops as low as it does in some polygynous ant societies, where new queens are adopted from outside the nest (reviewed in Ross 1988a).

The magnitude of direct reproduction by workers (uninseminated females in species lacking caste dimorphism) of swarm-founding polistines is not known. Some nonforaging workers of *Protopolybia acutiscutis* possess developed ovaries and oviposit frequently. Naumann (1970) speculated that a significant number of males may be derived from the eggs laid by such workers even though they generally were observed to eat their own eggs after oviposition. Ovarian-developed unmated females are common in nests of *Brachygastra scutellaris* (Richards 1978a, Carpenter and Ross 1984), whereas in several species of *Agelaia* worker ovarian development and oviposition are rare (Jeanne 1973, Richards 1978a).

In tropical regions with some seasonality, brood rearing, swarming, and the emergence of sexuals often are synchronized in a population (Naumann 1970, Jeanne 1975b, West-Eberhard 1982c); thus males may be present only during certain periods. The presence of males at newly established nests, reported for a number of swarming species (Richards 1978a, Castellón 1982, West-Eberhard 1982c), suggests that mating may occur at the time of colony founding in these wasps. West-Eberhard (1982c) speculated that males may follow pheromone trails of founding swarms, with a likely result being that males from a rather wide area are attracted to incipient nests. The mating system of *Polybia sericea* differs strikingly, however; males patrol open areas and mark substrates (West-Eberhard 1982c), reminiscent of the mating habits typical of independent-founding polistines. Both types of mating systems are likely to promote panmixis. In support of this, Queller et al. (1988) reported coefficients of inbreeding indistinguishable from zero in the three swarm-founding species they studied (including *P. sericea*) (see also West-Eberhard 1990).

## Vespinae

Wasps of the subfamily Vespinae traditionally have been considered to be homogeneous with respect to mode of colony establishment (haplometrosis) and queen number (monogyny) (Spradbery 1973a, this volume; Akre 1982; Fletcher and Ross 1985). Indeed, these highly eusocial wasps, together with honey bees (both of which have been well studied in north temperate areas), have stood as the archetypal eusocial Hymenoptera in terms of exhibiting the rather simple family structure seen to be conducive to kin selection. However, recent genetic data, as well as natural history studies of tropical and introduced species, suggest that colony genetic composition in vespines may often be more complex than was previously realized.

Species of *Vespula* and *Dolichovespula*, as well as most *Vespa*, invariably found colonies haplometrotically (Duncan 1939, Thomas 1960, Spradbery 1973a, Akre 1982, Matsuura 1984). Species in the tropical genus *Provespa* appear to found nests by means of swarms, but these swarms are unusual for eusocial wasps in that they contain only one queen (Matsuura 1985). Thus, in general, only a single queen is present during the period of colony establishment in the Vespinae.

In at least one tropical Asian species of *Vespa* (*V. affinis*) that generalization breaks down; nests are in many instances initiated by a group of several queens, perhaps in the company of workers (Matsuura 1983b, Spradbery 1986). Multiple inseminated queens with well-developed ovaries are present in mature nests of this species as well, and there is no evidence of overt dominance behaviors among them. These observations suggest that associative foundation leads to primary functional polygyny in this wasp (Matsuura 1983b, Spradbery 1986), which may also be true of the tropical colonies of *Vespa tropica leefmansi* from Sumatra and Malaysia (Matsuura, this volume). No data are available concerning genetic relatedness of co-foundress queens, nor is it known to what extent one or more queens may dominate reproduction in colonies of these tropical *Vespa*.

Factors demonstrated or suspected of conferring some complexity to the genetic composition of haplometrotic vespine colonies include usurpation of young colonies by conspecific queens, intercolony drift of workers, development of secondary polygyny in perennial colonies, direct production of males by workers, and multiple mating by queens. Usurpation is a well-established phenomenon in vespine wasps (Spradbery 1973a, Edwards 1980, MacDonald and Matthews 1981, Akre 1982, Greene, this volume), and, given that some proportion of the resident brood and workers survive the invasion of a foreign queen, relatedness of nestmate workers must be diluted somewhat during the period that the original residents survive. It is unlikely, however, that this period overlaps with the rearing of sexual brood in most instances.

A low level of intercolony worker drift (1–2%) in areas of high colony density has been reported by some authors (e.g., Lord et al. 1977), but such low levels will have no meaningful effect on colony genetic structure. An analysis of the genotypic composition of *Vespula squamosa* and *V. maculifrons* colonies, several of which were located in close proximity to other conspecific colonies, indicates that drift was not important in the study populations (Ross 1986, unpubl.). This conclusion is based on the following observations: (1) no workers were found to possess genotypes inconsistent with maternity by the resident foundress queen; (2) several colonies were monomorphic at loci that were highly polymorphic in the populations; and (3) arrays of genotypes of workers from a given colony were very similar to those of new queens, which are unlikely to move among colonies.

The development of secondary polygyny in originally monogynous colonies is now well established for *Vespula vulgaris* group species, and for the behaviorally convergent *V. squamosa*, in areas where these species have been introduced or in climatically favorable regions of their natural ranges (Duncan 1939; Spradbery 1973a,b, 1986; Ross and Visscher 1983). Such colonies commonly possess from several to many hundred egg-laying queens, which become functionally reproductive at the end of a single season of development of what look to be typical monogynous colonies. Newly recruited queens are usually presumed to be daughters of the original foundress queen (Spradbery 1973b), but this possibility has been ruled out for some nests (Ross and Matthews 1982, Spradbery, this volume). As in the swarming polistines, the origin of these queens is an important determinant of ensuing colony genetic structure, since at least a modest level of nestmate relatedness may be maintained through the second season if new queens take up egg laying in their natal nests. There are no data to suggest that one or relatively few queens dominate reproduction in polygynous *Vespula* colonies.

Nothing is known of local genetic structure in vespine populations, but considering that these are generally strong-flying wasps in which queen dispersal in temperate-zone forms occurs in two phases (pre- and posthibernation), it seems unlikely that genetic structure at a microgeographic scale is significant (see also Thomas 1960 for data on natural dispersal rates). Philopatry and reuse of natal nests are not known, although queens have infrequently been observed to hibernate in or near their natal nests (Spradbery 1973a, Edwards 1980). The large-scale migrations of spring queens of *Vespula rufa* and *Dolichovespula saxonica* that have been observed in Europe (e.g., Mikkola 1978) suggest the opportunity for periodic extensive gene flow among populations.

The possibility of direct reproduction by unmated vespine workers has received a good deal of attention (Montagner 1966b; Spradbery 1971; Ross 1985, 1986; Bourke 1988), but consensus on the reproductive role of workers in queenright colonies is yet to be reached. Worker production of males after loss or death of the foundress in monogynous colonies is well known (see Yamane 1974a; Matsuura 1984; Ross 1985, 1986), and worker oviposition may occur routinely in secondarily polygynous *Vespula* colonies (e.g., Spradbery 1973a, Ross and Visscher 1983). Several authors have concluded that *Vespula* workers also normally lay eggs that give rise to most or all of the males produced in queenright monogynous colonies, but others have failed to find evidence for such substantial worker reproduction (reviewed in Ross 1985, Bourke 1988). Recent results based on extensive behavioral and dissection data (e.g., Akre et al. 1976, Reed and Akre 1983d, Ross 1985) as well as on genetic analyses (Ross 1986) suggest that few, if any, males are the sons of workers in queenright *Vespula* colonies. In the latter study, over 2,300 male genotypes from 23 colonies of *V. maculifrons* and *V. squamosa* were examined, and none was found to be inconsistent with maternity by the foundress queen. Thus queen control over worker reproduction appears to be highly effective in at least these species, with the probable result that workers are able to directly reproduce only if the foundress dies prematurely. Premature death of the foundress may occur with significant frequency in most species, however (e.g., Yamane 1974a; Archer 1980a; Ross 1985, 1986).

Mating strategies of vespine males appear to be quite diverse. Males of *Vespula* and *Dolichovespula*, and of *Vespa crabro* in the United States, leave their nests to patrol prominent vegetation, where they mate (Schulz-Langner 1954, Schremmer 1962, MacDonald et al. 1974, Batra 1980, Post 1980). Competitive interactions among males are generally absent. Males of *Vespa mandarinia* in Japan are attracted to nests, where they mate with emerging queens at the entrance (Matsuura 1984, Ono et al. 1987). Mating in and around nests has also been reported for a number of vespine species in which males patrol (Schremmer 1962, Spradbery 1973a, MacDonald et al. 1974, Batra 1980, Ono et al. 1987, Bunn 1988), leading some authors to conclude that alternative male mating strategies may exist in some populations and that sibling mating may occur with some frequency at the natal nest (e.g., Reed and Akre 1983d, Ross 1983b). Given that most confirmed observations of sibling mating have involved captive colonies, the extent of its occurrence in nature remains conjectural.

A substantial body of behavioral evidence suggests that both male and queen vespines are capable of multiple mating (Schulz-Langner 1954, Thomas 1960, Schremmer 1962, MacDonald et al. 1974, Post 1980, Ross 1983b, Ono et al. 1987)—workers apparently are incapable of mating (Spradbery 1973a). Genetic markers have been used to study the frequency of mating by queens of *Vespula maculifrons* and *V. squamosa* (Ross 1986). The effective harmonic mean numbers of matings by queens were estimated to be 7.1 for *V. maculifrons* and 3.3 for *V. squamosa*, but these are minimum summary values because they do not take into account variance in paternity apportionment (see Page 1986). Estimates of paternity variance derived from sequential samples of worker genotypes from natural colonies were relatively high (Ross 1986); they can be used to estimate the actual harmonic mean numbers of matings by queens: 9.5 for *V. maculifrons* and 5.5 for *V. squamosa* (incorrectly reported in Ross 1986). Given that the harmonic mean is generally lower than the arithmetic mean, it is clear that queens of these two species are highly polyandrous. The result of these mating

habits is relatively low levels of nestmate relatedness in the typical monogynous colonies of the two species (Table 13.2).

The substantial variance in the apportionment of paternity among
the various mates of single queens that was evident in Ross's study
(1986) suggests that some males achieve a high proportion of egg
fertilizations through numerical superiority of sperm or some other
mechanism of sperm predominance. This variance in paternity apportionment in monogynous colonies results in a far more homogeneous
colony genetic composition (higher relatedness of female nestmates)
than would be the case if all of the mates of a foundress contributed
equally to egg fertilizations. Patterns of paternity apportionment appeared relatively stable through colony development for both *Vespula
maculifrons* and *V. squamosa*, and this low variance through time in male
parentage means that female nestmate relatedness remains relatively
unchanged as colonies grow (see Crozier and Brückner 1981). These
results further suggest that certain males cannot achieve a disproportionate number of fertilizations later in the colony cycle when fertilized
eggs give rise to new queens rather than to workers (Ross 1986).

## SOCIAL SPHECIDAE

Diverse forms of presocial behavior have been reported for the Sphecidae (Evans and West-Eberhard 1970, West-Eberhard 1978a, Eickwort
1981, Matthews, this volume), but eusocial species are known or suspected only in the genera *Microstigmus* and *Arpactophilus* (tribe Pemphredonini) (Matthews 1968a,b, this volume; Matthews and Naumann
1988; Ross and Matthews 1989b). The more advanced presocial and the
primitively eusocial species are important taxa for comparative studies
of social evolution because eusociality in sphecids originated independently of eusociality in other wasps (and bees).

In the nest-sharing Australian species *Cerceris antipodes*, female nestmate relatedness has been reported to be consistently high in one aggregation ($r = 0.51$–$0.64$) but relatively low and variable in three
others ($r = 0.25$–$0.34$), on the basis of electrophoretic data (McCorquodale 1988a) (see Table 13.2). Relatedness between wasps from
neighboring nests was found to be low and generally indistinguishable
from zero, indicating a lack of significant local genetic structure at a
level above the nest in the study population.

Among the presumed eusocial species, *Microstigmus comes* has received the most attention in terms of its population genetic structure
(Table 13.2) and social biology. Females of this Neotropical species
found nests singly or in groups. Average relatedness for foundresses in
pleometrotic groups approaches 0.50 (Ross and Matthews 1989a), sug-

gesting that some degree of philopatry and/or kin recognition may act to bring relatives together for cooperative nest initiation. Older multi-female nests with brood represent families of varying composition, although many appear to be simple families composed of a singly-mated female and her male and female offspring (Ross and Matthews 1989a,b). Relatedness of females in such nests is accordingly high ($r = 0.63$–$0.70$), with the relatedness of males to females also as expected for such simple families (Table 13.2).

A modest degree of local genetic structure occurs at levels above the individual nest in *M. comes*, so that females from different nests on the same or adjacent plants tend to be rather distantly related ($r \leq 0.20$) (Ross and Matthews 1989a). When this modest local structure is taken into account, the global estimates of female nestmate relatedness are reduced to local estimates in the range 0.60–0.65. This devaluing of global estimates for the effects of higher level genetic structure emphasizes the concept that, in structured populations with relatively reduced dispersal, the between-group component of kin selection is somewhat dampened because of more-limited opportunities for families to compete in their contributions to the population breeding pool (Pollock 1983, Wade 1985). From an inclusive fitness perspective, adjacent colonies likely to compete for local resources in structured populations share more genes identical by descent than do widely spaced noncompeting colonies (Crozier et al. 1984, Pamilo 1984a). Thus the high relatedness within families that favors kin selection through positive indirect fitness effects is offset to some extent by the negative indirect fitness consequences of competition between adjacent colonies.

Almost nothing is known of the mating behavior of eusocial sphecids, but some generalizations can be drawn for *M. comes* on the basis of population genetic data (Ross and Matthews 1989a). Subpopulation genotypic frequencies closely approximated those expected under Hardy-Weinberg conditions in the Costa Rican population studied, suggesting that significant inbreeding was unlikely. Furthermore, the simple genotype arrays observed in many colonies, coupled with frequent identification of a single putative mother female in such groups, suggests that females are effectively monandrous. Although the location of mating activities in this species is unknown, colony genotypic data indicate that males are unlikely to take up long-term residency on foreign nests to secure matings.

## CONCLUDING REMARKS

Our present knowledge of the diversity of social organization and reproductive behavior of social wasps confirms that enormous variability exists in the attributes that affect genetic structure in a wasp population, and thus, in the genetic structure itself. As work in this area continues, useful frameworks for organizing an expanding body of empirical data will be required. Two useful frameworks are one based on phylogenetic history and one based on complexity of social organization.

The objectives of comparative studies within a phylogenetic framework include using knowledge of phylogenetic relationships to infer primitive states for a character of interest in increasingly remote common ancestors of extant taxa. From the character "transformation series" that is generated from this exercise one can directly reconstruct an evolutionary pathway detailing the specific steps by which a trait has evolved in a given lineage (Carpenter 1989, this volume; Ross and Carpenter 1990). Use of this approach in analyses of hymenopteran social evolution unfortunately is hampered by the lack of robust phylogenetic hypotheses for many relevant groups as well as the lack of data for particular traits over a sufficient range of species. Nonetheless, early attempts to track the evolution of such important determinants of genetic structure as number of functional queens (Ward 1989, Ross and Carpenter 1990, Carpenter, this volume) and frequency of queen mating (Ross et al. 1988) illustrate the potential usefulness of the approach.

Complementary to a phylogenetic framework, and subsumed under it, is a framework based on social complexity. Because complex social organizations are derived from simpler ones, one should be able to survey determinants of genetic structure in lineages in which component taxa exhibit a range of social complexity and assess to what extent changes in those determinants are linked to changes in the level of social organization. Social wasps are ideal subjects for this approach because they include several independent lineages with species that differ profoundly in social organization. Particularly attractive candidates include the sphecid tribe Pemphredonini and the vespid subfamily Stenogastrinae, in which the transitions from solitary or temporary eusocial nesting (respectively) to permanent eusociality have occurred (Carpenter 1988a, Turillazzi, Matthews, this volume), as well as the polistine genus *Ropalidia*, in which the transition from independent founding/primitive eusociality to swarm founding/advanced eusociality has occurred (Jeanne 1980a, Kojima and Jeanne 1986). It is possible that these transitions occurred independently on more than one occasion in each of these taxa.

Comparative genetic studies undertaken within these frameworks may shed light on a particular hypothesis of the relationship between genetic structure and social organization that has been expressed frequently in the literature. That hypothesis is the idea that eusociality originates most often in populations structured into relatively simple families with high relatedness, whereas its subsequent maintenance and elaboration into more-complex forms can occur in the absence of high relatedness, for instance after the adoption of polyandry or polygyny (Hamilton 1964b, 1972; West-Eberhard 1975; Page and Metcalf 1982; Andersson 1984; Ross 1986). In this view, the elaboration of eusociality can take place under circumstances of low relatedness because increasing caste dimorphism makes workers less well equipped to reproduce directly but more efficient at helping to rear brood, thereby altering the ratio of their direct to indirect fitness effects for a given value of $r$. Carpenter (this volume) has concluded from a phylogenetic analysis that the advanced levels of eusociality attained by the typically monogynous Vespinae and the swarm-founding (polygynous) Polistinae are both derived from more-or-less short-term monogyny, such as characterizes extant independent-founding polistines. The shift from such short-term monogyny to polygyny that accompanied social advancement in polistines is expected to be paralleled by a decline in female nestmate relatedness, an expectation met on the basis of available data ($\bar{r} = 0.53$ for *Polistes* and *Mischocyttarus*; $\bar{r} = 0.25$ for the swarm-founding polistines: see Table 13.2). Interestingly, the habit of polyandry known or suspected for many of the highly eusocial vespines might also be expected to lead to a drop in relatedness in this group compared with the average for independent-founding polistines, in spite of the transition from short-term to long-term monogyny. Relatedness estimates available for two vespines are not at variance with this suggestion ($\bar{r} = 0.36$: Table 13.2).

In view of these early data, the hypothesis that eusociality in vespids originated under conditions of relatively high $r$, while its subsequent development is not dependent on such conditions, cannot be discarded, although we do not yet have sufficient data for conducting a robust, formal test. Moreover, rather high levels of $r$ have been reported also for the only primitively eusocial Hymenoptera outside of Vespidae to be studied genetically: *Microstigmus comes* ($r = 0.60$–$0.70$: Table 13.2), the halictid bee *Lasioglossum zephyrum* ($r = 0.43$–$0.70$: Crozier et al. 1987, Kukuk 1989), and the allodapine bee *Exoneura bicolor* ($r = 0.48$–$0.60$: Schwarz 1987). These species have no close highly eusocial relatives for comparison, however. Among all eusocial hymenopterans the vespid genus *Ropalidia* emerges as the most promising taxon for comparative studies of this hypothesis, because it includes both primitively and highly eusocial forms.

Aside from adopting a phylogenetic approach to the role of genetic structure in social evolution, future studies of relatedness in presocial or primitively eusocial wasps will profit from being combined with ecological studies that quantify the costs and benefits of group living— that is, the traditional approach to investigating inclusive fitness. All population genetic studies of social insects should take into account genetic structure above the level of the nest and its effects on within-colony relatedness, a consideration likely to figure prominently in populations with conspicuous spatial structure associated with local aggregations of nests. Statistical procedures for dealing with such hierarchical genetic structure in social insect populations are not well developed and will require further attention (see Crozier et al. 1984, 1987; Pamilo 1984a; Queller and Goodnight 1989; Ross and Matthews 1989a; Davis et al. 1990). Future studies also should not neglect inbreeding as a factor molding local genetic structure, although the role of inbreeding effects in constraining or favoring kin selection remains much in debate (e.g., Trivers and Hare 1976, Wade and Breden 1981, Pollock 1983, Uyenoyama 1984). Interestingly, significant levels of inbreeding have rarely been reported in genetic studies of any social wasps, or, for that matter, any other social Hymenoptera (Crozier 1980, Ross 1988a, Davis et al. 1990).

In addition to its importance to kin selection, the study of genetic structure and its determinants is intrinsically valuable because of the unique information it can provide on the natural history, social organization, and population biology of organisms. Such important population attributes as patterns of dispersal, levels of gene flow, and differential reproductive success of individuals, often difficult to study using behavioral methods alone, are amenable to analysis using the tools of population genetics. The description of genetic diversity and its distribution in natural populations, the traditional province of population genetics, remains a key to understanding a host of important evolutionary processes operating at levels above that where kin selection reigns. The complexity added by the existence of kinship structure, reproductive castes, and male haploidy in social Hymenoptera makes the task of describing the distribution of genetic diversity in these populations all the more challenging.

*Acknowledgments*

We were supported in part by grants from the National Science Foundation (BSR-8615238 to K. G. Ross, BSR-8508055 and BSR-8817608 to J. M. Carpenter) during the preparation of this chapter. We thank R. Owen for comments on an early draft.

# 14

# Evolution of Nest Architecture

JOHN W. WENZEL

The architecture of structures made by animals is an important source of information about behavioral evolution. Unlike most behaviors, construction leaves behind a physical trace that can be measured, stored, and later reexamined. By studying the products of construction behavior, one may gain unique and useful insights not easily achieved by studying other kinds of behaviors. Naturalists have long been fascinated with the nests of social insects because their architectural complexity contrasts with the builders' presumably limited intelligence; the origin and significance of this complexity remain major unsolved problems in understanding the evolution of nest architecture.

Within the diversity of social wasp nests, some structural features are evolutionarily conservative and serve as useful taxonomic characters. Indeed, the early taxonomy of eusocial wasps relied on architecture nearly as much as on morphology (Saussure 1853–1858), and this emphasis is at least partly maintained even in a recent major systematic revision of Polistinae (Richards 1978a). Distinct from this view of architecture faithfully recording phylogenetic relationships, Jeanne's (1975a) landmark paper on wasp architecture stressed adaptation to common selection pressures relating to economy and colony defense. In this chapter, both views are explored in an effort to determine what sort of variation is related to phylogenetic history and what sort is related to more recent adaptation.

The antiquity of eusocial wasps and their paper nests is revealed by a Cretaceous fossil from Utah, United States, that resembles the nest of

480

the cosmopolitan *Polistes* (Brown 1941, Wenzel 1990, USNM), suggesting that wasp eusociality and paper nests date back at least 63 million years. Perhaps because of such a long period of independent development of the various lineages, nests of modern eusocial vespids share only a few fundamental traits. For instance, to fabricate cells the wasps carry a foreign material to the nest site and masticate it, adding oral secretions. Also, larvae are isolated and supported individually by cells during development. Yet, there are exceptions even to these most basic traits, and nests of closely related species or higher level taxa may differ in virtually any conceivable architectural detail.

This chapter is the most comprehensive review of eusocial vespid nests to date, although brevity requires that it is still incomplete. For convenience, Fig. 14.3 supplies a simplified illustration of nest development in Vespinae and Polistinae, and my simple line drawings in Figs. 14.4–14.62 present much of what is known about nest ontogeny in social wasps. The reader should see Carpenter (this volume) for details of the phylogenetic relationships among the 39 genera treated in this chapter. Readers would also benefit from the well-illustrated introduction to wasp architecture in Berland and Grassé (1951). The appendixes at the end of the chapter define terms used, some in a sense different from that of other authors, and abbreviations.

The first of the three main sections in this chapter concerns the rather poorly known subfamily Stenogastrinae, and I hope new work will soon make this section obsolete. The second section concerns architectural variation found in the polistine genera with independent foundation and small colonies. These genera are generally regarded as primitive and it is appropriate to look to them for traces of the groundplan from which the more complex nests of other genera evolved. By examining in detail the variation among species in one of these genera, *Mischocyttarus*, I attempt to show that every architectural feature is subject to change between close relatives, while more distant relatives may converge on outwardly similar designs. The third section demonstrates the same lesson on a larger scale by examining the much greater nest variation observed among all the genera of Vespinae and Polistinae. I discuss examples of convergence in these genera, where similar characters have evolved many times, as well as the breadth of architectural diversity, ranging from symmetrical, continually expanding forms (e.g., Fig. 14.1) to those more modular and deterministically built (e.g., Fig. 14.2). I propose that a better understanding of architectural homology relies on the developmental stages of given designs, not just on the finished products. To support this position, I offer tests of such an ontogenetic model by examining the details of nest construction in genera that have modified the construction sequence by eliminating elements or building them precociously.

**Fig. 14.1.** Nest of *Agelaia flavipennis*. The brood comb grows in continuous spirals over itself, supported by pillars and serving temporarily as an envelope protecting the older comb within. Wasps may pass through the nest by means of several openings and ramps in the surface of the comb. (Courtesy of W. D. Hamilton.)

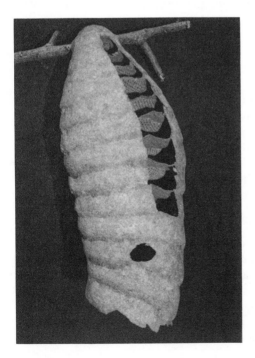

**Fig. 14.2.** Nest of *Epipona guerini*. The nest is built quickly and completed before the brood matures. Note the short height of uppermost (oldest) brood cells, and the decreasing diameter of the nest toward the bottom. The entrance is in the bottom on the far side of the nest, aligned with a single internal passageway through all levels.

## STENOGASTRINAE

The poorly known stenogastrines may prove to be the most versatile of insect architects. Alone and in groups they engineer variable, elegant, often ornately embellished structures. Despite thorough efforts (Pardi and Turillazzi 1982, Turillazzi, this volume), the evolution of diversity in stenogastrine architecture is still largely uninterpretable.

### Building Materials

Nest-building material is generally chewed vegetable matter, often mixed with soil (Pagden 1958a, Ohgushi et al. 1983a, Hansell 1987a), but may be primarily mud and inorganic particles in some species of *Stenogaster* (Spradbery 1975), *Liostenogaster* (Yoshikawa et al. 1969; Ohgushi et al. 1983a, 1985, 1986), and *Anischnogaster* (Vecht 1972, Spradbery 1989). Stenogastrine nests are generally far more fragile than nests built by other vespids, even those of noneusocial species such as *Calligaster*, *Eumenes*, and *Zethus*. Hansell (1984a,b, 1985) has shown that stenogastrine carton is looser and weaker than that of vespines or polistines (photographs in Hansell 1984b, 1987a), and he further hypothesizes that the mud or frail paper characteristic of the former group prohibited the evolution of large colonies. However, ideal colony size is a complex issue and certainly involves details of the species' predatory habits, larval development, and other factors, and I believe that "the failure in Stenogastrinae to evolve high-quality paper" (Hansell 1987a:179) may be no more responsible for limited colony size than any other single factor. Regardless of its relationship to sociality, I interpret architectural frailty to be a derived trait of this subfamily, not a primitive trait or a "failure" to evolve more durable designs.

### Nest Design

Nest design is highly variable both within genera and within species of Stenogastrinae. All nests are sessile (see Appendix 14.1) with no true pedicel, although they often are built on rootlets or on other slender supports (see also Turillazzi, this volume: Fig. 3.9). *Liostenogaster* may build cells scattered singly across the substrate (Ohgushi et al. 1983a). *Liostenogaster nitidipennis* builds mud cells that may lie on their sides on a substrate or on neighboring cells in double rows (Fig. 14.4), or the cells may form a loose comb joined by the side to a root (Iwata 1967, Yoshikawa et al. 1969). *Liostenogaster vechti* ( = *arcuata*) builds nests of carton with cells lying on their sides in curved double rows opening toward a central region (Pagden 1958a, Yoshikawa et al. 1969, Ohgushi et al. 1983a, SEM). *Liostenogaster variapicta* builds sessile mud cells "piled one on top of the other" (Reyes 1988:398). Species near *L. flavoli-neata* make large mud combs that are sessile at the base (Fig. 14.5); each cell may hold two immatures in tandem and a peripheral comb wall is drawn out far beyond the tops of the cells (Yoshikawa et al. 1969, Ohgushi et al. 1983a).

*Eustenogaster* builds bell-shaped nests with long downward entrance tubes (Fig. 14.6; BMNH, USNM, SEM) and often with elaborate fins or lacey carton (Pagden 1958a; Yoshikawa et al. 1969; Ohgushi et al. 1983a, 1986; see cover illustration). *Stenogaster concinna* (Fig. 14.7) has a nest similar to that of *Eustenogaster*, but without an entrance tube (Spradbery 1975). *Anischnogaster* builds barrel-shaped cells singly or with shared walls (Fig. 14.8; Vecht 1972, Spradbery 1989). *Metischnogaster* builds a long chain of cells end to end (Fig. 14.9) such that each cell must be built to full size before the next can be started (Pagden 1958a, 1962; Yoshikawa et al. 1969; Vecht 1972; Reyes 1988).

*Parischnogaster* probably has the most diverse architecture in the subfamily; species near *P. jacobsoni* and *P. nigricans* (Fig. 14.10) build cells singly or sometimes sharing walls (Iwata 1967, Yoshikawa et al. 1969, Pardi and Turillazzi 1982), whereas *P. mellyi* (Fig. 14.11) makes single cells or disorganized "combs" with cells of differing orientation (Saussure 1852; Pagden 1958a; Hansell 1981; Pardi and Turillazzi 1982; Ohgushi et al. 1983a, 1986; BPBM, SEM). The species near *P. alternata* (Fig. 14.12), *P. striatula* (Fig. 14.13), and *P. depressigaster* align the cells or extend their marginal walls such that access to the older cells of the long, stringlike or irregular comb is through an internal passageway beside the cells (Williams 1919, Pagden 1958a, Iwata 1967, Yoshikawa et al. 1969, Ohgushi and Yamane 1983, Ohgushi et al. 1983a, SEM). *Parischnogaster timida* builds a nest quite like *Metischnogaster* (Fig. 14.9), but less perfectly cylindrical (Williams 1919, Berland and Grassé 1951: fig. 1021).

Krombein (1976, USNM) found that nests of *Eustenogaster eximia* vary in shape, presence of external fins, and details of the entrance spout, thus demonstrating substantial intraspecific diversity. On the other hand, Sakagami and Yoshikawa (1968) used architectural differences to distinguish the sibling species *Eustenogaster calyptodoma* and *E. micans*, which are diagnosed primarily on the basis of whether the envelope is supported directly by the substrate or arises from marginal cell walls. There are still no studies of intraspecific architectural variation in either of these two "ethospecies," but diagnostic morphological characters have now been reported (C. Starr, unpubl.).

In addition to their notable variability, stenogastrine nests are also characterized as generally being cryptic (Pagden 1958a, Turillazzi

1989b), perhaps as an adaptation to avoid detection by *Vespa tropica*, a specialized predator of Southeast Asian social wasps (Matsuura 1984, this volume) and the main visual predator of Stenogastrinae (Turillazzi 1985d, Hansell 1987a). Williams (1919) recorded four attacks in eight days on *Parischnogaster depressigaster*, and attacks have also been recorded on *P. mellyi* (Ward 1965) and *Eustenogaster* (C. Starr, unpubl.). Hansell (1984b) proposed that construction of heavy mud nests (e.g., *Liostenogaster flavolineata* nests) is a defense mechanism against such predation.

The function of some stenogastrine nest structures, such as the putative ant guards, may be misinterpreted because they are still little studied. Many species build their nests on stems, hanging roots, or on other slender structures, with the ant guard between the nest and substrate or on both sides of the nest (Jacobson 1935, Turillazzi and Pardi 1981). The gelatinous guards of *Parischnogaster* (Fig. 14.10) do indeed have some ant-repellent properties (Turillazzi and Pardi 1981, Turillazzi 1985c), and the stickiness of the "guards" of an unidentified *Eustenogaster* (Pagden 1958a) suggests that they too may serve as ant deterrents. Williams interpreted the carton "umbrellas" of *Parischnogaster timida* (Williams 1919, Berland and Grassé 1951: fig. 1021) to be either ant guards or devices to keep water from running down the support onto the nest. The latter is a reasonable proposal, especially given the proclivity of many stenogastrines for nesting near cascades and in other humid sites. Furthermore, a *P. timida* nest in the Bishop Museum (collected by Williams and identified by Vecht) has five widely spaced umbrellas above it that probaby could be traversed easily by ants. Vecht (1977a) postulated that the longer, more conical carton structure on *Metischnogaster* nests (Fig. 14.9) would have to be made sticky to be an effective ant guard. Thus, these carton structures may serve a different function from the glandular material examined by Turillazzi and Pardi (1981) and Turillazzi (1985c).

## DIVERSIFICATION AND CONVERGENCE IN INDEPENDENT-FOUNDING POLISTINAE

It is appropriate to look to the so-called primitive genera of Polistinae for clues to the groundplan from which more complex nests of the other genera are derived. There clearly is less diversity in nest architecture among the independent-founding polistines (*Polistes, Mischocyttarus, Belonogaster, Parapolybia,* and most *Ropalidia*) than among swarm-founding wasps (all other Polistinae; Fig. 14.3). Considering the elements shared by the primitive groups, we can deduce that the com-

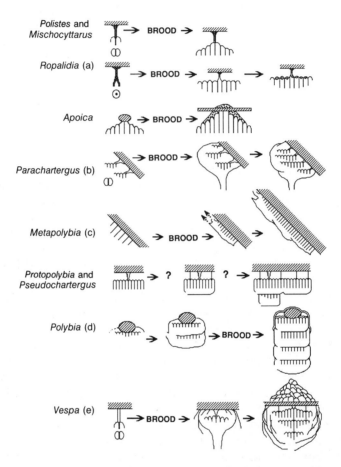

**Fig. 14.3.** Major designs in eusocial vespid nest architecture. Each row displays the nest development of one or two genera chosen to represent all members of the group; the groups are arranged to parallel Table 14.1. The left-most figure in a row illustrates initiation (pedicellate or sessile) of the nest on the substrate (shaded). Black pedicels (*Polistes*, *Mischocyttarus*, and *Ropalidia* groups) and shaded comb sections (*Ropalidia* group) represent oral secretion as the main material. Pale pedicels are primarily fibrous paper. *Apoica* builds a thick, light felt. Two ellipses broadly joined and beneath left-most figures represent cell-marginal pedicels with a cell on each side; one circle with a central spot represents cell-central pedicel with one cell at the end (see Ontogenetic Variation in Architecture: Polistinae for details). Initiation of brood rearing is designated by "BROOD." The second figure in each row illustrates the form of the nest built by the foundresses; the third figure shows the mature form after workers have enlarged the nest. A question mark indicates that the groundplan for brood initiation is ambiguous. Groups include other genera as follows: (a) *Belonogaster, Parapolybia, Polybioides*; (b) *Agelaia, Angiopolybia, Chartergellus, Leipomeles, Marimbonda, Nectarinella, Pseudopolybia*; (c) *Asteloeca, Clypearia, Occipitalia, Synoeca*; (d) *Brachygastra, Charterginus, Chartergus, Epipona, Protonectarina, Synoecoides*; (e) *Dolichovespula, Provespa, Vespula*.

486

mon ancestor of all Vespinae and Polistinae likely built simple, paper combs supported by a pedicel and without an envelope. Yet, if the nature of variation in these simple nests is any indication of which architectural features are subject to modification and adaptation, then we can conclude that wasps may change nearly any element of nest design—that is, no feature seems to be immutable.

## Variation within Species

Although the primitive genera predominantly build nests with a single comb, variation in comb number within species is widespread. Several separate combs may be built by single colonies of *Parapolybia* (Yamane 1984), many *Ropalidia* (Gadagkar and Joshi 1982b; Hook and Evans 1982; Kojima 1984b; Itô 1986b, 1987a), *Mischocyttarus insolitus* (Herre et al. 1986), and several species of *Polistes* (Jeanne 1979a, Gamboa 1981, Strassmann 1981b, Ono 1989). Perhaps a related phenomenon is the gregarious nesting of certain *Ropalidia*, such as *R. formosa* (Wenzel 1987b), *R. gregaria*, and *R. plebeiana* (Richards 1978b, Itô and Higashi 1987). These species build hundreds of nests within a few meters of each other or cut combs apart (Yamane et al. 1991), usually in protected sites such as beneath a bridge or overhanging rock.

Comb shape may vary nonrandomly within a given species; for example, *Ropalidia variegata* constructs round or linear combs, depending on the structure of vegetation at the nest site (Davis 1966a). *Ropalidia horni* may enlarge nests from their simple oval shape to a lobate form (Kojima 1982a, USNM), and *Ropalidia carinata* and *R. flavoviridis* from Madagascar build variable lobate nests, some of which may approach a pinnate form similar to palm leaves (Wenzel, unpubl., SEM). The mechanical and economic disadvantages of these nests, which depart from a more efficient oval design, may be balanced by increased crypsis, as proposed by Jeanne (1975a) for nests of *Mischocyttarus*.

Much of the intraspecific nest variation of independent-founding polistines may be ascribed to ontogenetic factors. Nests change in a more-or-less predictable fashion as they grow and age, and variation is usually greater between old nests than between young ones. Wenzel (1989) demonstrated that very young *Polistes annularis* combs have a stereotypical shape, while more mature combs vary considerably according in part to cues taken from the environment. Richards (1978b) used overall shape to classify *Ropalidia* nests, but apparently did not realize that the shape of young and old nests may differ. Many species of *Ropalidia* initiate the nest with a long comb only a few cells wide, and the emerging workers may then build evenly around the sides to make an oval nest (Wenzel, unpubl.). Workers of *Ropalidia ignobilis* in Madagascar not only broaden the nest, but also abruptly shift to building cells 70% larger by volume for the production of queens (Wenzel, unpubl.), somewhat like *Vespula* and *Dolichovespula*. *Parapolybia nodosa* has an ordinary pedicellate first comb, but later worker-initiated combs may be built directly from the substrate without pedicels (Yamane 1984).

## Variation among Species

Examination of the variation among species within a diverse genus should show whether some architectural features are prone to rapid evolution and others less so, or whether all features appear equally mutable. The large genera *Polistes, Ropalidia,* and *Mischocyttarus* might be important in this regard. In *Polistes,* both radial symmetry (Fig. 14.14) and eccentricity of the comb (Fig. 14.15) are common. Examples of more extreme departures from the basic *Polistes* nest structure can be found in *P. tepidus* (Yamane and Okazawa 1977), *P. olivaceus* (Wenzel, unpubl.), and probably other species of the Indo-Pacific subgenus *Megapolistes,* which build ordinary combs early, but subsequently lengthen the cells so that older nests often contain two immatures per cell, arranged with a pupa proximally and a young larva distally (Fig. 14.16). *Polistes riparius* (= *biglumis,* see Yamane and Yamane 1987) in Japan builds long cells and leaves empty the marginal cells of the comb, perhaps for thermoregulatory purposes (Yamane and Kawamichi 1975), although Lorenzi and Turillazzi (1986) dispute this interpretation for *P. biglumis bimaculatus* in the Alps. The Neotropical species *P. goeldii* builds a long comb two cells wide (Richards 1978a, AMNH, MCZ, SEM). In the most detailed study of *Polistes* construction to date, Downing and Jeanne (1986, 1987) quantify the more subtle architectural differences among several *Polistes* species. From the available data, we can conclude that the genus shows only relatively minor variation given its cosmopolitan distribution, presumed age, and number of species.

The paleotropical genus *Ropalidia* is the most architecturally diverse of these focal genera, and it is still quite poorly known. Some species build envelopes (Kojima and Jeanne 1986, Spradbery and Kojima 1989), some build ovoid combs, some build narrow or lobate combs (Kojima 1982a), and some build continuous spiral combs (Kojima and Jeanne 1986). As far as we know, all of the *Ropalidia* species with complex architecture, such as stacked combs or envelopes, are not independent-founding species but rather reproduce by swarming. This association of social complexity and nest architecture has an interesting parallel in the South American polistine genera. Understanding the evolutionary significance of *Ropalidia* architecture is a formidable task for future study.

Eight Elements in *Mischocyttarus*

Of the three largest social wasp genera, *Mischocyttarus* is currently the most informative with respect to variable nest architecture. Several designs appear to have evolved convergently in distantly related species, and among close species any of the following basic features of nest architecture may vary: (1) design of the pedicel, (2) shape of the cells, (3) orientation of the combs, (4) nature of the carton, (5) displacement of the cells relative to each other within a comb, (6) shape of the comb, (7) presence of noncellular carton structures, (8) number of cells per pedicel. Alone and in combination, these features are examined next.

*Design of the pedicel. Mischocyttarus* (subgenus *Mischocyttarus*) *drewseni* (Fig. 14.17; illustration in Jeanne 1975a) and closely related species build long pedicels. One *M. (M.) melanarius* nest had a pedicel over 38 mm long and about 0.7 mm wide (SEM). At the other extreme, *M. (Kappa) latior* builds a stout, buttresslike pedicel; the pedicel from a nest of 240 cells measured only 5 mm long and about 10 × 4 mm in cross-section (Richards 1978a, similar nest in CMNH).

*Cell shape.* Cells of polistine and vespine nests are typically hexagonal. *Mischocyttarus collaris* and the six other species of the subgenus *Megacanthopus* build either a long comb a few cells across with a pedicel at one end or a parallelogram-shaped comb with the pedicel at the acute vertex (Fig. 14.18). The marginal cells are flat-sided and without furrows marking the positions of walls between them, rather than hexagonal with furrows (see photos in Zikán 1935, 1949, 1951; BMNH; CMNH; MCZ).

*Cell shape and comb orientation. Mischocyttarus (Haplometrobius) weyrauchi* builds nests of a single row of cells, with the long axes of the cells and the row perpendicular to a stout pedicel (Fig. 14.19; MCZ, BMNH). Even a nest of 33 cells (CMNH) is only a single row of cells; Richards's (1978a) assertion that there may be a second row directed upward is without support to my knowledge. The lower surface of the cells (side opposite the pedicel) is often flat-sided and does not show furrows.

*Pedicel, comb orientation, and comb shape. Mischocyttarus (Monogynoecus) pelor* builds rows of cells back to back, pointing to either side of a central line of several short pedicels (Fig. 14.20; USNM). This species' near relatives *M. alienus* and *M. moralesi* suspend a single vertical row of cells from a long pedicel (Carpenter and Wenzel 1988: fig. 10).

*Cell displacement and shape, and carton.* Species of the subgenus *Omega* typically build oval nests, but *Mischocyttarus (Omega) punctatus* and *M. (O.) vaqueroi* build each cell to full size and initiate the next cell on the lip of the first (a displacement of 100% toward the cell mouth), resulting in a vertical string of tandem arrangement (Fig. 14.21) that may be as long as 32 cells (AMNH). Cells are cone-shaped and round in cross-section. Some nests have bifurcations such that one cell supports two strings below (BMNH, MCZ), and O. W. Richards (unpubl.) apparently believed these were built by undescribed species. A very smooth and uniform carton is reinforced by a layer of glossy secretion that can be peeled off both the inner and outer walls of the cells (SEM). In a different subgenus, *Mischocyttarus (Haplometrobius) stenoecus* builds a nest with cell displacement similar to that found in *M. punctatus* and *M. vaqueroi* (Richards 1978a).

*Cell displacement.* *Mischocyttarus (Haplometrobius) iheringi* builds a suboval comb with the long axes of the cells and comb perpendicular to the marginal pedicel. Successive rows of cells tend to be displaced toward the mouth of the comb by as much as 30% of the length of the cell (RMNH), and the nest back is covered with moss and other organic particles, giving it a cryptic and uneven appearance (photos in Zikán 1935, 1949, 1951).

*Cell displacement and comb shape.* While *Mischocyttarus (Haplometrobius) synoecus* builds ordinary suboval combs, other members of this subgenus (*M. artifex*, *M. mirificus*, *M. oecothrix*, and *M. ypiranguensis*) build a long comb with a single or double row of cells suspended at one end by a pedicel. The lower (newer) cell wall extends farther from the base of the cell than the upper (older) wall, allowing the next cell to be initiated with large displacement even if the one above is still incomplete (Fig. 14.22). Richards (1978a) measured a 32-cell *M. mirificus* nest that was 3.0 mm wide and 470 mm long, with the longer cell walls measuring about 18 mm.

*Comb shape and noncellular structure.* *Mischocyttarus cooperi*, which is closely related to *M. iheringi* mentioned above, builds clusters of cells hanging from the horizontal top row of cells, producing a comb that is roughly M-shaped (Richards 1978a). The legs of the M apparently were joined by a carton bridge in the specimen Richards saw.

*Noncellular structure.* One of the most interesting idiosyncracies in *Mischocyttarus* nest structure is that of *M. (Haplometrobius) decimus*, which builds a conical nest in which the margins of the comb foundation project beyond the brood cells to the extent that Richards (1978a) called this a "partial envelope" (see photo in Richards 1945). In three nests in the British Museum (Natural History) this circumferential wall is in about the same plane as the tops of the brood cells and does not restrict access to the face of the comb (Fig. 14.23). The inside surface of the wall later becomes the base of new cells, but the external surface of the nest is smooth and does not show the location of cell walls or bases. New cell walls are built between the older brood comb and the expanding outer wall by a process that must be at least initially different from ordinary peripheral comb expansion in the genus. Hopefully, future study will illuminate what function, if any, the wall serves.

*Cells per pedicel.* Several species spread cells over more than one pedicel. *Mischocyttarus (Monogynoecus) insolitus* builds a nest of several pedicels, each supporting one to four cells (Fig. 14.24; BMNH; photos in Zikán 1949, 1951; Herre et al. 1986). *Mischocyttarus (Kappa) latissimus* builds a line of contiguous cells, each on its own pedicel (BMNH). Although these designs compromise the defensive value of a single attachment (e.g., Gadagkar, this volume), both of these species nest on Melastomataceae leaves, where their nests are guarded by resident dolichoderine ants (Herre et al. 1986). *Mischocyttarus (Monogynoecus) pelor* and *Mischocyttarus (Monogynoecus) fraudulentus* initiate nests with single cells on separate pedicels, but as cells are added the "combs" may fuse to become one multipedicellate structure (Carpenter and Wenzel 1988). The bromeliad-nesting *Mischocyttarus (Phi) cryptobius* builds a more ordinary, perhaps initially unipedicellate comb that is later supported by multiple pedicels (Richards 1978a).

## Conclusions from *Mischocyttarus*

Considering all these modifications, it appears that in taxa such as *Mischocyttarus* all architectural features are subject to innovation, often resulting in dramatically different-looking nests among close relatives. On the other hand, different lineages may often converge on the same general design. From this perspective the question of what drives polistine architectural evolution becomes less one of how to explain the diversity of designs in groups such as *Mischocyttarus* or *Ropalidia* and more one of how to explain the relative uniformity apparent in *Polistes*. Independent derivation of similar structures is a theme repeated in the next section, which surveys the great variation of nest designs among the genera of Vespinae and Polistinae.

## VARIATION AMONG GENERA OF VESPINAE AND POLISTINAE

### Structural Variation

Saussure (1853–1858) described conspicuous features of mature nests of Vespinae and Polistinae and coined terms to describe classes of nests based on those features. Saussure's terms were later augmented, most notably by Richards and Richards (1951), to give major categories such as *calyptodomous* and *gymnodomous* (nests covered by an envelope vs. naked nests); *stelocyttarus*, *astelocyttarus*, or *phragmocyttarus* (comb supported by a pedicel, sessile on the substrate, or sessile on the previous envelope, respectively); and *laterinidal* versus *rectinidal* (pedicel peripherally supporting a comb perpendicular to the substrate vs. centrally supporting a comb parallel to the substrate). These categories often are not mutually exclusive; for example, *Protopolybia* species may have a stelocyttarus first comb and phragmocyttarus later combs.

Table 14.1 lists architectural characters that serve as the basis for this traditional terminology, grouping genera according to architectural ontogeny, which loosely corresponds to the presumed relationships (Carpenter, this volume). Given the widespread variation in these traditional traits, they are of dubious utility in inferring phylogenetic relatedness of the groups. Various ecological forces seem to have driven changes in choice of building materials in various groups. Convergence in design is widespread, as evidenced by the fact that several distantly related genera build sessile combs. Necessity of defense has no doubt independently produced envelopes several times, while common engineering problems have forced different lineages to converge on stacked or continuous combs and on similar techniques of reinforcement.

## Building Materials

The five categories of materials listed in Table 14.1 usually can be distinguished with low magnification. Nests of long woody fibers are usually gray; nests of plant hair are usually yellow, amber, or sometimes white; and nests of short chips of plant or inorganic material may

**Table 14.1.** Traditional characters of nest architecture in Polistinae and Vespinae

| Model genus for group[a] | Material[b] | Pedicellate combs parallel or perpendicular to substrate | Sessile combs | Envelope | Stacked combs |
|---|---|---|---|---|---|
| *Polistes* | 1,2 | Either | No | No | No |
| *Mischocyttarus* | 1,3,4 | Either | No | No | No |
| *Ropalidia*[c] | 1,2,3,4,5 | Either | Some | Some | Some |
| *Apoica* | 2 | No pedicel | Yes | No | No |
| *Parachartergus*[d] | 1,2 | Either | Some | Some | Some |
| *Metapolybia*[e] | 3,4,5 | No pedicel | Yes | Yes | No |
| *Protopolybia*[f] | 1,2,5 | Parallel | Some | Some | Some |
| *Polybia*[g] | 1,2,3,4 | No pedicel | Yes | Yes | Yes |
| *Vespa*[h] | 1,2,3,4 | Parallel | No | Yes | Yes |

[a]Genera are grouped by architectural ontogeny; groups are ordered roughly in phylogenetic sequence from *Polistes* to *Polybia* (see Carpenter, this volume).

[b]1, long woody fiber; 2, plant hairs; 3, short vegetable chips; 4, mud, inorganic particles; 5, pure glandular secretion, no vegetable pulp.

[c]Includes *Belonogaster, Parapolybia, Polybioides*.

[d]Includes *Agelaia, Angiopolybia, Chartergellus, Leipomeles, Marimbonda, Nectarinella, Pseudopolybia*.

[e]Includes *Asteloeca, Clypearia, Occipitalia, Synoeca*.

[f]Includes *Pseudochartergus*.

[g]Includes *Brachygastra, Charterginus, Chartergus, Epipona, Protonectarina, Synoecoides*.

[h]Includes *Dolichovespula, Provespa, Vespula*.

be any color depending on their origin. *Pure glandular secretion* refers to structural use of secretion in the envelope or comb (exclusive of the pedicel and adjacent region), without incorporation of the foreign matter found in ordinary carton. The origin of most secretions is poorly known (but see Downing, this volume). Most studies have focused on cuticular hydrocarbons of *Polistes* that, when applied to the nest, may play a role in repelling ants (Espelie and Hermann 1988, 1990; Henderson and Jeanne 1989) or in recognition (Espelie et al. 1990). Structural secretions may be chitinous (Schremmer et al. 1985) or proteinaceous (Espelie and Himmelssbach 1990). The glands producing them have not yet been identified definitively, but likely are associated with the oral region. In overview, it is apparent that none of the materials listed in Table 14.1 is restricted to any single taxon.

Among the Old World and nonswarming New World polistines, *Polistes* uses long woody fibers and plant hairs (Duncan 1928), the materials most widely used by eusocial wasps. *Mischocyttarus* uses long fiber and coarse chips, occasionally mixing in dirt (Richards 1978a). Many *Ropalidia* species use short chips, and all species I have studied rely on secretion more than does *Polistes*, many species making transparent windows in the bottoms of cells early in the construction process (Wenzel, unpubl.) or later to close the holes made to extract meconia (Kojima 1983c). *Ropalidia opifex* restricts gaps between supporting and adjacent substrates with a veil-like envelope of pure secretion (Vecht 1962). *Belonogaster* is unusual in that, although it starts with long fiber collected away from the nest, postemergence nests often are enlarged exclusively by completely removing old pupal cells and reusing this fiber elsewhere on the nest (Fig. 14.25; M. Keeping, unpubl.). To my knowledge, *Parapolybia* builds only with long fiber, plant hair, and generous secretion (Sekijima et al. 1980), often making nearly translucent amber-colored combs (AMNH, BMNH). The few *Polybioides melainus* and *P. tabidus* nests I have seen are of pale or gray long fiber (BMNH, MNHN).

The swarming Neotropical polistine genera exhibit great diversity in both building materials and their reliance on secretion. This variation is not congruent with the patterns of nest ontogeny or adult morphology; that is, genera may employ different materials despite their similarity in construction sequence (see Ontogenetic Variation in Architecture, below) or proximity in Carpenter's (this volume) proposed phylogeny. *Parachartergus, Chartergellus, Nectarinella*, and *Pseudopolybia* (all members of a single clade) build mostly gray nests with both plant hair and long fiber, while *Leipomeles* (from the same clade), and the more distantly related *Angiopolybia, Agelaia*, and *Apoica* generally build pale yellow or amber nests exclusively of plant hair and often generous secretion (see Schremmer 1983 for micrographs).

In yet another clade, *Metapolybia* (Rau 1943), *Asteloeca* (WDH), and probably *Occipitalia* (R. Jeanne, unpubl.) build nests of vegetable chips with numerous windows of pure secretion, yet the closely related *Synoeca* uses very little secretion and will sometimes build with mud. The last member of this clade, *Clypearia*, uses one type of chipped pulp to build most of the nest, varnishes this portion and makes an envelope with pure secretion, then adds to the outside distinctive particles, which may be stone cells from tree bark (Jeanne 1979c) or other small organic granules (WDH).

The sister genera *Pseudochartergus* (Jeanne 1970c) and *Protopolybia* (Wenzel, unpubl.) may build with both pale hairs and long fiber and commonly bridge gaps between leaves or join leaves to the envelope with a transparent film of pure chitinous oral secretion (Schremmer et al. 1985). These genera, segregated earlier on the basis of architecture, have recently been synonymized (Carpenter and Wenzel 1989).

The genera more closely related to *Polybia* also differ widely with respect to choice of building material. *Chartergus* builds heavy durable nests of long white hairs with little or no secretion matrix; *Charterginus* builds with short amber plant hairs and ample secretion. Two *Polybia* subgenera (*Pedothoeca* and *Furnariana*) build with heavy mud, yet they are not constrained to make nests smaller than their congeners: the *Polybia* (*Pedothoeca*) *singularis* nest figured by Berland and Grassé (1951, MNHN) is over 20 cm in diameter with walls 1 cm thick and may have weighed over 5 kg when it was intact. The remaining species of *Polybia*, as well as *Brachygastra*, *Epipona*, and *Synoecoides* (BMNH) build primarily with short chips, sometimes producing a brittle, hard, and dense material.

A notable feature of nests of *Polybia* and its nearest relatives is the lack of a conspicuous matrix of secretion binding together the chip material of the carton. *Polybia sericea* and other *Polybia* (subgenus *Trichinothorax*) species sometimes have a glossy carton, but none of the nests I have seen of *Polybia*, *Chartergus*, *Brachygastra*, *Synoecoides*, and *Epipona* has an amount of glandular secretion comparable to that of virtually all other Polistinae. It seems that in this group the paper-making process has changed such that trace amounts of secretion will suffice to make a cohesive (though brittle) paper of the irregular coarse chips (*Charterginus* is exceptional, see above). Interestingly, the honey-storing wasps *Protonectarina* (MCZ, CMNH) and *Brachygastra mellifica* (USNM) use more secretion than their near relatives listed above, making flexible, gray paper of long fiber.

The reduced role of secretion in the genera allied to *Polybia* may be due in part to the division of labor in nest construction, first mentioned by Schwarz (1931) and later measured by Jeanne (1986a) for *Polybia occidentalis*. Some wasps specialize in bringing pulp to the nest, and others specialize in building with pulp received from foragers. In large colonies, only 0.2% of pulp foragers retain part of their loads to build, and 95% of wasps that have just built with carton do so again as their next task (Jeanne 1986a, this volume: Chap. 11). If each worker participates in all duties, then the whole colony serves as a source or reservoir of glandular products, whereas increasing division of labor may shift the responsibility of secretion to progressively fewer wasps. Gland size, glandular chemistry, or the role of secretions required in paper making may have to change if fewer adults contribute directly to the tasks that require secretions. To my knowledge there is no study comparable to Jeanne's (1986a) study for any other swarm-founding wasp—it would be interesting to know if division of labor involving nest construction in *Polybia, Brachygastra, Synoecoides, Chartergus,* and *Epipona* is generally different from that of other genera that use more secretion. In different clades more distantly related to *Polybia, Synoeca* and *Pseudopolybia* rely on glandular secretion less than their closest relatives, so they may provide particularly useful information relevant to these questions.

Among the vespines, long fiber is the preferred material of *Dolichovespula* (Greene et al. 1976) and some *Vespula* (see Duncan 1939, Spradbery 1973a), and all members of Vespinae may use macerated long fiber occasionally. Plant hairs may be used by *Dolichovespula* (Laidlaw 1930) and *Provespa* (Matsuura and Yamane 1984), but many *Provespa* (Maschwitz and Hänel 1988), *Vespula,* and most *Vespa* nests are built primarily of short chips probably chewed from rotten wood and bark (MacDonald 1977). Dirt or mud may be incorporated into the nests of *Vespa* (Ishay et al. 1986) and used in repairs by *Vespula* (Spradbery 1973a).

*Provespa* (Matsuura and Yamane 1984), *Vespula* (SEM), *Dolichovespula* (FMNH), and probably *Vespa* sometimes paint a varnishlike secretion onto the upper surface of the envelope, presumably to improve water repellency. The paper of vespines is usually without an obvious matrix of secretion, and the adhesive is evident only under high magnification (Ishay and Ganor 1990). Akre et al. (1976) have reported that pulp foragers of *Vespula* always build with their load and do not transfer it, a trait apparently typical of all Vespinae (e.g., Matsuura, this volume; Jeanne, this volume: Chap. 11). It seems unlikely that the ancestors of modern Vespinae ever relied on secretions in making paper to the extent that the Polistinae do.

## Pedicels

A narrow, rodlike pedicel is easy for one wasp to build and serves to isolate the nest from the substrate, but it is often too weak to support a growing comb without modification. All genera with pedicellate nests reinforce the paper core of the pedicel with a lustrous secretion that serves in part to strengthen it. Many, if not all pedicellate genera are willing to build secondary pedicels, buttresses on the primary pedicel, or some other connection for better support of the growing nest or for maintaining space between the nest and nearby substrate (Downing and Jeanne 1990).

The orientation of the comb with respect to the pedicel or substrate varies continuously between perpendicular and parallel within many species (e.g., Schremmer 1978b on *Parachartergus*) and idiosyncratically among close relatives. Among the five independent-founding polistine genera, an elongate comb supported from the top margin (loosely, laterinidal) is the only form known for *Belonogaster* and *Parapolybia* (Figs. 14.25, 14.26) and is widely distributed in *Mischocyttarus*, *Polistes*, and *Ropalidia*. This may be the primitive form of comb orientation for Polistinae (Wenzel 1989), despite the traditional (and more broadly parsimonious) contention that a radially symmetrical (loosely, rectinidal) design is primitive.

Pedicels that initially are rodlike are later modified into wide, walllike structures in *Provespa* (Fig. 14.60, Matsuura and Yamane 1984, Maschwitz and Hänel 1988), the *Vespula rufa* group (Fig. 14.61c), and *Dolichovespula* (Fig. 14.62). *Polybia* (subgenus *Trichinothorax*) species have secondarily developed a suspended primary comb relying on a unique broad, buttressed sheet, which is analogous but not homologous to the narrow pedicel of *Polistes* (contrary to Vecht's [1967] statement). In mature nests, these wasps often build cells on the sides of this sheet above the primary comb (Fig. 14.51c), a habit unknown among builders of truly pedicellate nests.

Sessile Combs

By traditional definition (Saussure 1853–1858, Jeanne 1975a), sessile combs are those whose cell walls are built directly from the substrate with no paper foundation; examples are found in *Synoeca* and *Metapolybia*. Such combs are presumably cheaper to build than are freely hanging combs, so that one might expect many lineages to develop a sessile design for the sake of economy. Richards and Richards (1951) called such combs astelocyttarus and distinguished them from a comb built directly on a paper sheet (phragmocyttarus). This latter method of construction involves efficient reuse of old envelope to form new comb and is probably cheaper than building a free-standing pedicellate comb from the envelope, as *Ropalidia* (*Icarielia*) *flavopicta* does (Fig. 14.28; Vecht 1940). I prefer to call any cells sessile if they are built by adding straight walls directly to a flat surface (in contrast to adding cups at the margin of a pedicellate comb), regardless of whether the surface is paper or another substrate. This is an important departure from other authors.

Jeanne's (1975a) hypothesized ancestral economy nest, consisting of cells broadly joined by their bases to the substrate, differs little from nests of some *Liostenogaster* and *Parischnogaster* and from the sessile combs of *Agelaia cajennensis*. However, the primitive nest form of the common ancestor of Polistinae + Vespinae probably was a narrow pedicel supporting a one-sided paper comb of cells, like that of *Polistes*. Nevertheless, sessile cells occur frequently even in typically pedicellate taxa. I have seen occasional sessile cells built near natural nests of *Ropalidia carinata* and *Polistes annularis*, and anomalous sessile nest initiation (in the laboratory) by *Polistes dorsalis* (SEM), *P. instabilis*, and *Mischocyttarus mexicanus*. Yamane and Kawamichi (1975) have reported this phenomenon for *P. riparius* ( = *biglumis*) in Japan, and R. Gadagkar (unpubl.) has seen the same in Indian *Ropalidia marginata*. No *Polistes*, *Mischocyttarus*, *Belonogaster*, or independent-founding *Ropalidia* are known to build such combs by regular habit, however. To date no vespines are known to build sessile combs in nature, but *Vespula* and *Vespa* sometimes do so in the laboratory (Ross 1982b, Ishay et al. 1986).

There are at least nine independent origins of sessile cells as a regular part of nest design. *Parapolybia* seems to intitiate nests in ordinary pedicellate fashion, but large colonies may start secondary combs by building sessile cells on the substrate and then adding to these to produce a free-hanging comb that is ordinary in all other respects (Yamane 1984). In the *Ropalidia* subgenus *Icarielia*, *R. montana* (Carl 1934) and *R. nigrescens* (Kojima and Jeanne 1986, S. Reyes, unpubl.) swarms initiate nests with sessile cells supporting the spirally expanding comb, which stands mostly free of the substrate. Analogous nests occur in *Polybioides raphigastra* (Fig. 14.30; Pagden 1958b, Vecht 1966). *Polybioides* apparently has two kinds of sessile cells, those that I presume sometimes initiate free-hanging combs (like the secondary combs of *Parapolybia*), and those of *Polybioides tabidus* (MNHN) built on the outer sheets of paper, which hang between and perpendicular to the envelope halves (Fig. 14.31). These sheets are destined to become combs, and their surfaces are chewed to create dimples around which cell walls are constructed, much as in *Agelaia areata* (see below). *Apoica* initiates nests with a few basally sessile cells (Fig. 14.32), which may appear to be a fat pedicel since only the first cells actually reach the substrate (CMNH, FMNH). *Agelaia cajennensis* (Fig. 14.33; SEM, WDH) sometimes does the same. *Agelaia areata* has two kinds of sessile cells: those that sometimes serve as the point of initiation of the nest (Fig. 14.38; Jeanne 1973) and those that later are drawn from the dimples chewed in the envelope (SEM). *Nectarinella* (Fig. 14.42; Schremmer 1977) and its near relative *Marimbonda* (Richards 1978a) build a single sessile comb joined basally to the substrate.

The pedicellate primary combs of many *Protopolybia* species support on the envelope a sessile secondary comb (Fig. 14.44), as may also occur in nests of some species in the sister genus *Pseudochartergus*, especially *Pseudochartergus chartergoides* (Schremmer 1984) and *P. pallidibalteatus* (RMNH). *Protopolybia sedula* (CMNH) and *P. acutiscutis* (SEM) sometimes initiate nests with a sessile comb and will build sessile combs on leaves incorporated into the growing nest.

Nearly all nests of the genera near *Metapolybia* (Figs. 14.46, 14.47) and near *Polybia* (Figs. 14.48–14.50, 14.52–14.58) are initiated by a sessile comb, which may be drawn away from the substrate. *Polybia* (subgenus *Trichinothorax*) species have an exceptional primary comb (Fig. 14.51; see Pedicels, above).

Though not actually sessile, cells in the nests of *Agelaia lobipleura* (Fig. 14.35; Richards 1978a, BMNH) and *Mischocyttarus pelor* (Fig. 14.20; Carpenter and Wenzel 1988, USNM) are built back to back by adding cups to the comb margins. The resulting comb somewhat resembles that of *Apis* and uses approximately 7% less material than a standard one-sided comb (Jeanne 1975a). This design was erroneously reported for *Polybioides* (Bequaert 1918, Richards and Richards 1951), whose combs are in fact one-sided (Bequaert 1922, Vecht 1966)

Kojima and Jeanne (1986) point out that *Ropalidia* (subgenus *Icarielia*) is the only group that will build pedicellate combs from the envelope (e.g., *R. (I.) flavopicta*: Fig. 14.28). They hypothesize that *Ropalidia* species have not made use of sessile combs because this would interfere with the hygienic behavior of pulling meconia from the back of brood cells. That hypothesis is plausible, but small-colony *Ropalidia* and *Parapolybia* succeed in pulling the meconia from cells that are centered on the resinous pedicel, so other species probably could get through the 1 mm or less of envelope to accomplish the same task. There are no adequate studies of *Ropalidia* to show that sessile cells are not sometimes added to the envelope, and Vecht's (1940) illustration (redrawn in Fig. 14.28) of a *R. flavopicta* nest along with Spradbery and Kojima's (1989) photograph of a *R. bensoni* nest suggest that they may be. *Polybioides tabidus* (MNHN, BMNH) and *P. raphigastra* (Vecht 1966) apparently apply sessile cells to paper sheets, later stripping the backing sheet away so that the cells are open on both ends.

### The Envelope

One of the main points of Jeanne's (1975a) seminal paper on wasp nest architecture is that envelopes are primarily a response to ant predation. Nearly all Neotropical biologists agree that *Eciton* army ants are a major predator of many insects, and *Eciton hamatum* is known to specialize in raiding the nests of social insects. Recent information shows that all paper wasp taxa that typically lack envelopes (except perhaps *Agelaia*) use chemical ant repellents (Jeanne 1970a, Turillazzi and Ugolini 1979, Kojima 1983a,b, Keeping 1990), as do many Stenogastrinae (Turillazzi and Pardi 1981). For instance, *Polistes* incorporates into the pedicels fatty acids known to elicit necrophoric behavior in ants (Espelie and Hermann 1990), these compounds being similar to others with demonstrated ant-repellent properties (Henderson and Jeanne 1989). *Leipomeles* (Fig. 14.41; Williams 1928, SEM) and *Nectarinella* (Fig. 14.42; Schremmer 1977, SEM) employ an ant guard consisting of fiber pillars tipped with a sticky substance as well as an envelope as protection against ants. In the former genus most of these pillars are on the petiole of the leaf on which the nest is built, whereas in the latter they occur on and around the envelope. With a nest more chemically cryptic than repellent, *Parachartergus aztecus* lives with pugnacious *Pseudomyrmex* ants, apparently concealing its nest from them by smearing across the envelope a mixture of lipids like that of the ant's own cuticle (Espelie and Hermann 1988).

Some authors (e.g., Ohgushi et al. 1983a) use the term *pseudenvelope* for structures that arise directly from the cell walls, suggesting that they are not true envelopes, a conclusion with which I disagree. All envelopes may have been originally derived from extensions of the cell wall, later becoming displaced away from the comb. This transition presumably can be seen in the contemporary species pair *Eustenogaster micans* and *E. calyptodoma* (Sakagami and Yoshikawa 1968). Pseudenvelopes are no less effective in concealing or restricting access to the cells than are the true envelopes of, say, *Parachartergus*. The utility of an envelope for excluding small predators and parasites seems indisputable, and envelopes have evolved and have been lost several times, probably in accordance with their adaptive role.

Social vespids display at least seven independent origins of envelopes or wall-like structures that cover and restrict access to the brood comb. We find an envelope in *Eustenogaster* (Fig. 14.6) and one independently evolved in *Parischnogaster alternata* (Fig. 14.12). Vespine species typically build envelopes (Figs. 14.59–14.62). *Polybioides tabidus* (Richards 1969, BMNH, MNHN) and *P. melainus* (Fig. 14.31; Bequaert 1918) have distinctive envelopes of separate convex sheets resembling clam shells, which are eventually joined together. *Ropalidia* (subgenus *Icarielia*), but apparently not other *Ropalidia* groups, also may build envelopes (Vecht 1962, Richards 1978b). Virtually all the swarm-founding Neotropical polistines build envelopes (or probably did so primitively), though it is not clear whether they are homologous among all genera. For instance, *Protopolybia* draws the envelope from the margin of the pedicellate comb (Fig. 14.44), expanding the nest by building cells on the envelope or continuous with the original comb. In contrast, the genera allied to *Parachartergus* (Figs. 14.39–14.43) suspend the envelope from the substrate and remodel restrictive portions of envelope as pedicellate combs grow or are added. All cavity-dwelling *Agelaia* omit the envelope, with the exception of some *A. cajennensis* colonies (Fig. 14.33; Cooper 1986, MCZ). The arboreal *A. areata* (Fig. 14.38; Jeanne 1973, SEM) and *A. flavipennis* (Fig. 14.1) build an outwardly spiralling comb, the advancing edge of which provides protection to concentric older portions within. I interpret this as a modification of a fast-growing, curved brood comb that functions briefly as an envelope and so represents another independently derived envelope structure.

Envelopes presumably arose once in the ancestor of the Vespinae separately from those in Polistinae, and envelopes of modern species have acquired some important secondary functions. The best-known example is to provide insulation, which assists with homeothermy (see Spradbery 1973a, Makino and Yamane 1980). This may be a recent functional innovation based on an ancient design, assuming a Cretaceous origin of eusocial wasps with combs like those of modern Vespinae and Polistinae (Brown 1941, Wenzel 1990). Increasing evidence indicates an ice-free Mesozoic world with tropical trees and vertebrates persisting in the northernmost Arctic at least until about 50 million years ago (Dawson et al. 1986)—thermoregulation in such an environment seems of little importance. Vespines may have originally constructed multiple layers in their envelopes as an adaptation to heavy tropical rains (Fig. 14.59c), as Vecht (1957) suggested.

The loss or reduction of envelopes in cavity dwellers such as *Agelaia* is also known in *Pseudochartergus* (Fig. 14.45), *Metapolybia bromelicola* (between leaves: Richards 1978a), *Vespa* (Fig. 14.59d; Spradbery 1973a), *Dolichovespula* (Edwards 1980), *Vespula* (Duncan 1939), and *Ropalidia flavopicta ornaticeps* (Kojima and Jeanne 1986). The cavity-dwelling *Polybioides raphigastra* does not cover the brood comb with hanging sheets as do *P. melainus* and *P. tabidus*, but rather extends the comb in a downward helix and fixes the outer margin to the comb above (Fig. 14.30; Vecht 1966). Thus, social wasps commonly omit the envelope when a cavity wall serves to protect or insulate the nest.

## Engineering with Paper

Regardless of building material or nest form, different lineages sometimes evolve the same solution to problems of engineering. Aerial wasp nests can reach tremendous age and size (e.g., nests of *Polybia scutellaris* 30 years old and over 50 cm in diameter: Richards 1978a). Jeanne (1977a) points out that as such nests grow larger, an increasing fraction of the carton must be devoted to structural support as opposed to brood comb, probably imposing an economic limit on size unless branches are incorporated to support the weight. Species now known to use surrounding branches and leaves in this manner include *Polybia dimidiata* (Fig. 14.52), *Brachygastra mellifica*, *Ropalidia montana* (Fig. 14.29;

Matsuura and Yamane 1984), *Agelaia areata* (Fig. 14.38; Jeanne 1973), *A. flavipennis* (Fig. 14.1), *Protopolybia exigua* (CMNH), *Protopolybia acutiscutis* (SEM), *Provespa* species (Matsuura and Yamane 1984, Maschwitz and Hänel 1988), *Dolichovespula* species (Duncan 1939), and occasionally some *Parachartergus* (Wenzel, unpubl.).

Another engineering problem is how to arrange and expand brood combs efficiently. Stacking them vertically in discrete levels is a common solution (Table 14.1). Another solution is demonstrated by the helical combs of *Ropalidia (Icarielia) montana* (Fig. 14.29), once thought to be unique (Carl 1934). In fact, helically expanding combs are known in *R. (I.) flavobrunnea* and *R. (I.) nigrescens* (Kojima 1982a, Kojima and Jeanne 1986), and in *Polybioides raphigastra* (Fig. 14.30; Pagden 1958b, Vecht 1966). *Chartergus* builds several discrete stories of its nest initially, but in expanding mature nests additional stories may be interconnected to form a helix as opposed to discrete chambers. I have seen the specimen figured by Saussure (1853–1858: plate 33; MNHN), and the representation of spiral connections is accurate. This design has been overlooked despite frequent use of the original figure (Berland and Grassé 1951, Wilson 1971, Jeanne 1975a). Similarly, combs of large nests of *Protopolybia exigua* (CMNH) and *P. acutiscutis* (Naumann 1970 [as *P. pumila*], SEM) often grow in a downward helix or irregular spiral. *Agelaia areata* (Fig. 14.38; Jeanne 1973) and *A. flavipennis* (Fig. 14.1) build one or several interconnected spherical combs expanding spirally from a central point or axis, as do large colonies of *Brachygastra mellifica* and *B. lecheguana* (Saussure 1853–1858, Naumann 1968). Thus, no fewer than six genera have independently evolved an expanding spiral nest form, four times as a helical shape and twice as a growing sphere.

The most obvious engineering problem in wasp nest construction is how to strengthen a sheet of paper built by adding strips of material to the edge. The most common solutions are application of blots of material to the surface (probably used to some extent by all groups), lamination, imbrication, and corrugation (see Appendix 14.1). The laminate envelope of the Vespinae may have over a dozen sheets, which are mostly independent (photos in Hungerford 1930 and Duncan 1939 for *Dolichovespula*) or are interconnected (Kemper and Döhring 1967, fig. 39, for *Vespula*). *Pseudopolybia* (Fig. 14.43) shows remarkable convergence on the vespine design, building a laminate envelope similar in appearance, though more supple. Many species of the genera near *Polybia* reinforce a single sheet of paper by *Vespula*-like imbrication if the sheet is a side wall of the nest (Fig. 14.49), though the lower portion of envelope (which may later bear brood comb) is usually reinforced only by blots of material or not at all. *Agelaia areata* and large *Protopolybia acutiscutis* colonies build strong interconnected sheets of envelope, which will later bear cells (SEM). Nests of *A. areata* have a unique form of reinforcement; the surface of a sheet is chewed to create dimples, which are sometimes drawn out to form cells (SEM) or may be covered over later to make closed air spaces (Jeanne 1973).

Corrugation of a single sheet of paper, apparently for strength, has evolved more than once. *Parachartergus* often has an envelope stiffened by orderly ridges and furrows that run perpendicular to the lines of application of individual strips of pulp. The quite distantly related *Synoeca* also corrugates the envelope perpendicular to the lines of edge construction (Buysson 1906, Jeanne 1975a) (though apparently *S. virginea* does not: RMNH; nor does *S. chalybea*: W. Hamilton, unpubl.). Convergence on this design is understandable, since fibers applied by the edge-technique resist flexion on one axis of the plane and the corrugation resists it on the other axis. The distance from ridge to ridge is about the same as the cell diameter, but in neither *Synoeca* nor *Parachartergus* do the ripples invariably correspond to the contours of the walls of the brood cells (but see Castellón 1980b). *Nectarinella* builds irregular, fine corrugations in its domed envelope (SEM). Although the nests of *Protonectarina* (Fig. 14.54) are very different in overall design from those of *Nectarinella* (Fig. 14.42), the two sometimes have indistinguishably similar corrugations (MCZ, SEM).

*Leipomeles* occasionally builds widely spaced narrow ridges in its thin translucent envelope (Schremmer 1983). These ridges provide some stiffness, but because they are sometimes branched, resembling the veins of a leaf (Williams 1928, SEM, WDH), I do not think they are homologous to the corrugation in closely related genera such as *Parachartergus* or *Nectarinella*. The very distantly related *Charterginus nevermanni* has acute ridges and broad furrows rather similar to those of *Leipomeles* (Bequaert 1938), but these are parallel with the lines of edge construction of the envelope so that they approximate the successive shapes of the advancing edge of the incomplete envelope (SEM).

## Ontogenetic Variation in Architecture

Von Baer's laws regarding conservation of ontogenetic patterns survive as useful tools for interpreting evolution (Gould 1977, Nelson 1978). One useful principle is that characters expressed early in ontogeny often define higher taxonomic units, while characters expressed in later stages define lower units. Within both Vespinae and Polistinae this generalization seems to be valid with respect to nest architecture.

Fig. 14.3 shows several developmental steps of the major eusocial wasp nest types (Figs. 14.4–14.62 present more detail). The initiation of brood rearing serves as an important landmark for comparing sequences of nest development, since it is likely that the reproductive physiology of the wasps is more conservative than the occurrence or sequence of given building behaviors. The first few hours of construc-

tion are sufficient to distinguish Vespinae from Polistinae because vespine queens build a fibrous pedicel and initiate an envelope immediately after the first few comb cells are built (Matsuura 1971a,
Matsuura and Yamane 1984). Within the Polistinae, one finds a spectrum ranging from the simple development of the primitive *Polistes*-
type design to the structurally and ontogenetically more complex nests
of various South American swarm-founding genera. In the primitive
pattern, virtually the whole nest is built after brood rearing is initiated.
For instance, *Polistes* lays eggs in the first cell, and the comb is expanded gradually thereafter by repetition of the two steps, "initiate
cell" and "lay egg." In contrast, colonies of some swarm-founding genera, such as *Polybia*, complete most of the structure before or simultaneous with initiating brood production and later expand the nest
suddenly by adding identical modular units.

Several groups have evolved mechanisms that result in deterministic
nest growth. That is, once the wasps complete certain structures they
will not subsequently expand or remodel them. *Nectarinella* (Fig. 14.42;
Schremmer 1977), *Charterginus* (Fig. 14.55; Wenzel, unpubl.), and in
some cases *Clypearia* (Fig. 14.46; Ducke 1910, 1914; Jeanne 1979c; WDH)
apparently do not expand their nests beyond the confines of the envelope built during nest initiation. In the modular nests of *Polybia*, a
given comb may expand gradually to fill an area determined by the size
of the envelope, but it cannot be expanded beyond that. The envelope
is not remodeled to allow further comb growth, so to continue expansion of the nest a new module must be added. At the pinnacle of this
form of deterministic growth is *Epipona* (Fig. 14.57), which not only
uses modular units, but also sometimes builds the entire multilevel
nest at once, not expanding it later (Vesey-Fitzgerald 1938 [as *Tatua*],
Wenzel, unpubl.). For example, one *Epipona guerini* nest (Fig. 14.2;
SEM) was built to a size of 13 tiers before the brood cells in the oldest
levels were more than 1 or 2 mm deep, clearly showing that the founding swarm built the entire structure long before concentrating on rearing the larvae. The narrowed and asymmetrically sloped lower tiers
would seem to preclude the addition of more brood combs comparable
in size to those above (Figs. 14.2, 14.57), making the ultimate size of
the nest determined entirely during the period of initiation. This sort of
determinism is reminiscent of the acceleration of ontogeny through
evolutionary time, one of the classical mechanisms of morphological
evolution (Gould 1977).

Examples of more-specific ontogenetic differences in nest construction among related species are now considered.

### Vespinae

Although little is known about nest initiation in *Vespa* and *Provespa*
(but see Yamane and Makino 1977, Matsuura, this volume), one differ-

ence in queen nests between the two genera is that *Provespa* builds multiple pedicels on the primary comb, and later fuses them to form a single broad structure (Fig. 14.60; Matsuura and Yamane 1984, Maschwitz and Hänel 1988, BMNH). In contrast to both of these genera, queens of *Dolichovespula* and some *Vespula* build a distinctive hanging sheet of paper that supports a twisted pedicel attached to the small comb (Figs. 14.61, 14.62; Jeanne 1977c, Yamane et al. 1981). These latter two genera can be distinguished in that the arboreal *Dolichovespula* has a flexible pedicel (Jeanne 1977c), builds independent laminar sheets of envelope on the hanging sheet or substrate, and builds combs curved slightly upward at their margins, while the cavity-dwelling *Vespula* has a stiffer pedicel, initiates successive envelope layers on the preceeding sheets (Yamane et al. 1981), and builds more planar combs (Spradbery 1973a: fig. 47).

Within *Vespula*, several nest characters differ between the two major species groups, but which of these are primitive and which are derived is a complicated issue (Carpenter 1987a). The prominent differences are largely in traits evident only in mature nests, and therefore they are ontogenetically late characters. The pedicels of mature nests of the *V. rufa* group become buttresslike or ribbonlike between the envelope and the primary comb (Fig. 14.61c), whereas the numerous pedicels of the *V. vulgaris* group are postlike throughout (Fig. 14.61d; Akre et al. 1980), a condition Spradbery (1973a) argues is derived in Vespinae. The mature, worker-built envelope of the *V. rufa* group is partly laminar and has straighter applications of pulp than the more arcuate and imbricate (Appendix 14.1) applications in the *V. vulgaris* group (Fig. 14.61c,d; Akre et al. 1980). Finally, the secondary combs of the *V. rufa* group bear reproductive brood of both sexes. Species in the *V. vulgaris* group rear males in old worker cells and either build additional combs for queen production (MacDonald and Matthews 1976) or abruptly switch to queen-sized cells at the margin of older worker combs (Duncan 1939, RMNH; see also Greene, this volume).

Structures built at nest initiation (by foundresses) often differ sightly from homologous structures built later (by workers). This kind of variation through time is similar to that found among serially homologous morphological structures (e.g., front vs. hind legs). Such ontogenetic architectural variation is widespread throughout the family Vespidae, but the clearest examples seem to be found in the nests of Vespinae. The best known example is the discrepancy in cell size, with workers building larger cells than do foundresses (Duncan 1939). Also, the queen-built envelope differs from worker-built envelope; it is smoother and more uniform and built of straighter, more horizontal applications of pulp, and, in some groups, has a long entrance tube (Matsuura 1984). As a further example of variation through ontogeny, I have seen several *Vespula* nests (RMNH) in which the queen apparently built with long fiber, while her daughters used short vegetable chip.

## Polistinae

It is conventional to regard nests of the basal genus *Polistes* as being typical of the form that was ancestral to all other forms in Polistinae. Downing and Jeanne (1988, 1990) demonstrated that the behavioral program for nest construction in *Polistes fuscatus* is not a simple chain of events progressing in a strictly linear sequence. Rather, the program is flexible, although the end result is defined. Such flexibility may be generally true of nest construction in eusocial Vespidae, since progressive provisioning requires concurrent construction on regions of the nest where brood are in different stages of development. Furthermore, workers must be able to assume tasks on a partially completed nest rather than start from the first step.

Other than the coarser carton of *Mischocyttarus*, there is no single reliable criterion to distinguish nest structure of this genus from *Polistes* (Table 14.1). The remainder of Polistinae exemplify the principle that taxa at higher levels tend to differ early in the ontogeny of their nests, while lower level taxa differ in the later developmental stages.

The Old World genera *Ropalidia*, *Parapolybia*, *Belonogaster*, and *Polybioides* form a major clade in Carpenter's phylogenetic analysis (this volume) and can be distinguished from other groups by a feature appearing very early in nest construction, the arrangement of the first cells built on the pedicel. The primitive state for Polistinae is cell-marginal, having the pedicel support the first few cells from their margins, as is found also in the sister-group, the Vespinae (Janet 1895, Ishay et al. 1982). The first cell is often flat-sided where the pedicel attaches (Fig. 14.3; Downing and Jeanne 1987, 1988), indicating the impending addition of at least one other cell on the other side of the pedicel (*Mischocyttarus punctatus* is exceptional; see Fig. 14.21). In contrast, nests of *Ropalidia* and its three closest relatives are typified by a cell-central pedicel, where the pedicel is gently flared into a cup or cone, becoming the center of the base of the first cell (Fig. 14.3). The pedicel may be eccentric (Kojima 1988), but I have not seen in this group of genera any cells as flat-sided as those in genera that build cell-marginal pedicels. This character has been overlooked until now.

*Polybioides* may omit the pedicel, but it clearly is a member of this group because it tears away paper and removes the meconium through the back of the comb, as do *Belonogaster* (Marino Piccioli 1968), *Ropalidia*, and *Parapolybia* (Vecht 1966). The former two genera do not repair the comb (*Belonogaster* may later remove the comb completely: Fig. 14.25; illustration in Iwata 1966), but the latter two replace cell bottoms with windows of pure secretion (Figs. 14.26–14.28). This hygienic trait of removing meconia may reflect selection pressure by parasitic flies on the common ancestor of the Old World genera. Unlike Neotropical taxa, African nests in the field and museum are often laden with fly pupae (probably Tachinidae): for example, 58 flies, one per cell, occupied half the comb of a *Belonogaster discifera* nest (SEM). Tachinids are also reported to parasitize some Asian eusocial wasps heavily, with up to 80% of *Anischnogaster* cells in New Guinea parasitized (Spradbery 1989). The odor of meconia is often strong, and removing it may help reduce the probability of discovery by flies. The removal of meconia is not restricted to the *Ropalidia* group, but their method is unique, for other social vespids such as *Parischnogaster* (Turillazzi 1984), *Occipitalia*, *Asteloeca* (formerly *Occipitalia*), and one *Clypearia* species (Jeanne 1980b) apparently clean the cells through the open front end.

All the swarm-founding South American polistines, which form another major clade within the subfamily, are distinguished from *Polistes* and *Mischocyttarus* by loss or reduction of the narrow secretion-rich pedicel, another example of an early-developing feature. The *Parachartergus* group (Fig. 14.3; paraphyletic as used here, see Carpenter, this volume) builds a thicker, fibrous pedicel (Jeanne 1975a) painted with a tough polymer probably from the salivary glands. The apical clades represented by the *Metapolybia* and *Polybia* groups initiate the nest with sessile cells (Fig. 14.3; Jeanne 1975a). These latter two groups can be distinguished from each other by the tendency in the *Metapolybia* group to maintain upward-pointing entrances and to expand mature nests upward along the substrate (Figs. 14.46, 14.47), while genera near *Polybia* usually have peripheral entrances in mature nests and expand by adding modular tiers downward.

The individual genera (and often the species) of the polistine groups in Fig. 14.3 can usually be identified by details of the texture, reinforcement, or other characteristics of the mature envelope. Judging from photographs of new nests of *Parachartergus* (Schremmer 1978b) and *Angiopolybia* (WDH), these wasps apparently lay eggs immediately in the new combs and do not complete the primary envelope until several days later. Even closely related genera within the *Parachartergus* group of Fig. 14.3 build various envelope forms (Figs. 14.40–14.45) and expand and modify the primary envelope as the combs grow. The little known monotypic genus *Nectarinella* appears to be exceptional in the *Parachartergus* group since it does not remodel the primary envelope (Schremmer 1977). Given that envelopes and reinforcement characteristically appear relatively late in the construction of such nests, these distinctions support the idea that variation between close relatives should be more common in later stages than in early stages.

In contrast to the sequence described above, *Polybia* (Schwarz 1931; Richards 1978a: plate 1, figs. 1, 2; Jeanne 1986a), *Brachygastra* (Naumann 1968), and probably other closely related genera complete the primary envelope within a few hours of nest initiation, simultaneously with or before laying eggs in the brood cells. Envelopes are not remodeled to accommodate growing comb; nests are expanded by adding nearly identical modular units beneath the mature sections (Figs. 14.48–14.54, 14.56–14.58). The precocious envelopes of these genera are more similar to each other in design than are those of the *Parachartergus* group. Precocious completion of the envelope and more conservative evolution of envelope design in the *Polybia* group is consistent with the idea that variability is not focused on the envelope per se, but rather on whatever traits are expressed late in a building sequence.

Another similar test of the ontogenetic model of variability is supplied by the mature combs of *Agelaia*. Whereas all 35 species in the genera *Parachartergus*, *Chartergellus*, *Leipomeles*, *Pseudopolybia*, and *Angiopolybia* build ordinary combs enclosed in a variety of envelopes modified late in the building sequence, *Agelaia* typically omits the envelope so that comb expansion is the final stage of construction. Consistent with von Baer's law that embellishment should be concentrated in "late" traits, comb design is highly variable in the 22 species of *Agelaia* (Figs. 14.33–14.38). *Agelaia cajennensis* (Fig. 14.33) may build combs that are pedicellate (MCZ), sessile (SEM), or both (WDH); *A. vicina* builds straight parallel combs (Fig. 14.34); *A. testacea* builds gently curving combs (Fig. 14.36; Jeanne 1973); *A. angulata* builds a central, nearly conical primary comb with secondary arcs of concentric hemispheres outside it (Fig. 14.37; Evans and West-Eberhard 1970: fig. 91, WDH); *A. lobipleura* builds two-sided combs (Fig. 14.35; Richards 1978a, BMNH); and arboreal *A. areata* (Fig. 14.38; Jeanne 1973, SEM) and *A. flavipennis* (Fig. 14.1) initiate nests with a pedicellate or sessile comb, which expands outward and over itself in a spiral. The unusual diversity of comb designs in *Agelaia* may be due to embellishment of late ontogenetic stages (comb expansion here), possibly amplified by the preference for building in protected sites (e.g., cavities) where external forces such as wind and rain are less restrictive. In contrast to *Agelaia*, the cavity-nesting *Vespula* generally do not omit the envelope and do not differ much from one another in comb design. Rather, they differ in features of the secondary pedicels and mature envelope, which, again, are products of late stages in construction.

## CONCLUDING REMARKS

Some of the complex behavior of eusocial wasps is preserved in the architecture of their nests. Previous study of wasp nests shows that architectural information is useful both for inferring phylogenetic relationships (Ducke 1914) and also for demonstrating adaptive evolution in a suite of behaviors (Jeanne 1975a). Much of the work to date has focused on the challenge to determine which patterns of variation reflect historical relationships among species and which reflect recent adaptations. This chapter is a first effort to make sense of the diversity of nest designs among all eusocial Vespidae with respect to that challenge and to offer a mechanism that may explain much about the patterns observed in architectural evolution.

The relatively few species of wasps in the subfamily Stenogastrinae are confined to humid forest understory in the Indo-Pacific, yet they build a bewildering variety of nests. Here, one finds single scattered cells rather like the architecture of many solitary Eumeninae as well as organized combs sheltered within sometimes ornate envelopes. The nests are quite frail and cryptic and may include chemical defense barriers of glandular origin, but none has a narrow supportive pedicel built by the wasps. Our understanding of the origin and function of various nest structures remains largely speculative, since stenogastrine biology is still poorly known (but see Turillazzi, this volume). The Stenogastrinae present a promising frontier for future research.

The relatively simple nests of the five genera of primitively eusocial, independent-founding Polistinae share many common features that presumably were present also in the ancestor of the swarm-founding genera. Nests are suspended from a tough pedicel built largely of oral secretion applied over a core of vegetable matter, and all seem to be protected by a repellent secretion applied to the pedicel from abdominal glands. All other details of the nest vary in one or another group. The archetypal "primitive" genus *Polistes* is less variable in architecture than either *Mischocyttarus* or *Ropalidia*, despite its large number of species and cosmopolitan distribution. A survey of variation in *Mischocyttarus* suggests that in some lineages close species may have dramatically different nests, while more distant species may converge on similar designs. *Ropalidia* architecture is diverse and still poorly known, offering another challenge for future work.

Examination of variation among all genera of Polistinae and Vespinae shows that nearly every genus can be diagnosed by architectural details of the nest. Traditional classes of nests are polyphyletic, and different species within the same genus often build forms that fall into different classes. This finding is consistent with the conclusions drawn from studies of nests of *Mischocyttarus*, and probably reflects the importance of recent innovations superimposed on the ancestral groundplan. Furthermore, some intraspecific variation may be due to idiosyncratic features of colonies, such as differences in division of labor, colony reproductive phenology, or other emergent properties of the society as a whole.

Except for *Apoica*, all of the swarm-founding polistine genera include species that build an envelope, and some genera are known also to use a chemical defense against ants. Species of Vespinae generally build envelopes, whether they initiate nests independently or by swarms. Vespine nesting habits are little known in the tropics (see Matsuura, this volume), but from what is known of both tropical and temperate-zone species glandular secretions appear to be less important in both repellent and structural contexts than they are in Polistinae.

Ontogenetic analysis of nest construction seems to offer a better understanding of whether given architectural elements are homologous or merely analogous in different taxa. With few exceptions, differences in early stages of construction generally distinguish higher, more anciently diverged taxa, while differences in later stages separate lower, more recently diverged taxa. In accordance with classical principles of morphological ontogeny, comparison of homologous structures among genera shows that structural variation is more common when features appear late or have a protracted development, while greater uniformity marks features that appear early or precociously. However, this model of architectural evolution relies on a typological perspective, and future workers should not neglect the search for patterns in intraspecific or intrapopulational variation in nest design.

*Acknowledgments*

I am indebted to many people for their help and support. M. G. Naumann posthumously inspired me through his legacy at the University of Kansas. J. M. Carpenter, H. A. Downing, W. D. Hamilton, R. L. Jeanne, and J. van der Vecht generously offered to me their advice, photographs, and unpublished work. J. M. Carpenter, W. D. Hamilton, C. D. Michener, S. G. Reyes, K. G. Ross, C. K. Starr, and W. T. Wcislo improved the early drafts. I thank W. D. Hamilton and R. Jander for specific comments on the use of ontogenetic criteria. The curators and collection managers at the museums cited in the text allowed me to examine specimens in their care. Assistance with Dutch, German, and Japanese was provided respectively by G. De Boer, J. E. Baxter, and M. Shimada. Financial support was provided by the R. H. Beamer Fund of the Kansas University Endowment Association and NSF grant BNS-8200651 (C. D. Michener principal investigator).

## Major Architectural Forms

All nests are shown from the side in cross-section with the substrate shaded. Illustrations separated by a diagonal bar represent two views of the same structure. Alternative structures are separated by the word *or*. Most illustrations represent developmental series progressing as a, b, c. Black pedicels (Figs. 14.14–14.29) and shaded comb sections (Figs. 14.25–14.28) represent oral secretion as the main material. Pale pedicels are primarily fibrous paper. Two ellipses broadly joined and beneath a figure (e.g., Fig. 14.14a) represent cell-marginal pedicel with a cell on each side; one circle with a central spot (e.g., Figs. 14.25–14.28) represents cell-central pedicel with one cell at the end (see Ontogenetic Variation in Architecture: Polistinae for details). All nests are paper unless otherwise indicated. I have examined all specimens illustrated except where noted.

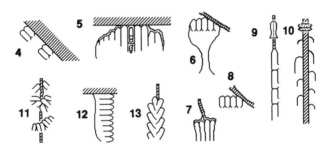

**14.4.** *Liostenogaster nitidipennis.*

**14.5.** *Liostenogaster flavolineata* mud nest with tandem brood in a cell.

**14.6.** *Eustenogaster.*

**14.7.** *Stenogaster concinna* (not examined).

**14.8.** *Anischnogaster iridipennis* (not examined).

**14.9.** *Metischnogaster cilipennis,* with conical carton structure above cells.

**14.10.** *Parischnogaster nigricans,* with gelatinous ant guard (stippled) above cells.

**14.11.** *Parischnogaster mellyi.*

**14.12.** *Parischnogaster alternata.*

**14.13.** *Parischnogaster striatula.*

510

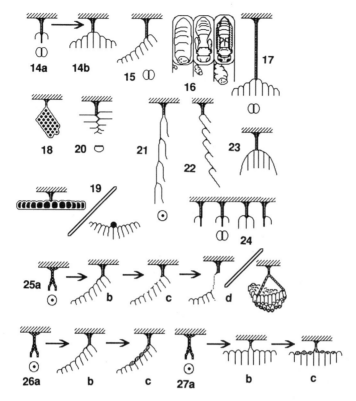

**14.14.** Some *Polistes* and *Mischocyttarus*. (a) Glandular secretion-reinforced, cell-marginal pedicel. (b) Mature, radially symmetrical comb.

**14.15.** Asymmetrical comb of some *Polistes* and *Mischocyttarus*.

**14.16.** Tandem brood of *Polistes olivaceus*.

**14.17.** *Mischocyttarus drewseni*, long pedicel.

**14.18.** *Mischocyttarus collaris*, view of comb face showing smooth exterior walls and parallelogram comb shape.

**14.19.** *Mischocyttarus weyrauchi*, face and top view of single row of cells constituting the comb.

**14.20.** *Mischocyttarus pelor*, cells back to back.

**14.21.** *Mischocyttarus punctatus*, single string of cells from cell-central pedicel.

**14.22.** *Mischocyttarus mirificus*, long lower cell walls and large displacement.

**14.23.** *Mischocyttarus decimus*, with peripheral wall.

**14.24.** *Mischocyttarus insolitus*, single nest of separate cells or combs on cell-marginal pedicels.

**14.25.** *Belonogaster*. (a) Initiation with cell-central pedicel. (b) Asymmetrical comb growth. (c) Cell bases removed and not repaired. (d) Carton removed and reused elsewhere, leaving glandular secretion-reinforced remnant of comb margin, side and oblique view.

**14.26.** *Parapolybia* and some *Ropalidia*. (a, b) Initiation of asymmetrical comb with cell-central pedicel. (c) Cell base replaced by secretion after removal of meconium, sometimes preceding removal.

**14.27.** Some *Ropalidia*. (a, b) Initiation of symmetrical comb with cell-central pedicel. (c) Cell base replaced by glandular secretion after removal of meconium, sometimes preceding removal.

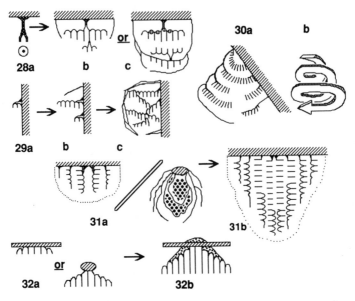

**14.28.** *Ropalidia* (subgenus *Icarielia*) spp. (a) Initiation with cell-central pedicel. (b) Envelope from substrate sometimes left unfinished, sometimes a secondary comb suspended beneath primary comb by a cell-central pedicel. (c) *Ropalidia flavopicta*, with secondary comb arising from envelope or sessile on primary envelope (not examined).

**14.29.** *Ropalidia* (subgenus *Icarielia*) *montana*. (a, b, c) With fused continuous comb and laminate envelope; removal of cell bases not illustrated (not examined).

**14.30.** *Polybioides raphigastra*. (a, b) Continuous downward spiral comb. Bases of older brood cells removed and not repaired; peripheral envelope possibly with imbricate reinforcement (not examined).

**14.31.** *Polybioides tabidus* and *P. melainus*. (a) Initial combs facing opposite directions, envelope sheets perpendicular to planes of combs and arising as separate sheets, one on each side in *P. tabidus*, multiple and loosely laminate in *P. melainus*, side and end view. (b) Mature *P. melainus* nest with cell bases removed and not repaired.

**14.32.** *Apoica*, felt nest. (a) Sessile initiation. (b) Expansion off substrate with fibrous dome (stippling) enclosing substrate.

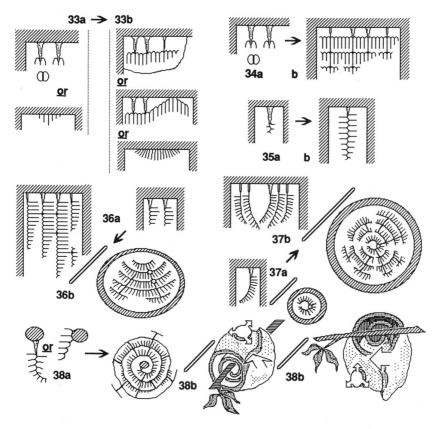

**14.33.** *Agelaia cajennensis.* (a) Pedicellate or sessile foundation in cavities, including between leaves. (b) Pedicellate comb, envelope restricting access to cavity, comb fusion; also sessile comb and combination comb.

**14.34.** *Agelaia vicina.* (a) In cavities. (b) Parallel planar combs, comb fusion.

**14.35.** *Agelaia lobipleura.* (a) In cavities. (b) Double-sided comb.

**14.36.** *Agelaia testacea.* (a) In cavities. (b) Concentric arcuate combs expanding in front or behind, comb fusion, side and top view.

**14.37.** *Agelaia angulata.* (a) In cavities, conical primary comb, side and top view. (b) Concentric arcuate combs expanding outward on all sides, comb fusion, side and top view.

**14.38.** *Agelaia areata* and *A. flavipennis.* (a) Arboreal, sessile or pedicellate initiation, sometimes initially double-sided comb. (b) Spiral comb expansion, rapidly growing comb margins (hollow arrows) form envelope, comb fusion; side view and alternative sections through spiral (also see Fig. 14.1).

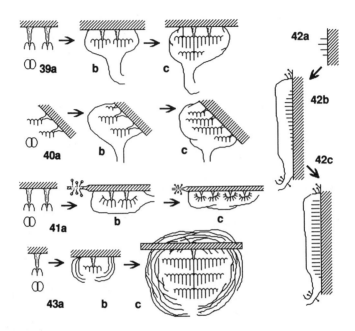

**14.39.** *Angiopolybia.* (a, b, c) Envelope modified as combs grow, comb fusion.

**14.40.** *Parachartergus* and *Chartergellus.* (a, b, c) Envelope modified as combs grow, fusion of combs rare or absent.

**14.41.** *Leipomeles.* (a, b) Ant guard of pillars and sticky secretion on leaf petiole, envelope with or without spout. (c) Envelope probably modified, no fusion of combs.

**14.42.** *Nectarinella* (possibly also *Marimbonda*, not examined). (a) Sessile initiation. (b) Envelope with ant guard of pillars and sticky secretion. (c) Envelope not later expanded.

**14.43.** *Pseudopolybia.* (a, b, c) Laminate envelope modified as combs grow, combs usually with single pedicel.

514

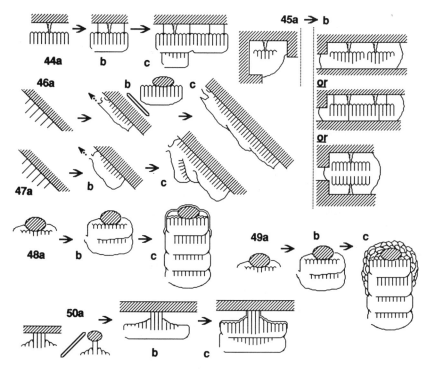

**14.44.** *Protopolybia* and some *Pseudochartergus*. (a) Comb walls fully extended. (b) Envelope from margin of comb of *Pseudochartergus pallidibalteatus, P. chartergoides,* and *Protopolybia*. (c) Nest expanded by construction contiguous with primary comb or on envelope.

**14.45.** *Pseudochartergus.* (a) In cavities, including between leaves, secretion envelope restricting access to cavity. (b) Separate combs, as in some *P. pallidibalteatus*; with pedicels to substrate from comb face, as in some *P. chartergoides*; with upward facing comb, as in *P. fuscatus.*

**14.46.** *Metapolybia* and *Clypearia* (*Occipitalia* and *Asteloeca* similar, not examined); some areas pure secretion with no vegetable pulp. (a) Sessile initiation. (b) Comb sometimes growing off narrow substrate, envelope usually with entrance pointing upward, side and end view. (c) Nest expanded upward, usually without cells on envelope.

**14.47.** *Synoeca.* (a) Sessile initiation. (b) Entrance pointing upward. (c) Nest expanded upward, sometimes with cells on envelope.

**14.48.** Some *Polybia* and *Brachygastra.* (a) Sessile initiation. (b) Precocious completion of several stories, entrance sometimes initially directed upward. (c) Reinforcement by blot-technique (double lines), expansion by addition of modular tiers below; primary comb may fuse with envelope.

**14.49.** Some *Polybia* and *Brachygastra* (as Fig. 14.48, but showing imbricate reinforcement).

**14.50.** *Polybia bistriata.* (a) Narrow, sessile initiation. (b) Precocious closure of envelope. (c) Modular expansion, envelope reinforced by blots (double lines).

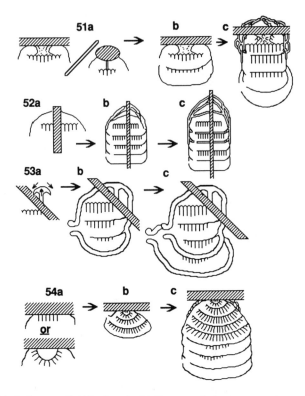

**14.51.** *Polybia* (subgenus *Trichinothorax*). (a) Buttressed sheet supporting primary comb, side and end view. (b) Precocious completion of envelope, secondary combs sessile and ordinary. (c) Imbricate reinforcement and blot-technique (double lines) sometimes used together, disorganized cells on surface of buttressed sheet; primary comb may fuse with envelope. New bottom envelope in mature nests often concealing one or more older tiers.

**14.52.** *Polybia dimidiata.* (a) Sessile initiation on upright trunk. (b) Nest constructed to include trunk through combs; precocious completion of envelope. (c) Imbricate reinforcement and blot-technique (double lines).

**14.53.** *Polybia emaciata*, mud nest. (a) Sessile initiation, post built upward from substrate and expanding downward like an umbrella. (b) Precocious completion of envelope, blot-reinforced exterior walls thick and heavy (double lines), internal septa thin and poorly reinforced. (c) Secondary entrance protruding.

**14.54.** *Protonectarina.* (a) Initiation sessile or with hemisphere of carton. (b) Precocious completion of envelope; early combs hemispherical, later combs flatter; entrance initially as upward spout. (c) Weak imbricate reinforcement above, weak corrugation on bottom envelope; later entrance simple.

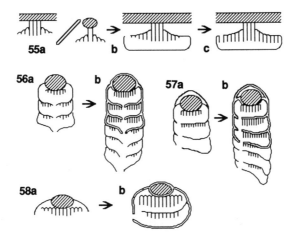

**14.55.** *Charterginus.* (a) Narrow sessile initiation, side and end view. (b) Precocious completion of envelope; entrance simple (peripheral or dorsal). (c) Comb grows, but envelope apparently not later expanded.

**14.56.** *Chartergus*, felt nest. (a) Sessile initiation. (b) Precocious completion of envelope; extensive blot-reinforcement (double lines); entrance central through combs and basal envelope.

**14.57.** *Epipona.* (a) Sessile initiation. (b) Precocious completion of envelope; extensive blot-reinforcement (double lines); later tiers often smaller and increasingly droopy opposite simple entrance. Some nests rapidly built to completion and not later expanded (see Fig. 14.2).

**14.58.** *Synoecoides.* (a) Possibly sessile initiation. (b) Blot-reinforcement (double lines). Nest possibly built to completion and not later expanded, or like Fig. 14.48.

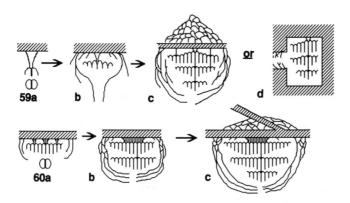

**14.59.** *Vespa.* (a) Cell-marginal initation of comb. (b) Queen nest with envelope. (c) Mature nest with imbricate roof (arboreal Southeast Asian species). (d) Nest in cavity.

**14.60.** *Provespa.* (a) Multipedicellate initiation. (b) Pedicels fused. (c) Mature nest, imbricate roof.

517

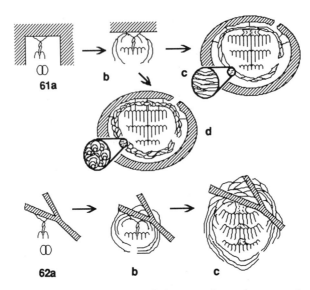

**14.61.** *Vespula*, in cavity. (a) Initiation with hanging sheet of paper and twisted, cell-marginal pedicel. (b) Queen nest with envelope. (c) Mature nest of *V. rufa* group with primary pedicel becoming secondarily buttresslike. Inset shows surface view of laminate envelope. (d) Mature nest of *V. vulgaris* group with all pedicels postlike. Inset shows surface view of imbricate envelope.

**14.62.** *Dolichovespula*. (a) Hanging sheet and twisted pedicel. (b) Queen nest with laminate envelope. (c) Mature nest with all pedicels secondarily buttresslike.

## APPENDIX 14.1 Definitions

Below are definitions of several terms used in this chapter, but often in a sense different from that of other authors.

*Arboreal*:   supported by the branches or trunk of a tree, bush, or other vegetation, in the open.

*Arcuate*:   archlike, curved.

*Blot-technique*:   spreading pulp across a surface, thereby producing a surface of many superimposed spots of pulp. Synonymous with Jeanne's (1975a) *surface building*.

*Cavity*:   any enclosed region (e.g., subterranean, in hollow trees, or between the walls of a house).

*Corrugation*:   alternating ridges and furrows.

*Displacement*:   following Downing and Jeanne (1986), relative positions of the cell bottoms on a comb. Displacement equals zero if cell bottoms are all in one plane; displacement is greater than zero if cell bottoms are initiated somewhere along the wall of older cells.

*Edge-technique*:   adding long strips of pulp to an edge, producing a sheet with the successive strips lying in roughly parallel lines or concentric arcs. Synonymous with Jeanne's (1975a) *edge building*.

*Envelope*:   any sheet of nest material constructed to cover and restrict access to the brood comb(s) or nest.

*Felt*:   material composed of interlocking plant hairs and containing little oral secretion.

*Foundation*:   the act of starting a nest. *Independent foundation* when one or a few females, sometimes later joined by others, choose a site and initiate a nest; *swarm foundation* when more than about 20 wasps move in an organized group to a chosen site and initiate a nest.

*Imbricate*:   structure organized like tiles of a roof or fish scales, usually built by adding strips of pulp in short concentric arcs of decreasing radius to produce a small shell of paper superimposed on other similar shells, sometimes making many closed air spaces. Synonymous with the more ambiguous terms *cellular* or *scalloping* of authors such as Spradbery (1973a).

*Laminate*:   close juxtaposition of separate broad sheets infrequently or not in contact with each other except at the margins.

*Pedicel*:   a strong narrow pillar, usually rodlike initially although often flatter later, between the nest and substrate or between parts of the nest. Synonymous with *petiole* or *peduncle*.

*Primary*:   the first-built of several homologous structures (such as the primary pedicel of *Dolichovespula*) or, alternatively, structures built before workers emerge (such as the primary combs of *Parapolybia*).

*Secondary*:   structures built after the primary structures.

*Secretion*:   oral secretion used in constructing nests, reported to be chitin (Schremmer et al. 1985) but possibly proteinaceous or some other polymer.

*Sessile*:   (1) a comb or nest built without a pedicel or base; (2) cells built by adding walls directly to a surface rather than by drawing out the edges of a basal cup added to the margin of the comb.


## APPENDIX 14.2  Abbreviations of Collections Containing Nest Specimens Cited in the Text

AMNH:   American Museum of Natural History, New York.

BMNH:   British Museum (Natural History), London.

BPBM:   Bernice P. Bishop Museum, Honolulu.

CMNH:   Carnegie Museum of Natural History, Pittsburgh.

FMNH:   Field Museum of Natural History, Chicago.

MCZ:   Museum of Comparative Zoology, Harvard University, Cambridge, Massachusetts.

MNHN:   Muséum National d'Histoire Naturelle, Paris.

RMNH:   Rijksmuseum van Natuurlijke Historie, Leiden.

SEM:   Snow Entomological Museum, University of Kansas, Lawrence, Kansas.

USNM:   National Museum of Natural History, Washington, D.C.

WDH:   Personal collection of W. D. Hamilton, Oxford, England.

## 15

# The Nest as the Locus
# of Social Life

CHRISTOPHER K. STARR

The nests of wasps and bees are among their most distinctive and studiable features. Any treatment of a species' basic biology will usually include a description of the nest, and at least among eusocial species there are now few genera or species groups whose nests are entirely unknown. The central importance of the nest is reflected in the common use of the term *nesting biology* for the behavioral ecology of these insects.

An understanding of the nests of social wasps must begin with those of solitary wasps. Since Evans's landmark paper (1958, expanded in Evans and West-Eberhard 1970), it is generally regarded as a precondition for the evolution of aculeate sociality that the female have a fixed nest to which she repeatedly returns. Behavioral mechanisms for such nest fidelity are well established in solitary aculeates, many of which show extraordinary capabilities of spatial orientation (Baerends 1941, Carthy 1958, Evans and West-Eberhard 1970). For spiders, the web has much the same social-evolutionary importance as a stationary physical nucleus around which a social group *can* form. The known colonial spiders are all web builders (Shear 1970, Buskirk 1981).

Nests of social insects are the locus of colony life in the most literal sense. Interactions between colony members mostly take place at the nest, and in a great many species the individual is effectively a solitary insect while away from the nest. This is certainly true of social wasps, which have no known territoriality, virtually no group foraging, and no demonstrated food-source recruitment, although many swarm-

520

founding polistines lay odor-spot trails to new nest sites (Jeanne 1981a,c).

The internal relations of any nonclonal society are necessarily a mixture of cooperation and conflict. On the one hand, the members have some overlap of interests on account of shared resources and, usually, high genetic relatedness. On the other hand, because they are not genetically identical, their overlap of fitness interests is incomplete. In my view, it is this very tension that makes sociobiology so interesting, and it is the great accomplishment of kin-selection theory (Hamilton 1972, West-Eberhard 1975, Michod 1982) to have shown how we can make some sense of it.

The nest is the stage on which this drama of cooperation and conflict is played, and what takes place offstage is rarely central to the drama. This chapter describes the uses to which nests are put and the interplay of nest structure with function. For a similar approach to the nests of social bees, see Michener (1974: chap. 7).

## FUNCTIONS OF THE NEST

> Wasp nests are constructs, at once light-weight and solid, which serve principally as nurseries for innumerable new generations and have the auxiliary purpose of sheltering the families of artisans which build them. (Saussure 1853–1858:LXIX)

The nest patently originated as a site of brood care and development, or nursery. In this respect, it is very different from the webs of spiders—mentioned above as having an analogous role in the evolution of sociality—which just as patently began as foraging devices. The nest's primary function, then, is to provide a protected microenvironment for the developing brood and its food provisions. Providing food, in turn, is the focus of foraging. Unlike members of many other social groups, such as social ungulates, flocking birds outside of the breeding season, or soldier crabs (*Mictyris* spp.), adult social insects do not forage mainly for themselves but for the brood held in the nest (see Hunt, this volume).

Animal architecture is by no means exempt from the general evolutionary tendency for biological structures to take on new functions (Hansell 1984a). Spider webs, for example, often serve as a substrate for vibratory sexual communication (Barth 1982, Foelix 1982). Comparable shifts of function are also found in the nests of solitary Hymenoptera. For example, while the female of the sphecid wasp genus *Trypoxylon* (subgenus *Trypargilum*) is away foraging, her consort male occupies and guards the newest cell, the physical focus of pair bonding

(Cross et al. 1975, Brockmann 1988). A single empty cell serves as a resting chamber for the female of the eumenine wasp *Calligaster williamsi* (Williams 1919).

Nonetheless, such secondary functions do not appear to be widespread among solitary Hymenoptera, and they probably are rarely important in the nesting economy. In contrast, sociality introduces a new aspect to the nest by virtue of the presence of at least a few interacting adults. Almost unavoidably, the nest thus takes on new roles that may be largely independent of the immediate presence of brood. This is analogous to the way bars, barbershops, and churches have assumed social functions having little to do with refreshment, haircuts, or prayer. There are hints of this point of view in Saussure's statement, quoted above, and in Hansell's (1984a) observation that among aculeates it is only in solitary, presocial, and some primitively eusocial species that the nest shelters only the brood; other species have added specialized structures to shelter the colony as a whole.

Jeanne (1977a) has briefly treated this subject as it applies to eusocial wasps. He defined the primary functions of the nest as those directly related to brood care, and identified four categories of secondary functions: (1) defense against enemies of the brood, (2) maintenance of favorable physical conditions, (3) serving as the main locus of social interactions, and (4) preservation of its own structural integrity. Jeanne's distinction is thus analogous to that between primary and secondary sexual characters. My own concept is much the same, but I find it more useful to treat Jeanne's functions 1, 2, and 4 as primary and to set secondary functions as nearly equivalent to his function 3. As with sexual characters, however, there can be no absolute division between primary and secondary nest functions.

In this chapter, I consider five broad types of secondary functions as they apply to the nests of eusocial wasps. Each of these is also relevant to other eusocial insects, in some of which we can identify particular structural features presumed to have evolved by reference to these or other secondary functions (e.g., Table 15.1).

*Defining colony membership.* As the colony's home, the nest is closely guarded against intrusion by foreign insects. Its perimeter thus sets a sharp boundary to the colony, and residency is nearly equivalent to colony membership. So confident are we of this general principle that in field studies we regularly associate insects collected from the same nest as members of the same colony, even if there is no separate confirmation of such status. Indeed, the term *nestmates* is in general use for what should strictly be called *colonymates*, and I use the latter term only where I wish to emphasize that what is shared is the social group. On the other hand, the frequent use by entomologists of *nest* to signify the colony is confusing and should be discontinued.

**Table 15.1.** Examples of eusocial insect nest features believed to have originated to serve secondary functions

| Taxon | Feature | Secondary Function | Reference |
|---|---|---|---|
| Some higher termites (Termitidae) | Thick-walled royal cell near the nest center | Queen's seat of influence: a sort of bunker | Noirot 1970 |
| Dwarf honey bees (*Apis florea*) | Broadened top of the comb | Communication: platform for communicative dances | Lindauer 1971 |
| Some stingless bees (Apidae: Meliponinae) | Chamber just proximal to entrance tube | Resting area: chamber for guard bees | Wille and Michener 1973 |
| Bumble bees (*Bombus* spp.) | Old cocoons retained in nest | Food storage: honey and pollen | Michener 1974, Alford 1975 |
| Giant honey bees (*Apis dorsata* group) | Deep cells in the upper part of the comb | Food storage: honey | Morse and Laigo 1969, Underwood 1986 |
| Leaf-cutter ants (Myrmicinae: Attini) | Fungus gardens | Food production | Batra and Batra 1979; Weber 1979, 1982 |
| Fungus-gardening termites (Termitidae: Macrotermitinae) | Fungus gardens | Food production | Howse 1970, Batra and Batra 1979 |
| Weaver ants (*Oecophylla smaragdina, Polyrhachis* spp.) | Auxiliary silk bowers | Shelters for symbiotic homopterans | Degen and Gersani 1989; Starr, unpubl. |
| Mound-building ants (*Formica montana*) | Auxiliary bowers of soil and plant matter | Shelters for symbiotic aphids | G. Henderson, unpubl. |
| Fire ants (*Solenopsis* spp.) | Large, temporary exit holes | Escape of sexuals for mating flights | Lofgren et al. 1975, Starr, unpubl. |
| Many higher termites (Termitidae) | Temporary tunnels to the surface, sometimes temporary waiting chambers | Escape of sexuals for mating flights | Nutting 1969 |
| *Macrotermes bellicosus* and some other termites | Peripheral tunnels, sometimes opening at the surface | Nest ventilation | Howse 1970, Noirot 1970 |

*The queen's seat of influence.* Second to being a nursery for the brood, the nest's most important role is perhaps as the seat of the queen. The importance of this function is reflected in studies of polyethism in polistine foundress groups (e.g., West-Eberhard 1969, Jeanne 1972, Yamane 1985, Röseler, this volume), in which the fraction of time spent on the nest is treated as a key variable. Dominant queens consistently spend less time away from the nest than do subordinates. This tendency may be so marked that it serves as a convenient index for recognizing the dominant queen. Similarly, in all eusocial insects it is the workers who undertake most or all off-nest tasks. We can thus regard the nest in some sense as a personal domain in which the queen holds sway.

*Communication.* The signals that pass among nestmates are almost entirely chemical and tactile, so that sight and (airborne) sound have little part in social insect communication (Hölldobler 1977). Given the prominence of tactile signals, we might expect to find that the nest is used to transmit vibrational signals to nestmates. Both adults and larvae of some social wasps use the nest in this way (see Communication and Nest Structure, below).

*A resting area.* Anyone who spends time watching social insects soon realizes that much of the time they do nothing (Wheeler 1957). It is mainly at the nest where they do nothing, or at least nothing in particular, so that the nest is a resting area as well as an area of activity.

*A mating site.* There are two fairly obvious foci for sexual interactions in aculeate hymenopterans, because there are two places where males could expect to find females: at the nest and at food sources. Compared with the sexual behavior of solitary wasps (e.g., Evans 1966a, Alcock et al. 1978, Thornhill and Alcock 1983), that of social wasps is still quite poorly known (e.g., note the paucity of references to the subject by Thornhill and Alcock 1983). Present knowledge suggests that most species mate away from the nest. There is some indirect evidence of at-nest mating in some polistines and within-nest mating in temperate vespines (reviewed by Ross and Carpenter, this volume), and it may be significant that males of tropical vespines and a few polistines and stenogastrines are rarely found away from the nest (Starr, unpubl.). Still, it remains to be seen in which groups, if any, the nest is the primary site of courtship and mating.

This listing by no means exhausts the range of known or putative secondary functions of social insect nests, as indicated in Table 15.1. Not all known secondary functions of nests appear to be significant for most social wasps, and I will comment on two that are not. First, the nests of many social insects serve as a food-storage facility in a manner quite different from the mass provisioning of most solitary aculeates and a few social species, in which the amount needed for complete larval development is served into a single cell along with the egg. Use of the nest as a food-storage facility is especially prominent in the highly eusocial bees (Michener 1974, Roubik 1989), honeypot ants (Wilson 1971: fig. 14-6), and fungus-gardening ants and termites (Sands 1969; Batra and Batra 1979; Weber 1979, 1982). *Polistes* and some other wasps often store droplets of honey in cells (Marchal 1896b, Rau 1928, Evans and West-Eberhard 1970, Strassmann 1979), but only the polistines *Protonectarina sylveirae* and two species of *Brachygastra* are reported to regularly stockpile large quantities of honey (Buysson 1905, Zikán 1951, Naumann 1968, Evans and West-Eberhard 1970, Richards 1978a). I have found *Polybia* nests with cells stuffed with prey, but that is exceptional among wasps (see also Gobbi 1984). The evident reason is that dismembered prey, unlike pollen and some other social insect staples, is not readily preserved.

Second, while the nest clearly has a key defensive role, especially in protecting the brood from predators (Jeanne 1975a), it does not stand as a refuge for individuals threatened away from the nest. When independent-founding polistines are disturbed at the nest, they often retreat to the far side of the comb. Similarly, in social insects with enclosed nests, individuals at the nest entrance or inside the nest often respond to outside disturbance by backing deeper into the nest (e.g., *Pseudochartergus fuscatus*: Jeanne 1970c). In this limited sense, the nest is a refuge. One might further expect a wasp threatened in the vicinity of her nest to flee into it, much as a marmot or fiddler crab runs to its burrow. However, I have never seen that happen, even in species that nest in impregnable hollows or burrows, nor have I found it mentioned in the literature.

## THE ELEMENTS OF NEST STRUCTURE

Here I name and briefly discuss the main features of the nests of social wasps. For detailed analyses of their materials, design, and evolution, see Jeanne (1975a) and Wenzel (this volume).

Unlike the nests of most termites and ants and many social bees, those of all social wasps are free-standing structures formed by the accumulation of material. Many vespines nest in ready-made cavities in the soil. Although they may enlarge the cavity (Spradbery 1973a, Akre et al. 1980, Edwards 1980), such excavation is not itself an enlargement of the nest, but simply gives room for nest building to proceed. As noted by Malyshev (1968), this structural independence of the nest from its environment should allow considerable freedom in the development of the overall structure and final size, and in fact social wasp nests seem relatively unconstrained by the substrate.

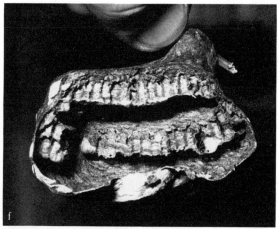

**Fig. 15.1.** Nests of eusocial wasps, illustrating main features and some of the gross variation. (a) Naked comb of *Mischocyttarus labiatus* suspended from a centric petiole (petiole central on comb). (b) Naked comb of *Polistes erythrocephalus* suspended from an eccentric petiole (petiole near edge of comb). (c) Nest of *Ropalidia horni* comprising three combs, each with an eccentric petiole. Although the two larger combs appear coalescent, they are discrete. (d) Nest of undescribed *Parischnogaster* species, a series of cells arranged approximately linearly along a substrate, with no true petiole. (e) Cross-section of *Vespa affinis* nest, showing the parallel combs connected by petioles and surrounded by an envelope (stelocyttarus calyptodomous arrangement). (f) Cross-section of *Brachygastra augusti* nest, without petioles between the parallel combs, which are extensions of the envelope (phragmocyttarus arrangement). Note single hole through the lower comb. (g) External aspect of *Synoeca septentrionalis* nest, the envelope perforated by a single entrance hole. See Jeanne (1975a: fig. 9) and Wenzel (this volume: Fig. 14.47) for the internal structure.

527

The accumulated nesting materials of most social insects can be grouped into four categories: (1) plant matter, (2) mud and feces, (3) silk, and (4) wax and resins. Social wasps build with plant material, mud, or a mixture of the two, often with some admixture of glandular material. Most social wasp species appear to use little or no mud. Rather, the nest consists of plant material chewed into a pulp and in most parts of the nest drawn into a thin sheet, which dries to form a variably strong, flexible material known as carton.

The fundamental unit of all wasp and most bee nests is the cell, the fixed chamber in which an individual develops from an egg into an adult. For present purposes, we can take this as a given. As Jeanne (1977a) has pointed out, though, the very interesting question of why wasps and bees segregate the brood in individual cells, while ants and termites do not, has yet to be answered.

Almost all social wasps compactly arrange a group of cells into a comb by orienting the cells in parallel, with a high degree of wall sharing between adjacent cells (Fig. 15.1a–c; see Fig. 15.1d for an exception). If the number of cells becomes very large, there comes a point at which a single comb is no longer the most compact arrangement. Most species with mature colonies of more than about 100–200 adults (Jeanne 1977a) and with roughly 1000 or more cells characteristically build two or more combs in parallel, most commonly one above the other (Fig. 15.1e–f; Wenzel, this volume: Figs. 14.2, 14.43, 14.54). A few polistines build concentrically arranged combs (Vecht 1966, Evans and West-Eberhard 1970: fig. 91; Wenzel, this volume: Figs. 14.1, 14.38). Adjacent combs are separated by a more-or-less regular wasp-space, a separation sufficient to allow a wasp to walk between them. Such a nest thus has a hierarchical structure, in which the cells are grouped into combs, which are grouped in turn into a parallel set.

It is conceivable that an additional layer might be added to this hierarchy through the construction and use by a single colony of several distinct nests. Many ants and some termites have such polydomous (= polycalic) colonies, but they are unknown among social bees. The nesting arrangements of some wasps could be regarded as polydomous, although none achieves the sort of dispersion that we usually associate with the term (see The Boundary between Colony and Noncolony, below). The flightlessness of ants and termites is probably sufficient to account for this difference. The apparent absence of accurate food-source communication in wasps and most social bees is undoubtedly due to the fact that they fly from one site to another, so that they cannot recruit by means such as odor trails and tandem running (see Hölldobler 1977). It is thus hard to see how social cohesion could be maintained among dispersed nests if newly emerged workers cannot easily learn their locations.

The first (or only) comb may be either sessile—i.e., sitting directly against the substrate—or attached by a slender petiole (stalk or pedicel) (Fig. 15.1a–c). If there are several combs, they often are connected across the wasp-spaces by comparable petioles (Fig. 15.1e). No species of Stenogastrinae builds a petiole, though many achieve a similar effect by building a sessile nest on a slender vine or twig (Fig. 15.1d; Wenzel, this volume: Fig. 14.7).

Vespines and most swarm-founding polistines nest in compact cavities and/or surround the comb(s) with a sheetlike envelope (Fig. 15.1e–g). The envelope is usually continuous except for a single narrow entrance hole (Fig. 15.1g). Many instances of nests with more than one hole are known (Starr 1989b), but only in the genus *Protopolybia* do multi-holed nests appear to be a regular architectural feature (J. Wenzel, unpubl.). An envelope thus sets a very definite perimeter to the nest, with the entrance hole as the point of communication between the exterior and interior. Many social insects alter or plug entrance holes in response to disturbance (e.g., Khoo and Yong 1987) or open temporary new holes for massed mating flights (e.g., Nutting 1969). However, social wasps have never been reported to make such alterations, so the nest's size and form can be treated as nearly constant in the short term.

Even very large nests are usually about as compact overall as substrate conditions will permit. That is to say, combs tend to be roughly circular and of such a number and arrangement that the total nest volume and outer surface area are not much beyond minimal for that number of cells. Such compactness is consistent both with economy of building materials (Jeanne 1975a, 1977a) and with maintenance of social cohesion. Where we find strong departures from this general rule of nest compactness, it draws attention to the social significance of within-colony dispersion of individuals. In a very elongate nest or one that separates into branches or lobes, some sets of individuals may have little physical contact with some others, with probable social consequences. Similarly, it is reasonable to think that any barriers between occupied parts of a nest may affect social cohesion, especially in matters of reproductive competition and other conflicts among nestmates.

## THE BOUNDARY BETWEEN COLONY AND NONCOLONY

As stated above, residency on the nest and colony membership are very nearly one and the same. Given such a distinct criterion, the dividing line between one colony and another should be plain to members of each, and we should expect stable membership, with little drifting of individuals between colonies. This expectation is for the most part upheld in eusocial insects (Wilson 1971). *Polistes* foundresses have an unruly tendency to switch nests (West-Eberhard 1969, Pratte 1980, Gamboa 1981, Noonan 1981, Itô 1984a), but nest switching is almost entirely confined to the earliest part of the colony cycle. The attention given to nest switching is testimony to its uncommonness among social insects, and evidently occasioned Gamboa's (1981:153) remark that "*Polistes metricus* . . . exhibits considerable nest infidelity."

Itô and Higashi (1987) studied a population of wasps under circumstances where a breakdown in colony boundaries might be expected. A very dense aggregation of the independent-founding species *Ropalidia plebeiana*, occupying some thousands of combs, is known to have persisted for several years in one place in New South Wales, Australia. Richards (1978b) had suggested that groups on different combs were not independent colonies but together constituted one or more "supercolonies," in the manner of some ants of the *Formica rufa* species group (e.g., Higashi 1978, Rosengren and Pamilo 1983). Nonetheless, observation of marked individuals showed very little drifting between combs from one day to another (Itô and Higashi 1987).

Given this general identity between the nest and its colony, tolerance of a new individual on the nest is the same as accepting her into the colony, and physical expulsion is a simple, unequivocal means of ostracism. A *Polistes* queen seeking to join an established foundress group, for example, may land nearby and remain at the margins of the nest, sometimes for days, awaiting an opportunity to insinuate or force herself into the colony by taking up physical residence (West-Eberhard 1969, Gamboa et al. 1978).

The converse of this process often takes place in pleometrotic *Polistes* colonies soon after the first workers emerge: active hostility among adult females escalates until the subordinate queens are driven off the nest and out of the colony (Pardi 1942a, 1947; West-Eberhard 1969; Gamboa et al. 1978; Reeve, Röseler, this volume). A comparable event happens in the colony cycle of the swarm-founding *Metapolybia aztecoides* and *M. docilis* (West-Eberhard 1973, 1978b). Fletcher and Blum (1983) suggest as a general rule in social Hymenoptera that it is the workers that do the culling in such situations. Unfortunately, little direct attention has been given to the question of which members of social wasp colonies are responsible for the ousting of supernumerary queens, but the implication has been that, for *Polistes*, the workers remain largely aloof from a battle between queens. The lack of worker involvement in the reduction of queen number would be consistent with our general conception of queen control as relatively direct and physical in these wasps. It seems likely, on the other hand, that the elimination of males from the colony is always carried out by workers.

The periodic violent expulsion of male honey bees by workers is well known (Morse et al. 1967, Free and Williams 1975, Winston 1987), and similar occurrences are known or suggested in some *Polistes* (West-Eberhard 1969, Kasuya 1983a).

Much the same processes of social acceptance and ostracism (through "peripheralization") are known from primates (e.g., Box 1984, Jolly 1985: chap. 15), the difference being that there is no unambiguous physical boundary such as the nest furnishes. Do social insects recognize the nest edge in particular as the boundary? It is hard to believe that they do not, yet I am not aware that the question has been addressed.

Consistent with this hypothesized relationship between nest and colony boundaries is the observation that the process of colony fission, as far as we know, always involves a change of site for at least one of the daughter colonies. We do not find two or more colonies of any species occupying a single nest. However, colony fission need not entail a shift into unfamiliar habitat for either group. For example, new colonies of stingless bees tend to be established well within the mother colony's foraging range (Lindauer 1971, Sakagami 1982, Wille 1983). The "satellite nests" of *Polistes exclamans* represent an elaboration of the colony cycle in which some females leave the nest to found daughter colonies before the usual end of the season (Strassmann 1981b,d). Although these are usually established nearby, it appears that they do not long remain auxiliaries of the mother colony but are soon independent.

Quite a different situation is presented by individual colonies of those *Polistes* and independent-founding *Ropalidia* species that often or characteristically construct a series of separate combs (Jeanne 1979a, Hook and Evans 1982, Kojima 1984b) (Fig. 15.1c). Although these are by definition unequivocal examples of polydomy, I doubt that they have much in common with the polydomy of many ants and termites. Combs within a group are "independent" (Kojima 1984b) in the physical sense that they neither support nor impede each other, but in the examples I have seen they are far from spatially independent. Rather, they are grouped close together and often show quite regular nearest-neighbor distances (Fig. 15.1c). More important, there is no indication that they are socially any more independent of each other than are different combs within a vespine nest, for example, so that the term *satellite combs* (Kojima 1984b) may be misleading. It is fortunate that Strassmann (1981b) has drawn a clear distinction between this sort of polydomy and a satellite nest situation.

Should such a division of the cells among several distinct combs affect social cohesion? If it is possible for wasps to be "associate members" of a colony, this seems like a good place to look for it. To date, this question has only been partially addressed in one species. Where

*Polistes canadensis* builds multiple combs, queens do not restrict their activities to one comb or another and seem to treat the entire nest much as if it consisted of a single comb (Jeanne 1979a). However, this species (where it builds multiple combs) has the anomalous tendency to utilize each comb for only one brood cycle and to keep moving down the line, like the Mad Hatter and his guests, so it makes little sense to generalize from it to other species. The possibility remains that fragmentation into several open combs breaks up the coincident out-lines—from the wasps' point of view—of the nest and the colony.

If this is so, might the extreme elongation of a single open comb have a similar effect? I very much doubt it, just because such nests seem always to be quite small, even where they consist of just one or two rows of cells. Accordingly, there is no part that could be consid-ered an outlying zone, nor is there opportunity for nestmates to be-come strangers to each other.

## COMMUNICATION AND NEST STRUCTURE

The durable cohesion of a social insect colony demands a flow of infor-mation from the queen and brood to the active workers, in order that the colony's status and requirements can be monitored. Except in the simplest colonies, it also requires considerable transfer of information among workers. On a simple level, such exchange is unavoidable, as workers respond to changes in nest features brought about by them-selves and their nestmates. There is no indication, though, that such stigmergic responses (Wilson 1971: chap. 11) go beyond just keeping the nest in order to the point of affecting social organization. It is thus more pertinent to ask whether there is a connection between variation in nest structure and the more direct forms of within-colony communi-cation.

Two types of chemical communication are known in eusocial wasps. At least some vespines have queen pheromones that seem to act much like those of honey bees (Ikan et al. 1969, Landolt et al. 1977, Edwards 1980), and chemical alarm is inferred in some polistines (Jeanne 1981b, 1982b) and vespines (Maschwitz 1964a,b, Edwards 1980, Heath and Landolt 1988). The relative development of such communication is evi-dently related to colony size and complexity, but I see two reasons to doubt any close connection with nest structure.

First, there is no indication that the slow-acting (primer) queen pher-omones cannot be distributed just as thoroughly in one kind of nest as another. If this were not the case we would expect poor queen control in the weaver ants *Oecophylla smaragdina* and *O. longinoda*, whose huge colonies are each dispersed among many discrete nests, often over several trees (Way 1954, Hölldobler and Wilson 1978, Starr, unpubl.). Yet all examined colonies of these species have been monogynous, and worker egg laying is unknown in queenright colonies.

Second, it is not yet known whether airborne alarm pheromones take significantly longer to reach *responsive* colony members in some kinds of nests than in others. The phragmocyttarous nests of *Polybia* and some other genera have more thorough internal partitioning (e.g., Fig. 15.1f; Wenzel, this volume: Fig. 14.48–14.58) than calyptodomous vespine nests (Fig. 15.1e; Wenzel, this volume: Figs. 14.59–14.62), so that a pheromone released at the entrance should take longer to reach the upper combs in the former type of nest. But this may be unimportant if wasps on the upper combs are mostly too young to be effective defenders.

Wenzel (this volume) has pointed out that in phragmocyttarous nests the lowest chamber is often without cells and may even be narrowed to such an extent that it appears unsuited for brood rearing. If this is the mature condition, then such a chamber would seem most likely to serve as a holding area for guard wasps, the most appropriate individuals to respond to alarm pheromone release. It is curious that in the many vespine nests that I have dissected I have never noticed any comparable elaboration of the area just inside the entrance hole.

Although we still know little more about the physical properties of carton and nests than did Henri de Saussure (as quoted above), they are plainly a very good substrate for producing vibrations, unlike mud, silk, or wax. In the course of their normal activities, females of any vigorous *Polistes* colony, for example, can be seen to shake the nest and occasionally to scrape or rub it audibly. Many types of such vibrations would seem well suited for communication among nestmates.

By playing back recordings of returning foragers of a species of stingless bee, Esch (1967) showed that workers at the nest respond to the particular sounds that returning foragers make. Inferences of responses by social wasps to airborne vibrations remain equivocal, in my view. Some polistines react strongly to human whistling (Overal 1985), but it is plausible that this induces vibrations in the nest material, which the wasps then feel. The response of *Vespa orientalis* workers to recordings of larval hunger signals (Ishay and Schwartz 1973) would seem comparable to those noted above for stingless bees, and I know of no direct objection to the evidence. However, the use of (poorly localizable) airborne sound to solicit food makes so little biological sense that I reluct to accept it.

On the other hand, the communication functions of several types of substrate-borne vibrations is now well established for social wasps. The two best studied classes of such signals are the lateral vibrations, or tail wagging, of *Polistes* females (Esch 1971, Gamboa and Dew 1981, Downing and Jeanne 1985, West-Eberhard 1986) and the hunger signals of vespine larvae (Ishay and Landau 1972, Ishay and Schwartz 1973, Es'kov 1977, Ishay 1977a).

Tail wagging (also known as abdominal wagging; see Reeve, this volume) may rattle the nest-comb and may be distinctly audible to a human observer. Because, within foundress groups, most tail wagging is done by the dominant queen (West-Eberhard 1969, 1986; Hermann and Dirks 1975; Gamboa et al. 1978; Strassmann 1981a; Hughes et al. 1987), some authors have inferred a dominance function for the behavior. I find this inference unconvincing. Tail wagging is also known from lone foundresses (Hermann et al. 1975, Gamboa et al. 1978), and foundresses of at least some species continue tail wagging after subordinate foundresses have disappeared (Starr, unpubl.). As several authors (e.g., Pardi 1942a) have noted, tail wagging is closely associated with the inspection of cells and food exchange with larvae, and my own experience with *Polistes annularis* and *P. exclamans* is that it very rarely occurs outside of this context. Present evidence points to a general function as an alerting signal to larvae, much like the antennal drumming of various species (Evans and West-Eberhard 1970, Pratte and Jeanne 1984), rather than as a means of communication between adults. This hypothesis is consistent with data showing that *P. annularis* tail wagging is a more usual part of cell inspection after the first larvae have hatched (Strassmann 1981a) and with the prediction that dominant foundresses inspect much more than others. Jeanne's (1972) observations of a very similar behavior in *Mischocyttarus drewseni* also agree with this interpretation.

This is not to suggest that tail wagging must have only one function or the same function(s) for all species. I merely contend that no strong evidence has yet been presented for any other. Tail wagging evidently transmits a clear signal to everyone on the nest, and it would be odd if it were utilized for only one kind of message. Still less is there reason to think that similar movements reported from vespines (Ishay 1977a, Ross 1982a) must have the same function(s). (For another view of tail wagging, see Reeve, this volume.)

The audible movements of vespine larvae bumping their heads and scraping their mandibles against cell walls unambiguously function in food solicitation from adults (Ishay and Landau 1972, Ishay and Schwartz 1973, Es'kov 1977, Ishay 1977a). These "hunger signals" are thus in apparent functional symmetry with some vibratory signals of adults, such that one conveys a demand for food and the other an offering.

Vibrations in carton would seem extremely well suited for transmitting alarm, and in a few swarm-founding polistines there is evidence for such a role (Chadab 1979b, West-Eberhard 1982b). I therefore find it odd that similar organized tapping, scraping, or buzzing reactions have not been reported from disturbed vespine colonies. West-Eberhard (1969) noted that *Polistes* females may scrape the comb in the course of taking off in a flight attack and suggested that this may serve in alarm communication. However, my observations of several *Polistes* species failed to show that either their threats or attack flight serve to alarm nestmates (Starr 1990).

Unlike chemical signals, vibratory communication cannot very well be indifferent to nest structure. At least in vespine nests, vibrations propagate well within a comb but hardly at all from one comb to another or to the envelope (Ishay 1977a). This property would seem to increase their value by ensuring against the production of noisy, hard-to-localize signals. There is no indication, though, that the nest is in any way elaborated to facilitate vibrational communication. At present the best hypothesis seems to be that such communication has evolved to utilize existing nest features.

We can take this point a step further by saying that no feature of any social wasp nest is yet suggested to have been shaped by the colony's communication needs, comparable to the top platform of an *Apis florea* comb, for example (Lindauer 1971). But let me emphasize that nests seem not to have been examined with this question in mind. Why, for example, do the nests of *Synoeca* species have such a sturdy, regularly ridged envelope (Fig. 15.1g)? I know of no evidence that it amplifies or regulates alarm vibrations (Overal 1982), nor has it been shown that the ridges are needed for structural support.

## CONFLICTS OF GENETIC INTEREST

I have considered above how nest features can influence the colony's struggles against outsiders. Many of these struggles are at least potentially mirrored in the uneasy tolerance among colony-mates with conflicting genetic interests.

A female wasp's most fundamental route to classical fitness (a measure of success in rearing her own offspring) is to monopolize egg laying in the colony. It is well known that co-foundresses of some species undergo an intense struggle over this prize early in the colony cycle (reviewed by Jeanne 1980a, Fletcher and Ross 1985, Röseler, this volume). In addition, a queen may be in conflict with unfertilized workers over the laying of male eggs (Bourke 1988).

The partitioning of the brood-rearing space into cells is unavoidably central to this struggle. In the first place, it strictly limits the number of brood reared at one time. If a queen can keep all cells occupied by her own brood, a nestmate has nowhere to reproduce. Second, the cellular comb keeps the brood stationary, which must give the dominant queen a considerable advantage. The habit among competing foundresses of patrolling the brood-comb and destroying each others' eggs has been noted in several *Polistes* species (Pardi 1942a, Gervet 1964a, West-Eberhard 1969, Hermann and Dirks 1975, Strassmann 1981a) and in *Mischocyttarus drewseni* (Jeanne 1972) (see also Röseler, Spradbery, this volume). Queens appear only to eat newly laid eggs (Heldmann 1936a, West-Eberhard 1969), which suggests that they do not maintain a detailed map of the brood but rely on constant vigilance to detect others' eggs soon after they have been laid.

Effective vigilance probably requires a great deal of patrolling, much of which may come about in the course of other tasks. We could reasonably expect every adult on an ordinary *Polistes* or *Mischocyttarus* nest, for example, to visit every part of the nest and to contact every nestmate in the course of the day's activities. It is difficult for us to keep track of the coverage of most nests, except very small ones, but the linear arrangement of cells in nests of *Parischnogaster jacobsoni* and related species (Stenogastrinae) (Fig. 15.1d) makes it relatively easy to do so. Turillazzi and Pardi (1982) found that a *Parischnogaster nigricans* queen periodically patrols the length of her nest. Turillazzi (1985a:124) further monitored the positions of females on a *P. nigricans* nest at one-minute intervals over five hours and reported that "four distinct zones of preference are evident." However, my own reading of the same data is that the wasps' attention was spread rather evenly along the nest.

It is plain that the reproductive monopoly of many primitively eusocial wasp queens is not easily gained or kept. Sôichi Yamane (cited in West-Eberhard 1986) has suggested that the species-characteristic pattern of comb growth may influence the completeness of the queen's reproductive monopoly. Fig. 15.2 illustrates Yamane's hypothesis in slightly expanded form. If we let the shaded area of new cells be the zone of reproductive opportunity, then the figure shows that two common variables in the nest structure of independent-founding polistines—shape of the comb and the centricity of the petiole—can affect the ease with which a queen can cover this zone. (That she tries to patrol the zone of reproductive opportunity is suggested by the observation that a *P. canadensis* queen—in a population where each nest has one comb [West-Eberhard 1986]—stays mainly in the "shelf" area of new cells and drives others from it.) For either a round or an elongate comb, an eccentric petiole (e.g., Fig. 15.1b; Wenzel, this volume: Fig. 14.26) renders the zone of reproductive opportunity more compact, and hence more amenable to queen vigilance, than does a centric petiole (e.g., Fig. 15.1a; Wenzel, this volume: Fig. 14.27). The most compact zone of all is found on an elongate comb with an eccentric petiole. This illustration is intended to demonstrate one way that structural modifications of the nest can have important social effects. It is perhaps not

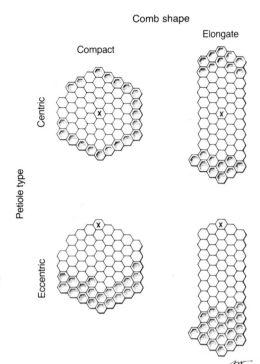

**Fig. 15.2.** Schematic single-comb nests of independent-founding polistines, illustrating the effects of comb shape and petiole placement on the shape of a zone of cells available to receive eggs. Each comb comprises 60 cells, with the 20 newest cells (shaded) empty or with new eggs. *x*, position of the petiole.

farfetched even to see the latter as candidates for multiplier effects (Wilson 1975), in which comb shape and petiole type affect the queen's relative reproductive hegemony, which in turn affects relatedness among workers, thence queen-worker and worker-worker conflicts of interest, the timetable of male production, length of the colony cycle, and even maximum colony size.

Colonies of the more highly eusocial insects become too large and complex to allow the queen to maintain reproductive hegemony by direct physical domination. A simple proof of this is seen in the many higher termites in which the queen is much too swollen with eggs to patrol the nest (e.g., Howse 1970, Wilson 1971: fig. 6-10), and in which she may even be inescapably held in a special chamber constructed by the workers. A less extreme example is found in the vespine wasps, queens of which remain mobile but are characteristically much larger than the workers. It would thus seem that the workers could exclude their queen from reproducing just by narrowing the wasp-space between combs to deny her access. The frequent assertion that in such a situation the queen exerts control by means of the subtler action of pheromones is at best an evasion of the essential question, which is not about communication but *enforcement*. It does little good for a queen to communicate her rank if she is powerless against insubordination. She can maintain control only by manipulating the workers' interests so that they act as enforcers against each other (Ratnieks 1988, Ratnieks and Visscher 1989), and it has been suggested (Starr 1984, Ratnieks 1990) that her own mating habits may serve this end. That is, by mating with several males, rather than just one, a queen decreases average worker-worker relatedness and thus their degree of common genetic interest without affecting her own relatedness to each worker.

It seems usually to be overlooked that larvae are also in competition with each other, as provisions are limited and a better nourished larva has a better expectation of reproductive success. Again, the division of the nest into cells is important. Unlike ant, termite, and some bumble bee larvae, wasp larvae have no direct contact with each other, nor do they have ad libitum access to a common food store, so all competition among them is necessarily mediated by adults. The tactics open to them apparently comprise exactly two behaviors that elicit visits from adults: (1) hunger signals, and (2) giving up attractive fluids in response to tactile solicitation (trophallaxis: see Hunt, this volume). It has yet to be experimentally confirmed that larvae that emit hunger signals less often or that give up less-attractive fluids are visited or fed less than others, and it will likely be difficult to do so. Nonetheless, it fits the known facts and makes biological sense that they should. Can especially demanding or generous larvae then gain a disproportionate share? The striking size uniformity in social wasps and bees among same-caste broodmates suggests that they cannot. More directly, Strassmann's (1981d) suggestion that larvae in the center of a large comb may receive more food than those on the periphery has not been upheld. Strassmann and Orgren (1983) found no such inequality of distribution of food among *Polistes exclamans* larvae.

## CONCLUDING REMARKS

The relationship between nest structure and the colony organization of eusocial insects, as presently understood, is extremely loose. In halictine bees, for example, closely related species usually have very similar nests, even if they are socially very different (Michener 1974). In contrast, closely related species of stenogastrine wasps often build widely divergent nests without any apparent divergence in social behavior (Pardi and Turillazzi 1982, Ohgushi 1986, Turillazzi 1986b). We are left, then, with only the very rough correlation between number of cells, colony size, and overall social complexity. The practical consequence of this—one that has often frustrated me a great deal—is that the social information to be gotten from even an unusual or ornate nest by itself is quite sparse. For social wasps, the metaphor of the nest as "frozen behavior" has limitations even beyond those that Noirot (1970) indicated for termites.

Michener (1974) suggested that physical constraints on nest size may have introduced social limitations in some lineages. Hansell (1987a) applied this general line of thinking to the Stenogastrinae, arguing that an inferior pulp preparation, compared with that of other social wasps, leads to structurally weak nests of sharply limited size, which in turn limits colony size and social complexity. While the known facts do not definitely dispute this thesis, I share Wenzel's (this volume) skepticism of the importance of this physical factor.

Except possibly in the Stenogastrinae—whose profuse architectural variety is most mysterious—the evolution of social wasp nests can be related fairly confidently to the primary function of providing a safe nursery for the brood. This relationship allows only a very limited role for secondary functions in shaping nest structure, in strong contrast to the situation in higher termites (examples in Table 15.1). Does any feature of any social wasp nest show loss of a (primitive) primary function and elaboration for a secondary function? A tentative example is suggested above (see The Elements of Nest Structure) in the empty bottom chamber in some phragmocyttarous nests. I am not aware of any other, though it should be noted that the question has not before been posed in quite this way. In particular, nest structure as the product of worker behavior gives no indication of being influenced by conflicts of interest between queens and workers or between different groups of workers. I believe it is shown that some nest features, especially cells, strongly influence secondary functions and that some other features can be suspected of doing so, but there is not yet any clear indication of feedback from secondary functions to nest structure. I find this conclusion remarkable.

### Acknowledgments

I prepared the first version of this chapter under a Smithsonian postdoctoral fellowship at the National Museum of Natural History. Criticism by George Eickwort and John Wenzel of an earlier version is much appreciated.

## 16

# The Function and Evolution of Exocrine Glands

HOLLY A. DOWNING

Many studies have demonstrated the important role of exocrine gland secretions in insect behavior (for reviews of the literature see Weaver 1978, Bell and Cardé 1984), and these glands and their secretions are prominently involved in many aspects of social wasp life (Landolt and Akre 1979a, Blum 1981, Akre 1982). Among various wasp species, glandular secretions are involved in a diverse array of behaviors, including dominance interactions and the maintenance of the queen's status, recruitment to new nest sites and to food sources, construction and defense of nests, and reproductive behavior. Unfortunately, little is known about the chemical nature or function of glandular secretions in social wasps (Blum 1981, Akre 1982).

The first section of this chapter provides background information about generalized gland structure and what is known of specific locations of exocrine glands in eusocial wasps. The second reviews the literature on the functions and possible evolution of social wasp glandular secretions.

## DESCRIPTION OF EXOCRINE GLANDS

### Gland Morphology

Three types of insect gland cells have been described (Noirot and Quennedey 1974). Class 1 cells are modified epidermal cells that are also involved in the secretion of cuticle (Fig. 16.1). The secretion from this type of gland cell must pass through the gland cell membrane and

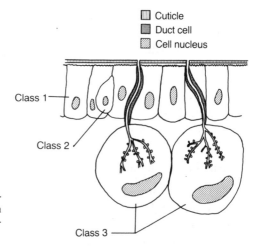

Fig. 16.1. Three classes of insect exocrine gland cells. (Modified from Noirot and Quennedey 1974, Delfino et al. 1979.)

the outer insect cuticle to be released outside the animal. Class 2 and class 3 gland cells are also modified epidermal cells but are not involved in the secretion of cuticle (Fig. 16.1). The secretion produced in a class 2 gland cell passes toward the outside through one or two epidermal cells and then passes through the cuticle in the same way as does secretion from a class 1 cell. Thus far the class 2 gland cell has been found only in termites (Noirot and Quennedey 1974).

Class 3 gland cells have complex branching tubular cavities that are directly connected to the outside surface of the cuticle by highly modified duct or canal cells (Delfino et al. 1979). The secretory cell's canals and the canal of the duct cell(s) are lined with cuticle, which is continuous with that on the outside of the body (Noirot and Quennedey 1974, Delfino et al. 1979). The glandular secretion passes into the lumen of the secretory cell and from there passes through the canal of the duct cell to the outside (Fig. 16.1). In more elaborate situations, there may be two secretory cells connected to the duct cell, and, presumably, each of these adds one or more components to the secreted product (Noirot and Quennedey 1974).

Class 3 gland cells with their associated duct cells may be found singly throughout many regions of the body, such as around joints and in the head, or they may be clustered in groups. Scattered individual gland cells may provide secretions that lubricate or in some other way maintain the integrity of the insect cuticle. The clusters of class 3 gland cells constitute exocrine glands that produce secretions with diverse functions. The cuticle associated with these glands is often modified in some way.

The degree of cuticular modification associated with a gland varies greatly (Landolt and Akre 1979a, Jeanne et al. 1983, Downing et al. 1985). For glands that open into the oral cavity or reproductive tract, the cuticle appears unmodified and simply lines the common ducts and reservior to the gland cells (Landolt and Akre 1979b, Hermann and Blum 1981). The cuticle associated with other exocrine glands, however, may be modified into ridges, scales, or hairs that presumably increase the surface area for volatilization or application of the secretion. Associated hairs and scales may be clustered into a brushlike structure that may be used by the wasp to apply the glandular secretion to the nest or other substrate surface.

## Gland Locations and Functions

The glands described in this section are known to occur in eusocial wasps, although they all do not occur in each species. Where possible, species or larger taxa that lack a particular gland are noted. Very little is known about the presence and function of glands in stenogastrines and eusocial sphecids (i.e., *Microstigmus* and, perhaps, *Arpactophilus*). Thus this chapter deals mainly with the Vespinae and Polistinae. As more researchers investigate the glands of stenogastrines and sphecids, some interesting differences from the patterns presented here may emerge.

### Glands of the Head

Eight pairs of exocrine glands occur in the head, all of them associated with the mouthparts or oral cavity (Fig. 16.2) (Landolt and Akre 1979a).

ECTAL MANDIBULAR GLANDS    In the paired ectal mandibular glands (= mandibular gland I: Spradbery 1973a) each member is associated with a mandible. Each of these glands consists of up to 70 ducted gland cells (Spradbery 1973a, Landolt and Akre 1979a) opening into a small reservoir, which in turn empties (via a sclerotized duct) just above a small brush of hairs on the anterior proximal surface of each mandible

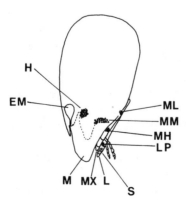

**Fig. 16.2.** Location of cephalic exocrine glands in eusocial wasps. A sagittal section through the head is shown, so that paired glands are represented singly. EM, ectal mandibular gland; MM, mesal mandibular gland; H, hypopharyngeal gland; LP, labial palp gland; ML, maxillary-labial gland; MH, maxillary-hypopharyngeal gland; S, sublingual gland. The endostipital gland is not shown. M, mandible; MX, maxilla; L, labium.

(Nedel 1960, Downing and Jeanne 1982). The ectal mandibular glands are found in at least nine *Vespula* species, three *Dolichovespula* species, *Vespa crabro*, five *Polistes* species, *Mischocyttarus flavitarsis*, three *Polybia* species, *Protopolybia minutissima*, *Apoica pallida*, and *Synoeca surinama* (Cruz-Landim and Pimenta Saenz 1972, Landolt and Akre 1979a, Downing and Jeanne 1982, Wenzel 1987a).

The average volume of reservoirs of these glands in *Vespula* queens is 0.03–0.10 mm$^3$, and in *Dolichovespula* queens, 0.01–0.04 mm$^3$ (Landolt and Akre 1979a). In a comparison between *Polistes fuscatus* queens and their subordinates at five times during the nesting season, no significant differences in reservoir size were found (Downing and Jeanne 1983). A survey of three independent-founding polistine species, five swarm-founding polistine species (*Polybia*, *Protopolybia*, and *Apoica* spp.), and two eumenine species revealed that these glands are very similar in appearance in all of these wasps (Cruz-Landim and Pimenta Saenz 1972).

Because these glands are relatively small compared with similar glands in bees and ants, and because they were thought to empty into the mouth, past authors have suggested that the ectal mandibular gland secretion has no social function in wasps (Hermann et al. 1971, Spradbery 1973a). Among *Polistes* females, however, the glands may be involved in dominance interactions (see Dominance and Queen Control, below; see also Downing and Jeanne 1982).

The ectal mandibular glands are large in males of *Polistes fuscatus* and *P. major* (Landolt and Akre 1979a, Wenzel 1987a). *Polistes major castaneicolor* males defend territories and mark perches within them by rubbing their clypeus and mandibles as well as their sterna on the substrate (Wenzel 1987a), suggesting that the ectal mandibular gland secretion has a possible role in territorial defense.

MESAL MANDIBULAR GLANDS The mesal mandibular glands (= mandibular gland II: Spradbery 1973a) each consist of a cluster of class 3 gland cells whose ducts are attached to the membrane connecting the mandible and maxilla along the mesal edge of the base of each mandible. Since these glands appear to be homologous to the postgenal glands in honey bees (Spradbery 1973a), past authors referring to the postgenal glands in wasps are probably referring to the mesal mandibular glands (Landolt and Akre 1979a). The mesal mandibular glands have been found in all female wasps surveyed to date (i.e., species of *Vespa*, *Vespula*, *Dolichovespula*, *Polistes*, and *Mischocyttarus*) and in *D. arctica* males (Landolt and Akre 1979a).

Landolt and Akre (1979a) found the cells of this gland to be larger on average in *Dolichovespula* queens (0.04–0.08 mm$^3$) than in *Vespula* queens (0.02–0.05 mm$^3$), which may simply be related to specific differences in overall body size. In *Polistes fuscatus*, the average gland cell size and color were found to vary with both the season and female rank in a dominance hierarchy (Downing and Jeanne 1983). Maximum cell size was found to average 0.038 mm in diameter and was significantly greater in dominant than in subordinate females in postemergence colonies (i.e., after the first workers had emerged).

Enlarged (and presumably active) gland cells in nesting females range from tan to black-brown, while in nonnesting females they range from white to tan (Downing and Jeanne 1983). Presumably the color changes because the secretory product accumulates in the cell vesicles during synthesis. The dark color of the cells matches that of the nest, which may indicate that the secretion is a component of building material.

HYPOPHARYNGEAL GLANDS   The hypopharyngeal glands are two relatively large clusters of class 3 gland cells (> 50 cells) located on the proximal end of the gnathal pouch of the hypopharynx (Landolt and Akre 1979a). The ducts open to the outer edge of the gnathal pouch. As with the previously described glands, the hypopharyngeal glands have been found in all eusocial wasp species surveyed thus far (Cruz-Landim and Pimenta Saenz 1972, Landolt and Akre 1979a, Downing and Jeanne 1983).

The average cell diameter for these glands in wasps surveyed by Landolt and Akre (1979a) was between 0.03 and 0.05 mm. *Vespula* queens tended to have larger gland cells than *Dolichovespula* queens. Among *Polistes fuscatus* females surveyed at five times during the nesting season, the average cell size in the hypopharyngeal glands (cell diameter = 0.04–0.05 mm) and, therefore, secretory activity showed little difference between dominant and subordinate individuals (Downing and Jeanne 1983). Landolt and Akre (1979a) found these glands to be active before colony founding in *Vespula* and *Dolichovespula* queens, suggesting that the glands do not necessarily function in a social context. However, the hypopharyngeal glands of honey bees produce food for larvae, and it has been suggested that during the nesting season these glands might serve a similar function among eusocial wasps (Spradbery 1973a).

LABIAL PALP GLANDS   A pair of glands containing 5–15 class 3 gland cells is located at the bases of the labial palps in females of all the wasp species studied by Landolt and Akre (1979a) (*Vespula, Dolichovespula, Vespa, Mischocyttarus,* and *Polistes* spp.). However, these glands were not found in *Dolichovespula arctica* males. The secretory cells are about 0.03 mm in diameter, with the duct cells opening into an area between the salivarium and labial prementum (Landolt and Akre 1979a). The glandular secretions probably pass out onto the anterior surface of the labium, but no function for these glands has yet been suggested.

MAXILLARY-LABIAL GLANDS    The maxillary-labial glands are two clusters of class 3 gland cells located along the maxillary-labial membrane that parallels the stipes (Landolt and Akre 1979a). These glands occur in all of the wasp queens Landolt and Akre (1979a) surveyed (see Labial Palp Glands) and in *D. arctica* males. Although the secretory cells of glands in queens collected during summer are yellow, these cells are white in the glands of queens collected at other times. Also, these glands are larger in *Dolichovespula* queens (0.5–0.7 mm long) than in *Vespula* queens (0.3–0.5 mm long) (Landolt and Akre 1979a). Again, no function has been postulated for these glands.

MAXILLARY-HYPOPHARYNGEAL GLANDS    The paired maxillary-hypopharyngeal glands consist of class 3 cells that are found in the membranous region between the maxillae and hypopharynx. Like the maxillary-labial glands, these two glands lie along the stipes (Landolt and Akre 1979a). All species surveyed by Landolt and Akre (1979a) have these glands, although they were not found in *Vespula pensylvanica* workers, in *D. arctica* males, or in all the individuals of the remaining species studied. The secretory cells are about 0.03 mm in diameter. No function for these glands has been suggested.

SUBLINGUAL GLANDS    The pair of small sublingual glands is found at the base of the ligula between the paraglossae of the labium in females of most of the species surveyed by Landolt and Akre (1979a). The only species found to lack these glands was *Vespula squamosa*. Each gland consists of 10–30 class 3 cells approximately 0.05 mm in diameter. No function is known for these glands.

ENDOSTIPITAL GLANDS    The endostipital glands are each composed of five to ten class 3 gland cells and are located at the proximal end of the stipes of each maxilla (Landolt and Akre 1979a). These glands have been found only in some *Vespula pensylvanica* and *Polistes fuscatus* queens (Landolt and Akre 1979a), and no function has been suggested.

## Thoracic Gland

The thorax contains only one gland. The thoracic gland (also known as the labial or salivary gland) has four lobes (Fig. 16.3); one pair of lobes lies laterally beneath both sides of the pronotum (prothoracic lobes), and the second pair lies ventrally among the midline (mesothoracic lobes). This gland has been found in all eusocial wasps surveyed (Cruz-Landim and Pimenta Saenz 1972, Landolt and Akre 1979a, Downing and Jeanne 1983).

The thoracic gland is made up of multicellular acini that empty into a network of collecting ductules, which then unite to form larger collecting ducts. These collecting ducts in turn unite to form a single duct that empties at the base of the salivarium (Deleurance 1955b, Landolt and

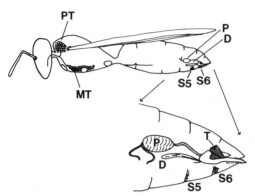

**Fig. 16.3.** Location of thoracic and abdominal exocrine glands in female eusocial wasps. Sagittal sections are shown, so that paired glands are represented singly (except for the poison gland tubules). PT, prothoracic lobes of the thoracic gland; MT, mesothoracic lobes of the thoracic gland; S5, fifth sternal gland; S6, sixth sternal gland; P, poison gland; D, Dufour's gland; T, seventh (gastral) tergal gland. Shaded area indicates the seventh tergum. (Portions based on Landolt and Akre 1979a, Billen 1987.)

Akre 1979b). Each acinus consists of a single large cell surrounded by six to nine parietal cells. The associated ductules are ringed by duct cells, which are in turn surrounded by another layer of cells, the extra-acinus ductule cells (Landolt and Akre 1979b).

The overall dimensions of the thoracic gland vary positively with wasp body size (Landolt and Akre 1979a). In a study of cell size differences in both pairs of thoracic gland lobes of *P. fuscatus*, Downing and Jeanne (1983) found that changes in average cell size indicate thoracic gland activity throughout the nesting cycle, with a decline in cell activity concurrent with the decline in nest construction behavior. These findings suggest that this gland's secretion may be used in nest construction. No significant trends in activity differences among individuals were found in this study, except that the acini of the mesothoracic lobe in dominant females were larger (indicating greater cell activity) than those of subordinate females in July and August. Because Downing and Jeanne (1983) found that the average cell size begins to decline before the peak in reproductive brood production, Deleurance's (1955b) suggestion that the gland provides critical nutrients for developing larvae is unsupported.

## Glands of the Abdomen

STERNAL GLANDS   Both male and female social wasps have glands along the anterior or posterior margins of various abdominal sterna and terga (Fig. 16.3) (Vecht 1968, Richards 1971, Landolt and Akre 1979a, Turillazzi 1979, Post and Jeanne 1980, Jeanne et al. 1983, Downing et al. 1985). These glands typically consist of clusters of class 3 gland cells but may also include an area of enlarged epidermal cells (class 1 gland cells) presumably involved in secretion (Jeanne et al. 1983, Downing et al. 1985). Although many of the sterna and terga in some wasps have small clusters of gland cells (Landolt and Akre 1979a, Turillazzi 1979, Post and Jeanne 1983c, Downing et al. 1985), most species have large, clearly discernible glands only on the last two (in females) or three (in males) sterna of the abdomen (Fig. 16.3). These larger glands are often associated with cuticular modifications such as brushes of hairs or scales.

The fifth gastral sternal gland (also known as Richards' gland or the sixth abdominal sternal gland) (Richards 1971) varies in occurrence and extent of development among the eusocial wasps. Whereas it is absent in *Angiopolybia, Agelaia* (= *Stelopolybia*), species of the *Vespula vulgaris* group, and all *Dolichovespula* species surveyed (Landolt and Akre 1979a), it is present and well developed (> 100 cells) in other species. Both males and females of most surveyed eusocial vespid species possess this gland in some form (Tables 16.1, 16.2). Among *Pseudopolybia* species the intersegmental membrane associated with this gland is greatly enlarged to form a reservoir (Jeanne and Post 1982, Jeanne et al. 1983).

In female wasps, the fifth sternal gland is a narrow to broad band of class 3 and, in some species of *Mischocyttarus* and *Polistes*, class 1 gland cells. In a study of average cell diameter for this gland in *P. fuscatus*, Downing and Jeanne (1983) found that at the start of the nesting season (May) there was little variation among 20 females from 14 nests ($\bar{X} \pm SD = 0.034 \pm 0.005$ mm). After the first workers emerged, however, the average cell dimensions were seen to vary with rank. Dominant queens' cell size remained relatively constant (less than 2% change from the average cell size in May), while cell size of subordinates dropped approximately 20% by August. Fall-collected females exhibited gland cell diameters similar to those of August-collected subordinates, indicating that the fifth sternal gland is relatively inactive in subordinates and overwintering females.

Four types of cuticular modifications (aside from simple pores) are associated with the duct openings of the fifth sternal gland (Tables 16.1, 16.2): (1) a bed of raised scalelike projections, (2) a narrow, transverse groove (*Polybia* spp. only: Jeanne and Post 1982), (3) a brush of hairs or setae arranged in a broad band across the fifth gastral sternum (*Apoica*: Jeanne et al. 1983; male *Mischocyttarus mexicanus*: Post and Jeanne 1982b), (4) reticulated ridges surrounding pits into which groups of gland ducts open (*Occipitalia sulcata* only: Jeanne et al. 1983). The scalelike projections and setae may occur in combination.

Males and females of the same species are not always equivalent in terms of the presence (Tables 16.1, 16.2) or exact location of the fifth sternal gland. In particular, males of the independent-founding polistine genera *Polistes, Belonogaster*, and *Parapolybia* have gland cells on

**Table 16.1.**  Sternal glands found in females or males of swarm-founding species of Polistinae

| Species[a] | Gastral sternum no.[b] | | | | | Reference[c] |
|---|---|---|---|---|---|---|
| | III | IV | V | VI | VII | |
| *Agelaia areata*[d,e] (similar for three congeners) | −,? | −,− | −,− | p,3 | | 2 |
| *Angiopolybia pallens* | −,− | −,− | −,− | −,− | | 2 |
| *Apoica pallida* (similar for a congener) | −,? | −,− | B,3 | B,3 | | 2 |
| male: *A. pallida* | | −,− | B,3 | B,3 | −,? | 3 |
| *Brachygastra augusti* (similar for two congeners) | −,− | −,− | S,3 | −,− | | 2 |
| male: *B. augusti* | | −,− | p,3 | p,3 | p,3 | 3 |
| *Chartergellus communis* | −,? | −,? | −,? | −,? | | 2 |
| *Charterginus fulvus* | −,? | −,? | S,? | −,? | | 2 |
| *Chartergus metanotalis* (similar for a congener) | ?,? | −,? | p,? | −,? | | 2 |
| male: *C. chartarius* | | −,? | p,− | −,? | −,? | 3 |
| *Clypearia apicipennis* | ?,? | −,− | S,3 | −,− | | 2 |
| *Epipona tatua* | −,? | −,? | p,? | −,? | | 2 |
| *Leipomeles dorsata* | −,− | −,− | −,− | −,− | | 2 |
| *Metapolybia docilis* | −,− | −,− | S,3 | −,− | | 2 |
| male: *M. docilis* | | −,− | S,3 | −,− | −,− | 3 |
| *Nectarinella championi* | −,? | −,? | −,? | −,? | | 2 |
| *Occipitalia sulcata* | ?,? | −,− | R,3 | −,− | | 2 |
| *Parachartergus fraternus* | −,− | −,− | −,− | −,− | | 2 |
| *Polybia chrysothorax* (similar for 11 congeners) | ?,? | −,− | G/S,3 | −,− | | 1 |
| male: *P. simillima* (similar for two congeners) | | −,− | S,3 | −,− | −,− | 3 |
| *Polybioides tabidus* | −,? | −,? | −,? | −,? | | 2 |
| *Protopolybia alvarengai* (similar for three congeners) | −,− | −,− | S,3 | −,− | | 2 |
| *Pseudochartergus fuscatus* | ?,? | −,? | S,? | −,? | | 2 |
| *Pseudopolybia compressa*[d] (similar for a congener) | −,− | −,− | S,3 | −,− | | 2 |
| *Ropalidia montana* (similar for three other swarming congeners) | −,− | −,− | −,− | B,− | | 2 |
| male: *R. montana* | | −,− | p,3 | p,3 | −,1 | 3 |
| *Synoeca surinama*[d] (similar for a congener) | −,? | −,− | p,3 | −,− | | 2 |

[a]Unless otherwise noted, the information pertains to females.

[b]The first entry for each gastral sternum indicates the type of cuticular modification associated with the gland; the second gives the type of gland cells present. The types of cuticular modification are: B, brush of hairs or setae; G, transverse groove into which ducts open; p, pores only; R, reticulated ridges; S, raised scales. A slash (/) indicates that both types are present within a genus. 1, class 1 gland cells; 3, class 3 gland cells. −, either no cuticular modification or no gland cells are present; ?, not examined.

[c]1, Jeanne and Post 1982; 2, Jeanne et al. 1983; 3, Downing et al. 1985.

[d]Males of at least one species in the genus have been studied, and no sternal glands were found.

[e]Formerly *Stelopolybia areata*.

**Table 16.2.** Sternal glands found in females or males of independent-founding species of eusocial Vespidae

| Species[a] | Gastral sternum no.[b] | | | | | Reference[c] |
|---|---|---|---|---|---|---|
| | III | IV | V | VI | VII | |
| **Polistinae** | | | | | | |
| Belonogaster grisea (similar for a congener) | −,? | −,− | −,− | B,b | | 4 |
| male: B. grisea | | ?,? | p,b | p,b | p,b | 6 |
| Mischocyttarus mexicanus (similar for two congeners) | ?,− | ?,− | −,b | B,b | | 4 |
| male: M. mexicanus (similar for two congeners) | | ?,? | B/p,b | B/p,b | p,1 | 3 |
| Parapolybia indica (similar for a congener) | ?,? | −,− | −,− | B,b | | 4 |
| male: P. indica | | ?,? | p,3 | p,3 | p,b | 6 |
| Polistes fuscatus (similar for four congeners) | −,− | −,− | p,b | B,b | | 4, 2 |
| male: P. (subgenus Fuscopolistes) fuscatus (similar for nine other subgenus Fuscopolistes and subgenus Polistes spp.) | | p,b | p,b | p,b | C/p,b | 5, 2 |
| male: P. (subgenus Aphanilopterus) annularis (similar for six other subgenus Aphanilopterus spp.) | | −,− | −,− | −,− | B,b | 6 |
| male: P. cinerascens | | p,? | p,? | p,? | p,? | 6 |
| male: P. jadwigae | | ?,? | p,b | p,b | F,b | 6 |
| male: P. mandarinus | | ?,? | −,− | p,b | p,b | 6 |
| Ropalidia horni | −,− | −,− | −/p,3 | B,b | | 4 |
| **Vespinae** | | | | | | |
| Dolichovespula arctica (similar for two congeners) | −,− | −,− | −,− | B,3 | | 1 |
| Vespa crabro (similar for a congener) | −,− | −,− | p,3 | B,3 | | 1 |
| Vespula squamosa (similar for four congeners) | −,− | −,− | p,3 | B,3 | | 1 |
| Vespula vulgaris (similar for three congeners within the V. vulgaris species group) | −,− | −,− | −,− | B,3 | | 1 |

[a]Unless otherwise noted, the information pertains to females.

[b]The first entry for each gastral sternum indicates the type of cuticular modification associated with the gland; the second gives the type of gland cells present. The types of cuticular modification are: B, brush of hairs or setae; C, combination of scales and setae; F, fingerlike projections; p, pores only. A slash (/) indicates that both types are present within a genus. 1, class 1 gland cells; 3, class three gland cells; b, both types are present. −, either no cuticular modification or no gland cells are present; ?, not examined.

[c]1, Landolt and Akre 1979a; 2, Turillazzi 1979; 3, Post and Jeanne 1982b; 4, Jeanne et al. 1983; 5, Post and Jeanne 1983c; 6, Downing et al. 1985.

the posterior one-half to two-thirds of the sternum rather than along the anterior margin, as in females (Post and Jeanne 1982b, 1983c; Downing et al. 1985). In addition, male sternal glands usually lack the cuticular modifications typical of female glands. In only a few cases are the glands associated with a brush of setae or scales in males (Tables 16.1, 16.2).

The sixth gastral sternal gland (also known as van der Vecht's gland or seventh abdominal sternal gland) is found in females of all vespine species surveyed by Landolt and Akre (1979a) (*Vespula* spp., *Dolichovespula* spp., and *Vespa crabro*), as well as in all independent-founding polistine species (*Polistes* spp., *Mischocyttarus* spp., *Parapolybia* spp., *Ropalidia* spp., and *Belonogaster* spp.) (Post and Jeanne 1980, Jeanne et al. 1983). Most swarm-founding polistines lack this gland, except species of *Agelaia* and *Apoica*, in which the gland appears identical to that on the fifth sternum (Jeanne et al. 1983).

In most species, the sixth gastral sternal gland, when present, consists of two clusters of gland cells on either side of the midline of the anterior margin of the sixth sternum (Delfino et al. 1979, Landolt and Akre 1979a, Post and Jeanne 1980, Jeanne and Post 1982). *Apoica, Belonogaster,* and *Agelaia* have glands that form a band across the sternum (Jeanne et al. 1983). In other species, Landolt and Akre (1979a) have found considerable individual variation, with rare specimens having a band of cells rather than two clusters.

Males of *Polistes* (except subgenus *Aphanilopterus*), *Mischocyttarus,* and about half of the swarm-founding polistine wasps surveyed to date possess a sixth sternal gland (Tables 16.1, 16.2). As with the fifth sternal gland, males often have glands similar to those of conspecific females. On the other hand, only males of *Brachygastra* have the sixth sternal gland. Again, males of independent-founding polistine species, except *Mischocyttarus,* have glands that cover the posterior one-half to two-thirds of the sternum.

The actively secreting class 3 gland cells of the sixth sternal gland have well-developed rough endoplasmic reticulum filling much of the cytoplasm, suggesting that proteins constitute at least part of this gland's secretion (Delfino et al. 1982). This gland's secretory activity remains high in females throughout the nesting season, but there are no signs of activity in overwintering females (Delfino et al. 1982, Downing and Jeanne 1983). Solitary foundresses of *P. fuscatus* have significantly larger average cell diameters than do co-foundresses in multifoundress nests during the late preemergence stage (before workers emerge), and queens have larger cell diameters than their subordinates during early postemergence (Downing and Jeanne 1983).

Setae are the only type of cuticular modification aside from simple pores that is associated with the sixth sternal gland (in both sexes the setae are usually arranged in the form of a brush; Tables 16.1, 16.2) (Downing et al. 1985). In males of most species this gland, if present, lacks any kind of cuticular modification.

Males have one more gastral sternum visible externally than do females, and this frequently has a gland associated with it (Turillazzi 1979; Post and Jeanne 1982b, 1983c; Downing et al. 1985). Interestingly, with the exception of *Ropalidia montana* and *Brachygastra augusti*, males of swarm-founding species lack the seventh gastral sternal gland (eighth abdominal sternal gland) whereas those of independent-founding species appear without exception to have it (Downing et al. 1985). In all species, this gland is located along the posterior margin of the sternum and usually consists of both class 1 and class 3 gland cells (Turillazzi 1979; Post and Jeanne 1982b, 1983c; Downing et al. 1985). Males of the subgenus *Aphanilopterus* (genus *Polistes*) have setae associated with this gland, while males of other studied *Polistes* subgenera have the seventh sterna concave posteriorly and frequently devoid of setae. Class 3 gland cells are clustered in the adjacent lateral areas. *Polistes jadwigae* is exceptional in that the lateral regions each have a fingerlike, outward projection of cuticle. The ducts to gland cells open at the bases of these projections (Downing et al. 1985).

TERGAL GLAND    Most, but not all, female Vespinae have a pair of gland cell clusters on the seventh gastral tergum, but little is known about this gland (Landolt and Akre 1979a).

POISON GLAND    The poison gland is a modified accessory gland to the reproductive system (Hermann and Blum 1981). Found in all female aculeate hymenopterans, it has been described in detail for honey bees and various species of ants and wasps (Snodgrass 1956, Landolt and Akre 1979a, Billen 1987). Typically, gland cells are found along two long tubules (filaments) and along a common duct with which the filaments connect. The common duct in turn empties into a large musculated reservoir (Fig. 16.3) (Hunt and Hermann 1970, Landolt and Akre 1979a, Hermann and Blum 1981). The reservoir pumps venom out through a duct to the base of the sting bulb (Billen 1987).

Venom consists of, among other things, phospholipases that act as antigens, lyse cells, and cause pain and toxicity. The enzyme hyaluronidase breaks down connective tissue and thus enhances the spread of venom through the tissue. Vespid venoms also contain neurotoxins, which can induce rapid, short-term paralysis by reducing the sodium current in the axon and hence stopping the transmission of nerve action potentials, as well as kinin peptides, which increase vascular permeability and affect smooth muscle contraction. Various venom micromolecules such as histamine, 5-hydroxytryptamine, acetylcholine, and dopamine cause itching, vasodilation, pain, and increased vascular permeability. For a more detailed account of these and other venom components in wasps, see Schmidt (1982).

Venoms are highly diverse and thus broad generalizations about hymenopteran taxonomic groups and the constituents of their venoms are difficult to make (Schmidt et al. 1980). Yet in a detailed study of specific component markers, genetic relationships between two species and their hybrid have been confirmed in fire ants by analyzing their venom constituents using gas chromatography (Ross et al. 1987). Thus it is probable that further detailed analyses of the components of this gland (and others) will contribute to our understanding of genetic relationships between and within species, and perhaps among higher taxonomic groups of Hymenoptera as well.

DUFOUR'S GLAND   The Dufour's gland is found in all eusocial wasps (Landolt and Akre 1979a). It lines a thin-walled sac (1.5–3.2 mm long and approximately 0.2 mm wide) into which a yellow oily substance is secreted (Landolt and Akre 1979a, Downing and Jeanne 1983). This reservoir empties into a duct, which opens onto the dorsal vaginal wall, near the base of the sting (Billen 1987). When the gland is inactive, the reservoir is opaque, white, and empty. This condition was observed by Landolt and Akre (1979a) in *Vespula, Dolichovespula,* and *Polistes* queens just emerging from winter diapause, but with the eclosion of workers the glands in queens became active (increasing in width and filling with the yellow secretion). In queens collected in late summer, the glands again appeared to be inactive, with only a minor residue of yellow material remaining (Landolt and Akre 1979a). In *P. fuscatus,* Downing and Jeanne (1983) found that the gland became deepest yellow in the dominant females of co-foundress associations during the late preemergence stage. At other times this gland showed little indication of secretory activity, regardless of the rank of an individual. Glands in nonnesting wasps maintained in pairs secrete more actively than do those of solitary wasps (Downing and Jeanne 1983).

The function of Dufour's gland varies considerably in ants (Wilson 1971, Billen 1987) and bees (Batra 1969; Lello 1971, 1976); little is known of its function in wasps. Spradbery (1973a) has suggested that it produces a sting lubricant, and Wigglesworth (1972) has suggested that it provides a coating or glue for attaching eggs to cells in the nest. The activity cycle of the gland does not support either of these hypotheses (Landolt and Akre 1979a, Downing and Jeanne 1983). In Stenogastrinae, where it is well developed, the Dufour's gland is involved in care of the brood (Jacobson 1935, Turillazzi, this volume).

In summary, there is a great deal of variability in gland presence and location among species and between the sexes in eusocial wasps. Generally, most wasps have at least ectal and mesal mandibular glands, a hypopharyngeal gland, a thoracic gland, and various sternal glands, and females invariably have both the Dufour's and poison glands (which are associated with the female reproductive tract). The remaining five oral glands occur in only certain species. For a summary of the glands, their distributions, and possible functions in eusocial wasps and other eusocial Hymenoptera, see Tables 16.3 and 16.4.

## PHEROMONES AND GLAND EVOLUTION

### Dominance and Queen Control

Most research on queen pheromones in social wasps has been done on the Vespinae, because among these species the effect that the queen's presence has on workers is quite striking (Spradbery, this volume). In some species of *Vespa*, workers gather around the queen and lick her body (Ishay 1964, Ishay and Schwarz 1965, Edwards 1980, Matsuura 1984). If the queen is removed, the workers of *Vespa*, *Vespula*, and *Dolichovespula* grow restless and cease foraging and nest construction, but calm down and resume work if the queen is replaced (Hölldobler 1977, Jeanne 1980a, Matsuura 1984). Since these behavioral changes extend beyond those wasps that come in direct contact with the queen, it is probable that the queens release pheromones that affect worker behavior. Indeed, a queen pheromone of *Vespa orientalis* has been isolated and identified as δ-*n*-hexadecalactone, which has the following chemical structure:

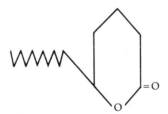

(Ikan et al. 1969). There are about 6 μg of the pheromone in the head of each queen, but as yet the glandular source has not been identified (Blum and Brand 1972). In honey bees, the ectal mandibular gland is the major source of queen pheromone; this gland may also be the source of queen pheromone in *V. orientalis*, because it is similar in its location and general structure to that of honey bees (Brian 1979). δ-*n*-Hexadecalactone is highly attractive to *V. orientalis* workers, appears to mediate the initiation of reproductive cells, and affects other aspects of

**Table 16.3.** Summary of exocrine glands and gland functions in eusocial vespids and sphecids

| Gland | Type[a] | Distribution | Function[b] | Reference[c] |
|---|---|---|---|---|
| Ectal mandibular | 3,R | All taxa examined | Dominance interactions(?), male calling behavior(?) | 7,14,22 |
| Mesal mandibular | 3 | All taxa examined | Construction(?), digestion(?) | 7,14 |
| Hypopharyngeal | 3 | All taxa examined | — | 7 |
| Labial palp | 3 | Females: all taxa examined | — | 7 |
| Maxillary | | | | |
| labial | 3 | All taxa examined | — | 7 |
| hypopharyngeal | 3 | Some females: all taxa examined | — | 7 |
| Sublingual | 3 | All taxa examined except *Vespula squamosa* | — | 7 |
| Endostipital | 3 | Some females of *Vespula pensylvanica* and *Polistes fuscatus* | — | 7 |
| Thoracic | 3 | All taxa examined | Construction(?), digestion(?) | 1,7,14 |
| Fifth sternal | 3,b | Most taxa[d] | Trail pheromone, dominance interactions(?) | 8,9,13 |
| Sixth sternal | 3,b | Most taxa[d] | Ant repellent | 4,11,15 |
| Seventh sternal | 3 or 1,b | Males: most taxa[d] | Perch marking, calling females(?) | 10,19 |
| Third tergal | 3 | Male Stenogastrinae | Territorial display(?) | 17 |
| Sixth tergal | — | *Microstigmus* | Construction | 3 |
| Seventh tergal | 3 | Most female Vespinae | — | 7 |
| Poison | 1,R | Females: all taxa | Defense, female sex attractant | 2,7,16,18,20,23 |
| Dufour's | 1,R | Females: all taxa | Dominance in *Polistes*(?), egg coating, nest usurpation, larval care in Stenogastrinae | 5,6,7,12,14,21 |

[a]1, class 1 gland cells present; 3, class 3 gland cells present; b, both class 1 and class 3 gland cells present; R, a reservoir is associated with the gland.

[b]?, hypothetical involvement of glands; —, unknown.

[c]1, Deleurance 1955b; 2, Maschwitz 1964b; 3, Matthews 1968b; 4, Jeanne 1970a; 5, MacDonald and Matthews 1975; 6, Jeanne 1977b; 7, Landolt and Akre 1979a; 8, Gamboa and Dew 1981; 9, Jeanne 1981a; 10, Kasuya 1981d; 11, Post and Jeanne 1981; 12, Turillazzi and Pardi 1981; 13, West-Eberhard 1982c; 14, Downing and Jeanne 1983; 15, Kojima 1983a; 16, Post and Jeanne 1983d; 17, Turillazzi and Calloni 1983; 18, Veith et al. 1984; 19, Downing et al. 1985 (literature review); 20, Keeping et al. 1986; 21, Billen 1987; 22, Wenzel 1987a; 23, Heath and Landolt 1988.

[d]See Tables 16.1, 16.2.

**Table 16.4.**  Summary of exocrine glands in ants and eusocial bees that are similar in location to those found in wasps

| Gland | Type[a] | Distribution | Function[b] | Reference[c] |
|---|---|---|---|---|
| Ectal mandibular | 3,R | All taxa examined | Queen control, attraction, sex attractant, stimulation of comb building, trail pheromone, alarm, defense | 1,2,3,5,6,7,8 |
| Mesal mandibular (postgenal) | 3 | Some bees | — | 2 |
| Hypopharyngeal | 3 | All taxa examined | Brood food | 2,8 |
| Thoracic | 3 | All taxa examined | — | 2,8, |
| Sternal | 1 | All bees | Construction | 2,8 |
|  | 3 | Some ants | Trail marking | 4 |
| Tergal | 3 | *Apis*, some ants | Queen control, attraction, trail marking | 4,8 |
| Poison | 1,R | All taxa | Defense, alarm, trail marking (ants) | 2,8 |
| Dufour's | 1,R | All taxa | Attraction, trail marking | 2,8 |

[a]1, class 1 gland cells present; 3, class 3 gland cells present; R, a reservoir is associated with the gland.
[b]—, unknown.
[c]1, Butler 1959; 2, Wilson 1971 (and cited references); 3, Avitabile et al. 1975; 4, Hölldobler 1980; 5, Honk et al. 1980; 6, Röseler et al. 1981; 7, Smith and Roubik 1983; 8, Free 1987 (and cited references).

construction behavior (Ikan et al. 1969, Akre 1982). Similarly acting pheromones seem to be produced by *Vespula atropilosa* and *V. pensylvanica* queens (Landolt et al. 1977, Akre and Reed 1983).

Transmission of queen pheromones to workers in at least three *Vespula* species appears to involve direct contact either by trophallaxis or by contact with the substrate (see Greene, this volume). When colonies of *V. atropilosa* were divided in half by a double screen, workers on the queenless side developed functional ovaries and began behaving as if they were uninfluenced by the queen (Landolt et al. 1977), indicating that the queen pheromone in this species is not airborne. Furthermore, experiments with both *V. pensylvanica* and *V. vulgaris* have suggested that queens deposit pheromone on the comb that inhibits worker ovarian development (Akre and Reed 1983). Among some *Vespula* species the queen's excrement is imbibed by workers and presumably is distributed among them by trophallaxis, suggesting that this might be a means of dispersing a queen pheromone (Akre et al. 1976, but see also Spradbery, this volume). Evidence that queen quality and age influence colony productivity may be explained by differences in endocrine and exocrine activity among queens (Brian 1979, Archer 1981b).

Little is known about the method of queen control (inhibition of subordinate reproduction) among species of swarm-founding polistines (Jeanne, this volume: Chap. 6; Spradbery, this volume). West-Eberhard (1977a) demonstrated that queens of *Metapolybia aztecoides* are recognized by their nestmates and that the source of the cue originates in the head (presumably from exocrine glands located there). As West-Eberhard (1977a) points out, however, queen recognition and control are not necessarily the same thing. Richards (1971) found that the fifth sternal glands of *Polybia* queens usually had more secretion than those of workers of the same species, and that the secretion was often of a different color. He suggested that these glands may produce a queen control pheromone. However, among species of *Polybia* this gland appears to be the source of trail pheromone used to direct colony movement to new nest sites (Jeanne 1981a; see also Recruitment, below). Because queens are older individuals and rarely exhibit trail-marking behavior, this secretion presumably would accumulate on their sterna; this accumulation alone may explain Richards's (1971) observations. Glands may have multiple functions, however, so it is possible that the fifth sternal gland may also be involved in queen control or may be used by queens to attract workers during swarming (Jeanne 1981a).

Among independent-founding polistine species, queen inhibition of worker ovarian development is apparently accomplished primarily by means of aggressive behavior (Spradbery, this volume). From their behavior, it is clear that subordinate and dominant nestmates recognize one another before contact, and probably in this way avoid potentially costly aggressive exchanges (West-Eberhard 1977a, Barnard and Burk 1979, Reeve and Gamboa 1983, Downing and Jeanne 1985). If a dominant wasp approaches a subordinate, the latter often will move out of the way or lower its head toward the nest (Pardi 1948, West-Eberhard 1969, Jeanne 1972). It has also been demonstrated that dominant *Polistes fuscatus* females attack higher ranked subordinates more frequently than lower ranked ones. In *P. fuscatus*, odors from both the head and the ovaries provide recognition cues (Downing and Jeanne 1985).

In a study of the influence of seasonal and rank differences on the development of exocrine glands in *P. fuscatus*, two glands—the Dufour's and fifth sternal glands—were larger and appeared to be more active in dominant than in subordinate wasps (Downing and Jeanne 1983). The Dufour's gland secretion may be added to eggs during oviposition, and may function to communicate status and prevent egg eating by subordinates during times of intense reproductive competition among nestmates.

Because abdominal wagging (also called tail wagging) and the more violent abdominal lateral vibrations described in many *Polistes* species (West-Eberhard 1969, Esch 1971, Hermann and Dirks 1974, Gamboa and Dew 1981) are primarily performed by the queen of a colony, they may serve to add a secretion from the sternal glands to the nest substrate and thus may mediate domination of subordinates (Hermann et al. 1975, Hermann and Dirks 1975, Gamboa and Dew 1981). However, both abdominal wagging and lateral vibrations are also performed by solitary foundresses (Gamboa and Dew 1981, Downing and Jeanne 1985). Abdominal wagging is initiated when the queen's developing larvae enter the third instar, suggesting that this behavior may be a form of communication with the brood (Gamboa and Dew 1981, see also Reeve, Starr, this volume).

Because the ectal mandibular gland is a source of queen pheromone in honey bees and bumble bees (Butler 1959, Honk et al. 1980), it is possible that this gland is also involved in dominance interactions in at least some eusocial wasps. It opens to the front of the face in *Polistes*, and during the face-to-face aggressive jabbing that is observed among females the mandibles are held slightly apart and the opening and its associated brush are thus exposed (Downing and Jeanne 1982).

Because physical aggression is an important method of queen control among primitively eusocial wasps and chemical control characterizes highly eusocial wasps, chemical control may be derived from physical aggression (Wilson 1971, Reeve and Gamboa 1983). The energy waste and potential danger from physically determined and maintained dominance would select for systems that avoid conflict. Dominance rank is established initially by aggressive interactions in *Polistes, Mischocyttarus*, and other independent-founding polistine species (Röseler, this volume), but subordinates learn the odor cues of a dominant wasp and thus can avoid or appease the dominant and avoid further conflict (West-Eberhard 1967, 1977a; Jeanne 1972). Queen pheromones that influence egg laying in subordinates may have evolved concurrently with or subsequent to the recognition cues, either through the extension of recognition cues into this new role or by the development of a separate set of odor cues. The development of these pheromones would further reduce physical aggression over reproductive roles among nestmates (as in the Vespinae).

## Social Parasitism

The queen of a socially parasitic species usurps the nest of a queen of another species of eusocial wasp, eventually killing or driving off the host queen. The young of the parasitic wasp are then reared by the host queen's worker offspring (Spradbery 1973a). The parasitic species may be an obligate parasite, unable to initiate a colony on its own and, in some cases, produce workers, or it may be a facultative parasite, opportunistically initiating a nest under some conditions (Spradbery 1973a, Matthews 1982).

Odor cues may be involved in parasite-host recognition and in the ability of the parasite queen to recognize her most obstinate rivals (Scheven 1958). In addition, pheromones appear to be involved in the domination of the host queen and her workers (Jeanne 1977b, Greene et al. 1978). Although no one has experimentally tested the role of the Dufour's gland in the usurpation process, two lines of evidence suggest that this gland may be involved. First, parasitic queens drag their abdomen over the nest envelope around the nest entrance, while host queens and workers do not. Second, the Dufour's gland of queens of both the obligate social parasite *Dolichovespula arctica* (Jeanne 1977b, Greene et al. 1978, Landolt and Akre 1979a) and the facultative social parasite *Vespula squamosa* (MacDonald and Matthews 1975) is 4–18 times as large as the Dufour's gland of nonparasitic species.

There are several possible ways in which Dufour's gland secretion (or that of other glands) may be involved in the usurpation process. The gland(s) of the parasite queen may produce a super stimulus, and thus to the host workers she may appear pheromonally superior to, and therefore more acceptable than, the host queen (Jeanne 1977b, Fletcher and Blum 1983). The secretion could also mask or alter the odor of the host colony and thus confuse the host system for recognizing nest and queen (Schmidt et al. 1984). In addition, the secretion might alter worker behavior in some way to make the usurpation process easier (Schmidt et al. 1984).

Concurrent with selection on the parasitic species, selection must also be acting on the host species, leading to a dynamic co-evolutionary system. Workers able to detect a parasite queen before she has usurped the nest would benefit in terms of their inclusive fitness, as would their egg-laying queen. Evidence that parasite queens do not easily invade nests comes from the presence of dead parasite queens around some colonies (Akre 1982).

## Recruitment

In social wasps, four types of recruitment have been found: (1) to an aggregation, (2) to a food source, (3) to a new nest site, and (4) to a site of defense (alarm recruitment).

### To an Aggregation

Little is known about female aggregation pheromones in social wasps (Akre 1982), but aggregation or assembly pheromones appear to be deposited by queens and workers of vespine and *Polistes* species when they rest in one spot for a period of time (Ishay and Perna 1979). When moving to a new nest or temporary swarm location, individuals of some swarm-founding polistine species either drag their gasters or lick the substrate at the new site (Chadab and Rettenmeyer 1979, Forsyth 1981, Jeanne 1981a, West-Eberhard 1982c). These behaviors appear to function in applying glandular secretions to the substrate; such secretions may act as aggregation pheromones.

### To a Food Source

Recruitment to a food source is rare among social wasp species and does not appear to involve trail laying, as it does in many ants (Hölldobler 1977). Recruitment of naive wasps appears to occur by returning foragers carrying food odors back to the nest, as has been suggested for two *Vespula* species (Maschwitz et al. 1974) and several swarm-founding polistine species (Jeanne 1980a). However, not all species in these groups recruit to a food source. In tests with *V. pensylvanica*, Akre (1982) could find no evidence of such recruitment.

When attacking honey bee colonies, *Vespa mandarinia* appears to exhibit specific phases: hunting (killing bees individually and carrying them back to the nest), slaughter (killing bees rapidly without carrying them back to the nest), and occupation of the honey bee hive (Matsuura 1984, this volume). The shift from hunting to slaughter appears to be rapid, and increasing numbers of nestmates appear to be involved, indicating that individuals switching from a hunting to a slaughter phase may emit some stimulant that attracts nestmates to the attack. Further support for this hypothesis comes from the finding that hornets killed in a slaughter are attractive to other hornets (Matsuura 1984).

### To a New Nest Site

Recruitment to a new nest site occurs in two situations. The first is when the original nest is destroyed by a natural disaster or predator. In this situation, the absconding wasps temporarily cluster in a safe location while scouts search for a new nest site. In the second situation, the colony moves to a new location as a natural part of the colony cycle. The behavior associated with recruitment to the new nest site is similar in the two cases.

Gaster dragging (rubbing the fifth and/or sixth sterna on the substrate) associated with swarming to a new nest location was first observed in the four swarm-founding polistine genera *Angiopolybia*, *Leipomeles*, *Polybia*, and *Agelaia* (Naumann 1975), and was subsequently observed also in the swarm-founding genera *Metapolybia*, *Parachartergus*, and *Synoeca* (Jeanne 1975c, 1981a; Chadab and Rettenmeyer 1979; Forsyth 1981; West-Eberhard 1982c). Females of most swarm-founding species have either a fifth or sixth (*Agelaia* only) sternal gland, but females of *Apoica* have both (Jeanne et al. 1983). Jeanne (1981a) demonstrated experimentally that *Polybia sericea* uses the fifth sternal gland secretion as a trail pheromone to communicate the location of a new nest site (see Jeanne, this volume: Chap. 6). The association of these sternal glands (in particular the fifth) with swarm-founding habits suggests that most swarm-founding species use sternal gland secretions to direct swarm movement (Jeanne et al. 1983). Swarms moving over short distances do not appear to use a trail pheromone, however, since dragging is not observed in this situation (Chadab and Rettenmeyer 1979, West-Eberhard 1982c).

No evidence of glandular tissue on sterna III–VI has been found in species of *Angiopolybia*, *Leipomeles*, or *Parachartergus* (Jeanne et al. 1983; cf. Vecht 1968), although these species exhibit dragging behavior during migrations. At least two explanations are possible. First, secretion from some other gland source(s) may be applied to the gaster, which may then be used as an applicator. This seems unlikely since some indication of this behavior should have been observed but has not (Naumann 1975, West-Eberhard 1982c). A second explanation is that the sternal glands of these species may consist only of class 1 cells. This type of gland cell has no duct associated with it, and its secretory activity might be closely regulated so that it becomes enlarged only during the time swarming typically takes place. Unless these wasps were in a swarming phase at the time of collection, their glands might be difficult to identify even using histological techniques.

Sternal glands also have not been found in swarm-founding species of four genera in which dragging behavior has not been recorded (*Ropalidia*, *Chartergellus*, *Nectarinella*, and *Polybioides*), although one *Ropalidia* species has a brush of hairs on the sixth gastral sternum (Jeanne et al. 1983) and *Polybioides* apparently has external modifications to this area (Vecht 1968). How these species communicate the location of new nest sites remains unknown.

The use of trail/aggregation pheromones during swarm movement greatly reduces the chance of individuals getting lost during the migration, and thus presumably increases the distances that can be traveled by a swarm. But this system also limits dispersal, because emigrations are impossible across relatively narrow water barriers (50 m or more) (Jeanne 1981a). The effect of this limitation is clear when the numbers of swarm-founding and independent-founding species on islands are compared (Richards 1978a, Jeanne 1981a).

Some behavioral observations suggest that an abdominal secretion may also be used by returning scouts to induce flying in a group of wasps about to emigrate. This suggestion was first made by Naumann (1970) when he observed preswarming behavior in *Protopolybia*. Scouts ran through the wasp swarm, buzzing their wings and extending their abdomens as if releasing a secretion. This activity appeared to stimulate nestmates to fly.

The pattern of occurrence of sternal glands among the swarm-founding polistine species suggests at least three possible pathways for the evolution of these glands (Jeanne et al. 1983). First, the glands of extant species may all be derived from a common swarm-founding ancestor with small glands on every gastral sternum. Since small glands produce only a small amount of glandular secretion, multiple glands might have been advantageous. With selection for greater glandular production and ease of application, glands became concentrated on sterna near the tip of the gaster and became associated with specialized sensory receptors and muscular control. Whether the reduction in the number of gland-bearing sterna resulted in the presence of glands on the fifth, sixth, or both sterna might simply reflect morphological differences among species affecting the ease with which certain sterna can be flexed and extended during glandular application. Two other hypotheses are that the trail-marking glands have separate origins from undifferentiated fifth and sixth sternal glands present in the common ancestor of swarm-founding polistines, or that glands on these two different sterna arose de novo completely independently (do not exhibit serial homology) and have different functions (Jeanne et al. 1983).

When the nests of independent-founding species are destroyed, nestmates will often reinitiate the nest in the same location or nearby. Litte (1981) has observed nest relocation in *Mischocyttarus labiatus* and found that queens drag their abdomens over leaves and stems in the area between the old and new nest sites, presumably depositing a glandular secretion. Nestmates flying in the area antennate these spots and may use the odor cues to locate the new nest location. Litte also observed queens dragging their gasters over the nest when co-foundresses disappeared or were removed. Queens of both *Polistes* and *Mischocyttarus* frequently rub their abdomens on the nest, which may serve to leave identifying cues (see Dominance and Queen Control). A simple extension of this behavior may be the rubbing of these same identifying cues on the substrate to communicate a new nest location.

## To a Site of Defense

Alarm pheromones need to be volatile to disperse rapidly and persist only a short time. For high volatility the predicted molecular weight of alarm pheromones is 100–200 (Bossert and Wilson 1963). Such pheromones have been reported in *Vespula vulgaris*, *V. germanica*, *V. squamosa*, and *Vespa crabro*, and the glandular source appears to be the poison gland (Maschwitz 1964b, Moritz and Bürgin 1987, Heath and Landolt 1988). Alarm pheromones have not been found in *Vespula atropilosa* or

*V. pensylvanica* (Hermann and Blum 1981, Akre 1982), nor have Maschwitz's results been confirmed with North American *V. vulgaris* (Akre 1982).

Chemical components of the alarm pheromones of two species have been identified. N-3-methylbutylacetamide, a component of venom in *Vespula squamosa*, elicits the same alarm response as the venom (Heath and Landolt 1988). A major component of the venom alarm pheromone of *Vespa crabro* is 2-methyl-3-butene-2-ol (Veith et al. 1984). However, this component does not elicit as great a response as pure venom, suggesting that there may be one or more other active components.

Maschwitz (1964b) found no evidence of an alarm pheromone in the polistine wasp *Polistes biglumis bimaculatus*. In contrast, Jeanne (1981b, 1982b) has demonstrated that the venom acts as an alarm pheromone in both *Polistes canadensis* and *Polybia occidentalis* by lowering the threshold of attack and, in *Polistes canadensis*, by acting as an attractant. The venom of *Polybia occidentalis* induces wasps to exit the nest and spread out on the nest envelope, but the release of flying to attack a disturbance requires a visual stimulus such as a moving dark object (Jeanne 1981b).

Although venom may alarm wasps and lower their threshold for attack, we still do not know whether the wasps actually release venom at the nest independent of stinging behavior in order to communicate alarm to nestmates. Jeanne (1982b) observed that *Polistes canadensis* opens the sting chamber while on the nest, but release of venom was not observed. More recent work by Post et al. (1984b) demonstrated that the venom of both *Polistes exclamans* and *P. fuscatus* could induce an alarm response, but when colonies placed in a wind tunnel were alarmed, downwind conspecific colonies did not respond, indicating that venom was not released by alarmed wasps at the nest in either species. It is possible that the sting chambers are opened as a part of the defensive display, which involves bending the gaster toward the source of disturbance, or possibly that the sting chamber is opened in preparation for stinging (Post et al. 1984b). On the other hand, *P. canadensis* is unusual in that it regularly builds nests consisting of many separate but closely spaced combs (Jeanne 1979a). Such nests may hinder communication of alarm among nestmates by vibrations in the nest substrate. Therefore, a chemically induced alarm response may have evolved in *P. canadensis* as an effective method of rapidly spreading alarm between separated combs (Post et al. 1984b).

## Nest Construction

Wasps typically build their nests by gathering small loads of some sort of building material such as wood pulp or mud and adding them load by load to their developing nest. That most wasps add glandular secretions to the building material either while it is gathered or shortly afterward is obvious. The change in cohesive properties of the material as it dries and the behavior of the wasps during construction both indicate that some sort of oral secretion is used to cement the building material together. The type of nesting material and the glandular secretions added to that material have probably influenced the evolution of nest architecture and thus the evolution of social organization among wasps (Hansell 1985). However, little is known about the source(s) and nature of the cement added to wood pulp or mud to create the nest-building material. Recent studies indicate that the cement is water resistant and that it may contain N-acetyl glucosamine, a component of chitin (Schremmer et al. 1985), as well as various amino acids, some similar to the components of raw silk (McGovern et al. 1988).

The thoracic gland is the largest gland to open into the oral cavity in vespid wasps, and, therefore, it has been suggested as the source of the building cement (Janet 1903, Spradbery 1973a, MacDonald 1977, Wenzel, this volume). This hypothesis is supported by the fact that this gland appears to be active only during the nesting period, when wasps are actively building their nests (Downing and Jeanne 1983).

Observations of the eusocial sphecid wasp *Microstigmus comes* indicate that an abdominal gland on the apical tergum, and not an oral gland, is the source of a smooth, silklike material added to the nest during construction (Matthews 1968b, this volume; Matthews and Starr 1984). In this species, one or more females scrape up plant fibers and glue them together with their silklike glandular secretion. The ball is suspended from the leaf by a ropelike petiole, which is gradually lengthened and shaped by the wasps using the secretion. The cells are then constructed by burrowing and adding more glandular secretion in the center of the suspended ball. This behavior is unique among wasps, particularly with respect to the addition of a conspicuous glandular secretion.

In addition to adding glandular secretions while applying construction materials to a growing nest, many social vespids also lick the comb and supporting petioles (referred to as "mouthing" in Jeanne 1972). Over time, the comb and petiole(s) are darkened with a shellaclike material, presumably the glandular secretion added during licking and rubbing (West-Eberhard 1969; Jeanne 1972, 1977c; Litte 1981; Downing and Jeanne 1987). Thoracic gland secretions may be added by licking, but the only oral gland that produces a dark secretion is the mesal mandibular gland, suggesting that this gland may produce at least one of the components of the oral secretion (Downing and Jeanne 1983). Among the Vespinae, the oral secretion added to the original nest petiole facilitates its flexibility (Jeanne 1977c). The oral secretion of polistine wasps with open-combed nests hardens, and it may strengthen the comb and petiole and help to make them water-repellent (Hermann and Blum 1981) as well as inhibit fungal growth (MacDonald 1977).

Evidence for pheromonal regulation of building behavior in vespids has been reported. *Vespa orientalis* queen pheromone mediates queen-cell construction at the end of the nesting season (Ikan et al. 1969), and initiation of nest-building activity of *V. orientalis* appears to be stimulated by a "building-initiating pheromone" applied to the substrate by workers (Ishay and Perna 1979).

## Recognition of Nestmates and Nests

Recognition of nestmates (and, therefore, presumably genetically related individuals), both adult and immature, has been demonstrated in *Polistes* and *Dolichovespula* (Klahn 1979, Ross and Gamboa 1981, Post and Jeanne 1982a, Klahn and Gamboa 1983, Ryan et al. 1985). In addition, *Polistes fuscatus* can discriminate between unrelated nonnestmates and nonnestmate kin, clearly suggesting a genetic component to the recognition cues (Gamboa et al. 1987). Recognition cues appear to be chemical and are produced and acquired during both early development and later adult life (Gamboa et al. 1986b). Possible glandular involvement in these recognition cues has not been investigated.

Various recognition cues also appear to be associated with the nest. Butler et al. (1969) discovered a "footprint" pheromone in *Vespula vulgaris* that is apparently left on the nest by workers as they walk over the surface. It is believed that this pheromone aids returning foragers in locating their nest entrance. In another study, discrimination between broodless natal and nonnatal nest fragments was found to occur in both *Dolichovespula maculata* and *P. fuscatus* (Ferguson et al. 1987). The nature of the nest recognition cues is uncertain, however, and may or may not involve the glandular secretions added to the nest.

## Brood Care

Females of a number of vespine species warm pigmented pupae even when they are removed from the nest. The pupae emit an odor that attracts and stimulates warming behavior in adults (Ishay 1972).

A glandular (possibly Dufour's gland) secretion has also been implicated in the care of brood in stenogastrine wasps (Turillazzi and Pardi 1981, Hansell 1982a). This secretion is drawn in a long thread from the tip of the abdomen and wound up into a ball by the front legs and mouthparts, during which oral secretions may be added. Larvae do not appear to feed directly on the secretion in cells, rather the secretion serves as a container for the larval food and for reserves of liquid food for adults (Turillazzi 1985c, this volume).

## Nest Defense

The architecture of a wasp nest is its primary defense mechanism (for a review of the literature see Jeanne 1975a). In addition, wasps that build stalked combs without an enclosing envelope add glandular secretions to the petiole of the nest, which prevent ants and perhaps other small ambulatory predators from crossing to the comb. Among the Vespidae, this defensive allomone is produced by the sixth sternal gland (Jeanne 1970a, Post and Jeanne 1981, Kojima 1983a). A wasp applying this substance rubs its gastral sterna laterally over the petiole and surrounding comb (see Gadagkar, this volume: Fig. 5.10). While rubbing, the wasp extends its abdomen, thus exposing the entire brush associated with the sixth sternal gland. Rubbing behavior has been observed in the polistine genera *Mischocyttarus* (Jeanne 1970a, Litte 1981), *Polistes* (Corn 1972, Turillazzi and Ugolini 1979, Post and Jeanne 1981), *Ropalidia* (Darchen 1976a; Kojima 1982b, 1983a), *Parapolybia* (Kojima 1983b), and *Belonogaster* (Marino Piccioli and Pardi 1970). The sixth sternal gland secretion of *Polistes fuscatus* has two active components that repel ants; one has been identified as methyl palmitate ($C_{17}H_{34}O_2$) (Post et al. 1984a).

The sphecid wasp *Microstigmus comes* exhibits a similar behavior of rubbing its abdomen over the nest (Matthews 1968b, Matthews and Starr 1984). In this case it rubs the apical tergum, which has a small brush of setae associated with a gland. This wasp appears to apply a silklike secretion, which functions as a building cement over the outer surface of its nest and may repel ants as well.

Vespine species have a sixth sternal gland with a brush of hairs, but rubbing behavior on the petioles or outside of the nest has not been observed in these wasps (Landolt and Akre 1979a).

Some species of stenogastrine wasps apply a gelatinous secretion around the branch or other narrow support to which the nest is attached (Turillazzi and Pardi 1981, Turillazzi, this volume). Females exude the secretion from the tip of their abdomen and then collect the small drops on the hind legs. The collected secretion is passed to the mouth and applied to the substrate. Ants appear to have difficulty crossing the secretion, and if offered an alternative path they will take it. The source of this ant guard material may be the Dufour's gland, but it is possible that oral gland secretions are also added to the material (Turillazzi and Pardi 1981).

Studies of both nest architecture (Jeanne 1975a, Downing and Jeanne

1986) and chemical defenses (Jeanne 1970a, Post and Jeanne 1981, Kojima 1983a) indicate that ant predation has strongly influenced the evolution of social wasp building and defensive behavior.

## Reproductive Behavior

Two major categories of pheromones are associated with reproductive behavior in social wasps: (1) sex attractants and recognition pheromones produced by females and (2) perch- and possibly territory-marking pheromones produced by males (Fig. 16.4).

Females are highly attractive to conspecific males, and this attractiveness appears to be chemically induced. Components of the venom of *Polistes* (Post and Jeanne 1983d, 1984b) and of *Belonogaster petiolata* (Keeping et al. 1986) have been shown to attract males and to induce copulatory behavior. In addition, male *Polistes* also respond to a contact pheromone on the female's cuticle that apparently communicates spe-

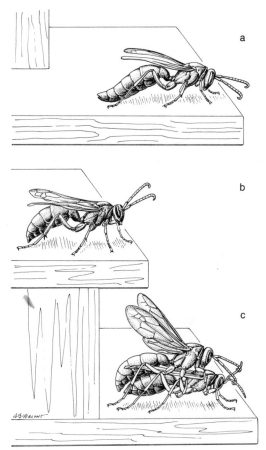

**Fig. 16.4.** Pheromones involved in social wasp reproductive behavior. (a) A male *Polistes fuscatus* rubs his sternal glands on his perch site. (b) A male alert but not rubbing at a perch site. (c) Female venom attracts the male and releases copulatory behavior. Upon contact, cuticular pheromones identify the species of the female.

cies identity and inhibits inappropriate mating attempts (Post and Jeanne 1984b). Observations of *Mischocyttarus flavitarsis* (Litte 1979), *Vespa crabro* (Batra 1980), and *Vespula maculifrons* (Ross 1983b) suggest that these species have a similar contact or close-range pheromone that may be involved in the recognition of conspecific females.

Queen but not worker solvent extracts induce copulatory responses in males of six *Vespa* species, with little apparent species specificity (Ono and Sasaki 1987). The interspecific attractiveness of the extracts raises questions about the mechanism of reproductive isolation in these wasps. The fact that only the extracts of queens excite males indicates a caste distinction in these highly eusocial wasps that has not been found in the primitively eusocial genus *Polistes* (Post and Jeanne 1985).

The males of some vespines and stenogastrines form clusters around single females (Sandeman 1938, MacDonald et al. 1974, Turillazzi and Pardi 1982). This observation only suggests the possibility of a female-released sex attractant, because males joining these clusters could be orienting either to the cluster itself or to the odor of the visually obscured female.

Among the Vespinae, queen attractiveness appears to vary from day to day, possibly depending on age and when the individual last mated (MacDonald et al. 1974). Thus care should be taken when interpreting negative results in studies on female sex attractants.

Social vespid males exhibit a wide range of mate-searching behaviors, from defending territories around female hibernacula to nonterritorial patrolling of routes around vegetation (Kasuya 1981d, Litte 1981, Turillazzi and Cervo 1982, Post and Jeanne 1983a, Wenzel 1987a, Ross and Carpenter, this volume). Dragging of abdominal sterna on territory perch sites or on leaves along a patrol route has been observed in *Polistes* (Turillazzi and Cervo 1982, Post and Jeanne 1983a, Wenzel 1987a) (Fig. 16.4), *Mischocyttarus* (Litte 1979, 1981), and *Polybia* (West-Eberhard 1982c). Glands on one or more abdominal sterna are well developed in males of *Polistes* (Turillazzi 1979, Post and Jeanne 1983c, Downing et al. 1985), *Mischocyttarus* (Post and Jeanne 1982b), *Belonogaster, Parapolybia, Ropalidia, Brachygastra, Polybia, Apoica, Metapolybia,* and *Chartergus* (Downing et al. 1985) (Tables 16.1, 16.2).

Only males of *Polistes jadwigae* are known to exhibit what could be female-calling behavior. Territorial males bend their gasters while resting in their territories. Associated with their apical sternal gland are two fingerlike projections that may aid in dispersing the gland's secretion (Kasuya 1981d, Downing et al. 1985). Males of *Polistes major* have a unique behavior of rubbing the front of the face over perches along patrol routes, which suggests that they rub secretions from their ectal mandibular glands on the surface. The volatility of the secretion suggests its use in long-range communication (Wenzel 1987a).

Although the likely gland sources of male pheromones involved in mating and associated behaviors have been identified, the exact functions of these secretions are unknown. Females do not appear to be particularly attracted by either the ectal mandibular gland secretion of *P. major* (Wenzel 1987a) or the sternal gland secretion of *Mischocyttarus flavitarsis* (Litte 1979), and the secretions produced by males of other species have not been studied in detail. Since these secretions appear to be used in a territory or along a patrol route, they may be involved in advertising or maintaining ownership of these areas.

Males of the stenogastrine *Parischnogaster nigricans serrei* hover in front of landmarks and extend their abdomens, thus exposing white strips on their gastral terga as well as a gland located on their third gastral tergum (Turillazzi 1982, Turillazzi and Calloni 1983). As with other social wasp species the exact function of the secretion of this gland is unknown, but presumably it plays a role in the male wasp's display.

Male mating behavior has been observed in some nine Vespinae species (Post 1980 and cited references), yet no evidence for territorial marking or for pheromones used to attract females has been found.

## CONCLUDING REMARKS

Although our understanding of the role of pheromones and other exocrine secretions in social wasp behavior has greatly increased since the 1960s, there remain many unanswered questions (e.g., see Table 16.3). For many glands no function(s) has been identified for the product. For those glands producing substances with an identified function, researchers may have overlooked secondary or tertiary functions. Most glandular secretions are a mix of many chemicals, and not all cells within a gland become activated at once, which may indicate that glands often serve more than one role in communication (Delfino et al. 1979, Downing and Jeanne 1983). This is the case for the venom of *Polistes*, which induces alarm response in conspecific females and sexual attraction in conspecific males (Post and Jeanne 1983d, Post et al. 1984b). Whenever a gland produces a large amount of material, there is the potential for the evolution of multiple communicative roles for that glandular secretion.

There are many species of tropical social wasps about which virtually nothing is known, even regarding what glands they possess and whether these show differences in secretory activity through the nesting cycle or with social rank. In addition, possible differences between the sexes in glands have not been studied for most species. The importance of individual differences in exocrine gland development also has not been explored extensively, even though such differences may be important in individual recognition, attraction, or ability to regulate the behavior of others. It should not be assumed that such differences do not have some biological significance (Barrows et al. 1975).

The development of advanced levels of eusociality correlates with selection for greater pheromonal involvement in communication and regulation of social organization within the colony (Blum 1977, Reeve and Gamboa 1983, Spradbery, this volume). In order to piece together the evolution of chemical communication in wasp societies we need to (1) learn more about the distribution and development of various glands among social wasp species, (2) identify the product(s) and function(s) of more glands, (3) learn more about glands with multiple roles (e.g., how different secretory products are produced and how these are altered with time, individual social rank, and colony developmental stage), and (4) determine the evolutionary relationships of glands and gland products in various species by reference to well-established phylogenies.

*Acknowledgments*

I thank Art and Naomi Moss for their hospitality and David Post for valuable suggestions on improving this chapter. Work on the manuscript was supported in part by the University of Wisconsin–Whitewater.

# 17

# Evolution of Social Behavior in Sphecid Wasps

ROBERT W. MATTHEWS

The family Sphecidae is a diverse group of about 8,000 species, the vast majority of which seem to be strictly solitary in their nesting habits. Yet repeatedly within this group there occur apparently independently evolved instances of cooperative nesting, ranging from temporary nest sharing to full eusociality with reproductive division of labor. My intent here is to examine factors that may have facilitated the evolution of varying degrees of social behavior in the Sphecidae. As for all social species, combinations of many factors are required to "explain" the evolution of social behavior, the relative importance of each being determined by the particular ecological setting.

At the outset, four points need to be emphasized. First, the Sphecidae are not phylogenetically closely related to other social wasps, all of which belong to the family Vespidae. Rather, the sphecid wasps are allied to the bees (Brothers 1975), both having been derived from a common ancestor not shared with the vespid wasps (Michener 1944, Königsmann 1978, Lomholdt 1982) (Fig. 17.1). Thus, for comparisons of evolutionary trends in the behavioral patterns of sphecids, it is most appropriate to look to the various bees, rather than to the vespids. There is some consensus as to which group of sphecids is closest to the bees. On the basis of morphology, Bohart and Menke (1976:31) state that "it is our impression that the subfamily Pemphredoninae contains some of the most beelike forms in the family." Malyshev (1968) argues a similar thesis primarily on the basis of behavioral traits. Fossil evi-

This chapter is dedicated to Howard E. Evans in honor of his 70th birthday.

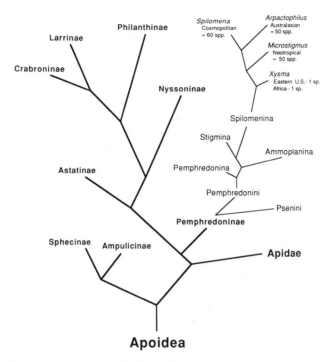

## Apoidea

**Fig. 17.1.** Dendogram of suggested relationships among component taxa of the Apoidea (see Michener 1986), which includes the eight sphecid subfamilies together with the bees (modified from Bohart and Menke 1976, Lomholdt 1982, A. Menke, unpubl.). The bees are placed together in the single family Apidae as has been suggested by Lomholdt (1982) and Gauld and Bolton (1988). Because of their relatively well-developed social behavior, the subfamily Pemphredoninae has been amplified to show hypothesized relationships between the tribes and subtribes and the included genera of the Spilomenina (Budrys 1988, Menke 1989). *Note:* The validity of retaining the traditional separation between the Larrinae and Crabroninae has recently been questioned by Lomholdt (1985) and Menke (1988), who propose uniting these under the Larrinae. Such unification concurs with a similar arrangement based on larval characters proposed by Evans (1964a).

dence (Evans 1969, 1973b; Lomholdt 1982), though fragmentary, shows the oldest sphecids (from the Cretaceous) to be very similar to living Pemphredoninae. (There is no corresponding "missing link" among the bees, for the oldest bee fossils [from Cretaceous amber] belong not to a bee group judged to be basal [e.g., Colletidae], but to one of the most derived genera, *Trigona* [Michener and Grimaldi 1988]). Thus, the Pemphredoninae appear to constitute a relatively ancient lineage among extant sphecids, particularly if the subfamilies Sphecinae and Ampulicinae are recognized as separate from the remaining sphecids as some authors suggest (Evans 1964a, Lomholdt 1982). Lomholdt (1982) advocates recognizing Apidae (in the broad sense) as a subordinate group of the Sphecoidea, possibly most closely related to the Pemphre-doninae. Alexander (1990) provides further details.

Second, behavioral plasticity in the Sphecidae is remarkable (e.g., Ohgushi 1954, Evans 1966b, Evans and West-Eberhard 1970)—so much so that generalizations based on available data are best regarded as tentative. Brockmann's (1980) record of five different nesting patterns for the pipe-organ mud dauber (*Trypoxylon politum*) illustrates that the range of variation present within a single species may encompass most of the variation found within a subfamily of sphecids. Behavioral plasticity also has increasingly been recognized in the presocial and primitively eusocial bees (e.g., Michener 1985, Eickwort 1986, Yanega 1988).

Although it is convenient and often useful to describe various species as representing a particular social level, this should not be meant to imply that the more primitive species are "striving" toward "higher" social organization, and there is no implication that the more primitive species are somehow less well adapted. The past emphasis on characterizing a species as representing a particular category of social development has given way to increasing recognition of the importance of ecological factors that influence the expression of facultative social traits (Evans 1977a, Michener 1985, Strassmann and Queller 1989). In addition, there is an increasing appreciation of the fact that colonies of a given species may often progress through different stages of social development (Eickwort 1986, Yanega 1989). Eickwort and Kukuk (1987) have pointed out that individual roles in halictid bees are defined more by behavior than by morphology, and they have suggested a caste terminology that could be applied to sphecids as additional information accumulates.

Third, social level among sphecid wasps does not progress in orderly parallel with the current phylogenetic arrangement of these groups. Although the most socially advanced sphecids currently known belong to the Pemphredoninae, many other members of that subfamily are strictly solitary, and the group appears to have arisen relatively early in the diversification of the family. The Nyssoninae and Crabroninae are regarded as the most specialized groups on the basis of morphology (Bohart and Menke 1976), but in overall behavioral complexity the Nyssoninae and Philanthinae appear to be the most advanced subfamilies (Evans 1966a, Evans and O'Neill 1988). This sphecid pattern contrasts with that in the vespid wasps, in which eusociality characterizes the most recently diverged subfamilies (Carpenter, this volume).

Fourth, absolutely no biological information exists for many genera of sphecids (Bohart and Menke 1976). Thus, current interpretations may be expected to change as new discoveries are made, particularly for tropical groups. The recent discovery of relatively complex social behavior in *Arpactophilus mimi* (Matthews and Naumann 1988) is a case in point.

The literature on sphecid biology is vast; much of the descriptive literature is briefly summarized by Bohart and Menke (1976) in their catalog of the sphecid wasps of the world. Other noteworthy general sources are Iwata (1942, 1976), Olberg (1959), Evans (1966b), Krombein (1967), and Evans and West-Eberhard (1970). Scientific names mentioned here follow usage by Bohart and Menke (1976).

## AN OVERVIEW OF SPHECID SOCIAL BEHAVIOR

Relatively low fecundity seems to be a general phenomenon in the Sphecidae. The entire group consists of classic $K$-selected organisms with high levels of parental investment. In general, the reproductive groundplan for the Sphecidae appears to be relatively large egg size and slow maturation of oocytes (Iwata 1955, O'Neill 1985). Where fecundity has been measured over an individual's lifetime, it is seldom greater than 15 eggs. For example, females of *Sceliphron assimile* in Jamaica have an average lifetime fecundity of 8.86 eggs (Freeman 1982).

In every sphecid subfamily, the vast majority of species are solitary nesters that nevertheless exhibit considerable parental care in preparing and provisioning their nests. The relatively short adult life span and absence of direct contact with other adult conspecifics or with their own developing young probably obviate further social opportunities in these species, for sociality presupposes cohabitation of adults.

Scattered through six subfamilies are many reports of adult cohabitation (Table 17.1). These encompass a broad range of behaviors from accidental joint occupancy to eusociality. Because of the anecdotal nature of most of the reports, meaningful categorization of social organization is all but impossible until more information becomes available. Many early records of nest cohabitation were viewed as "aberrant" behavior on the part of one or a few individuals in an otherwise strictly solitary population. However, among better studied species, different populations vary in the extent to which nest sharing occurs (Brockmann and Dawkins 1979, McCorquodale 1989a). Such variation may prove to be characteristic of most nest-sharing species. As Michener (1985) has suggested for bees, shifts between solitary and group nesting may occur even between generations in some of these wasps.

Variation in social development between closely related species is at least as great as intraspecific variation, and further investigation promises to be fruitful. For example, females of 25 of the approximately 900 known species in the philanthine genus *Cerceris* have been reported to

**Table 17.1.** Summary of published records of nest sharing among sphecid wasps

| Species | Reference |
|---|---|
| Sphecinae | |
| *Ammophila heydeni* | Molitor 1933 |
| *A. procera* | Criddle 1924 |
| *Sphex argentatus* | Olberg 1959, Tsuneki 1963 |
| *S. ichneumoneus* | Brockmann and Dawkins 1979, Brockmann et al. 1979 |
| *S. rufocinctus* | Olberg 1959 |
| *Trigonopsis cameronii*[a] | Eberhard 1972, 1974 |
| Pemphredoninae | |
| *Arpactophilus mimi*[a] | Naumann 1983, Matthews and Naumann 1988 |
| *Carinostigmus monstrosus*[a] | Iwata 1964a |
| *Diodontus tristis* | Nielson 1933 |
| *Microstigmus comes*[a] | Matthews 1968a,b |
| *M. theredii*[a] | Myers 1934 |
| *M. thripoctenus*[a] | Matthews 1970 |
| *Microstigmus* spp.[a] | Richards 1972, West-Eberhard 1977b, Matthews, unpubl. |
| *Spilomena subterranea*[a] | McCorquodale and Naumann 1988 |
| *Spilomena* sp. | West-Eberhard 1978a |
| *Stigmus solskyi* | Bristowe 1948 |
| Larrinae | |
| *Dalara mandibularis* | Williams 1919, 1928 |
| *Dicranorhina ruficornis* | Iwata 1976 |
| *Liris aurulenta* | Williams 1919 |
| *L. haemorrhoidalis* | Williams 1928 |
| *Sericophorus sydneyi*[a] | Rayment 1955b |
| *S. victoriensis* | Rayment 1955a |
| *Tachytes distinctus* | Lin and Michener 1972 |
| *Trypoxylon clavatum* | Iwata 1976 |
| *T. fabricator* (= *rugifrons*)[a] | Rau 1933, Richards 1934, Sakagami and Zucchi 1978 |
| *T. politum*[a] | Brockmann 1980 |
| *T. rugiceps* | Iwata 1976 |
| Crabroninae | |
| *Crabro cribrellifer* | Wcislo et al. 1985 |
| *C. hilaris* | Miller and Kurczewski 1973 |
| *Crossocerus dimidiatus* | Peters 1973 |
| *C. elongatulus* | Bristowe 1948 |
| *C. maculiclypeus* | Miller and Kurczewski 1973 |
| *C. megacephalus* (= *leucostoma*) | Saunders 1896, Hamm and Richards 1926 |
| *Dasyproctus kibonotensis* | Bowden 1964 |
| *Ectemnius sexcinctus* (= *quadricinctus*) | Hamm and Richards 1926, Bristowe 1948 |
| *Entomognathus memorialis* | Miller and Kurczewski 1973 |

**Table 17.1**—*continued*

| Species | Reference |
|---|---|
| *Lindenius armaticeps; L. buccadentis;* *L. columbianus* | Miller and Kurczewski 1973 |
| *Moniaecera asperta* | Evans 1964b |
| *Rhopalum longinodum* | Claude-Joseph 1928 |
| Nyssoninae | |
| *Bembix bubalus* | Gess and Gess 1989 |
| *Mellinus arvensis* | Bristowe 1948, Huber 1961 |
| *Sphecius speciosus* | Dambach and Good 1943, Lin and Michener 1972 |
| Philanthinae | |
| *Cerceris acanthopila; C. californica; C.* *dilatata; C. echo; C. femurrubrum* | Hook 1987 |
| *C. anthicivora; C. armigera; C. goddardi;* *C. minuscula; C. unispinosa; C.* *windorum; C. xanthura* | Evans and Hook 1982b, 1986 |
| *C. antipodes* | Evans and Matthews 1970, Alcock 1980, Evans and Hook 1986, McCorquodale 1988b |
| *C. arenaria* | Willmer 1985 |
| *C. australis* | Evans and Hook 1982a, 1986; Evans 1989 |
| *C. fumipennis* | Kurczewski and Miller 1984 |
| *C. gilberti* | Evans and Hook 1986 |
| *C. hortivaga* | Tsuneki 1965 |
| *C. intricata* | Alcock 1975a |
| *C. japonica* | Tsuneki 1947 |
| *C. rubida* | Grandi 1944, 1961 |
| *C. rufimana* | Evans et al. 1976 |
| *C. storeyi* | Evans 1989 |
| *C. watlingensis* | Salbert and Elliott 1979, Elliott et al. 1986 |
| *C. zonata* | Elliott et al. 1981 |
| *Eucerceris bitruncata* | Krombein 1960 |
| *Philanthus crabroniformis* | Alcock 1974 |
| *P. gibbosus*[a] | Peckham and Peckham 1905, Evans 1973c |
| *P. ventrilabris*[a] | Alcock 1975b |
| *Trachypus denticollis*[a] | Claude-Joseph 1928 |
| *T. patagonensis* (= *magnificus*); *T.* *romandi* | Bertoni 1911 |
| *T. petiolatus*[a] | Evans and Matthews 1973b |

*Note*: Brockmann and Dawkins's (1979) table 7, and West-Eberhard's (1978a) table 1, provide additional details for many of these species. The order of the subfamilies listed here follows that of Bohart and Menke (1976).

[a]Males are also regularly present in some nests.

share nests at least occasionally (18 reported since 1981); these nest-sharing species belong to various species groups, indicating that the phenomenon has evolved independently more than once even within this single genus. An instance of interspecific nest sharing between two species of *Cerceris* has also been reported (McCorquodale and Thomson 1988).

With regular nest sharing, the stage is clearly set for eusociality, which is usually considered to originate with either association of daughters with their mothers (the *subsocial* or *familial route*) or association of females of the same generation who cooperate to construct, defend, and provision a nest (the *parasocial* or *communal route*) (Wilson 1971, Brockmann 1984, Spradbery, this volume). This dichotomy is not absolute, because colonies could arise through both routes under certain conditions. Both types of associations are represented among the Sphecidae—for example, accidental (parasitic?) communal associations (e.g., *Sphex ichneumoneus*) or, more commonly, adult offspring remaining on their natal nests (e.g., many *Cerceris*). The term *primitively social* has been proposed by West-Eberhard (1978a) to describe species that regularly share nests but lack a reproductive division of labor (cf. *primitively eusocial* in Cowan, this volume: Table 2.1).

On the whole, how social are the sphecid nest sharers? Because it is difficult to observe activities that occur inside the nest, it is rarely known whether cohabiting wasps maintain separate brood cells in a common nest (communal nesting) or jointly share each brood cell as well as the nest (quasisocial or semisocial nesting) (see Wilson 1971, Cowan, this volume, for definitions of types of nesting behavior). The best data on this point are provided by Brockmann and Dawkins (1979) for *Sphex ichneumoneus*. When two females shared a nest, they provisioned a common brood cell in 88% of the nests but maintained separate cells in a shared burrow in the remaining 12%.

Nearly all sphecids build individual nests that contain separated cells, each cell being provided with paralyzed prey and a single egg, but there are occasional exceptions. Certain species of *Isodontia* construct a single large brood chamber in which 2–12 larvae develop together apparently amicably (Krombein 1967, Bohart and Menke 1976). *Isodontia mexicana* constructs both multicellular and brood chamber nests (Bohart and Menke 1976); because of mass provisioning there is little contact between mother and offspring. Iwata (1964a) mentions a Chilean species, *Rhopalum* ( = *Crabro*) *claudii* (Bohart and Menke 1976), in which six to ten eggs are laid in an empty cell; after the eggs hatch, the female provisions progressively with aphids. Partitions between cells are sometimes omitted in *Pemphredon lethifer*, with the result that more than one larva develops in a cell (Ohgushi 1954). Although there is apparently no overlap of adult generations, these species are noteworthy because of their convergence to the situation found in many allodapine bees that have brood chamber nests. In these bees, sociality has been achieved through the subsocial route because the mother's life span overlaps that of her offspring (Michener 1974, 1990a).

Considerable variation in individual reproductive condition, as indicated by relative ovarian development, is characteristic of nest-sharing sphecid females (Matthews 1968a,b, Evans and Hook 1986, Matthews and Naumann 1988, McCorquodale 1988b, McCorquodale and Naumann 1988). Given this variation, there exists the potential for partitioning of nesting tasks, including egg laying, among cohabiting individuals. Whether such an incipient reproductive division of labor is realized may depend on several factors, both intrinsic and extrinsic.

The only sphecid presently known to be eusocial (i.e., with confirmed reproductive division of labor) is *Microstigmus comes* (Matthews 1968a, Ross and Matthews 1989a,b), a member of the subfamily Pemphredoninae. All of the approximately 50 species of *Microstigmus* known from the Neotropics appear to have more than one adult present in mature nests (Matthews, unpubl.). In addition, similarly advanced levels of sociality appear to have been achieved in two closely related pemphredonine genera, *Arpactophilus* (Matthews and Naumann 1988, Menke 1989), a genus restricted to the Australasian region, and *Spilomena* (McCorquodale and Naumann 1988), which is distributed throughout the world. Thus the most socially advanced sphecids all belong to a single phyletic branch, the subtribe Spilomenina of the tribe Pemphredonini in the subfamily Pemphredoninae (Budrys 1988, Menke 1989) (Fig. 17.1).

Another Old World genus, *Carinostigmus*, which belongs to the sister subtribe of Spilomenina, Stigmina, has at least one species, *C. monstrosus*, reported to be subsocial (Iwata 1964a). *Diodontus tristis*, belonging to a third subtribe, Pemphredonina, has been reported to have more than one female provisioning a common cell (Nielson 1933). The remaining pemphredonine genera are apparently strictly solitary, but the nesting habits of a majority of genera are entirely unknown (Bohart and Menke 1976), and several new genera have only recently been described (Budrys 1988).

## FACTORS FAVORING SOCIAL BEHAVIORS

Contexts favoring the evolution of aggregative nesting and nest sharing in presocial insects are discussed by Lin and Michener (1972), Evans (1977a), West-Eberhard (1978a), Brockmann and Dawkins (1979), Eickwort (1981, 1986), Andersson (1984), Brockmann (1984), Michener (1985), Gadagkar (1985a), Velthuis (1987), and Rosenheim (1990). As a

general rule, nest sharing tends to be more common in species occurring in warmer parts of the globe (e.g., Philanthinae: Evans and O'Neill
1988). In the following sections, life history factors that have been regarded as significant in social evolution in other groups of Hymenoptera are discussed, as are some factors that may be peculiar to the
sphecids.

## Pressure from Parasites and Predators

Evans (1977a) has suggested that natural enemies play a major role
in all stages of socal evolution, and Michener (1974) considered para-
site-predator pressure to be one of the main factors promoting social
evolution in bees. Quantitative studies of the bionomics of various
sphecids (e.g., Freeman and Parnell 1973, Peckham 1977, Wcislo et al.
1985) have demonstrated that nesting success is often greatly reduced
by parasites and predators (see also Cowan, this volume). McCorquodale (1989b) showed that multifemale *Cerceris antipodes* nests were
better defended against parasitic mutillid wasps than were nests tended by solitary females. However, there was no consistent division of
labor between guarding and foraging or other tasks (McCorquodale
1988b).

Many gregarious species of burrowing sphecid wasps construct one
or more "accessory burrows" (Evans 1966c), shallow empty tunnels a
short distance from the true nest entrance. In effect mimicking the entrances of conspecifics, these tunnels are made at various times in the
nest cycle as a fixed part of nest construction behavior. A major postulated function of accessory burrows is to decoy parasitic flies and mutillid wasps that search out the nest entrances to gain access to hosts.
Wcislo (1986) has shown that the sarcophagid fly *Metopia* is attracted to
artificial burrows punched in the soil. Comparing two sympatric species of *Sphex*, Tsuneki (1963) found that the species that made accessory
burrows (*Sphex argentatus*) suffered consistently lower levels of parasitism from miltogrammine flies than the species that never made accessory burrows (*S. flammitrichus*).

Wcislo (1984) has suggested that accessory burrows be interpreted in
a "selfish herd" context (Hamilton 1971) as improving the female's geo-
metric position in an aggregation by increasing the apparent density of
nearby nests and thereby decreasing the probability of being parasitized. However, an alternative (though not mutually exclusive) hypothesis is that accessory burrows may be a ploy to deter competing
conspecific females from overly close nest spacing, thus reducing their
opportunities for prey theft or usurpation of cells or nests. The fact that
accessory burrows are made most often by gregariously nesting species

(Evans 1966c) supports this view. If the burrows' function were solely to decoy parasites, then they should be made equally often by wasps that nest in isolation.

Ant predation has been identified as a major selective force on social evolution and nest architecture of many nest-sharing wasps and bees (Jeanne 1975a, 1979b; Kukuk et al. 1989), and ant-wasp interactions appear to be much more common among sphecids than generally realized (Eberhard 1974; Cane and Miyamoto 1979; Salbert and Elliott 1979; Matthews et al. 1981; Kurczewski and Miller 1984; Evans and Hook 1986; Larsson 1986; Rosenheim 1987, 1988; Brockmann and Grafen 1989; McCorquodale 1989b). Some adult sphecids respond to foraging ants by actively pursuing them, biting at them, and displaying overt aggression (Rosenheim 1988), or even airlifting individual ants away (Cane and Miyamoto 1979). Many solitary ground-nesting species "level" or disperse the accumulated burrow soil (Evans 1966c). In addition to altering visual cues to the nests' presence, this behavior would also disrupt ant forager odor trails that might have previously existed at the nest site and would disperse any telltale wasp odor(s) that could serve as a cue for marauding ants or other parasites. An equally effective alternative behavior would be to cover odor sources, much as some animals scrape soil over their feces. Perhaps this explains the enigmatic mound-building (or reverse-leveling) behavior of the Australian *Bembix littoralis* (Evans and Matthews 1973a).

Glandular secretions applied to the nests by females of many eusocial vespids serve to repel ants (see Downing, this volume). Among sphecids, glands serving similar functions have been identified only in *Chalybion californicum* (Obin 1983), which constructs mud nests. Extracts of sternal glands of *C. californicum* were found to be strongly repellent to two species of ants and to *Melittobia* parasites that search nest sites by crawling. *Chalybion californicum* also is unusual in being one of the few sphecids in which females maintain nest territories by scent marking, and the same sternal gland products appear to serve as these nest-marking pheromones (Obin 1983).

Aggregations are likely to be more vulnerable than scattered individuals to discovery and exploitation by parasitoid flies or wasps. For two species of mud-daubing *Sceliphron* in Jamaica in which nest placement ranges from isolated mud cells to relatively dense aggregations, Freeman (1982) found a significant positive correlation between nest density and brood mortality due to *Melittobia* parasitic wasps. On the other hand, individuals nesting in an aggregation may benefit indirectly from the activities of neighboring individuals in their defensive reactions toward threats of predation or parasitism. Genise (1979) observed that females of the gregarious sand wasp *Rubrica nasuta* chase parasitic bombyliid flies hovering over neighbors' nests.

Wcislo (1984) has shown that more centrally situated nests in an aggregation of the soil-nesting *Crabro cribrellifer* have lower levels of parasitism than peripheral nests, and hence wasps in central positions enjoy higher reproductive success. The fact that nests of *C. cribrellifer* in the centers of aggregations are more closely spaced than those close to the periphery suggests that members of a nesting population may be competing for the central positions. Thus placement of nests in an aggregation, like use of accessory burrows, may be a selfish herd strategy, with individual females employing other individuals as living shields against parasites. Individual *Bembix rostrata* digger wasps likewise respond to increased parasite/predator pressure by decreasing the distance between neighboring nests (Larsson 1986). Opposing this, Rubink (1982) has found that pressure from conspecifics causes nests of *Bembix pruinosa* to be less densely spaced than expected.

## Pressure from Conspecifics: Usurpation

Aggregations of simultaneously nesting conspecifics provide increased opportunities for individuals to incidentally enter nests of their neighbors, a common occurrence among all kinds of nesting animals. The literature contains numerous anecdotal references to nest supersedure by conspecifics among sphecids, which suggests that facultative intraspecific nest parasitism is probably widespread and deserves more detailed study (see, e.g., Alcock 1982, Elliott and Elliott 1987).

The existence of such behavior provides the essential variation upon which natural selection can act, in any of several different ways. If the entered nest turns out to be empty or abandoned, selection may favor its reuse since construction of a nest often requires a considerable expenditure of time and energy. Alternatively, if the nest is occupied, usurpation or nest sharing may result; if the founding and joining females each enjoy greater reproductive success on the average than if each nested alone, then the basis for continued social evolution may be established.

Once two females find themselves in a common nest, mutual tolerance would be required for further social evolution to occur. Many examples of such tolerance among sphecids have been summarized by Brockmann and Dawkins (1979) (see also Wcislo et al. 1985), but the degree of tolerance or aggression may be context-dependent (see Reeve 1989).

Females of many nonsocial species throughout the Hymenoptera routinely exploit the efforts of conspecifics. As Alexander (1987) points out, the propensity for nest usurpation and stealing of nest provisions is almost exclusively the province of aculeate hymenopterans, even though nesting behavior (including provisioning and parental care) is widespread over four orders of insects. Cleptoparasitism, in which a female seeks out the provisions of another female (usually of a different species) and appropriates it for rearing her own offspring, has been recorded in many species in over 80 genera of Hymenoptera, including many sphecids, but outside of the Hymenoptera it occurs only in a few species of dung beetles. Social parasitism is unknown outside of the Hymenoptera. Alexander suggests that this distribution of exploitive intra- and interspecific behaviors may be related in some way to the frequent evolution of eusociality in Hymenoptera. Stubblefield and Charnov (1986) have described a scenario for such exploitive social evolution occurring in a subsocial context.

## Philopatry and Construction of Persistent Nests

Aggregations of many burrowing sphecids tend to persist at the same site over long periods of time, often for several years (reviewed by Evans 1966b; see also Rubink 1982). (Related to this is the matter of what factors influence nest site selection in the first place. Several studies on both bees and wasps have attempted to quantify various physical and edaphic factors regarded as possibly important, with varying success [e.g., Michener et al. 1958, Brockmann 1979, Rubink 1979, Wcislo 1984, Weaving 1989]). Such site tenacity or philopatry is probably the result of the tendency of individual females to nest close to their natal nest. For example, Yanega (1988) observed a marked population of the primitively eusocial bee *Halictus rubicundus* over five years in New York State and found that returning marked foundresses nearly always nested within 0.5 m of their natal nest. Wcislo et al. (1985) recorded five marked females of *Crabro cribellifer* to return to nest within 1 m of their emergence site; essentially the same phenomenon was reported by Eberhard (1972, 1974) for *Trigonopsis cameronii*, in which three of eight marked females returned to nest at their mother's nest. Such site fidelity is probably widespread; its significance for social evolution, like gregarious nesting, is that it brings individuals who are likely to be relatives together in one place, where other ecological factors and/or kin selection may tip the balance in favor of nest sharing. For example, McCorquodale (1989c) suggests that nest sharing in *Cerceris antipodes* results in part from the limited opportunities for newly emerged females to excavate nests in hard surface soils.

Individual nests of some sphecid wasps persist for a considerable time before they disintegrate (e.g., mud dauber nests: Naumann 1983), an attribute that may facilitate nest sharing among females and/or nest reuse by subsequent generations. McCorquodale (1988b) has documented that many nests of the well-studied *Cerceris antipodes* are used for two consecutive seasons; at one study site some were reused for three successive summers. Likewise, there is suggestive evidence that soil nests may be maintained over several generations by the larrine wasp *Dicranorhina cavernicola* (Dupont and Franssen 1937, Iwata and Yoshikawa 1964).

Nest reuse, like philopatry, gives emerging females the opportunity to associate with close relatives, a situation that can pay off in terms of inclusive fitness effects under certain conditions. West-Eberhard (1987c) suggests that this single trait, the widespread tendency of solitary wasp females to reuse existing cells, could provide sufficient raw material for the development of a whole suite of behaviors, including cell usurpation and prey theft, which, though often expressed facultatively as a consequence of contextual changes, may serve as precursors to further social evolution.

## Nest Site and Type: The Opportunity to Be Useful

The nest site and type of nest made also may have an effect on whether a given species is predisposed to nest sharing. Brockmann and Dawkins (1979) found a greater tendency for nest sharing in species that fabricate nests from mud or plant materials than in species that dig burrows in the soil or nest in twigs. It may be more efficient for two or more individuals to collaborate in nest construction than to build alone when repeated trips must be made to obtain the raw materials for nest fabrication, whereas it may be equally efficient for a lone wasp to simply excavate excess soil or pith away from the burrow entrance as for two or more. In other words, the significance of the association of fabricated nests with the transition from solitary to social nesting may be that there is increased opportunity for individuals to provide useful help when nests are to be built from raw materials that must be gathered and fashioned. It may also be a matter of working space; with excavation, extra individuals may simply get in the way.

Nesting on flat substrates such as walls or the ground requires that parasites, predators, and even potential usurpers need only search in two dimensions, whereas twigs and stems add a third dimension, so that populations of species nesting in these latter substrates tend to be more dispersed. Freeman (1982) reported that parasitism of the mud dauber *Sceliphron fistularum* in Jamaica differed significantly depending on substrate. *Melittobia* parasitism of mud nests constructed on vines was 5% versus 49% in nests on flat substrates at the same locality. Thus nest site differences may be another factor that can change the cost-benefit equation in favor of two or more cooperating individuals, because some types of nest sites may offer greater opportunities for individuals to provide useful assistance in the form of guarding behaviors. Michener (1985) has suggested that nest guarding, in particular, will have a higher payoff in two-dimensional settings.

Nest guarding is common in presocial and primitively eusocial halictids (Michener 1974, 1990b), and females typically perform the task. However, males of several genera of sphecid wasps often are associated with their mates' nests (see Table 17.1). Males of some species have been shown to perform guard tasks, whereas males of other species remain closely associated with nests for mating purposes, but have not been observed to actually guard. Hook and Matthews (1980) and Brockmann and Grafen (1989) have reviewed the occurrence of sphecid male guarding and point out that it has evolved at least four times. Of course, female sphecids also undertake guard duty (Fig. 17.2a), but overall, female guards appear to be far less common among sphecids than in the presocial and primitively eusocial bees.

For some birds, such as the scrub jay (Woolfenden and Fitzpatrick 1984), availability of suitable nest sites may limit opportunities for young individuals to successfully establish nesting territories, leading to nest sharing. In general, lack of suitable nest sites is probably not a major limiting factor for most sphecids. But for the Australian *Arpactophilus mimi*, intense competition for empty mud cells may be important in promoting nest sharing (Matthews and Naumann 1988). In such a situation, the only way for many individuals to achieve any fitness may be indirectly, by helping relatives to rear additional progeny.

**Fig. 17.2.** (a) External view of a nest of the Australian sphecid wasp *Arpactophilus mimi* constructed in a reused mud dauber wasp cell. The extensive silk produced by adults has closed the damaged space, leaving only a small circular entrance, which is always guarded by an adult wasp, (from Matthews and Naumann 1988, courtesy of CSIRO Editorial and Publishing Unit). (b) Interior view of an *A. mimi* nest showing the extensive silk lining and the five adult inhabitants. Cells are silk-enshrouded at the left side of the nest. Adults are rather tenacious and reluctant to leave the nest, even after considerable disturbance.

The type of nest material may also channel opportunities for social evolution. For example, Hansell (1987a) postulates that the ability to produce high-quality building materials in the form of high-grade nest paper was the key factor that allowed development of the polistine and vespine petiolate nest suspension system, and this in turn opened virtually unlimited opportunities for nest site choice for these groups. In contrast, the primitively eusocial stenogastrine wasps remained restricted to suspending their nests from naturally occurring pendant objects, such as rootlets, and their poorer quality paper seems to have limited nest size as well (but see Wenzel, this volume). Apparently no sphecids use masticated paper for nest construction, and except for *Microstigmus* (see *Microstigmus comes*: Case Study of a Eusocial Sphecid), none has a petiolate suspension system. Thus, if paper manufacture and/or petiolate nest suspension somehow predispose wasps to sociality, sphecids as a group must be largely excluded from opportunities for the development of advanced social organization.

However, other aspects of the nest itself may facilitate or constrain individual reproductive success and group living (see also Starr, this volume). For instance, *Microstigmus* nests are unique among sphecids not only because of their petiolate suspension, but also because of the use of plant fibers embedded in a silk matrix. The related pemphredonine genera that have social species nest in the soil (*Spilomena*: McCorquodale and Naumann 1988) or in reused mud nests (*Arpactophilus*: Matthews and Naumann 1988), neither of which is particularly distinctive for sphecids. What is distinctive in all three genera is that they use adult-manufactured silk to construct or line cells (Figs. 17.2, 17.3). Acquisition of the ability to produce silk may have been an important factor that preadapted adults in these groups for a social existence. If increased amounts of silk conferred increased protection against pathogens of brood, or made it less likely that a nest would fail, or made it relatively easy to reuse cells, then shared silk production by females might be selectively advantageous, especially if the silk production ability of individuals was limited in these relatively small wasps. An analogous idea has been suggested for communally living spiders (Reichert 1985). To date, no data are available concerning the energetics of silk production in these wasps.

Other features of nest architecture in these three pemphredonine genera also may have facilitated social evolution. Nests of *Microstigmus comes* (Fig. 17.3), and to a lesser degree those of *Arpactophilus mimi* (Fig. 17.2b), characteristically include a large open area in which the adults congregate. Called the vestibule in *Microstigmus* nests (see West-Eberhard 1977b: fig. 1), this area is a high-use space between the cells and the entrance, so that such an architecture tends to necessitate in-

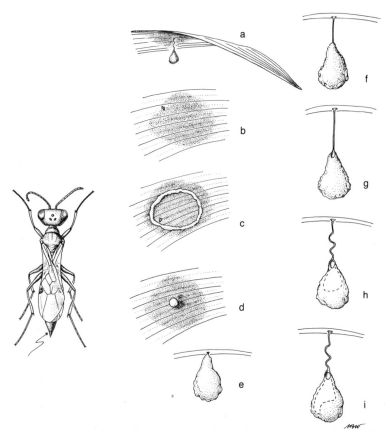

**Fig. 17.3.** Completed nest (a) and nest construction of *Microstigmus comes*. All material to be used is first loosened from the underside of the *Chryosophila* palm leaf surface (b), then gathered together in a single undifferentiated mass (c, d), bound and wrapped with silk (e), and finally lowered on a silken thread (f, g). The suspended mass is then partially hollowed and the entrance is made (g). Finally, the vestibule is made (h, dotted line), and the first pouchlike cell is excavated in the bottom of the nest and heavily lined with silk (i), after which provisioning begins. Also shown is an *M. comes* female (left); the arrow shows the location of the silk gland and application brush at the tip of the abdomen.

creased interaction among adults. In contrast, soil-nesting sphecids such as *Cerceris* typically construct a relatively long main burrow with short lateral burrows leading to cells, an architecture that requires only minimal interactions among nest occupants. Many ground-nesting presocial and primitively eusocial halictids also seem prone to construct an enlarged communally used space just inside the entrance (Kukuk and Eickwort 1987).

586                                    Robert W. Matthews

## Female Longevity, Multivoltinism, and Adult Overwintering

Attributes discussed in this section tend to reflect the regional orientation of most biologists. Among sphecids, it is not the case that there are more species in temperate regions than in the tropics, nor is it true that group living occurs relatively more often in cooler latitudes. Rather, since most biologists live in temperate regions, they have tended to study temperate species disproportionately, and their interpretations of phenomena tend to reflect this bias. Thus, it has sometimes been difficult to resist the tendency to overgeneralize from data gathered largely from higher latitudes.

Clearly, in temperate regions at least, if there are to be overlapping adult generations (a requisite feature of eusociality), there must be more than one generation per year. Thus life history traits that allow generational overlap require scrutiny. Potentially significant preadaptations for such overlap, especially in extratropical regions, include increased female longevity, multivoltinism, and adult overwintering. Evans and West-Eberhard (1970) concluded that an important factor explaining the lack of significant social evolution in the family Sphecidae is the relatively short female life span, such that mothers rarely overlap with daughters. Relatively few data on female longevity are available for presocial or primitively eusocial sphecids. Eberhard (1974) recorded one marked female of *Trigonopsis cameronii* continuously provisioning her nest over a period of 63 days. Adult females of *Cerceris antipodes* overwinter, and some maintain residency in nests for over 50 days (McCorquodale 1989a); nonprovisioning females lived up to 69 days in the population studied by Evans and Hook (1986). Strictly solitary female sphecids seem to have shorter lives as adults, rarely more than 30 days (e.g., Toft 1987). Females of many species of sphecids do not live long enough to overlap their young even when there are two generations in a single summer.

The proximate basis of female longevity may be relative body fat content. For example, polistine queens tend to have much more fat than workers (see Hunt, this volume). The physiology of fat deposition and the role(s) that differential individual fat content may play are aspects of sphecid wasp sociality that have been largely ignored, but whose investigation may prove fruitful.

Among temperate-zone sphecid wasps, the most widespread condition is univoltinism—only one generation per year—throughout a species' range. Why this should be is a bit of an enigma, for often there is ample time for a second and even third generation. For example, the solitary sand wasp *Oxybelus uniglumis* has only a single generation per year in Athens, Georgia, United States (Matthews, unpubl.), though it preys on muscid flies, which are always abundant, and there is no obvious developmental or other reason why it should not have multiple generations.

At the other extreme, among a number of sphecids, especially in tropical species, nesting continues year-round with neither interruption nor synchrony. Provided the mother is long-lived and ecological factors make helping a rewarding option, such multivoltinism offers the option for females to assist their mothers in a subsocial nest association.

By overwintering as inseminated adults rather than as immatures that must complete development, emerge, and mate, female wasps can begin nesting earlier in the favorable season, a factor that theoretically might favor development of at least partial multivoltinism. The fact that all primitively eusocial halictid bees (Michener 1974, 1990b) and the eusocial vespids overwinter as adults suggests that such behavior might be a significant preadaptation for social development. In contrast, relatively few sphecids overwinter as adults (see O'Brien and Kurczewski 1982). All of the temperate-zone species in Table 17.1 apparently overwinter as prepupae, with the exception of several of the *Cerceris* species (Tsuneki 1965, Evans et al. 1976, Salbert and Elliott 1979, Evans and Hook 1982a,b, Kurczewski and Miller 1984, McCorquodale 1989a). Thus, in the temperate-zone Sphecidae, the general absence of winter hibernation by adults may be an important factor limiting the extent of social evolution. *Cerceris antipodes* may be the closest sphecid approximation to the primitively eusocial halictid pattern (McCorquodale 1988b).

## Maternal Control over Sex Ratio, Size, and Fecundity

Sex investment ratios in Hymenoptera are highly plastic and often differ from 50:50, with a bias toward excess female production commonplace. Such biased sex ratios can result from several causes, including inbreeding, within-sex competition among relatives for resources, and group selection (Frank 1986). Because sex investment ratios are clearly related to fitness and can be measured with relatively great precision, and because of the peculiar asymmetries of relatedness resulting from the haplodiploid method of sex determination, hymenopteran sex investment ratios have received a great deal of theoretical and empirical attention (Hamilton 1967, Joshi and Gadagkar 1985, Grafen 1986, Frank and Crespi 1989).

Sex ratio biases can become important preadaptations for the origin and maintenance of eusociality in those species whose generations overlap and where offspring are cooperatively produced. Seger (1983) has described a model in which sex ratio biases may preadapt a species for sociality. Part of a population of some species may be univoltine while another subset develops directly to produce a second generation during the same season. If spring males from the first generation overlap females from the second and successfully mate with those females, then these males make a disproportionate genetic contribution to the next generation compared with their sisters, whose reproductive success is unaffected by the overlap (Seger 1983). In addition, if spring males are able to mate with summer-generation females, summer males become less necessary. The predicted outcome of such a life history is that the second (summer) generation will be female-biased. The significance for sociality is that daughters will be more likely to help mothers when they can rear sisters in preference to brothers because of the potentially greater relatedness between sisters with male haploidy, assuming other factors are also favorable.

Among solitary bees and wasps, there is a striking tendency for nonrandom patterns of sex allocation within a nest (Krombein 1967, Longair 1981). Seasonal swings in population sex ratio also are being increasingly documented (see also Cowan, this volume). Data from the most extensively documented example of seasonal sex ratio swings among sphecids, the pipe-organ mud dauber (*Trypoxylon politum*), fail to support Seger's model, perhaps because of the relatively infrequent overlap between spring and summer adults and the unique role of males as nest guards in this species (J. Brockmann and A. Grafen, unpubl.). A similarly low probability of spring-emerging males overlapping to participate in summer matings has been reported for *Megachile rotundata* (Tepedino and Parker 1988), although the summer generation of this solitary bee exhibits a significantly female-biased numerical sex ratio, consistent with Seger's model. Clearly, further data are needed for species that appear to better fit the model's criteria. One possibility is *Cerceris antipodes* of temperate Australia, in which inseminated females overwinter and may share nests with their daughters (McCorquodale 1989a). Unfortunately, no sex ratio data are yet available for *C. antipodes*. Sex investment ratios of the tropical *Microstigmus comes*, which does not conform to Seger's scenario, are consistently female-biased (see *Microstigmus comes*: Case Study of a Eusocial Sphecid).

In addition to determining what sex her offspring will be, a mother can influence her offspring's size (and, for daughters, presumably fecundity) by the amount of food she provides them as larvae. Earlier in the season, prey are smaller and/or less abundant, and there would seem to be a premium placed on rapid initial colony growth. In accordance with the expectation that these conditions favor smaller individuals, wasps in a first brood of eusocial vespids are commonly smaller than those produced later (e.g., West-Eberhard 1969, Spradbery 1972).

Small size and reduced fecundity due to maternal investment decisions affect the options available to individuals and the payoffs resulting from alternative behavioral choices. The critical consideration for social evolution is that other (ecological) conditions must be favorable to make even a relatively diminutive daughter's help pay off in terms of inclusive fitness benefits. Because larger amounts of food tend to correlate with production of daughters rather than sons in solitary or presocial wasps (Brockmann and Grafen 1989 and references therein), and helpers would enable more food to be provisioned per unit of time (and hence per offspring), mothers with helpers would be expected to produce a higher proportion of daughters. As the brood becomes increasingly female-biased the helpers' inclusive fitness inevitably rises (Frank and Crespi 1989), assuming that the mother has fewer than two effective matings (so that sisters are more closely related to each other than brothers are to their sisters) and that the reproductive value of the sexes is equal. However, as a population becomes increasingly female-biased the reproductive value of females is diminished relative to males (Andersson 1984).

## Provisioning with Multiple Small Prey

All the species in Table 17.1 characteristically provision each brood cell with more than a single prey item. This requires them to repeatedly come and go to a fixed nest. Such a situation may also be regarded as a preadaptation that favors helpers. This may account for the fact that few members of the family Pompilidae, all of which provision brood cells with a single spider, are known to share nests (for exceptions [all of which are tropical species] see Williams 1919, Kimsey 1980b, Wcislo et al. 1988). The same reasoning may apply also to many members of the subfamilies Sphecinae and Ampulicinae, which characteristically provide only a single prey item for each larva.

Evans (1958) considered the transition from mass provisioning (all prey stocked before oviposition) to progressive provisioning (prey brought in over several days after the egg hatches) to be a critical behavioral change promoting social behavior in wasps. He was perhaps strongly influenced by the fact that eusocial vespids universally provision their nests progressively after ovipositing into the empty cell. Progressive provisioning by bees occurs primarily in the allodapines (Anthophoridae), bumble bees, and honey bees, while the eusocial halictid species are mass provisioners (Michener 1974). Progressive and mass provisioning behaviors of sphecids often occur in different species of the same genus, indicating great plasticity in this aspect of the nesting behavior of this group. For example, Weaving (1989) found both progressive- and mass-provisioning species among eight species of African

*Ammophila*. Most species of the cosmopolitan genus *Bembix* are mass provisioners, but at least one species in North America, in Africa, and in Australia is known to be a progressive provisioner (Evans and Matthews 1973a, Gess and Gess 1989). *Microstigmus comes* is a mass-provisioning species (Matthews 1968b), yet apparently all members of the *M. bicolor* group lay the egg in the empty cell and provision progressively (West-Eberhard 1977b). Thus the form of provisioning, per se, seems to be less relevent as a preadaptation for sociality than was previously thought.

Provisioning behavior may be significant for social evolution in sphecids in a different way, however. Malyshev (1968) pointed out that the most behaviorally advanced sphecids (Pemphredoninae, in his opinion) have switched from true "hunting" to essentially "collecting" provisions by preying on abundant, more-or-less defenseless groups of insects such as aphids and psocids, which are little more than minute sacks of protein, easily caught and transported. In general, wasps using these prey kill them by biting rather than stinging, and the prey mass putrifies and/or dehydrates relatively rapidly if isolated from the nest. Large numbers of such small prey are collected (the record may be 171 individual thrips found in a single cell provisioned by *Microstigmus thripoctenus*: Matthews 1970), and these are packed into a compact ball that may be masticated before oviposition. The most socially advanced sphecids (which were unknown to Malyshev) also belong to the Pemphredoninae, and they similarly prey on tiny but abundant insects—e.g., Collembola, Thysanoptera, and Cicadellidae in *Microstigmus* (Matthews 1968b, 1970; West-Eberhard 1977b), Psyllidae in *Arpactophilus mimi* (Matthews and Naumann 1988), and Psyllidae and Eulophidae in *Spilomena subterranea* (McCorquodale and Naumann 1988). *Microstigmus comes* and *M. thripoctenus* oviposit on the mass of prey (see Matthews 1968b: fig. 10) rather than on a specific place on a single paralyzed prey's body as do most other sphecids. In this respect the manner of egg deposition is similar to the situation in most mass-provisioning bees.

Rapid collection of small prey can be shared easily among several females, whereas a single female would not only require an inordinately long time to collect the equivalent amount, but would incur the additional costs of an extended period of nest vulnerability to parasites and predators, prey deterioration, and prey theft by conspecifics. Paradoxically, small prey size could also serve to minimize theft, since such individual prey items that are easily gathered may not be "worth" the cost of stealing them. Prey mastication may serve a similar function by rendering prey relatively more difficult to steal, transport, or use (Eberhard 1974).

A requirement for numerous small prey provides a situation in which a helper could in fact provide an enormous boost to the fitness of its mother, perhaps more than compensating for its loss of personal fitness by forgoing reproduction. This singular fact may help to explain why some pemphredonines have progressed so far toward eusociality. Most other sphecids take larger prey and stock far fewer prey items per cell; the more separate prey items to be collected, the greater the number of individual adults that can be productively involved in the colony work. Interestingly, workers of the highly eusocial vespids characteristically collect masticated prey one jawful at a time, thereby effectively partitioning the food resource in a manner analogous to the socially advanced pemphredonines, although many primitively eusocial polistines retain the habit of solitary vespids of transporting whole caterpillars (R. Jeanne, unpubl.).

In addition to being small, the prey used by the social pemphredonines in the genera *Microstigmus*, *Arpactophilus*, and *Spilomena* probably represent a food resource that lacks much seasonal fluctuation in abundance. A constant and relatively predictable food supply reduces the chance that larval development will fail and provides continuous work opportunities for potential helpers. Among Australian birds such uniformity of prey abundance tends to favor cooperative breeding (Ford et al. 1988). Whether it is important for sphecid sociality would be an interesting avenue to explore further.

## Nestmate or Kin Recognition

For incipient division of labor to develop to any degree, it might be expected that individuals would be able to recognize and discriminate among nest associates so that altruistic aid can be dispensed nonrandomly to more closely related individuals. Nestmate recognition is linked with population structure in that philopatry makes it likely that relatives will be nearby. Although evidence for nestmate or kin recognition and discrimination is rather extensive for primitively eusocial bees (e.g., Kukuk et al. 1977, Greenberg 1979, Smith and Ayasse 1987) and *Polistes* wasps (Gamboa et al. 1986b), to date there exist no similar data for any sphecid (see also Gadagkar 1985b, Venkataraman et al. 1988). However, it seems likely that such an ability exists among sphecids, and in fact, nestmate recognition may be widespread. It seems likely that a guard wasp, such as occurs in *Arpactophilus mimi* (Fig. 17.2a) or *Cerceris antipodes* (McCorquodale 1989b), must be able to discriminate between residents and nonresidents, indirect evidence for the existence of nestmate or kin recognition. Nesting females of the solitary cicada-killer wasp (*Sphecius speciosus*) are more aggressive toward nonneighbors than toward neighbors, indicating some ability to discriminate (Pfennig and Reeve 1989). Male guards of *Trypoxylon pol-*

*itum* almost surely possess the ability to distinguish their female mates from other females. Smith (1983) has suggested that kin recognition may have initially evolved as a mechanism for choosing nonrelated mates and preventing inbreeding. The ability to discriminate between kin and nonkin, or among kin of varying degrees of relatedness, would then seem to be but a small shift.

## *MICROSTIGMUS COMES*: CASE STUDY OF A EUSOCIAL SPHECID

The fundamental discoveries of *Microstigmus* biology were made by Myers (1934), who determined that the curious baglike nest on a spiral stalk was made by the wasp itself, thereby correcting the original interpretation of Ducke (1907), who supposed that the nest bags were spider egg sacs and the wasps were parasites of a spider. (The specific epithet of the type species, *Microstigmus theridii*, was given in reference to the supposition that it parasitized theridiid spiders.) Myers discovered that the nests were constructed from hairs of the supporting plant leaf. He further recorded the presence of two adult females from one of the nests in his sample, but the significance of this finding was overlooked. (Actually, Ducke had reported the same thing in his original description of the genus in 1907, saying that he had observed five nests, all of which contained several females.)

In 1968, I presented evidence suggesting that *M. comes* from Costa Rica was eusocial, with most nests having three to eight females, one of which always had better developed ovaries than the others, a situation indicative of a reproductive division of labor (Matthews 1968a,b). Some females in such nests were clearly teneral (newly emerged), suggesting overlap of adult generations. Females were also found to cooperate in provisioning cells with collembolans and to cooperatively respond to artificial nest disturbance or the presence of a parasitic braconid wasp. These observations seemed to satisfy all of the criteria for eusociality (see Cowan, this volume: Table 2.1), thereby earning *M. comes* a permanent niche in the literature on insect societies as the only eusocial sphecid (Wilson 1971).

Because the nature of the evidence used to support the contention of eusociality was not conclusive, the hypothesized level of *Microstigmus* sociality was not regarded as firm (West-Eberhard 1978a, Eickwort 1981). Knowledge of individual females' behavior has proved frustratingly difficult to obtain because of the delicate nature of the tiny nests and the impossibility of directly observing activities within the nest. The relatively small size of the adults (< 4 mm long) makes them difficult to mark for individual recognition. Nevertheless, the large numbers of nests available for studying colony development and the use of techniques for determining nestmate genotypes have provided important insights into social organization in this species. Recent studies (see Reproductive Division of Labor) have made the eusocial status of *M. comes* unequivocal (Ross and Matthews 1989a,b).

## Nest Construction

*Microstigmus* nest construction is unusual for at least four reasons. First, the petiolate pendant fabricated nests made by *Microstigmus* wasps are unique in the family Sphecidae. All other members of the family burrow in some type of substrate (e.g., sand, logs, twigs) or construct nests of mud plastered to a substrate or "rent" preexisting cavities. In species such as *M. comes*, the spiraled petiolate nest suspension system seems to have a buffering effect against strong winds that sometimes accompany tropical rains (Matthews, unpubl.), though other functions are also possible.

Second, several *Microstigmus* species, including *M. comes*, are host-plant specific, and always construct their nests from the pubescence found on the leaves of their host plant, an unusual commensal type of insect-plant interaction. Other species use bits of punky wood or stone fragments for their nests, which may be suspended from rootlets or from the underside of sheltered rock faces. A few species regularly nest under overhanging roofs of man-made structures (Matthews, unpubl.). Although several sphecid lineages excavate their nests in pithy plant stems, only *Microstigmus* uses plant fibers as the raw material for nest construction in a fashion similar to eusocial vespids.

The third unique aspect of *Microstigmus* nesting is the fact that the nests are given shape and form by silk secreted by glands located in the tip of the female's abdomen (see Downing, this volume). As noted above, silk production by an adult hymenopteran is notable; it occurs only in *Microstigmus* and two closely related genera, *Spilomena* and *Arpactophilus*. (Another pemphredonine, *Psenulus*, also produces a silklike material to line its nest, and may possess a similar gland: see Malyshev 1968: fig. 115, W. Rubink, unpubl.).

The fourth unique aspect is the manner of nest construction, first described by Matthews and Starr (1984). We originally had supposed that nests were constructed piecemeal, one mouthful at a time, as is the case for the pendant nests of familiar social wasps such as *Polistes*. However, the *Microstigmus* method is radically different from that known for any other social insect (Fig. 17.3). Nest size depends on the amount of raw material gathered by the founding wasps at the outset; no opportunity for subsequent expansion or alteration exists. This contrasts sharply with the situation in virtually all other eusocial insects, in which the nest typically gradually increases in size with age (e.g., Wenzel, this volume).

## Nest Establishment

Starting a new colony is a risky venture for social insects (Wilson 1971). The extremely delicate nature of *M. comes* nests is a fact that cannot be ignored; I know of no other sphecid whose nests are so potentially vulnerable to environmental effects. In Costa Rica, frequent brief downpours accompanied by strong winds have been observed to tear incipient *M. comes* nests from the underside of the palm leaves before the nests have been sufficiently well anchored with silk.

Nests can be constructed entirely by a single foundress (haplometrotic founding) or by several females who work cooperatively to construct a shared nest (pleometrotic founding). Groups of cooperating females are able to establish a nest more rapidly than are solitary foundresses. In one population in Costa Rica, groups of two to six females required an average of 36.8 working hours (about four days) to construct a new nest, whereas a solitary female was observed to require 16 days for the same task (Matthews and Starr 1984). Thus, compared with pleometrotically established nests, single-foundress nests clearly are at risk to the vagaries of weather over a much longer period and hence have a correspondingly higher chance of being destroyed. Furthermore, assuming female life span and fecundity are finite, delays necessitated by prolonged time spent in haplometrotic nest construction lessen an individual female's chance of realizing any reproductive success. Queller (1989) has suggested that the overall head start enjoyed by a female associating with a viable nest may be an important factor favoring female offspring remaining with their parent rather than independently starting a new nest.

Average nest longevity for *Microstigmus comes* was 3.7 months in the Costa Rican study population (SD = 3.3 months, $N$ = 801), with five nests having continuous documented activity for 20 months (Matthews, unpubl.). Thus nests may persist over relatively long periods of time where successive generations can inhabit them and cells can be reused.

## Parasitism, Predation, and Usurpation

Another factor that could favor cooperative nesting is the benefit that accrues to each individual derived from improved protection against natural enemies (see Factors Favoring Social Behaviors). Such benefits are possible regardless of the genetic relationship of the cohabiting females. To establish whether such benefits might apply in the case of *M. comes* requires knowledge of the extent of parasite/predation/usurpation pressures, and the effectiveness of a single female versus a group in nest defense. Data exist concerning the former but not the latter.

The most common parasite of *M. comes* is a braconid wasp, *Heterospilus microstigmi*. (The only other known parasite of *M. comes* is a chalcid wasp, *Trichokaleva microstigmi* [Bouček 1972], which was extremely rare at my Costa Rican site.) Large samples of nests from a single locality in Costa Rica from three years have revealed parasitism levels to be rather low—three of 44 nests in 1981, four of 52 nests in 1982, and 34 of 102 nests in 1987 (Matthews, unpubl.). Few parasites were reared from nests that had fewer than four cells; in the 1987 sample, parasitized nests almost always had eight or more cells (Fig. 17.4a). While it may be that larger nests are more active and thereby attract more parasites, a combination of parasite morphology and host nest structure, more than *M. comes* behavior, may account for the observed pattern of parasitism. Because the braconid's ovipositor is rela-

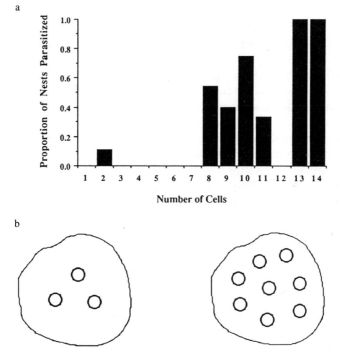

**Fig. 17.4.** (a) The incidence of parasitism of *Microstigmus comes* by the braconid *Heterospilus microstigmi* in relation to nest size, based on a sample of 102 nests from Corcovado National Park, Costa Rica, collected in March 1987. (b) The placement of cells in a young *M. comes* nest (left) compared with that in an older nest (right). The thick layer of unconsolidated plant fibers surrounding each cell in young nests is hypothesized to account for the lower incidence of parasitism of younger nests by *H. microstigmi*, which must drill through the nest with its ovipositor from the outside.

tively short, it is physically unable to penetrate deep enough through
the nest bag to reach the cell contents when there are fewer than four
cells, each of which tends to be centrally placed and to have relatively
thick walls (Fig. 17.4b). However, as the number of cells increases, the
bag is stretched to accommodate them, thinning the outer wall and
consequently making cells more accessible to the parasite's ovipositor.
Since the rate of parasitism per group member may be increased over
that sustained by solitary individuals (Lin 1964), parasitism has been
suggested also as a disadvantage of group life. However, parasitism by
H. microstigmi appears not to occur often enough to be a major current
selection pressure either for or against group life in M. comes.

Predation likewise seems not to be a major mortality source to M.
comes nests. Nests with torn bags have seldom been found, and predation by birds has never been observed. The trunks of the mature host
Cryosophila palms have long sharp spines, which appear to deter most
climbing vertebrates. In other groups of tropical eusocial wasps, army
ants constitute a major predation threat (Jeanne 1970a); nests under
siege from an advancing column of army ants are simply abandoned
and the resident adults abscond to rebuild elsewhere. Only once in
hundreds of hours of field observation did I observe Eciton army ants
swarming over a palm with an active M. comes nest. After the column
departed, I removed the nest and upon dissection found three cells,
their provisions and immature stages undisturbed even though ants
had crawled over the nest. A fully provisioned cell contains a mass of
collembolan prey tightly packed into a ball about the size of the head of
a pin, a very small payoff to a plundering ant in any case.

Workers of a variety of other ant species commonly forage on the
leaf surface in the vicinity of nests. The petiolate suspension of M.
comes nests clearly narrows the point of contact by which ant foragers
may chance upon the nest. Given that the petiole of nests of the polistine Mischocyttarus drewseni is coated with an ant repellent (Jeanne
1970a), it seems possible that Microstigmus comes nests might be similarly protected. In several hundred hours of field observations, C.
Starr and I never observed foraging ants (other than army ants) to
crawl onto the nest petiole, but in choice tests similar to Jeanne's
(1970a), Solenopsis ants were quite willing to crawl down petioles of
active M. comes nests and showed no hesitancy or avoidance behaviors.
Thus we have concluded that the petioles are not chemically protected
(Matthews and C. Starr, unpubl.). However, it may be that the scraped
area of the leaf undersurface from which the nest raw material is gathered is in some way made repellent to ants. Only rarely do ants tra-
verse the scraped area surrounding the petiolar attachment. Instead
they generally detour around nests.

Interspecific usurpation appears to be more prevalent than either
parasitism or predation. Perhaps the greatest threat to nesting M. comes

is another sphecid wasp, *Trypoxylon latro*, which is prone to use *M. comes* nests for its own nesting activities. Incidence of *T. latro* takeovers in the Costa Rican nest samples was 7.8% in 1981 and 9.2% in 1982 (Matthews 1983). A more recent sample (March 1987) revealed a somewhat lower 3.7% ($N = 194$ nests) takeover rate by *T. latro*. Because most other members of the genus *Trypoxylon* typically reuse a variety of empty cavities for nesting, it is possible that many *T. latro* nests are made in abandoned *M. comes* nests. However, *M. comes* nests apparently persist on leaves for only a short time after they have been abandoned (only seven of 201 nests collected in March 1987 were empty), and *T. latro* wasps are much larger than *M. comes*, observations that lead me to speculate that usurpation of active nests may be the more common habit of *T. latro*. Analysis of *T. latro* nest incidence during monthly censuses of all *Microstigmus* nests in my study site over a 15-month period revealed that nests of all ages and sizes were equally vulnerable, with an average of 5% of active nests converting to *T. latro* nests each month (Matthews 1983). Thus, neither nest size nor number of resident wasps seems to be related to the likelihood of a takeover by *T. latro*, suggesting that nests with more females are not able to better defend themselves against usurpation.

Other occasional occupants of *M. comes* nests included three spider species: *Myrmarachne centralis* (Salticidae), a *Trachelas* species (Clubionidae), and a *Eustala* species (Araneidae). A species of *Tapinoma* ant claimed one nest, the only time that an ant colony has been observed to use any *Microstigmus* nest. Whether these occasional nest occupants displace nesting *M. comes* adults or simply reuse an abandoned nest is not known.

Finally, the extent to which conspecific usurpation occurs has not been determined because it is so difficult to mark these small wasps. Because intraspecific takeovers are commonplace among all groups of nesting Hymenoptera it would be surprising if it did not also occur in *M. comes*; such data, when available, will be interesting for comparing single- and multifemale nest vulnerability.

In summary, evidence for any important role of parasitism, predation, or usurpation in the evolution of eusociality in *M. comes* is at best equivocal. Perhaps the more important point that emerges from these data is the relatively high immature survival rate in *M. comes*, as reflected in high nest survivorship, compared with that reported for solitary sphecids (e.g., Danks 1971, Freeman and Parnell 1973).

## Genetic Relatedness

Average genetic relatedness ($r$) for female nestmates of *M. comes* ranges from 0.63 to 0.70 (Ross and Matthews 1989a,b). These high levels of relatedness apparently result from simple family (mother-daughter) groups constituting colonies. Indeed, in nearly two-thirds of all nests having five or more females ($N = 64$ nests), genotype distributions were consistent with a matrifilial monogynous society involving a single-mated "queen." Colonies that were inconsistent with this interpretation still had average $r$ values greater than 0.60. Similarly high relatedness values have been reported for some aggregations of *Cerceris antipodes* (McCorquodale 1988a, see also Ross and Carpenter, this volume).

Populations of *M. comes* appear on the basis of observational data to be moderately "viscous" (Hamilton 1964a,b), so that some measurable level of relatedness at the microgeographic level (i.e., between nests on the same host plant or group of plants) is to be expected. In the field, there appears to be a pronounced tendency for particular plants to host nests over several years (Matthews, unpubl.), other nearby host plants that appear perfectly suitable for nesting by *M. comes* not having been recorded to have any nests. Orphaned colonies mentioned in Matthews and Starr (1984) rebuilt on a different palm frond on the same host plant in five of six cases monitored, and the sixth colony moved to an immediately adjacent (leaves touching) plant. Possibly the clustering of nests on particular plants over long periods of time reflects some as yet undetected chemical factor(s) or physical microenvironmental differences. Regardless of the underlying causal factors, these behavioral observations suggest that *M. comes* fits Hamilton's predictions in having relatively low dispersal and high philopatry. Consistent with this is the discovery of a moderate level of genetic relatedness among nonnestmate wasps from the same or from neighboring host plants (Ross and Matthews 1989a). An important additional inference from the genetic studies is that *M. comes* is likely to be outbreeding, because observed subpopulation genotypic ratios closely approximated those expected under Hardy-Weinberg conditions.

Despite the unusually high degrees of genetic relatedness among cohabiting females, as yet there is no direct evidence of nestmate or kin discrimination for *M. comes*. Indirect evidence is provided by the observation that orphaned nestmates appear to remain together and establish new nests close to the original nest (Matthews and Starr 1984). The genetic data on multifoundress groups, for which relatedness is high (Ross and Matthews 1989a), can be taken as strong inferential evidence that individuals discriminate kin from nonkin or former nestmates from nonnestmates.

## Sex Ratio

I (1968b) reported an apparent numerically female-biased sex ratio for *M. comes*, and, more recently, K. Ross and I (1989a) confirmed that females outnumber males by nearly four to one. Among newly reared progeny, the female bias increased to 6.7:1. Allowing for the differential weight of males and females (to get a better idea of investment levels) and separating putative reproductive females from the total female group (see Reproductive Division of Labor) yielded an estimated population-wide sex investment ratio of 1.78:1.

The pattern of production of the two sexes (Fig. 17.5) is particularly significant in *M. comes* in that it resembles the pattern characteristic of

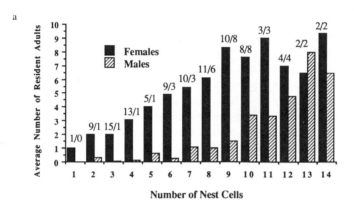

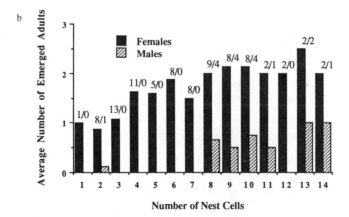

**Fig. 17.5.** Production of females and males of *Microstigmus comes* in relation to nest size in nests collected from Corcovado National Park, Costa Rica, in March 1987. (a) Average number of adults of each sex present at time of collection (all 102 nests collected at night). (b) Average number of adults of each sex reared from pupae following collection ($N = 87$ nests). In only one case was a male reared from a nest that contained fewer than eight cells. For both parts of the figure numbers above bars refer to the number of nests that contained individuals of each sex, i.e., "8/4" means that eight nests of that size contained females and four nests had males. The first number is also the total number of nests of that size sampled.

other eusocial wasps and bees, where male production typically is de-layed until late in the colony cycle (when the nests are larger). Invariably, the first several progeny in a given nest are females, so early-emerging daughters who remain on their mother's nest will be helping to rear siblings that are quite closely related to them (Ross and Matthews 1989a).

## Reproductive Division of Labor

The putative mothers of the colonies of *M. comes* identified on the basis of genotype distributions were found consistently to weigh more than other females in their nests (Ross and Matthews 1989b). These "queen" individuals proved to be significantly heavier on average than the female population at large ($P < 0.001$), as were females from young nests containing solitary (hence known to be reproductive) found-resses. These data suggest that lower-weight females probably are not reproducing, giving rise to a size-structured reproductive division of labor. This size differentiation probably results from differential mater-nal investment in larvae, which may effectively limit the reproductive options of lower-weight individuals, forcing them to become workers.

A dominance hierarchy that reflects different reproductive roles among females in a nest is a possibility. While scraping, rasping, and rolling the plant material during early stages of nest construction, indi-vidual females respond to each other in characteristic dominance/sub-ordinance postures when they meet. Furthermore, one individual ap-pears to "take the lead" in deciding where the new nest will ultimately be suspended; this female makes a disproportionate number of visits to the spot and probes repeatedly there with her abdomen (Matthews, unpubl.) The occasional occurrence of nests with double petioles (Fig. 17.6) suggests that female competition and conflict may exist at the nest-founding stage.

## Evolution of Eusociality in *Microstigmus comes*

The achievement of eusociality in *M. comes* seems best explained by a combination of attributes. These include a unique nest type and mode of construction, sufficiently long adult life spans to allow continuously overlapping generations, use of small but abundant prey, high genetic relatedness among females within individual nests, female-biased sex investment ratios, and delayed male production. From an economic standpoint, it seems clear that for daughter offspring in *M. comes*, stay-ing and helping at the parental nest is a viable option, especially since there is ample work daughters can do in provisioning, nest mainte-nance, and defense that is of great value for nest survival and produc-tivity, thereby increasing their mother's personal fitness as well as their

**Fig. 17.6.** A nest of *Microstigmus comes* with a double petiole. Similar anomalies occur in about 1% of all nests and suggest that cooperating females may have been in conflict concerning the point of nest suspension during construction.

own inclusive fitness. The unusually low immature mortality in *M. comes* is striking compared with that recorded for solitary sphecids. The recent discovery of an apparent size-based reproductive division of labor indicates an elaboration of sociality unprecedented elsewhere in the Sphecidae. In many respects, *M. comes* probably very nearly fits the paradigm of social evolution facilitated by kin selection originally envisioned by Hamilton.

## CONCLUDING REMARKS

The Sphecidae is by far the largest family of wasps, and displays perhaps the greatest range of life histories of any group of aculeate Hymenoptera. All but a very few species construct nests and provide significant amounts of maternal care for their offspring, yet relatively few have evolved beyond a strictly solitary existence. Studies of socially advanced sphecids lag considerably behind those of primitively eusocial bees and polistine wasps, in part because it is difficult to observe intranest activity and nearly impossibile to rear most sphecids in the laboratory (but see Steiner 1984b). Data on individual lifetime fecundity, sex investment ratios, longevity, kin discrimination ability, individual survivorship, and mating frequency are virtually nonexistent.

Quantitative natural history data on nest-sharing sphecids are more sorely needed than ever before in order to test specific evolutionary and mechanistic hypotheses about sociality. Detailed field studies on key species (especially those tropical species with more-or-less continuous overlap of generations) are necessary to obtain data on the relative reproductive success of individuals that adopt alternative strategies as solitary foundresses or as members of different-sized groups, and on the effects of different strategies on inclusive fitness at different seasons and under various environmental conditions. We know next to nothing about the effects of variable group size on efficiency of brood rearing. Finally, we must better understand those aspects of the ecology that make sociality pay in terms of realized inclusive fitness. A focus on causes and extent of nest failure in both social and solitary situations may reveal unsuspected advantages and costs of cooperative behavior, so this aspect of natural history must not be neglected.

The transition from solitary to social living in sphecids clearly involves a number of factors. Parasite pressure, widely thought to be a major selective factor for group living, does not in itself appear to make sociality pay for sphecids; all species seem to have their parasites. Likewise, haplodiploidy is universal in Hymenoptera, so that in the absence of other "enabling" factors it is inadequate for explaining the origins of social living. Maternal investment decisions, increased difficulty of independent nest initiation, and the opportunity to help a long-lived mother may tip the balance from solitary to social nesting. There may be other factors in the sphecid equation as well. For example, what is the significance of adult-produced silk? Does predation on relatively abundant, innocuous small prey in some way facilitate female cooperation? Do nest structural attributes play a role? Has some physical or physiological trait been a major enabling factor? It must be more than mere coincidence that all of the presently known socially advanced sphecid genera are on the same phyletic branch of the tribe Pemphredonini.

*Acknowledgments*

I thank Jane Brockmann, James Carpenter, Janice Matthews, David McCorquodale, Arnold Menke, Kenneth Ross, and Christopher Starr for their critical comments on various versions of this chapter. Jerry Freilich assisted with statistical analysis and preparation of the graphics. H. W. Levi kindly identified the spider inhabitants of the *Microstigmus* nests. National Science Foundation grants BNS-7925708 and BSR-8407840 supported my studies of *Microstigmus*.

# Literature Cited

Abbas, N. D., Sô. Yamane, and M. Matsuura. 1983. Nest architecture and some biological aspects of three *Polistes* species in Sumatera Barat. Ecological Study on Social Insects in Central Sumatra with Special Reference to Wasps and Bees—Sumatra Nature Study (Entomology), Kanazawa University, pp. 29–36.

Abrams, J., and G. C. Eickwort. 1981. Nest switching and guarding by the communal sweat bee *Agapostemon virescens* (Hymenoptera, Halictidae). Insectes Soc. 28: 105–116.

Abugov, R., and R. E. Michod. 1981. On the relation of family structured models and inclusive fitness models for kin selection. J. Theor. Biol. 88: 743–754.

Akre, R. D. 1982. Social wasps. Pp. 1–105 *in* H. R. Hermann (ed.), Social Insects, vol. 4. Academic Press, New York.

Akre, R. D., and D. P. Bleicher. 1985. Nests of *Dolichovespula norwegica* and *D. norvegicoides* in North America (Hymenoptera: Vespidae). Entomol. News 96: 29–35.

Akre, R. D., and J. F. MacDonald. 1986. Biology, economic importance and control of yellow jackets. Pp. 353–412 *in* S. B. Vinson (ed.), Economic Impact and Control of Social Insects. Praeger, New York.

Akre, R. D., and H. C. Reed. 1981a. A polygynous colony of *Vespula pensylvanica* (Saussure) (Hymenoptera: Vespidae). Entomol. News 92: 27–31.

Akre, R. D., and H. C. Reed. 1981b. Population cycles of yellowjackets (Hymenoptera: Vespinae) in the Pacific Northwest. Environ. Entomol. 10: 267–274.

Akre, R. D., and H. C. Reed. 1983. Evidence for a queen pheromone in *Vespula* (Hymenoptera: Vespidae). Can. Entomol. 115: 371–377.

Akre, R. D., and H. C. Reed. 1984. Vespine defense. Pp. 59–94 *in* H. R. Hermann (ed.), Defensive Mechanisms in Social Insects. Praeger, New York.

Akre, R. D., W. B. Garnett, J. F. MacDonald, A. Greene, and P. Landolt. 1976. Behavior and colony development of *Vespula pensylvanica* and *V. atropilosa* (Hymenoptera: Vespidae). J. Kansas Entomol. Soc. 49: 63–84.

Akre, R. D., C. F. Roush, and P. J. Landolt. 1977. A *Vespula pensylvanica/Vespula vulgaris* nest (Hymenoptera: Vespidae). Environ. Entomol. 6: 525–526.

Akre, R. D., A. Greene, J. F. MacDonald, P. J. Landolt, and H. G. Davis. 1980. The Yellowjackets of America North of Mexico. United States Department of Agriculture, Agric. Handbook no. 552.

Akre, R. D., H. C. Reed, and P. J. Landolt. 1982. Nesting biology and behavior of the blackjacket, *Vespula consobrina* (Hymenoptera: Vespidae). J. Kansas Entomol. Soc. 55: 373–405.

Akre, R. D., C. Ramsay, A. Grable, C. Baird, and A. Stanford. 1989. Additional range extension by the German yellowjacket, *Paravespula germanica* (Fabricius), in North America (Hymenoptera: Vespidae). Pan-Pac. Entomol. 65: 79–88.

Alcock, J. 1974. The behaviour of *Philanthus crabroniformis* (Hymenoptera: Sphecidae). J. Zool. Soc. London 173: 233–246.

Alcock, J. 1975a. Social interactions in the solitary wasp *Cerceris simplex* (Hymenoptera: Sphecidae). Behaviour 54: 142–152.

Alcock, J. 1975b. The nesting behavior of some sphecid wasps of Arizona, including *Bembix, Microbembex*, and *Philanthus*. J. Arizona Acad. Sci. 10: 160–165.

Alcock, J. 1978. Notes on male mate-locating behavior in some bees and wasps of Arizona. Pan-Pac. Entomol. 54: 215–225.

Alcock, J. 1980. Communal nesting in an Australian solitary wasp, *Cerceris antipodes* Smith (Hymenoptera: Sphecidae). J. Aust. Entomol. Soc. 19: 223–228.

Alcock, J. 1982. Nest usurpation and sequential nest occupation in the digger wasp *Crabro monticola* (Hymenoptera: Sphecidae). Can. J. Zool. 60: 921–925.

Alcock, J. 1985. Hilltopping behavior in the wasp *Pseudomasaris maculifrons* (Fox) (Hymenoptera: Masaridae). J. Kansas Entomol. Soc. 58: 162–166.

Alcock, J. 1987. Leks and hilltopping insects. J. Nat. Hist. 21: 319–328.

Alcock, J. 1989. Animal Behavior: An Evolutionary Approach, 4th ed. Sinauer, Sunderland, MA.

Alcock, J., E. M. Barrows, G. Gordh, L. J. Hubbard, L. L. Kirkendall, D. Pyle, T. L. Ponder, and F. G. Zalom. 1978. The ecology and evolution of male reproductive behaviour in the bees and wasps. Zool. J. Linn. Soc. (London) 64: 293–326.

Alexander, B. 1986. Alternative methods of nest provisioning in the digger wasp *Clypeadon laticinctus* (Hymenoptera: Sphecidae). J. Kansas Entomol. Soc. 59: 59–63.

Alexander, B. 1987. Eusociality and parasitism in nest-provisioning insects. P. 387 *in* J. Eder and H. Rembold (eds.), Chemistry and Biology of Social Insects. J. Peperny, Munich.

Alexander, B. 1990. A preliminary phylogenetic analysis of sphecid wasps and bees. Sphecos (20): 7–16.

Alexander, R. D. 1974. The evolution of social behavior. Annu. Rev. Ecol. Syst. 5: 325–383.

Alexander, R. D., and G. Borgia. 1978. Group selection, altruism, and the levels of organization of life. Annu. Rev. Ecol. Syst. 9: 449–474.

Alexander, R. D., and P. W. Sherman. 1977. Local mate competition and parental investment in social insects. Science 196: 494–500.

Alexander, R. D., J. L. Hoogland, R. D. Howard, K. M. Noonan, and P. W. Sherman. 1979. Sexual dimorphism and breeding systems in pinnipeds, ungulates, primates and humans. Pp. 402–435 *in* N. A. Chagnon and W. Irons (eds.), Evolutionary Biology and Human Social Behavior: An Anthropological Perspective. Wadsworth, Belmont, CA.

Alford, D. V. 1975. Bumblebees. Davis-Poynter, London.

Allen, J. L., K. Schulze-Kellman, and G. J. Gamboa. 1982. Clumping patterns during overwintering in the paper wasp, *Polistes exclamans*: effects of relatedness. J. Kansas Entomol. Soc. 55: 97–100.

Anderson, R. H. 1963. The laying worker in the Cape honey bee, *Apis mellifera capensis*. J. Apic. Res. 2: 85–92.

Andersson, M. 1984. The evolution of eusociality. Annu. Rev. Ecol. Syst. 15: 165–189.

Araujo, R. L. 1940. Contribuição para o conhecimento de "*Gymnopolybia meridionalis*" (Iher., 1904) (Hym.). Arch. Inst. Biol. São Paulo 2: 11–16.

Archer, M. E. 1972a. Studies of the seasonal development of *Vespula vulgaris* (L.) (Hymenoptera: Vespidae) with special reference to queen production. J. Entomol. (A) 47: 45–59.

Archer, M. E. 1972b. The significance of worker size in the seasonal development of the wasps *Vespula vulgaris* (L.) and *Vespula germanica* (F.). J. Entomol. (A) 46: 175–183.

Archer, M. E. 1972c. The importance of worker behaviour in the colony organization of the social wasps, *Vespula vulgaris* (L.) and *V. germinaca* [*sic*] (F.) (Vespidae; Hymenoptera). Insectes Soc. 19: 227–242.

Archer, M. E. 1976. Tree wasp workers (Hym., Vespidae) excavating soil from underground nests. Entomol. Mon. Mag. 112: 88.

Archer, M. E. 1977. The weights of forager loads of *Paravespula vulgaris* (Linn.) (Hymenoptera: Vespidae) and the relationship of load weight to forager size. Insectes Soc. 24: 95–102.

Archer, M. E. 1980a. Population dynamics. Pp. 172–207 *in* R. Edwards, Social Wasps: Their Biology and Control. Rentokil, East Grinstead, W. Sussex, England.

Archer, M. E. 1980b. Numerical characteristics of nests of *Vespa crabro* L. (Hym., Vespidae). Entomol. Mon. Mag. 116: 117–121.

Archer, M. E. 1980c. Possible causes of the yearly fluctuations in wasp numbers. Brit. Isles Bee Breed. Assoc. News 18: 26–29.

Archer, M. E. 1981a. A simulation model for the colonial development of *Paravespula vulgaris* (Linnaeus) and *Dolichovespula sylvestris* (Scopoli) (Hymenoptera: Vespidae). Melanderia 36: 1–59.

Archer, M. E. 1981b. Successful and unsuccessful development of colonies of *Vespula vulgaris* (Linn.) (Hymenoptera: Vespidae). Ecol. Entomol. 6: 1–10.

Archer, M. E. 1984a. Secondary nests of *Vespa crabro* L. (Hym., Vespidae). Entomol. Mon. Mag. 120: 125.

Archer, M. E. 1984b. Life and fertility tables for the wasp species *Vespula vulgaris* and *Dolichovespula sylvestris* (Hymenoptera: Vespidae) in England. Entomol. Gener. 9: 181–188.

Archer, M. E. 1985a. Secondary nests of the hornet, *Vespa crabro* (L.) (Hymenoptera: Vespidae), produce queens. Proc. Trans. British Entomol. Nat. Hist. Soc. 18: 35–36.

Archer, M. E. 1985b. Population dynamics of the social wasps *Vespula vulgaris* and *Vespula germanica* in England. J. Anim. Ecol. 54: 473–485.

Avitabile, A., R. A. Morse, and R. Boch. 1975. Swarming honey bees guided by pheromones. Ann. Entomol. Soc. Amer. 68: 1079–1082.

Bacardit, R., and M. Moreno-Manas. 1983. Synthesis of δ-lactonic pheromones of *Xylocopa hirsutissima* and *Vespa orientalis* and an allomone of some ants of genus *Camponotus*. J. Chem. Ecol. 9: 703–714.

Baerends, G. P. 1941. Fortpflanzungsverhalten und Orientierung der Grabwespe *Ammophila campestris* Jur. Tijdschr. Entomol. 84: 68–208.

Baker, H. G. 1975. Studies of nectar-constitution and pollinator-plant co-evolution. Pp. 100–140 *in* L. E. Gilbert and P. H. Raven (eds.), Co-evolution of Animals and Plants. University of Texas Press, Austin.

Baker, H. G. 1978. Chemical aspects of the pollination biology of woody plants in the tropics. Pp. 57–82 *in* P. B. Tomlinson and M. H. Zimmerman (eds.), Tropical Trees as Living Systems. Cambridge University Press, New York.

Baker, H. G., and I. Baker. 1973a. Amino acids in nectar and their evolutionary significance. Nature 241: 543–545.

Baker, H. G., and I. Baker. 1973b. Some anthecological aspects of the evolution of nectar-producing flowers, particularly amino acid production in nectar. Pp. 243–264 *in* V. H. Heywood (ed.), Taxonomy and Ecology. Academic Press, New York.

Baker, H. G., and P. Hurd. 1968. Intrafloral ecology. Annu. Rev. Entomol. 13: 385–414.

Balduf, W. V. 1954. Observations on the white-faced wasp *Dolichovespula maculata* (Linn.) (Vespidae, Hymenoptera). Ann. Entomol. Soc. Amer. 47: 455–458.

Barnard, C. J. 1983. Animal Behaviour. Croom Helm, London.

Barnard, C. J., and T. Burk. 1979. Dominance hierarchies and the evolution of "individual recognition." J. Theor. Biol. 81: 65–73.

Barrows, E. M., W. J. Bell, and C. D. Michener. 1975. Individual odor differences and their social functions in insects. Proc. Natl. Acad. Sci. USA 72: 2824–2828.

Barth, F. G. 1982. Spiders and vibratory signals: sensory reception and behavioral significance. Pp. 67–122 *in* P. N. Witt and J. S. Rovner (eds.), Spider Communication: Mechanisms and Ecological Significance. Princeton University Press, Princeton, NJ.

Barth, R. H., L. J. Lester, P. Sroka, T. Kessler, and R. Hearn. 1975. Juvenile hormone promotes dominance behavior and ovarian development in social wasps (*Polistes annularis*). Experientia 31: 691–692.

Batra, L. R., and S. W. T. Batra. 1979. Termite-fungus mutualism. Pp. 117–163 *in* L. R. Batra (ed.), Insect-Fungus Symbiosis: Nutrition, Mutualism and Commensalism. Allanheld, Osmun, Montclair, NJ.

Batra, S. W. T. 1968. Behavior of some social and solitary halictine bees within their nests: a comparative study. J. Kansas Entomol. Soc. 41: 120–133.

Batra, S. W. T. 1980. Sexual behavior and pheromones of the European hornet, *Vespa crabro germana* (Hymenoptera: Vespidae). J. Kansas Entomol. Soc. 53: 461–469.

Beani, L., and S. Turillazzi. 1988a. An experiment on the relationship between spatial behaviour and mating success in the male *Polistes gallicus* (L.) (Hymenoptera, Vespidae). Monit. Zool. Ital. (N.S.) 22: 323–330.

Beani, L., and S. Turillazzi. 1988b. Alternative mating tactics in males of *Polistes dominulus* (Hymenoptera, Vespidae). Behav. Ecol. Sociobiol. 22: 257–264.

Beebe, W., G. I. Hartley, and T. G. Howes. 1917. Tropical Wild Life in British Guiana, vol. 1. New York Zoological Society, New York.

Beirne, B. P. 1944. The causes of the occasional abundance or scarcity of wasps (*Vespula* spp.) (Hym., Vespidae). Entomol. Mon. Mag. 80: 121–124.

Belavadi, V. V., and R. Govindan. 1981. Nesting habits and behaviour of *Ropalidia* (*Icariola*) *marginata* (Hymenoptera: Vespidae) in south India. Colemania 1: 95–101.

Bell, W. J., and R. T. Cardé. 1984. Chemical Ecology of Insects. Sinauer, Sunderland, MA.

Bendegem, J. P. van, D. L. Gibo, and T. M. Alloway. 1981. Effects of colony division on foundress associations in *Polistes fuscatus* (Hymenoptera: Vespidae). Can. Entomol. 113: 551–556.

Bequaert, J. 1918. A revision of the Vespidae of the Belgian Congo based on the collection of the American Museum Congo Expedition, with a list of Ethiopian diplopterous wasps. Bull. Amer. Mus. Nat. Hist. 39: 1–384.

Bequaert, J. 1922. Études sur les Hyménoptères Diploptères d'Afrique. I. À propos des moeurs et de la distribution géographique du genre *Polybioides*. Rev. Zool. Afric. 10: 309–317.

Bequaert, J. 1928. A study of certain types of diplopterous wasps in the collection of the British Museum. Ann. Mag. Nat. Hist. (Ser. 10) 2: 138–176.

Bequaert, J. 1938. A new *Charterginus* from Costa Rica, with notes on *Charterginus*, *Pseudochartergus*, *Pseudopolybia*, *Epipona*, and *Tatua* (Hymenoptera, Vespidae). Rev. Entomol. 9: 99–116.

Bequaert, J. 1940. An introductory study of *Polistes* in the United States and Canada with descriptions of some new North and South American forms (Hymenoptera: Vespidae). J. New York Entomol. Soc. 48: 1–31.

Berland, L., and P. P. Grassé. 1951. Super-famille des Vespoidea Ashmead. Pp. 1127–1174 in P. P. Grassé (ed.), Traité de Zoologie, vol. 10. Masson, Paris.

Bertoni, A. 1911. Contribución a la biología de las avispas y abejas del Paraguay. Ann. Mus. Nac. Buenos Aires 22: 97–146.

Billen, J. P. J. 1987. New structural aspects of the Dufour's and venom glands in social insects. Naturwissenschaften 74: 340–341.

Blackith, R. E. 1958. An analysis of polymorphism in social wasps. Insectes Soc. 5: 263–272.

Blum, M. S. 1969. Alarm pheromones. Annu. Rev. Entomol. 14: 57–80.

Blum, M. S. 1977. Behavioral responses of Hymenoptera to pheromones and allomones. Pp. 149–167 in H. H. Shorey and J. J. McKelvey (eds.), Chemical Control of Insect Behavior: Theory and Application. Wiley-Interscience, New York.

Blum, M. S. 1981. Sex pheromones in social insects: chemotaxonomic potential. Pp. 163–174 in P. E. Howse and J.-L. Clément (eds.), Biosystematics of Social Insects. Academic Press, New York.

Blum, M. S., and J. M. Brand. 1972. Social insect pheromones: their chemistry and function. Amer. Zool. 12: 553–576.

Blüthgen, P. 1938. Systematisches Verzeichnis der Faltenwespen Mitteleuropas, Skandinaviens und Englands. Konowia 16: 270–295.

Bodenheimer, F. S. 1951. Citrus Entomology in the Middle East. Dr. W. Junk, The Hague.

Bohart, G. E., F. D. Parker, and V. J. Tepedino. 1982. Notes on the biology of *Odynerus dilectus* (Hym.: Eumenidae), a predator of the alfalfa weevil, *Hypera postica* (Col.: Curculionidae) Entomophaga 27: 23–31.

Bohart, R. M. 1940. A revision of the North American species of *Pterocheilus* and notes on related genera (Hymenoptera, Vespidae). Ann. Entomol. Soc. Amer. 33: 162–208.

Bohart, R. M., and A. S. Menke. 1976. Sphecid Wasps of the World: A Generic Revision. University of California Press, Berkeley.

Bohart, R. M., and L. A. Stange. 1965. A revision of the genus *Zethus* Fabricius in the Western Hemisphere (Hymenoptera: Eumenidae). Univ. California Publ. Entomol. 40: 1–208.

Bohm, M. F. K. 1972a. Reproduction in *Polistes metricus*. Ph.D. dissertation, University of Kansas, Lawrence.

Bohm, M. F. K. 1972b. Effects of environment and juvenile hormone on ovaries of the wasp *Polistes metricus*. J. Insect Physiol. 18: 1875–1883.

Bohm, M. F. K., and K. A. Stockhammer. 1977. The nesting cycle of a paper wasp, *Polistes metricus*. J. Kansas Entomol. Soc. 50: 275–286.

Bonelli, B., L. Bullini, and R. Cianchi. 1980. Paralyzing behavior of the wasp *Rynchium oculatum* Scop. (Hymenoptera, Eumenidae). Monit. Zool. Ital. (N.S.) 14: 95–96.

Boomsma, J. J. 1989. Sex-investment ratios in ants: has female bias been systematically overestimated? Amer. Nat. 133: 517–532.

Bornais, K. M., C. M. Larch, G. J. Gamboa, and R. B. Daily. 1983. Nestmate discrimination among laboratory overwintered foundresses of the paper wasp *Polistes fuscatus* (Hymenoptera: Vespidae). Can. Entomol. 115: 655–658.

Bossert, W. H., and E. O. Wilson. 1963. The analysis of olfactory communication among animals. J. Theor. Biol. 5: 443–469.

Bouček, Z. 1972. A new genus and species of Pteromalidae (Hym.) parasitic on sphecids in South America. Mitt. Schweiz. Entomol. Gesell. 45: 113–116.

Bourdon, V., J. Lecomte, M. Leclercq, and J. Leclercq. 1975. Présence de nor-adrenaline conjuguée dans l'enveloppe du nid de "*Vespula vulgaris* Linné." Bull. Soc. R. Sci. Liège 44: 474–476.

Bourke, A. F. G. 1988. Worker reproduction in the higher eusocial Hymenoptera. Quart. Rev. Biol. 63: 291–311.

Bowden, J. 1964. Notes on the biology of two species of *Dasyproctus* Lep. and Br. in Uganda. J. Entomol. Soc. Sth. Africa 26: 425–437.

Box, H. O. 1984. Primate Behaviour and Social Ecology. Chapman and Hall, London.

Bradbury, J. W. 1981. The evolution of leks. Pp. 138–169 *in* R. D. Alexander and D. W. Tinkle (eds.), Natural Selection and Social Behavior: Recent Research and New Theory. Chiron, New York.

Bradley, J. C. 1922. The taxonomy of the masarid wasps, including a monograph on the North American species. Univ. California Publ. Entomol. 1: 369–464.

Brauns, H. 1910. Biologisches über südafrikanische Hymenopteren. Z. Wiss. Insektenbiol. 6: 384–387, 445–447.

Brian, M. V. 1965. Social Insect Populations. Academic Press, London.

Brian, M. V. 1979. Caste differentiation and division of labor. Pp. 121–222 *in* H. R. Hermann (ed.), Social Insects, vol. 1. Academic Press, New York.

Brian, M. V. 1980. Social control over sex and caste in bees, wasps and ants. Biol. Rev. (Cambridge) 55: 379–415.

Brian, M. V. 1983. Social Insects: Ecology and Behavioural Biology. Chapman and Hall, London.

Brian, M. V. 1985. Comparative aspects of caste differentiation in social insects. Pp. 385–398 *in* J. A. L. Watson, B. M. Okot-Kotber, and C. Noirot (eds.), Caste Differentiation in Social Insects. Pergamon, New York.

Brian, M. V., and A. D. Brian. 1948. Nest construction by queens of *Vespula sylvestris* Scop. (Hym., Vespidae). Entomol. Mon. Mag. 84: 193–198.

Brian, M. V., and A. D. Brian. 1952. The wasp, *Vespula sylvestris* Scopoli: feeding, foraging and colony development. Trans. R. Entomol. Soc. London 103: 1–26.

Bristowe, W. S. 1948. Notes on the habits and prey of twenty species of British hunting wasp. Proc. Linn. Soc. London 160: 12–37.

Brockmann, H. J. 1979. Nest-site selection in the great golden digger wasp, *Sphex ichneumoneus* L. (Sphecidae). Ecol. Entomol. 4: 211–224.

Brockmann, H. J. 1980. Diversity in the nesting behavior of mud-daubers (*Trypoxylon politum* Say; Sphecidae). Florida Entomol. 63: 53–64.

Brockmann, H. J. 1984. The evolution of social behaviour in insects. Pp. 340–361 *in* J. R. Krebs and N. B. Davies (eds.), Behavioural Ecology: An Evolutionary Approach, 2nd ed. Sinauer, Sunderland, MA.

Brockmann, H. J. 1988. Father of the brood. Nat. Hist. 97(7): 32–37.

Brockmann, H. J., and R. Dawkins. 1979. Joint nesting in a digger wasp as an evolutionarily stable preadaption to social life. Behaviour 71: 203–245.

Brockmann, H. J., and A. Grafen. 1989. Mate conflict and male behaviour in a solitary wasp, *Trypoxylon (Trypargilum) politum* (Hymenoptera: Sphecidae). Anim. Behav. 37: 232–255.

Brockmann, H. J., A. Grafen, and R. Dawkins. 1979. Evolutionarily stable nesting strategy in a digger wasp. J. Theor. Biol. 77: 473–496.

Brooke, M. de L. 1981. The nesting biology and population dynamics of the Seychelles potter wasp *Eumenes alluaudi* Perez. Ecol. Entomol. 6: 365–377.

Brooks, R. W., and D. B. Wahl. 1987. Biology and mature larva of *Hemipimpla pulchripennis* (Saussure), a parasite of *Ropalidia* (Hymenoptera: Ichneumonidae, Vespidae). J. New York Entomol. Soc. 95: 547—552.

Brothers, D. J. 1975. Phylogeny and classification of the aculeate Hymenoptera, with special reference to Mutillidae. Univ. Kansas Sci. Bull. 50: 483–648.

Brothers, D. J. 1976. Modifications of the metapostnotum and origin of the "propodeal triangle" in Hymenoptera Aculeata. Syst. Entomol. 1: 177–182.

Brown, A. W. 1941. The comb of a wasp nest from the upper Cretaceous of Utah. Amer. J. Sci. 239: 54–56.

Bruch, C. 1936. Notas sobre el "Camuatí" y las avispas que lo construyen. Physis 12: 125–135.

Budrys, E. R. 1988. Digging wasps of the subfamily Pemphredoninae (Hymenoptera: Sphecidae) in the fauna of the USSR. Ph.D. dissertation abstract, Zoological Institute, Academy of Sciences of the USSR, Leningrad (in Russian).

Bulmer, M. G. 1981. Worker-queen conflict in annual social Hymenoptera. J. Theor. Biol. 93: 239–251.

Bulmer, M. G. 1983. The significance of protandry in social Hymenoptera. Amer. Nat. 121: 540–551.

Bunn, D. S. 1982a. Notes on the nesting cycle of Dolichovespula sylvestris Scop. (Hym., Vespidae). Entomol. Mon. Mag. 118: 213–218.

Bunn, D. S. 1982b. Usurpation in wasps (Hym., Vespidae). Entomol. Mon. Mag. 118: 171–173.

Bunn, D. S. 1983. A further observation of usurpation in Vespula vulgaris L. (Hym., Vespidae). Entomol. Mon. Mag. 119: 27.

Bunn, D. S. 1988. The nesting cycle of the hornet Vespa crabro L. (Hym., Vespidae). Entomol. Mon. Mag. 124: 117–122.

Buskirk, R. 1981. Sociality in the Arachnida. Pp. 281–368 in H. R. Hermann (ed.), Social Insects, vol. 2. Academic Press, New York.

Butani, D. K. 1979. Insects and Fruits. Khosla and Pragati, Delhi.

Butler, C. G. 1959. The source of the substance produced by a queen honey-bee (Apis mellifera L.) which inhibits development of the ovaries of the workers of her colony. Proc. R. Entomol. Soc. London (A) 34: 137–138.

Butler, C. G., D. J. C. Fletcher, and D. Watler. 1969. Nest-entrance marking with pheromones by the honeybee–Apis mellifera L., and by a wasp, Vespula vulgaris L. Anim. Behav. 17: 142–147.

Buysson, R. du. 1905. Monographie des Vespides du genre Nectarina. Ann. Soc. Entomol. France 74: 537–564.

Buysson, R. du. 1906. Monographie des Vespides appartenant aux genres Apoica et Synoeca. Ann. Soc. Entomol. France 75: 333–362.

Caïnadas, E. 1987. [Vespula germanica populations in Greece.] Sphecos (14): 3.

Calmbacher, C. W. 1977. The nest of Zethus otomitus (Hymenoptera: Eumenidae). Florida Entomol. 60: 135–137.

Cameron, S. A. 1985. Brood care by male bumble bees. Proc. Natl. Acad. Sci. USA 82: 6371–6373.

Cameron, S. A. 1986. Brood care by males of Polistes major (Hymenoptera: Vespidae). J. Kansas Entomol. Soc. 59: 183–185.

Cane, J. H., and M. M. Miyamoto. 1979. Nest defense and foraging ethology of a Neotropical sand wasp, Bembix multipicta (Hymenoptera: Sphecidae). J. Kansas Entomol. Soc. 52: 667–672.

Carl, J. 1934. Ropalidia montana n. sp. et son nid. Un type nouveau d'architecture vespienne. Rev. Suisse Zool. 41: 675–691.

Carpenter, F. M., and H. R. Hermann. 1979. Antiquity of sociality in insects. Pp. 81–89 in H. R. Hermann (ed.), Social Insects, vol. 1. Academic Press, New York.

Carpenter, J. M. 1982. The phylogenetic relationships and natural classification of the Vespoidea (Hymenoptera). Syst. Entomol. 7: 11–38.

Carpenter, J. M. 1986. A synonymic generic checklist of the Eumeninae (Hymenoptera: Vespidae). Psyche 93: 61–90.

Carpenter, J. M. 1987a. Phylogenetic relationships and classification of the Vespinae (Hymenoptera: Vespidae). Syst. Entomol. 12: 413–431.

Carpenter, J. M. 1987b. On "The evolutionary genetics of social wasps" and the phylogeny of the Vespinae (Hymenoptera: Vespidae). Insectes Soc. 34: 58–64.

Carpenter, J. M. 1988a. The phylogenetic system of the Stenogastrinae (Hymenoptera: Vespidae). J. New York Entomol. Soc. 96: 140–175.

Carpenter, J. M. 1988b. Choosing among multiple equally parsimonious cladograms. Cladistics 4: 291–296.

Carpenter, J. M. 1989. Testing scenarios: wasp social behavior. Cladistics 5: 131–144.

Carpenter, J. M. 1990. On Brothers' aculeate phylogeny. Sphecos (19): 9–10.

Carpenter, J. M., and J. M. Cumming. 1985. A character analysis of the North American potter wasps (Hymenoptera: Vespidae; Eumeninae). J. Nat. Hist. 19: 877–916.

Carpenter, J. M., and M. C. Day. 1988. Nomenclatural notes on Polistinae (Hymenoptera: Vespidae). Proc. Entomol. Soc. Washington 90: 323–328.

Carpenter, J. M., and K. G. Ross. 1984. Colony composition in four species of Polistinae from Suriname, with a description of the larva of *Brachygastra scutellaris* (Hymenoptera: Vespidae). Psyche 91: 237–250.

Carpenter, J. M., and J. W. Wenzel. 1988. A new species and nest type of *Mischocyttarus* from Costa Rica (Hymenoptera: Vespidae; Polistinae), with descriptions of nests of three related species. Psyche 95: 89–99.

Carpenter, J. M., and J. W. Wenzel. 1989. Synonymy of the genera *Protopolybia* and *Pseudochartergus* (Hymenoptera: Vespidae; Polistinae). Psyche 96: 177–186.

Carthy, J. D. 1958. An Introduction to the Behaviour of Invertebrates. Allen and Unwin, London.

Castellón, E. G. 1980a. Reprodução e dinâmica de população em *Synoeca surinama* (Hymenoptera: Vespidae). Acta Amaz. 10: 679–690.

Castellón, E. G. 1980b. Orientação, arquitetura e construção dos ninhos de *Synoeca surinama* (L.) (Hymenoptera; Vespidae). Acta Amaz. 10: 883–896.

Castellón, E. G. 1982. Evidências sôbre o comportamento de cópula dos machos de *Synoeca surinama* L. (Hymenoptera: Vespidae) num ninho em construção. Acta Amaz. 12: 665–670.

Cervo, R., and S. Turillazzi. 1985. Associative foundation and nesting sites in *Polistes nimpha*. Naturwissenschaften 72: 48–49.

Chadab, R. 1979a. Army-ant Predation on Social Wasps. Ph.D. dissertation, University of Connecticut, Storrs.

Chadab, R. 1979b. Early warning cues for social wasps attacked by army ants. Psyche 86: 115–123.

Chadab, R., and C. W. Rettenmeyer. 1979. Observations on swarm emigrations and dragging behavior by social wasps (Hymenoptera: Vespidae). Psyche 86: 347–352.

Chandrashekara, K., and R. Gadagkar. 1990. Evolution of eusociality: lessons from social organization in *Ropalidia marginata* (Lep.) (Hymenoptera: Vespidae). Pp. 73–74 *in* G. K. Veeresh, B. Mallik, and C. A. Viraktamath (eds.), Social Insects and the Environment. Oxford & IBH, New Delhi.

Chandrashekara, K., S. Bhagavan, S. Chandran, P. Nair, and R. Gadagkar. 1990. Perennial indeterminate colony cycle in a primitively eusocial wasp. P. 81 *in* G. K. Veeresh, B. Mallik, and C. A. Viraktamath (eds.), Social Insects and the Environment. Oxford & IBH, New Delhi.

Chapman, R. F. 1959. Some observations on *Pachyophthalmus africa* Curran (Diptera: Calliphoridae), a parasite of *Eumenes maxillosus* De Geer (Hymenoptera: Eumenidae). Proc. R. Entomol. Soc. London (A) 34: 1–6.

Charnley, H. W. 1973. The value of the propodeal orifice and the phallic capsule in vespid taxonomy (Hymenoptera, Vespidae). Bull. Buffalo Soc. Nat. Sci. 26: 1–79.

Charnov, E. L. 1978a. Evolution of eusocial behavior: offspring choice or parental parasitism? J. Theor. Biol. 75: 451–465.

Charnov, E. L. 1978b. Sex-ratio selection in eusocial Hymenoptera. Amer. Nat. 112: 317–326.

Charnov, E. L., R. L. Los-Den Hartogh, W. Jones, and J. van den Assem. 1981. Sex ratio evolution in a variable environment. Nature 289: 27–33.

Clark, A. H., and G. A. Sandhouse. 1936. The nest of *Odynerus tempeferus* var. *macio* Bequaert, with notes on the habits of the wasps. Proc. U.S. Natl. Mus. 84: 89–95.

Claude-Joseph, F. (H. Janvier). 1928. Recherches biologiques sur les prédateurs du Chili. Ann. Sci. Nat. Zool. (Sér. 10) 11: 67–207.

Claude-Joseph, F. (H. Janvier). 1930. Recherches biologiques sur les prédateurs du Chili. Ann. Sci. Nat. Zool. (Sér. 10) 13: 235–354.

Clement, S. L. 1972. Notes on the biology and larval morphology of *Stenodynerus canus canus*. Pan-Pac. Entomol. 48: 271–275.

Clement, S. L., and E. E. Grissell. 1968. Observations of the nesting habits of *Euparagia scutellaris* Cresson (Hymenoptera: Masaridae). Pan-Pac. Entomol. 44: 34–37.

Coddington, J. A. 1988. Cladistic tests of adaptational hypotheses. Cladistics 4: 3–22.

Coke, J. L., and A. B. Richon. 1976. Synthesis of optically active δ-*n*-hexadecalactone, the proposed pheromone from *Vespa orientalis*. J. Org. Chem. 41: 3516–3517.

Cole, B. J. 1983. Multiple mating and the evolution of social behavior in the Hymenoptera. Behav. Ecol. Sociobiol. 12: 191–201.

Cooper, K. W. 1952. Records and flower preferences of masarid wasps. II. Polytropy or oligotropy in *Pseudomasaris*? Amer. Midl. Nat. 48: 103–110.

Cooper, K. W. 1953. Biology of eumenine wasps. I. The ecology, predation, nesting and competition of *Ancistrocerus antilope* (Panzer). Trans. Amer. Entomol. Soc. 79: 13–35.

Cooper, K. W. 1966. Ruptor ovi, the number of moults in development, and method of exit from masoned nests. Biology of eumenine wasps VII. Psyche 73: 238–250.

Cooper, K. W. 1979. Plasticity in nesting behavior of a renting wasp, and its evolutionary implications. Studies on eumenine wasps, VIII (Hymenoptera, Aculeata). J. Washington Acad. Sci. 69: 151–158.

Cooper, M. 1986. Nests of *Stelopolybia cajennensis* (F.) (Vespidae-Polistinae). Sphecos (11): 17.

Corn, M. L. 1972. Notes on the biology of *Polistes carnifex* (Hymenoptera, Vespidae) in Costa Rica and Colombia. Psyche 79: 150–157.

Cory, E. N. 1931. Notes on the European hornet. J. Econ. Entomol. 24: 50–52.

Cottam, R. 1948. On new methods of controlling ground wasps. North Western Nat. 23: 127–129.

Cowan, D. P. 1979. Sibling matings in a hunting wasp: adaptive inbreeding? Science 205: 1403–1405.

Cowan, D. P. 1981. Parental investment in two solitary wasps, *Ancistrocerus adiabatus* and *Euodynerus foraminatus* (Eumenidae: Hymenoptera). Behav. Ecol. Sociobiol. 9: 95–102.

Cowan, D. P. 1984. Life history and male dimorphism in the mite *Kennethiella trisetosa* (Acarina: Winterschmidtiidae), and its symbiotic relationship with the wasp *Ancistrocerus antilope* (Hymenoptera: Eumenidae). Ann. Entomol. Soc. Amer. 77: 725–732.

Cowan, D. P. 1986. Sexual behavior of eumenid wasps (Hymenoptera: Eumenidae). Proc. Entomol. Soc. Washington 88: 531–541.

Cowan, D. P., and G. P. Waldbauer. 1984. Seasonal occurrence and mating at flowers by *Ancistrocerus antilope* (Hymenoptera: Eumenidae). Proc. Entomol. Soc. Washington 86: 930–934.

Craig, R. 1979. Parental manipulation, kin selection, and the evolution of altruism. Evolution 33: 319–334.

Craig, R. 1980. Sex ratio changes and the evolution of eusociality in the Hymenoptera: simulation and games theory studies. J. Theor. Biol. 87: 55–70.

Craig, R. 1983. Subfertility and the evolution of eusociality by kin selection. J. Theor. Biol. 100: 379–397.

Criddle, N. 1924. Observations on the habits of *Sphex procera* in Manitoba. Can. Field-Nat. 38: 121–123.

Cross, E. A., M. G. Stith, and T. R. Bauman. 1975. Bionomics of the organ-pipe mud-dauber, *Trypoxylon politum* (Hymenoptera: Sphecidae). Ann. Entomol. Soc. Amer. 60: 901–916.

Crozier, R. H. 1970. Coefficients of relationship and the identity of genes by descent in the Hymenoptera. Amer. Nat. 104: 216–217.

Crozier, R. H. 1975. Animal Cytogenetics 3. Insecta 7. Hymenoptera. Borntraeger, Berlin.

Crozier, R. H. 1977. Evolutionary genetics of the Hymenoptera. Annu. Rev. Entomol. 22: 263–288.

Crozier, R. H. 1979. Genetics of sociality. Pp. 223–286 *in* H. R. Hermann (ed.), Social Insects, vol. 1. Academic Press, New York.

Crozier, R. H. 1980. Genetical structure of social insect populations. Pp. 129–146 *in* H. Markl (ed.), Evolution of Social Behavior: Hypotheses and Empirical Tests. Verlag Chemie, Weinheim, Federal Republic of Germany.

Crozier, R. H. 1982. On insects and insects: twists and turns in our understanding of the evolution of eusociality. Pp. 4–9 *in* M. D. Breed, C. D. Michener, and H. E. Evans (eds.), The Biology of Social Insects. Westview, Boulder, CO.

Crozier, R. H., and D. Brückner. 1981. Sperm clumping and the population genetics of Hymenoptera. Amer. Nat. 117: 561–563.

Crozier, R. H., and P. Pamilo. 1980. Asymmetry in relatedness: who is related to whom? Nature 283: 604.

Crozier, R. H., P. Pamilo, and Y. C. Crozier. 1984. Relatedness and microgeographic genetic variation in *Rhytidoponera mayri*, an Australian arid-zone ant. Behav. Ecol. Sociobiol. 15: 143–150.

Crozier, R. H., B. H. Smith, and Y. C. Crozier. 1987. Relatedness and population structure of the primitively eusocial bee *Lasioglossum zephyrum* (Hymenoptera: Halictidae) in Kansas. Evolution 41: 902–910.

Cruz-Landim, C. da, and M. H. Pimenta Saenz. 1972. Estudo comparativo de algumas glândulas dos Vespoidea (Hymenoptera). Pap. Avul. Zool. São Paulo 25: 251–263.

Cumber, R. A. 1951. Some observations on the biology of the Australian wasp *Polistes humilis* Fabr. (Hymenoptera: Vespidae) in North Auckland (New Zealand), with special reference to the nature of the worker caste. Proc. R. Entomol. Soc. London (A) 26: 11–16.

Cumming, J. M., and F. L. Leggett. 1985. Cephalic foveae of eumenine wasps (Hymenoptera: Vespidae). J. Nat. Hist. 19: 1197–1207.

Dambach, C. A., and F. Good. 1943. Life history and habits of the cicada killer in Ohio. Ohio J. Sci. 43: 32–41.

Danks, H. V. 1971. Nest mortality factors in stem-nesting aculeate Hymenoptera. J. Anim. Ecol. 40: 79–82.

Dantas de Araújo, C. Z. 1982. Bionomia comparada de *Myschocyttarus* [sic] *drewseni drewseni* das regiões subtropical (Curitiba, PR) e tropical (Belém, PA) do Brasil (Hymenoptera, Vespidae). Dusenia 13: 165–172.

Darchen, R. 1964. Biologie de *Vespa orientalis*. Les premiers stades de développement. Insectes Soc. 11: 141–157.

Darchen, R. 1976a. *Ropalidia cincta*, guêpe sociale de la savane de Lamto (Côte-d'Ivoire) (Hym. Vespidae). Ann. Soc. Entomol. France (N.S.) 12: 579–601.

Darchen, R. 1976b. La formation d'une nouvelle colonie de *Polybioides tabidus* Fab. (Vespidae, Polybiinae). C. R. Acad. Sci. Paris (D) 282: 457–459.

Darwin, C. 1964. On the Origin of Species [facsimile of 1st ed.]. Harvard University Press, Cambridge, MA.

Das, B. P., and V. K. Gupta. 1989. The social wasps of India and the adjacent countries (Hymenoptera: Vespidae). Oriental Ins. Monogr. no. 11: 1–292.

Davis, S. K., J. E. Strassmann, C. Hughes, L. S. Pletscher, and A. R. Templeton. 1990. Population structure and kinship in *Polistes* (Hymenoptera, Vespidae): an analysis using ribosomal DNA and protein electrophoresis. Evolution 44: 1242–1253.

Davis, T. A. 1966a. Nest structure of a social wasp (*Ropalidia variegata*) varying with siting of leaves. Nature 210: 966–967.

Davis, T. A. 1966b. Observations on *Ropalidia variegata* (Smith) (Hymenoptera: Vespidae). Entomol. News 77: 271–277.

Davis, W. T. 1919. A remarkable nest of *Vespa maculata*, with notes on some other wasps' nests. Bull. Brooklyn Entomol. Soc. 14: 119–123.

Davis, W. T. 1925. *Cicada tibicen*, a South American species, with records and descriptions of North American cicadas. J. New York Entomol. Soc. 33: 35–51.

Dawkins, R. 1980. Good strategy or evolutionarily stable strategy? Pp. 331–367 *in* G. W. Barlow and J. Silverberg (eds.), Sociobiology: Beyond Nature/Nurture? Reports, Definitions and Debate. Westview, Boulder, CO.

Dawson, M. R., L. J. Hickey, K. Johnson, and C. J. Morrow. 1986. Dermopteran skull from the Paleogene of Arctic Canada. Natl. Geogr. Res. 2: 112–115.

Day, M. C. 1979. The species of Hymenoptera described by Linnaeus in the genera *Sphex, Chrysis, Vespa, Apis* and *Mutilla*. Biol. J. Linn. Soc. (London) 12: 45–84.

Day, M. C., G. R. Else, and D. Morgan. 1981. The most primitive Scoliidae (Hymenoptera). J. Nat. Hist. 15: 671–684.

Deevey, E. S. 1947. Life tables for natural populations of animals. Quart. Rev. Biol. 22: 283–314.

Degen, A. A., and M. Gersani. 1989. Environmental effects on activity and honeydew collection by the weaver ant *Polyrhachis simplex* (Hymenoptera: Formicidae) when attending the mealybug *Trabutina* sp. (Homoptera: Pseudococcidae). J. Zool., London 218: 421–432.

Deleurance, É.-P. 1946. Une régulation sociale à base sensorielle périphérique: l'inhibition de la ponte des ouvrières par la présence de la fondatrice chez les *Polistes* (Hyménoptères-Vespidae). C. R. Acad. Sci. Paris 223: 871–872.

Deleurance, É.-P. 1947. Le cycle évolutif du nid de *Polistes* (Hyménoptères-Vespidae). C. R. Acad. Sci. Paris 224: 228–230.

Deleurance, É.-P. 1948a. Sur le cycle biologique de *Polistes* (Hyménoptères Vespides). C. R. Acad. Sci. Paris 226: 601–603.

Deleurance, É.-P. 1948b. Le comportement reproducteur est indépendant de la présence des ovaires chez *Polistes* (Hyménoptères Vespides). C. R. Acad. Sci. Paris 227: 866–867.

Deleurance, É.-P. 1950. Sur le mécanisme de la monogynie fonctionnelle chez les *Polistes* (Hyménoptères-Vespides). C. R. Acad. Sci. Paris 230: 782–784.

Deleurance, É.-P. 1952a. Le polymorphisme social et son déterminisme chez les guêpes. Struct. Physiol. Soc. Anim., Colloq. Cent. Natl. Rech. Sci. 34: 141–155.

Deleurance, É.-P. 1952b. Etude du cycle biologique du couvain chez *Polistes*. Les phases "couvain normal" et "couvain abortif." Behaviour 4: 104–115.

Deleurance, É.-P. 1955a. L'influence des ovaires sur l'activité de construction chez les Polistes (Hyménoptères Vespides). C. R. Acad. Sci. Paris 241: 1073–1075.

Deleurance, É.-P. 1955b. Contribution à l'étude biologique des Polistes (Hyménoptères Vespides). II. Le cycle évolutif du couvain. Insectes Soc. 2: 285–302.

Deleurance, É.-P. 1957. Contribution à l'étude biologique des Polistes (Hyménoptères-Vespides). I. L'activité de construction. Behaviour 11: 67–84.

Delfino, G., M. T. Marino Piccioli, and C. Calloni. 1979. Fine structure of the glands of van der Vecht's organ in Polistes gallicus (L.) (Hymenoptera Vespidae). Monit. Zool. Ital. (N.S.) 13: 221–247.

Delfino, G., S. Turillazzi, and C. Calloni. 1982. Secretory cycle of the van der Vecht organ glands in foundresses of Polistes foederatus (Kohl) (Hymenoptera Vespidae). Monit. Zool. Ital. (N.S.) 16: 161–175.

Delfino, G., S. Turillazzi, and C. Calloni. 1988. A light and electron microscope study of the Dufour's gland in Parischnogaster (Hymenoptera: Stenogastrinae). Z. Mikrosk. Anat. Forsch. (Leipzig) 102: 627–644.

DeMarco, B. B. 1982. Population Studies of a Paper Wasp, Polistes metricus Say. M.S. thesis, University of Missouri-St. Louis, St. Louis.

Dew, H. E. 1983. Division of labor and queen influence in laboratory colonies of Polistes metricus (Hymenoptera: Vespidae). Z. Tierpsychol. 61: 127–140.

Dew, H. E., and C. D. Michener. 1978. Foraging flights of two species of Polistes wasps (Hymenoptera: Vespidae). J. Kansas Entomol. Soc. 51: 380–385.

Dew, H. E., and C. D. Michener. 1981. Division of labor among workers of Polistes metricus (Hymenoptera: Vespidae): laboratory foraging activities. Insectes Soc. 28: 87–101.

Döhring, E. 1960. Zur Häufigkeit, hygienischen Bedeutung und zum Fang sozialer Faltenwespen in einer Großstadt. Z. Angew. Entomol. 47: 69–79.

Doorn, A. van. 1989. Factors influencing dominance behaviour in queenless bumblebee workers (Bombus terrestris). Physiol. Entomol. 14: 211–221.

Downing, H. A., and R. L. Jeanne. 1982. A description of the ectal mandibular gland in the paper wasp Polistes fuscatus (Hymenoptera: Vespidae). Psyche 89: 317–320.

Downing, H. A., and R. L. Jeanne. 1983. Correlation of season and dominance status with activity of exocrine glands in Polistes fuscatus (Hymenoptera: Vespidae). J. Kansas Entomol. Soc. 56: 387–397.

Downing, H. A., and R. L. Jeanne. 1985. Communication of status in the social wasp Polistes fuscatus (Hymenoptera: Vespidae). Z. Tierpsychol. 67: 78–96.

Downing, H. A., and R. L. Jeanne. 1986. Intra- and interspecific variation in nest architecture in the paper wasp Polistes (Hymenoptera, Vespidae). Insectes Soc. 33: 422–443.

Downing, H. A., and R. L. Jeanne. 1987. A comparison of nest construction behavior in two species of Polistes paper wasps (Insecta, Hymenoptera: Vespidae). J. Ethol. 5: 53–66.

Downing, H. A., and R. L. Jeanne. 1988. Nest construction by the paper wasp Polistes: a test of stigmergy theory. Anim. Behav. 36: 1729–1739.

Downing, H. A., and R. L. Jeanne. 1990. The regulation of complex building behaviour in the paper wasp, Polistes fuscatus (Insecta, Hymenoptera, Vespidae). Anim. Behav. 39: 105–124.

Downing, H. A., D. C. Post, and R. L. Jeanne. 1985. Morphology of sternal glands in male polistine wasps (Hymenoptera: Vespidae). Insectes Soc. 32: 186–197.

Dropkin, J. A., and G. J. Gamboa. 1981. Physical comparisons of foundresses of the paper wasp Polistes metricus (Hymenoptera: Vespidae). Can. Entomol. 113: 457–461.

Ducke, A. 1905. Sobre as vespidas sociaes do Pará. (I. Supplemento). Bol. Mus. Para. Emilio Goeldi 4: 652–698.

Ducke, A. 1907. Nouveau genre de Sphégides. Ann. Soc. Entomol. France 76: 28–30.

Ducke, A. 1910. Révision des guêpes sociales polygames d'Amérique. Ann. Hist.-Nat. Mus. Natl. Hungarici 8: 449–544.

Ducke, A. 1914. Über Phylogenie und Klassifikation der sozialen Vespiden. Zool. Jahrb. Abt. Syst. Geogr. Biol. Tiere 36: 303–330.

Duncan, C. D. 1928. Plant hairs as building material for *Polistes* (Hymenoptera: Vespidae). Pan-Pac. Entomol. 5: 90.

Duncan, C. D. 1939. A contribution to the biology of North American vespine wasps. Stanford Univ. Publ. Biol. Sci. 8: 1–272.

Dupont, F., and C. J. H. Franssen. 1937. Aanteekeningen over de levenswijze van *Piagetia cavernicola* v. d. Vecht. Entomol. Meded. Ned.-Indië 3: 27–29.

Eberhard, W. G. 1972. Altruistic behavior in a sphecid wasp: support for kin-selection theory. Science 175: 1390–1391.

Eberhard, W. G. 1974. The natural history and behaviour of the wasp *Trigonopsis cameronii* Kohl (Sphecidae). Trans. R. Entomol. Soc. London 125: 295–328.

Eck, R. 1979. Biometrische Untersuchung zur Klärung der Artunterschiede bei sozialen Faltenwespen (Hymenoptera, Vespinae). Entomol. Abh. 42: 315–344.

Edwards, R. 1980. Social Wasps: Their Biology and Control. Rentokil, East Grinstead, W. Sussex, England.

Eickwort, G. C. 1975. Gregarious nesting of the mason bee *Hoplitis anthocopoides* and the evolution of parasitism and sociality among megachilid bees. Evolution 29: 142–150.

Eickwort, G. C. 1981. Presocial insects. Pp. 199–280 *in* H. R. Hermann (ed.), Social Insects, vol. 2. Academic Press, New York.

Eickwort, G. C. 1986. First steps into eusociality: the sweat bee *Dialictus lineatulus*. Florida Entomol. 69: 742–754.

Eickwort, G. C., and P. F. Kukuk. 1987. Reproductive castes in primitively eusocial halictid bees. Pp. 261–262 *in* J. Eder and H. Rembold (eds.), Chemistry and Biology of Social Insects. J. Peperny, Munich.

Eickwort, K. 1969. Separation of the castes of *Polistes exclamans* and notes on its biology (Hym.: Vespidae). Insectes Soc. 16: 67–72.

Eisner, T. 1957. A comparative morphological study of the proventriculus of ants (Hymenoptera: Formicidae). Bull. Mus. Comp. Zool. Harvard Univ. 116: 439–490.

Eldredge, N., and J. Cracraft. 1980. Phylogenetic Patterns and the Evolutionary Process. Columbia University Press, New York.

Elliott, N. B., and W. M. Elliott. 1987. Nest usurpation by females of *Cerceris cribrosa* (Hymenoptera: Sphecidae). J. Kansas Entomol. Soc. 60: 397–402.

Elliott, N. B., W. M. Elliott, and P. Salbert. 1981. Nesting behavior of *Cerceris zonata* (Hymenoptera: Philanthinae). Ann. Entomol. Soc. Amer. 74: 127–129.

Elliott, N. B., T. Shlotzhauer, and W. M. Elliott. 1986. Nest use by females of the presocial wasp *Cerceris watlingensis* (Hymenoptera: Sphecidae). Ann. Entomol. Soc. Amer. 79: 994–998.

Emerson, A. E. 1959. Social insects. Encyclopaedia Britannica 20: 871–878.

Emlen, S. T. 1982a. The evolution of helping. I. An ecological constraints model. Amer. Nat. 119: 29–39.

Emlen, S. T. 1982b. The evolution of helping. II. The role of behavioral conflict. Amer. Nat. 119: 40–53.

Emlen, S. T., and L. W. Oring. 1977. Ecology, sexual selection, and the evolution of mating systems. Science 197: 215–223.

Esch, H. 1967. Die Bedeutung der Lauterzeugung für die Verständigung der stachellosen Bienen. Z. Vergl. Physiol. 56: 199–220.

Esch, H. 1971. Wagging movements in the wasp *Polistes versicolor vulgaris* Bequaert. Z. Vergl. Physiol. 72: 221–225.

Es'kov, E. K. 1977. The structure of larval acoustic signals in the wasp *Dolichovespula sylvestris* and the hornet *Vespa crabro*. Zhur. Evol. Biokhim. Fiziol. 13: 371–375 (in Russian).

Espelie, K. E., and H. R. Hermann. 1988. Congruent cuticular hydrocarbons: biochemical convergence of a social wasp, an ant and a host plant. Biochem. Syst. Ecol. 16: 505–508.

Espelie, K. E., and H. R. Hermann. 1990. Surface lipids of the social wasp *Polistes annularis* (L.) and its nest and nest pedicel. J. Chem. Ecol. 16: 1841–1852.

Espelie, K. E., and D. S. Himmelssbach. 1990. Characterization of pedicel, paper, and larval silk from nest of *Polistes annularis* (L.). J. Chem. Ecol. 16: 3467–3477.

Espelie, K. E., J. W. Wenzel, and G. Chang. 1990. Surface lipids of the social wasp *Polistes metricus* Say and of its nest and nest pedicel and their relation to nestmate recognition. J. Chem. Ecol. 16: 2229–2241.

Evans, H. E. 1956. Notes on the biology of four species of ground-nesting Vespidae (Hymenoptera). Proc. Entomol. Soc. Washington 58: 265–270.

Evans, H. E. 1958. The evolution of social life in wasps. Proc. 10th Int. Cong. Entomol., pp. 449–457.

Evans, H. E. 1962. The evolution of prey-carrying mechanisms in wasps. Evolution 16: 468–483.

Evans, H. E. 1964a. The classification and evolution of digger wasps as suggested by larval characters (Hymenoptera: Sphecoidea). Entomol. News 75: 225–237.

Evans, H. E. 1964b. Observations on the nesting behavior of *Moniaecera asperata* (Fox) with comments on the communal nesting in solitary wasps. Insectes Soc. 11: 71–78.

Evans, H. E. 1966a. The Comparative Ethology and Evolution of the Sand Wasps. Harvard University Press, Cambridge, MA.

Evans, H. E. 1966b. The behavior patterns of solitary wasps. Annu. Rev. Entomol. 11: 123–154.

Evans, H. E. 1966c. The accessory burrows of digger wasps. Science 152: 465–471.

Evans, H. E. 1969. Three new Cretaceous aculeate wasps. Psyche 76: 251–261.

Evans, H. E. 1973a. Notes on the nests of *Montezumia* (Hymenoptera, Eumenidae). Entomol. News. 84: 285–290.

Evans, H. E. 1973b. Cretaceous aculeate wasps from Taimyr, Siberia. Psyche 80: 166–178.

Evans, H. E. 1973c. Burrow sharing and nest transfer in the digger wasp *Philanthus gibbosus* (Fabricius). Anim. Behav. 21: 302–308.

Evans, H. E. 1975. Social parasitism of a common yellowjacket. Ins. World Dig. 2(1): 6–13.

Evans, H. E. 1977a. Extrinsic versus intrinsic factors in the evolution of insect sociality. Bioscience 27: 613–617.

Evans, H. E. 1977b. Observations on the nests and prey of eumenid wasps (Hymenoptera, Eumenidae). Psyche 83: 255–259.

Evans, H. E. 1989. Further studies on the systematics and nesting behavior of eastern Australian *Cerceris* wasps (Hymenoptera: Sphecidae). Trans. Amer. Entomol. Soc. 114: 1–13.

Evans, H. E., and A. W. Hook. 1982a. Communal nesting in the digger wasp *Cerceris australis* (Hymenoptera: Sphecidae). Aust. J. Zool. 30: 557–568.

Evans, H. E., and A. W. Hook. 1982b. Communal nesting in Australian *Cerceris* digger wasps. Pp. 159–163 *in* M. D. Breed, C. D. Michener, and H. E. Evans (eds.), The Biology of Social Insects. Westview, Boulder, CO.

Evans, H. E., and A. W. Hook. 1986. Nesting behavior of Australian *Cerceris* digger wasps, with special reference to nest reutilization and nest sharing (Hymenoptera: Sphecidae). Sociobiology 11: 275–302.

Evans, H. E., and R. W. Matthews. 1970. Notes on the nests and prey of Australian wasps of the genus *Cerceris* (Hymenoptera: Sphecidae). J. Aust. Entomol. Soc. 9: 153–156.

Evans, H. E., and R. W. Matthews. 1973a. Systematics and nesting behavior of Australian *Bembix* sand wasps (Hymenoptera: Sphecidae). Mem. Amer. Entomol. Inst. (20): 1–387.

Evans, H. E., and R. W. Matthews. 1973b. Observations on the nesting behavior of *Trachypus petiolatus* (Spinola) in Colombia and Argentina. J. Kansas Entomol. Soc. 46: 165–175.

Evans, H. E., and R. W. Matthews. 1974. Notes on nests and prey of two species of ground-nesting Eumenidae from South America (Hymenoptera). Entomol. News 85: 149–153.

Evans, H. E., and K. M. O'Neill. 1988. The Natural History and Behavior of North American Beewolves. Cornell University Press, Ithaca, NY.

Evans, H. E., and M. J. West-Eberhard. 1970. The Wasps. University of Michigan Press, Ann Arbor.

Evans, H. E., R. W. Matthews, J. Alcock, and M. A. Fritz. 1976. Notes on the nests and prey of two subspecies of *Cerceris rufimana* Taschenberg (Hymenoptera: Sphecidae: Cercerini). J. Kansas Entomol. Soc. 49: 126–132.

Farris, J. S. 1969. A successive approximations approach to character weighting. Syst. Zool. 18: 374–385.

Farris, J. S. 1970. Methods for computing Wagner trees. Syst. Zool. 19: 83–92.

Farris, J. S. 1979. The information content of the phylogenetic system. Syst. Zool. 28: 483–519.

Farris, J. S. 1983. The logical basis of phylogenetic analysis. Pp. 7–36 *in* N. I. Platnick and V. A. Funk (eds.), Advances in Cladistics 2. Columbia University Press, New York.

Ferguson, I. D., G. J. Gamboa, and J. K. Jones. 1987. Discrimination between natal and non-natal nests by the social wasps *Dolichovespula maculata* and *Polistes fuscatus* (Hymenoptera: Vespidae). J. Kansas Entomol. Soc. 60: 65–69.

Fisher, R. A. 1958. The Genetical Theory of Natural Selection, 2nd ed. Dover, New York.

Fisher, R. M. 1983. Inability of the social parasite *Psithyrus ashtoni* to suppress ovarian development in workers of *Bombus affinis* (Hymenoptera: Apidae). J. Kansas Entomol. Soc. 56: 69–73.

Fitzgerald, D. V. 1950. Notes on the genus *Ropalidia* (Hymenoptera: Vespidae) from Madagascar. Proc. R. Entomol. Soc. London (A) 25: 81–86.

Fletcher, D. J. C., and M. S. Blum. 1983. Regulation of queen number by workers in colonies of social insects. Science 219: 312–314.

Fletcher, D. J. C., and C. D. Michener (eds.). 1987. Kin Recognition in Animals. Wiley, New York.

Fletcher, D. J. C., and K. G. Ross. 1985. Regulation of reproduction in eusocial Hymenoptera. Annu. Rev. Entomol. 30: 319–343.

Foelix, R. F. 1982. Biology of Spiders. Harvard University Press, Cambridge, MA.

Forbes, H. O. 1885. A Naturalist's Wanderings in the Eastern Archipelago. Harper, New York.

Ford, H. A., H. Bell, R. Nias, and R. Noske. 1988. The relationship between ecology and the incidence of cooperative breeding in Australian birds. Behav. Ecol. Sociobiol. 22: 239–249.

Forsyth, A. B. 1975. Usurpation and dominance behavior in the polygynous social wasp *Metapolybia cingulata* (Hymenoptera: Vespidae: Polybiini). Psyche 82: 299–303.

Forsyth, A. B. 1978. Studies on the Behavioral Ecology of Polygynous Social Wasps. Ph.D. dissertation, Harvard University, Cambridge, MA.

Forsyth, A. B. 1981. Swarming activity of polybiine social wasps (Hymenoptera: Vespidae: Polybiini). Biotropica 13: 93–99.

Fox-Wilson, G. 1946. Factors affecting populations of social wasps, *Vespula* species, in England (Hymenoptera). Proc. R. Entomol. Soc. London (A) 21: 17–27.

Frank, S. A. 1986. Hierarchical selection theory and sex ratios. I. General solutions for structured populations. Theor. Popul. Biol. 29: 312–342.

Frank, S. A., and B. J. Crespi. 1989. Synergism between sib-rearing and sex ratio in Hymenoptera. Behav. Ecol. Sociobiol. 24: 155–162.

Free, J. B. 1987. Pheromones of Social Bees. Cornell University Press, Ithaca, NY.

Free, J. B., and I. H. Williams. 1975. Factors determining the rearing and rejection of drones by the honey bee colony. Anim. Behav. 23: 650–675.

Freeman, B. E. 1982. The comparative distribution and population dynamics in Trinidad of *Sceliphron fistularium* (Dahlbom) and *S. asiaticum* (L.) (Hymenoptera: Sphecidae). Biol. J. Linn. Soc. (London) 17: 343–360.

Freeman, B. E., and D. B. Jayasingh. 1975a. Factors controlling the distribution in Jamaica of *Pachodynerus nasidens* (Latr.) (Hymenoptera: Eumenidae). Biol. J. Linn. Soc. (London) 7: 231–241.

Freeman, B. E., and D. B. Jayasingh. 1975b. Population dynamics of *Pachodynerus nasidens* (Hymenoptera) in Jamaica. Oikos 26: 86–91.

Freeman, B. E., and J. R. Parnell. 1973. Mortality of *Sceliphron assimile* Dahlbom (Sphecidae) caused by the eulophid *Melittobia chalybii* Ashmead. J. Anim. Ecol. 42: 779–784.

Freeman, B. E., and C. A. Taffe. 1974. Population dynamics and nesting behaviour of *Eumenes colona* (Hymenoptera) in Jamaica. Oikos 25: 388–394.

Freisling, J. 1938. Die Bauinstinkte der Wespen (Vespidae). Z. Tierpsychol. 2: 81–98.

Fye, R. E. 1965. The biology of the Vespidae, Pompilidae, and Sphecidae (Hymenoptera) from trap nests in northwestern Ontario. Can. Entomol. 97: 716–744.

Gadagkar, R. 1980. Dominance hierarchy and division of labour in the social wasp *Ropalidia marginata* (Lep.) (Hymenoptera: Vespidae). Curr. Sci. 49: 772–775.

Gadagkar, R. 1985a. Evolution of insect sociality—a review of some attempts to test modern theories. Proc. Indian Acad. Sci. (Anim. Sci.) 94: 309–324.

Gadagkar, R. 1985b. Kin recognition in social insects and other animals—a review of recent findings and a consideration of their relevance for the theory of kin selection. Proc. Indian Acad. Sci. (Anim. Sci.) 94: 587–621.

Gadagkar, R. 1987. Social structure and the determinants of queen status in the primitively eusocial wasp *Ropalidia cyathiformis*. Pp. 377–378 *in* J. Eder and H. Rembold (eds.), Chemistry and Biology of Social Insects. J. Peperny, Munich.

Gadagkar, R. 1990a. The haplodiploidy threshold and social evolution. Curr. Sci. 59: 374–376.

Gadagkar, R. 1990b. Social biology of *Ropalidia*: investigations into the origins of eusociality. Pp. 9–11 *in* G. K. Veeresh, B. Mallik, and C. A. Viraktamath (eds.), Social Insects and the Environment. Oxford & IBH, New Delhi.

Gadagkar, R. 1990c. Evolution of eusociality: the advantage of assured fitness returns. Phil. Trans. R. Soc. London (B) 329: 17–25.

Gadagkar, R. 1990d. Origin and evolution of eusociality: a perspective from studying primitively eusocial wasps. J. Genet. 69: 113–125.

Gadagkar, R., and N. V. Joshi. 1982a. A comparative study of social structure in colonies of *Ropalidia*. Pp. 187–191 *in* M. D. Breed, C. D. Michener, and H. E. Evans (eds.), The Biology of Social Insects. Westview, Boulder, CO.

Gadagkar, R., and N. V. Joshi. 1982b. Behaviour of the Indian social wasp *Ropalidia cyathiformis* on a nest of separate combs (Hymenoptera: Vespidae). J. Zool., London 198: 27–37.

Gadagkar, R., and N. V. Joshi. 1983. Quantitative ethology of social wasps: time-activity budgets and caste differentiation in *Ropalidia marginata* (Lep.) (Hymenoptera: Vespidae). Anim. Behav. 31: 26–31.

Gadagkar, R., and N. V. Joshi. 1984. Social organisation in the Indian wasp *Ropalidia cyathiformis* (Fab.) (Hymenoptera: Vespidae). Z. Tierpsychol. 64: 15–32.

Gadagkar, R., and N. V. Joshi. 1985. Colony fission in a social wasp. Curr. Sci. 54: 57–62.

Gadagkar, R., M. Gadgil, N. V. Joshi, and A. S. Mahabal. 1982a. Observations on the natural history and population ecology of the social wasp *Ropalidia marginata* (Lep.) from peninsular India (Hymenoptera: Vespidae). Proc. Indian Acad. Sci. (Anim. Sci.) 91: 539–552.

Gadagkar, R., M. Gadgil, and A. S. Mahabal. 1982b. Observations on population ecology and sociobiology of the paper wasp *Ropalidia marginata* (Lep.) (Family Vespidae). Proc. Symp. Ecol. Anim. Popul. Zool. Surv. India 4: 49–61.

Gadagkar, R., C. Vinutha, A. Shanubhogue, and A. P. Gore. 1988. Pre-imaginal biasing of caste in a primitively eusocial insect. Proc. R. Soc. London (B) 233: 175–189.

Gadagkar, R., K. Chandrashekara, S. Chandran, and S. Bhagavan. 1990a. Serial polygyny in *Ropalidia marginata*: implications for the evolution of eusociality. Pp. 227–228 *in* G. K. Veeresh, B. Mallik, and C. A. Viraktamath (eds.), Social Insects and the Environment. Oxford & IBH, New Delhi.

Gadagkar, R., S. Bhagavan, R. Malpe, and C. Vinutha. 1990b. On reconfirming the evidence for pre-imaginal caste bias in a primitively eusocial wasp. Proc. Indian Acad. Sci. (Anim. Sci.) 99: 141–150.

Gadagkar, R., S. Bhagavan, K. Chandrashekara, and C. Vinutha. 1990c. The role of larval nutrition in pre-imaginal biasing of caste in the primitively eusocial wasp *Ropalidia marginata* (Lep.). Submitted.

Gadgil, M., and A. S. Mahabal. 1974. Caste differentiation in the paper wasp *Ropalidia marginata* (Lep.). Curr. Sci. 43: 482.

Gambino, P. 1986. Winter prey collection at a perennial colony of *Paravespula vulgaris* (L.) (Hymenoptera: Vespidae). Psyche 93: 331–340.

Gamboa, G. J. 1978. Intraspecific defense: advantage of social cooperation among paper wasp foundresses. Science 199: 1463–1465.

Gamboa, G. J. 1980. Comparative timing of brood development between multiple- and single-foundress colonies of the paper wasp, *Polistes metricus*. Ecol. Entomol. 5: 221–225.

Gamboa, G. J. 1981. Nest sharing and maintenance of multiple nests by the paper wasp, *Polistes metricus* (Hymenoptera: Vespidae). J. Kansas Entomol. Soc. 54: 153–155.

Gamboa, G. J. 1988. Sister, aunt-niece, and cousin recognition by social wasps. Behav. Genet. 18: 409–423.

Gamboa, G. J., and H. E. Dew. 1981. Intracolonial communication by body oscillations in the paper wasp, *Polistes metricus*. Insectes Soc. 28: 13–26.

Gamboa, G. J., and J. A. Dropkin. 1979. Comparisons of behaviors in early vs. late foundress associations of the paper wasp, *Polistes metricus* (Hymenoptera: Vespidae). Can. Entomol. 111: 919–926.

Gamboa, G. J., B. D. Heacock, and S. L. Wiltjer. 1978. Division of labor and subordinate longevity in foundress associations of the paper wasp, *Polistes metricus* (Hymenoptera: Vespidae). J. Kansas Entomol. Soc. 51: 343–352.

Gamboa, G. J., H. K. Reeve, I. D. Ferguson, and T. L. Wacker. 1986a. Nestmate recognition in social wasps: the origin and acquisition of recognition odours. Anim. Behav. 34: 685–695.

Gamboa, G. J., H. K. Reeve, and D. W. Pfennig. 1986b. The evolution and ontogeny of nestmate recognition in social wasps. Annu. Rev. Entomol. 31: 431–454.

Gamboa, G. J., J. E. Klahn, A. O. Parman, and R. E. Ryan. 1987. Discrimination between nestmate and non-nestmate kin by social wasps (*Polistes fuscatus*, Hymenoptera: Vespidae). Behav. Ecol. Sociobiol. 21: 125–128.

García A., R. J. 1974. Observaciones sobre *"Polistes peruvianus"* Bequaert (Hym., Vespidae) en los alrededores de Lima. Biota (Lima) 10: 11–27.

Gaul, A. T. 1941. An unrecorded sleeping habit of *Dolichovespula arenaria*, Fab. (Hymenoptera, Vespidae). Bull. Brooklyn Entomol. Soc. 36: 49.

Gaul, A. T. 1947. Additions to vespine biology III: notes on the habits of *Vespula squamosa* Drury (Hymenoptera, Vespidae). Bull. Brooklyn Entomol. Soc. 42: 87–96.

Gaul, A. T. 1948. Additions to vespine biology—V: the distribution of labor in the colonies of hornets and yellowjackets. Bull. Brooklyn Entomol. Soc. 43: 73–79.

Gaul, A. T. 1952. Additions to vespine biology IX: temperature regulation in the colony. Bull. Brooklyn Entomol. Soc. 47: 79–82.

Gaul, A. T. 1953. Additions to vespine biology XI: defense flight. Bull. Brooklyn Entomol. Soc. 48: 35–37.

Gauld, I., and B. Bolton (eds.). 1988. The Hymenoptera. British Museum (Natural History), London.

Genise, J. F. 1979. Interacción social en una agrupación de *Rubrica nasuta* (Christ) (Hymenoptera: Sphecidae: Nyssoninae). Physis (C) 38: 8–30.

Gervet, J. 1956. L'action des températures différentielles sur la monogynie fonctionelle chez les *Polistes* (Hyménoptères Vespides). Insectes Soc. 3: 159–176.

Gervet, J. 1962. Etude de l'effet de groupe sur la ponte dans la société polygyne de *Polistes gallicus* (Hymén. Vesp.). Insectes Soc. 9: 231–263.

Gervet, J. 1964a. Le comportement d'oophagie différentielle chez *Polistes gallicus* L. (Hymén. Vesp.). Insectes Soc. 11: 343–382.

Gervet, J. 1964b. La ponte et sa régulation dans la société polygyne de *Polistes gallicus* L. (Hyménoptère Vespidé). Ann. Sci. Nat. Zool. Biol. Anim. (Sér. 12) 6: 601–778.

Gervet, J., M. Pratte, S. Semenoff, and D. Gabouriaut. 1986. Pattern of offspring production in colonies of paper wasps (*Polistes gallicus* L.) II. Demographic data hypotheses concerning fitness implications. Insectes Soc. 33: 375–387.

Gess, F. W. 1981. Some aspects of an ethological study of the aculeate wasps and bees of a karroid area in the vicinity of Grahamstown, South Africa. Ann. Cape Prov. Mus. Nat. Hist. 14: 1–80.

Gess, F. W., and S. K. Gess. 1976. An ethological study of *Parachilus insignis* (Saussure) (Hymenoptera: Eumenidae) in the Eastern Cape Province of South Africa. Ann. Cape Prov. Mus. Nat. Hist. 11: 83–102.

Gess, F. W., and S. K. Gess. 1980. Ethological studies of *Jugurtia confusa* Richards, *Ceramius capicola* Brauns, *C. linearis* Klug and *C. lichtensteinii* (Klug) (Hymenoptera: Masaridae) in the Eastern Cape Province of South Africa. Ann. Cape Prov. Mus. Nat. Hist. 13: 63–83.

Gess, S. K., and F. W. Gess. 1989. Notes on nesting behaviour in *Bembix bubalus* Handlirsch in southern Africa, with the emphasis on nest sharing and reaction to nest parasites (Hymenoptera: Sphecidae). Ann. Cape Prov. Mus. Nat. Hist. 18: 151–160.

Gibo, D. L. 1974. A laboratory study on the selective advantage of foundress associations in *Polistes fuscatus*. Can. Entomol. 106: 101–106.

Gibo, D. L. 1978. The selective advantage of foundress associations in *Polistes fuscatus* (Hymenoptera: Vespidae): a field study of the effects of predation on productivity. Can. Entomol. 110: 519–540.

Gibo, D. L., and R. A. Metcalf. 1978. Early survival of *Polistes apachus* (Hymenoptera: Vespidae) colonies in California: a field study of an introduced species. Can. Entomol. 110: 1339–1343.

Gibo, D. L., H. E. Dew, and A. S. Hajduk. 1974a. Thermoregulation in colonies of *Vespula arenaria* and *Vespula maculata* (Hymenoptera: Vespidae). II. The relation between colony biomass and calorie production. Can. Entomol. 106: 873–879.

Gibo, D. L., R. M. Yarascavitch, and H. E. Dew. 1974b. Thermoregulation in colonies of *Vespula arenaria* and *Vespula maculata* (Hymenoptera: Vespidae) under normal conditions and under cold stress. Can. Entomol. 106: 503–507.

Gibo, D. L., A. Temporale, T. P. Lamarre, B. M. Soutar, and H. E. Dew. 1977. Thermoregulation in colonies of *Vespula arenaria* and *Vespula maculata* (Hymenoptera: Vespidae). III. Heat production in queen nests. Can. Entomol. 109: 615–620.

Gibson, G. A. P. 1985. Some pro- and mesothoracic structures important for phylogenetic analysis of Hymenoptera, with a review of terms used for the structures. Can. Entomol. 117: 1395–1443.

Gobbi, N. 1977. Ecologia de *Polistes versicolor*. Ph.D. dissertation, Universidade de São Paulo, Ribeirão Prêto.

Gobbi, N. 1984. Contribuição ao estudo do ciclo basico de espécies do gênero *Polybia*, com especial referência à *Polybia (Myrapetra) paulista* (Ihering, 1896) e *Polybia occidentalis occidentalis* (Olivier, 1791) (Hymenoptera, Vespidae). Rev. Brasil. Entomol. 28: 451–457.

Gobbi, N., and V. L. L. Machado. 1985. Material capturado e utilizado na alimentação de *Polybia (Myrapetra) paulista* Ihering, 1896 (Hymenoptera-Vespidae). An. Soc. Entomol. Brasil 14: 189–195.

Gobbi, N., and V. L. L. Machado. 1986. Material capturado e utilizado na alimentação de *Polybia (Trichothorax) ignobilis* (Haliday, 1836) (Hymenoptera, Vespidae). An. Soc. Entomol. Brasil 15: 117–124 (suppl.).

Gobbi, N., V. L. L. Machado, and J. A. Tavares Filho. 1984. Sazonalidade das presas utilizadas na alimentação de *Polybia occidentalis occidentalis* (Olivier, 1791) (Hym., Vespidae). An. Soc. Entomol. Brasil 13: 63–69.

Gorton, R. E. 1978. Observations on the nesting behavior of *Mischocyttarus immarginatus* (Rich.) (Vespidae: Hymenoptera) in a dry forest in Costa Rica. Insectes Soc. 25: 197–204.

Gould, S. J. 1977. Ontogeny and Phylogeny. Harvard University Press, Cambridge, MA.

Gould, S. J. 1987. Freudian slip. Nat. Hist. 96(2): 14–21.

Grafen, A. 1984. Natural selection, kin selection and group selection. Pp. 62–89 *in* J. R. Krebs and N. B. Davies (eds.), Behavioural Ecology: An Evolutionary Approach, 2nd ed. Sinauer, Sunderland, MA.

Grafen, A. 1986. Split sex ratios and the evolutionary origins of eusociality. J. Theor. Biol. 122: 95–121.

Grandi, G. 1944. Un interessante caso di socialità negli sfecidi. Mem. Acad. Sci. Ist. Bologna (Ser. 10) 1: 63–66.

Grandi, G. 1961. Studi di un entomologo sugli imenotteri superiori. Boll. Ist. Entomol. Univ. Bologna 25: 1–659.

Grechka, E. O. 1986. The regulation of seasonal development and caste determination in *Polistes* wasps. Pp. 43–73 *in* V. B. Zinchuk (ed.), Essays in Memory of N. A. Kholodkovskii. Nauka, Leningrad (in Russian).

Greenberg, L. 1979. Genetic component of bee odor in kin recognition. Science 206: 1095–1097.

Greene, A. 1979. Behavioral characters as indicators of yellowjacket phylogeny (Hymenoptera: Vespidae). Ann. Entomol. Soc. Amer. 72: 614–619.

Greene, A. 1982. Comparative early growth and foraging of two naturally established vespine wasp colonies. Pp. 85–89 *in* M. D. Breed, C. D. Michener, and H. E. Evans (eds.), The Biology of Social Insects. Westview, Boulder, CO.

Greene, A. 1984. Production schedules of vespine wasps: an empirical test of the bang-bang optimization model. J. Kansas Entomol. Soc. 57: 545–568.

Greene, A., and D. M. Caron. 1980. Entomological etymology: the common names of social wasps. Bull. Entomol. Soc. Amer. 26: 126–130.

Greene, A., R. D. Akre, and P. J. Landolt. 1976. The aerial yellowjacket, *Dolichovespula arenaria* (Fab.): nesting biology, reproductive production, and behavior (Hymenoptera: Vespidae). Melanderia 26: 1–34.

Greene, A., R. D. Akre, and P. J. Landolt. 1978. Behavior of the yellowjacket social parasite, *Dolichovespula arctica* (Rohwer) (Hymenoptera: Vespidae). Melanderia 29: 1–28.

Grissell, E. E. 1975. Ethology and larva of *Pterocheilus texanus* (Hymenoptera: Eumenidae). J. Kansas Entomol. Soc. 48: 244–253.

Grogan, D. E., and J. H. Hunt. 1977. Digestive proteases of two species of wasps of the genus *Vespula*. Insect Biochem. 7: 191–196.

Gwynne, D. T. 1984. Male mating effort, confidence of paternity, and insect sperm competition. Pp. 117–149 *in* R. L. Smith (ed.), Sperm Competition and the Evolution of Animal Mating Systems. Academic Press, New York.

Haeseler, V. 1986. Attachment to the nest and worker-like activities in young queens: effects on reproduction in *Dolichovespula media* (Retzius 1783) (Hymenoptera, Vespidae). Zool. Anz. 217: 145–155.

Haggard, C. M., and G. J. Gamboa. 1980. Seasonal variation in body size and reproductive condition of a paper wasp, *Polistes metricus* (Hymenoptera: Vespidae). Can. Entomol. 112: 239–248.

Hamilton, W. D. 1963. The evolution of altruistic behavior. Amer. Nat. 97: 354–356.

Hamilton, W. D. 1964a. The genetical evolution of social behaviour. I. J. Theor. Biol. 7: 1–16.

Hamilton, W. D. 1964b. The genetical evolution of social behaviour. II. J. Theor. Biol. 7: 17–52.

Hamilton, W. D. 1967. Extraordinary sex ratios. Science 156: 477–478.

Hamilton, W. D. 1971. Geometry for the selfish herd. J. Theor. Biol. 31: 295–311.

Hamilton, W. D. 1972. Altruism and related phenomena, mainly in social insects. Annu. Rev. Ecol. Syst. 3: 193–232.

Hamilton, W. D. 1975. Innate social aptitudes of man: an approach from evolutionary genetics. Pp. 133–155 *in* R. Fox (ed.), Biosocial Anthropology. Wiley, New York.

Hamm, A. H., and O. W. Richards. 1926. The biology of the British Crabronidae. Trans. Entomol. Soc. London 74: 297–331.

Hammerstein, P. 1981. The role of asymmetries in animal conflicts. Anim. Behav. 29: 193–205.

Hansell, M. H. 1977. Social behaviour of a three wasp colony: Stenogastrinae Vespidae. Proc. 15th Int. Ethol. Conf., p. 30.

Hansell, M. H. 1981. Nest construction in the subsocial wasp *Parischnogaster mellyi* (Saussure) Stenogastrinae (Hymenoptera). Insectes Soc. 28: 208–216.

Hansell, M. H. 1982a. Brood development in the subsocial wasp *Parischnogaster mellyi* (Saussure) (Stenogastrinae, Hymenoptera). Insectes Soc. 29: 3–14.

Hansell, M. H. 1982b. Colony membership in the wasp *Parischnogaster striatula* (Stenogastrinae). Anim. Behav. 30: 1258–1259.

Hansell, M. H. 1983. Social behaviour and colony size in the wasp *Parischnogaster mellyi* (Saussure), Stenogastrinae (Hymenoptera, Vespidae). Proc. K. Nederl. Akad. Wetens. (Ser. C) 86: 167–178.

Hansell, M. H. 1984a. Animal Architecture and Building Behaviour. Longman, London.

Hansell, M. H. 1984b. How to build a social life. New Scientist 102: 16–18.

Hansell, M. H. 1985. The nest material of Stenogastrinae (Hymenoptera, Vespidae) and its effect on the evolution of social behaviour and nest design. Actes Coll. Insectes Soc. 2: 57–63.

Hansell, M. H. 1986a. Colony biology of the stenogastrine wasp *Holischnogaster gracilipes* (van der Vecht) (Hym.) on Mount Kinabalu (Borneo). Entomol. Mon. Mag. 122: 31–36.

Hansell, M. H. 1986b. The nest of *Holischnogaster gracilipes* (van der Vecht) (Hym., Vespidae, Stenogastrinae). Entomol. Mon. Mag. 122: 185–188.

Hansell, M. H. 1987a. Nest building as a facilitating and limiting factor in the evolution of eusociality in the Hymenoptera. Pp. 155–181 *in* P. H. Harvey and L. Partridge (eds.), Oxford Surveys in Evolutionary Biology, vol. 4. Oxford University Press, Oxford.

Hansell, M. H. 1987b. Elements of eusociality in colonies of *Eustenogaster calyptodoma* (Sakagami & Yoshikawa) (Stenogastrinae, Vespidae). Anim. Behav. 35: 131–141.

Hansell, M. H. 1989. Les nids des insectes sociaux. La Recherche 20: 14–22.

Hansell, M. H., C. Samuel, and J. I. Furtado. 1982. *Liostenogaster flavolineata*: social life in the small colonies of an Asian tropical wasp. Pp. 192–195 *in* M. D. Breed, C. D. Michener, and H. E. Evans (eds.), The Biology of Social Insects. Westview, Boulder, CO.

Hase, A. 1936. Über das Leben der staatenbildenden Wespe *Polybia atra* Sauss. Sitzungsber. Ges. Naturforsch. Freunde (Berlin) 1936: 1–51.

Haskins, C. P. 1966. [Discussion of E. O. Wilson, "Behaviour of social insects."] Symp. R. Entomol. Soc. London 3: 93–94.

Haskins, C. P., and R. M. Whelden. 1965. "Queenlessness," worker sibship, and colony versus population structure in the formicid genus *Rhytidoponera*. Psyche 72: 87–112.

Heath, R. R., and P. J. Landolt. 1988. The isolation, identification and synthesis of the alarm pheromone of *Vespula squamosa* (Drury) (Hymenoptera: Vespidae) and associated behavior. Experientia 44: 82–83.

Heinrich, B. 1984. Strategies of thermoregulation and foraging in two vespid wasps, *Dolichovespula maculata* and *Vespula vulgaris*. J. Comp. Physiol. (B) 154: 175–180.

Heithaus, E. R. 1979. Community structure of Neotropical flower visiting bees and wasps: diversity and phenology. Ecology 60: 190–202.

Heldmann, G. 1936a. Über das Leben auf Waben mit mehreren überwinterten Weibchen von *Polistes gallica* L. Biol. Zentralbl. 56: 389–400.

Heldmann, G. 1936b. Über die Entwicklung der polygynen Wabe von *Polistes gallica* L. Arb. Physiol. Angew. Entomol. Berlin-Dahlem 3: 257–259.

Henderson, G., and R. L. Jeanne. 1989. Response of aphid tending ants to a repellent produced by wasps (Hymenoptera: Formicidae, Vespidae). Ann. Entomol. Soc. Amer. 82: 516–519.

Hennig, W. 1966. Phylogenetic Systematics. University of Illinois Press, Urbana.

Hensen, R. V., and L. H. M. Blommers. 1987. Review of the Malagasy species of *Belonogaster* Saussure (Hymenoptera, Vespidae). Tijdschr. Entomol. 130: 11–31.

Hermann, H. R., and M. S. Blum. 1981. Defensive mechanisms in the social Hymenoptera. Pp. 77–197 *in* H. R. Hermann (ed.), Social Insects, vol. 2. Academic Press, New York.

Hermann, H. R., and J. T. Chao. 1984a. Nesting biology and defensive behavior of *Mischocyttarus (Monocyttarus) mexicanus cubicola* (Vespidae: Polistinae). Psyche 91: 51–65.

Hermann, H. R., and J. T. Chao. 1984b. Distribution of *Mischocyttarus (Monocyttarus) mexicanus cubicola* in the United States. Florida Entomol. 67: 516–520.

Hermann, H. R., and J. T. Chao. 1984c. Morphology of the venom apparatus of *Mischocyttarus mexicanus cubicola* (Hymenoptera: Vespidae: Polistinae). J. Georgia Entomol. Soc. 19: 339–344.

Hermann, H. R., and T. F. Dirks. 1974. Sternal glands in polistine wasps: morphology and associated behavior. J. Georgia Entomol. Soc. 9: 1–8.

Hermann, H. R., and T. F. Dirks. 1975. Biology of *Polistes annularis* (Hymenoptera: Vespidae). I. Spring behavior. Psyche 82: 97–108.

Hermann, H. R., A. N. Hunt, and W. F. Buren. 1971. Mandibular gland and mandibular groove in *Polistes annularis* (L.) and *Vespula maculata* (L.) (Hymenoptera: Vespidae). Int. J. Insect Morphol. Embryol. 1: 43–49.

Hermann, H. R., D. Gerling, and T. F. Dirks. 1974. The cohibernation and mating activity of five polistine wasp species (Hymenoptera: Vespidae: Polistinae). J. Georgia Entomol. Soc. 9: 203–204.

Hermann, H. R., R. Barron, and L. Dalton. 1975. Spring behavior of *Polistes exclamans* (Hymenoptera: Vespidae: Polistinae). Entomol. News 86: 173–178.

Herre, E. A., D. M. Windsor, and R. B. Foster. 1986. Nesting associations of wasps and ants on lowland Peruvian ant-plants. Psyche 93: 321–330.

Hicks, C. H. 1929. *Pseudomasaris edwardsii* Cresson, another pollen-provisioning wasp, with further notes on *P. vespoides* (Cresson). Can. Entomol. 61: 121–125.

Higashi, S. 1978. Analysis of internest drifting in a supercolonial ant, *Formica (Formica) yessensis* by individually marked workers. Kontyû 46: 176–191.

Himmer, A. 1927. Ein Beitrag zur Kenntnis des Wärmehaushalts im Nestbau sozialer Hautflügler. Z. Vergl. Physiol. 5: 375–389.

Himmer, A. 1932. Die Temperaturverhältnisse bei den sozialen Hymenopteren. Biol. Rev. (Cambridge) 7: 224–253.

Hingston, R. W. G. 1926. The mason wasp (*Eumenes conica*). J. Bombay Nat. Hist. Soc. 31: 241–247, 754–761, 890–896.

Hingston, R. W. G. 1929. Instinct and Intelligence. Macmillan, New York.

Hirose, Y., and M. Yamasaki. 1984a. Dispersal of females for colony founding in *Polistes jadwigae* Dalla Torre (Hymenoptera, Vespidae). Kontyû 52: 65–71.

Hirose, Y., and M. Yamasaki. 1984b. Foundress association in *Polistes jadwigae* Dalla Torre (Hymenoptera, Vespidae): relatedness among co-foundresses and colony productivity. Kontyû 52: 172–174.

Höfling, J. C., and V. L. L. Machado. 1985. Análise populacional de colonias de *Polybia ignobilis* (Haliday, 1836) (Hymenoptera, Vespidae). Rev. Brasil. Entomol. 29: 271–284.

Hölldobler, B. 1977. Communication in social Hymenoptera. Pp. 418–471 *in* T. A. Sebeok (ed.), How Animals Communicate. Indiana University Press, Bloomington.

Hölldobler, B. 1980. The pygidial gland and chemical recruitment communication in *Pachycondyla (= Termitopone) laevigata*. J. Chem. Ecol. 6: 883–893.

Hölldobler, B., and E. O. Wilson. 1977. The number of queens: an important trait in ant evolution. Naturwissenschaften 64: 8–15.

Hölldobler, B., and E. O. Wilson. 1978. The multiple recruitment systems of the African weaver ant *Oecophylla longinoda* (Latreille) (Hymenoptera: Formicidae). Behav. Ecol. Sociobiol. 3: 19–60.

Hölldobler, B., and E. O. Wilson. 1990. The Ants. Harvard University Press, Cambridge, MA.

Holmes, H. B. 1974. Patterns of sperm competition in *Nasonia vitripennis*. Can. J. Genet. Cytol. 16: 789–795.

Honk, C. G. J. van, H. H. W. Velthuis, P.-F. Röseler, and M. E. Malotaux. 1980. The mandibular glands of *Bombus terrestris* queens as a source of queen pheromones. Entomol. Exp. Appl. 28: 191–198.

Hook, A. W. 1982. Observations on a declining nest of *Polistes tepidus* (F.) (Hymenoptera: Vespidae). J. Aust. Entomol. Soc. 21: 277–278.

Hook, A. W. 1987. Nesting behavior of Texas *Cerceris* digger wasps with emphasis on nest reutilization and nest sharing (Hymenoptera: Sphecidae). Sociobiology 13: 93–118.

Hook, A. W., and H. E. Evans. 1982. Observations on the nesting behaviour of three species of *Ropalidia* Guérin-Méneville (Hymenoptera: Vespidae). J. Aust. Entomol. Soc. 21: 271–275.

Hook, A. W., and R. W. Matthews. 1980. Nesting biology of *Oxybelus sericeus* with a discussion of nest guarding by male sphecid wasps (Hymenoptera). Psyche 87: 21–37.

Hoshikawa, T. 1979. Observations on the polygynous nests of *Polistes chinensis antennalis* Pérez (Hymenoptera: Vespidae) in Japan. Kontyû 47: 239–243.

Houston, T. F. 1984. Bionomics of a pollen-collecting wasp, *Paragia tricolor* (Hymenoptera: Masarinae), in Western Australia. Rec. West. Aust. Mus. 11: 141–151.

Howse, P. E. 1970. Termites: A Study in Social Behaviour. Hutchinson, London.

Huber, A. 1961. Zur Biologie von *Mellinus arvensis* L. (Hymenoptera: Sphecidae). Zool. Jahrb. Abt. Syst. Ökol. Geogr. Tiere 89: 43–118.

Hughes, C. R. 1987. Group nesting and reproductive conflict in primitively eusocial wasps. Ph.D. dissertation, Rice University, Houston, TX.

Hughes, C. R., and J. E. Strassmann. 1988a. Foundress mortality after worker emergence in social wasps (*Polistes*). Ethology 79: 265–280.

Hughes, C. R., and J. E. Strassmann. 1988b. Age is more important than size in determining dominance among workers in the primitively eusocial wasp, *Polistes instabilis*. Behaviour 107: 1–15.

Hughes, C. R., M. O. Beck, and J. E. Strassmann. 1987. Queen succession in the social wasp, *Polistes annularis*. Ethology 76: 124–132.

Hung, A. C. F., H. C. Reed, and S. B. Vinson. 1981. Chromosomes of four species of *Polistes* wasps (Hymenoptera: Vespidae). Caryologia 34: 225–230.

Hungerford, H. B. 1930. An unusual nest of *Vespula* (*Dolichovespula*) *arenaria* Fabr. (= *V. diabolica* de Saussure) (Hym: Vespidae). Entomol. News 41: 329–330.

Hunt, A. N., and H. R. Hermann. 1970. The hymenopterous poison apparatus. X. *Polistes annularis* (Hymenoptera: Vespidae). J. Georgia Entomol. Soc. 5: 210–216.

Hunt, J. H. 1982. Trophallaxis and the evolution of eusocial Hymenoptera. Pp. 201–205 *in* M. D. Breed, C. D. Michener, and H. E. Evans (eds.), The Biology of Social Insects. Westview, Boulder, CO.

Hunt, J. H. 1984. Adult nourishment during larval provisioning in a primitively eusocial wasp, *Polistes metricus* Say. Insectes Soc. 31: 452–460.

Hunt, J. H. 1988. Lobe erection behavior and its possible social role in larvae of *Mischocyttarus* paper wasps. J. Insect Behav. 1: 379–386.

Hunt, J. H., and G. J. Gamboa. 1978. Joint nest use by two paper wasp species. Insectes Soc. 25: 373–374.

Hunt, J. H., and K. C. Noonan. 1979. Larval feeding by male *Polistes fuscatus* and *Polistes metricus* (Hymenoptera: Vespidae). Insectes Soc. 26: 247–251.

Hunt, J. H., I. Baker, and H. G. Baker. 1982. Similarity of amino acids in nectar and larval saliva: the nutritional basis for trophallaxis in social wasps. Evolution 36: 1318–1322.

Hunt, J. H., R. L. Jeanne, I. Baker, and D. E. Grogan. 1987. Nutrient dynamics of the swarm-founding social wasp species, *Polybia occidentalis* (Hymenoptera: Vespidae). Ethology 75: 291–305.

Ihering, H. von. 1896a. Zur Biologie der sozialen Wespen Brasiliens. Zool. Anz. 19: 449–453.

Ihering, H. von. 1896b. L'état des guêpes sociales du Brésil. Bull. Soc. Zool. France 21: 159–162.

Ihering, R. von. 1903. Zur Frage nach dem Ursprung der Staatenbildung bei den sozialen Hymenopteren. Zool. Anz. 27: 113–118.

Ikan, R., and J. Ishay. 1966. Larval wasps secretions and honeydew of the aphids,

*Chaitophorus populi*, feeding on *Populus euphratica* as sources of sugars in the diet of the Oriental hornet, *Vespa orientalis* F. Israel J. Zool. 15: 64–68.

Ikan, R., R. Gottlieb, E. D. Bergmann, and J. Ishay. 1969. The pheromone of the queen of the Oriental hornet, *Vespa orientalis*. J. Insect Physiol. 15: 1709–1712.

Isely, D. 1913. The biology of some Kansas Eumenidae. Univ. Kansas Sci. Bull. 8: 235–309.

Ishay, J. 1964. Observations sur la biologie de la guêpe orientale, *Vespa orientalis* F. Insectes Soc. 11: 193–206.

Ishay, J. 1965. Observations and experiments on colonies of the Oriental wasp. Twentieth Int. Beekeep. Jubilee Cong., pp. 140–145.

Ishay, J. 1972. Thermoregulatory pheromones in wasps. Experientia 28: 1185–1187.

Ishay, J. 1973a. The influence of cooling and queen pheromone on cell building and nest architecture by *Vespa orientalis* (Vespinae, Hymenoptera). Insectes Soc. 20: 243–252.

Ishay, J. 1973b. Thermoregulation by social wasps: behavior and pheromones. Trans. New York Acad. Sci. (Ser. 2) 35: 447–462.

Ishay, J. 1975a. Caste determination by social wasps: cell size and building behaviour. Anim. Behav. 23: 425–431.

Ishay, J. 1975b. Frequencies of the sounds produced by the Oriental hornet. J. Insect Physiol. 21: 1737–1740.

Ishay, J. 1977a. Acoustical communication in wasp colonies (Vespinae). Proc. 15th Int. Cong. Entomol., pp. 406–435.

Ishay, J. 1977b. Filial attachment in *Vespa orientalis* colonies. Proc. 8th Int. Cong. Int. Union Study Soc. Insects, p. 152.

Ishay, J. 1981. On the filial attachment phenomenon in colonies of the Oriental hornet, *Vespa orientalis* (Hymenoptera: Vespidae). Entomol. Gener. 7: 167–174.

Ishay, J., and M. B. Brown. 1975. Patterns in the sounds produced by *Paravespula germanica* wasps. J. Acoust. Soc. Amer. 57: 1521–1525.

Ishay, J. S., and E. Ganor. 1990. Comb cells and puparial silk in the Oriental hornet nest: structure and function. J. Morphol. 203: 11–19.

Ishay, J., and E. M. Landau. 1972. *Vespa* larvae send out rhythmic hunger signals. Nature 237: 286–287.

Ishay, J., and D. Nachshen. 1975. On the nature of the sounds produced within the nest of the wasp *Paravespula germanica* F. Insectes Soc. 22: 213–218.

Ishay, J., and B. Perna. 1979. Building pheromones of *Vespa orientalis* and *Polistes foederatus*. J. Chem. Ecol. 5: 259–272.

Ishay, J., and A. Schwartz. 1973. Acoustical communication between the members of the Oriental hornet (*Vespa orientalis*) colony. J. Acoust. Soc. Amer. 53: 640–649.

Ishay, J., and J. Schwarz. 1965. On the nature of the sounds produced within the nest of the Oriental hornet, *Vespa orientalis* F. Insectes Soc. 12: 383–388.

Ishay, J., R. Ikan, and E. D. Bergmann. 1965. The presence of pheromones in the Oriental hornet, *Vespa orientalis* F. J. Insect Physiol. 11: 1307–1309.

Ishay, J., H. Bytinski-Salz, and A. Shulov. 1968. Contributions to the bionomics of the Oriental hornet (*Vespa orientalis* F.). Israel J. Entomol. 2: 45–106.

Ishay, J., A. Motro, S. Gitter, and M. B. Brown. 1974. Rhythms in acoustical communication by the Oriental hornet, *Vespa orientalis*. Anim. Behav. 22: 741–744.

Ishay, J. S., E. Rosenzweig, and V. A. Paniry. 1982. The genesis of hexagonal cells and the frugality in comb and cell building by social Vespinae (Hymenoptera). Insectes Soc. 29: 34–43.

Ishay, J. S., D. Levanon, N. Avidor, and O. S. Shechter. 1983a. Photoperiodicity stimulates cell size and caste determination in the Oriental hornet *Vespa orientalis* Linné, 1775. Insectes Soc. 30: 332–346.

Ishay, J. S., Z. A. Dotan, and A. Pinshasov. 1983b. Combativeness among Oriental hornet queens. Insectes Soc. 30: 57–69.

Ishay, J. S., E. Rosenzweig, and H. Pechhaker. 1986. Comb building by worker groups of *Vespa crabro* L., *V. orientalis* L. and *Paravespula germanica* Fabr. (Hymenoptera Vespinae). Monit. Zool. Ital. (N.S.) 20: 31–51.

Itino, T. 1986. Comparison of life tables between the solitary eumenid wasp *Anterhynchium flavomarginatum* and the subsocial eumenid wasp *Orancistrocerus drewseni* to evaluate the adaptive significance of maternal care. Res. Popul. Ecol. 28: 185–199.

Itô, Y. 1983. Social behaviour of a subtropical paper wasp, *Ropalidia fasciata* (F.): field observations during founding stage. J. Ethol. 1: 1–14.

Itô, Y. 1984a. Shifts of females between adjacent nests of *Polistes versicolor* (Hymenoptera: Vespidae) in Panama. Insectes Soc. 31: 103–111.

Itô, Y. 1984b. Social behaviour and social structure of neotropical paper wasps, *Mischocyttarus angulatus* Richards and *M. basimacula* (Cameron). J. Ethol. 2: 17–29.

Itô, Y. 1985a. A comparison of frequency of intra-colony aggressive behaviours among five species of polistine wasps (Hymenoptera: Vespidae). Z. Tierpsychol. 68: 152–167.

Itô, Y. 1985b. Colony development and social structure in a subtropical paper wasp, *Ropalidia fasciata* (F.) (Hymenoptera: Vespidae). Res. Popul. Ecol. 27: 333–349.

Itô, Y. 1985c. Social behaviour of an Australian paper wasp, *Ropalidia plebeiana*, with special reference to the process of acceptance of an alien female. J. Ethol. 3: 21–25.

Itô, Y. 1986a. On the pleometrotic route of social evolution in the Vespidae. Monit. Zool. Ital. (N.S.) 20: 241–262.

Itô, Y. 1986b. Social behaviour of *Ropalidia fasciata* (Hymenoptera: Vespidae) females on satellite nests and on a nest with multiple combs. J. Ethol. 4: 73–80.

Itô, Y. 1986c. Observations on the social behaviour of three polistine wasps (Hymenoptera: Vespidae). J. Aust. Entomol. Soc. 25: 309–314.

Itô, Y. 1986d. Spring behaviour of an Australian paper wasp, *Polistes humilis synoecus*: colony founding by haplometrosis and utilization of old nests. Kontyû 54: 191–202.

Itô, Y. 1987a. Social behaviour of the Australian paper wasp, *Ropalidia revolutionalis* (de Saussure) (Hymenoptera: Vespidae). J. Ethol. 5: 115–124.

Itô, Y. 1987b. Role of pleometrosis in the evolution of eusociality in wasps. Pp. 17–34 *in* Y. Itô, J. L. Brown, and J. Kikkawa (eds.), Animal Societies: Theories and Facts. Japan Scientific Societies Press, Tokyo.

Itô, Y., and S. Higashi. 1987. Spring behaviour of *Ropalidia plebeiana* (Hymenoptera: Vespidae) within a huge aggregation of nests. Appl. Entomol. Zool. 22: 519–527.

Itô, Y., and Sk. Yamane. 1985. Early male production in a subtropical paper wasp *Ropalidia fasciata* (Hymenoptera: Vespidae). Insectes Soc. 32: 403–410.

Itô, Y., O. Iwahashi, Sô. Yamane, and Sk. Yamane. 1985. Overwintering and nest reutilization in *Ropalidia fasciata* (Hymenoptera, Vespidae). Kontyû 53: 486–490.

Itô, Y., Sô. Yamane, and J. P. Spradbery. 1988. Population consequences of huge nesting aggregations of *Ropalidia plebeiana* (Hymenoptera: Vespidae). Res. Popul. Ecol. 30: 279–295.

Iwata, K. 1938a. Habits of eight species of Eumeninae (*Rhynchium, Lionotus*, and *Symmorphus*). Mushi 11: 110–132.

Iwata, K. 1938b. Habits of four species of *Odynerus* (*Ancistrocerus*) in Japan. Tenthredo 2: 19–32.

Iwata, K. 1939. Habits of a paper making potter wasp (*Eumenes architectus* Smith) in Japan. Mushi 12: 83–85.

Iwata, K. 1942. Comparative studies on the habits of solitary wasps. Tenthredo 4: 1–142.

Iwata, K. 1953. Biology of *Eumenes* in Japan. Mushi 25: 25–47.

Iwata, K. 1955. The comparative anatomy of the ovary in Hymenoptera. Part I. Aculeata. Mushi 29: 17–37.

Iwata, K. 1964a. Bionomics of nonsocial wasps in Thailand. Nat. Life Southeast Asia 3: 323–383.

Iwata, K. 1964b. Egg giantism in subsocial Hymenoptera, with ethological discussion on tropical bamboo carpenter bees. Nat. Life Southeast Asia 3: 399–435.

Iwata, K. 1966. Description of the nests of so called *Belonogaster griseus* var. *menelikii* Gribodo [sic] collected by Dr. K. Yamashita in Ethiopia, with a general consideration on the life of the genus (Hymenoptera, Vespidae). Mushi 39: 57–64.

Iwata, K. 1967. Report of the fundamental research on the biological control of insect pests in Thailand. II. The report on the bionomics of aculeate wasps. Bionomics of subsocial wasps of Stenogastrinae (Hymenoptera, Vespidae). Nat. Life Southeast Asia 5: 259–293.

Iwata, K. 1969. On the nidification of *Ropalidia (Anthreneida) taiwana koshuensis* Sonan in Formosa (Hymenoptera, Vespidae). Kontyû 37: 367–372.

Iwata, K. 1971. Ethological notes on four Japanese species of *Vespa* (Hymenoptera). Pp. 219–223 *in* S. Asahina, J. L. Gressitt, Z. Hidaka, T. Nishida, and K. Nomura (eds.), Entomological Essays to Commemorate the Retirement of Professor K. Yasumatsu. Hokuryukan, Tokyo.

Iwata, K. 1976. Evolution of Instinct: Comparative Ethology of Hymenoptera. Amerind, New Delhi.

Iwata, K., and K. Yoshikawa. 1964. Biological records on two Saltatoria-hunters of the genera *Dicranorhina* and *Gastrosericus* in Thailand. Nature Life Southeast Asia 3: 385–390.

Jacobson, E. 1922. *Vespa cincta* F. Trop. Natuur. 11: 182–187 (in Dutch).

Jacobson, E. 1935. Aanteekeningen over Stenogastrinae (Hym., Vespidae). Entomol. Meded. Ned.-Indië 1: 15–19. [English translation in: C. K. Starr. 1988. Sphecos (16): 17–19.]

Jacobson, R. S., R. W. Matthews, and J. F. MacDonald. 1978. A systematic study of the *Vespula vulgaris* group with a description of a new yellowjacket species in eastern North America (Hymenoptera: Vespidae). Ann. Entomol. Soc. Amer. 71: 299–312.

Janet, C. 1895. Etudes sur les fourmis, les guêpes et les abeilles. Note 9. Sur *Vespa crabro* L. Histoire d'un nid depuis son origine. Mém. Soc. Zool. France 8: 1–140.

Janet, C. 1903. Observations sur les guêpes. C. Naud, Paris.

Jarvis, J. U. M. 1981. Eusociality in a mammal: cooperative breeding in naked mole-rat colonies. Science 212: 571–573.

Jayakar, S. D. 1963. "Proterandry" in solitary wasps. Nature 198: 208–209.

Jayakar, S. D. 1966. Sexual behavior in solitary eumenid wasps. J. Bombay Nat. Hist. Soc. 63: 760–763.

Jayakar, S. D., and H. Spurway. 1965a. Normal and abnormal nests of *Eumenes emarginatus conoideus* (Gmelin) including notes on crepissage in this and other members of the genus (Vespoidea, Hymenoptera). J. Bombay Nat. Hist. Soc. 62: 193–200.

Jayakar, S. D., and H. Spurway. 1965b. Winter diapause in the squatter wasps *Antodynerus flavescens* (Fabr.) and *Chalybion bengalense* (Dahlb.) (Vespoidea and Sphecoidea). J. Bombay Nat. Hist. Soc. 61: 662–667.

Jayakar, S. D., and H. Spurway. 1966a. Re-use of cells and brother-sister mating in the Indian species *Stenodynerus miniatus* (Sauss.) (Vespidae: Eumenidae). J. Bombay Nat. Hist. Soc. 63: 378–398.

Jayakar, S. D., and H. Spurway. 1966b. Sex ratios of some mason wasps (Vespoidea and Sphecoidea). Nature 212: 306–307.

Jayakar, S. D., and H. Spurway. 1967. The nesting activities of the vespoid potter wasp *Eumenes campaniformis esuriens* (Fabr.) compared with the ecologically similar sphecoid *Sceliphron madraspatanum* (Fabr.) (Hymenoptera). J. Bombay Nat. Hist. Soc. 64: 307–332.

Jayakar, S. D., and H. Spurway. 1968. The nesting activities of the vespoid potter wasp *Eumenes campaniformis esuriens* (Fabr.) compared with the ecologically similar sphecoid *Sceliphron madraspatanum* (Fabr.) (Hymenoptera). J. Bombay Nat. Hist. Soc. 65: 148–181.

Jayasingh, D. B., and C. A. Taffe. 1982. The biology of the eumenid mud-wasp *Pachodynerus nasidens* in trapnests. Ecol. Entomol. 7: 283–289.

Jeanne, R. L. 1970a. Chemical defense of brood by a social wasp. Science 168: 1465–1466.

Jeanne, R. L. 1970b. Note on a bat (*Phylloderma stenops*) preying upon the brood of a social wasp. J. Mammal. 51: 624–625.

Jeanne, R. L. 1970c. Descriptions of the nests of *Pseudochartergus fuscatus* and *Stelopolybia testacea*, with a note on a parasite of *S. testacea* (Hymenoptera, Vespidae). Psyche 77: 54–69.

Jeanne, R. L. 1972. Social biology of the Neotropical wasp *Mischocyttarus drewseni*. Bull. Mus. Comp. Zool. Harvard Univ. 144: 63–150.

Jeanne, R. L. 1973. Aspects of the biology of *Stelopolybia areata* (Say) (Hymenoptera: Vespidae). Biotropica 5: 183–198.

Jeanne, R. L. 1975a. The adaptiveness of social wasp nest architecture. Quart. Rev. Biol. 50: 267–287.

Jeanne, R. L. 1975b. Social biology of *Stelopolybia areata* (Say) in Mexico (Hymenoptera: Vespidae). Insectes Soc. 22: 27–34.

Jeanne, R. L. 1975c. Behavior during swarm movement in *Stelopolybia areata* (Hymenoptera: Vespidae). Psyche 82: 259–264.

Jeanne, R. L. 1977a. Ultimate factors in social wasp nesting behavior. Proc. 8th. Int. Cong. Int. Union Study Soc. Insects, pp. 164–168.

Jeanne, R. L. 1977b. Behavior of the obligate social parasite *Vespula arctica* (Hymenoptera: Vespidae). J. Kansas Entomol. Soc. 50: 541–557.

Jeanne, R. L. 1977c. A specialization in nest petiole construction by queens of *Vespula* spp. (Hymenoptera: Vespidae). J. New York Entomol. Soc. 85: 127–129.

Jeanne, R. L. 1978. Intraspecific nesting associations in the Neotropical social wasp *Polybia rejecta* (Hymenoptera: Vespidae). Biotropica 10: 234–235.

Jeanne, R. L. 1979a. Construction and utilization of multiple combs in *Polistes canadensis* in relation to the biology of a predaceous moth. Behav. Ecol. Sociobiol. 4: 293–310.

Jeanne, R. L. 1979b. A latitudinal gradient in rates of ant predation. Ecology 60: 1211–1224.

Jeanne, R. L. 1979c. Nest of the wasp *Clypearia weyrauchi* (Hymenoptera, Vespidae). J. New York Entomol. Soc. 87: 78–84.

Jeanne, R. L. 1980a. Evolution of social behavior in the Vespidae. Annu. Rev. Entomol. 25: 371–396.

Jeanne, R. L. 1980b. Observações sobre limpeza e reutilização de células em ninhos de vespas sociais (Hymenoptera: Vespidae). Bol. Mus. Para. Emilio Goeldi (N.S.) Zoologia) 101: 1–8.

Jeanne, R. L. 1981a. Chemical communication during swarm emigration in the social wasp *Polybia sericea* (Olivier). Anim. Behav. 29: 102–113.

Jeanne, R. L. 1981b. Alarm recruitment, attack behavior, and the role of the alarm pheromone in *Polybia occidentalis* (Hymenoptera: Vespidae). Behav. Ecol. Sociobiol. 9: 143–148.

Jeanne, R. L. 1981c. Migração de enxames de vespas sociais: demonstração de um caminho químico. Acta Amaz. 11: 153–156.

Jeanne, R. L. 1982a. Predation, defense, and colony size and cycle in the social wasps. Pp. 280–284 *in* M. D. Breed, C. D. Michener, and H. E. Evans (eds.), The Biology of Social Insects. Westview, Boulder, CO.

Jeanne, R. L. 1982b. Evidence for an alarm substance in *Polistes canadensis*. Experientia 38: 329–330.

Jeanne, R. L. 1986a. The organization of work in *Polybia occidentalis*: the costs and benefits of specialization in a social wasp. Behav. Ecol. Sociobiol. 19: 333–341.

Jeanne, R. L. 1986b. The evolution of the organization of work in social insects. Monit. Zool. Ital. (N.S.) 20: 119–133.

Jeanne, R. L. 1987. Do water foragers pace nest construction activity in *Polybia occidentalis?* Pp. 241–251 *in* J. M. Pasteels and J.-L. Deneubourg (eds.), From Individual to Collective Behavior in Social Insects. Birkhäuser, Basel.

Jeanne, R. L., and E. G. Castellón Bermúdez. 1980. Reproductive behavior of a male Neotropical social wasp, *Mischocyttarus drewseni* (Hymenoptera: Vespidae). J. Kansas Entomol. Soc. 53: 271–276.

Jeanne, R. L., and D. W. Davidson. 1984. Population regulation in social insects. Pp. 559–590 *in* C. B. Huffaker and R. L. Rabb (eds.), Ecological Entomology. Wiley, New York.

Jeanne, R. L., and R. Fagen. 1974. Polymorphism in *Stelopolybia areata* (Hymenoptera, Vespidae). Psyche 81: 155–166.

Jeanne, R. L., and D. C. Post. 1982. Richards' gland and associated cuticular modifications in social wasps of the genus *Polybia* Lepeletier (Hymenoptera, Vespidae, Polistinae, Polybiini). Insectes Soc. 29: 280–294.

Jeanne, R. L., H. A. Downing, and D. C. Post. 1983. Morphology and function of sternal glands in polistine wasps (Hymenoptera: Vespidae). Zoomorphology 103: 149–164.

Jeanne, R. L., H. A. Downing, and D. C. Post. 1988. Age polyethism and individual variation in *Polybia occidentalis*, an advanced eusocial wasp. Pp. 323–357 *in* R. L. Jeanne (ed.), Interindividual Behavioral Variability in Social Insects. Westview, Boulder, CO.

Jolly, A. 1985. The Evolution of Primate Behavior, 2nd ed. Macmillan, New York.

Jorgensen, P. 1942. Biological observations on some solitary vespids. Entomol. Medd. 22: 299–335.

Joshi, N. V., and R. Gadagkar. 1985. Evolution of sex ratios in social Hymenoptera: kin selection, local mate competition, polyandry and kin recognition. J. Genet. 64: 41–58.

Kasuya, E. 1980. Behavioral ecology of Japanese paper wasps, *Polistes* spp. (Hymenoptera: Vespidae). I. Extranidal activities of *Polistes chinensis antennalis*. Res. Popul. Ecol. 22: 242–254.

Kasuya, E. 1981a. Polygyny in the Japanese paper wasp, *Polistes jadwigae* Dalla Torre (Hymenoptera, Vespidae). Kontyû 49: 306–313.

Kasuya, E. 1981b. Nest foundation by a single worker of the Japanese paper wasp, *Polistes chinensis antennalis* (Hymenoptera: Vespidae). Insectes Soc. 28: 341–342.

Kasuya, E. 1981c. Internidal drifting of workers in the Japanese paper wasp, *Polistes chinensis antennalis* (Vespidae: Hymenoptera). Insectes Soc. 28: 343–346.

Kasuya, E. 1981d. Male mating territory in a Japanese paper wasp, *Polistes jadwigae* Dalla Torre (Hymenoptera, Vespidae). Kontyû 49: 607–614.

Kasuya, E. 1982. Take-over of nests in a Japanese paper wasp, *Polistes chinensis antennalis* (Vespidae: Hymenoptera). Appl. Entomol. Zool. 17: 427–431.

Kasuya, E. 1983a. Social behavior of early emerging males of a Japanese paper wasp, *Polistes chinensis antennalis* (Hymenoptera: Vespidae). Res. Popul. Ecol. 25: 143–149.

Kasuya, E. 1983b. Behavioral ecology of Japanese paper wasps, *Polistes* spp. (Hymenoptera: Vespidae). II. Ethogram and internidal relationship in *Polistes chinensis antennalis* in the founding stage. Z. Tierpsychol. 63: 303–317.

Kasuya, E. 1983c. Behavioral ecology of Japanese paper wasps, *Polistes* spp. IV. Comparison of ethograms between queens and workers of *P. chinensis antennalis* in the ergonomic stage. J. Ethol. 1: 34–45.

Kayes, B. M. 1978. Digestive proteases in four species of *Polistes* wasps. Can. J. Zool. 56: 1454–1459.

Keeping, M. G. 1984. A beetle predacious on the brood of a social wasp. J. Entomol. Soc. Sth. Africa 47: 355–358.

Keeping, M. G. 1990. Rubbing behavior and morphology of van der Vecht's gland in *Belonogaster petiolata* (Hymenoptera: Vespidae). J. Insect Behav. 3: 85–104.

Keeping, M. G., and R. M. Crewe. 1983. Parasitoids, commensals and colony size in nests of *Belonogaster* (Hymenoptera: Vespidae). J. Entomol. Soc. Sth. Africa 46: 309–323.

Keeping, M. G., and R. M. Crewe. 1987. The ontogeny and evolution of foundress associations in *Belonogaster petiolata* (Hymenoptera: Vespidae). Pp. 383–384 *in* J. Eder and H. Rembold (eds.), Chemistry and Biology of Social Insects. J. Peperny, Munich.

Keeping, M. G., D. Lipschitz, and R. M. Crewe. 1986. Chemical mate recognition and release of male sexual behavior in polybiine wasp, *Belonogaster petiolata* (Degeer) (Hymenoptera: Vespidae). J. Chem. Ecol. 12: 773–779.

Kemper, H. 1961. Nestunterschiede bei den sozialen Faltenwespen Deutschlands. Z. Angew. Zool. 48: 31–85.

Kemper, H., and E. Döhring. 1967. Die sozialen Faltenwespen Mitteleuropas. Paul Parey, Berlin.

Kerr, W. E. 1952. A variação do número de cromosomas na evolução dos Hymenoptera. Scientia Genet. 4: 182–190.

Keyel, R. E. 1983. Some aspects of niche relationships among yellowjackets (Hymenoptera: Vespidae) of the northeastern United States. Ph.D. dissertation, Cornell University, Ithaca, NY.

Khoo, S. G., and H. S. Yong. 1987. Nest structure and colony defence in the stingless bee *Trigona terminata* Smith. Nature Malaysiana 12: 4–15.

Kimsey, L. S. 1980a. The behaviour of male orchid bees (Apidae, Hymenoptera, Insecta) and the question of leks. Anim. Behav. 28: 996–1004.

Kimsey, L. S. 1980b. Notes on the biology of some Panamanian Pompilidae, with a description of a communal nest (Hymenoptera). Pan-Pac. Entomol. 56: 98–100.

Kirkpatrick, M. 1987. Sexual selection by female choice in polygynous animals. Annu. Rev. Ecol. Syst. 18: 43–70.

Klahn, J. E. 1979. Philopatric and nonphilopatric foundress associations in the social wasp *Polistes fuscatus*. Behav. Ecol. Sociobiol. 5: 417–424.

Klahn, J. E. 1981. Alternate reproductive tactics of single foundresses of a social wasp, *Polistes fuscatus*. Ph.D. dissertation, University of Iowa, Iowa City.

Klahn, J. E. 1988. Intraspecific comb usurpation in the social wasp *Polistes fuscatus*. Behav. Ecol. Sociobiol. 23: 1–8.

Klahn, J. E., and G. J. Gamboa. 1983. Social wasps: discrimination between kin and nonkin brood. Science 221: 482–484.

Kluge, A. G., and J. S. Farris. 1969. Quantitative phyletics and the evolution of anurans. Syst. Zool. 18: 1–32.

Knisley, C. D. 1985. Utilization of tiger beetle larval burrows by a nest-provisioning wasp, *Leucodynerus russatus* (Bohart) (Hymenoptera: Eumenidae). Proc. Entomol. Soc. Washington 87: 481.

Kojima, J. 1982a. Nest architecture of three *Ropalidia* species (Hymenoptera: Vespidae) on Leyte Island, the Philippines. Biotropica 14: 272–280.

Kojima, J. 1982b. Notes on rubbing behavior in *Ropalidia gregaria* (Hymenoptera, Vespidae). New Entomol. 31: 17–19.

Kojima, J. 1983a. Defense of the pre-emergence colony against ants by means of a chemical barrier in *Ropalidia fasciata* (Hymenoptera, Vespidae). Jpn. J. Ecol. 33: 213–223.

Kojima, J. 1983b. Occurrence of the rubbing behavior in a paper wasp, *Parapolybia indica* (Hymenoptera, Vespidae). Kontyû 51: 158–159.

Kojima, J. 1983c. Peritrophic sac extraction in *Ropalidia fasciata* (Hymenoptera, Vespidae). Kontyû 51: 502–508.

Kojima, J. 1984a. *Ropalidia* wasps in the Philippines (Hymenoptera, Vespidae). (I) Subgenus *Icariola*. Kontyû 52: 522–532.

Kojima, J. 1984b. Construction of multiple independent combs in *Ropalidia fasciata* (Hymenoptera: Vespidae). Jpn. J. Ecol. 34: 233–234.

Kojima, J. 1984c. Division of labor and dominance interaction among co-foundresses on pre-emergence colonies of *Ropalidia fasciata* (Hymenoptera, Vespidae). Biol. Mag. Okinawa 22: 27–35.

Kojima, J. 1988. Nest construction of *Ropalidia fasciata* (Hymenoptera, Vespidae). Kontyû 56: 169–179.

Kojima, J. 1989. Growth and survivorship of preemergence colonies of *Ropalidia fasciata* in relation to foundress group size in the subtropics (Hymenoptera: Vespidae). Insectes Soc. 36: 197–218.

Kojima, J., and R. L. Jeanne. 1986. Nests of *Ropalidia (Icarielia) nigrescens* and *R. (I.) extrema* from the Philippines, with reference to the evolutionary radiation in nest architecture within the subgenus *Icarielia* (Hymenoptera: Vespidae). Biotropica 18: 324–336.

Kojima, J., and M. G. Keeping. 1985. Larvae of *Belonogaster juncea colonialis* Kohl and *B. petiolata* (Degeer) (Hymenoptera: Vespidae). J. Entomol. Soc. Sth. Africa 48: 233–239.

Kojima, J., and T. Suzuki. 1986. Timing of mating in five Japanese polistine wasps (Hymenoptera: Vespidae): anatomy of fall females. J. Kansas Entomol. Soc. 59: 401–404.

Kojima, J., and Sk. Yamane. 1980. Biological notes on *Vespa luctuosa luzonensis* from Leyte Island, the Philippines, with descriptions of adults and larvae (Hymenoptera: Vespidae). Insecta Matsumurana (N.S.) 19: 79–87.

Königsmann, E. 1978. Das phylogenetische System der Hymenoptera. Teil 4: Aculeata (Unterordnung Apocrita). Dtsch. Entomol. Z. (N.F.) 25: 365–435.

Krispyn, J. W. 1979. Colony productivity and survivorship of the paper wasp *Polistes annularis*. Ph.D. dissertation, University of Georgia, Athens.

Krombein, K. V. 1960. Biological notes on several southwestern ground-nesting wasps. Bull. Brooklyn Entomol. Soc. 55: 75–79.

Krombein, K. V. 1967. Trap-nesting Wasps and Bees: Life Histories, Nests, and Associates. Smithsonian Institution Press, Washington, D.C.

Krombein, K. V. 1976. *Eustenogaster*, a primitive social Sinhalese wasp. Loris 14: 303–306.

Krombein, K. V. 1978. Biosystematic studies of Ceylonese wasps. III. Life history, nest, and associates of *Paraleptomenes mephitis* (Cameron) (Hymenoptera: Eumenidae). J. Kansas Entomol. Soc. 51: 721–734.

Krombein, K. V., P. D. Hurd, D. R. Smith, and B. D. Burks. 1979. Catalog of Hymenoptera in America North of Mexico. Smithsonian Institution Press, Washington, D.C.

Kugler, J., T. Orion, and J. Ishay. 1976. The number of ovarioles in the Vespinae (Hymenoptera). Insectes Soc. 23: 525–533.

Kukuk, P. F. 1989. Evolutionary genetics of a primitively eusocial halictine bee, *Dialictus zephyrus*. Pp. 183–202 *in* M. D. Breed and R. E. Page (eds.), The Genetics of Social Evolution. Westview, Boulder, CO.

Kukuk, P. F., and G. C. Eickwort. 1987. Alternative social structures in halictine bees. Pp. 555–556 *in* J. Eder and H. Rembold (eds.), Chemistry and Biology of Social Insects. J. Peperny, Munich.

Kukuk, P. F., M. D. Breed, A. Sobti, and W. J. Bell. 1977. The contributions of kinship and conditioning to nest recognition and colony member recognition in a primitively eusocial bee, *Lasioglossum zephyrum* (Hymenoptera: Halictidae). Behav. Ecol. Sociobiol. 2: 319–327.

Kukuk, P. F., G. C. Eickwort, M. Raveret-Richter, B. Alexander, R. Gibson, R. A. Morse, and F. Ratnieks. 1989. Importance of the sting in the evolution of sociality in the Hymenoptera. Ann. Entomol. Soc. Amer. 82: 1–5.

Kulike, H. 1987. Zur Biologie der Hornisse *Vespa crabro*: Stachelaufbau und Funktion, Gift und Stichwirkung, Hornissen im Umfeld des Menschen. Ph.D. dissertation, Freie Universität Berlin, Berlin.

Kurczewski, F. E., and R. C. Miller. 1984. Observations on the nesting of three species of *Cerceris* (Hymenoptera: Sphecidae). Florida Entomol. 67: 146–155.

Laidlaw, W. B. R. 1930. Notes on some humble bees and wasps in Scotland. The social wasps. Scottish Nat. 185: 135–136.

Landolt, P. J., and R. D. Akre. 1979a. Occurrence and location of exocrine glands in some social Vespidae (Hymenoptera). Ann. Entomol. Soc. Amer. 72: 141–148.

Landolt, P. J., and R. D. Akre. 1979b. Ultrastructure of the thoracic gland of queens of the western yellowjacket *Vespula pensylvanica* (Hymenoptera: Vespidae). Ann. Entomol. Soc. Amer. 72: 586–590.

Landolt, P. J., and R. R. Heath. 1987. Alarm pheromone behavior of *Vespula squamosa* (Hymenoptera: Vespidae). Florida Entomol. 70: 222–225.

Landolt, P. J., R. D. Akre, and A. Greene. 1977. Effects of colony division on *Vespula atropilosa* (Sladen) (Hymenoptera: Vespidae). J. Kansas Entomol. Soc. 50: 135–147.

Larch, C. M., and G. J. Gamboa. 1981. Investigation of mating preference for nestmates in the paper wasp *Polistes fuscatus* (Hymenoptera: Vespidae). J. Kansas Entomol. Soc. 54: 811–814.

Larsson, F. K. 1986. Increased nest density of the digger wasp *Bembix rostrata* as a response to parasites and predators (Hymenoptera: Sphecidae). Entomol. Gener. 12: 71–75.

Lashomb, J. H., and A. L. Steinhauer. 1975. Observations of *Zethus spinipes* Say (Hymenoptera: Eumenidae). Proc. Entomol. Soc. Washington 77: 164.

Lecomte, J., V. Bourdon, J. Damas, M. Leclercq, and J. Leclercq. 1976. Présence de noradrénaline conjuguée dans les parois du nid de *Vespula germanica* Linné. C. R. Soc. Biol. (Paris) 170: 212–215.

Lello, E. de. 1971. Adnexal glands of the sting apparatus in bees: anatomy and histology, I (Hymenoptera: Colletidae and Andrenidae). J. Kansas Entomol. Soc. 44: 5–13.

Lello, E. de. 1976. Adnexal glands of the sting apparatus in bees: anatomy and histology, V (Hymenoptera: Apidae). J. Kansas Entomol. Soc. 49: 85–99.

Lester, L. J., and R. K. Selander. 1981. Genetic relatedness and the social organization of *Polistes* colonies. Amer. Nat. 117: 147–166.

Lin, N. 1964. Increased parasitic pressure as a major factor in the evolution of social behavior in halictine bees. Insectes Soc. 11: 187–192.

Lin, N. 1972. Territorial behavior among males of the social wasp *Polistes exclamans* Viereck (Hymenoptera: Vespidae). Proc. Entomol. Soc. Washington 74: 148–155.

Lin, N. 1978. Sequential hypermalaxation in the digger wasp *Diodontus franclemonti* Krombein (Hymenoptera: Sphecidae). J. Kansas Entomol. Soc. 51: 235–238.

Lin, N., and C. D. Michener. 1972. Evolution of sociality in insects. Quart. Rev. Biol. 47: 131–159.

Lindauer, M. 1952. Ein Beitrag zur Frage der Arbeitsteilung im Bienenstaat. Z. Vergl. Physiol. 34: 299–345.

Lindauer, M. 1971. Communication among Social Bees, 2nd ed. Harvard University Press, Cambridge, MA.

Litte, M. 1977. Behavioral ecology of the social wasp, *Mischocyttarus mexicanus*. Behav. Ecol. Sociobiol. 2: 229–246.

Litte, M. 1979. *Mischocyttarus flavitarsis* in Arizona: social and nesting biology of a polistine wasp. Z. Tierpsychol. 50: 282–312.

Litte, M. 1981. Social biology of the polistine wasp *Mischocyttarus labiatus*: survival in a Colombian rain forest. Smithsonian Contr. Zool. 327: 1–27.

Lofgren, C. S., W. A. Banks, and B. M. Glancey. 1975. Biology and control of imported fire ants. Annu. Rev. Entomol. 20: 1–30.

Loher, W., I. Ganjian, I. Kubo, D. Stanley-Samuelson, and S. Tobe. 1981. Prostaglandins: their role in egg-laying of the cricket *Teleogryllus commodus*. Proc. Natl. Acad. Sci. USA 78: 7835–7838.

Lomholdt, O. 1982. On the origin of the bees (Hymenoptera: Apidae, Sphecidae). Entomol. Scand. 13: 185–190.

Lomholdt, O. 1985. A reclassification of the larrine tribes with a revision of the Miscophini of southern Africa and Madagascar (Hymenoptera: Sphecidae). Entomol. Scand. 24: 1–183 (suppl.).

Long, G. E., C. F. Roush, and R. D. Akre. 1979. A linear model of development for colonies of *Vespula pensylvanica* (Hymenoptera: Vespidae) collected from Pullman, Washington. Melanderia 31: 27–36.

Longair, R. W. 1981. Sex-ratio variations in xylophilous aculeate Hymenoptera. Evolution 35: 597–600.

Longair, R. W. 1984. Male mating behavior in solitary wasps (Hymenoptera: Vespidae). Ph.D. dissertation, Colorado State University, Fort Collins.

Longair, R. W. 1985. Male behavior in *Euparagia richardsi* Bohart (Hymenoptera: Vespidae). Pan-Pac. Entomol. 61: 318–320.

Longair, R. W. 1987. Mating behavior at floral resources in two species of *Pseudomasaris* (Hymenoptera: Vespidae: Masarinae). Proc. Entomol. Soc. Washington 89: 759–769.

Lord, W. D. 1979. Foraging, colony productivity, and competition in *Vespula maculifrons* (Buysson). M.S. thesis, University of Delaware, Newark.

Lord, W. D., and R. R. Roth. 1978. Is there intraspecific competition in *Vespula maculifrons* (Hymenoptera: Vespidae)? J. New York Entomol. Soc. 86: 304–305.

Lord, W. D., and R. R. Roth. 1985. Density, distribution and reproductive success in *Vespula maculifrons* (Hymenoptera: Vespidae). Amer. Midl. Nat. 113: 353–360.

Lord, W. D., D. A. Nicolson, and R. R. Roth. 1977. Foraging behavior and colony drift in *Vespula maculifrons* (Buysson) (Hymenoptera: Vespidae). J. New York Entomol. Soc. 85: 186.

Lorenzi, M. C., and S. Turillazzi. 1986. Behavioural and ecological adaptations to the high mountain environment of *Polistes biglumis bimaculatus*. Ecol. Entomol. 11: 199–204.

Lövgren, B. 1958. A mathematical treatment of the development of colonies of different kinds of social wasps. Bull. Math. Biophys. 20: 119–148.

Lucas, M. H. 1867. Quelques remarques sur les nids des *Polybia scutellaris* et *liliacea*, Hyménoptères sociaux de la tribu des Vespides. Ann. Soc. Entomol. France (Sér. 4) 7: 365–370.

Lucas, M. H. 1885. [Remarks on a nest of *Myraptera scutellaris*]. Ann. Soc. Entomol. France (Sér. 6) 5: LIV (in French).

MacCormack, A. T. 1982. Foundress associations and early colony failure in *Polistes exclamans*. Ph.D. dissertation, University of North Carolina, Chapel Hill.

MacDonald, J. F. 1977. Comparative and adaptive aspects of vespine nest construction. Proc. 8th Int. Cong. Int. Union Study Soc. Insects, pp. 169–172.

MacDonald, J. F., and R. D. Akre. 1984. Range extension and emergence of subterranean nesting by the German yellowjacket, *Vespula germanica*, in North America (Hymenoptera: Vespidae). Entomol. News 95: 5–8.

MacDonald, J. F., and R. W. Matthews. 1975. *Vespula squamosa*: a yellow jacket wasp evolving toward parasitism. Science 190: 1003–1004.

MacDonald, J. F., and R. W. Matthews. 1976. Nest structure and colony composition of *Vespula vidua* and *V. consobrina* (Hymenoptera: Vespidae). Ann. Entomol. Soc. Amer. 69: 471–475.

MacDonald, J. F., and R. W. Matthews. 1981. Nesting biology of the eastern yellowjacket, *Vespula maculifrons* (Hymenoptera: Vespidae). J. Kansas Entomol. Soc. 54: 433–457.

MacDonald, J. F., and R. W. Matthews. 1984. Nesting biology of the southern yellowjacket, *Vespula squamosa* (Hymenoptera: Vespidae): social parasitism and independent founding. J. Kansas Entomol. Soc. 57: 134–151.

MacDonald, J. F., R. D. Akre, and W. B. Hill. 1974. Comparative biology and behavior of *Vespula atropilosa* and *V. pensylvanica* (Hymenoptera: Vespidae). Melanderia 18: 1–66.

MacDonald, J. F., R. D. Akre, and W. B. Hill. 1975a. Locations and structure of nests of *Vespula atropilosa* and *V. acadica* (Hymenoptera: Vespidae). J. Kansas Entomol. Soc. 48: 114–121.

MacDonald, J. F., R. D. Akre, and W. B. Hill. 1975b. Nest associates of *Vespula atropilosa* and *V. pensylvanica* in southeastern Washington state (Hymenoptera: Vespidae). J. Kansas Entomol. Soc. 48: 53–63.

MacDonald, J. F., R. D. Akre, and R. E. Keyel. 1980a. The German yellowjacket (*Vespula germanica*) problem in the United States (Hymenoptera: Vespidae). Bull. Entomol. Soc. Amer. 26: 436–442.

MacDonald, J. F., R. W. Matthews, and R. S. Jacobson. 1980b. Nesting biology of the yellowjacket, *Vespula flavopilosa* (Hymenoptera: Vespidae). J. Kansas Entomol. Soc. 53: 448–458.

Macevicz, S., and G. F. Oster. 1976. Modeling social insect populations. II: optimal reproductive strategies in annual eusocial insect colonies. Behav. Ecol. Sociobiol. 1: 265–282.

Machado, V. L. L. 1977a. Estudos biológicos de *Polybia occidentalis occidentalis* (Olivier, 1791) (Hym.-Vespidae). An. Soc. Entomol. Brasil 6: 7–24.

Machado, V. L. L. 1977b. Aspectos da biologia de *Protopolybia pumila* (Saussure, 1863) (Hym., Vespidae). Rev. Brasil. Biol. 37: 771–784.

Machado, V. L. L. 1982. Plants which supply "hair" material for nest building of *Protopolybia sedula* (Saussure, 1854). Pp. 189–192 *in* P. Jaisson (ed.), Social Insects in the Tropics, vol. 1. Université de Paris-Nord, Paris.

Machado, V. L. L. 1985. Análise populacional de colónias de *Polybia (Myrapetra) paulista* (Ihering, 1896) (Hymenoptera, Vespidae). Rev. Brasil. Zool. 2: 187–201.

Machado, V. L. L., and N. J. Hebling. 1972. Polimorfismo das castas femininas de *Chartergus chartarius* (Olivier, 1791) (Hymenoptera-Vespidae). Pp. 199–204 *in* C. da Cruz-Landim, N. J. Hebling, E. de Lello, and C. S. Takahashi (eds.), Homenagem à Warwick E. Kerr. Faculdade de Ciências Médicas e Biológicas de Botucatu, Rio Claro, São Paulo.

Machado, V. L. L., and V. M. Rodrigues. 1975. Anatomia comparativa das glândulas salivares das diferentes castas de *Protopolybia pumila* (Saussure, 1863) e *Protopolybia exigua exigua* (Saussure, 1854) (Hymenoptera-Vespidae). Rev. Agric. (Piracicaba) 50: 95–102.

Machado, V. L. L., N. Gobbi, and D. Simões. 1987. Material capturado e utilizado na alimentação de *Stelopolybia pallipes* (Olivier, 1791) (Hymenoptera-Vespidae). An. Soc. Entomol. Brasil 16: 73–79.

Machida, J. 1934. The spermatogenesis of the three species of *Polistes* (Hymenoptera). Proc. Imp. Acad. Japan 10: 515–518.

MacLean, B. K., L. Chandler, and D. B. MacLean. 1978. Phenotypic expression in the paper wasp *Polistes fuscatus* (Hymenoptera: Vespidae). Great Lakes Entomol. 11: 105–116.

Madden, J. L. 1981. Factors influencing the abundance of the European wasp (*Paravespula germanica* [F.]). J. Aust. Entomol. Soc. 20: 59–65.

Maher, G. D. 1976. Some notes on social interactions in *Polistes exclamans* (Hymenoptera: Vespidae). Entomol. News 87: 218–222.

Makino, S. 1982. Nest structure, colony composition and productivity of *Dolichovespula media media* and *D. saxonica nipponica* in Japan (Hymenoptera, Vespidae). Kontyû 50: 212–224.

Makino, S. 1983. Larval feeding by *Polistes biglumis* males. Kontyû 51: 487.

Makino, S. 1985a. Solitary life of the foundress: field observations of behavior of *Dolichovespula media media* and *D. saxonica nipponica* in the pre-emergence stage (Hymenoptera, Vespidae). Kontyû 53: 475–485.

Makino, S. 1985b. Unwelcome conspecific visitors to *Dolichovespula* early nests (Hymenoptera, Vespidae). Kontyû 53: 578–579.

Makino, S. 1985c. Foundress-replacement on nests of the monogynic paper wasp *Polistes biglumis* in Japan (Hymenoptera, Vespidae). Kontyû 53: 143–149.

Makino, S., and S. Aoki. 1982. Observations on two polygynic colonies of *Polistes biglumis* in Hokkaido, northern Japan (Hymenoptera, Vespidae). Kontyû 50: 175–182.

Makino, S., and Sk. Yamane. 1980. Heat production by the foundress of *Vespa simillima*, with description of its embryo nest (Hymenoptera: Vespidae). Insecta Matsumurana (N.S.) 19: 89–101.

Makino, S., Sk. Yamane, T. Ban, and I. Kunou. 1981. The Japanese hornet *Vespa simillima* Smith, an important nuisance pest in urban areas (Hymenoptera: Vespidae). Jpn. J. Sanit. Zool. 32: 203–213.

Makino, S., Sk. Yamane, T. Sunose, and S. Aoki. 1987. Dispersion distance of queens from natal sites in the two haplometrotic paper wasps *Polistes riparius* and *P. snelleni* (Hymenoptera: Vespidae). Res. Popul. Ecol. 29: 111–117.

Malyshev, S. I. 1968. Genesis of the Hymenoptera and the Phases of Their Evolution. Translated by B. Haigh; edited by O. W. Richards and B. Uvarov. Methuen, London.

Marchal, P. 1896a. La reproduction et l'évolution des guêpes sociales. Arch. Zool. Expér. Gén. (Sér. 3) 4: 1–100.

Marchal, P. 1896b. Observations sur les *Polistes*. Cellule primitive et première cellule du nid. Provision de miel. Association de reines fondatrices. Bull. Soc. Zool. France 21: 15–21.

Marchal, P. 1897. La castration nutriciale chez les Hyménoptères sociaux. C. R. Soc. Biol. (Paris) 1897: 556–557. [English translation in: C. K. Starr. 1982. Sphecos (5): 26–27.]

Marino Piccioli, M. T. 1968. The extraction of the larval peritrophic sac by the adults in *Belonogaster*. Monit. Zool. Ital. (N.S.) 2: 203–206 (suppl.).

Marino Piccioli, M. T., and L. Pardi. 1970. Studi sulla biologia di *Belonogaster* (Hymenoptera, Vespidae). I. Sull'etogramma di *Belonogaster griseus* (Fab.) [sic]. Monit. Zool. Ital. (N.S.) 3: 197–225 (suppl.).

Marino Piccioli, M. T., and L. Pardi. 1978. Studies on the biology of *Belonogaster* (Hymenoptera Vespidae). 3. The nest of *Belonogaster griseus* (Fab.) [sic]. Monit. Zool. Ital. (N.S.) 10: 179–228 (suppl.).

Markin, G. P., and A. R. Gittins. 1967. Biology of *Stenodynerus claremontensis* (Cameron) (Hymenoptera: Vespidae). Univ. Idaho Coll. Agric. Res. Bull. 74: 1–25.

Martin, S. J. 1988. Thermoregulation in *Vespa simillima xanthoptera* (Hymenoptera, Vespidae). Kontyû 56: 674–677.

Maschwitz, U. 1964a. Alarm substances and alarm behaviour in social Hymenoptera. Nature 204: 324–327.

Maschwitz, U. 1964b. Gefahrenalarmstoffe und Gefahrenalarmierung bei sozialen Hymenopteren. Z. Vergl. Physiol. 47: 596–655.

Maschwitz, U. 1966a. Das Speichelsekret der Wespenlarven und seine biologische Bedeutung. Z. Vergl. Physiol. 53: 228–252.

Maschwitz, U. 1966b. Alarm substances and alarm behavior in social insects. Vitam. Horm. 24: 267–290.

Maschwitz, U. 1984. Alarm pheromone in the long-cheeked wasp *Dolochovespula* [*sic*] *saxonica* (Hym. Vespidae). Dtsch. Entomol. Z. (N.F.) 31: 33–34.

Maschwitz, U., and H. Hänel. 1988. Biology of the Southeast Asian nocturnal wasp, *Provespa anomala* (Hymenoptera: Vespidae). Entomol. Gener. 14: 47–52.

Maschwitz, U., W. Beier, I. Dietrich, and W. Keidel. 1974. Futterverständigung bei Wespen der Gattung *Paravespula*. Naturwissenschaften 61: 506.

Matsuura, M. 1966. Notes on the hibernating habits of the genus *Vespa* in Japan (Hymenoptera, Vespidae). Kontyû 34: 52–67 (in Japanese).

Matsuura, M. 1968. Life of hornets. Part 11. Royal court around queens of *Vespa* species. Jpn. Bee J. 21: 21–25 (in Japanese).

Matsuura, M. 1969. Behaviour of post-hibernating female hornets, *Vespa*, in the prenesting stage, with special reference to intra- and interspecific dominance relationships. Jpn. J. Ecol. 19: 196–203.

Matsuura, M. 1970a. Data for the nests of the giant paper-wasp, *Polistes gigas* Kirby, in Formosa with special consideration on colony size. Life Study (Fukui) 14: 35–40.

Matsuura, M. 1970b. Intraspecific invasion behavior of *Vespa crabro flavofasciata* Cameron in the early stage of nesting. Life Study (Fukui) 14: 21–26 (in Japanese).

Matsuura, M. 1971a. Nest foundation by the female wasps of the genus *Vespa* (Hymenoptera, Vespidae). Kontyû 39: 99–105.

Matsuura, M. 1971b. Nesting sites of the Japanese *Vespa* species. Kontyû 39: 43–54 (in Japanese).

Matsuura, M. 1973a. Colony development of *Vespa crabro flavofasciata* Cameron in the early stage of nesting. Life Study (Fukui) 17: 1–12 (in Japanese).

Matsuura, M. 1973b. Intracolonial polyethism in *Vespa*. III. Foraging activities. Life Study (Fukui) 17: 81–99 (in Japanese).

Matsuura, M. 1974. Intracolonial polyethism in *Vespa*. I. Behaviour and its change of the foundress of *Vespa crabro flavofasciata* in early nesting stage in relation to worker emergence. Kontyû 42: 333–350.

Matsuura, M. 1983a. Discovery of polygynous colonies in *Vespa affinis indosinensis*. Ecological Study on Social Insects in Central Sumatra with Special Reference to Wasps and Bees—Sumatra Nature Study (Entomology), Kanazawa University, pp. 25–27.

Matsuura, M. 1983b. Preliminary report on the polygynous colonies of *Vespa affinis indosinensis* (Hymenoptera, Vespidae) in Sumatra. Kontyû 51: 80–82.

Matsuura, M. 1983c. Some biological aspects of the nocturnal vespine genus *Provespa*. Ecological Study on Social Insects in Central Sumatra with Special Reference to Wasps and Bees—Sumatra Nature Study (Entomology), Kanazawa University, pp. 27–29.

Matsuura, M. 1984. Comparative biology of the five Japanese species of the genus *Vespa* (Hymenoptera, Vespidae). Bull. Fac. Agric. Mie Univ. 69: 1–131.

Matsuura, M. 1985. Life history of the nocturnal vespine, *Provespa anomalla* [*sic*]. Evolutionary Ecology of Insects in Humid Tropics, Especially in Central Sumatra—Sumatra Nature Study (Entomology), Kanazawa University, pp. 27–36.

Matsuura, M. 1988. Ecological study on vespine wasps (Hymenoptera: Vespidae) attacking honeybee colonies. I. Seasonal changes in the frequency of visits to apiaries by vespine wasps and damage inflicted, especially in the absence of artificial protection. Appl. Entomol. Zool. 23: 428–440.

Matsuura, M., and S. F. Sakagami. 1973. A bionomic sketch of the giant hornet, *Vespa mandarinia*, a serious pest for Japanese apiculture. J. Fac. Sci. Hokkaido Univ. (Ser. VI) 19: 125–162.

Matsuura, M., and Sk. Yamane. 1984. Comparative Ethology of the Vespine Wasps. Hokkaido University Press, Sapporo (in Japanese).

Matthews, R. W. 1968a. *Microstigmus comes*: sociality in a sphecid wasp. Science 160: 787–788.

Matthews, R. W. 1968b. Nesting biology of the social wasp *Microstigmus comes* (Hymenoptera: Sphecidae, Pemphredoninae). Psyche 75: 23–45.

Matthews, R. W. 1970. A new thrips-hunting *Microstigmus* from Costa Rica (Hymenoptera: Sphecidae, Pemphredoninae). Psyche 77: 120–126.

Matthews, R. W. 1982. Social parasitism in yellowjackets (*Vespula*). Pp. 193–202 *in* P. Jaisson (ed.), Social Insects in the Tropics, vol. 1. Université de Paris-Nord, Paris.

Matthews, R. W. 1983. Biology of a new *Trypoxylon* that utilizes nests of *Microstigmus* in Costa Rica (Hymenoptera: Sphecidae). Pan-Pac. Entomol. 59: 152–162.

Matthews, R. W., and J. R. Matthews. 1979. War of the yellowjacket queens. Nat. Hist. 88(8): 56–65.

Matthews, R. W., and I. D. Naumann. 1988. Nesting biology and taxonomy of *Arpactophilus mimi*, a new species of social sphecid (Hymenoptera: Sphecidae) from northern Australia. Aust. J. Zool. 36: 585–597.

Matthews, R. W., and C. K. Starr. 1984. *Microstigmus comes* wasps have a method of nest construction unique among social insects. Biotropica 16: 55–58.

Matthews, R. W., R. A. Saunders, and J. R. Matthews. 1981. Nesting behavior of the sand wasp *Stictia maculata* (Hymenoptera: Sphecidae) in Costa Rica. J. Kansas Entomol. Soc. 54: 249–254.

Matthews, R. W., K. G. Ross, and R. A. Morse. 1982. Comparative development of queen nests of four species of yellowjackets (Hymenoptera: Vespidae) reared under identical conditions. Ann. Entomol. Soc. Amer. 75: 123–129.

Maynard Smith, J. 1982. Evolution and the Theory of Games. Cambridge University Press, Cambridge.

Maynard Smith, J. 1988. The evolution of recombination. Pp. 106–125 *in* R. E. Michod and B. R. Levin (eds.), The Evolution of Sex. Sinauer, Sunderland, MA.

McCorquodale, D. B. 1988a. Relatedness among nestmates in a primitively social wasp, *Cerceris antipodes* (Hymenoptera: Sphecidae). Behav. Ecol. Sociobiol. 23: 401–406.

McCorquodale, D. B. 1988b. Nest sharing in the sphecid wasp, *Cerceris antipodes* Smith. Ph.D. dissertation, Australian National University, Canberra.

McCorquodale, D. B. 1989a. Nest sharing, nest switching, longevity and overlap of generations in *Cerceris antipodes* (Hymenoptera: Sphecidae). Insectes Soc. 36: 42–50.

McCorquodale, D. B. 1989b. Nest defense in single- and multifemale nests of *Cerceris antipodes* (Hymenoptera: Sphecidae). J. Insect Behav. 2: 267–276.

McCorquodale, D. B. 1989c. Soil softness, nest initiation and nest sharing in the wasp, *Cerceris antipodes* (Hymenoptera: Sphecidae). Ecol. Entomol. 14: 191–196.

McCorquodale, D. B., and I. D. Naumann. 1988. A new Australian species of communal ground nesting wasp, in the genus *Spilomena* (Hymenoptera: Sphecidae: Pemphredoninae). J. Aust. Entomol. Soc. 27: 221–231.

McCorquodale, D. B., and C. E. Thomson. 1988. A nest shared by the solitary wasps, *Cerceris antipodes* and *C. australis* Saussure (Hymenoptera: Sphecidae). J. Aust. Entomol. Soc. 27: 9–10.

McGovern, J. N., R. L. Jeanne, and M. J. Effland. 1988. The nature of wasp nest paper. Tappi J. 71: 133–139.

Meade-Waldo, G. 1911. New species of Diploptera in the collection of the British Museum. III. Ann. Mag. Nat. Hist. (Ser. 8) 7: 98–113.

Medler, J. T. 1964. Biology of *Rygchium foraminatum* in trap-nests in Wisconsin (Hymenoptera: Vespidae). Ann. Entomol. Soc. Amer. 57: 56–60.

Menke, A. S. 1988. *Pison* in the New World: a revision (Hymenoptera: Sphecidae: Trypoxylini). Contrib. Amer. Entomol. Inst. 24(3): 1–171.

Menke, A. S. 1989. *Arpactophilus* reassessed, with three bizarre new species from New Guinea (Hymenoptera: Sphecidae: Pemphredoninae). Invert. Taxon. 2: 737–747.

Metcalf, R. A. 1980. Sex ratios, parent-offspring conflict, and local competition for mates in the social wasps *Polistes metricus* and *Polistes variatus*. Amer. Nat. 116: 642–654.

Metcalf, R. A., and G. S. Whitt. 1977a. Intra-nest relatedness in the social wasp *Polistes metricus*. A genetic analysis. Behav. Ecol. Sociobiol. 2: 339–351.

Metcalf, R. A., and G. S. Whitt. 1977b. Relative inclusive fitness in the social wasp *Polistes metricus*. Behav. Ecol. Sociobiol. 2: 353–360.

Metcalf, R. A., J. C. Marlin, and G. S. Whitt. 1984. Genetics of speciation within the *Polistes fuscatus* species complex. J. Heredity 75: 117–120.

Michener, C. D. 1944. Comparative external morphology, phylogeny, and a classification of the bees. Bull. Amer. Mus. Nat. Hist. 82: 151–326.

Michener, C. D. 1958. The evolution of social behavior in bees. Proc. 10th Int. Cong. Entomol., pp. 441–447.

Michener, C. D. 1964. Reproductive efficiency in relation to colony size in hymenopterous societies. Insectes Soc. 11: 317–341.

Michener, C. D. 1969. Comparative social behavior of bees. Annu. Rev. Entomol. 14: 299–342.

Michener, C. D. 1974. The Social Behavior of the Bees: A Comparative Study. Harvard University Press, Cambridge, MA.

Michener, C. D. 1985. From solitary to eusocial: need there be a series of intervening species? Pp. 293–305 *in* B. Hölldobler and M. Lindauer (eds.), Experimental Behavioral Ecology and Sociobiology. Sinauer, Sunderland, MA.

Michener, C. D. 1986. Family-group names among bees. J. Kansas Entomol. Soc. 59: 219–234.

Michener, C. D. 1990a. Castes in xylocopine bees. Pp. 123–146 *in* W. Engels (ed.), Social Insects: An Evolutionary Approach to Castes and Reproduction. Springer Verlag, New York.

Michener, C. D. 1990b. Reproduction and castes in social halictine bees. Pp. 77–121 *in* W. Engels (ed.), Social Insects: An Evolutionary Approach to Castes and Reproduction. Springer Verlag, New York.

Michener, C. D., and D. J. Brothers. 1974. Were workers of eusocial Hymenoptera initially altruistic or oppressed? Proc. Natl. Acad. Sci. USA 71: 671–674.

Michener, C. D., and D. A. Grimaldi. 1988. The oldest fossil bee: apoid history, evolutionary stasis, and antiquity of social behavior. Proc. Natl. Acad. Sci. USA 85: 6424–6426.

Michener, C. D., and B. H. Smith. 1987. Kin recognition in primitively eusocial insects. Pp. 209–242 *in* D. J. C. Fletcher and C. D. Michener (eds.), Kin Recognition in Animals. Wiley, New York.

Michener, C. D., R. B. Lange, J. J. Bigarella, and R. Salamuni. 1958. Factors influencing the distribution of bees' nests in earth banks. Ecology 39: 207–217.

Michod, R. E. 1979. Genetical aspects of kin selection: effects of inbreeding. J. Theor. Biol. 81: 223–233.

Michod, R. E. 1982. The theory of kin selection. Annu. Rev. Ecol. Syst. 13: 23–55.

Mickevich, M. F., and J. S. Farris. 1981. The implications of congruence in *Menidia*. Syst. Zool. 30: 351–370.

Mikkola, K. 1978. Spring migrations of wasps and bumble bees on the southern coast of Finland (Hymenoptera, Vespidae and Apidae). Ann. Entomol. Fenn. 44: 10–26.

Miller, R. C., and F. E. Kurczewski. 1973. Intraspecific interaction in aggregations of *Lindenius* (Hymenoptera: Sphecidae, Crabroninae). Insectes Soc. 20: 365–378.

Miotk, P. 1979. Zur Biologie und Ökologie von *Odynerus spinipes* (L.) und *O. reniformis* (Gmel.) an den Lösswänden des Kaiserstuhls (Hymenoptera: Eumenidae). Zool. Jahrb. Abt. Syst. Ökol. Geogr. Tiere 106: 374–405.

Misra, J. S., and M. D. L. Srivastava. 1971. Chromosomal changes correlated with differentiation during embryonic development of *Polistes hebraeus* (Family Vespidae: Order Hymenoptera). Proc. Natl. Acad. Sci. India 41: 97–112.

Miyamoto, M. M. 1985. Consensus cladograms and general classifications. Cladistics 1: 186–189.

Miyano, S. 1980. Life tables of colonies and workers in a paper wasp, *Polistes chinensis antennalis* (Hymenoptera: Vespidae). Res. Popul. Ecol. 22: 69–88.

Miyano, S. 1983. Number of offspring and seasonal changes of their body weight in a paper wasp, *Polistes chinensis antennalis* Pérez (Hymenoptera: Vespidae), with reference to male production by workers. Res. Popul. Ecol. 25: 198–209.

Miyano, S. 1986. Colony development, worker behavior and male production in orphan colonies of a Japanese paper wasp, *Polistes chinensis antennalis* Pérez (Hymenoptera: Vespidae). Res. Popul. Ecol. 28: 347–361.

Moebius, A. 1856. Die Nester der geselligen Wespen. Abh. Naturwiss. Ver. Hamburg 3: 121–171.

Molitor, A. 1933. Neue Beobachtungen und Experimente mit Grabwespen. IV. Biol. Zentralbl. 53: 496–512.

Montagner, H. 1963. Etude préliminaire des relations entre les adultes et le couvain chez les guêpes sociales du genre *Vespa* au moyen d'un radio-isotope. Insectes Soc. 10: 153–165.

Montagner, H. 1964. Etude du comportement alimentaire et des relations trophallactiques des mâles au sein de la société de guêpes, au moyen d'un radio-isotope. Insectes Soc. 11: 301–316.

Montagner, H. 1966a. Le mécanisme et les conséquences des comportements trophallactiques chez les guêpes du genre *Vespa*. Bull. Biol. France Belg. 100: 187–323.

Montagner, H. 1966b. Sur l'origine des mâles dans les sociétés de guêpes du genre *Vespa*. C. R. Acad. Sci. Paris (D) 263: 785–787.

Montagner, H. 1967. Contribution à l'étude du déterminisme des castes chez les guêpes du genre *Vespa*. Ph.D. dissertation, Université de Nancy, Nancy.

Moore, W. S. 1975. Observations on the egg laying and sleeping habits of *Euparagia scutellaris* Cresson (Hymenoptera: Vespoidea). Pan-Pac. Entomol. 51: 286.

Mori, K., and T. Otsuka. 1985. Synthesis of the enantiomers of 5–hexadecanolide, the pheromone of the queen of the Oriental hornet, *Vespa orientalis*, employing enzymic resolution of ($\pm$)-2–aminotridecanoic acid as the key-step. Tetrahedron 41: 547–551.

Morimoto, R. 1961a. On the dominance order in *Polistes* wasps. I. (Studies on the social Hymenoptera of Japan, XII). Sci. Bull. Fac. Agric. Kyushu Univ. 18: 339–351 (in Japanese).

Morimoto, R. 1961b. On the dominance order in *Polistes* wasps. II. (Studies on the social Hymenoptera of Japan, XIII). Sci. Bull. Fac. Agric. Kyushu Univ. 19: 1–17 (in Japanese).

Moritz, R. F. A., and H. Bürgin. 1987. Group response to alarm pheromones in social wasps and the honeybee. Ethology 76: 15–26.

Morse, R. A., and F. M. Laigo. 1969. *Apis dorsata* in the Philippines (including an annotated bibliography). Monogr. Phil. Assoc. Entomol. 1: 1–96.

Morse, R. A., G. E. Strang, and J. Nowakowski. 1967. Fall death rates of drone honey bees. J. Econ. Entomol. 60: 1198–1202.

Morse, R. A., G. C. Eickwort, and R. S. Jacobson. 1977. The economic status of an immigrant yellowjacket, *Vespula germanica* (Hymenoptera: Vespidae), in northeastern United States. Environ. Entomol. 6: 109–110.

Motro, M., U. Motro, J. S. Ishay, and J. Kugler. 1979. Some social and dietary prerequisites of oocyte development in *Vespa orientalis* L. workers. Insectes Soc. 26: 155–164.

Muralidharan, K., M. S. Shaila, and R. Gadagkar. 1986. Evidence for multiple mating in the primitively eusocial wasp *Ropalidia marginata* (Lep.) (Hymenoptera: Vespidae). J. Genet. 65: 153–158.

Myers, J. G. 1934. Two Collembola-collecting crabronids in Trinidad. Trans. R. Entomol. Soc. London 82: 23–26.

Myers, J. G. 1935. Nesting associations of birds with social insects. Trans. R. Entomol. Soc. London 83: 11–22.

Nakahara, L. M. 1980. Western yellowjacket (*Vespula pensylvanica*)—first record of aerial nest in state. United States Department of Agriculture, APHIS Coop. Plant Pest Rep. 5: 270.

Nakahara, L. M., and P. Lai. 1981. Western yellowjacket, *Vespula pensylvanica* (Saussure). Hawaii Pest Rep. 1(3): 1–4.

Naumann, I. D. 1983. The biology of mud-nesting Hymenoptera (and their associates) and Isoptera in rock shelters of the Kakadu region, Northern Territory. Pp. 127–189 *in* D. Gillespie (ed.), The Rock Art Sites of Kakadu National Park—Some Preliminary Research Findings for Their Conservation and Management (Special Publication 10). Australian National Parks and Wildlife Service, Canberra.

Naumann, M. G. 1968. A revision of the genus *Brachygastra* (Hymenoptera: Vespidae). Univ. Kansas Sci. Bull. 47: 929–1003.

Naumann, M. G. 1970. The nesting behavior of *Protopolybia pumila* in Panama (Hymenoptera: Vespidae). Ph.D. dissertation, University of Kansas, Lawrence.

Naumann, M. G. 1975. Swarming behavior: evidence for communication in social wasps. Science 189: 642–644.

Nedel, J. O. 1960. Morphologie und Physiologie der Mandibeldrüse einiger Bienen-Arten (Apidae). Z. Morphol. Ökol. Tiere 49: 139–183.

Neff, J. L., and B. B. Simpson. 1985. Hooked setae and narrow tubes: foretarsal pollen collection by *Trimeria buyssoni* (Hymenoptera: Masaridae). J. Kansas Entomol. Soc. 58: 730–732.

Nei, M. 1987. Molecular Evolutionary Genetics. Columbia University Press, New York.

Nelson, G. 1978. Ontogeny, phylogeny, paleontology, and the biogenetic law. Syst. Zool. 27: 324–345.

Nelson, G. 1979. Cladistic analysis and synthesis: principles and definitions, with a historical note on Adanson's *Familles des Plantes* (1763–1764). Syst. Zool. 28: 1–21.

Nelson, J. M. 1968. Parasites and symbionts of nests of *Polistes* wasps. Ann. Entomol. Soc. Amer. 61: 1528–1539.

Nielson, E. T. 1932. Sur les habitudes des Hyménoptères aculéates solitaires. II. (Vespidae, Chrysididae, Sapygidae et Mutillidae). Entomol. Medd. 18: 84–174.

Nielson, E. T. 1933. Sur les habitudes des Hyménoptères aculéates solitaires. III. (Sphegidae). Entomol. Medd. 18: 259–348.

Nijhout, H. F., and D. E. Wheeler. 1982. Juvenile hormone and the physiological basis of insect polymorphisms. Quart. Rev. Biol. 57: 109–133.

Nixon, G. E. J. 1935. Notes on wasps. IV. Entomol. Mon. Mag. 71: 106–111.

Nixon, G. E. J. 1936. Notes on wasps. V. Entomol. Mon. Mag. 72: 6–8.

Nixon, G. E. J. 1955. The World of Bees. Philosophical Library, New York.

Nixon, G. E. J. 1981. Colony failure in the hornet (*Vespa crabro* L.) (Hym., Vespidae). Entomol. Mon. Mag. 117: 241–244.

Nixon, G. E. J. 1983a. A note on intercolonial behaviour in the hornet, *Vespa crabro* L. (Hym., Vespidae). Entomol. Mon. Mag. 119: 253–256.

Nixon, G. E. J. 1983b. Notes on colony failure and the phenomenon of usurpation in the hornet, *Vespa crabro* L. (Hym., Vespidae). Entomol. Mon. Mag. 119: 1–11.

Nixon, G. E. J. 1984. An attempt to rear in captivity an incipient nest of the hornet, *Vespa crabro* L. (Hym., Vespidae). Entomol. Mon. Mag. 120: 93–96.

Nixon, G. E. J. 1985a. Secondary nests in the hornet, *Vespa crabro* L. (Hym., Vespidae). Entomol. Mon. Mag. 121: 189–198.

Nixon, G. E. J. 1985b. The phenomenon of usurpation in certain Vespidae, especially *Vespula vulgaris* L. (Hym.). Entomol. Mon. Mag. 121: 145–148.

Noirot, C. 1970. The nests of termites. Pp. 73–125 *in* K. Krishna and F. M. Weesner (eds.), Biology of Termites, vol. 2. Academic Press, New York.

Noirot, C., and A. Quennedey. 1974. Fine structure of insect epidermal glands. Annu. Rev. Entomol. 19: 61–80.

Nonacs, P. 1988. Queen number in colonies of social Hymenoptera as a kin-selected adaptation. Evolution 42: 566–580.

Noonan, K. C. 1986. Recognition of queen larvae by worker honey bees (*Apis mellifera*). Ethology 73: 295–306.

Noonan, K. M. 1978. Sex ratio of parental investment in colonies of the social wasp *Polistes fuscatus*. Science 199: 1354–1356.

Noonan, K. M. 1981. Individual strategies of inclusive-fitness-maximizing in *Polistes fuscatus* foundresses. Pp. 18–44 *in* R. D. Alexander and D. W. Tinkle (eds.), Natural Selection and Social Behavior: Recent Research and New Theory. Chiron, New York.

Norden, B., S. W. T. Batra, H. M. Fales, A. Hefetz, and G. J. Shaw. 1980. *Anthophora* bees: unusual glycerides from maternal Dufour's glands serve as larval food and cell lining. Science 207: 1095–1097.

Nutting, W. L. 1969. Flight and colony foundation. Pp. 233–282 *in* K. Krishna and F. M. Weesner (eds.), Biology of Termites, vol. 1. Academic Press, New York.

Obin, M. S. 1983. Behavior and nesting biology of the blue mud-dauber *Chalybion californicum* (Saussure) (Hymenoptera: Sphecidae). M.S. thesis, University of Florida, Gainesville.

O'Brien, M. F., and F. E. Kurczewski. 1982. Ethology and overwintering of *Polalonia luctuosa* (Hymenoptera: Sphecidae). Great Lakes Entomol. 15: 261–275.

Ohgushi, R. 1954. On the plasticity of the nesting habit of a hunting wasp, *Pemphredon lethifer fabricii* Müller. Mem. College Sci. Univ. Kyoto (B) 21: 45–48.

Ohgushi, R. 1986. A review of the biology and ecology of stenogastrine wasps. Hymenopterists' Comm. (25/26): 1–76 (in Japanese).

Ohgushi, R., and Sô. Yamane. 1983. Supplementary notes on the nest architecture and biology of some *Parischnogaster* species in Sumatera Barat (Hymenoptera, Vespidae). Sci. Rep. Kanazawa Univ. 28: 69–78.

Ohgushi, R., S. F. Sakagami, Sô. Yamane, and N. D. Abbas. 1983a. Nest architecture and related notes of stenogastrine wasps in the province of Sumatera Barat, Indonesia (Hymenoptera, Vespidae). Sci. Rep. Kanazawa Univ. 28: 27–58.

Ohgushi, R., S. F. Sakagami, Sô. Yamane, and N. D. Abbas. 1983b. Nest architecture of Sumatran species of Stenogastrinae. Ecological Study on Social Insects in Central Sumatra with Special Reference to Wasps and Bees—Sumatra Nature Study (Entomology), Kanazawa University, pp. 2–12.

Ohgushi, R., Sô. Yamane, and N. D. Abbas. 1985. Descriptions and a re-description of 5 types of stenogastrine nests collected in Sumatera Barat, Indonesia, with some biological notes (Hymenoptera, Vespidae). Evolutionary Ecology of Insects in Humid Tropics, Especially in Central Sumatra—Sumatra Nature Study (Entomology), Kanazawa University, pp. 1–12.

Ohgushi, R., Sô. Yamane, and N. D. Abbas. 1986. Additional description and records of stenogastrine nests collected in Sumatera Barat, Indonesia, with some biological notes (Hymenoptera, Vespidae). Kontyû 54: 1–11.

Olberg, G. 1959. Das Verhalten der solitären Wespen Mitteleuropas (Vespidae, Pompilidae, Sphecidae). Deutscher Verlag der Wissenschaften, Berlin.

O'Neill, K. M. 1985. Egg size, prey size, and sexual size dimorphism in digger wasps (Hymenoptera: Sphecidae). Can. J. Zool. 63: 2187–2193.

Ono, M. 1989. Multiple-comb nest foundation by a single inseminated worker of the temperate paper wasp, *Polistes snelleni* Saussure (Hymenoptera: Vespidae). J. Ethol. 7: 57–58.

Ono, M., and M. Sasaki. 1987. Sex pheromones and their cross-activities in six Japanese sympatric species of the genus *Vespa*. Insectes Soc. 34: 252–260.

Ono, M., M. Sasaki, and I. Okada. 1987. Reproductive biology of the giant hornet, *Vespa mandarinia* Smith. Pp. 299–300 *in* J. Eder and H. Rembold (eds.), Chemistry and Biology of Social Insects. J. Peperny, Munich.

O'Rourke, M. E., and F. E. Kurczewski. 1984. Nest usurpation of *Vespula vulgaris* by *Dolichovespula arenaria* with successive parasitism of *D. arenaria* by *D. arctica* (Hymenoptera: Vespidae). Entomol. News 95: 212–214.

Osten, T. 1982. Vergleichend-funktionsmorphologische Untersuchungen der Kopfkapsel und der Mundwerkzeuge ausgewählter "Scolioidea" (Hymenoptera, Aculeata). Stuttgarter Beitr. Naturk. (A) 354: 1–60.

Oster, G. F., and E. O. Wilson. 1978. Caste and Ecology in the Social Insects. Princeton University Press, Princeton, NJ.

Overal, W. L. 1982. Acoustical behavior and variable nest architecture in *Synoeca virginea* (Hymenoptera, Vespidae). J. Georgia Entomol. Soc. 17: 1–4.

Overal, W. L. 1985. Wasp-whistling. Sphecos (10): 18–19.

Overal, W. L., D. Simões, and N. Gobbi. 1981. Colony defense and sting autotomy in *Polybia rejecta* (Hymenoptera: Vespidae). Rev. Brasil. Entomol. 25: 41–47.

Owen, J. 1962. The behavior of a social wasp *Polistes fuscatus* (Vespidae) at the nest, with special reference to differences between individuals. Ph.D. dissertation, University of Michigan, Ann Arbor.

Pagden, H. T. 1958a. Some Malayan social wasps. Malayan Nat. J. 12: 131–148.

Pagden, H. T. 1958b. A little known wasp's nest, *Polybioides raphigastra* (Sauss.) (Hymenoptera: Vespoidea). Malayan Nat. J. 12: 112–114.

Pagden, H. T. 1962. More about *Stenogaster*. Malayan Nat. J. 16: 95–102.

Pagden, H. T. 1976. A note on colony founding by *Ropalidia (Icarielia) timida* van der Vecht. Proc. K. Nederl. Akad. Wetens. (Ser. C) 79: 508–509.

Page, R. E. 1986. Sperm utilization in social insects. Annu. Rev. Entomol. 31: 297–320.

Page, R. E., and E. H. Erickson. 1988. Reproduction by worker honey bees. Behav. Ecol. Sociobiol. 23: 117–126.

Page, R. E., and H. H. Laidlaw. 1988. Full sisters and super sisters: a terminological paradigm. Anim. Behav. 36: 944–945.

Page, R. E., and R. A. Metcalf. 1982. Multiple mating, sperm utilization, and social evolution. Amer. Nat. 119: 263–281.

Page R. E., D. C. Post, and R. A. Metcalf. 1989a. Satellite nests, early males, and plasticity of reproductive behavior in a paper wasp. Amer. Nat. 134: 731–748.

Page, R. E., G. E. Robinson, and M. K. Fondrk. 1989b. Genetic specialists, kin recognition and nepotism in honey-bee colonies. Nature 338: 576–579.

Pallett, M. J. 1984. Nest site selection and survivorship of *Dolichovespula arenaria* and *Dolichovespula maculata* (Hymenoptera: Vespidae). Can. J. Zool. 62: 1268–1272.

Pallett, M. J., R. C. Plowright, and D. L. Gibo. 1983. A method of capturing colonies of aerial nesting vespid wasps (Hymenoptera: Vespidae). Can. Entomol. 115: 351–355.

Pamilo, P. 1983. Genetic differentiation within subdivided populations of *Formica* ants. Evolution 37: 1010–1022.

Pamilo, P. 1984a. Genotypic correlation and regression in social groups: multiple alleles, multiple loci and subdivided populations. Genetics 107: 307–320.

Pamilo, P. 1984b. Genetic relatedness and evolution of insect sociality. Behav. Ecol. Sociobiol. 15: 241–248.

Pamilo, P. 1987. Sex ratios and the evolution of eusociality in the Hymenoptera. J. Genet. 66: 111–122.

Pamilo, P., and R. H. Crozier. 1982. Measuring genetic relatedness in natural populations: methodology. Theor. Popul. Biol. 21: 171–193.

Pardi, L. 1941. Ricerche sui Polistini. 3. Ancora sulla poliginia iniziale di *Polistes gallicus* (L.) e sul comportamento delle femmine associate fino alla schiusa delle prime operaie. Proc. Verb. Soc. Toscana Sci. Nat. (Pisa) 50: 3–15.

Pardi, L. 1942a. Ricerche sui Polistini. V. La poliginia iniziale di *Polistes gallicus* (L.). Boll. Ist. Entomol. Univ. Bologna 14: 1–106.

Pardi, L. 1942b. Ricerche sui Polistini. 8. La spermatogenesi di *Polistes gallicus* (L.) e di *Polistes (Leptopolistes) omissus* (Weyrauch). Scientia Genet. 3: 14–22.

Pardi, L. 1946. Ricerche sui Polistini. VII. La "dominazione" e il ciclo ovarico annuale in *Polistes gallicus* (L.). Boll. Ist. Entomol. Univ. Bologna 15: 25–84.

Pardi, L. 1947. Beobachtungen über das interindividuelle Verhalten bei *Polistes gallicus* (Untersuchungen über die Polistini, no. 10). Behaviour 1: 138–172.

Pardi, L. 1948. Dominance order in *Polistes* wasps. Physiol. Zool. 21: 1–13.

Pardi, L. 1950. Recenti ricerche sulla divisione di lavoro negli imenotteri sociali. Boll. Zool. 17: 17–66 (suppl.).

Pardi, L. 1977. Su alcuni aspetti della biologia di *Belonogaster* (Hymenoptera, Vespidae). Boll. Ist. Entomol. Univ. Bologna 33: 281–299.

Pardi, L., and M. T. Marino Piccioli. 1970. Studi sulla biologia di *Belonogaster* (Hymenoptera, Vespidae). 2. Differenziamento castale incipiente in *B. griseus* (Fab.) [sic]. Monit. Zool. Ital. (N.S.) 3: 235–265 (suppl.).

Pardi, L., and M. T. Marino Piccioli. 1981. Studies on the biology of *Belonogaster* (Hymenoptera Vespidae). 4. On caste differences in *Belonogaster griseus* (Fab.) [sic] and the position of this genus among social wasps. Monit. Zool. Ital. (N.S.) 14: 131–146 (suppl.).

Pardi, L., and S. Turillazzi. 1981. Behaviour and social organization of *Parischnogas-*

*ter nigricans serrei* (du Buysson) (Hymenoptera Vespoidea). Monit. Zool. Ital. (N.S.) 15: 322–323.

Pardi, L., and S. Turillazzi. 1982. Biologia delle Stenogastrinae (Hymenoptera, Vespoidea). Atti Accad. Naz. Ital. Entomol. Rendic. 30: 1–21.

Parker, G. A. 1970. Sperm competition and its evolutionary consequences in the insects. Biol. Rev. (Cambridge) 45: 525–567.

Parker, G. A. 1984. Sperm competition and the evolution of animal mating strategies. Pp. 1–60 *in* R. L. Smith (ed.), Sperm Competition and the Evolution of Animal Mating Systems. Academic Press, New York.

Peckham, D. J. 1977. Reduction of miltogrammine cleptoparasitism by male *Oxybelus sublatus* (Hymenoptera: Sphecidae). Ann. Entomol. Soc. Amer. 70: 823–828.

Peckham, G. W., and E. G. Peckham. 1905. Wasps: Social and Solitary. Houghton Mifflin, New York.

Perrott, D. C. F. 1975. Factors affecting use of mirex-poisoned protein baits for control of European wasp (*Paravespula germanica*) in New Zealand. New Zealand J. Zool. 2: 491–508.

Peters, D. S. 1973. *Crossocerus dimidiatus* (Fabricius, 1781), eine weitere soziale Crabroninen-Art. Insectes Soc. 20: 103–108.

Pfennig, D. W., and J. E. Klahn. 1985. Dominance as a predictor of cofoundress disappearance order in social wasps (*Polistes fuscatus*). Z. Tierpsychol. 67: 198–203.

Pfennig, D. W., and H. K. Reeve. 1989. Neighbor recognition and context-dependent aggression in a solitary wasp, *Sphecius speciosus* (Hymenoptera, Sphecidae). Ethology 80: 1–18.

Pfennig, D. W., G. J. Gamboa, H. K. Reeve, J. Shellman-Reeve, and I. D. Ferguson. 1983a. The mechanism of nestmate discrimination in social wasps (*Polistes*, Hymenoptera: Vespidae). Behav. Ecol. Sociobiol. 13: 299–305.

Pfennig, D. W., H. K. Reeve, and J. S. Shellman. 1983b. Learned component of nestmate discrimination in workers of a social wasp, *Polistes fuscatus* (Hymenoptera: Vespidae). Anim. Behav. 31: 412–416.

Pianka, E. R. 1988. Evolutionary Ecology, 4th ed. Harper and Row, New York.

Piek, T. 1987. A toxinological argument in favour of the close relationship of the Vespidae and the Scoliidae (Hymenoptera). Entomol. Ber. (Amsterdam) 47: 96–98.

Piel, P. O. 1935. Biologie de *Pareumenes quadrispinosus* Saussure (Hyménoptères: Vespidae) et de ses parasites, en particulier: *Calosota sinensis* Ferriere. Notes Entomol. Chinoise 2: 105–138.

Pollock, G. B. 1983. Population viscosity and kin selection. Amer. Nat. 122: 817–829.

Pompolo, S. das G., and C. S. Takahashi. 1986. Karyotype of two species of wasps of the genus *Polistes* (Polistinae, Vespidae, Hymenoptera). Insectes Soc. 33: 142–148.

Post, D. C. 1980. Observations on male behavior of the eastern yellowjacket, *Vespula maculifrons* (Hymenoptera: Vespidae). Entomol. News 91: 113–116.

Post, D. C., and R. L. Jeanne. 1980. Morphology of the sternal glands of *Polistes fuscatus* and *P. canadensis* (Hymenoptera: Vespidae). Psyche 87: 49–58.

Post, D. C., and R. L. Jeanne. 1981. Colony defense against ants by *Polistes fuscatus* (Hymenoptera: Vespidae) in Wisconsin. J. Kansas Entomol. Soc. 54: 599–615.

Post, D. C., and R. L. Jeanne. 1982a. Recognition of former nestmates during colony founding by the social wasp *Polistes fuscatus* (Hymenoptera: Vespidae). Behav. Ecol. Sociobiol. 11: 283–285.

Post, D. C., and R. L. Jeanne. 1982b. Sternal glands in three species of male social wasps of the genus *Mischocyttarus* (Hymenoptera: Vespidae). J. New York Entomol. Soc. 90: 8–15.

Post, D. C., and R. L. Jeanne. 1983a. Male reproductive behavior of the social wasp *Polistes fuscatus* (Hymenoptera: Vespidae). Z. Tierpsychol. 62: 157–171.

Post, D. C., and R. L. Jeanne. 1983b. Relatedness and mate selection in *Polistes fuscatus* (Hymenoptera: Vespidae). Anim. Behav. 31: 1260–1261.

Post, D. C., and R. L. Jeanne. 1983c. Sternal glands in males of six species of *Polistes* (*Fuscopolistes*) (Hymenoptera: Vespidae). J. Kansas Entomol. Soc. 56: 32–39.

Post, D. C., and R. L. Jeanne. 1983d. Venom: source of a sex pheromone in the social wasp *Polistes fuscatus* (Hymenoptera: Vespidae). J. Chem. Ecol. 9: 259–266.

Post, D. C., and R. L. Jeanne. 1984a. Recognition of conspecifics and sex by territorial males of the social wasp *Polistes fuscatus*. Behaviour 91: 78–92.

Post, D. C., and R. L. Jeanne. 1984b. Venom as an interspecific sex pheromone, and species recognition by a cuticular pheromone in paper wasps (*Polistes*, Hymenoptera: Vespidae). Physiol. Entomol. 9: 65–75.

Post, D. C., and R. L. Jeanne. 1985. Sex pheromone in *Polistes fuscatus* (Hymenoptera: Vespidae): effect of age, caste and mating. Insectes Soc. 32: 70–77.

Post, D. C., M. A. Mohamed, H. C. Coppel, and R. L. Jeanne. 1984a. Identification of ant repellent allomone produced by social wasp *Polistes fuscatus* (Hymenoptera: Vespidae). J. Chem. Ecol. 10: 1799–1807.

Post, D. C., H. A. Downing, and R. L. Jeanne. 1984b. Alarm response to venom by the social wasps *Polistes exclamans* and *P. fuscatus* (Hymenoptera: Vespidae). J. Chem. Ecol. 10: 1425–1433.

Post, D. C., R. L. Jeanne, and E. H. Erickson. 1988. Variation in behavior among workers of the primitively social wasp *Polistes fuscatus variatus*. Pp. 283–321 *in* R. L. Jeanne (ed.), Interindividual Behavioral Variability in Social Insects. Westview, Boulder, CO.

Potter, N. B. 1964. A study of the biology of the common wasp, *Vespula vulgaris* L., with special reference to the foraging behaviour. Ph.D. dissertation, University of Bristol, Bristol, England.

Pratte, M. 1980. Mobilité des fondatrices de *Polistes gallicus*. Pp. 133–140 *in* D. Cherix (ed.), Ecologie des Insectes Sociaux. Musée Zoologique, Lausanne.

Pratte, M. 1982. Relations antérieures et association de fondation chez *Polistes gallicus* L. Insectes Soc. 29: 352–357.

Pratte, M. 1989. Foundress association in the paper wasp *Polistes dominulus* Christ (Hymen. Vesp.). Effects of dominance hierarchy on the division of labour. Behaviour 111: 208–219.

Pratte, M., and J. Gervet. 1980. Le modèle sociobiologique, ses conditions de validité dans le cas des sociétés d'hyménoptères. Année Biol. 19: 163–201.

Pratte, M., and R. L. Jeanne. 1984. Antennal drumming behavior in *Polistes* wasps (Hymenoptera: Vespidae). Z. Tierpsychol. 66: 177–188.

Pratte, M., J. Gervet, and S. Semenoff. 1984. L'évolution de la production de descendance dans le guêpier de Poliste (*Polistes gallicus* L.). Insectes Soc. 31: 34–50.

Queller, D. C. 1985. Kinship, reciprocity and synergism in the evolution of social behaviour. Nature 318: 366–367.

Queller, D. C. 1989. The evolution of eusociality: reproductive head starts of workers. Proc. Natl. Acad. Sci. USA 86: 3224–3226.

Queller, D. C., and K. F. Goodnight. 1989. Estimation of genetic relatedness using allozyme data. Evolution 43: 258–275.

Queller, D. C., and J. E. Strassmann. 1988. Reproductive success and group nesting in the paper wasp, *Polistes annularis*. Pp. 76–96 *in* T. H. Clutton-Brock (ed.), Reproductive Success: Studies of Individual Variation in Contrasting Breeding Systems. University of Chicago Press, Chicago.

Queller, D. C., and J. E. Strassmann. 1989. Measuring inclusive fitness in social wasps. Pp. 103–122 *in* M. D. Breed and R. E. Page (eds.), The Genetics of Social Evolution. Westview, Boulder, CO.

Queller, D. C., J. E. Strassmann, and C. R. Hughes. 1988. Genetic relatedness in colonies of tropical wasps with multiple queens. Science 242: 1155–1157.

Queller, D. C., C. R. Hughes, and J. E. Strassmann. 1990. Wasps fail to make distinctions. Nature 344: 388.

Qureshi, M. Y., and N. Ahmad. 1978. A study of the mud dauber nest (Hymenoptera). Pakistan J. Sci. 30: 79–83.

Rabb, R. L. 1960. Biological studies of *Polistes* in North Carolina (Hymenoptera: Vespidae). Ann. Entomol. Soc. Amer. 53: 111–121.

Rasnitsyn, A. P. 1980. The origin and evolution of Hymenoptera. Trudy Paleontol. Inst. Akad. Nauk SSSR 174: 1–190 (in Russian).

Rasnitsyn, A. P. 1988. An outline of evolution of the hymenopterous insects (Order Vespida). Oriental Ins. 22: 115–145.

Ratnieks, F. L. W. 1988. Reproductive harmony via mutual policing by workers in eusocial Hymenoptera. Amer. Nat. 132: 217–236.

Ratnieks, F. L. W. 1990. Assessment of queen mating frequency by workers in eusocial Hymenoptera. J. Theor. Biol. 142: 87–94.

Ratnieks, F. L. W., and P. K. Visscher. 1989. Worker policing in the honeybee. Nature 342: 796–797.

Rau, P. 1928. The honey-gathering habits of *Polistes* wasps. Biol. Bull. (Woods Hole) 54: 503–519.

Rau, P. 1929. The nesting habits of the bald-faced hornet, *Vespa maculata*. Ann. Entomol. Soc. Amer. 22: 659–675.

Rau, P. 1933. The Jungle Bees and Wasps of Barro Colorado Island (with Notes on Other Insects). Phil Rau, Kirkwood, MO.

Rau, P. 1935. The courtship and mating of the wasp *Monobia quadridens* (Hymenoptera: Vespidae). Entomol. News 46: 57–58.

Rau, P. 1940. The nesting habits of several species of Mexican social wasps. Ann. Entomol. Soc. Amer. 33: 81–93.

Rau, P. 1943. The nesting habits of Mexican social and solitary wasps of the family Vespidae. Ann. Entomol. Soc. Amer. 36: 516–536.

Rau, P., and N. Rau. 1918. Wasp Studies Afield. Princeton University Press, Princeton, NJ.

Raw, A. 1985. *Asteloeca*, a new Neotropical wasp genus (Hymenoptera, Vespidae). Rev. Brasil. Entomol. 29: 185–188.

Rayment, T. 1955a. Biology of two hunting wasps. The specific descriptions of a new species and one allotype of *Sericophorus* and a new blowfly *Pollenia*. Aust. Zool. 12: 132–141.

Rayment, T. 1955b. Taxonomy, morphology and biology of sericophorine wasps: with diagnoses of two new genera and descriptions of forty new species and six sub-species. Mem. Natl. Mus. Victoria 19: 11–105.

Réaumur, R. A. F. de. 1719 (1721). Histoire des guêpes. Mém. Acad. R. Sci. Paris 21(1719): 302–364.

Réaumur, R. A. F. de. 1740. Mémoires pour servir à l'histoire des insectes. Imprimerie Royale, Paris.

Reed, H. C., and R. D. Akre. 1982. Morphological comparisons between the obligate social parasite, *Vespula austriaca* (Panzer), and its host, *Vespula acadica* (Sladen) (Hymenoptera: Vespidae). Psyche 89: 183–195.

Reed, H. C., and R. D. Akre. 1983a. Nesting biology of a forest yellowjacket *Vespula acadica* (Sladen) (Hymenoptera: Vespidae), in the Pacific Northwest. Ann. Entomol. Soc. Amer. 76: 582–590.

Reed, H. C., and R. D. Akre. 1983b. Colony behavior of the obligate social parasite *Vespula austriaca* (Panzer) (Hymenoptera: Vespidae). Insectes Soc. 30: 259–273.

Reed, H. C., and R. D. Akre. 1983c. Usurpation behavior of the yellowjacket social parasite, *Vespula austriaca* (Panzer) (Hymenoptera: Vespidae). Amer. Midl. Nat. 110: 419–432.

Reed, H. C., and R. D. Akre. 1983d. Comparative colony behavior of the forest yellowjacket, *Vespula acadica* (Sladen) (Hymenoptera: Vespidae). J. Kansas Entomol. Soc. 56: 581–606.

Reed, H. C., and S. B. Vinson. 1979. Nesting ecology of paper wasps (*Polistes*) in a Texas urban area (Hymenoptera: Vespidae). J. Kansas Entomol. Soc. 52: 673–689.

Reed, H. C., R. D. Akre, and W. B. Garnett. 1979. A North American host of the yellowjacket social parasite *Vespula austriaca* (Panzer) (Hymenoptera: Vespidae). Entomol. News 90: 110–113.

Reed, H. C., J. Gallego, and J. Nelson. 1988. Morphological evidence for polygyny in post-emergence colonies of the red paper wasp, *Polistes perplexus* Cresson (Hymenoptera: Vespidae). J. Kansas Entomol. Soc. 61: 453–463.

Reeve, H. K. 1989. The evolution of conspecific acceptance thresholds. Amer. Nat. 133: 407–435.

Reeve, H. K., and G. J. Gamboa. 1983. Colony activity integration in primitively eusocial wasps: the role of the queen (*Polistes fuscatus*, Hymenoptera: Vespidae). Behav. Ecol. Sociobiol. 13: 63–74.

Reeve, H. K., and G. J. Gamboa. 1987. Queen regulation of worker foraging in paper wasps: a social feedback control system (*Polistes fuscatus*, Hymenoptera: Vespidae). Behaviour 102: 147–167.

Reichert, S. E. 1985. Why do some spiders cooperate? *Agelena consociata*, a case study. Florida Entomol. 68: 105–116.

Reid, J. A. 1942. On the classification of the larvae of the Vespidae. Trans. R. Entomol. Soc. London 93: 285–331.

Reyes, S. 1988. A review of Philippine Stenogastrinae (Hymenoptera: Vespidae). Philipp. Entomol. 7: 387–434.

Richards, O. W. 1927. The specific characters of the British humblebees (Hymenoptera). Trans. Entomol. Soc. London 75: 233–268.

Richards, O. W. 1934. The American species of the genus *Trypoxylon* (Hymenoptera: Sphecoidea). Trans. R. Entomol. Soc. London 82: 173–362.

Richards, O. W. 1945. A revision of the genus *Mischocyttarus* de Saussure (Hymen., Vespidae). Trans. R. Entomol. Soc. London 95: 295–462.

Richards, O. W. 1953. The Social Insects. MacDonald, London.

Richards, O. W. 1962. A Revisional Study of the Masarid Wasps (Hymenoptera, Vespoidea). British Museum (Natural History), London.

Richards, O. W. 1963. The species of *Pseudomasaris* Ashmead (Hymenoptera: Masaridae). Univ. California Publ. Entomol. 27: 283–304.

Richards, O. W. 1969. The biology of some W. African social wasps (Hymenoptera: Vespidae, Polistinae). Mem. Soc. Entomol. Ital. 48: 79–93.

Richards, O. W. 1971. The biology of the social wasps (Hymenoptera, Vespidae). Biol. Rev. (Cambridge) 46: 483–528.

Richards, O. W. 1972. The species of the South American wasps of the genus *Microstigmus* Ducke (Hymenoptera: Sphecoidea, Pemphredoninae). Trans. R. Entomol. Soc. London 124: 123–148.

Richards, O. W. 1978a. The Social Wasps of the Americas, Excluding the Vespinae. British Museum (Natural History), London.

Richards, O. W. 1978b. The Australian social wasps (Hymenoptera: Vespidae). Aust. J. Zool. 61: 1–132 (suppl.).

Richards, O. W. 1982. A revision of the genus *Belonogaster* de Saussure (Hymenoptera: Vespidae). Bull. Brit. Mus. (Nat. Hist.) Entomol. Ser. 44: 31–114.

Richards, O. W., and M. J. Richards. 1951. Observations on the social wasps of South America (Hymenoptera Vespidae). Trans. R. Entomol. Soc. London 102: 1–170.

Ricklefs, R. E. 1987. Ecology, 3rd ed. Chiron, New York.

Rivnay, E., and H. Bytinski-Salz. 1949. The Oriental hornet (*Vespa orientalis* F.). Its biology in Israel. Bull. Agric. Res. Sta. Rehovoth 52: 1–34 (in Hebrew).

Rocha, I., and A. Raw. 1982. Dinâmica das populações da vespa solitária *Zeta argillacea*. An. Soc. Entomol. Brasil 11: 57–78.

Rodrigues, V. M. 1968. Estudo sôbre as vespas sociais do Brasil (Hymenoptera-Vespidae). Ph.D. dissertation, Universidade de Campinas, Campinas.

Rodrigues, V. M., and B. B. dos Santos. 1974. Vespídeos sociais: estudo de uma colônia de *Polybia dimidiata* (Olivier, 1791) (Hymenoptera, Polistinae). Rev. Brasil. Entomol. 18: 37–42.

Roland, C. 1976. Approche éco-éthologique et biologique des sociétés de *Paravespula vulgaris* et *germanica*. Ph.D. dissertation, Université de Nancy, Nancy.

Rösch, G. A. 1925. Untersuchungen über die Arbeitsteilung im Bienenstaat. 1. Teil. Die Tätigkeiten im normalen Bienenstaate und ihre Beziehungen zum Alter der Arbeitsbienen. Z. Vergl. Physiol. 2: 571–631.

Rösch, G. A. 1930. Untersuchungen über die Arbeitsteilung im Bienenstaat. 2. Teil. Die Tätigkeiten der Arbeitsbienen unter experimentell veränderten Bedingungen. Z. Vergl. Physiol. 12: 1–71.

Röseler, P.-F. 1985. Endocrine basis of dominance and reproduction in polistine paper wasps. Pp. 259–272 *in* B. Hölldobler and M. Lindauer (eds.), Experimental Behavioral Ecology and Sociobiology. Sinauer, Sunderland, MA.

Röseler, P.-F., and I. Röseler. 1989. Dominance of ovariectomized foundresses of the paper wasp *Polistes gallicus*. Insectes Soc. 36: 219–234.

Röseler, P.-F., I. Röseler, and A. Strambi. 1980. The activity of corpora allata in dominant and subordinated females of the wasp *Polistes gallicus*. Insectes Soc. 27: 97–107.

Röseler, P.-F., I. Röseler, and C. G. J. van Honk. 1981. Evidence for inhibition of corpora allata activity in workers of *Bombus terrestris* by a pheromone from the queen's mandibular glands. Experientia 37: 348–351.

Röseler, P.-F., I. Röseler, A. Strambi, and R. Augier. 1984. Influence of insect hormones on the establishment of dominance hierarchies among foundresses of the paper wasp, *Polistes gallicus*. Behav. Ecol. Sociobiol. 15: 133–142.

Röseler, P.-F., I. Röseler, and A. Strambi. 1985. Role of ovaries and ecdysteroids in dominance hierarchy establishment among foundresses of the primitively social wasp, *Polistes gallicus*. Behav. Ecol. Sociobiol. 18: 9–13.

Röseler, P.-F., I. Röseler, and A. Strambi. 1986. Studies of the dominance hierarchy in the paper wasp, *Polistes gallicus* (L.) (Hymenoptera, Vespidae). Monit. Zool. Ital. (N.S.) 20: 283–290.

Rosengren, R., and P. Pamilo. 1983. The evolution of polygyny and polydomy in mound-building *Formica* ants. Acta Entomol. Fenn. 42: 65–77.

Rosenheim, J. A. 1987. Nesting behavior and bionomics of a solitary ground-nesting wasp, *Ammophila dysmica* (Hymenoptera: Sphecidae): influence of parasite pressure. Ann. Entomol. Soc. Amer. 80: 739–749.

Rosenheim, J. A. 1988. Parasite presence acts as a proximate cue in the nest-site selection process of the solitary digger wasp, *Ammophila dysmica* (Hymenoptera: Sphecidae). J. Insect Behav. 1: 333–342.

Rosenheim, J. A. 1990. Density dependent parasitism and the evolution of aggregated nesting in the solitary Hymenoptera. Ann. Entomol. Soc. Amer. 83: 277–286.

Ross, K. G. 1982a. Gastral vibrations in laboratory *Vespula vulgaris* and *V. maculifrons* colonies (Hymenoptera: Vespidae). Florida Entomol. 65: 187–188.

Ross, K. G. 1982b. Atypical yellowjacket nests initiated in the laboratory (Hymenoptera: Vespidae). J. Kansas Entomol. Soc. 55: 563–567.

Ross, K. G. 1983a. Studies of the foraging and feeding behavior of yellowjacket foundresses, *Vespula* (*Paravespula*) (Hymenoptera: Vespidae), in the laboratory. Ann. Entomol. Soc. Amer. 76: 903–912.

Ross, K. G. 1983b. Laboratory studies of the mating biology of the eastern yellowjacket, *Vespula maculifrons* (Hymenoptera: Vespidae). J. Kansas Entomol. Soc. 56: 523–537.

Ross, K. G. 1985. Aspects of worker reproduction in four social wasp species (Insecta: Hymenoptera: Vespidae). J. Zool., London (A) 205: 411–424.

Ross, K. G. 1986. Kin selection and the problem of sperm utilization in social insects. Nature 323: 798–800.

Ross, K. G. 1988a. Population and colony-level genetic studies of ants. Pp. 189–215 *in* J. C. Trager (ed.), Advances in Myrmecology. E. J. Brill, New York.

Ross, K. G. 1988b. Differential reproduction in multiple-queen colonies of the fire ant *Solenopsis invicta* (Hymenoptera: Formicidae). Behav. Ecol. Sociobiol. 23: 341–355.

Ross, K. G., and J. M. Carpenter. 1990. Phylogenetic analysis and the evolution of queen number in eusocial Hymenoptera. J. Evol. Biol. (in press).

Ross, K. G., and R. W. Matthews. 1982. Two polygynous overwintered *Vespula squamosa* colonies from the southeastern U.S. (Hymenoptera: Vespidae). Florida Entomol. 65: 176–184.

Ross, K. G., and R. W. Matthews. 1989a. Population genetic structure and social evolution in the sphecid wasp *Microstigmus comes*. Amer. Nat. 134: 574–598.

Ross, K. G., and R. W. Matthews. 1989b. New evidence for eusociality in the sphecid wasp *Microstigmus comes*. Anim. Behav. 38: 613–619.

Ross, K. G., and P. K. Visscher. 1983. Reproductive plasticity in yellowjacket wasps: a polygynous, perennial colony of *Vespula maculifrons*. Psyche 90: 179–191.

Ross, K. G., R. W. Matthews, and R. A. Morse. 1981. Laboratory culture of four species of yellowjackets, *Vespula* spp.; foundress nest initiation. Ann. Entomol. Soc. Amer. 74: 247–254.

Ross, K. G., R. K. Vander Meer, D. J. C. Fletcher, and E. L. Vargo. 1987. Biochemical phenotypic and genetic studies of two introduced fire ants and their hybrid (Hymenoptera: Formicidae). Evolution 41: 280–293.

Ross, K. G., E. L. Vargo, and D. J. C. Fletcher. 1988. Colony genetic structure and queen mating frequency in fire ants of the subgenus *Solenopsis* (Hymenoptera: Formicidae). Biol. J. Linn. Soc. (London) 34: 105–117.

Ross, N. M., and G. J. Gamboa. 1981. Nestmate discrimination in social wasps (*Polistes metricus*, Hymenoptera: Vespidae). Behav. Ecol. Sociobiol. 9: 163–165.

Rossi, A. M., and J. H. Hunt. 1988. Honey supplementation and its developmental consequences: evidence for food limitation in a paper wasp, *Polistes metricus*. Ecol. Entomol. 13: 437–442.

Roth, R. R., and W. D. Lord. 1987. Cycles of queen size and abundance in a population of *Vespula maculifrons* (Hymenoptera: Vespidae). Environ. Entomol. 16: 649–652.

Rothney, G. A. J. 1877. Notes on the habits of *Chlorion lobatum* and two species of wasps in India. Entomol. Mon. Mag. 14: 91–92.

Roubaud, E. 1911. The natural history of the solitary wasps of the genus *Synagris*. Smithsonian Inst. Ann. Rep. 1910: 507–525.

Roubaud, E. 1916. Recherches biologiques sur les guêpes solitaires et sociales d'Afrique. La genèse de la vie sociale et l'évolution de l'instinct maternel chez les vespides. Ann. Sci. Nat. Zool. (Sér. 10) 1: 1–160.

Roubik, D. W. 1989. Ecology and Natural History of Tropical Bees. Cambridge University Press, Cambridge.

Roush, C. F., and R. D. Akre. 1978. Nesting biologies and seasonal occurrence of yellowjackets in northeastern Oregon forests (Hymenoptera: Vespidae). Melanderia 30: 57–94.

Rowland, C. M., and A. R. McLellan. 1979. A model of a wasp colony population, *Paravespula vulgaris* (L.). Ecol. Modelling 7: 151–162.

Rubink, W. L. 1979. The use of edaphic factors as cues for nest-site selection by sand wasps. Ph.D. dissertation, Colorado State University, Fort Collins.

Rubink, W. L. 1982. Spatial patterns in a nesting aggregation of solitary wasps: evidence for the role of conspecifics in nest-site selection. J. Kansas Entomol. Soc. 55: 52–56.

Ruiter, B. A. de. 1916. De *Vespa cincta* als poppenroover. Trop. Natuur. 5: 190–191.

Ryan, R. E., and G. J. Gamboa. 1986. Nestmate recognition between males and gynes of the social wasp *Polistes fuscatus*. Ann. Entomol. Soc. Amer. 79: 572–575.

Ryan, R. E., T. C. Cornell, and G. J. Gamboa. 1985. Nestmate recognition in the bald-faced hornet, *Dolichovespula maculata* (Hymenoptera: Vespidae). Z. Tierpsychol. 69: 19–26.

Sailer, R. I. 1950. Nest temperature of the common yellow jacket *Vespula arenaria* (F.). J. Kansas Entomol. Soc. 23: 134–137.

Sakagami, S. F. 1982. Stingless bees. Pp. 361–423 *in* H. R. Hermann (ed.), Social Insects, vol. 3. Academic Press, New York.

Sakagami, S. F., and K. Fukushima. 1957. *Vespa dybowskii* André as a facultative temporary social parasite. Insectes Soc. 4: 1–12.

Sakagami, S. F., and Sô. Yamane. 1983. Behavior inventory of *Parischnogaster mellyi*. Ecological Study on Social Insects in Central Sumatra with Special Reference to Wasps and Bees—Sumatra Nature Study (Entomology), Kanazawa University, pp. 12–18.

Sakagami, S. F., and K. Yoshikawa. 1968. A new ethospecies of *Stenogaster* wasps from Sarawak, with a comment on the value of ethological characters in animal taxonomy. Annot. Zool. Jpn. 41: 77–84.

Sakagami, S., and R. Zucchi. 1978. Nests of *Hylaeus (Hylaepsis) tricolor*: the first record of non-solitary life in colletid bees, with notes on communal and quasisocial colonies (Hymenoptera: Colletidae). J. Kansas Entomol. Soc. 51: 597–614.

Salbert, P., and N. Elliott. 1979. Observations on the nesting behavior of *Cerceris watlingensis* (Hymenoptera: Sphecidae, Philanthinae). Ann. Entomol. Soc. Amer. 72: 591–595.

Samuel, C. T. 1987. Factors affecting colony size in the stenogastrine wasp *Liostenogaster flavolineata*. Ph.D. dissertation, University of Malaya, Kuala Lumpur.

Samuel, C. T., and M. H. Hansell. 1987. Fully social life in the small colonies of *Liostenogaster flavolineata* (Cameron) (Stenogastrinae, Vespidae). Abstr. 20th Int. Ethol. Conf., p. 156.

Sandeman, R. G. 1938. The swarming of the males of *Vespula sylvestris* (Scop.) around a queen. Proc. R. Entomol. Soc. London (A) 13: 87–88.

Sands, W. A. 1969. The association of termites and fungi. Pp. 495–524 *in* K. Krishna and F. M. Weesner (eds.), Biology of Termites, vol. 1. Academic Press, New York.

Santamour, F. S., and A. Greene. 1986. European hornet damage to ash and birch trees. J. Arboriculture 12: 273–279.

Saunders, E. 1896. The Hymenoptera Aculeata of the British Isles. L. Reeve, London.

Saussure, H. de. 1852. Description du genre *Ischnogaster*. Ann. Soc. Entomol. France 10: 19–27.

Saussure, H. de. 1853–1858. Etudes sur la famille des vespides. 2. Monographie des guêpes sociales, ou de la tribu des vespiens. Masson, Paris.

Schaal, B. A., and W. J. Leverich. 1981. The demographic consequences of two-stage life cycles: survivorship and the time of reproduction. Amer. Nat. 118: 135–138.

Scheven, J. 1958. Beitrag zur Biologie der Schmarotzerfeldwespen *Sulcopolistes atrimandibularis* Zimm., *S. semenowi* F. Morawitz und *S. sulcifer* Zimm. Insectes Soc. 5: 409–437.

Schmidt, G. H. (ed.). 1974. Sozialpolymorphismus bei Insekten. Wissenschaftliche Verlagsgesellschaft, Stuttgart.

Schmidt, J. O. 1982. Biochemistry of insect venoms. Annu. Rev. Entomol. 27: 339–368.

Schmidt, J. O., M. S. Blum, and W. L. Overal. 1980. Comparative lethality of venoms from stinging Hymenoptera. Toxicon 18: 469–474.

Schmidt, J. O., H. C. Reed, and R. D. Akre. 1984. Venoms of a parasitic and two nonparasitic species of yellowjackets (Hymenoptera: Vespidae). J. Kansas Entomol. Soc. 57: 316–322.

Schremmer, F. 1962. Wespen und Hornissen. Neue Brehm-Bücherei, Wittenberg, German Democratic Republic.

Schremmer, F. 1972. Beobachtungen zur Biologie von *Apoica pallida* (Olivier, 1791), einer neotropischen sozialen Faltenwespe (Hymenoptera, Vespidae). Insectes Soc. 19: 343–357.

Schremmer, F. 1977. Das Baumrinden-Nest der neotropischen Faltenwespe *Nectarinella championi*, umgeben von einem Leimring als Ameisen-Abwehr (Hymenoptera: Vespidae). Entomol. Germ. 3: 344–355.

Schremmer, F. 1978a. Das bisher unbekannte Nest von *Charterginus carinatus*, einer neotropischen sozialen Faltenwespe (Hymenoptera: Vespidae). Entomol. Gener. 5: 17–24.

Schremmer, F. 1978b. Zum Einfluss verschiedener Nestunterlagen-Neigungen auf Nestform und Wabengrösse bei zwei neotropischen Arten sozialer Faltenwespen der Gattung *Parachartergus* (Hymenoptera: Vespidae). Entomol. Germ. 4: 356–367.

Schremmer, F. 1983. Das Nest der neotropischen Faltenwespe *Leipomeles dorsata*. Ein Beitrag zur Kenntnis der Nestarchitektur der sozialen Faltenwespen (Vespidae, Polistinae, Polybiini). Zool. Anz. 211: 95–107.

Schremmer, F. 1984. Das Lehmnest der sozialen Faltenwespe *Polybia emaciata* und Untersuchungen über seine Regenfestigkeit (Hymenoptera, Vespidae, Polybiinae). Sitzungsber. Österr. Akad. Wiss. Mathem.-Naturwiss. Kl. (Abt. 1) 193: 7–22.

Schremmer, F., L. März, and P. Simonsberger. 1985. Chitin im Speichel der Papierwespen (soziale Faltenwespen, Vespidae): Biologie, Chemismus, Feinstruktur. Mikroskopie 42: 52–56.

Schulz-Langner, E. 1954. Beobachtungen zur Copulation gefangen gehaltener Wespen. Zool. Anz. 152: 39–42.

Schwarz, H. F. 1929. Honey wasps. Nat. Hist. 29: 421–426.

Schwarz, H. F. 1931. The nest habits of the diplopterous wasp *Polybia occidentalis* variety *scutellaris* (White) as observed at Barro Colorado, Canal Zone. Amer. Mus. Novit. 471: 1–27.

Schwarz, M. P. 1987. Intra-colony relatedness and sociality in the allodapine bee *Exoneura bicolor*. Behav. Ecol. Sociobiol. 21: 387–392.

Schwarz, M. P. 1988. Intra-specific mutualism and kin-association of cofoundresses in allodapine bees (Hymenoptera; Anthophoridae). Monit. Zool. Ital. (N.S.) 22: 245–254.

Seeley, T. D. 1979. Queen substance dispersal by messenger workers in honeybee colonies. Behav. Ecol. Sociobiol. 5: 391–415.

Seeley, T. D. 1982. Adaptive significance of the age polyethism schedule in honeybee colonies. Behav. Ecol. Sociobiol. 11: 287–293.

Seeley, T. D. 1985. Honeybee Ecology: A Study of Adaptation in Social Life. Princeton University Press, Princeton, NJ.

Seeley, T. D. 1989. Social foraging in honey bees: how nectar foragers assess their colony's nutritional status. Behav. Ecol. Sociobiol. 24: 181–199.

Seeley, T. D., and B. Heinrich. 1981. Regulation of temperature in the nests of social insects. Pp. 159–234 in B. Heinrich (ed.), Insect Thermoregulation. Wiley, New York.

Seeley, T. D., and R. H. Seeley. 1980. A nest of a social wasp, Vespa affinis, in Thailand (Hymenoptera: Vespidae). Psyche 87: 299–304.

Seger, J. 1983. Partial bivoltinism may cause alternating sex-ratio biases that favour eusociality. Nature 301: 59–62.

Sekijima, M., M. Sugiura, and M. Matsuura. 1980. Nesting habits and brood development of Parapolybia indica Saussure (Hymenoptera: Vespidae). Bull. Fac. Agric. Mie Univ. 61: 11–23.

Shear, W. A. 1970. The evolution of social phenomena in spiders. Bull. Brit. Arachnol. Soc. 1: 65–76.

Shellman, J. S., and G. J. Gamboa. 1982. Nestmate discrimination in social wasps: the role of exposure to nest and nestmates (Polistes fuscatus, Hymenoptera: Vespidae). Behav. Ecol. Sociobiol. 11: 51–53.

Shellman-Reeve, J. S., and G. J. Gamboa. 1985. Male social wasps (Polistes fuscatus, Hymenoptera: Vespidae) recognize their male nestmates. Anim. Behav. 33: 331–332.

Sherman, P. W., and W. G. Holmes. 1985. Kin recognition: issues and evidence. Pp. 437–460 in B. Hölldobler and M. Lindauer (eds.), Experimental Behavioral Ecology and Sociobiology. Sinauer, Sunderland, MA.

Shida, T. 1959. Vespula lewisii in Musashino. Nippon-Kontyuki 1: 77–145 (in Japanese).

Siew, Y. S., and K. I. Sudderuddin. 1982. Malaysian hornets—some interesting facts. Nature Malaysiana 7: 18–21.

Silva, M. N. da. 1981. Ciclo de desenvolvimento das colônias de Mischocyttarus (Kappa) atramentarius Zikán, 1949 (Hym.-Vespidae): fase de pré-emergência. M.S. thesis, Universidade Estadual Paulista "Júlio de Mesquita Filho," Rio Claro, Brasil.

Simmons, G. E. 1973. Patterns of life in woodland communities. 12. The papermaking wasps. Quart. J. Forestry 67: 13–20.

Simões, D. 1977. Etologia e diferenciação de casta em algumas vespas sociais (Hymenoptera, Vespidae). Ph.D. dissertation, Universidade de São Paulo, Ribeirão Prêto.

Skutch, A. F. 1971. A Naturalist in Costa Rica. University of Florida Press, Gainesville.

Smith, A. P. 1978. An investigation of the mechanisms underlying nest construction in the mud wasp Paralastor sp. (Hymenoptera: Eumenidae). Anim. Behav. 26: 232–240.

Smith, A. P., and J. Alcock. 1980. A comparative study of the mating systems of Australian eumenid wasps (Hymenoptera). Z. Tierpsychol. 53: 41–60.

Smith, B. H. 1983. Recognition of female kin by male bees through olfactory signals. Proc. Natl. Acad. Sci. USA 80: 4551–4553.

Smith, B. H., and M. Ayasse. 1987. Kin-based male mating preferences in two species of halictine bee. Behav. Ecol. Sociobiol. 20: 313–318.

Smith, B. H., and D. W. Roubik. 1983. Mandibular glands of stingless bees (Hymenoptera: Apidae): chemical analysis of their contents and biological function in two species of *Melipona*. J. Chem. Ecol. 9: 1465–1472.

Smith, R. L. 1979. Repeated copulation and sperm precedence: paternity assurance for a male brooding water bug. Science 205: 1029–1031.

Snelling, R. R. 1974. Changes in the status of some North American *Polistes* (Hymenoptera: Vespidae). Proc. Entomol. Soc. Washington 76: 476–479.

Snelling, R. R. 1983. Taxonomic and nomenclatural studies on American polistine wasps (Hymenoptera: Vespidae). Pan-Pac. Entomol. 59: 267–280.

Snodgrass, R. E. 1956. Anatomy of the Honeybee. Cornell University Press, Ithaca, NY.

Spieth, H. T. 1948. Notes on a colony of *Polistes fuscatus hunteri* Bequaert. J. New York Entomol. Soc. 56: 155–169.

Spradbery, J. P. 1965. The social organization of wasp communities. Symp. Zool. Soc. London 14: 61–96.

Spradbery, J. P. 1971. Seasonal changes in the population structure of wasp colonies (Hymenoptera: Vespidae). J. Anim. Ecol. 40: 501–523.

Spradbery, J. P. 1972. A biometric study of seasonal variation in worker wasps (Hymenoptera: Vespidae). J. Entomol. (A) 47: 61–69.

Spradbery, J. P. 1973a. Wasps: An Account of the Biology and Natural History of Solitary and Social Wasps. University of Washington Press, Seattle.

Spradbery, J. P. 1973b. The European wasp, *Paravespula germanica* (F.) (Hymenoptera: Vespidae) in Tasmania, Australia. Proc. 7th Int. Cong. Int. Union Study Soc. Insects, pp. 375–380.

Spradbery, J. P. 1975. The biology of *Stenogaster concinna* van der Vecht, with comments on the phylogeny of Stenogastrinae (Hymenoptera: Vespidae). J. Aust. Entomol. Soc. 14: 309–318.

Spradbery, J. P. 1986. Polygyny in the Vespinae with special reference to the hornet *Vespa affinis picea* Buysson (Hymenoptera, Vespidae) in New Guinea. Monit. Zool. Ital. (N.S.) 20: 101–118.

Spradbery, J. P. 1989. The nesting of *Anischnogaster iridipennis* (Smith) (Hymenoptera: Vespidae) in New Guinea. J. Aust. Entomol. Soc. 28: 225–228.

Spradbery, J. P., and J. Kojima. 1989. Nest descriptions and colony populations of eleven species of *Ropalidia* (Hymenoptera, Vespidae) in New Guinea. Jpn. J. Entomol. 57: 632–653.

Starr, C. K. 1976. Nest reutilization by *Polistes metricus* (Hymenoptera: Vespidae) and possible limitation of multiple foundress associations by parasitoids. J. Kansas Entomol. Soc. 49: 142–144.

Starr, C. K. 1978. Nest reutilization in North American *Polistes* (Hymenoptera: Vespidae): two possible selective factors. J. Kansas Entomol. Soc. 51: 394–397.

Starr, C. K. 1979. Origin and evolution of insect sociality: a review of modern theory. Pp. 35–79 *in* H. R. Hermann (ed.), Social Insects, vol. 1. Academic Press, New York.

Starr, C. K. 1984. Sperm competition, kinship, and sociality in the aculeate Hymenoptera. Pp. 427–464 *in* R. L. Smith (ed.), Sperm Competition and the Evolution of Animal Mating Systems. Academic Press, New York.

Starr, C. K. 1985a. A simple pain scale for field comparison of hymenopteran stings. J. Entomol. Sci. 20: 225–232.

Starr, C. K. 1985b. Enabling mechanisms in the origin of sociality in the Hymenoptera—the sting's the thing. Ann. Entomol. Soc. Amer. 78: 836–840.

Starr, C. K. 1987. A colony of the hornet *Vespa philippinensis* (Hymenoptera: Vespidae). Insectes Soc. 34: 1–9.

Starr, C. K. 1989a. In reply, is the sting the thing? Ann. Entomol. Soc. Amer. 82: 6–8.

Starr, C. K. 1989b. The ins and outs of a tropical social wasp nest. Proc. Entomol. Soc. Washington 91: 26–28.

Starr, C. K. 1990. Holding the fort: colony defense in some primitively social wasps. Pp. 421–463 in D. L. Evans and J. O. Schmidt (eds.), Insect Defenses, Adaptive Mechanisms and Strategies of Prey and Predators. State University of New York Press, Stony Brook.

Stein, K. J. 1986. The ecology and pest status of yellowjackets in Ohio (Hymenoptera: Vespidae). M.S. thesis, Ohio State University, Columbus.

Steiner, A. L. 1983. Predatory behavior of solitary wasps. V. Stinging of caterpillars by Euodynerus foraminatus (Hymenoptera: Eumenidae). Biol. Behav. 8: 11–26.

Steiner, A. L. 1984a. Observations on the possible use of habitat cues and token stimuli by caterpillar-hunting wasps: Euodynerus foraminatus (Hymenoptera, Eumenidae). Quaest. Entomol. 20: 25–33.

Steiner, A. L. 1984b. L'organisation spatio-temporelle des activités nidificatrices de la guêpe solitaire holarctique Oxybelus uniglumis (Hym., Sphecidae: Crabroninae) en conditions de laboratoire. Ann. Sci. Nat. Zool. Biol. Anim. (Sér. 13) 6: 33–55.

Strambi, A. 1969. La fonction gonadotrope des organes neuroendocrines des guêpes femelles du genre Polistes (Hyménoptères). Influence du parasite Xenos vesparum Rossi (Strepsiptères). Ph.D. dissertation, Université de Paris, Paris.

Strambi, A. 1985. Physiological aspects of caste differentiation in social wasps. Pp. 371–384 in J. A. L. Watson, B. M. Okot-Kotber, and C. Noirot (eds.), Caste Differentiation in Social Insects. Pergamon, Oxford.

Strambi, A., C. Strambi, and M. L. de Reggi. 1977. Ecdysones and ovarian physiology in the adult wasp Polistes gallicus. Proc. 8th Int. Cong. Int. Union Study Soc. Insects, pp. 19–20.

Strambi, C., A. Strambi, and R. Augier. 1982. Protein level in the haemolymph of the wasp Polistes gallicus L. at the beginning of imaginal life and during overwintering. Action of the strepsiterian [sic] parasite Xenos vesparum Rossi. Experientia 38: 1189–1190.

Strassmann, J. E. 1979. Honey caches help female paper wasps (Polistes annularis) survive Texas winters. Science 204: 207–209.

Strassmann, J. E. 1981a. Wasp reproduction and kin selection: reproductive competition and dominance hierarchies among Polistes annularis foundresses. Florida Entomol. 64: 74–88.

Strassmann, J. E. 1981b. Evolutionary implications of early male and satellite nest production in Polistes exclamans colony cycles. Behav. Ecol. Sociobiol. 8: 55–64.

Strassmann, J. E. 1981c. Parasitoids, predators and group size in the paper wasp, Polistes exclamans. Ecology 62: 1225–1233.

Strassmann, J. E. 1981d. Kin selection and satellite nests in Polistes exclamans. Pp. 45–58 in R. D. Alexander and D. W. Tinkle (eds.), Natural Selection and Social Behavior: Recent Research and New Theory. Chiron, New York.

Strassmann, J. E. 1983. Nest fidelity and group size among foundresses of Polistes annularis (Hymenoptera: Vespidae). J. Kansas Entomol. Soc. 56: 621–634.

Strassmann, J. E. 1984. Female-biased sex ratios in social insects lacking morphological castes. Evolution 38: 256–266.

Strassmann, J. E. 1985a. Relatedness of workers to brood in the social wasp, Polistes exclamans (Hymenoptera: Vespidae). Z. Tierpsychol. 69: 141–148.

Strassmann, J. E. 1985b. Worker mortality and the evolution of castes in the social wasp, Polistes exclamans. Insectes Soc. 32: 275–285.

Strassmann, J. E. 1989. Early termination of brood rearing in the social wasp Polistes annularis (Hymenoptera: Vespidae). J. Kansas Entomol. Soc. 62: 353–362.

Strassmann, J. E., and C. R. Hughes. 1986. Latitudinal variation in protandry and protogyny in polistine wasps. Monit. Zool. Ital. (N.S.) 20: 87–100.

Strassmann, J. E., and D. C. Meyer. 1983. Gerontocracy in the social wasp, *Polistes exclamans*. Anim. Behav. 31: 431–438.

Strassmann, J. E., and M. C. F. Orgren. 1983. Nest architecture and brood development times in the paper wasp *Polistes exclamans* (Hymenoptera: Vespidae). Psyche 90: 237–248.

Strassmann, J. E., and D. C. Queller. 1989. Ecological determinants of social evolution. Pp. 81–101 *in* M. D. Breed and R. E. Page (eds.), The Genetics of Social Evolution. Westview, Boulder, CO.

Strassmann, J. E., D. C. Meyer, and R. L. Matlock. 1984a. Behavioral castes in the social wasp, *Polistes exclamans*. Sociobiology 8: 211–224.

Strassmann, J. E., R. E. Lee, R. R. Rojas, and J. G. Baust. 1984b. Caste and sex differences in cold-hardiness in the social wasps *Polistes annularis* and *P. exclamans* (Hymenoptera: Vespidae). Insectes Soc. 31: 291–301.

Strassmann, J. E., D. C. Queller, and C. R. Hughes. 1987. Constraints on independent nesting by *Polistes* foundresses in Texas. Pp. 379–380 *in* J. Eder and H. Rembold (eds.), Chemistry and Biology of Social Insects. J. Peperny, Munich.

Strassmann, J. E., D. C. Queller, and C. R. Hughes. 1988. Predation and the evolution of sociality in the paper wasp *Polistes bellicosus*. Ecology 69: 1497–1505.

Strassmann, J. E., C. R. Hughes, D. C. Queller, S. Turillazzi, R. Cervo, S. K. Davis, and K. F. Goodnight. 1989. Genetic relatedness in primitively eusocial wasps. Nature 342: 268–269.

Strassmann, J. E., D. C. Queller, C. R. Solís, and C. R. Hughes. 1991. Relatedness and queen number in the Neotropical wasp, *Parachartergus colobopterus*. Anim. Behav. (in press).

Stubblefield, J. W., and E. L. Charnov. 1986. Some conceptual issues in the origin of eusociality. Heredity 57: 181–187.

Sugiura, M., M. Sekijima, and M. Matsuura. 1983a. Intracolonial polyethism in *Parapolybia indica* (Hymenoptera, Vespidae) I. Behaviour and its changes in the foundress and worker wasps in relation to colony development. Bull. Fac. Agric. Mie Univ. 66: 27–43 (in Japanese).

Sugiura, M., M. Sekijima, and M. Matsuura. 1983b. Life cycle of *Parapolybia indica* (Hymenoptera, Vespidae), with special reference to colony development. Bull. Fac. Agric. Mie Univ. 66: 11–25 (in Japanese).

Sullivan, J. D., and J. E. Strassmann. 1984. Physical variability among nest foundresses in the polygynous social wasp, *Polistes annularis*. Behav. Ecol. Sociobiol. 15: 249–256.

Suzuki, H., and M. Murai. 1980. Ecological studies of *Roparidia* [*sic*] *fasciata* in Okinawa Island. I. Distribution of single- and multiple-foundress colonies. Res. Popul. Ecol. 22: 184–195.

Suzuki, T. 1980. Flesh intake and production of offspring in colonies of *Polistes chinensis antennalis* (Hymenoptera, Vespidae) I. Flesh intake and worker production by solitary foundresses. Kontyû 48: 149–159.

Suzuki, T. 1981a. Flesh intake and production of offspring in colonies of *Polistes chinensis antennalis* (Hymenoptera, Vespidae) II. Flesh intake and production of reproductives. Kontyû 49: 283–301.

Suzuki, T. 1981b. Male production by foundresses of *Polistes chinensis antennalis* (Hymenoptera, Vespidae) in the founding stage. Kontyû 49: 519–520.

Suzuki, T. 1981c. Effect of photoperiod on male egg production by foundresses of *Polistes chinensis antennalis* (Hymenoptera, Vespidae). Jpn. J. Ecol. 31: 347–351.

Suzuki, T. 1982. Cessation and resumption of laying of female-producing eggs by foundresses of a polistine wasp, *Polistes chinensis antennalis* (Hymenoptera, Vespidae) under experimental conditions. Kontyû 50: 652–655.

Suzuki, T. 1983. Resting posture at night in the lone foundress of *Polistes chinensis antennalis* (Hymenoptera, Vespidae). Kontyû 51: 488–489.

Suzuki, T. 1985. Mating and laying of female-producing eggs by orphaned workers of a paper wasp, *Polistes snelleni* (Hymenoptera: Vespidae). Ann. Entomol. Soc. Amer. 78: 736–739.

Suzuki, T. 1986. Production schedules of males and reproductive females, investment sex ratios, and worker-queen conflict in paper wasps. Amer. Nat. 128: 366–378.

Suzuki, T. 1987. Egg-producers in the colonies of a polistine wasp, *Polistes snelleni* (Hymenoptera: Vespidae), in central Japan. Ecol. Res. 2: 185–189.

Taffe, C. A. 1978. Temporal distribution of mortality in a field population of *Zeta abdominale* (Hymenoptera) in Jamaica. Oikos 31: 106–111.

Taffe, C. A. 1979. The ecology of two West Indian species of mud-wasps (Eumenidae: Hymenoptera). Biol. J. Linn. Soc. (London) 11: 1–17.

Taffe, C. A. 1983. The biology of the mud-wasp *Zeta abdominale* (Drury) (Hymenoptera: Eumenidae). Zool. J. Linn. Soc. (London) 77: 385–393.

Taffe, C. A., and K. Ittyeipe. 1976. Effect of nest substrata on the mortality of *Eumenes colona* Saussure (Hymenoptera) and its inquilines. J. Anim. Ecol. 45: 303–311.

Taylor, L. H. 1939. Observations on social parasitism in the genus *Vespula* Thomson. Ann. Entomol. Soc. Amer. 32: 304–315.

Tepedino, V. J. 1979. Notes on the flower visiting habits of *Pseudomasaris vespoides* (Hymenoptera: Masaridae). Southwestern Nat. 24: 380–381.

Tepedino, V. J., and F. D. Parker. 1988. Alternation of sex ratio in a partially bivoltine bee, *Megachile rotundata* (Hymenoptera: Megachilidae). Ann. Entomol. Soc. Amer. 81: 467–476.

Tepedino, V. J., L. L. McDonald, and R. Rothwell. 1979. Defense against parasitization in mud-nesting Hymenoptera: can empty cells increase net reproductive output? Behav. Ecol. Sociobiol. 6: 99–104.

Thomas, C. R. 1960. The European wasp (*Vespula germanica* Fab.) in New Zealand. New Zealand Dept. Sci. Ind. Res. Inf. Ser. 27: 1–74.

Thornhill, R., and J. Alcock. 1983. The Evolution of Insect Mating Systems. Harvard University Press, Cambridge, MA.

Tinbergen, N. 1932. Über die Orientierung des Bienenwolfes (*Philanthus triangulum* Fabr.). Z. Vergl. Physiol. 16: 305–334. [Translated into English, pp. 103–127 *in* N. Tinbergen. 1972. The Animal in Its World; Explorations of an Ethologist. Vol. 1, Field Studies. Harvard University Press, Cambridge, MA.]

Tinbergen, N. 1935. Über die Orientierung des Bienenwolfes (*Philanthus triangulum* Fabr.). II. Die Bienenjagd. Z. Vergl. Physiol. 21: 699–716. [Translated into English, pp. 128–145 *in* N. Tinbergen. 1972. The Animal in Its World; Explorations of an Ethologist. Vol. 1, Field Studies. Harvard University Press, Cambridge, MA.]

Tinbergen, N., and W. Kruyt. 1938. Über die Orientierung des Bienenwolfes (*Philanthus triangulum* Fabr.). III. Die Bevorzugung bestimmter Wegmarken. Z. Vergl. Physiol. 25: 292–334. [Translated into English, pp. 146–196 *in* N. Tinbergen. 1972. The Animal in Its World; Explorations of an Ethologist. Vol. 1, Field Studies. Harvard University Press, Cambridge, MA.]

Tissot, A. N., and F. A. Robinson. 1954. Some unusual insect nests. Florida Entomol. 37: 73–92.

Toft, C. A. 1987. Population structure and survival in a solitary wasp (*Microbembix cubana*: Hymenoptera, Sphecidae, Nyssoninae). Oecologia 73: 338–350.

Torchio, P. F. 1970. The ethology of the wasp, *Pseudomasaris edwardsii* (Cresson), and a description of its immature forms (Hymenoptera: Vespoidea, Masaridae). Los Angeles County Mus. Contr. Sci. 202: 1–32.

Trivers, R. L. 1971. The evolution of reciprocal altruism. Quart. Rev. Biol. 46: 35–57.

Trivers, R. L., and H. Hare. 1976. Haplodiploidy and the evolution of the social insects. Science 191: 249–263.

Tsuneki, K. 1947. On the primitive sociality found in a dense aggregation of *Cerceris harmandi* Pérez. Kagaku 7: 119–120.

Tsuneki, K. 1963. Comparative studies on the nesting biology of the genus *Sphex* (s.l.) in East Asia (Hymenoptera: Sphecidae). Mem. Fac. Lib. Arts Fukui Univ. (Ser. 2 Nat. Sci.) 13: 13–78.

Tsuneki, K. 1965. The biology of East-Asiatic *Cerceris* (Hym., Sphecidae) with special reference to the peculiar social relationships and return to the nest in *Cerceris hortivaga* Kohl. Etizenia 9: 1–46.

Turillazzi, S. 1979. Tegumental glands in the abdomen of some European *Polistes* (Hymenoptera Vespidae). Monit. Zool. Ital. (N.S.) 13: 67–70.

Turillazzi, S. 1980. Seasonal variation in the size and anatomy of *Polistes gallicus* (L.) (Hymenoptera: Vespidae). Monit. Zool. Ital. (N.S.) 14: 63–75.

Turillazzi, S. 1982. Patrolling behaviour in males of *Parischnogaster nigricans serrei* (Du Buysson) and *P. mellyi* (Saussure) (Hymenoptera, Stenogastrinae). Accad. Naz. Lincei Rendic. Classe Sci. Fis. Mat. Nat. 72: 153–157.

Turillazzi, S. 1983. Extranidal behaviour of *Parischnogaster nigricans serrei* (Du Buysson) (Hymenoptera, Stenogastrinae). Z. Tierpsychol. 63: 27–36.

Turillazzi, S. 1984. Larval development and larva-adult interactions in a stenogastrine wasp (Hymenoptera Vespidae). Monit. Zool. Ital. (N.S.) 18: 180.

Turillazzi, S. 1985a. Brood rearing behaviour and larval development in *Parischnogaster nigricans serrei* (Du Buysson) (Hymenoptera Stenogastrinae). Insectes Soc. 32: 117–127.

Turillazzi, S. 1985b. Egg deposition in the genus *Parischnogaster* (Hymenoptera: Stenogastrinae). J. Kansas Entomol. Soc. 58: 749–752.

Turillazzi, S. 1985c. Function and characteristics of the abdominal substance secreted by wasps of the genus *Parischnogaster* (Hymenoptera Stenogastrinae). Monit. Zool. Ital. (N.S.) 19: 91–99.

Turillazzi, S. 1985d. Colonial cycle of *Parischnogaster nigricans serrei* (Du Buysson) in West Java (Hymenoptera Stenogastrinae). Insectes Soc. 32: 43–60.

Turillazzi, S. 1985e. Associative nest foundation in the wasp *Parischnogaster alternata*. Naturwissenschaften 72: 100–102.

Turillazzi, S. 1986a. Colony composition and social behaviour of *Parischnogaster alternata* Sakagami (Hymenoptera: Stenogastrinae). Monit. Zool. Ital. (N.S.) 20: 333–347.

Turillazzi, S. 1986b. Les Stenogastrinae: un groupe clé pour l'étude de l'évolution du comportement social chez les guêpes. Actes Coll. Insectes Soc. 3: 7–32.

Turillazzi, S. 1987a. Distinguished features of the social behaviour of stenogastrine wasps. Pp. 492–495 *in* J. Eder and H. Rembold (eds.), Chemistry and Biology of Social Insects. J. Peperny, Munich.

Turillazzi, S. 1987b. Social biology of the Stenogastrinae: temporary dynamic reproductive strategy in the wet tropics. Pp. 381–382 *in* J. Eder and H. Rembold (eds.), Chemistry and Biology of Social Insects. J. Peperny, Munich.

Turillazzi, S. 1987c. Colony defence in stenogastrine wasps (Hymenoptera). Monit. Zool. Ital. (N.S.) 21: 205.

Turillazzi, S. 1988a. Social biology of *Parischnogaster jacobsoni* (Du Buysson) (Hymenoptera, Stenogastrinae). Insectes Soc. 35: 133–143.

Turillazzi, S. 1988b. Adults and nest of *Liostenogaster vechti* n. sp. (Hymenoptera, Stenogastrinae). Tropical Zool. 1: 193–201.

Turillazzi, S. 1989a. The origin and evolution of social life in the Stenogastrinae (Hymenoptera, Vespidae). J. Insect Behav. 2: 649–661.

Turillazzi, S. 1989b. Hover wasps, amazing inhabitants of the jungle. Nature Malaysiana 14: 52–57.

Turillazzi, S. 1990. Social biology of *Liostenogaster vechti* Turillazzi 1989 (Hymenoptera Stenogastrinae). Tropical Zool. 3: 69–87.

Turillazzi, S., and L. Beani. 1985. Mating behaviour in the paper wasp *Polistes gallicus* (L.) (Hymenoptera Polistinae). Monit. Zool. Ital. (N.S.) 19: 169–170.

Turillazzi, S., and C. Calloni. 1983. Tegumental glands in the third gastral tergite of male *Parischnogaster nigricans serrei* (du Buysson) and *P. mellyi* (Saussure) (Hymenoptera Stenogastrinae). Insectes Soc. 30: 455–460.

Turillazzi, S., and R. Cervo. 1982. Territorial behaviour in males of *Polistes nimpha* (Christ) (Hymenoptera, Vespidae). Z. Tierpsychol. 58: 174–180.

Turillazzi, S., and A. Conte. 1981. Temperature and caste differentiation in laboratory colonies of *Polistes foederatus* (Kohl) (Hymenoptera Vespidae). Monit. Zool. Ital. (N.S.) 15: 275–297.

Turillazzi, S., and E. Francescato. 1989. Observations on the behaviour of male stenogastrine wasps (Hymenoptera, Vespidae, Stenogastrinae). Actes Coll. Insectes Soc. 5: 181–187.

Turillazzi, S., and E. Francescato. 1990. Patrolling behaviour and related secretory structures in the males of some stenogastrine wasps (Hymenoptera Vespidae). Insectes Soc. 37: 146–157.

Turillazzi, S., and C. Marucelli Turillazzi. 1985. Notes on the social behaviour of *Ropalidia fasciata* (F.) in West Java (Hymenoptera Vespidae). Monit. Zool. Ital. (N.S.) 19: 219–230.

Turillazzi, S., and L. Pardi. 1977. Body size and hierarchy in polygynic nests of *Polistes gallicus* (L.) (Hymenoptera, Vespidae). Monit. Zool. Ital. (N.S.) 11: 101–112.

Turillazzi, S., and L. Pardi. 1981. Ant guards on nests of *Parischnogaster nigricans serrei* (Buysson) (Stenogastrinae). Monit. Zool. Ital. (N.S.) 15: 1–17.

Turillazzi, S., and L. Pardi. 1982. Social behavior of *Parischnogaster nigricans serrei* (Hymenoptera: Vespoidea) in Java. Ann. Entomol. Soc. Amer. 75: 657–664.

Turillazzi, S., and A. Ugolini. 1979. Rubbing behaviour in some European *Polistes* (Hymenoptera Vespidae). Monit. Zool. Ital. (N.S.) 13: 129–142.

Turillazzi, S., M. T. Marino-Piccioli, L. Hervatin, and L. Pardi. 1982. Reproductive capacity of single foundress and associated foundress females of *Polistes gallicus* (L.) (Hymenoptera Vespidae). Monit. Zool. Ital. (N.S.) 16: 75–88.

Turillazzi, S., R. Cervo, and I. Cavallari. 1990. Invasion of the nest of *Polistes dominulus* by the social parasite *Sulcopolistes sulcifer* (Hymenoptera, Vespidae). Ethology 84: 47–59.

Underwood, B. A. 1986. The natural history of *Apis laboriosa* in Nepal. M.S. thesis, Cornell University, Ithaca, NY.

Uyenoyama, M. K. 1984. Inbreeding and the evolution of altruism under kin selection: effects on relatedness and group structure. Evolution 38: 778–795.

Varvio-Aho, S. L., P. Pamilo, and A. Pekkarinen. 1984. Evolutionary genetics of social wasps (Hymenoptera, Vespidae, *Vespula*). Insectes Soc. 31: 375–386.

Vawter, L., and J. E. Strassmann. 1982. Juvenile hormone production in polistine wasps. P. 222 *in* M. D. Breed, C. D. Michener, and H. E. Evans (eds.), The Biology of Social Insects. Westview, Boulder, CO.

Vecht, J. van der. 1936. Some further notes on *Provespa*, Ashm. (Hym., Vespidae). J. Fed. Malay States Mus. 18: 159–166.

Vecht, J. van der. 1940. The nesting habits of *Ropalidia flavopicta* (Sm.) (Hym., Vespidae). Entomol. Meded. Ned.-Indië 6: 47–50.

Vecht, J. van der. 1941. The Indo-Australian species of the genus *Ropalidia* ( = *Icaria*) (Hym., Vespidae). Treubia 18: 103–190.

Vecht, J. van der. 1957. The Vespinae of the Indo-Malayan and Papuan areas (Hymenoptera, Vespidae). Zool. Verhand. (Leiden) 34: 1–83.

Vecht, J. van der. 1959. Notes on Oriental Vespinae, including some species from China and Japan (Hymenoptera, Vespidae). Zool. Meded. (Leiden) 36: 205–232.

Vecht, J. van der. 1962. The Indo-Australian species of the genus *Ropalidia* (*Icaria*) (Hymenoptera, Vespidae) (2nd part). Zool. Verhand. (Leiden) 57: 1–72.

Vecht, J. van der. 1963. Studies on Indo-Australian and east-Asiatic Eumenidae. Zool. Verhand. (Leiden) 60: 1–116.

Vecht, J. van der. 1965. The geographical distribution of the social wasps (Hymenoptera, Vespidae). Proc. 12th Int. Cong. Entomol., pp. 440–441.

Vecht, J. van der. 1966. The east Asiatic and Indo-Australian species of *Polybioides* Buysson and *Parapolybia* Saussure (Hym., Vespidae). Zool. Verhand. (Leiden) 82: 1–42.

Vecht, J. van der. 1967. Bouwproblemen van sociale wespen. Versl. Gewone Vergad. Afd. Natuur. K. Nederl. Akad. Wetens. 76: 59–68.

Vecht, J. van der. 1968. The terminal gastral sternite of female and worker social wasps (Hymenoptera, Vespidae). Proc. K. Nederl. Akad. Wetens. (Ser. C) 71: 411–422.

Vecht, J. van der. 1972. A review of the new genus *Anischnogaster* in the Papuan region (Hymenoptera, Vespidae). Zool. Meded. (Leiden) 47: 240–256.

Vecht, J. van der. 1975. A review of the genus *Stenogaster* Guerin (Hymenoptera: Vespoidea). J. Aust. Entomol. Soc. 14: 283–308.

Vecht, J. van der. 1977a. Studies of Oriental Stenogastrinae (Hymenoptera: Vespoidea). Tijdschr. Entomol. 120: 55–75.

Vecht, J. van der. 1977b. Important steps in the evolution of nest construction in social wasps. Proc. 8th Int. Cong. Int. Union Study Soc. Insects, p. 319.

Veenendaal, R. L., and T. Piek. 1988. Predatory behavior of *Discoelius zonalis* (Hymenoptera: Eumenidae). Entomol. Ber. (Amsterdam) 48: 8–12.

Vehrencamp, S. L. 1983a. A model for the evolution of despotic versus egalitarian societies. Anim. Behav. 31: 667–682.

Vehrencamp, S. L. 1983b. Optimal degree of skew in cooperative societies. Amer. Zool. 23: 327–335.

Veith, H. J., and N. Koeniger. 1978. Identifizierung von cis-9-Pentacosen als Auslöser für das Wärmen der Brut bei der Hornisse. Naturwissenschaften 65: 263.

Veith, H. J., N. Koeniger, and U. Maschwitz. 1984. 2-Methyl-3-butene-2-ol, a major component of the alarm pheromone of the hornet *Vespa crabro*. Naturwissenschaften 71: 328–329.

Velthuis, H. H. W. 1987. The evolution of sociality: ultimate and proximate factors leading to primitive social behavior in carpenter bees. Pp. 405–430 *in* J. M. Pasteels and J.-L. Deneubourg (eds.), From Individual to Collective Behavior in Social Insects. Birkhäuser, Basel.

Venkataraman, A. B., V. B. Swarnalatha, P. Nair, and R. Gadagkar. 1988. The mechanism of nestmate discrimination in the tropical social wasp *Ropalidia marginata* and its implications for the evolution of sociality. Behav. Ecol. Sociobiol. 23: 271–279.

Vesey-Fitzgerald, D. 1938. Social wasps (Hym. Vespidae) from Trinidad, with a note on the genus *Trypoxylon* Latreille. Trans. R. Entomol. Soc. London 87: 181–191.

Vesey-Fitzgerald, D. 1950. Notes on the genus *Ropalidia* (Hymenoptera: Vespidae) from Madagascar. Proc. R. Entomol. Soc. London (A) 25: 81–86.

Visscher, P. K. 1989. A quantitative study of worker reproduction in honey bee colonies. Behav. Ecol. Sociobiol. 25: 247–254.

Vuillaume, M., J. Schwander, and C. Roland. 1969. Note préliminaire sur l'existence de colonies pérennes et polygynes de *Paravespula germanica*. C. R. Acad. Sci. Paris (D) 269: 2371–2372.

Wade, M. J. 1979. The evolution of social interactions by family selection. Amer. Nat. 113: 399–411.

Wade, M. J. 1980. Kin selection: its components. Science 210: 665–667.

Wade, M. J. 1982. The effect of multiple inseminations on the evolution of social behaviors in diploid and haplo-diploid organisms. J. Theor. Biol. 95: 351–368.

Wade, M. J. 1985. The influence of multiple inseminations and multiple foundresses on social evolution. J. Theor. Biol. 112: 109–121.

Wade, M. J., and F. Breden. 1981. Effect of inbreeding on the evolution of altruistic behavior by kin selection. Evolution 35: 844–858.

Wafa, A. K., and S. G. Sharkawi. 1972. Contribution to the biology of *Vespa orientalis* Fab. Bull. Soc. Entomol. Egypte 56: 219–226.

Wafa, A. K., F. M. El-Borolossy, and S. G. Sharkawi. 1968. Studies on *Vespa orientalis* F. Bull. Soc. Entomol. Egypte 52: 9–27.

Waldbauer, G. P., and D. P. Cowan. 1985. Defensive stinging and Müllerian mimicry among eumenid wasps (Hymenoptera: Vespoidea: Eumenidae). Amer. Midl. Nat. 113: 198–199.

Ward, P. 1965. A raid on a stenogastrine nest by a hornet *Vespa tropica*. Malayan Nat. J. 19: 152.

Ward, P. S. 1989. Genetic and social changes associated with ant speciation. Pp. 123–148 *in* M. D. Breed and R. E. Page (eds.), The Genetics of Social Evolution. Westview, Boulder, CO.

Waser, P. M., S. M. Austad, and B. Keane. 1986. When should animals tolerate inbreeding? Amer. Nat. 128: 529–537.

Way, M. J. 1954. Studies of the life history and ecology of the ant *Oecophylla longinoda* Latreille. Bull. Entomol. Res. 45: 93–112.

Wcislo, W. T. 1984. Gregarious nesting of a digger wasp as a "selfish herd" response to a parasitic fly (Hymenoptera: Sphecidae; Diptera: Sarcophagidae). Behav. Ecol. Sociobiol. 15: 157–160.

Wcislo, W. T. 1986. Host discrimination by a cleptoparasitic fly, *Metopia campestris* (Fallén) (Diptera: Sarcophagidae: Miltogramminae). J. Kansas Entomol. Soc. 59: 82–88.

Wcislo, W. T. 1987. The roles of seasonality, host synchrony, and behaviour in the evolution and distributions of nest parasites in Hymenoptera (Insecta), with special reference to bees (Apoidea). Biol. Rev. (Cambridge) 62: 515–543.

Wcislo, W. T., B. S. Low, and C. J. Karr. 1985. Parasite pressure and repeated burrow use by different individuals of *Crabro* (Hymenoptera: Sphecidae; Diptera: Sarcophagidae). Sociobiology 11: 115–125.

Wcislo, W. T., M. J. West-Eberhard, and W. G. Eberhard. 1988. Natural history and behavior of a primitively social wasp, *Auplopus semialatus*, and its parasite, *Irenangelus eberhardi* (Hymenoptera: Pompilidae). J. Insect Behav. 1: 247–260.

Weaver, N. 1978. Chemical control of behavior—intraspecific. Pp. 359–389 *in* M. Rockstein (ed.), Biochemistry of Insects. Academic Press, New York.

Weaving, A. J. S. 1989. Nesting strategies in some southern African species of *Ammophila* (Hymenoptera: Sphecidae). J. Nat. Hist. 23: 1–16.

Weber, N. A. 1979. Fungus-culturing by ants. Pp. 77–116 *in* L. R. Batra (ed.), Insect-Fungus Symbiosis: Nutrition, Mutualism and Commensalism. Allanheld, Osmun, Montclair, NJ.

Weber, N. A. 1982. Fungus ants. Pp. 255–385 *in* H. R. Hermann (ed.), Social Insects, vol. 4. Academic Press, New York.

Weir, B. S., and C. C. Cockerham. 1984. Estimating *F*-statistics for the analysis of population structure. Evolution 38: 1358–1370.

Wenzel, J. W. 1987a. Male reproductive behavior and mandibular glands in *Polistes major* (Hymenoptera: Vespidae). Insectes Soc. 34: 44–57.

Wenzel, J. W. 1987b. *Ropalidia formosa*, a nearly solitary paper wasp from Madagascar (Hymenoptera: Vespidae). J. Kansas Entomol. Soc. 60: 549–556.

Wenzel, J. W. 1989. Endogenous factors, external cues and eccentric construction in *Polistes annularis* (Hymenoptera: Vespidae). J. Insect Behav. 2: 679–699.

Wenzel, J. W. 1990. A social wasp's nest from the Cretaceous period, Utah, USA, and its biogeographic significance. Psyche 97: 21–29.

Wernstedt, F. L. 1972. World Climatic Data. Climatic Data Press, Lemont, PA.

Werren, J. H., and E. L. Charnov. 1978. Facultative sex ratios and population dynamics. Nature 272: 349–350.

West-Eberhard, M. J. 1967. Foundress associations in polistine wasps: dominance hierarchies and the evolution of social behavior. Science 157: 1584–1585.

West-Eberhard, M. J. 1968. Range extension and solitary nest founding in *Polistes exclamans* (Hymenoptera: Vespidae). Psyche 75: 118–123.

West-Eberhard, M. J. 1969. The social biology of polistine wasps. Misc. Publ. Mus. Zool. Univ. Michigan 140: 1–101.

West-Eberhard, M. J. 1973. Monogyny in "polygynous" social wasps. Proc. 7th Int. Cong. Int. Union Study Soc. Insects, pp. 396–403.

West-Eberhard, M. J. 1975. The evolution of social behavior by kin selection. Quart. Rev. Biol. 50: 1–33.

West-Eberhard, M. J. 1977a. The establishment of reproductive dominance in social wasp colonies. Proc. 8th Int. Cong. Int. Union Study Soc. Insects, pp. 223–227.

West-Eberhard, M. J. 1977b. Morphology and behavior in the taxonomy of *Microstigmus* wasps. Proc. 8th Int. Cong. Int. Union Study Soc. Insects, pp. 164–168.

West-Eberhard, M. J. 1978a. Polygyny and the evolution of social behavior in wasps. J. Kansas Entomol. Soc. 51: 832–856.

West-Eberhard, M. J. 1978b. Temporary queens in *Metapolybia* wasps: nonreproductive helpers without altruism? Science 200: 441–443.

West-Eberhard, M. J. 1979. Sexual selection, social competition, and evolution. Proc. Amer. Phil. Soc. 123: 222–234.

West-Eberhard, M. J. 1981. Intragroup selection and the evolution of insect societies. Pp. 3–17 *in* R. D. Alexander and D. W. Tinkle (eds.), Natural Selection and Social Behavior: Recent Research and New Theory. Chiron, New York.

West-Eberhard, M. J. 1982a. Diversity of dominance displays in *Polistes* and its possible evolutionary significance. P. 222 *in* M. D. Breed, C. D. Michener, and H. E. Evans (eds.), The Biology of Social Insects. Westview, Boulder, CO.

West-Eberhard, M. J. 1982b. Communication in social wasps: predicted and observed patterns, with a note on the significance of behavioral and ontogenetic flexibility for theories of worker "altruism." Pp. 13–36 *in* A. de Haro and X. Espalader (eds.), La communication chez les sociétés d'insectes. Universidad Autónoma de Barcelona, Bellaterra.

West-Eberhard, M. J. 1982c. The nature and evolution of swarming in tropical social wasps (Vespidae, Polistinae, Polybiini). Pp. 97–128 *in* P. Jaisson (ed.), Social Insects in the Tropics, vol. 1. Université de Paris-Nord, Paris.

West-Eberhard, M. J. 1982d. Introduction: symposium on the evolution and ontogeny of eusociality. Pp. 185–186 *in* M. D. Breed, C. D. Michener, and H. E. Evans (eds.), The Biology of Social Insects. Westview, Boulder, CO.

West-Eberhard, M. J. 1983. Sexual selection, social competition, and speciation. Quart. Rev. Biol. 58: 155–183.

West-Eberhard, M. J. 1986. Dominance relations in *Polistes canadensis* (L.), a tropical social wasp. Monit. Zool. Ital. (N.S.) 20: 263–281.

West-Eberhard, M. J. 1987a. The epigenetical origins of insect sociality. Pp. 369–372 *in* J. Eder and H. Rembold (eds.), Chemistry and Biology of Social Insects. J. Peperny, Munich.

West-Eberhard, M. J. 1987b. Observations of *Xenorhynchium nitidulum* (Fabricius) (Hymenoptera, Eumeninae), a primitively social wasp. Psyche 94: 317–323.

West-Eberhard, M. J. 1987c. Flexible strategy and social evolution. Pp. 35–51 *in* Y. Itô, J. L. Brown, and J. Kikkawa (eds.), Animal Societies: Theories and Facts. Japan Scientific Societies Press, Tokyo.

West-Eberhard, M. J. 1989. Scent-trail diversion, a novel defense against ants by tropical social wasps. Biotropica 21: 280–281.

West-Eberhard, M. J. 1990. The genetic and social structure of polygynous social wasp colonies (Vespidae: Polistinae). Pp. 254–255 *in* G. K. Veeresh, B. Mallik, and C. A. Viraktamath (eds.), Social Insects and the Environment. Oxford & IBH, New Delhi.

Weyrauch, W. 1935. *Dolichovespula* und *Vespa*. Vergleichende Übersicht über zwei wesentliche Lebenstypen bei sozialen Wespen. Mit Bezugnahme auf die Frage nach der Fortschrittlichkeit tierischer Organisation. I. Teil. Biol. Zentralbl. 55: 484–524.

Weyrauch, W. 1936a. Wie entsteht ein Wespennest? I. Teil. Beobachtungen und Versuche über den Papierbereitungsinstinkt bei *Vespa*, *Dolichovespula* und *Macrovespa*. Z. Morphol. Ökol. Tiere 30: 401–431.

Weyrauch, W. 1936b. *Dolichovespula* und *Vespa*. Vergleichende Übersicht über zwei wesentliche Lebenstypen bei sozialen Wespen. Mit Bezugnahme auf die Frage nach der Fortschrittlichkeit tierischer Organisation. II. Teil. Biol. Zentralbl. 56: 287–301.

Weyrauch, W. 1936c. Das Verhalten sozialer Wespen bei Nestüberhitzung. Z. Vergl. Physiol. 23: 51–63.

Weyrauch, W. 1939. Wie ein Wespennest entsteht (nach Versuchen und Beobachtungen an *Vespa vulgaris* L. und *V. germanica* F.). Naturwissenschaften 27: 49–57, 73–77.

Wheeler, D. E. 1986. Developmental and physiological determinants of caste in social Hymenoptera: evolutionary implications. Amer. Nat. 128: 13–34.

Wheeler, G. C. 1957. Don't go to the ant. Bios 28: 94–103.

Wheeler, W. M. 1911. The ant-colony as an organism. J. Morphol. 22: 307–325.

Wheeler, W. M. 1918. Study of some ant larvae with a consideration of the origin and meaning of social habits among insects. Proc. Amer. Phil. Soc. 57: 293–343.

Wheeler, W. M. 1922. Social life among the insects. II. Wasps solitary and social. Scient. Mon. 15: 68–88, 119–131.

Wheeler, W. M. 1923. Social Life among the Insects. Harcourt, Brace, New York.

Wheeler, W. M. 1928. The Social Insects. Harcourt, Brace, New York.

Wigglesworth, V. B. 1972. The Principles of Insect Physiology. Wiley, New York.

Wilkes, A. 1965. Sperm transfer and utilization by the arrhenotokous wasp *Dahlbominus fuscipennis* (Zett.) (Hymenoptera: Eulophidae). Can. Entomol. 97: 647–657.

Wilkes, A. 1966. Sperm utilization following multiple insemination in the wasp *Dahlbominus fuscipennis*. Can. J. Genet. Cytol. 8: 451–461.

Wilkinson, G. S., and G. F. McCracken. 1985. On estimating relatedness using genetic markers. Evolution 39: 1169–1174.

Wille, A. 1983. Biology of the stingless bees. Annu. Rev. Entomol. 28: 41–64.

Wille, A., and C. D. Michener. 1973. The nest architecture of stingless bees, with special reference to those of Costa Rica (Hymenoptera: Apidae). Rev. Biol. Trop. 21: 1–278 (suppl.).

Willer, D. E., and H. R. Hermann. 1989. Multiple foundress associations and nest switching among females of *Polistes exclamans* (Hymenoptera: Vespidae). Sociobiology 16: 197–216.

Williams, F. X. 1919. Philippine wasp studies. II. Descriptions of new species and life history studies. Bull. Exp. Sta. Hawaii Sugar Planters' Assoc. (Entomol.) 14: 19–186.

Williams, F. X. 1927. *Euparagia scutellaris* Cresson, a masarid wasp that stores its cells with the young of a curculionid beetle. Pan-Pac. Entomol. 4: 38–39.

Williams, F. X. 1928. Studies in tropical wasps—their hosts and associates (with descriptions of new species). Bull. Exp. Sta. Hawaii Sugar Planters' Assoc. (Entomol.) 19: 1–179.

Williams, G. C. 1957. Pleiotropy, natural selection, and the evolution of senescence. Evolution 11: 398–411.

Williams, G. C. 1966. Adaptation and Natural Selection. Princeton University Press, Princeton, NJ.

Willmer, P. G. 1985. Thermal ecology, size effects, and the origins of communal behavior in *Cerceris* wasps. Behav. Ecol. Sociobiol. 17: 151–160.

Wilson, D. S., and E. Sober. 1989. Reviving the superorganism. J. Theor. Biol. 136: 337–356.

Wilson, E. O. 1966. Behaviour of social insects. Symp. R. Entomol. Soc. London 3: 81–96.

Wilson, E. O. 1971. The Insect Societies. Harvard University Press, Cambridge, MA.

Wilson, E. O. 1975. Sociobiology: The New Synthesis. Harvard University Press, Cambridge, MA.

Wilson, E. O. 1976. Which are the most prevalent ant genera? Studia Entomol. 19: 187–200.

Wilson, E. O. 1984. The relation between caste ratios and division of labor in the ant genus *Pheidole* (Hymenoptera: Formicidae). Behav. Ecol. Sociobiol. 16: 89–98.

Wilson, E. O. 1985. The sociogenesis of insect colonies. Science 228: 1489–1495.

Wilson, E. O. 1989. The coming pluralization of biology and the stewardship of systematics. Bioscience 39: 242–245.

Windsor, D. M. 1972. Nesting association between two neotropical polybiine wasps (Hymenoptera, Vespidae). Biotropica 4: 1–3.

Windsor, D. M. 1976. Birds as predators on the brood of *Polybia* wasps (Hymenoptera: Vespidae: Polistinae) in a Costa Rican deciduous forest. Biotropica 8: 111–116.

Winston, M. L. 1978. Intra-colony demography and reproductive rate of the Africanized honeybee in South America. Behav. Ecol. Sociobiol. 4: 279–292.

Winston, M. L. 1987. The Biology of Honey Bees. Harvard University Press, Cambridge, MA.

Woolfenden, G. E., and J. W. Fitzpatrick. 1984. The Florida Scrub Jay: Demography of a Cooperatively Breeding Bird (Monogr. Popul. Biol. 20). Princeton University Press, Princeton, NJ.

Woyciechowski, M., and A. Łomnicki. 1987. Multiple mating of queens and the sterility of workers among eusocial Hymenoptera. J. Theor. Biol. 128: 317–327.

Wray, D. L. 1954. Nesting notes on the European hornet in North Carolina. Entomol. News 65: 1–5.

Yamane, Sk. 1973. Discovery of a pleometrotic association in *Polistes chinensis* Pérez (Hymenoptera: Vespidae). Life Study 17: 3–4.

Yamane, Sk. 1974a. Observations on an orphan nest of *Vespa simillima* Smith (Hymenoptera, Vespidae). Kontyû 42: 404–415.

Yamane, Sk. 1974b. On the genus *Vespa* (Hymenoptera: Vespidae) from Nepal. Kontyû 42: 29–39.

Yamane, Sk. 1976. Morphological and taxonomic studies on vespine larvae, with reference to the phylogeny of the subfamily Vespinae (Hymenoptera: Vespidae). Insecta Matsumurana (N.S.) 8: 1–45.

Yamane, Sk. 1982. A new subspecies of *Dolichovespula adulterina* Buysson from Japan, with a brief note on its host (Hymenoptera: Vespidae). Trans. Shikoku Entomol. Soc. 16: 109–115.

Yamane, Sk., and J. Kojima. 1982. *Pseudonomadina* new genus, with a description of a new species from the Philippines (Hymenoptera, Trigonalidae). Kontyû 50: 183–188.

Yamane, Sk., and Y. Maeta. 1985. Notes on the hibernation of *Parapolybia indica* (Hymenoptera, Vespidae). Kontyû 53: 576–577.

Yamane, Sk., and S. Makino. 1977. Bionomics of *Vespa analis insularis* and *V. mandarinia latilineata* in Hokkaido, northern Japan, with notes on vespine embryo nests (Hymenoptera: Vespidae). Insecta Matsumurana (N.S.) 12: 1–33.

Yamane, Sk., and M. Terayama. 1983. Description of a new subspecies of *Bakeronymus typicus* Rohwer parasitic on the social wasp *Parapolybia varia* Fabricius in Taiwan (Hymenoptera: Trigonalidae). Mem. Kagoshima Univ. Res. Ctr. South Pacific 3: 169–173.

Yamane, Sk., and Sô. Yamane. 1987. A new species and new synonymy in the subgenus *Polistes* of eastern Asia (Hymenoptera, Vespidae). Kontyû 55: 215–219.

Yamane, Sk., S. Makino, and M. J. Toda. 1980. Nests of *Dolichovespula albida* from the Arctic Canada (Hymenoptera: Vespidae). Low Temp. Sci. (B) 38: 61–68.

Yamane, Sk., S. Makino, and R. P. MacFarlane. 1981. Embryo nest architecture in three *Vespula* species (Hymenoptera, Vespidae). Kontyû 49: 491–497.

Yamane, Sô. 1969. Preliminary observations on the life history of two polistine wasps, *Polistes snelleni* and *P. biglumis* in Sapporo, northern Japan. J. Fac. Sci. Hokkaido Univ. (Ser. 6) 17: 78–105.

Yamane, Sô. 1972. Life cycle and nest architecture of *Polistes* wasps in the Okushiri Island, northern Japan (Hymenoptera: Vespidae). J. Fac. Sci. Hokkaido Univ. (Ser. 6) 18: 440–459.

Yamane, Sô. 1980. Social biology of the *Parapolybia* wasps in Taiwan. Ph.D. dissertation, Hokkaido University, Sapporo.

Yamane, Sô. 1984. Nest architecture of two Oriental paper wasps, *Parapolybia varia* and *P. nodosa*, with notes on its adaptive significance (Vespidae, Polistinae). Zool. Jahrb. Abt. Syst. Ökol. Geogr. Tiere 111: 119–141.

Yamane, Sô. 1985. Social relations among females in pre- and postemergence colonies of a subtropical paper wasp, *Parapolybia varia* (Hymenoptera: Vespidae). J. Ethol. 3: 27–38.

Yamane, Sô. 1986. The colony cycle of the Sumatran paper wasp *Ropalidia (Icariola) variegata jacobsoni* (Buysson), with reference to the possible occurrence of serial polygyny (Hymenoptera Vespidae). Monit. Zool. Ital. (N.S.) 20: 135–161.

Yamane, Sô., and T. Kawamichi. 1975. Bionomic comparison of *Polistes biglumis* (Hymenoptera: Vespidae) at two different localities in Hokkaido, northern Japan, with reference to its probable adaptation to cold climate. Kontyû 43: 214–232.

Yamane, Sô., and T. Okazawa. 1977. Some biological observations on a paper wasp, *Polistes (Megapolistes) tepidus malayanus* Cameron (Hymenoptera, Vespidae) in New Guinea. Kontyû 45: 283–299.

Yamane, Sô., and Sk. Yamane. 1979. Polistine wasps from Nepal (Hymenoptera, Vespidae). Insecta Matsumurana (N.S.) 15: 1–37.

Yamane, Sô., S. F. Sakagami, and R. Ohgushi. 1983a. Multiple behavioral options in a primitively social wasp, *Parischnogaster mellyi*. Insectes Soc. 30: 412–415.

Yamane, Sô., J. Kojima, and Sk. Yamane. 1983b. Queen/worker size dimorphism in

666 Literature Cited

an Oriental polistine wasp, *Ropalidia montana* Carl (Hymenoptera: Vespidae). Insectes Soc. 30: 416–422.

Yamane, Sô., S. F. Sakagami, R. Ohgushi, N. D. Abbas, and M. Matsuura. 1983c. Life history and social organization in *Parischnogaster mellyi*. Ecological Study on Social Insects in Central Sumatra with Special Reference to Wasps and Bees—Sumatra Nature Study (Entomology), Kanazawa University, pp. 18–24.

Yamane, Sô., Y. Itô, and J. P. Spradbery. 1991. Comb cutting in *Ropalidia plebeiana*: a new process of colony fission in social wasps (Hymenoptera: Vespidae). Insectes Soc. (in press).

Yanega, D. 1988. Social plasticity and early-diapausing females in a primitively social bee. Proc. Natl. Acad. Sci. USA 85: 4374–4377.

Yanega, D. 1989. Caste determination and differential diapause within the first brood of *Halictus rubicundus* in New York (Hymenoptera: Halictidae). Behav. Ecol. Sociobiol. 24: 97–107.

Yoshikawa, K. 1955. A polistine colony usurped by a foreign queen. Ecological studies of *Polistes* wasps, II. Insectes Soc. 2: 255–260.

Yoshikawa, K. 1956. Compound nest experiments in *Polistes fadwigae* Dalla Torre [*sic*]. Ecological studies of *Polistes* wasps. IV. J. Inst. Polytech. Osaka City Univ. (D) 7: 229–243.

Yoshikawa, K. 1957. A brief note on the temporary polygyny in *Polistes fadwigae* Dalla Torre [*sic*], the first discovery in Japan (Ecological studies of *Polistes* wasps, III). Mushi 30: 37–39.

Yoshikawa, K. 1962. Introductory studies on the life economy of polistine wasps. I. Scope of problems and consideration on the solitary stage. Bull. Osaka Mus. Nat. Hist. 15: 3–27.

Yoshikawa, K. 1963. Introductory studies on the life economy of polistine wasps. II. Superindividual stage. 2. Division of labor among workers. Jpn. J. Ecol. 13: 53–57.

Yoshikawa, K., R. Ohgushi, and S. F. Sakagami. 1969. Preliminary report on entomology of the Osaka City University 5th Scientific Expedition to Southeast Asia 1966. With descriptions of two new genera of stenogasterine [*sic*] wasps by J. van der Vecht. Nature Life Southeast Asia 6: 153–182.

Young, A. M. 1986. Natural history notes on the social paper wasp *Polistes erythrocephalus* Laetreille (Hymenoptera, Vespidae, Polistinae) in Costa Rica. J. Kansas Entomol. Soc. 59: 712–722.

Zikán, J. F. 1935. Die sozialen Wespen der Gattung *Mischocyttarus* Saussure, nebst Beschreibung von 27 neuen Arten (Hym. Vespidae). Arch. Inst. Biol. São Paulo 1: 143–203.

Zikán, J. F. 1949. O gênero *Mischocyttarus* Saussure (Hymenoptera, Vespidae), com a descrição de 82 espécies novas. Bol. Parq. Nac. Itatiáia (Rio de Janeiro) 1: 1–251.

Zikán, J. F. 1951. Polymorphismus und Ethologie der sozialen Faltenwespen (Vespidae Diploptera). Acta Zool. Lilloana 11: 5–51.

Zucchi, R., Sô. Yamane, and S. F. Sakagami. 1976. Preliminary notes on the habits of *Trimeria howardi*, a Neotropical communal masarid wasp, with a description of the mature larva (Hymenoptera: Vespoidea). Studies on the vespoid larvae. 2. Insecta Matsumurana (N.S.) 8: 47–57.

# Subject Index

Page numbers in italic type indicate the locations of figures and tables. The abbreviation "def." following a page number denotes that the entry is defined on that page.

Abdominal lobe, 439–440, *439*
Abdominal substance, of stenogastrines, 77–82, 96, 97
Abdominal wagging, 319, 533–534, 556–557
Abundance. *See* Diversity; Numbers, relative
Abundance index, 196–198, *196, 197, 198*
"Accessory burrows," 578–579
Age polyethism. *See* Polyethism
Aggregations, nesting, 457, 459, 530–531, 579–581
Aggression, 88–89, 173–174, 280–281, *280*, 295–296, 320–322, *321, 322*, 327, 333, 407, 418, 530–531, 556–557. *See also* Dominance; Mauling
Alarm recruitment. *See* Recruitment
Allomones, and usurpation, 280–281
Altruism, 451 (def.)
Altruist, 378–379 (def.)
Ant guards, 81–82, *81*, 91–94, *91*, 485, 499, 565–566
Ants: defenses against, 150, 157, 166–170, *168, 190*, 220–223, 392, *393–395*, 485, 493, 498–499, 508; predation by, 57, *190*, 211, *211*, 227, 230–231, 579
Apomorphies, *19–21*, 25–26. *See also* Autapomorphies; Synapomorphies

Associative foundation. *See* Pleometrosis
Autapomorphies, 8 (def.), *9, 14*, 30

Behavior: extranidal vs. intranidal, 246 (def.); larval, 79–80, 170–171, 534–538 (*see also* Trophallaxis); male, 59–62, *60, 66*, 88–90, *88*, 172, 464 (see also Mating); queen, 524, 536. *See also entries for individual behaviors*
Behavioral castes. *See* Caste; Polyethism
Behavioral profiles, *176, 178*
Blot-technique, 518 (def.)
Breeding system, and genetic relatedness, 451–479. *See also* Mating systems; Monogyny; Polygyny
Brood cannibalism, 422, 434, 437–438, *438*
Brood care, 30, 77–82, 96–97, 287, 449, 564–565
Brood maps, 437, *438*

Caste, 22, 24, 30, 31, 175–179, *176, 178*, 215–216, 245–248, 281–284, 293, 345, 443–444, 449–450. *See also* Division of labor; Polyethism; Queen control; Task partitioning
Caste bias, preimaginal, 179–181, 184, 440–441

667

Casteless nest sharing, 27. *See also* Nest sharing: intraspecific
Castration, trophic (alimentary, nutricial, nutritional), 293, 432
Cavity, 518 (def.)
Cell construction, 286, 342, 365, 383–384, 393–395
Character weighting, 15
Chromosome numbers, in *Polistes*, 101, 102
Cladistic analysis: of polistines, 15–21, 16, 17; of vespids, 8–21, 25–29
Cladograms, 25–26; for polistine genera, 18; for stenogastrine genera, 10; for vespid subfamilies, 10; for vespine genera, 13
Classification. *See* Cladistic analysis; Cladograms; Dendrogram; Diversity; Generic relationships; Phylogeny
Cleptoparasitism, 45 (def.), 581
Colony, 208 (def.); polydomous, 528 (def.). *See also* Perennial colonies, in *Vespula*
Colony characteristics: in *Belonogaster, Parapolybia,* and *Mischocyttarus,* 355–356; in independent-founding polistines, 373–374; in *Polistes,* 339–340; in *Ropalidia,* 350–353; in stenogastrines, 346–347
Colony cycle, 208 (def.); factors influencing, 161–162, 210, 421–423, 440–441; and nesting cycle, compared, 207–210, 209; and queen control, 421, 422; in specific taxa, *see* Life history
Colony founding. *See* Nest founding
Colony growth, 82–87, 132–135, 199–206, 274–281, 284–290
Colony phases, 159–161 (def.)
Colony productivity, 212 (def.), 282–284
Colony reproduction, 204–206, 208 (def.)
Colony size: correlates of, 31, 214–231, 227, 291, 403, 405, 423; in independent-founding polistines, 150; in *Polistes,* 104; in stenogastrines, 83–84; in swarm-founding polistines, 214–231; in vespines, 234, 251–252, 269–274, 282–284
Colonymates, 522 (def.)
Comb size, limits to, 528
Comb type: astelocyttarus, 491 (def.), 496 (def.); gymnodomous, 153 (def.); phragmocyttarus, 491 (def.), 496 (def.); sessile, 496 (def.), 496–498, 519 (def.); stelocyttarus, 153 (def.), 491 (def.), 496 (def.)
Communal behaviors, 34 (def.), 52–55, 54, 70–71
Communication, acoustic/vibrational, 249–250, 288–290, 296, 297, 532–535, 556–

557 (*see also* Abdominal wagging); chemical, *see* Glands; Pheromones; Recruitment
Consensus tree, 16, 18
"Contact dominance," 417 (def.). *See also* Dominant-subordinate behaviors; Trophallaxis
Copulation. *See* Mating
Corpora allata, 324, 327, 328, 329, 344
Corrugation, 518 (def.)
Courtship. *See* Mating

Defense, 93–95, 165–170, 289–290, 565–566. *See also* Aggression; Ant guards; Sting
Demographics, of social wasp populations, model for, 444–448
Dendrogram: of Apoidea, 571; of polistine genera, 14–16
Development: of *Polistes,* 127; of solitary and presocial vespids, 45–46; of stenogastrines, 79–80. *See also* Life history
Direct fitness effect, 451 (def.)
Discrimination, nestmate/kin, 95, 183, 295, 326, 459–460, 591–592
Dispersion. *See* Aggregations, nesting
Displacement, cell, 518 (def.)
Distribution: of polistines, 10, 14, 101, 102, 151–153, 152; of stenogastrines, 9, 10, 75; of subfamilies of Vespidae, 10; of vespines, 10, 232, 233, 265
Diversity: of polistines, 10, 14–21, 101, 151–153; of stenogastrines, 9–11, 10, 75; of subfamilies of Vespidae, 10; of vespines, 10, 232, 266
Division of labor, 22–24, 86, 86, 87, 245–248, 284–290, 494–495; in independent-founding polistines, 396–397; in *Microstigmus,* 600; reproductive, 390 (def.), 391–396, 393–395; in stenogastrines, 11, 22, 85–87; in swarm-founding polistines, 400–402; in vespines, 397–400. *See also* Caste; Polyethism; Queen control; Task partitioning
Dominance: confrontational vs. nonconfrontational, 330–331; and nest architecture, 535–538, 537; physiological basis of, 293–294, 296–299, 344, 543–547, 550, 553–557; reproductive, characteristics of, 322–325; and reproductive success, 331–333; reproductive vs. social, 310, 328; role in polyethism, 414–420, 416; social, characteristics of, 328–330. *See also* Pheromones; Queen control
Dominance hierarchies, 11, 31, 159, 318–334, 321, 414–420; in *Belonogaster,* 354–357, 357; in independent-founding polistines, 173–174, 391–392; linear, 318

Dominance hierarchies (*cont.*)
  (def.); in *Microstigmus*, 600, *601*; in *Mischocyttarus*, *355–356*, 358–359; in *Parapolybia*, 356, 358; in *Polistes*, 120–126, 128–129, 337–340, *338, 339–340*, 408–411; in *Polybioides*, 357–360; in *Ropalidia*, 173–174, 349–354; in stenogastrines, 85–87, 344–349; in swarm-founding polistines, 359–360
Dominant-subordinate behaviors, 293–296, 333–334, *341*, 358, 360, 367, 368, 383, 417. *See also* Aggression; Mauling; Submissive behavior, role in evolution of dominance hierarchies

Ecdysteroids, 327–328
Ecotrophobiosis, 431 (def.). *See also* Trophallaxis
Edge-technique, 518 (def.)
Egg eating. *See* Oophagy
"Emery's rule," 279–280
Endocrine activity, 324–325, 327–330, 331, 333. *See also* Corpora allata; Juvenile hormone
Ethocline, 35 (def.)
Eusociality, 7 (def.), 24 (def.), *34* (def.)
—antiquity of, 480–481, 500
—benefits of, 181–185, 372–376
—genetic basis of, 376–380, 451–456, 476–479
—importance of haplodiploidy for origin of, 182–183, 452–455, *453*
—importance of pleometrosis for origin of, 372, 435
—in independent-founding polistines, 181–185
—in *Microstigmus*, 600–601
—models and hypotheses for evolution of, 23–24, 185–187, 376, 377, 477–479; "dominance inhibition hypothesis," 415; "hopeful reproductive" model, 338, 379; nourishment model, generalized, 447–448; "nutritional castration," 432–433, 441; parental manipulation hypothesis, 184 (def.), 379–380, 448; polygynous family hypothesis, 21–29, 380–382, 435; subfertility hypothesis, 184 (def.); "superorganism" concept, 407, 423. *See also* Haplodiploidy; Inclusive fitness; Kin selection; Sociality, stages (levels) of
—predisposing factors for origin of, 69–73, 96, 182–184, 371–372, 376–380, 408, 577–592
—role of ecology in origin of, 183–184
Even-age cohorts, 172 (def.)
Evolutionarily stable strategy, 411

Fecundity, and provisioning method, 51–52

Feeding, 427–430, 436, 440, 443; in independent-founding polistines, 170–171; in stenogastrines, 79–81; in swarm-founding polistines, 204–205; in vespines, 242–245, *244*, 270–271. *See also* Brood care; Foraging; Provisioning
Female choice, in solitary and presocial vespids, 63–64
Fisher's Principle, 67 (def.), 68. *See also* Sex ratios
Food materials, flow in colony, *430*. *See also* Trophallaxis
Food sources, of adult Aculeata, 428–430
Food/nectar storage, 80–81, 170, 224, 524–525
Foraging, 46–47, 392, *393–395*, 402–407, *406*, 415, 420–421, 520–521, 528, 559. *See also* Feeding
Foundation, nest, 519 (def.)
Foundress associations. *See* Pleometrosis

Generations, overlap of, 69, 70, 72–73
Generic relationships: in Polistinae, 14–21; in Stenogastrinae, 9–12; in Vespinae, 12–14
Genetic relatedness, 108, 182–185, 451–479, *453, 454*; electrophoretic estimates of, *461–462*. *See also* Hamilton's rule; Haplodiploidy; Population genetic structure
Genetic structure. *See* Population genetic structure
Glands
—abdominal: Dufour's, 77, 299, 546, 552, *554, 555*, 556, 558, 564; poison, 551–552, *554, 555*, 562; Richard's, 151, 547; silk, 563, 593; sternal, 150, 167, 299, 386–387, 546–553, *546, 548, 549, 554, 555*; tergal, 552, *546, 554, 555*
—cephalic, 542–545, *542, 554, 555*, 563
—function and evolution of, 540–569
—thoracic, 545–546, *546, 555*, 563

Hamilton's rule, 113, 145–148, 378, 451 (def.), 452
Haplodiploidy, 71–72, 372, 452–455, *453*. *See also* Population genetic structure
Haplometrosis, 11 (def.), 82–83, 106, 158, 311, 375, 435 (def.)
Hexadecalactone, 386, 553. *See also* Pheromones: queen
Hibernation, 329–330, 344. *See also* Overwintering, adaptations to
Homeostasis, and colony size, 223–224

Inbreeding, 59–61, 63–64, 67–68, 141, 468, 471, 479

Inclusive fitness, 182–185, 378–379 (def.), 451 (def.)
Independent founding, of nests, 150 (def.), 151, 192 (def.), 193
Indirect fitness effect, 451 (def.), 455
Inquilinism, 279–281, *280. See also* Social parasitism
Intermediate females, 201 (def.)

Juvenile hormone, 324, 327–328, *329*, 344

Kin recognition. *See* Discrimination, nest-mate/kin
Kin selection: exact genetic models of, 452–456; predictions and correlates of, 378–379, 412, 451–479. *See also* Genetic relatedness

Lateral vibrations. *See* Abdominal wagging; Communication: acoustic/vibrational
Latitude: and colony size, 225–227; and seasonal synchrony, 227–229
Life cycle. *See* Colony cycle; Life history; Nesting cycle
Life history: in independent-founding polistines, 149–190, 457–469, 463–464, 467–468; in *Microstigmus*, 592–597; in *Polistes*, 99–148; in social sphecids, 475–476; in solitary and presocial vespids, 33–73; in stenogastrines, 74–98, 456–457; in swarm-founding polistines, 191–231, 471; in vespines, 232–262, 263–305, 471–475
Longevity, and nourishment, 436–437

Maternal manipulation, 174, 423, 587–589. *See also* Queen control
Mating, 89, 524; in *Dolichovespula* and *Vespula*, 302–304; in independent-founding polistines, 139–143, 171–172; in solitary and presocial vespids, 58–66, *66*; in stenogastrines, 88–90; in swarm-founding polistines, 206
Mating systems: in independent-founding polistines, 463–464, 467–468; in stenogastrines, 457; in swarm-founding polistines, 471; in vespines, 474. *See also* Monandry; Polyandry; Polygyny (mating system)
Matrilines, 455 (def.)
Mauling, 295–296, 417, 419
Meconium extraction, 153, 498, 505
Monandry, 456, 476. *See also* Mating systems
Monogyny, 7 (def.), 24, 255 (def.), 460; matrifilial (or long-term monogyny), 28, 29; periodic, 201 (def.); short-term (or

serial polygyny), 28, 29. *See also* Population genetic structure
Mortality, causes of, 51, 55–58, 212, 275–281, 594–595. *See also* Parasites; Predation; Survival
"Mouthing," 563
Multiple-foundress associations. *See* Pleometrosis
Multivoltinism, role in evolution of eusociality, 586–587

Nest architecture, 31–32, 480–519, 520–539; in *Agelaia*, 483; in *Brachygastra*, 526–527; in *Epipona*, 483; in independent-founding polistines, 103, 153–158, *154–155, 156, 157, 159*, 485–491, 508; major forms of, *510–518*; major variations in vespids, *486*; in *Mischocyttarus*, 489–491, *526–527*; in *Parischnogaster*, *526–527*; in *Polistes*, *526–527*; in polistines, 505–507; primary structure, 519 (def.); in *Ropalidia*, *526–527*; secondary structure, 519 (def.); in solitary and presocial vespids, 36–43, *38, 39, 40, 50*; in stenogastrines, 90–93, *91*, 483–485, 508; in swarm-founding polistines, 199–200, 222–223, 508; in *Synoeca*, *526–527*; terminology, traditional, 491 (def.); in *Vespa*, 257, *526–527*; in vespines, 503–504; in *Vespula* and *Dolichovespula*, 267, 268–270, *268, 273, 301. See also* Nest construction behaviors; Nests: queen
Nest construction behaviors, 36–43, 157–158, *157*, 284–287, 421, 562–564, *585*, 593–594. *See also* Nest architecture; Nest founding; Nests: queen
Nest cycle. *See* Nesting cycle
Nest envelope, 498–500, 519 (def.)
Nest founding: associative, 83, 315–318 (*see also* Pleometrosis); correlates of mode of, 191–199, *194, 195*, 210–214; in *Microstigmus*, 592–597; in *Polistes*, 106–127, 408–411; in *Ropalidia*, 349, 354; solitary, *see* Haplometrosis; swarm vs. independent, 191–199, 392–396; in vespines, 396
Nest growth, deterministic, 503. *See also* Colony growth
Nest guarding, 167–168, 582–583, *583*
Nest material, 40–43, 96–97, 236–237, 402–403, 483, 492–495, 500–502, 528, 584; mud, 40–41; paper, 500–502
Nest ontogeny, 501–506, *509*
Nest pedicel, 495–496, 505, 519 (def.); laterinidal, 491–492 (def.); rectinidal, 491–492 (def.)
Nest relocation, 240–242, *241*
Nest "renting," 37, *38, 39*, 42

Nest sharing: interspecific, 11, 27, 30, 83, 166, 222, 499, 576; intraspecific, 11, 27, 30, 573–592, *574–575*

Nest structure: and communication, 532–535; and social evolution, 584–585

Nest switching, in *Polistes*, 314, 530

Nest type: calyptodomous, 491 (def.); gymnodomous, 491 (def.)

Nest usurpation, 53, 57–58, 87, 235–236, 276–281, 312–313, 348–349, 463, 580–581, 596–597. *See also* Social parasitism

Nesting, group. *See* Pleonetrosis

Nesting biology. *See* Colony cycle; Life history; Nesting cycle

Nesting cycle, 208 (def.), *209*; and colony cycle compared, 207–210, *209*; determinate, 210 (def.), *209*; factors influencing, 225–229, 234–235; in independent-founding polistines, 158–165, *162*; indeterminate, 164–165, *209*, 210 (def.), 229–231; in *Polistes*, 103–104; in solitary and presocial wasps, 36–37, 41–43, 520–522; in swarm-founding polistines, 229–231; in vespines, 236–242, 258–260, *259*

Nestmate, 522 (def.)

Nestmate recognition. *See* Discrimination, nestmate/kin

Nests, 208 (def.), 522 (def.); arboreal, 518 (def.); fidelity to, 529–532; as locus of social life, 520–539; queen, 238–240, *238*, *239*, 274–275; satellite (or combs), 131, 313–314, 531; secondary functions of, 521–525, *523*

Nocturnal habits, of *Provespa*, 257–258

Numbers, relative, of social wasps, 193–199, *194*, *196*, *197*, *198*. *See also* Diversity

Nutrition, larval, 48, 293, 426–450. *See also* Trophallaxis

Ontogenetic worker, 432 (def.), 433–444, 447

Oogenesis, inhibition of, 324–325

Oophagy, 89, 293–294, 316–317, *393–395*, 419, 430, 463, 536, 556; differential, 323 (def.), 342, *343*, 383; nutritional, 323–324

Ovarian development, 299–300, 327, 338, 391–392, 412–413, 577

Overwintering, adaptations to, 24, 29, 586–587. *See also* Hibernation

Oviposition, 44–45, 78–79, 204, *393–395*; in solitary and presocial vespids, 44–45; in stenogastrines, 78–79; in swarm-founding polistines, 204

Parasites: of independent-founding polistines, 588; of *Microstigmus*, 595–596,

595; of polistines, 165–170, *188–189*, 222–223, 505–506; of solitary and presocial vespids, 50, 55–57; of stenogastrines, 94–95, 506

Parental manipulation. *See* Maternal manipulation

Patrilines, 455 (def.)

Perennial colonies, in *Vespula*, 271–274, 317–318, 362–363

Petiole, abdominal, 428, *429*

Pheromones, 192, 286–287, 319, 365, 419, 553–568, *566*; alarm, 290, 532–533, 561–562; in polistines, 172, 199, 216–220, *219*; queen, 291, 296, 299, 303, 366, 369, 370, 384–387, 419, 532–533, 537–538, 553–556, 564; recognition of, 564; in *Vespa*, 248–249

Philopatry, 326, 334, 459 (def.), 465, 581–582

Phoresy, 56–57

Phylogeny: of *Polistes*, 101; of polistines, 14–21; of sphecids, 570–573, *571*; of stenogastrines, 9–11, *10*, 75, 95–97; of vespids, 8–21, *10*; of vespines, 12–14, *13*. *See also* Cladistic analysis; Cladograms; Diversity

Pleometrosis, *83*, 106, 158, 315–318, 360–362, *361*, *373*, 375, 435

Pollen foraging: oligolectic, 47 (def.), 62; polylectic, 47 (def.)

Polyandry, 464, 468, 474. *See also* Mating systems

Polybiinae, 14

Polybiini, 14, 153

Polydomy, 531 (def.)

Polyethism, 215–216, 247–248, 389 (def.), 390–425, 524; worker, 390 (def.), 396–402, *398*, *401*, 413–414, *413*

Polygyny (mating system): "lek," 89, 140, 468; "resource-defense," 139; "scramble competition," 140, 468. *See also* Mating systems

Polygyny (queen number), 7 (def.), 24, 27, 28, 30, 32, 183, 200, 255–256, 271–274, 336–388, 451–479; long-term, 28, 211–212; secondary, in Vespinae, 271–274, 362–364, *363*, 473; serial (or short-term monogyny), 28, 29, 359, 463. *See also* Queen supersedure; Swarm founding, of nests

Polymorphism. *See* Caste

Population dynamics, in *Vespa*, 250–254

Population genetic structure, 451–479; in independent-founding polistines, 457–469; role in social evolution, 451–455, 476–479; in social Sphecidae, 475–476, 534–535; in stenogastrines, 456–457; in vespines, 471–475

Population structure, and sex ratio, 66–69
Predation: on independent-founding polistines, 165–170, *190*; on *Microstigmus*, 594–596, *595*; role in evolution of eusociality, 578–580; by social wasps, *see* Feeding; on solitary and presocial vespids, 55–57; on stenogastrines, 93–95; on swarm-founding polistines, 221–223, *221, 225, 227. See also* Ants: predation by
"Prevalence," as measure of relative dominance, 196–198, *198. See also* Diversity
Principal components analysis, *176, 178*
Productivity, of haplometrotic and pleometrotic colonies, 375, *375*
Provisioning, 51, 58, 436; delayed, 34 (def.), 49 (def.), 50–52; fully progressive, 49 (def.); mass, 33 (def.), 44, 47–48, 70, 589 (def.), 590; progressive, 30, 33 (def.), 44, 49–52, 70, 71, 589 (def.), 590; truncated progressive, 49 (def.). *See also* Feeding; Foraging
Pseudenvelope, 92, 93

Queen control, 31, 279, 248–250, 290–302, *291*, 336–388, 418–420, 448, 449, 553–557. *See also* Dominance hierarchies; Pheromones: queen
Queen number, 200–204, 336–388, 460
Queen recognition, 385
Queen removal, 365–366, 530. *See also* Split colony experiments
Queen supersedure, 84, 129–132, 365–366

Receptivity, female, 63, 65, 66
Reciprocal altruism, 185
Recognition. *See* Discrimination, nestmate/kin
Recruitment, 216–217, *217*, 534–535, 558–562
Releasers, behavioral, 420–421
Reproduction, direct, 331 (def.), 332
Reproductive behavior. *See* Breeding system; Mating
Reproductive competition, 309–335
Reproductive dominance. *See* Dominance; Queen control
"Reproductive efficiency," 215
Reproductive fitness, Schaal-Leverich model of, 445–446, 449
Reproductive potential, and supplemental feeding, 440
Reproductive swarms, 208 (def.)
Reproductive value, 409–414, *409, 413, 423, 424. See also* Inclusive fitness
Reproductives, production of, 205, 251–252, *254*, 281, 300–302, 414, 422, 473–474, 599 (*see also* Caste); removal of,

530–531 (*see also* Queen removal; Split colony experiments)
Requeening, and relatedness in polygynous societies, 470–471
"Resting circle." *See* Royal court
Royal court, 248–249, *248*, 260–261, *261*, 298, 302, 364–365, 385–386

Saliva, larval, 437–440, 447–448. *See also* Trophallaxis
Sanitation, in *Dolichovespula* and *Vespula* nests, 285
Seasonality, in social wasps, 227–229, 444–446, *444*
Secretion, glandular, 493 (def.), 519 (def.); and nest construction, 492–495, *492*, 509
Sex ratios, 63, 66–69, *68, 449, 453*, 455, 587–588, 598–599, *599*
Sexual dimorphism, 66
Sibling mating. *See* Inbreeding
Silk production, 563, 593
Social parasitism, 235–236, 278–281, *280*, 367, 557–558, 58l. *See also* Nest usurpation
Sociality, stages (levels) of, 22–24, *34* (def.), 185–186; casteless polygynous family group, 22 (def.); communal, 34 (def.); highly eusocial stage, 35 (def.); highly social stage, 24 (def.); parasocial level, *34* (def.), 97, 432, 433; presocial level, 33 (def.); primitively eusocial level, 22 (def.), 35 (def.); primitively social level, 22 (def.), *34* (def.); quasisocial level, 34 (def.), 52–55; rudimentary-caste-containing stage, 23 (def.), 97; semisocial level, 24, *34* (def.); solitary nesting stage, 22 (def.), 33 (def.); subsocial level, 33 (def.), 432–435
Species diversity. *See* Diversity; Numbers, relative
Sperm competition, 59, 62
Split colony experiments, 367, 369–370, 555
Stigmergic responses, 103, 532 (def.)
Sting, 168–170, 427, 435. *See also* Venom
Submissive behavior, role in evolution of dominance hierarchies, 330, 333–334
Subordinate worker, 433 (def.)
Supernumerary brood, 274 (def.)
Supersedure. *See* Queen supersedure
Survival, 51, 252, 372–376, *373–375. See also* Mortality, causes of
Survivorship curve, 250–251, 444–446, *445, 446*
Swarm, 208 (def.); types of, 206–207
Swarm founding, of nests, 26, 30, 32, 150 (def.), 151, 192 (def.), *211*, 217–218, *219*, 258, 317, 531, 560–561

Symplesiomorphies, 14 (def.)
Synapomorphies, 8 (def.), 9, 13, 32
Systematics. *See* Cladistic analysis; Clado-
grams; Diversity; Phylogeny

Tail wagging. *See* Abdominal wagging
Task partitioning, 392–396, *393–395*, 402–
403, *402, 404–405*, 411–413, *411, 413*,
425. *See also* Caste; Division of labor;
Polyethism
Thermoregulation, 287, 420, 500
Time activity budgets, 176–178. *See also*
Caste
Trophallaxis, 79–80, 170–171, 181, 241,

260, 288–289, 319–320, 338, *370*, 417–
418, 428, *430*, 431 (def.), 432–433, 434,
437–440, 447–448, 538, 544, 546

Usurpation. *See* Nest usurpation

Variability, individual, 403, 406–407, *406*
Venom, 551–552, 562, 568. *See also* Sting

Worker oviposition, 128–132, 293, 412,
466–467, 471, 473–474

Zethinae, 9

# Taxonomic Index

This index contains the arthropod genera mentioned in the text. Page numbers in italic type indicate the locations of figures and tables.

*Abispa*, 37, 62–65
*Agelaia*, 19, 21, *194, 196, 197, 198,* 199, 201, 202, 205–207, 212, 215, 218, 220, 222, 224, 225, 227, 229, 230, 360, 385, 400, 417, *462,* 470, 471, *482, 486, 492,* 493, 497, 498, 500, 501, 507, *513,* 547, *548,* 550, 559
*Allomerus*, 166
*Allorhynchium*, 37, 66
*Ammophila*, *574,* 590
*Anacamptomyia*, 188
*Ancistroceroides*, 37, *39*
*Ancistrocerus*, 37, 60, 62, 63, *68*
*Angiopolybia*, 19, *194, 196, 197, 198,* 199, 202, 225, *486, 492,* 493, 506, 507, *514,* 547, *548,* 559, 560
*Anischnogaster*, 27, 75, *91, 92, 345, 346,* 483, 506, *510*
*Antepipona*, 37, 52
*Anterhynchium*, 37, 51
*Anthophora*, 79
*Apis*, 218, 220, 243, 412, 440, 498, *523,* 535, *555*
*Apoica*, 19, 29, *194, 196, 197, 198,* 202, 215, 217, 224, 257, *486, 492,* 493, 497, 508, *512,* 543, *548,* 550, 567
*Arpactophilus*, 475, 542, 573, *574,* 577, *583, 584,* 590, 591, *593*
*Arthula*, 189

*Asteloeca*, 15, *194, 486, 492, 494,* 506, *515*
*Azteca*, 22, 230, 231

*Bakeronymus*, 188
*Belonogaster*, 3, 19, 25, 29, 149–194, *152, 154, 160, 188, 194,* 315, 319, 343, *354–357, 355, 373, 379, 381, 382, 393, 397,* 458–464, 485, *486, 492,* 493, *505,* 511, 547, *549,* 550, 565–567
*Bembix*, 51, *575,* 579, 580, 590
*Bombus*, 367, *523*
*Brachygastra*, 19–21, *194, 196, 197, 198,* 202, 210, 212, 218, 220, 224, 471, *486, 492, 494, 495,* 501, 506, *515, 525, 526, 548,* 550, 551, 567

*Calligaster*, 37, 40, 50, 51, 53, 72, 483, 522
*Camponotus*, 190, 221
*Camptotypus*, 188
*Carinostigmus*, *574,* 577
*Ceramius*, 37, 52
*Cerceris*, *462,* 475, 573, *575,* 576, 578, 581, 585, 587, 588, 591, 598
*Chalcoela*, 111, 188, 223
*Chalybion*, 579
*Chartergellus*, 20, *194, 196, 197, 198,* 202, 227, *486, 492,* 493, 507, *514, 548,* 560
*Charterginus*, 15, 20, *194, 197, 486, 492, 494,* 502, 503, *517, 548*

*Chartergus*, 20, 21, *194, 196, 197, 198, 202,*
    222, 230, *486, 492,* 494, 495, 501, *517,*
    *548,* 567
*Chrysis,* 55
*Clypearia,* 15, 20, *194, 197, 202, 486, 492,*
    494, 503, 506, *515, 548*
*Crabro, 574,* 576, 580, 581
*Crematogaster,* 57
*Crossocerus, 574*

*Dalara, 574*
*Dasyproctus, 574*
*Delta, 37,* 40, 44, 46, 48, 56, 66
*Dicranorhina, 574,* 582
*Diodontus, 574,* 577
*Dolichovespula,* 12, 191, 232, 234–237, 250,
    252, 261, 263–305, 312, 313, 326, 366,
    367, 385, 396, 398, 399, 472–474, *486,*
    488, *492,* 495, 500, 501, 504, *518,* 519,
    543–545, 547, *549,* 550, 552, 553, 558,
    564

*Eciton,* 221, 222, 498, 596
*Ectemnius, 574*
*Elasmus,* 111
*Entomognathus, 574*
*Epipona,* 20, *194, 198,* 206, 210, 222, 228,
    230, *482, 486, 492,* 495, 503, 517, *548*
*Epsilon, 37,* 60, 61
*Eucerceris, 575*
*Eumenes, 37, 39, 41,* 483
*Euodynerus, 37, 38, 39,* 55, 58, 60, 63, 65–
    68, *66, 68*
*Euparagia, 37,* 63
*Eustala,* 597
*Eustenogaster,* 11, 12, 27, 74–83, 87, 90–97,
    *91, 346,* 348, *404,* 485, 499, *510*
*Exoneura,* 478

*Formica,* 523, 530

*Gorytes,* 51

*Halictus,* 449, 581
*Hemipimpla,* 189
*Heterospilus,* 595–596, *595*
*Holischnogaster,* 9, 77
*Hoplostomus,* 190

*Isodontia,* 576

*Koralliomyia,* 189

*Lasioglossum,* 478
*Leipomeles,* 20, *194, 197, 202,* 222, 223, *486*
    *492, 493,* 499, 502, 507, *514, 548,* 559,
    560
*Leucodynerus,* 37

*Lindenius, 575*
*Liostenogaster,* 11, 27, 75–78, 80, 82–85, 87,
    89–98, *91,* 315, *346, 404,* 483–485, 497,
    *510*
*Liris, 574*

*Macrotermes,* 523
*Marimbonda,* 19, *194, 196, 197, 198, 202,*
    225, *486, 492,* 497, *514*
*Megachile,* 588
*Megaselia,* 188
*Melittobia,* 44, 56, 579, 582
*Mellinus, 575*
*Metapolybia,* 15, 20, *194, 196, 197, 198,*
    199–207, *202,* 211, 212, 215, 218, 220,
    222, 223, 228, 229, 317, 319, 323, 334,
    *341,* 360, 383–385, 400, 406, 417, 418,
    469, 470, *486, 492,* 496, 498, 500, 506,
    *515,* 530, *548,* 556, 559, 567
*Metischnogaster,* 11, 27, 75, 76, 88, 89, *91,*
    93, 94, 97, 485, *510*
*Metopia, 578*
*Microstigmus, 462,* 475, 476, 478, 542, *554,*
    563, 565, *574,* 577, 584, *585,* 588, 590–
    601, *595, 600, 601*
*Mictyris,* 521
*Mischocyttarus,* 19, 24, 29, 149–190, *152,*
    *154, 160, 163, 168, 188, 190,* 192–198,
    *194, 196, 197, 198,* 315, 316, 319, 321–
    323, 326, 332, *341,* 355–359, *355–356,*
    373, 381, 392, *393, 404,* 406, 407, 415,
    416, 421, 434, 436, 437, 439, *439,* 443,
    458–464, *461,* 478, 481, 485, *486,* 487–
    498, *492,* 505, 506, 508, *511, 526,* 534,
    536, 543, 544, 547, *549,* 550, 557, 561,
    565, 567, 568, *595, 596, 600, 601*
*Moniaecera, 575*
*Monobia, 37,* 65
*Monodontomerus,* 188
*Monomorium,* 190
*Montezumia,* 30, *37,* 53
*Myrmarachne,* 597

*Nectarinella,* 19, *194, 197,* 222, *486, 492,*
    493, 497, 499, 502, 503, 506, *514, 548,*
    560
*Nothomyrmecia,* 426

*Occipitalia,* 15, 20, *194, 197, 202,* 211, 222,
    *486, 492,* 494, 506, *515,* 547, *548*
*Odynerus, 37,* 41, 52, 65
*Oecophylla,* 523, 532
*Orancistrocerus, 37, 38,* 51
*Oreumenes, 37,* 41
*Osprynchotus,* 50
*Oxybelus,* 62, 586

*Pachodynerus, 37,* 41, 69
*Parachartergus,* 19, *194, 198, 202, 206, 462, 470, 486, 492,* 494, 496, 499, 501, 502, 506, 507, *514, 548,* 559, 560
*Parachilus, 37,* 47, 57, 66
*Paralastor, 37,* 61, 63
*Paraleptomenes, 37,* 51, 60, 63, 65, 67
*Parancistrocerus, 37,* 65
*Parapolybia,* 19, 29, 149–190, *152, 155,* 160, *162, 163, 188, 190,* 192, 194, 234, 243, *356,* 358, *373, 381, 394,* 458–464, 485–488, *486, 492,* 493, 496–498, 505, *511,* 519, 547, 549, 550, 565, 567
*Pararrhynchium, 37,* 51
*Paravespula,* 264
*Pareumenes, 37,* 44
*Parischnogaster* 9, 11, 74–98, *76, 77, 78, 81, 91,* 310, 315, 323, 345–349, *346–347, 348, 381, 393,* 457, 497, 499, 506, *510, 526,* 536, 568
*Pediobius,* 188
*Pemphredon,* 576
*Pernis,* 252
*Pheidole,* 166
*Philanthus, 575*
*Pison,* 62
*Polistes,* 9, 11, 14, 24, 25, 27, *37,* 41, 87, 95, 97, 99–148, *102, 105, 115, 116, 117, 133, 137, 142,* 151, 192–198, *194, 196, 197, 198,* 207, 223, 234, 243, 299, 310, 313–344, *321, 322, 328, 329, 338, 339– 340, 343,* 359, *373–374,* 375, 380–385, 390, 392, *394–395,* 396, 397, *404,* 407, 408, 414–423, 432–444, *438, 444,* 458, 461, 465–469, 478, 481, 485–488, *486, 492,* 493, 496, 497, 499, 503, 505, 508, *511,* 525, *526,* 530–538, 543–551, *549,* 556–558, *559,* 561, 562, 564–568, *566,* 591, 593
*Polybia,* 15, 19, 20, 21, 29, 166, 171, 194– 199, *194, 196, 197, 198, 202,* 204–230, *209, 217, 219, 221, 226,* 360, 385, 400– 407, *401, 404, 405, 406,* 420–421, *430, 462, 470,* 471, *486, 492,* 494–507, *515, 516,* 525, 533, 543, 547, *548,* 556, 559, 560, 562, 567
*Polybioides,* 19, 29, 193, *194,* 199, 200, *486, 492,* 493, 497–501, 505, *512, 548,* 560
*Protonectarina,* 20, *194, 197,* 220, 224, *486, 492,* 494, 502, *516,* 525
*Protopolybia,* 20, 21, 194–199, *194, 196, 197, 198, 203, 204,* 207, 212, 217, 220, 221, 224, 228, 383–385, 400, 406, 417, 471, *486, 492,* 494, 497–499, 501, *515,* 529, 543, *548,* 560
*Provespa,* 12, 192, 232–262, *233, 259, 261,* 364–366, 472, *486, 492,* 495, 496, 501, 503, *517*

*Psenulus,* 593
*Pseudochartergus,* 15, 20, *194, 197,* 200, 399, *486, 492,* 494, 498, 500, *515, 548*
*Pseudomasaris, 37,* 41, 60–62, 64
*Pseudomyrmex,* 499
*Pseudonomadina,* 189
*Pseudopolybia,* 15, 19, *194, 196, 197, 198, 203, 486, 492,* 493, 495, 501, 507, *514, 547, 548*
*Psithyrus,* 367
*Pterocheilus, 37,* 41, 42

*Rhopalum, 575*
*Ropalidia,* 14, 19, 25, 29, 97, 149–190, *152, 155, 156, 157, 159,* 160, 165, 169, 176, 178, 189, *190,* 192–194, *194, 198, 199, 203,* 225, 243, 313–315, 319, 321, 326, 332, 343, 349–354, *350–353,* 372, 374, 379, 381, 383, *395, 404,* 415, 419, 441, 443, 458–464, 469, 485–488, *486, 492,* 493, 496–501, 505, 506, 508, *511, 512, 526,* 530, 531, 548–551, *548, 549,* 560, 565, 567
*Rubrica, 579*

*Sceliphron, 573, 579, 582*
*Sericophorus, 574*
*Solenopsis,* 221, *523,* 596
*Sphecius, 575,* 591
*Sphecophaga,* 275, 283
*Sphex, 574, 576, 578*
*Spilomena, 574,* 584, 590, 591, 593
*Stenodynerus, 37,* 41, 52, 57, 58, 61, 63
*Stenogaster,* 22, 27, 74, 75, 80, 82, 83, 85, 88, 91, 92, 345, 347, 483, *510*
*Stigmus, 574*
*Sulcopolistes,* 14
*Synagris,* 3, *37,* 49, *50,* 60, 66, 72
*Synoeca,* 15, 20, *194, 196, 197, 198, 199, 203, 206,* 216, 220, 222, 230, *486, 492,* 494–496, 502, *526,* 543, *548,* 559
*Synoecoides,* 21, *194, 486, 492,* 495, *515, 517*
*Syntomosphyrum,* 95

*Tachytes, 574*
*Tapinoma,* 597
*Technomyrmex,* 94
*Trachelas,* 597
*Trachypus, 575*
*Trichokaleva, 595*
*Trigona, 571*
*Trigonopsis, 581, 586*
*Trimeria, 37,* 53, 62
*Trypoxylon,* 35, 62, 521, *572, 574,* 588, 591, 597

*Vespa*, 28, 110, 164, 166, *190*, 192, 232–
  262, 298, 302, 305, 312–314, 317, 360,
  *361*, 364–366, 381, 382, 385, *395*, 396,
  397, *404*, 414, 472, 474, 485, *486*, *492*,
  495, 497, 503, *517*, 533, 543, 544, *549*,
  550, 553, 559, 561, 564, 567
*Vespula*, 26, 28, *37*, 191, 232–237, *233*, *236*,
  *238*, *239*, *241*, 243, 244, *248*, 250, 252,
  *254*, 261, 263–305, 312, 313, 317, 344,
  360–365, *361*, *363*, 367–371, 381, 386,

387, *395*, 398, 399, 403, *405*, 417, 419,
420, *429*, 437, 441, *462*, 472–475, *486*,
488, *492*, 496, 497, 500, 501, 504, *518*,
543–545, 547, *549*, 550, 552, 553, *554*,
555, 558, 559, 561, 562, 564, 567

*Xenorhynchium*, 30, *37*, 41, 53

*Zeta*, *37*, 44, 51
*Zethus*, 30, *37*, 41, 44, 53, *54*, 55, 72, 483

# Challenge
# Exam

# Introduction

One of the least attractive aspects of pursuing an education is the necessity of being examined on what has been learned. Instructors do not like to prepare tests, and students do not like to take them.

However, students are required to take many examinations during their learning careers, and little if any time is spent acquainting them with the positive aspects of tests and with systematic and successful methods for approaching them. Students perceive tests as punitive and sometimes feel that they are merely opportunities for the instructor to discover what the student has forgotten or has never learned. Students need to view tests as opportunities to display their knowledge and to use them as tools for developing prescriptions for further study and learning.

A brief history and discussion of the National Board of Medical Examiners (NBME) examinations (i.e., Parts I, II, and III and FLEX) are presented in this preface, along with ideas concerning psychological preparation for the examinations. Also presented are general considerations and test-taking tips as well as how practice exams can be used as educational tools. (The literature provided by the various examination boards contains detailed information concerning the construction and scoring of specific exams.)

## National Board of Medical Examiners Examinations

Before the various NBME exams were developed, each state attempted to license physicians through its own procedures. Differences between the quality and testing procedures of the various state examinations resulted in the refusal of some states to recognize the licensure of physicians licensed in other states. This made it difficult for physicians to move freely from one state to another and produced an uneven quality of medical care in the United States.

To remedy this situation, the various state medical boards decided they would be better served if an outside agency prepared standard exams to be given in all states, allowing each state to meet its own needs and have a common standard by which to judge the educational preparation of individuals applying for licensure.

One misconception concerning these outside agencies is that they are licensing authorities. This is not the case; they are examination boards only. The individual states retain the power to grant and revoke licenses. The examination boards are charged with designing and scoring valid and reliable tests. They are primarily concerned with providing the states with feedback on how examinees have performed and with making suggestions about the interpretation and usefulness of scores. The states use this information as partial fulfillment of qualifications upon which they grant licenses.

Students should remember that these exams are administered nationwide and, although the general medical information is similar, educational methodologies and faculty areas of expertise differ from institution to institution. It is unrealistic to expect that students will know all the material presented in the exams; they may face questions on the exams in areas that were only superficially covered in their classes. The testing authorities recognize this situation, and their scoring procedures take it into account.

### Scoring the Exams

The diversity of curriculum necessitates that these tests be scored using a criteria-based normal curve. An individual score is based not only on how many questions were answered correctly by a specific student but also on how this one performance relates to the distribution of all scores of the criteria group. In the case of NBME, Part I, the criteria group consists of those students who have completed 2 years of medical training in the United States and are taking the test for the first time and those students who took the test during the previous four June sittings.

Since this test has been constructed to measure a wide range of educational situations, the mean, or average, score generally can be achieved by answering 64% to 68% of the questions correctly. Passing the exam requires answering correctly 55% to 60% of the questions. The competition for acceptance into medical school and the performance levels necessary to stay in school are so high that many students who have always achieved these high levels naturally assume they must perform in a similar fashion and attain equivalent scores on the NBME exams. This is not the case. In fact, among students who are accustomed to performing at levels exceeding 80% to 90%, fewer than 4% taking these tests perform at that high level. Unrealistically high personal expectations leave students psychologically unprepared for these tests, and the anxiety of the moment renders them incapable of doing their best work.

Actually, **most students have learned quite well**, but they fail to display this learning when they are tested because they do not understand the construction, purpose, or scoring procedures of board exams. It is imperative that they understand that they are **not** expected to score as well as they have in the past and that the measurement criteria is group performance, not only individual performance.

While preparing for an exam, it is important that students learn as much as they can about the subject they will be tested on as well as prepare to discover just how much they may not know. Students should study to acquire knowledge, not just to prepare for tests. **For the well-prepared candidate, the chances of passing far exceed the chances of failing.**

### Materials Needed for Test Preparation

In preparation for a test, many students collect far too much study material only to find that they simply do not have the time to go through all of it. They are defeated before they begin because either they cannot get through all the material leaving areas unstudied, or they race through the material so quickly that they cannot benefit from the activity.

It is generally more efficient for the student to use materials already at hand; that is, class notes, one good outline to cover or strengthen areas not locally stressed and for quick review of the whole topic, and one good text as a reference for looking up complex material needing further explanation.

Also, many students attempt to memorize far too much information, rather than learning and understanding less material and then relying on that learned information to determine the answers to questions at the time of the examination. Relying too heavily on memorized material causes anxiety, and the more anxious students become during a test, the less learned knowledge they are likely to use.

## Positive Attitude

A positive attitude and a realistic approach are essential to successful test taking. If concentration is placed on the negative aspects of tests or on the potential for failure, anxiety increases and performance decreases. A negative attitude generally develops if the student concentrates on "I must pass" rather than on "I can pass." "What if I fail?" becomes the major factor motivating the student to **run from failure rather than toward success**. This results from placing too much emphasis on scores rather than understanding that scores have only slight relevance to future professional performance.

The score received is only one aspect of test performance. Test performance also indicates the student's ability to use information during evaluation procedures and reveals how this ability might be used in the future. For example, when a patient enters the physician's office with a problem, the physician begins by asking questions, searching for clues, and seeking diagnostic information. Hypotheses are then developed, which will include several potential causes for the problem. Weighing the probabilities, the physician will begin to discard those hypotheses with the least likelihood of being correct. Good differential diagnosis involves the ability to deal with uncertainty, to reduce potential causes to the smallest number, and to use all learned information in arriving at a conclusion.

This same thought process can and should be used in testing situations. It might be termed **paper-and-pencil differential diagnosis**. In each question with five alternatives, of which one is correct, there are four alternatives that are incorrect. If deductive reasoning is used, as in solving a clinical problem, the choices can be viewed as having possibilities of being correct. The elimination of wrong choices increases the odds that a student will be able to recognize the correct choice. Even if the correct choice does not become evident, the probability of guessing correctly increases. Just as differential diagnosis in a clinical setting can result in a correct diagnosis, eliminating incorrect choices on a test can result in choosing the correct answer.

Answering questions based on what is incorrect is difficult for many students since they have had nearly 20 years experience taking tests with the implied assertion that knowledge can be displayed only by knowing what is correct. It must be remembered, however, that students can display knowledge by knowing something is wrong, just as they can display it by knowing something is right. **Students should begin to think in the present as they expect themselves to think in the future**.

## Paper-and-Pencil Differential Diagnosis

The technique used to arrive at the answer to the following question is an example of the paper-and-pencil differential diagnosis approach.

A recently diagnosed case of hypothyroidism in a 45-year-old man may result in which of the following conditions?

**(A)** Thyrotoxicosis

**(B)** Cretinism

**(C)** Myxedema

**(D)** Graves' disease

**(E)** Hashimoto's thyroiditis

It is presumed that all of the choices presented in the question are plausible and partially correct. If the student begins by breaking the question into parts and trying to discover what the question is attempting to measure, it will be possible to answer the question correctly by using more than memorized charts concerning thyroid problems.

- The question may be testing if the student knows the difference between "hypo" and "hyper" conditions.
- The answer choices may include thyroid problems that are not "hypothyroid" problems.
- It is possible that one or more of the choices are "hypo" but are not "thyroid" problems, that they are some other endocrine problems.
- "Recently diagnosed in a 45-year-old man" indicates that the correct answer is not a congenital childhood problem.
- "May result in" as opposed to "resulting from" suggests that the choices might include a problem that **causes** hypothyroidism rather than **results from** hypothyroidism, as stated.

By applying this kind of reasoning, the student can see that choice **A**, thyroid toxicosis, which is a disorder resulting from an overactive thyroid gland ("hyper") must be eliminated. Another piece of knowledge, that is, Graves' disease is thyroid toxicosis, eliminates choice **D**. Choice **B**, cretinism, is indeed hypothyroidism, but it is a childhood disorder. Therefore, **B** is eliminated. Choice **E** is an inflammation of the thyroid gland—here the clue is the suffix "itis." The reasoning is that thyroiditis, being an inflammation, may **cause** a thyroid problem, perhaps even a hypothyroid problem, but there is no reason for the reverse to be true. Myxedema, choice **C**, is the only choice left and the obvious correct answer.

## Preparing for Board Examinations

1. **Study for yourself.** Although some of the material may seem irrelevant, the more you learn now, the less you will have to learn later. Also, do not let the fear of the test rob you of an important part of your education. If you study to learn, the task is less distasteful than studying solely to pass a test.

2. **Review all areas.** You should not be selective by studying perceived weak areas and ignoring perceived strong areas. This is probably the last time you will have the time and the motivation to review **all** of the basic sciences.

3. **Attempt to understand, not just to memorize, the material.** Ask yourself: To whom does the material apply? When does it apply? Where does it apply? How does it apply? Understanding the connections among these points allows for longer retention and aids in those situations when guessing strategies may be needed.

4. Try to **anticipate questions that might appear on the test.** Ask yourself how you might construct a question on a specific topic.

5. **Give yourself a couple days of rest before the test.** Studying up to the last moment will increase your anxiety and cause potential confusion.

### Taking Board Examinations

1. In the case of NBME exams, be sure to **pace yourself** to use the time optimally. As soon as you get your test booklet, go through and circle the questions numbered 40, 80, 120, and 160. The test is constructed so that you will have approximately 45 seconds for each question. If you are at a circled number every 30 minutes, you will be right on schedule. A 2-hour test will have 150–170 questions and a 2½-hour test will have approximately 200 questions. You should use all of your allotted time; if you finish too early, you probably did so by moving too quickly through the test.

2. **Read each question and all the alternatives carefully** before you begin to make decisions. Remember the questions contain clues, as do the answer choices. As a physician, you would not make a clinical decision without a complete examination of all the data; the same holds true for answering test questions.

3. **Read the directions for each question set carefully.** You would be amazed at how many students make mistakes in tests simply because they have not paid close attention to the directions.

4. It is not advisable to leave blanks with the intention of coming back to answer the questions later. Because of the way board examinations are constructed, you probably will not pick up any new information that will help you when you come back, and the chances of getting numerically off on your answer sheet are greater than your chances of benefiting by skipping around. If you feel that you must come back to a question, mark the best choice and place a note in the margin. Generally speaking, it is best not to change answers once you have made a decision, unless you have learned new information. Your intuitive reaction and first response are correct more often than changes made out of frustration or anxiety. **Never turn in an answer sheet with blanks**. Scores are based on the number that you get correct; you are not penalized for incorrect choices.

5. **Do not try to answer the questions on a stimulus–response basis.** It generally will not work. Use all of your learned knowledge.

6. **Do not let anxiety destroy your confidence.** If you have prepared conscientiously, you know enough to pass. Use all that you have learned.

7. **Do not try to determine how well you are doing as you proceed.** You will not be able to make an objective assessment, and your anxiety will increase.

8. **Do not expect a feeling of mastery** or anything close to what you are accustomed. Remember, this is a nationally administered exam, not a mastery test.

9. **Do not become frustrated or angry** about what appear to be bad or difficult questions. You simply do not know the answers; you cannot know everything.

### Specific Test-Taking Strategies

Read the entire question carefully, regardless of format. Test questions have multiple parts. Concentrate on picking out the pertinent key words that might help you begin to problem solve. Words such as "always," "all," "never," "mostly," "primarily," and so forth play significant roles. In all types of questions, distractors with terms such as "always" or "never" most often are incorrect. Adjectives and adverbs can completely change the meaning of questions—pay close attention to them. Also, medical prefixes and suffixes (e.g., "hypo-," "hyper-," "-ectomy," "-itis") are sometimes at the root of the question. The knowledge and application of everyday English grammar often is the key to dissecting questions.

### Multiple-Choice Questions

Read the question and the choices carefully to become familiar with the data as given. Remember, in multiple-choice questions there is one correct answer and there are four distractors, or incorrect answers. (Distractors are plausible and possibly correct or they would not be called distractors.) They are generally correct for part of the question but not for the entire question. Dissecting the question into parts aids in discerning these distractors.

If the correct answer is not immediately evident, begin eliminating the distractors. (Many students feel that they must always start at option A and make a decision before they move to B, thus forcing decisions they are not ready to make.) Your first decisions should be made on those choices you feel the most confident about.

Compare the choices to each part of the question. **To be wrong**, a choice needs to be incorrect for only part of the question. **To be correct**, it must be **totally** correct. If you believe a choice is partially incorrect, tentatively eliminate that choice. Make notes next to the choices regarding tentative decisions. One method is to place a minus sign next to the choices you are certain are incorrect and a plus sign next to those that potentially are correct. Finally, place a zero next to any choice you do not understand or need to come back to for further inspection. Do not feel that you must make final decisions until you have examined all choices carefully.

When you have eliminated as many choices as you can, decide which of those that are left has the highest probability of being correct. Remember to use paper-and-pencil differential diagnosis. Above all, be honest with yourself. If you do not know the answer, eliminate as many choices as possible and choose reasonably.

### Multiple True–False Questions

Multiple true–false questions are not as difficult as some students make them. These are the questions in which you must mark:

> **A** if **1, 2, and 3** are correct,
> **B** if **1 and 3** are correct,
> **C** if **2 and 4** are correct,
> **D** if only **4** is correct, or
> **E** if **all** are correct.

Remember that the name for this type of question is multiple true–false and then use this concept. Become familiar with each choice and make notes. Then concentrate on the one choice you feel is definitely incorrect. If you can find one incorrect alternative, you can eliminate three choices immediately and be down to a fifty–fifty probability of guessing the correct answer. In this format, if choice 1 is incorrect, so is choice 3; they go together. Alternatively, if 1 is correct, so is 3. The combinations of alternatives are constant; they will not be mixed. You will not find a situation where choice 1 is correct, but 3 is incorrect.

After eliminating the choices you are sure are incorrect, concentrate on the choice that will make your final decision. For instance, if you discard choice 1, you have eliminated alternatives A, B, and E. This leaves C (2 and 4) and D (4 only). Concentrate on choice 2, and decide if it is true or false. Rereading and concentrating on choice 4 only wastes time; choice 2 will be the decision maker. (Take the path of least resistance and concentrate on the smallest possible number of items while making a decision.) Obviously, if none of the choices is found to be incorrect, the answer is E (all).

## Comparison-Matching Questions

Comparison-matching questions are also easier to address if you concentrate on one alternative at a time. Choose option:

    **A**  if the question is associated with **(A) only**,
    **B**  if the question is associated with **(B) only**,
    **C**  if the question is associated with **both (A) and (B)**, or
    **D**  if the question is associated with **neither (A) nor (B)**.

Here again, the elimination of obvious wrong alternatives helps clear away needless information and can help you make a clearer decision.

## Single Best Answer–Matching Sets

Single best answer–matching sets consist of a list of words or statements followed by several numbered items or statements. Be sure to pay attention to whether the choices can be used more than once, only once, or not at all. Consider each choice individually and carefully. Begin with those with which you are the most familiar. It is important always to break the statements and words into parts, as with all other question formats. **If a choice is only partially correct, then it is incorrect.**

## Guessing

Nothing takes the place of a firm knowledge base, but with little information to work with, even after playing paper-and-pencil differential diagnosis, you may find it necessary to guess at the correct answer. A few simple rules can help increase your guessing accuracy. Always guess consistently if you have no idea what is correct; that is, after eliminating all that you can, make the choice that agrees with your intuition or choose the option closest to the top of the list that has not been eliminated as a potential answer.

When guessing at questions that present with choices in numerical form, you will often find the choices listed in an ascending or descending order. It is generally not wise to guess the first or last alternative, since these are usually extreme values and are most likely incorrect.

## Using the Challenge Exam to Learn

All too often, students do not take full advantage of practice exams. There is a tendency to complete the exam, score it, look up the correct answers to those questions missed, and then forget the entire thing.

In fact, great educational benefits can be derived if students would spend more time using practice tests as learning tools. As mentioned earlier, incorrect choices in test questions are plausible and partially correct or they would not fulfill their purpose as distractors. This means that it is just as beneficial to look up the incorrect choices as the correct choices to discover specifically why they are incorrect. In this way, it is possible to learn better test-taking skills as the subtlety of question construction is uncovered.

Additionally, it is advisable to go back and attempt to restructure each question to see if all the choices can be made correct by modifying the question. By doing this, four times as much will be learned. By all means, look up the right answer and explanation. Then, focus on each of the other choices and ask yourself under what conditions they might be correct? For example, the entire thrust of the sample question concerning hypothyroidism could be altered by changing the first few words to read:

"Hyperthyroidism recently discovered in. . . ."

"Hypothyroidism prenatally occurring in. . . ."

"Hypothyroidism resulting from. . . ."

This question can be used to learn and understand thyroid problems in general, not only to memorize answers to specific questions.

The Challenge Exam that follows contains 180 questions and explanations. Every effort has been made to simulate the types of questions and the degree of question difficulty in the various licensure and qualifying exams (i.e., NBME Parts I, II, and III and FLEX). While taking this exam, the student should attempt to create the testing conditions that might be experienced during actual testing situations. Approximately 1 minute should be allowed for each question, and the entire test should be finished before it is scored.

## Summary

Ideally, examinations are designed to determine how much information students have learned and how that information is used in the successful completion of the examination. Students will be successful if these suggestions are followed:

- Develop a positive attitude and maintain that attitude.
- Be realistic in determining the amount of material you attempt to master and in the score you hope to attain.
- Read the directions for each type of question and the questions themselves closely and follow the directions carefully.
- Guess intelligently and consistently when guessing strategies must be used.
- Bring the paper-and-pencil differential diagnosis approach to each question in the examination.
- Use the test as an opportunity to display your knowledge and as a tool for developing prescriptions for further study and learning.

National Board examinations are not easy. They may be almost impossible for those who have unrealistic expectations or for those who allow misinformation concerning the exams to produce anxiety out of proportion to the task at hand. They are manageable if they are approached with a positive attitude and with consistent use of all the information the student has learned.

Michael J. O'Donnell

# QUESTIONS

**Directions:** Each question below contains five suggested answers. Choose the **one best** response to each question.

1. Endometriosis may involve the presence of all of the following signs or symptoms EXCEPT

(A) cul-de-sac nodularity
(B) endometriomas of the ovaries
(C) endometrial glands and stroma outside the uterine cavity
(D) functioning endometrium within the myometrium
(E) pelvic adhesions

2. A mother brings her 5-year-old daughter to the emergency room and relates a 4-day history of sore throat and a painless but persistent serosanguineous discharge from the child's vagina. Pelvic examination confirms the history, and the saline/potassium chloride preparations are negative. The most likely etiology is

(A) *Candida*
(B) foreign body
(C) *Neisseria gonorrhoeae*
(D) *Streptococcus*
(E) *Trichomonas*

3. All of the following dietary instructions are appropriate for a pregnant woman EXCEPT

(A) restrict salt intake
(B) take 1200 mg of calcium a day
(C) take 800 $\mu$g of folic acid a day
(D) take supplemental iron
(E) gain at least 15 lbs during the pregnancy

4. What is the recurrence risk of trisomy 21 for a 25-year-old woman who has one child with trisomy 21?

(A) 1.5%
(B) 3%
(C) 4.5%
(D) 6%
(E) None of the above

5. All of the following statements about menopausal osteoporosis are true EXCEPT

(A) most bone loss occurs to trabecular bone
(B) bone loss is most rapid after bilateral oophorectomy in a 35-year-old woman
(C) estrogen therapy can retard osteoporosis
(D) osteoporosis is more common in black than in white women
(E) about one-third of American women can expect a hip fracture

6. Which of the following bacteria has been found almost exclusively in women with pelvic inflammatory disease who use an intrauterine device?

(A) *Actinomyces israelii*
(B) *Bacteroides fragilis*
(C) *Chlamydia trachomatis*
(D) *Mycoplasma hominis*
(E) *Neisseria gonorrhoeae*

7. Characteristics of a hydatidiform mole include all of the following EXCEPT

(A) enlargement of the villi
(B) absence of fetal tissue
(C) proliferation of the lining trophoblast
(D) presence of villous blood vessels
(E) edema of the villi

8. A patient presents in her thirty-second week of pregnancy with a below average estimated fetal weight. Her only significant history is smoking over a pack of cigarettes a day. She is worried about fetal movement. Which of the following hormone levels should be measured at this time?

(A) Estriol
(B) Progesterone
(C) Prolactin
(D) Human placental lactogen
(E) Human chorionic gonadotropin

9. A woman who is 16-weeks pregnant has an α-fetoprotein level of 2.8 MOM. All of the following conditions could explain this abnormal finding EXCEPT

(A) anencephaly
(B) Down's syndrome
(C) duodenal atresia
(D) omphalocele
(E) twins

10. A prolonged gestation is associated with which of the following fetal abnormalities?

(A) Meningomyelocele
(B) Spina bifida
(C) Anencephaly
(D) Omphalocele
(E) None of the above

11. Criteria for good prognosis metastatic gestational trophoblastic neoplasia include all of the following EXCEPT

(A) short duration of disease
(B) no prior chemotherapy
(C) pretreatment human chorionic gonadotropin values of less than 40,000 mIU/ml
(D) pregnancy 6 months previously
(E) no brain metastases

12. *Early* signs of premature labor include all of the following EXCEPT

(A) increased vaginal discharge
(B) increase in uterine contractions
(C) low back pain
(D) cervical dilation to 4 cm
(E) worsening pelvic pressure

13. The most accurate method of dating a pregnancy using ultrasound is to obtain

(A) a crown–rump length at 8 weeks gestation
(B) a biparietal diameter at 24 weeks gestation
(C) a femur length at 24 weeks gestation
(D) an abdominal circumference and femur length at 26 weeks gestation
(E) two biparietal diameters at 32 and 36 weeks gestation

14. A 23-year-old woman presents to the physician's office complaining of a mucopurulent vaginal discharge, lower abdominal pain, and a fever, which began towards the end of her menstrual period. Which sexually transmitted disease is she most likely to have?

(A) *Gardnerella vaginalis*
(B) *Chlamydia trachomatis*
(C) *Neisseria gonorrhoeae*
(D) Chancroid
(E) Lymphogranuloma venereum

15. All of the following statements concerning the function of progesterone are correct EXCEPT

(A) it prepares the endometrium for nidation
(B) it relaxes the myometrium
(C) it elevates serum binding proteins
(D) it stimulates aldosterone production
(E) it has natriuretic actions

16. A woman with ruptured membranes is in the active phase of labor and is 5 cm dilated with sustained, deep variable decelerations. The decision is made to perform a cesarean section. All of the following are appropriate intrauterine resuscitative measures that should be performed before the cesarean section EXCEPT

(A) increase the intravenous fluids
(B) place the patient in the supine position
(C) start nasal oxygen
(D) start amnioinfusion
(E) administer subcutaneous terbutaline

17. Hypertension from oral contraceptive use is associated with all of the following EXCEPT

(A) increased plasma renin activity
(B) increased aldosterone secretion
(C) elevated levels of angiotensin
(D) increased renal excretion of sodium
(E) prolonged use (more than 5 years)

18. Which of the following methods of contraception is associated with the lowest incidence of pelvic inflammatory disease?

(A) Condom
(B) Diaphragm
(C) Foam
(D) Intrauterine device
(E) Oral contraceptives

19. All of the following drugs are used to inhibit premature labor EXCEPT

(A) ethanol
(B) magnesium sulfate
(C) phenobarbital
(D) ritodrine
(E) terbutaline

20. A fractional dilatation and curettage for post-menopausal bleeding in a 51-year-old woman reveals a uterine depth of 7 mm, negative endocervical curettings, and a well-differentiated adenocarcinoma. A metastatic workup is negative. Which of the following treatments would be most appropriate for this woman?

(A) Total abdominal hysterectomy and bilateral salpingo-oophorectomy
(B) Radical hysterectomy
(C) Pelvic exenteration
(D) Radiation therapy
(E) Chemotherapy

21. A mediolateral episiotomy is associated with all of the following advantages and disadvantages EXCEPT

(A) more space for delivery
(B) more blood loss
(C) occasional faulty anatomic repair
(D) occasional dyspareunia
(E) anal sphincter tear

22. The advantage of cordocentesis over amniocentesis in evaluating Rh disease is

(A) its lower complication rate
(B) its ability to measure fetal hemoglobin levels directly
(C) the ease with which it is performed
(D) the fact that it does not have to be performed as often
(E) because it is not followed by an anamnestic response as often as after amniocentesis

23. The hot flush that is characteristic of the menopause is thought to be due to

(A) a surge of follicle-stimulating hormone
(B) a surge of luteinizing hormone
(C) an acute drop in estrogen
(D) an acute drop in progesterone
(E) none of the above

24. The biophysical profile for the risk assessment of a postdates fetus consists of all of the following evaluations EXCEPT

(A) fetal breathing
(B) amniotic fluid volume
(C) fetal tone
(D) contraction stress test
(E) fetal motion

25. Which of the following terms best describes menses with a 100 ml blood loss that occur every 35 days?

(A) Menometrorrhagia
(B) Metrorrhagia
(C) Polymenorrhea
(D) Menorrhagia
(E) Oligomenorrhea

26. Correct statements regarding the respiratory distress syndrome include all of the following EXCEPT

(A) there is a deficiency of surfactant in the lung due to immaturity
(B) hypoxia and acidosis stimulate the production of lung phospholipids
(C) it is the most common problem of the preterm neonate
(D) there is a "ground-glass" appearance on x-ray
(E) positive end expiratory pressures are useful treatment modalities

27. A patient begins labor prematurely, and despite all the appropriate measures and medications, a 3 lb infant is born at 32 weeks gestation. The infant has respiratory distress syndrome but eventually is discharged with his mother. At her 6-week postpartum visit, the patient asks about future pregnancies, specifically the risk of recurrence of her premature labor. The physician should respond that the risk is

(A) 0% (no increase)
(B) 5%–15%
(C) 25%–40%
(D) 45%–60%
(E) 65%–80%

28. Motile spermatozoa found on a wet mount of vaginal secretions is indicative of intercourse within the past

(A) 6 hours
(B) 12 hours
(C) 24 hours
(D) 48 hours
(E) 72 hours

29. Prerequisites for a forceps delivery include all of the following EXCEPT

(A) a completely dilated cervix
(B) an empty bladder
(C) the vertex in the occiput anterior position
(D) ruptured membranes
(E) the known position of the vertex

30. After having identified a fetus at risk in prolonged pregnancy, management should consist of

(A) amniocentesis for maturity studies
(B) delivery regardless of the status of the cervix
(C) fetal sampling of scalp pH
(D) measurement of human placental lactogen
(E) repeat antepartum studies in 1 week

31. Falling levels of estriol could be anticipated in all of the following clinical presentations EXCEPT

(A) preeclampsia and eclampsia
(B) Rh isoimmunization
(C) intrauterine growth retardation
(D) maternal renal disease
(E) pregnancy-induced hypertension

32. Which of the following historical features or physical findings confirm the diagnosis of endometriosis?

(A) Bilaterally enlarged ovaries
(B) Cul-de-sac nodularity
(C) Increasingly severe dysmenorrhea
(D) Infertility
(E) None of the above

33. A class II cardiac patient at 38 weeks gestation presents to the hospital in labor with dyspnea on exertion and chest rales. Proper management of this pregnant patient includes all of the following EXCEPT

(A) epidural anesthesia
(B) diuretics
(C) digitalis
(D) oxygen therapy
(E) cesarean delivery

34. The level of human chorionic gonadotropin at which vaginal ultrasound is useful is

(A) 1500–2000 mIU/ml
(B) 2500–3000 mIU/ml
(C) 3500–4000 mIU/ml
(D) 4500–5000 mIU/ml
(E) 5500–6000 mIU/ml

35. Follicle-stimulating hormone (FSH) stimulates all of the following actions EXCEPT

(A) growth and maturation of granulosa cells
(B) aromatase activity
(C) luteinizing hormone (LH) release
(D) increase in FSH receptors on the follicle
(E) development of LH receptors

36. The fetal cardiac output is defined as

(A) output of the foramen ovale and left ventricle
(B) output of the aorta and ductus arteriosus
(C) output from the left ventricle alone
(D) output from the right ventricle alone
(E) being equal to total umbilical blood flow

37. Indications for surgery on a myomatous uterus include all of the following EXCEPT

(A) hypermenorrhea with anemia
(B) 1 year of infertility
(C) uterine enlargement to an 18-week pregnancy
(D) rapid enlargement of the myomas
(E) hydronephrosis

38. A woman reports to the physician's office on June 12, 1988 with a positive pregnancy test. She reports regular 35-day cycles and a last menstrual period that began on April 1, 1988 and ended April 4, 1988. Her estimated due date is

(A) January 1, 1989
(B) January 8, 1989
(C) January 15, 1989
(D) January 22, 1989
(E) January 29, 1989

39. Exclusion of the diagnosis of ectopic pregnancy is supported by

(A) culdocentesis revealing unclotted blood
(B) proliferative endometrium on dilatation and curettage
(C) absence of an intrauterine sac at 6 weeks on pelvic ultrasound
(D) negative urine pregnancy test
(E) negative serum human chorionic gonadotropin β-subunit levels

40. All of the following statements about sexual abuse of children are true EXCEPT

(A) it is underreported
(B) it is seen in all ethnic groups
(C) it occurs among all socioeconomic groups
(D) most sexually victimized children do not know their abuser
(E) no age-group among children is spared

41. Characteristics of active-phase uterine contractions include all of the following EXCEPT

(A) it creates 40 mm Hg of pressure
(B) it causes dilation of the cervix
(C) it causes thickening of the lower uterine segment
(D) it occurs every 2–4 minutes
(E) it lasts for 45 seconds

42. A 30-year-old woman has a 2-year history of infertility. During this time, the diagnosis of moderate endometriosis was made. She decides to have surgery. Requirements of conservative surgery for moderate endometriosis include all of the following EXCEPT

(A) reperitonealization of all raw surface areas
(B) suspension of the uterus
(C) meticulous hemostasis
(D) postoperative danazol
(E) instillation of 3% dextran 70

43. Characteristics of the androgen insensitivity syndrome include all of the following EXCEPT

(A) an XY gonad
(B) a vaginal pouch
(C) breast development
(D) pubic hair
(E) the presence of müllerian-inhibiting factor

44. After birth, the intra-abdominal portion of the umbilical vein becomes the

(A) lateral umbilical ligament
(B) ligamentum teres
(C) urachus
(D) right hepatic ligament
(E) ligamentum venosum

45. Factors that are important in the pathophysiology of pelvic inflammatory disease include all of the following EXCEPT

(A) intrauterine device use
(B) intercourse
(C) menstruation
(D) uterine contractions
(E) uterine fibroids

46. An infertility patient reports that her menses, which have been irregular for 2 years, occur anywhere from 30–50 days apart. She has been unable to conceive for a year. She is a healthy appearing, thin woman who states that she has been competing in distance races for 18 months. Which of the following procedures are most likely to reveal information regarding her infertility?

(A) Laparoscopy
(B) Semen analysis
(C) Basal body temperature record
(D) Postcoital test
(E) Hysteroscopy

47. Complications of a pregnancy in which an intrauterine device has been left in place include all of the following EXCEPT

(A) congenital anomalies
(B) ectopic pregnancy
(C) pelvic infection
(D) prematurity
(E) spontaneous abortion

48. The critical period of organogenesis when the fetus is most susceptible to malformations from teratogenic agents is how many days from the last menstrual period?

(A) $< 5$
(B) 5–17
(C) 7–57
(D) 57–100
(E) 100–200

49. At 28 weeks gestation, amniocentesis reveals a $\Delta OD_{450}$ of 0.20, which is at the top third of zone II on the Liley curve. Appropriate timing for the next amniocentesis would be

(A) 1 day
(B) 1 week
(C) 2 weeks
(D) 3 weeks
(E) 4 weeks

50. At what point is a pelvic examination indicated in suspected placenta previa?

(A) At 27 weeks at the time of hospital admission
(B) After the patient is admitted to the hospital and the bleeding has stopped
(C) At any time in the third trimester when continued vaginal bleeding occurs
(D) Prior to localization of the placenta by ultrasound
(E) Only after the decision has been made to perform a cesarean section

51. A 35-year-old woman who has just had radical endometriosis surgery but takes no estrogen replacement might expect all of the following signs or symptoms EXCEPT

(A) hot flashes
(B) osteoporosis
(C) delayed aging of the cardiovascular system
(D) atrophic vaginitis
(E) decreased libido

52. Cesarean section is indicated in all of the following situations EXCEPT

(A) a woman with active herpes lesions whose membranes ruptured 2 hours prior to admission
(B) a woman who had a positive herpes culture 1 week prior to the onset of labor
(C) a woman who had a negative herpes culture 2 weeks previously but who currently has active herpes lesions and is in labor
(D) a woman with active herpes lesions whose membranes ruptured 24 hours prior to admission
(E) a woman who had a positive herpes culture 10 days previously and who is now in active labor but whose membranes have not ruptured

53. Blood levels of which of the following substances correlate best with the excess androgen state?

(A) Free testosterone
(B) Androstenedione
(C) Total testosterone
(D) Dehydroepiandrosterone sulfate
(E) 17-Hydroxyprogesterone

54. A woman presents with an invasive squamous cancer of the cervix, which extends to the lower third of the vagina. A metastatic workup reveals a right hydronephrosis on intravenous pyelogram. The correct clinical stage of this woman's cancer is most likely to be

(A) state IB
(B) stage IIB
(C) stage IIIA
(D) stage IIIB
(E) stage IVA

55. The appearance of late decelerations would not be a surprise in any of the following situations EXCEPT

(A) intrauterine growth retardation
(B) preeclampsia
(C) chronic abruptio placentae
(D) chronic hypertension
(E) placenta previa

56. A false-normal fetal scalp pH reflects a disparity between predicted good Apgar scores and fetal pH. One could expect a normal scalp pH and Apgar scores of 2 and 4 at 1 and 5 minutes, respectively, in all of the following clinical presentations EXCEPT

(A) prematurity
(B) sedation of the mother
(C) intrauterine growth retardation
(D) terminal abruptio placentae
(E) fetal infection

57. All of the following therapeutic procedures are recommended for ectopic pregnancies EXCEPT

(A) salpingectomy
(B) salpingo-oophorectomy
(C) linear salpingostomy
(D) segmental resection of the portion of the tube containing the ectopic pregnancy
(E) milking the unruptured pregnancy from the tube

58. A 38-year-old Rh-negative woman had amniocentesis at 18 weeks gestation. Because she had a negative indirect Coombs' test, she received 300 $\mu$g of Rh$_o$ (anti-D) immune globulin. Eight weeks later, an indirect Coombs' test was positive at a titer of 1:4. The most likely explanation for this is

(A) not enough immune globulin was administered at the time of amniocentesis
(B) the woman was sensitized after amniocentesis
(C) the titer represents Rh immune globulin that has not been cleared from the maternal circulation
(D) the indirect Coombs' test of 1:4 is false-positive and may be ignored
(E) none of the above

59. When the classical uterine incision is used in cesarean section, the major concern is

(A) rupture of the scar in a subsequent pregnancy and labor
(B) postoperative adhesion formation
(C) discomfort to the patient during healing
(D) injury to the uterine vessels
(E) inability to reach the fetal head for delivery

60. The presence of meconium in amniotic fluid is associated with all of the following conditions EXCEPT

(A) postdates pregnancy
(B) placental insufficiency
(C) polyhydramnios
(D) full-term gestation
(E) intrauterine growth retardation

61. Gestational trophoblastic neoplasia is characterized by all of the following statements EXCEPT

(A) it may follow a normal pregnancy
(B) it is more common in women older than 45 years of age
(C) it develops in 3%–5% of molar pregnancies
(D) less than 50% occur within a year of the preceding pregnancy, mole, or abortion
(E) it may present as abnormal uterine bleeding

62. The advantage of the vaginal cystourethropexy over other surgical approaches in the operative treatment of stress urinary incontinence is that

(A) vaginal surgery is accompanied by fewer risks than abdominal surgery
(B) symptoms of pelvic relaxation can be corrected simultaneously
(C) the cure rate is greater than in retropubic surgery
(D) there is less chance of postoperative urinary retention
(E) a suprapubic catheter is easier to place in the bladder

63. The luteal phase of the menstrual cycle is characterized by

(A) a variable length
(B) growth and development of ovarian follicles
(C) secretion of estrogen
(D) a low basal body temperature
(E) secretion of progesterone

64. All of the following statements concerning uterine myomas are true EXCEPT

(A) malignant degeneration occurs in less than 1% of uterine myomas
(B) myomas can be found in the fallopian tubes and the vagina
(C) myomas rarely appear or grow after menopause
(D) hyaline degeneration is the least common form of myomatous degeneration
(E) although myomas appear encapsulated, no real capsule exists

65. Appropriate forms of management for the preeclamptic patient (blood pressure of 140/95 and 1 + proteinuria) include all of the following EXCEPT

(A) bed rest
(B) contraction stress tests
(C) hospitalization
(D) serial sonography of the fetus
(E) diuretics

**Questions 66 and 67**

Each phase of the menstrual cycle is accompanied by bodily signs and symptoms, the observation of which is necessary for women who practice natural family planning.

66. Which phase of the cycle is characterized by peak estrogen secretion with abundant, clear cervical mucus and a sensation of vulvar wetness?

(A) Relatively infertile phase (phase I)
(B) Fertile phase (phase II)
(C) Absolutely infertile phase (phase III)
(D) Menstrual phase (phase IV)
(E) None of the above

67. Which phase of the cycle is characterized by a rise in the basal body temperature; thick, scant cervical mucus; and the presence of progesterone?

(A) Relatively infertile phase (phase I)
(B) Fertile phase (phase II)
(C) Absolutely infertile phase (phase III)
(D) Menstrual phase (phase IV)
(E) None of the above

68. What is the most accurate method of diagnosing an ectopic pregnancy?

(A) Culdocentesis
(B) Endometrial biopsy
(C) Laparoscopy
(D) Measurement of serial human chorionic gonadotropin levels
(E) Pelvic ultrasound

69. Amenorrhea in a 16-year-old girl may result from all of the following conditions EXCEPT

(A) imperforate hymen
(B) androgen insensitivity syndrome
(C) Turner's syndrome
(D) cystic fibrosis
(E) granulosa–theca cell tumor

70. A woman comes to the office for the first time complaining of mild pinkish vaginal spotting. An ultrasound is performed and a fetus is revealed as well as fluid-filled vesicles in the uterus. The most likely presumptive diagnosis would be

(A) impending abortion
(B) normal pregnancy
(C) hydatidiform mole
(D) twin pregnancy with death of one twin
(E) an ovarian cyst

71. The menopausal vagina has all of the following characteristics EXCEPT

(A) a pale, dry epithelium
(B) a reduction in the size of the upper vagina
(C) the appearance of superficial cells
(D) a loss of vaginal tone
(E) an increase in parabasal cells

72. A 24-year-old primigravida comes for her first obstetric visit. From her menstrual dating and pelvic examination, it is ascertained that she is 10 weeks pregnant. She tells you that 4 weeks ago she had a chest x-ray performed for a possible rib fracture. You advise her that

(A) she has probably endangered the fetus and should abort the pregnancy
(B) her risk of a malformed fetus is 20%
(C) her risk of a malformed fetus is probably not increased over the baseline risk for all pregnancies
(D) it is impossible to cite a risk since the x-ray was probably performed during the critical period of organogenesis
(E) there is no risk to a fetus from an x-ray of the chest area

73. Ovulation is associated with all of the following processes EXCEPT

(A) reduction division in the oocyte
(B) depression of follicle-stimulating hormone
(C) luteinizing hormone surge
(D) prostaglandin synthesis
(E) progesterone secretion

74. A 25-year-old woman presents with 3 days of vaginal spotting. She states that her previous menstrual period was 6 weeks prior to the spotting and that she has had regular cycles prior to that. She is having no pain. Pelvic examination reveals a uterus that is at the upper limit of normal size, and there are no adnexal masses. The physician should

(A) perform a hysteroscopy
(B) order a dilatation and curettage
(C) chart a basal body temperature
(D) measure human chorionic gonadotropin levels
(E) prescribe a progestin, such as medroxyprogesterone

75. Luteinizing hormone (LH) stimulates all of the following actions EXCEPT

(A) luteinization of the granulosa cells
(B) reduction division in the oocyte
(C) progesterone secretion by the corpus luteum
(D) production of LH receptors on the granulosa cells
(E) androgen synthesis by the theca cells

76. Genetic counseling is advisable in each of the following clinical situations EXCEPT for

(A) a 35-year-old woman who plans to begin a family
(B) a single, nonpregnant woman who is a carrier of the Tay-Sachs gene
(C) a 39-year-old divorced woman who has one child with a neural tube defect and would like to have more children
(D) a couple who has two children with unbalanced translocations
(E) a pregnant 21-year-old woman who has no family history of genetic disease

77. The most common causes of maternal mortality include all of the following EXCEPT

(A) premature labor
(B) hemorrhage
(C) hypertension
(D) infection
(E) pulmonary embolism

**Directions:** Each question below contains four suggested answers of which **one or more** is correct. Choose the answer

**A**  if **1, 2, and 3** are correct
**B**  if **1 and 3** are correct
**C**  if **2 and 4** are correct
**D**  if **4** is correct
**E**  if **1, 2, 3, and 4** are correct

78. Smoking during pregnancy is associated with

(1) functional inactivation of fetal and maternal hemoglobin
(2) low-birth-weight infants
(3) reduced perfusion of the placenta
(4) intrauterine growth retardation

79. Human papillomavirus (HPV) infection is characterized by which of the following features?

(1) HPV types 16 and 18 are found in cervical cancer
(2) Genital warts are caused by this virus
(3) The koilocyte or "halo" cell is pathognomonic
(4) Sexual intercourse is the only possible means of transmission

80. Correct statements regarding the normal pressure relationship between the urethra and bladder in continent patients include which of the following?

(1) Urethral pressure decreases during the Valsalva maneuver
(2) Urethral pressure is greater than intravesical pressure when standing
(3) Increased urethral length lowers urethral pressure
(4) The pressures are related to the anatomic position of the bladder neck

81. After 20 weeks gestation, the effects of ionizing radiation on the fetus include

(1) hair loss
(2) skin lesions
(3) bone marrow suppression
(4) mental retardation

## SUMMARY OF DIRECTIONS

| A | B | C | D | E |
|---|---|---|---|---|
| 1, 2, 3 only | 1, 3 only | 2, 4 only | 4 only | All are correct |

82. Surgery to remove an ovary in the ninth week of pregnancy was necessary in a woman who presented with torsion of the ovary containing the corpus luteum of pregnancy. Which of the following outcomes could be expected in this pregnancy?

(1) Uterine bleeding
(2) Uterine cramping
(3) Spontaneous abortion
(4) No change in the pregnancy

83. Myomas are associated with which of the following clinical conditions?

(1) Anemia
(2) Pyelonephritis
(3) Urinary frequency
(4) Dysmenorrhea

84. Fetal lung maturation is assured by measurement of

(1) lecithin
(2) prostaglandin
(3) sphingomyelin
(4) phosphatidylglycerol

85. A woman who has had a first-trimester abortion, a 22-week spontaneous loss, an ectopic pregnancy, and a term twin gestation is described as a

(1) multigravida
(2) nullipara
(3) primipara
(4) multipara

86. The distressed fetus reveals which of the following acid–base characteristics?

(1) The accumulation of lactic acid
(2) An increase in $P_{CO_2}$
(3) A fall in fetal pH
(4) Respiratory alkalosis

87. Potential fetal and neonatal complications of the uncontrolled diabetic mother include

(1) congenital anomalies
(2) stillbirth
(3) macrosomia
(4) hypoglycemia

88. A woman in active labor (4–5 cm dilated) requests pain relief. She is having regular contractions, and the vertex is at 0 station. Appropriate anesthesia includes

(1) paracervical block
(2) caudal anesthesia
(3) lumbar epidural anesthesia
(4) pudendal block

89. Tests taken at 5 and 6 weeks after the last menstrual period that suggest an ectopic pregnancy include which of the following?

| | Five weeks | Six weeks |
|---|---|---|
| (1) | Positive serum (HCG) | Negative urine HCG |
| (2) | 175 mIU/ml HCG | 2000 mIU/ml HCG |
| (3) | 45 mIU/ml HCG | 400 mIU/ml HCG |
| (4) | Negative urine HCG | Negative serum HCG |

90. A 1-hour glucose tolerance test in a woman with a previous stillborn infant resulted in the following values: fasting, 110 and 1 hour, 140. Follow-up for this patient should include

(1) nothing further as the 1-hour test was normal
(2) a 2000-calorie diet
(3) home glucose urine testing
(4) a standard glucose tolerance test

91. The follicular phase of the menstrual cycle is characterized by

(1) a variable length
(2) growth and development of ovarian follicles
(3) secretion of estrogen by the ovary
(4) vascular growth of the endometrium

92. Birth control pills have a positive effect on cancer prevention in which of the following organs?

(1) Breast
(2) Ovary
(3) Endometrium
(4) Cervix

93. Hormones that are elevated during pregnancy include

(1) progesterone
(2) thyroxine
(3) human placental lactogen
(4) cortisol

94. A 30-year-old woman (gravida 2, para 0010) has been treated for premature labor for a total of 15 hours with magnesium sulfate (3 g/hr). She complains to the nurse caring for her that she cannot catch her breath. The physician should

(1) perform a brief neurologic examination of her deep tendon reflexes
(2) administer 40 mg of furosemide intravenously.
(3) perform a chest and cardiac examination
(4) administer 1 L of normal saline over 1 hour

95. Abnormal luteal phase function in a non-pregnant woman could result from depressed

(1) follicle-stimulating hormone stimulation of the follicle
(2) estradiol levels
(3) progesterone production
(4) human chorionic gonadotropin production

96. The diagnosis of placental abruption should be suspected with which of the following signs or symptoms?

(1) Late fetal heart rate decelerations
(2) Coupling of contractions
(3) Loss of fetal heart rate variability
(4) Unresponsive premature contractions

97. A 35-year-old woman (gravida 1, para 0) presents to a physician at 35 weeks gestation, complaining of the abrupt onset of frequent, painful abdominal contractions, severe back pain, and moderate vaginal bleeding. On examination, a firm uterus, which is moderately tender, is noted. The physician should

(1) administer 0.3 mg of ritodrine as an intravenous bolus
(2) accompany the patient to labor and delivery to rule out a placental abruption
(3) assure the patient that this often happens towards the end of pregnancy
(4) prepare for immediate delivery

98. In a case of severe placental abruption with fetal demise, which of the following would be appropriate management?

(1) Rupture of membranes
(2) Oxytocin administration
(3) Platelet determination
(4) Watchful waiting

99. Correct statements concerning lymphogranuloma venereum include which of the following?

(1) It is caused by *Chlamydia trachomatis*
(2) It can cause extensive anogenital destruction
(3) The "groove sign" is found in the secondary disease stage
(4) The organism is exquisitely sensitive to penicillin

100. Fetal heart rate decelerations associated with fetal head compression are characterized by

(1) systemic hypoxia
(2) poor fetal outcome
(3) an association with oligohydramnios
(4) decelerations beginning at the onset of the contraction

101. A 40-year-old woman delivered an infant at 22 weeks gestation after experiencing abdominal pressure but no cramping pain. When she arrived at the hospital, she was completely dilated, and both feet were palpable vaginally through the membranes. When advising her about future pregnancies, the physician should state that

(1) she probably has an incompetent cervix
(2) she is likely to have another preterm delivery
(3) she may have a structural uterine abnormality
(4) hysterosalpingography in 6 months could help to identify structural cervical and uterine problems

102. Physiologic characteristics of menstruation include

(1) constriction of spiral arteries
(2) desquamation of endometrial tissue
(3) ischemia of endometrial tissue
(4) estrogen secretion

SUMMARY OF DIRECTIONS

| A | B | C | D | E |
|---|---|---|---|---|
| 1, 2, 3 only | 1, 3 only | 2, 4 only | 4 only | All are correct |

103. Correct statements concerning vulvar anatomy include which of the following?

(1) The nerve supply includes the perineal, genitofemoral, and ilioinguinal nerves
(2) The lymphatic drainage is to the superficial inguinal lymph node group
(3) Both the Skene's and Bartholin's glands drain into the vestibule of the vulva
(4) The vulva is covered exclusively by keratinized epithelium

104. An Rh-negative woman comes to the office as a new obstetric patient. Her blood type, antibody screen, complete blood count, serology, and rubella titer are drawn. Other important initial studies include

(1) antinuclear antibody screen
(2) reticulocyte count
(3) bilirubin level
(4) blood type from the infant's father

105. Appropriate therapy for a luteal phase defect includes

(1) clomiphene citrate
(2) human chorionic gonadotropin
(3) postovulatory progesterone supplementation
(4) low-dose estrogen

106. An 18-year-old woman enters the emergency room with acute lower abdominal pain, which started 1 week after a period that was 7 days late. She is sexually active and not using birth control. Her temperature is 99.4°F, and her white blood cell count is 12.4. On examination, her pain extends to the upper abdomen on the right side. The differential diagnosis should include which of the following conditions?

(1) Ectopic pregnancy
(2) Fitz-Hugh–Curtis syndrome
(3) Appendicitis
(4) Pelvic tuberculosis

107. Barrier methods of contraception are associated with which of the following advantages?

(1) They protect the user against sexually transmitted diseases
(2) Regular office visits are not required
(3) They are relatively free from side effects
(4) They prevent cervical cancer

108. Cigarette smoking during pregnancy has been shown to be particularly harmful to the fetus. Noted deleterious effects include the development of

(1) low birth weight
(2) bleeding during pregnancy
(3) premature rupture of membranes
(4) intrauterine fetal demise

109. *Gardnerella vaginalis* vaginitis is associated with which of the following features?

(1) A pH of 4.5
(2) "Clue" cells
(3) Marked inflammatory background in vaginal secretions
(4) Positive "whiff" test

110. A 24-year-old woman reports that it has been 7 weeks since her last menstrual period. She states that her menstrual cycles have always been irregular, but she is now experiencing some vaginal spotting and mild right lower quadrant pain. On examination, she has a normal sized uterus with a soft top and mild tenderness in the right lower quadrant. A human chorionic gonadotropin (HCG) β-subunit level of 1000 mIU/ml from the day prior to the examination is reported. The physician should do which of the following?

(1) Recommend diagnostic laparoscopy
(2) Perform an ultrasound examination of the pelvis
(3) Perform a culdocentesis
(4) Repeat the HCG β-subunit measurement in 24 hours

111. A poor postcoital test can be reflective of

(1) low sperm count
(2) poor sperm motility
(3) poor cervical mucus
(4) poor coital technique

112. The hyper- or hypogonadotropic syndromes that demand excision of the gonad before or after puberty include which of the following?

(1) Androgen insensitivity syndrome
(2) Turner's syndrome
(3) Mixed gonadal dysgenesis
(4) Kallmann's syndrome

113. An 18-year-old woman complains of a long history of extremely painful menses, which last 5–7 days. She states that her last menstrual period was 5 weeks ago and was very heavy. Examination reveals a fairly normal pelvis. The physician should treat this patient by

(1) performing a dilatation and curettage
(2) ruling out pregnancy
(3) ruling out thyroid dysfunction
(4) administering oral contraceptives and anti-prostaglandins

114. Evidence of ejaculation at the time of a sexual assault can be determined by

(1) positive Wood's light examination
(2) positive acid phosphatase test
(3) wet mount of vaginal secretions showing motile spermatozoa
(4) high human chorionic gonadotropin levels

115. The dependence of myomas on estrogen is demonstrated by the fact that they

(1) stop growing after menopause
(2) often grow rapidly during pregnancy
(3) are unusual prior to menarche
(4) may be found along with endometrial hyperplasia

116. Persistent fetal circulation is a life-threatening complication of meconium aspiration. Factors in its pathophysiology include which of the following?

(1) Meconium plugs in the bronchial tree cause atelectasis
(2) Bacteria from the meconium causes inoculation of the lung tissue contributing to pneumonia and sepsis
(3) Meconium causes spasm of the pulmonary arterioles, resulting in pulmonary hypertension
(4) Meconium passes from the lungs into the general circulation initiating disseminated intravascular coagulation

117. Appropriate therapy for a 21-year-old single college student with proven mild endometriosis and dysmenorrhea is

(1) oral contraceptives
(2) danazol
(3) nonsteroidal anti-inflammatory drugs
(4) gonadotropin-releasing hormone agonists

118. Conditions considered relative emergencies in a teenager with amenorrhea include

(1) vaginal agenesis
(2) transverse vaginal septum
(3) uterine anomaly
(4) imperforate hymen

119. Which of the following drug combinations would be appropriate in the treatment of acute pelvic inflammatory disease?

(1) Clindamycin/metronidazole
(2) Doxycycline/cefoxitin
(3) Ampicillin/cefoxitin
(4) Clindamycin/gentamicin

120. Treatment options for a 49-year-old woman with a 12-week asymptomatic myomatous uterus include

(1) hysterectomy
(2) pelvic ultrasound examinations
(3) gonadotropin-releasing hormone agonists
(4) myomectomy

121. Prophylactic measures used to prevent aspiration during induction of general anesthesia include

(1) cricoid pressure
(2) corticosteroids
(3) antacids
(4) antibiotics

122. Endometrial hyperplasia is discovered in a 23-year-old nulligravid woman with irregular menstrual cycles who is anxious to start a family. Appropriate therapy includes which of the following?

(1) Cyclic combination estrogen/progestin therapy for 3 months
(2) Repeat endometrial biopsy after therapy
(3) Clomiphene stimulation to induce ovulation
(4) Cyclic progestin therapy

---

### SUMMARY OF DIRECTIONS

| A | B | C | D | E |
|---|---|---|---|---|
| 1, 2, 3 only | 1, 3 only | 2, 4 only | 4 only | All are correct |

---

123. Which of the following tests should be part of the workup of a woman who presents with a postdates pregnancy?

(1) Nonstress test
(2) Contraction stress test
(3) Measurement of maternal urinary or plasma estriol
(4) Measurement of human placental lactogen

124. An increased incidence of ectopic pregnancy has been associated with

(1) endometritis
(2) chronic salpingitis
(3) adenomyosis
(4) use of the intrauterine device

125. A pregnant woman comes to the physician's office for her first prenatal visit. She states that her first child was healthy but that her second and third pregnancies ended in intrauterine fetal demises at 28 and 30 weeks gestation, respectively. The pathology reports of both fetuses showed evidence of erythroblastosis fetalis. She states that she was given $Rh_0$ (anti-D) immune globulin after the first delivery. Which of the following statements explains why she became sensitized?

(1) Her antibodies may not be anti-D
(2) She was most likely sensitized during the pregnancy before the immune globulin was given
(3) Not enough immune globulin was administered
(4) Her first child was Rh negative

126. The pathophysiology of hirsutism involves which of the following hormone or enzyme activities?

(1) Increased concentration of bound testosterone
(2) Decreased $5\alpha$-reductase activity
(3) Increased levels of luteinizing hormone
(4) Decreased levels of sex hormone binding globulin

**Directions:** The groups of questions below consist of lettered choices followed by several numbered items. For each numbered item select the **one** lettered choice with which it is **most** closely associated. Each lettered choice may be used once, more than once, or not at all.

### Questions 127 and 128

For each of the following clinical presentations, select the treatment that would be most appropriate.

(A) Fractional dilatation and curettage
(B) Unilateral salpingo-oophorectomy
(C) Conization of the cervix
(D) Cyclic estrogen/progesterone
(E) Laser vaporization of the cervix

127. A 23-year-old woman underwent colposcopy for the evaluation of a class III Pap smear. The squamocolumnar junction was seen in its entirety, and the endocervical curettage was negative. A directed biopsy of the cervix revealed a 1 mm focus of invasion.

128. A 37-year-old woman has heavy, painless bleeding every 3–6 months. She is currently interested in contraception. Inspection of her cervix is normal, and her Pap smear is class I.

### Questions 129–131

For each of the drugs listed below, select the malformation most commonly caused by it.

(A) Chondrodysplasia punctata
(B) Neural tube defect
(C) Discolored teeth
(D) Hutchinson's molars
(E) Limb reduction

129. Valproic acid
130. Warfarin
131. Tetracycline

**Questions 132–134**

Each laboratory value listed below is diagnostic of a particular clinical condition associated with hirsutism. For each laboratory value, select the therapy that would be most useful for the suppression of the androgen excess associated with that condition.

(A) Spironolactone
(B) Prednisone
(C) Clomiphene
(D) Birth control pills
(E) Cimetidine

132. Elevated dihydroepiandrosterone sulfate

133. Elevated serum androstenedione

134. Elevated 17-hydroxyprogesterone

**Questions 135–138**

For each of the following clinical situations, select the form of treatment that would be most appropriate.

(A) Progestin
(B) Nonsteroidal anti-inflammatory drugs
(C) Oral conjugated estrogen
(D) Dilatation and curettage
(E) Oral contraceptive agents

135. A 30-year-old woman presents with a chief complaint of menorrhagia. She states that her cycles are regular at 28 or 29 days apart. Physical examination reveals a blood pressure of 140/90 mm Hg and a uterus of normal size.

136. A 35-year-old woman presented 3 months ago to another physician with irregular (every 30–40 days) heavy menses and some intermenstrual spotting. Her doctor prescribed two packs of oral contraceptives. She is returning now to a new physician, reporting that she bled almost every day for the past 2 months, during which time she was taking the birth control pills.

137. A 23-year-old nulligravid woman presents with a 3-week history of bleeding, the last 3 days of which were heavy bleeding with clots. She feels weak. Her previous menstrual period was 3 months prior to this bleeding episode. Her hemoglobin is 9.0 g.

138. A 47-year-old woman complains of menometrorrhagia. She states that this type of bleeding has been occurring for the past 18 months. Physical examination reveals a mildly obese woman with a blood pressure of 140/95 mm Hg and a uterus of normal size.

**Questions 139–142**

For each legal term listed below, select the statement that defines it.

(A) An action brought by a child, alleging that because of a physician's negligence in genetic counseling, he or she was born with congenital defects of such magnitude that no life (through mother's choice of abortion) would have been preferable
(B) An action brought by a child, alleging that his or her cerebral palsy was proximately caused by the negligent monitoring of his or her mother's pregnancy by the obstetrician
(C) An action brought by parents of a normal child, alleging that because of a physician's negligence in the performance of a sterilization procedure or an abortion, an unwanted child was born
(D) An action brought by the parents of a congenitally defective child, alleging that because of inadequate or absent genetic counseling, the option of abortion was taken away from them; thus, they seek compensation for the harm arising from the child's defects
(E) An action brought by a woman against her obstetrician stating that if she had known that there was a possibility of cesarean section in the delivery of her child, she would have sought the assistance of another obstetrician

139. Wrongful conception

140. Wrongful life

141. Wrongful birth

142. Birth injury

## Questions 143–145

For each clinical presentation listed below, select the postpartum clinical entity most likely to be associated with it.

(A) Pelvic thrombophlebitis
(B) Infected hematoma
(C) Urinary tract infection
(D) Parametritis
(E) Episiotomy breakdown

143. A woman spikes a temperature to 102.6°F (39.1°C) on her first postpartum day. She had ruptured membranes for 36 hours prior to delivery, and the cervical culture was group B β-hemolytic *Streptococcus*. She has lower abdominal tenderness.

144. A woman had a cesarean section for fetal distress 5 days previously. Her clinical course has been marked by a spiking fever for 4 days that has not responded to triple antibiotic therapy.

145. A woman had a vaginal delivery with epidural anesthesia. She had to be catheterized once in the recovery room for urinary retention. One day postpartum she complains of shaking chills and back pain.

## Questions 146–148

For each case presented below, select the treatment modality that is most appropriate.

(A) Monthly human chorionic gonadotropin (HCG) titers
(B) Chest x-ray
(C) Chemotherapy
(D) Pelvic ultrasound
(E) Quantitative HCG titer

146. A woman presents with vaginal bleeding and a positive serum pregnancy test. She states that she had a spontaneous abortion 4 months ago. Her uterus is at the level of the umbilicus, and the hand-held ultrasound instrument picks up no fetal heart tones.

147. A woman with metastatic gestational trophoblastic neoplasia has been on chemotherapy for 1 year. She had her third negative monthly HCG titer and a normal chest x-ray 3 months ago. Since then, her monthly HCG titers have been negative.

148. A woman had a suction dilatation and evacuation for a molar pregnancy. Her postoperative HCG titers dropped steadily until 3 weeks ago. For the past 3 weeks, her HCG titer has been around 6500 mIU/ml.

## Questions 149–151

For each obstetric situation listed below, select the sign or symptom with which it is most likely to be associated.

(A) Breech presentation
(B) Oliguria
(C) Occiput posterior position
(D) Transient fetal distress
(E) Hyperreflexia

149. A multiparous woman presents at 33 weeks gestation, complaining of vaginal bleeding in the absence of both contractions and ruptured membranes.

150. A woman at term is admitted with a tender uterus and a tense abdomen and no audible fetal heart tones.

151. A woman presents with moderate vaginal bleeding and uterine contractions that do not completely relax between contractions.

## Questions 152–154

For each case presented below, select the diagnosis that best describes the patient's clinical condition.

(A) False labor
(B) Hypertonic uterine dysfunction
(C) Hypotonic uterine dysfunction
(D) Active phase of labor
(E) Latent phase of labor

152. A woman presents to the labor floor complaining of painful contractions that occur every 2 minutes. She is 2 cm dilated. Two hours later, she continues to complain of frequent painful contractions, but she is still only 2 cm dilated.

153. A woman presents to the labor floor 3 cm dilated with contractions every 5–7 minutes. About 2 hours later she is having contractions every 3 minutes and is 6 cm dilated. She is 8 cm dilated 1 hour later.

154. A woman presents to the labor floor with contractions 8–12 minutes apart. She complains of lower abdominal discomfort with her contractions, which last for only 20 seconds each. Sedation causes the contractions to space out to intervals of 15–20 minutes.

## Questions 155–157

For each case presented below, select the treatment that would be most appropriate.

(A) Medroxyprogesterone
(B) Estrogen cream
(C) Sequential estrogen/progestin therapy
(D) Endometrial biopsy
(E) Hysterectomy

155. A 55-year-old woman presents with complaints of hot flashes and night sweats. Her last menstrual period was 3 years ago. Her only other medical complaint is migraine headaches. Family history includes breast cancer in her maternal aunt.

156. A 49-year-old woman presents with irregular cycles, intermenstrual bleeding, and hot flashes. She is very bothered by the hot flashes and insists that she needs medication. She has had the irregular bleeding for 18 months.

157. A 54-year-old woman presents complaining of vaginal spotting, discharge, and dyspareunia. She has been menopausal for 3 years. She had an endometrial biopsy 3 months ago, which revealed an atrophic endometrium.

## Questions 158–160

For each description of amenorrhea listed below, select the type of amenorrhea that is described by it.

(A) Eugonadotropic amenorrhea
(B) Physiologic amenorrhea
(C) Hypergonadotropic amenorrhea
(D) Androgen excess amenorrhea
(E) Hypogonadotropic amenorrhea

158. A 24-year-old nulligravid woman had a normal menstrual history until 8 months ago when she began intensive long-distance running. She has not had a menstrual flow since her first marathon 4 months ago.

159. An 18-year-old woman with well-developed secondary sexual characteristics complains of amenorrhea. She has a vaginal pouch and XX chromosomes.

160. A 15-year-old adolescent with normal sexual development complains of 5 months of amenorrhea. She states that her first menses were 9 months ago after which she had three menses.

## Questions 161–164

For each diagnosis listed below, select the underlying etiology.

(A) Arrhenoblastoma
(B) Immature hypothalamic–pituitary axis
(C) Epidermolysis bullosum
(D) 21-Hydroxylase defect
(E) Maternal ingestion of diethylstilbestrol

161. Dysfunctional uterine bleeding in an adolescent girl

162. Heterosexual precocious puberty

163. Congenital adrenal hyperplasia

164. Clear cell adenocarcinoma

## Questions 165 and 166

For each of the following clinical situations involving a primary infertility problem, select the procedure that would be most appropriate to evaluate it.

(A) Laparoscopy
(B) Hysteroscopy
(C) Postcoital test
(D) Endometrial biopsy
(E) Hysterosalpingography

165. A woman states that she has been infertile for 3 years with the exception of a pregnancy 10 months previously that aborted between the fifth and sixth weeks. She says she knows that she ovulates. She has a record of basal body temperatures, which shows an average luteal phase temperature rise of 9–10 days.

166. A woman states that she has been infertile for 3 years. An operation for acute appendicitis at 16 years of age was the only event of any significance in her medical history. Her postcoital test and her husband's semen analysis were good. Her basal body temperature charts show consistent 12–14-day luteal phases.

**Questions 167-169**

For each congenital condition listed below, select the diagnostic procedure that would be most helpful in identifying it.

(A) Maternal serum $\alpha$-fetoprotein
(B) Chorionic villi sampling
(C) Maternal chromosome analysis
(D) Amniocentesis
(E) Real-time ultrasonography

167. Meningomyelocele

168. Tay-Sachs disease

169. Sickle cell anemia

**Questions 170 and 171**

For each of the following clinical situations, select the management possibility that would be most appropriate.

(A) Strict bed rest at home
(B) Hospitalization
(C) Cesarean section
(D) Magnesium sulfate
(E) Induction of labor

170. A woman with a 32-week pregnancy presents to the office for the second week in a row with a blood pressure of 140/90, 1 + proteinuria, peripheral edema, and normal reflexes.

171. A woman with a 32-week pregnancy is admitted to the hospital with a blood pressure of 160/110, 3 + proteinuria, facial edema, hyperactive reflexes, and upper abdominal pain.

**Questions 172-175**

Match each description with the appropriate incision or procedure.

(A) Kerr
(B) Shirodkar
(C) Sellheim
(D) McDonald
(E) Sanger

172. Performed in the lower (noncontractile) segment of the uterus

173. Simple, easy to perform, and incurs minimal trauma to the cervix

174. Can obviate the need for repeating the procedure in a subsequent pregnancy

175. Dehiscence is more likely secondary to the scar extending into the uterine corpus

**Questions 176-178**

For each of the following obstetric presentations, select the analgesic or anesthetic modality that would be most appropriate.

(A) Pudendal block
(B) Spinal anesthesia
(C) Intramuscular morphine
(D) Epidural anesthesia
(E) Intravenous meperidine

176. A woman presents with painful uterine contractions that occur every 2-3 minutes. On examination, she is 2 cm dilated and 60% effaced. Three hours later, she is even more uncomfortable with the same contraction pattern, but her cervix is still only 2 cm dilated.

177. A woman is in active labor having contractions every 2-3 minutes. She is 3 cm dilated with the vertex at the - 1 station. Two hours later, she is 5-6 cm dilated with the vertex at the +1 station; she requests pain relief.

178. A woman is in the delivery room ready for delivery. She has had no analgesia or anesthesia up to this point. The vertex is on the perineum; the infant's head is visible at the perineum with each push.

**Questions 179 and 180**

For each clinical presentation listed below, select the test or therapy that would be most appropriate.

(A) Weekly human chorionic gonadotropin levels
(B) Repeat serum test for syphilis
(C) Dilatation and evacuation
(D) Diethylstilbestrol therapy
(E) Intrauterine device insertion

179. A 21-year-old woman was raped approximately 2 weeks after her last menstrual period. She was not using any birth control because she was not sexually active. One week after her assault she went to the physician's office for help.

180. A 21-year-old woman was raped approximately 2 weeks after her last menstrual period. She was not using any birth control because she was not sexually active. She was taken to the emergency room immediately after the assault and started on diethylstilbestrol prophylaxis therapy.

## ANSWERS AND EXPLANATIONS

**1. The answer is D.** (*Chapter 25 I A*) Endometriosis involves the presence of functioning endometrium, both glands and stroma, outside the uterine cavity. This endometrium can implant anywhere in the pelvis, causing nodularity of the cul-de-sac, pelvic adhesions, or endometriomas of the ovaries. All of this is due to the monthly bleeding that occurs within the implants, which then creates the nodularity or cystic ovarian enlargement. Functioning endometrium within the myometrium is called adenomyosis not endometriosis.

**2. The answer is D.** [*Chapter 20 II F 1 b (2)*] A streptoccocal infection elsewhere is almost always the predisposing factor for streptococcal vaginitis. Infections with *Candida, Trichomonas*, and *Neisseria gonorrhoeae* are rarely associated with bloody vaginal discharge. A foreign body should always be suspected, but in the absence of pain, it is an unlikely causative factor.

**3. The answer is A.** (*Chapter 3 V A 2, D 1 a, 2, 3, E 1 a*) Although it is not necessary to be restrictive about weight gain during pregnancy, the usual weight gain is 20–30 lbs. A weight gain of less than 15 lbs, even in an obese woman, is a cause for concern. Daily supplements of iron, calcium (1200 mg), and folic acid (800 $\mu$g) are recommended. Because of the natriuretic effect of progesterone, it is unwise to restrict the sodium intake of a pregnant woman. Thus, the use of diuretics in pregnancy can be harmful, leading to a hyponatremic state.

**4. The answer is A.** (*Chapter 12 III C 1; Table 12–1*) Of children with Down's syndrome, 50% are born to mothers younger than age 35 years. The risk for recurrence for the mother in the question is the same as the general population up to age 30 years (1%–2%), after which time recurrence is that expected based on maternal age.

**5. The answer is D.** (*Chapter 36 III C 1–5; V A 2 a*) Osteoporosis is the main health hazard associated with the menopause. Bone loss occurs in the axial skeleton, most of which involves trabecular bone with thinning of the cortex. With the current low numbers of postmenopausal women on estrogen replacement, about 32% can expect a hip fracture at some time, and this involves significant mortality in the postmenopausal population. Bone loss is most rapid in women under 40 years of age who undergo castration by surgical oophorectomy. At any time, estrogen can prevent or retard bone loss, depending on when the replacement is started with respect to the menopause. Osteoporosis is more common in white than black women.

**6. The answer is A.** (*Chapter 24 III A 1–6, B 1–5*) *Neisseria gonorrhoeae* may be responsible for as many as 40% of cases of pelvic inflammatory disease (PID), and penicillinase-producing strains, should they continue to proliferate, will make it difficult to eradicate. *Chlamydia trachomatis* may be responsible for up to 40% of cases of PID, but rigid laboratory growth requirements have made it difficult to link *C. trachomatis* with PID. *Mycoplasma hominis* may be responsible for 20% of cases of PID, and *Actinomyces israelii* is found in 15% of IUD-associated cases of PID. *A. israelii* is rarely found in women who do not use an IUD.

**7. The answer is D.** (*Chapter 19 II A 1, 2*) The microscopic characteristics of a hydatidiform mole include marked edema and enlargement of the villi. In addition, there is proliferation of the lining trophoblast of the villi and an absence of any fetal tissue. Another important feature is the disappearance of the villous blood vessels so that their presence would not be compatible with the diagnosis of a mole.

**8. The answer is A.** (*Chapter 1 VI D 1–5, E 1–3*) In a woman who smokes heavily, the suspected diagnosis of below average fetal weight at 32 weeks gestation must be intrauterine growth retardation. If the fetus is distressed, there may be an actual reduction in fetal movement. One way to assess and follow this situation is by serial estriol determinations.

**9. The answer is B.** (*Chapter 3 IV A 3 a, b*) $\alpha$-Fetoprotein levels are measured at 16 weeks and are reported as the multiple of the mean (MOM). Elevated levels (more than 2.5 MOM) can be associated with neural tube defects (i.e., anencephaly), omphalocele, multiple gestation, and duodenal atresia. Approximately 15%–20% of cases of Down's syndrome are associated with very low (0.5 MOM and below) $\alpha$-fetoprotein levels.

**10. The answer is C.** (*Chapter 4 I C*) In one theory about the initiation of labor, fetal cortisol is thought to play an important part. In anencephalic infants, there is faulty brain–pituitary–adrenal function, and fetal cortisol levels are low. A naturally prolonged gestation often results in an anencephalic fetus, perhaps because of the low levels of cortisol. None of the other abnormalities listed in the question are associated with low fetal cortisol levels.

**11. The answer is D.** (*Chapter 19 II D 1 a*) Good prognosis metastatic gestational trophoblastic neoplasia (GTN) is associated with a low pretreatment human chorionic gonadotropin titer (less than 40,000 mIU/ml), no significant prior chemotherapy, no metastatic spread to the brain or liver, and a short duration (i.e., the last pregnancy occurred less than 4 months prior to the diagnosis of GTN). Therefore, a pregnancy that occurred 6 months previously is no longer of short duration and is not included in the criteria for good prognosis metastatic GTN.

**12. The answer is D.** (*Chapter 10 IV A 1 a–e, B 3 a, b*) One of the *late* signs of premature labor is cervical dilation past 2 cm. Pharmacologic inhibition of premature labor is less effective when cervical effacement is advanced beyond 80% or dilation has progressed beyond 3 or 4 cm. The early determination of premature labor, therefore, improves the chance that labor can be halted and that the pregnancy will continue until a gestational age that is safe for delivery of the fetus.

**13. The answer is A.** (*Chapter 13 III A 1–3; IV A, B 1–3, E 3*) The crown–rump length is the most accurate method for dating a pregnancy: It is accurate to within 3 or 4 days. Biparietal diameter measurements are also accurate to within 7 days if done before 20 weeks gestation; however, from 20–26 weeks gestation, the biparietal diameter is accurate to only ± 11 days. The femur length is accurate to ± 14 days in the second trimester. The ratio of the abdominal circumference to femur length is valuable in assessing fetal growth, not in dating a pregnancy. Two biparietal diameters in the third trimester are notoriously *unreliable* for dating a pregnancy.

**14. The answer is C.** (*Chapter 30 II A 1, 2, 4, C 1, 2*) The symptoms with which the woman described in the question presents are most commonly seen in *Neisseria gonorrhoeae*–related acute pelvic inflammatory disease (PID). *Gardnerella vaginalis* produces a thin, greyish vaginal discharge associated with symptomatic vulvovaginitis, but not acute PID. *Chlamydia trachomatis*, serotypes D–K, are associated with PID, but the clinical course is of an indolent nature and not distinctly related to the menstrual cycle. Chancroid and lymphogranuloma venereum are part of the ulcerative genital diseases and are not associated with acute PID.

**15. The answer is C.** (*Chapter 1 I A 3 a; V E 1–5*) Progesterone has a number of important functions during pregnancy. It prepares the endometrium for implantation of the embryo and maintains the uterus in a quiescent state during pregnancy by relaxing the myometrium. It has natriuretic actions and, as a result of this, stimulates an increased production of aldosterone. It is the elevated levels of circulating estrogen, not progesterone, that cause an increase in the serum binding proteins, which, in turn, leads to spuriously elevated levels of certain hormones, such as thyroxine and cortisol.

**16. The answer is B.** (*Chapter 5 VI A 1–3, B 1, C, D*) In an effort to get the fetus in the best possible condition before delivery, it is advisable to perform some intrauterine resuscitative measures before the cesarean section. Increasing intravenous fluids increases the intravascular volume, which assures maximum blood flow to the uterus. Nasal oxygen to the patient can result in small increments in fetal $PO_2$. Amnioinfusion takes the pressure off the compressed cord, and tocolysis, the use of subcutaneous terbutaline, decreases contractions and allows maximal placental blood flow. The patient should be placed in a lateral recumbent position, not in the supine position; the supine position may lead to decreased venous return and cardiac output due to compression of the vena cava by the pregnant uterus.

**17. The answer is D.** (*Chapter 22 IV D 2 a–d*) Renin substrate, plasma renin activity, and angiotensin are elevated in oral contraceptive users. There is an increase in aldosterone secretion and renal retention of sodium. The resulting hypertension, which is usually seen in users of at least 5 years, may represent the failed suppression of renin substrate and plasma renin activity with elevated levels of angiotensin.

**18. The answer is E.** (*Chapter 24 II C 1–3*) Women who are sexually active and use no contraception develop 3.42 cases of pelvic inflammatory disease (PID) per 100 woman years. An intrauterine device is linked to a higher incidence of PID than any other method of contraception. Oral contraceptive use is linked to the lowest incidence of PID, even lower than the barrier methods, perhaps as a result of a decreased menstrual flow, a decreased ability of pathogenic bacteria to attach to endometrial cells, and the presence of progesterone.

**19. The answer is C.** (*Chapter 10 V B 3 a–c*) Ethanol, administered by intravenous infusion, was one of the most popular methods for the treatment of premature labor. Although it is an effective inhibitor of premature labor, it has many untoward side effects, including inebriation, nausea, vomiting, headache, lactic acidosis, respiratory depression, and aspiration. There are no significant fetal side effects, except sedation.

Ritodrine is the only $\beta_2$-mimetic agent to be approved by the Food and Drug Administration for the inhibition of premature labor. Its mechanism of action is stimulation of $\beta_2$-adrenergic receptors, producing relaxation of the uterus and bronchi, vasodilatation, and muscle glycogenolysis. Because of cross-reactivity with $\beta_2$ receptors throughout the body, there are significant cardiovascular side effects.

Magnesium sulfate is a neuromuscular blocking agent which inhibits myometrial contractions by competing with free calcium. It is the drug of choice for stopping premature labor in women with medical problems, such as diabetes mellitus.

Terbutaline is a selective $\beta_2$ agonist that is used as a bronchodilator when given orally; however, it has been used successfully to inhibit premature labor. Side effects, such as headache, tremor, hypotension, nausea, vomiting, and maternal and fetal tachycardia, are common with $\beta_2$ agonists.

Phenobarbital is an antiseizure medication and has never been shown to be an effective tocolytic agent.

**20. The answer is A.** [*Chapter 37 III D 1, 3 a, 4 a (1)*] The 51-year-old woman described in the question who presents with postmenopausal bleeding has a stage IA, grade 1 endometrial cancer, which is characterized by a well-differentiated adenocarcinoma (grade 1) and a uterine depth of 8 cm or less (stage IA). There is a very low risk of deep myometrial penetration or pelvic lymph node metastasis. The incidence of recurrent disease following a total hysterectomy and bilateral salpingo-oophorectomy is very low, making this the treatment of choice for this woman.

**21. The answer is E.** [*Chapter 4 IV C 2 b (1)*] The mediolateral episiotomy has several disadvantages when compared with the median episiotomy. It is more difficult to repair and occasionally has a poor anatomic result. There is more blood loss, and with resumption of intercourse, there is occasional dyspareunia due to the repair. However, if there is an extension of the episiotomy, it does not extend to or through the anal sphincter as does the median episiotomy in a similar situation. The mediolateral episiotomy gives more space at the introitus for difficult deliveries, such as the breech and shoulder dystocia presentations.

**22. The answer is B.** (*Chapter 11 IV C 1, 2*) Cordocentesis is advantageous since it measures direct fetal hemoglobin levels, blood type, bilirubin levels, and reticulocyte count; however, it has higher rates of complications and of fetal losses and is much more difficult to perform. It is performed as often as amniocentesis. Usually, it is a better predictor for timing of an intrauterine transfusion since in many cases the fetal hematocrit falls at 1% per day. Most infants become hydropic when their hemoglobin is 7 g/dl lower than that anticipated for their gestational age.

**23. The answer is B.** (*Chapter 36 III A 1–2*) The hot flushes that are characteristic of the menopause do not appear to be related to any changes in the estrogen or progesterone levels, even though most of the other menopausal symptoms are related to the hypoestrogenic state. The flush appears to coincide with a surge of luteinizing hormone, not follicle-stimulating hormone, and is more frequent and severe at night and during times of stress.

**24. The answer is D.** (*Chapter 6 III C 2, D 1*) The biophysical profile for risk assessment of a postdates fetus evaluates fetal breathing, tone, and motion; fetal heart rate, using the nonstress test (NST); and amniotic fluid volume. The contraction stress test could be used as a substitute for the NST but only if the NST is nonreactive.

**25. The answer is D.** (*Chapter 27 I C–G*) Menses that are characterized by prolonged and excessive bleeding at regular intervals is called menorrhagia. Menometrorrhagia and metrorrhagia are characterized by bleeding that occurs at irregular intervals, while polymenorrhea is normal uterine bleeding that occurs at frequent regular intervals. Oligomenorrhea is infrequent uterine bleeding that occurs at intervals greater than 40 days.

**26. The answer is B.** [*Chapter 16 III A 2, 4 a, 5 a (1)–(3)*] The respiratory distress syndrome is the most common and most serious problem of the preterm neonate. The more premature the infant, the more likely it is to occur. The cause of the respiratory distress syndrome is pulmonary immaturity and, thus, a deficiency of surfactant, which is composed of phospholipids that increase in concentration as the fetus matures. Surfactant acts to decrease the surface tension of the alveoli, allowing them to stay open, without which there would be alveolar collapse, hypoventilation, atelectasis, and poor gas exchange. Hypoxia leads to acidosis, which further inhibits surfactant production. X-ray findings are a "ground-glass" appearance of the lung. Treatment involves ventilatory support with a respirator capable of providing positive end expiratory pressure, continuous positive airway pressure, and intermittent mandatory ventilation.

**27. The answer is C.** (*Chapter 9 III B 3*) Even with no identifiable cause or associated abnormality responsible for preterm labor, the recurrence is 25%–40% of all cases. In the next pregnancy, subtle signs of premature labor, such as low back pain, increased vaginal discharge, pelvic pressure or abdominal crampiness, should be monitored. If diagnosed early and treated with bed rest and medication, premature labor may be successfully controlled and a premature delivery avoided or at least delayed.

**28. The answer is C.** (*Chapter 28 III D 1 a*) Motile spermatozoa found on a wet mount of vaginal secretions indicates that the spermatozoa have been deposited there within the previous 24 hours. Spermatozoa cannot live for a longer period of time because of the vaginal acidity, which irreversibly inactivates them.

**29. The answer is C.** [*Chapter 17 III A 1, 2 c (1)–(4)*] The known position of the infant's head, a completely dilated cervix, an empty bladder, and ruptured membranes are absolute prerequisites for a forceps delivery. The vertex does not have to be in the occiput anterior position prior to forceps application. The vertex can be posterior or transverse and still have a safe forceps application.

**30. The answer is B.** (*Chapter 6 IV B*) Delivery of a fetus at risk in prolonged pregnancy should be accomplished without regard to the status of the cervix. Delivery can, however, be accomplished through induction of labor, which is always accompanied by electronic intrapartum monitoring. One week is commonly the interval used for repetitive testing, but only if testing remains negative. Fetal scalp testing cannot be done with intact membranes.

**31. The answer is B.** (*Chapter 1 VI D 4, 5*) In such clinical situations as preeclampsia, eclampsia, hypertension in pregnancy, intrauterine growth retardation, and maternal renal disease, there is often a decline in estriol production due to placental and fetal compromise. Twenty-four hour urinary estriol values have been helpful in monitoring these high-risk pregnancies as they approach term. In Rh isoimmunization, large quantities of estriol are produced.

**32. The answer is E.** (*Chapter 25 III A, B*) The only way to diagnose endometriosis is by visualization (through the laparoscope) or biopsy of the disease. Cul-de-sac nodularity could be metastatic carcinoma. Any benign or neoplastic lesion can cause ovarian enlargement. Infertility and dysmenorrhea are symptoms associated with endometriosis, but neither is diagnostic because there are other reasons for each of those symptoms.

**33. The answer is E.** (*Chapter 7 VI C 2, D 1 a–d*) The patient described in the question who presents with cardiac failure due to the stress of the pregnancy on her cardiorespiratory system must be stabilized while preparations are made for delivery. The usual methods of therapy in a patient in cardiac failure include the administration of oxygen, diuretics, and digitalis. The delivery should be accomplished vaginally unless there is an overriding reason for a cesarean section as there is less morbidity and mortality with a vaginal delivery. Epidural anesthesia effectively relieves pain, allowing for a safe forceps or vacuum extraction delivery to shorten the second stage of delivery and to eliminate or minimize the stress associated with this stage. A cardiac patient should not be permitted to push in the second stage of labor.

**34. The answer is A.** (*Chapter 29 IV C 1–3*) Ultrasonography has proved to be a useful tool in diagnosing normal and abnormal pregnancies. For pelvic ultrasound to demonstrate a gestational sac in the uterus, the human chorionic gonadotropin (HCG) titer should be about 6000 mIU/ml. Vaginal ultrasound can pick up the presence of a gestational sac 4–6 days before pelvic ultrasound with an HCG titer in the 1500–2000 mIU/ml range. The presence of a gestational sac in the uterus at the expected time is helpful in ruling out an ectopic pregnancy. The absence of a gestational sac, depending on the HCG level, by either vaginal or pelvic ultrasound, is suggestive of an ectopic pregnancy.

**35. The answer is C.** (*Chapter 21 III A 1–4*) Follicle-stimulating hormone (FSH), a glycoprotein secreted by the pituitary, stimulates the growth and maturation of granulosa cells and aromatase activity, which is necessary for the synthesis of estradiol. In the follicular phase of the menstrual cycle, FSH and estrogen promote increased numbers of FSH receptors on the follicle, and FSH induces the development of luteinizing hormone (LH) receptors on granulosa cells. LH release is stimulated by elevated levels of estradiol, not by FSH.

**36. The answer is B.** (*Chapter 8 II A 3, 4, B*) The ductus arteriosus blood flow and the aortic blood flow constitute the cardiac output of the fetal heart. This enables adequate tissue oxygenation despite the low oxygen saturation of fetal blood. The fetal cardiac output of 200 ml/kg/min is higher than that of the adult.

**37. The answer is B.** (*Chapter 31 IV C 1–5*) Since the bleeding associated with myomas is not hormonally controlled, the presence of a profound anemia that cannot be corrected with iron is an indication for surgery. Rapid enlargement may be an indication of a malignant change in the myoma, and the hydronephrosis caused by pressure of a myomatous uterus on a ureter can be dangerous. A uterus the size of an 18-week pregnancy that fills the entire pelvis and makes examination of the ovaries very difficult is an indication for surgery. A year of infertility would not be a reason for surgery because the infertility may have nothing to do with the myomas. Myomas are usually not associated with infertility but rather with problems in maintaining a pregnancy once it is established.

**38. The answer is C.** (*Chapter 2 I D*) The calculated due date of 9 months plus 7 days from the last menstrual period is based on the assumption that a woman has 28-day cycles with ovulation on day 14. The calculation is made from the first day of the last menstrual period, assuming that a woman ovulates 14 days prior to her menses. In a 35-day cycle, ovulation occurs about day 21 instead of day 14. Thus, the due date of this patient would be January 15, 1989 because 7 days must be added to the regular calculation because of ovulation on day 21.

**39. The answer is E.** (*Chapter 29 IV C 1–6*) All (100%) ectopic pregnancies have a positive serum human chorionic gonadotropin β-subunit pregnancy test. Only 50% of urine pregnancy tests are positive. Unclotted blood on culdocentesis could represent a bleeding corpus luteum. A gestational sac on pelvic ultrasound cannot be expected before 7 weeks. The finding of a proliferative endometrium means only that ovulation has not occurred.

**40. The answer is D.** (*Chapter 28 V A 1–4*) Like rape, sexual abuse of children is underreported often because the abuser is a family member or member of the community; in most cases, sexually abused children are victims of someone they know. Children of all ages, including infants and toddlers, are abused. There is no racial, ethnic, educational, or economic barrier to sexual abuse of children; it occurs without regard to social status.

**41. The answer is C.** (*Chapter 4 II D 1 a, d, E 1, 3 a, b*) Uterine contractions in the active phase of labor cause progressive effacement and dilation of the cervix, occur every 2–4 minutes, last 30–90 seconds, and attain pressures of 20–50 mm Hg. During the active phase, the uterus differentiates into two parts. The active upper segment or uterine fundus thickens and contracts down. The passive lower uterine segment thins out with the contractions, thus promoting effacement.

**42. The answer is D.** (*Chapter 25 IV C 1 a–f*) As with all infertility surgery, gentle handling of tissue, lysis of adhesions, and meticulous hemostasis are important. Because of the adhesions with endometriosis and the raw surface areas that remain after dissection, reperitonealization helps to prevent reformation of the adhesions. The dextran is a useful adjunct because it is slowly absorbed from the peritoneal cavity and actually separates surfaces long enough to allow mesothelial cells to spontaneously reperitonealize the areas. Uterine suspension prevents refixation of the retroverted uterus. Danazol is helpful after surgery for severe endometriosis, but not for mild to moderate endometriosis, particularly because the woman described in the question wants to attempt pregnancy as soon after surgery as possible.

**43. The answer is D.** (*Chapter 34 III B 3 a–e*) The androgen insensitivity syndrome is characterized by a phenotypic female with an XY gonad and a vaginal pouch. The presence of müllerian-inhibiting factor inhibits uterine development. Because the cytosol receptors for testosterone are defective, there will be no evidence of circulating testosterone, such as pubic and axillary hair.

**44. The answer is B.** (*Chapter 8 V D*) The umbilical vein is patent for some time after birth and is used for intravenous administration and exchange transfusions in the early neonatal period. The lumen never disappears entirely but is not normally functional and, thus, is termed the ligamentum teres.

**45. The answer is E.** (*Chapter 24 IV A 1–3*) Degenerating endometrium provides a good culture medium, allowing bacteria to ascend the uterus to the tubal lumen; thus, two-thirds of acute cases of pelvic inflammatory disease (PID) begin just after the menses. Sperm can act as mobile transporters of bacteria into the uterus and tubes, and uterine contractions during sexual intercourse may assist their ascent. Intrauterine devices (IUDs) have been linked to an increased risk of PID (5.21 cases/ 100 woman years) as compared to sexually active women who use no contraception (3.42 cases/100 woman years).

**46. The answer is C.** [*Chapter 33 III F 2 b (1), 4*] All of the tests listed in the question—laparoscopy, semen analysis, basal body temperature record, postcoital test, and hysteroscopy—are important in an infertility workup, but only a basal body temperature record would be helpful in identifying a potential ovulatory problem. Well-trained female distance runners are known to have infrequent or absent menses. Cycles lasting between 30 and 50 days are suggestive of irregular or absent ovulations. A basal body temperature record would help to determine whether or not ovulation was occurring, as a thermal shift indicates ovulation.

**47. The answer is A.** (*Chapter 22 III E 1–4*) Whenever a pregnancy is associated with an intrauterine device (IUD) left in situ, the IUD should be removed immediately as there is an increased likelihood of a spontaneous abortion. In addition, serious infection can result, causing a loss of reproductive capacity or death. Ectopic pregnancy is higher than normal in IUD users than in nonusers, and prematurity, if the IUD remains throughout the pregnancy, may result from the irritative influence of the IUD on the endometrium in the third trimester. Since the IUD is always extra-amniotic in a pregnancy, there is no chance of the IUD causing pressure damage or anomalies of the fetus.

**48. The answer is C.** (*Chapter 15 I B 2*) During the embryonic period, 7–57 days after conception, the rudiments of all major structures are developing. The three primary germ layers differentiate into various tissues and organs. By the end of the eighth week postconception, the main organ systems have been established. Because the foundations of all essential internal and external structures are formed during this period, developmental disturbances may give rise to major congenital malformations.

**49. The answer is B.** (*Chapter 11 IV B 4*) One week is the appropriate time to repeat the amniocentesis. It is most important, however, to determine whether the $\Delta OD_{450}$ is rising or falling. A rising trend should be watched very closely since the present $OD_{450}$ is near zone III on the Liley curve, and the fetus may be developing hydrops fetalis. It should be safe to wait 1 week, since at the time of amniocentesis, the ultrasound presumably indicated good fetal condition. There is a very low risk for fetal death in zone II within 7 days of the study.

**50. The answer is C.** (*Chapter 14 I E 4*) A pelvic examination in suspected placenta previa is only done when delivery of the infant is necessary or indicated. This examination is contraindicated at the time of admission, after the bleeding has stopped, or prior to placental localization because there is no need to deliver the infant at that point and a pelvic examination could precipitate massive bleeding and the need for instant delivery. The pelvic examination is performed prior to any decision regarding a cesarean section; the palpation of a low-lying placenta through the cervix demands a cesarean section. However, a pelvic examination is indicated regardless of gestational age whenever there is enough bleeding to be a threat to maternal well-being.

**51. The answer is C.** (*Chapter 25 IV C 2*) Radical endometriosis surgery means a bilateral oophorectomy. Castration in a woman in her reproductive years induces an immediate hypoestrogenic state, which has short-term and long-term consequences. Hot flashes, atrophic vaginitis, and a decreased libido are symptoms women notice days to weeks following surgery. Osteoporosis and premature aging of the cardiovascular system are long-term sequelae of early castration that does not involve estrogen replacement.

**52. The answer is D.** [*Chapter 26 V H 5 a (1)–(3)*] All of the possibilities listed in the question are indications for cesarean section in the interest of preventing fetal contact with the herpesvirus with the exception of the woman with active herpes lesions whose membranes ruptured 24 hours prior to admission. In this case, the fetus has already been infected and nothing is gained by performing a cesarean section. With ruptured membranes and active herpes, it is important to deliver the fetus within 4 hours of the rupture.

**53. The answer is A.** (*Chapter 35 II A, B; IV B 1–4*) Dehydroepiandrosterone sulfate and androstenedione are preandrogens that measure activity from the adrenal and ovary, respectively; they are weak androgens, which do not accurately indicate androgen excess. Total testosterone includes testosterone bound to protein, much of which is inactive. It is the free testosterone, which is only a small portion of the total testosterone, that reflects a woman's androgen activity. 17-Hydroxyprogesterone is elevated in congenital adrenal hyperplasia because of enzyme deficiencies in the steroid pathway.

**54. The answer is D.** [*Chapter 37 II E 2 d (3) (b)*] The lesion of invasive cervical cancer is either clinically inapparent and found on cone biopsy or is clinically obvious in which cases a simple biopsy confirms the diagnosis. In stage III disease, the carcinoma extends onto the pelvic sidewall and to the lower third of the vagina. Hydronephrosis detected on intravenous pyelogram is a feature of stage IIIB disease for invasive cervical cancer.

**55. The answer is E.** (*Chapter 5 III C 2 e, f*) Late decelerations are suggestive of a problem with uteroplacental perfusion, which is commonly seen in such clinical states as preeclampsia, chronic abruptio placentae, and chronic hypertension. Intrauterine growth retardation is a reflection of poor placental perfusion; late decelerations are seen during the labor of a growth-retarded infant. Placenta previa is an acute bleeding event, which usually occurs in the absence of labor and, thus, is not associated with decelerations.

**56. The answer is C.** (*Chapter 5 IV C 2 a–c*) A false-normal fetal scalp pH value is seen in infants who are flaccid at birth with poor respiratory efforts in such clinical situations as prematurity, sedation or anesthesia in the mother, fetal infection, and events, such as terminal placental abruption, that occur after the pH determination and prior to delivery. With intrauterine growth retardation, pH values and Apgar scores are low and complementary.

**57. The answer is B.** (*Chapter 29 V A 2*) The goal in treating an ectopic pregnancy is to conserve as much of the reproductive organs as possible. Linear salpingostomy, segmental resection of the tube, and milking of the tube all preserve potential future tubal function. The salpingectomy removes the damaged tube but preserves the nearby ovary. The salpingo-oophorectomy is rarely used today because that procedure removes an ovary that might be of use at a later date as a source of ova for in vitro fertilization.

**58. The answer is C.** (*Chapter II V B 1–9, C 2 b*) After $Rh_o$ (anti-D) immune globulin is given in an adequate dose, there should be a positive indirect Coombs' test. This indicates that an excess of free immunoglobulin remains in the maternal serum and was not bound by fetal Rh-negative cells. Since $Rh_o$ (anti-D) immune globulin is an IgG molecule, it may persist in maternal serum for 8–12 weeks after the initial injection. A low titer of indirect Coombs' test (1:2–1:4) is usually indicative of persistent immune globulin that has not been cleared from the maternal circulation. These low titers are seldom a cause for concern since they rarely represent maternal sensitization. However, if low titers persist for longer than 3 months after the initial dosage, then serial titers should be obtained to exclude maternal sensitization.

**59. The answer is A.** [*Chapter 17 I D 4 b (3)*] Because the classical incision is entirely in the fundus, which is the most contractile portion of the uterus, it is the incision that poses the greatest risk for rupture during subsequent pregnancy and labor; thus, it carries the greatest risk of maternal and fetal mortality. Adhesions and healing discomfort are additional problems of this incision, but these are of less consequence than rupture of the scar.

**60. The answer is C.** (*Chapter 16 II E 1 a; IV C 4 a*) The significance of meconium in amniotic fluid has long been controversial. Although it is present often in normal pregnancies, it may indicate intermittent or chronic hypoxia. Its presence with a fetus who demonstrates fetal distress mandates prompt delivery with suctioning of the oropharynx and intubation to remove meconium below the vocal cords. Polyhydramnios simply indicates a greater than usual amount of amniotic fluid. It may be secondary to diabetes, Rh isoimmunization, congenital anomalies, duodenal atresia, or it may be idiopathic.

**61. The answer is D.** (*Chapter 19 I B 1 b, d, 2 a–c; III A 1–3*) Gestational trophoblastic neoplasia (GTN) usually presents with abnormal uterine bleeding at some point following a normal pregnancy, a molar pregnancy, or an abortion. It may appear within months or years of the preceding pregnancy. GTN develops in 3%–5% of molar pregnancies. Women at the end of their childbearing years have the highest incidence of moles and, therefore, of GTN.

**62. The answer is B.** (*Chapter 32 III B 1–3*) Although the degree of cystocele does not correlate with the severity of stress urinary incontinence, cystocele is a symptom of pelvic relaxation. Pelvic relaxation tends to be a function of age as well as obstetric experience and is often found in combination with stress urinary incontinence. The vaginal approach does make the simultaneous correction of both disorders easier.

**63. The answer is E.** (*Chapter 21 I C 1–4*) The luteal or secretory phase of the menstrual cycle extends from ovulation to the onset of the next menses. The length of the luteal phase is fairly constant at 12–16 days. Prior to ovulation, there is growth and development of the ovarian follicles with secretion of estrogen; the basal body temperature is low (under 98°F) during the preovulatory or follicular phase of the cycle. With ovulation, the corpus luteum secretes progesterone, which has a thermogenic effect on the body, raising the basal body temperature to above 98°F for the length of the luteal phase.

**64. The answer is D.** (*Chapter 31 I A 4, B 2 a, b; II B 1, C 1–5*) A leiomyoma (i.e., myoma or fibroid) is a localized proliferation of smooth muscle cells. Most of these appear in the uterus, but they may also occur elsewhere, including the fallopian tubes, vagina, round ligament, uterosacral ligaments, vulva, and the gastrointestinal tract. Myomas do not have a true capsule; the apparent capsule or pseudocapsule is composed of fibrous and muscle tissue that has been flattened by the tumor. It is believed that myomas are dependent on estrogen for growth because new myomas rarely appear and existing myomas stop growing after menopause. Hyaline degeneration is the *most common* form of degeneration as it is present in almost all leiomyomas. Malignant or sarcomatous degeneration occurs in less than 1% of leiomyomas.

**65. The answer is E.** (*Chapter 7 II E 1 a, b, 2 a–c*) When a pregnant patient has a blood pressure of 140/90 or above with proteinuria, she is preeclamptic and must be hospitalized. Outpatient management is inappropriate for this type of patient. She should be confined to bed in an effort to lessen the blood pressure and improve placental perfusion. Contraction stress tests and serial sonography of the fetus (for continued fetal growth) are necessary to assess fetal well-being. Diuretics in the preeclamptic patient are contraindicated because they can decrease the already contracted intravascular volume and reduce renal perfusion.

**66. The answer is B.** (*Chapter 22 I A 2 a–c, B 2 a–e*) Phase II of the cycle is the fertile phase, the time during which a woman can get pregnant. It is characterized by high levels of estrogen, which stimulate the cervix to secrete clear, abundant cervical mucus. When the estrogen level peaks, ovulation occurs. Since spermatozoa can live in the cervical mucus for several days, birth control is necessary whenever a woman notices vulvar wetness or mucus that stretches out to several inches when held between two fingers.

**67. The answer is C.** (*Chapter 22 I B 3 a, b, C 1*) Phase III of the cycle is the absolutely infertile phase when a woman cannot get pregnant. Ovulation has occurred, and the corpus luteum secretes progesterone. The progesterone causes the basal body temperature to rise 0.4°F–1.0°F and the cervical mucus to become thick and hostile to sperm penetration and migration.

**68. The answer is C.** (*Chapter 29 IV C 5*) Laparoscopy allows direct visualization of the pelvis, including both tubes and ovaries, in which case a tubal pregnancy is obvious. None of the other tests are absolutely diagnostic of an ectopic pregnancy. For example, a culdocentesis that reveals unclotting blood could be due to a bleeding corpus luteum, and serial human chorionic gonadotropin levels that do not rise at the expected rate could be due to an impending abortion.

**69. The answer is E.** (*Chapter 20 II D; V E; VI C 1–6, E 2 e*) Imperforate hymen is a vaginal anomaly, which does not allow menstrual blood to escape. The androgen insensitivity syndrome is characterized by an absent uterus, and a vagina that ends in a blind pouch. Systemic illnesses, such as cystic fibrosis, can be associated with amenorrhea. Turner's syndrome contains only one X chromosome and is associated with amenorrhea. On the other hand, the granulosa–theca cell tumor, which may secrete estrogen, can present as precocious puberty with early (under 9 years of age) uterine bleeding.

**70. The answer is C.** (*Chapter 13 III A 1–3, B 1, 3, 5, 7*) A hydatidiform (hydatid) mole classically presents with mild uterine bleeding and an ultrasound appearance of a snowstorm due to multiple fluid-filled small cystic spaces in the uterus. First-trimester molar pregnancies may mimic a blighted ovum or a threatened abortion, but certainly not a normal pregnancy. If there is a coexistent fetus, hydropic degeneration or an incomplete mole, which are often associated with fetal triploidy, should be excluded.

**71. The answer is C.** (*Chapter 36 II A 1–4*) One of the most sensitive end organs to estrogen is the vagina. Consequently, changes are seen in and around the vagina after the menopause with the natural decline in estrogen. The epithelium becomes pale, thin, and dry. The vagina becomes more narrow with a diminution in the size of the upper vagina. There is a loss of muscle tone, which can predispose to prolapse. Because of the estrogen deficiency, there is a change in the cellular maturation of the vaginal epithelium. There is an increase in the less mature parabasal cells and an absence of the most mature superficial cells. Thus, one would not expect to see superficial cells in a cytologic evaluation of the menopausal vagina.

**72. The answer is C.** (*Chapter 15 II A 1, 2; Table 15-1*) Prenatal exposure to therapeutic radiation in low doses (less than 5 rads) is not believed to be teratogenic. In fact, from a practical standpoint, serious risk to the fetus does not occur until the absorbed dose is 10 rads or more. Since a chest x-ray delivers between 1 and 8 mrads to the ovary and uterus, it is highly unlikely that the risk of a malformation is increased above the baseline (2%–4%). In all pregnancies, there is a 2%–4% risk of congenital malformations regardless of exposure to teratogens in utero. Patients exposed to x-rays in the first trimester may be reassured, after the radiologist computes the exact dose of radiation that could have been delivered to the fetus.

**73. The answer is B.** (*Chapter 21 IV E 1–4*) The luteinizing hormone (LH) surge is central to the ovulation process; it is triggered by an estradiol peak, which exerts a positive feedback effect on LH secretion. Along with the LH surge, there is a similar but smaller rise in follicle-stimulating hormone levels, a response that is progesterone dependent. The LH surge stimulates completion of reduction division in the oocyte, luteinization of the granulosa cells, and synthesis of progesterone and prostaglandins within the follicle. Prostaglandins play a part in the actual discharge of the ovum from the follicle.

**74. The answer is D.** (*Chapter 27 IV A 1–3*) In a woman of childbearing age, pregnancy or a pregnancy-related complication, such as an ectopic pregnancy, must always be ruled out with a presentation of either amenorrhea or unusual bleeding (i.e., spotting). Therefore, measurement of human chorionic gonadotropin (HCG) levels is necessary to rule out or confirm a pregnancy. Hysteroscopy, dilatation and curettage, and basal body temperatures are inappropriate ways of diagnosing a pregnancy. A pregnancy diagnosis should always be excluded in a woman of childbearing age prior to prescribing a progestin, which should not be taken in early pregnancy.

**75. The answer is D.** (*Chapter 21 III B 1–3; IV E 2*) Luteinizing hormone (LH), a glycoprotein secreted by the pituitary gland, stimulates germ cell maturation and completion of reduction division in the oocyte. LH stimulates the theca cells to produce androgens, which are subsequently converted to estradiol in the granulosa cells. LH effects luteinization of the granulosa cells with subsequent progesterone secretion. Follicle-stimulating hormone, not LH, induces the production of LH receptors on the granulosa cells.

**76. The answer is E.** (*Chapter 12 I A; Table 12–1*) A normal, 21-year-old pregnant woman has a 1%–2% chance of having a child with a chromosomal anomaly or a congenital anomaly. Genetic counseling, therefore, is not indicated in this low-risk age-group. The risk is the same in this group even with a previous child with Down's syndrome, for instance. The risk for neural tube defect, however, would change with a family history.

**77. The answer is A.** (*Chapter 9 II B 1–4*) All of the choices listed in the question—hemorrhage, hypertension, infection, and pulmonary embolism—except for premature labor are well-known causes of maternal death. Premature labor is rarely life-threatening unless it is complicated by cardiotoxic effects, caused by β-mimetic agents or an overdose of magnesium, or by side effects of certain drugs, such as steroids, which can cause pulmonary edema.

**78. The answer is E (all).** (*Chapter 3 VI E 1–3*) There is nothing good that results from smoking during pregnancy. The infants are small, and there is an increased perinatal mortality. Carbon monoxide is present in the maternal and fetal circulation, which may cause a functional inactivation of hemoglobin. The nicotine in cigarette smoke causes vasoconstriction and reduced perfusion of the placenta; this, in turn, can lead to intrauterine growth retardation because of the compromised intrauterine environment.

**79. The answer is A (1, 2, 3).** [*Chapter 30 II D 1 a, c (2)*] Human papillomavirus (HPV), a double-stranded DNA virus, is one of the most common viral sexually transmitted diseases (genital warts). HPV types 16 and 18 are associated with squamous cell cancer of the cervix. The koilocyte or "halo" cell is an exfoliated squamous cell with a wrinkled pyknotic nucleus surrounded by a perinuclear clear zone. This virus can be transmitted to the neonate at the time of vaginal birth and is responsible for neonatal/juvenile respiratory papillomatosis.

**80. The answer is C (2, 4).** (*Chapter 32 I B 1–3*) When intravesical pressure exceeds urethral pressure, incontinence will occur. The reverse of the normal pressure differential is commonly a consequence of anatomic distortion mainly from trauma to the bladder neck. Urethral length has no bearing on the mechanism of continence. In continent women, the urethra is an intra-abdominal organ. The Valsalva maneuver increases intra-abdominal pressure, which is equal to bladder and proximal urethral pressure.

**81. The answer is A (1, 2, 3).** (*Chapter 15 II A 1, 2*) Mental retardation frequently results when the fetus is exposed to ionizing radiation early in gestation (between 4 and 12 weeks). This period is one of maximum susceptibility to the adverse effects of radiation as the organs are differentiating and the cells are multiplying; thus, the problems of microcephaly and mental retardation are common. After 20 weeks gestation, however, the effects of ionizing radiation are the same as with postnatal exposure—that is, hair loss, skin lesions, and bone marrow suppression—but not mental retardation.

**82. The answer is D (4).** (*Chapter 1 V B 2*) The corpus luteum of pregnancy is the primary source of progesterone until the seventh to eighth week of pregnancy. Removal of the corpus luteum before that time would produce uterine bleeding, cramping, and spontaneous abortion. By the ninth week of pregnancy, the placenta is the primary source of progesterone, and the corpus luteum contributes only insignificant amounts. Removal of the corpus luteum in the ninth week would, therefore, cause no change in the status of the pregnancy.

**83. The answer is E (all).** (*Chapter 31 III A 1–3*) Depending on the location of the myomas within the uterus, a number of different clinical conditions can exist. The submucous myoma can lead to hypermenorrhea, which results in anemia. The posterior subserous myoma can put pressure on the ureters—leading to hydronephrosis, stasis of urine, and pyelonephritis. An anterior subserous myoma—growing beneath the bladder—can compromise bladder capacity, leading to urinary frequency. Intramural myomas can result in the reappearance of dysmenorrhea after years of pain-free menses.

**84. The answer is D (4).** (*Chapter 2 III A*) Fetal lung maturity is indicated by measuring the surface active lipids (i.e., lecithin and phosphatidylglycerol), which are secreted by fetal lung alveoli. For example, the lecithin/sphingomyelin (L/S) ratio is usually mature at 2/1 at 35–36 weeks. However, lecithin and sphingomyelin are present in both immature and mature fetal lungs, so neither by itself is assurance of fetal lung maturity. The presence of phosphatidylglycerol in the amniotic fluid is assurance of lung maturity. Prostaglandin is involved with the initiation of labor and has nothing to do with fetal lung maturity.

**85. The answer is B (1, 3).** (*Chapter 3 I B 1–5*) The woman in the question is a multigravida because she has been pregnant a number of times. She is a primipara because she has had one pregnancy that resulted in a viable fetus. She is not a multipara (as might be expected with twins) because it is the number of pregnancies reaching viability—in this case just one—not the number of fetuses delivered that determines parity.

**86. The answer is A (1, 2, 3).** (*Chapter 5 II B 1–3*) In the hypoxic fetus, there is a decreased transfer of oxygen to the fetus with a decreased $Po_2$, an increased $CO_2$ level, and an increased $Pco_2$. With an inability on the part of the fetus to sustain an aerobic metabolism, there is a shift to an anaerobic metabolism with an accumulation of organic acids (lactic and pyruvic) and a drop in fetal pH. All of this is analogous to adult respiratory or metabolic acidosis, not alkalosis.

**87. The answer is E (all).** (*Chapter 9 III C 5 a–k*) There are a vast number of potential complications in the uncontrolled diabetic. The cornerstone of treatment, therefore, consists of early rigid control so that early complications (e.g., congenital anomalies), late complications (e.g., stillbirth), and neonatal complications (e.g., hypoglycemia and macrosomia) can be avoided.

**88. The answer is A (1, 2, 3).** (*Chapter 18 III A–C, E 1, 2*) The woman described in the question is in active labor and needs pain relief. She needs something that will take away the pain of cervical dilation (the parasympathetic fibers of the second, third, and fourth sacral nerves) and the pain of uterine contractions (the eleventh and twelfth thoracic nerves). The paracervical block and epidural anesthesia (either lumbar or caudal) successfully block the pain fibers associated with labor. The pudendal block is used for delivery as it provides for perineal anesthesia.

**89. The answer is B (1, 3).** (*Chapter 29 IV C 1 a*) The level of serum human chorionic gonadotropin (HCG) and the rise of the serum HCG are very important in determining the health of a pregnancy. A normal pregnancy exhibits an HCG value that doubles every 2 days. An HCG rise showing a slow rise, one that does not double every 2 days, is suggestive of an abnormal pregnancy, such as an ectopic pregnancy. There should be three and a half doublings in HCG titers tested a week apart. Answer (1), a negative urine HCG taken a week after a positive serum HCG, is suggestive of an ectopic pregnancy because the lower than expected serum HCG level will be too low in the urine to register positive. Answer (2), 175 mIU/ml HCG and 2000 mIU/ml HCG, represents the appropriate doubling of a normal pregnancy. Answer (3), 45 mIU/ml HCG and 400 mIU/ml HCG, does not represent the normal doubling, which is suggestive of an ectopic pregnancy. Answer (4), a negative urine and serum HCG, indicates no pregnancy.

**90. The answer is D (4).** (*Chapter 7 I A 1, 3 a, b*) The screening 1-hour glucose tolerance test was ordered because of the woman's history of a stillborn infant. In a woman with a history of unexplained fetal losses, there should be a high index of suspicion of diabetes. Although the 1-hour test was abnormal with a fasting blood sugar level of 110 (normal: 105 or below), a diagnosis of gestational diabetes cannot be made at this time; thus, there is no need for dietary control or home urine testing. However, the abnormal 1-hour test should be followed up by the standard 3-hour glucose tolerance test.

**91. The answer is E (all).** (*Chapter 21 I B 1–5*) The follicular, or the proliferative, phase of the menstrual cycle extends from day 1 of menstrual bleeding to ovulation. Characteristics of this phase are a variable length, low basal body temperature, growth and development of the ovarian follicles, vascular growth of the endometrium, proliferation of endometrial glands, and secretion of estrogen from the ovary.

**92. The answer is A (1, 2, 3).** (*Chapter 22 IV F 1–4*) Because of the progestogen component in the birth control pill, there is an antiestrogen effect on the normal proliferative, stimulatory influence of estrogen on the breast and endometrium; this is thought to be protective in lessening the chance of the development of cancer in those organs. Because the pill suppresses ovarian activity and the number of ovulatory cycles, there may be protection against the incidence of ovarian cancer. There is no positive effect of the pill on the cervix. Some of the carcinogenic influences on the cervix are viral in nature, and only a barrier method of contraception is, therefore, a positive influence. Since the mode of action of the pill provides no barrier protection, the pill is not protective against cervical cancer.

**93. The answer is E (all).** (*Chapter 1 I A 3 a; III D; V D 1–5*) Progesterone is elevated many times above the normal luteal phase level because of the large amounts of the hormone produced by the placenta; progesterone levels continue to rise during pregnancy, reaching 250 mg/day at term. Human placental lactogen is only found in pregnancy, as it is manufactured by the placenta; it rises steadily during the first and second trimesters and disappears rapidly after delivery. Both thyroxine and cortisol are elevated during pregnancy because their respective binding proteins are elevated due to elevated levels of circulating estrogen. The measured amounts of these two hormones are only artificially elevated (due to the elevated binding proteins) because the free fraction of both thyroxine and cortisol changes little. Thus, the metabolic processes dependent on these hormones are usually unaltered.

**94. The answer is B (1, 3).** (*Chapter 10 V B 3 b*) The patient described in the question may be suffering from decreased respiratory drive secondary to hypermagnesemia. An initial evaluation should check for the presence or absence of deep tendon reflexes and a chest and cardiac examination to rule out pulmonary edema. An arterial blood gas and serum magnesium levels should be obtained. If reflexes are absent, the intravenous magnesium should be discontinued. If necessary, 10 ml of a 10% solution of calcium gluconate can be administered intravenously over 30 minutes.

**95. The answer is A (1, 2, 3).** (*Chapter 21 IV F 2 a–c, 3*) An abnormal luteal phase function is important because it can be the cause of both infertility and pregnancy wastage. Since oogenesis is sequential, it is logical that poor follicle-stimulating hormone stimulation of the follicle can result in depressed estradiol levels, which can affect the midcycle luteinizing hormone (LH) surge; the depressed LH surge can, in turn, affect the formation of the corpus luteum with a small luteal cell mass and a low midluteal progesterone production. Human chorionic gonadotropin is not involved in corpus luteum maintenance unless conception has occurred.

**96. The answer is E (all).** (*Chapter 14 II E 1–5*) The diagnosis of placental abruption is basically a clinical diagnosis. Ultrasound is not much help in defining the separation or a retroplacental clot. As the placenta separates, there are signs of placental insufficiency, that is, late decelerations and loss of variability on the fetal heart monitor. The contraction monitor may show a coupling of contractions with poor uterine relaxation between the contractions. A more subtle clue suggesting abruption is continued premature labor that tocolytic agents cannot quiet.

**97. The answer is C (2, 4).** (*Chapter 10 II B 4, 5; V B 3 c, 4; Chapter 14 II E, F*) Given the abrupt onset of symptoms with vaginal bleeding, a placental abruption is likely. Until the etiology of bleeding is clarified, it is important not to administer tocolytic agents, which may cause vasodilation with hypotension. Since this patient is 35 weeks pregnant, significant bleeding or fetal compromise is best treated by immediate delivery.

**98. The answer is A (1, 2, 3).** (*Chapter 14 II F 3, G 2 a, b*) Even though the fetus is dead and there may seem to be no urgency in getting the delivery accomplished, there is a risk of developing consumptive coagulopathy with delay in emptying the uterus. The thromboplastin that enters the circulation will continue to enter the circulation as long as the retroplacental clot remains in situ. Watchful waiting is inappropriate. While stimulating labor with rupture of membranes and an oxytocin infusion, platelet levels must be monitored to pick up early consumptive coagulopathy.

**99. The answer is A (1, 2, 3).** (*Chapter 30 II C 1 b, d*) Lymphogranuloma venereum is caused by *Chlamydia trachomatis,* serotypes $L_{1-3}$. The initial lesion is a painless vesicular or papular lesion in the genital region. The secondary stage is accompanied by lymphadenopathy of the inguinal lymph nodes, which become matted and retract the overlying subinguinal skin, creating the "groove" sign. In advanced or untreated states, ulceration, fistulization, and destruction of recognizable anogenital structures occur. The treatments of choice are either erythromycin or tetracycline. *C. trachomatis* is resistant to penicillin.

**100. The answer is D (4).** (*Chapter 5 III C 1 a-f, 3 d*) Fetal heart rate decelerations associated with fetal head compression are classified as early decelerations and are benign. The decelerations are mirror images of the contractions, beginning at the onset of the contraction. There is no fetal hypoxia and no association with a poor fetal outcome. It is the variable deceleration that is associated with oligohydramnios.

**101. The answer is E (all).** (*Chapter 10 II A 1-5; IV B 2-4*) The woman described in the question suffers from an incompetent cervix. The incompetent cervix may be acquired congenitally or may be secondary to a previous gynecologic procedure, such as dilatation and evacuation. Once suspected, it should be ruled out by hysterosalpingography, which would show dilation of the internal os. This study would also serve to exclude an associated uterine anomaly, which may increase the risk for preterm delivery and abnormal presentation of the fetus.

**102. The answer is E (all).** (*Chapter 21 V A 1-2, B*) When pregnancy does not occur in a cycle, steroid production from the corpus luteum declines, leading to increased coiling and constriction of the spiral arteries in the endometrium. The decreased blood flow leads to ischemia and degradation of the endometrial tissue with subsequent desquamation or shedding of that tissue, which results in bleeding. Soon after the menses begin, regeneration of the surface epithelium under the influence of estrogen occurs at the same time that the endometrium is shedding.

**103. The answer is A (1, 2, 3).** (*Chapter 26 II A 1, 2*) The sensory nerve supply of the vulva includes contributions from the perineal, genitofemoral, and ilioinguinal nerves. These sensory nerves are responsible for most of the "symptoms" of vulvovaginitis; that is, pruritus, burning, and soreness. Many times in acute vulvovaginitis, superficial inguinal adenopathy can be seen. The vestibule, which many physicians mistake as being part of the vagina, is a nonkeratinized mucus membrane in between the inner aspect of the labia minor and the hymenal rings. The secretory glands (Skene's and Bartholin's) empty into the vestibule.

**104. The answer is D (4).** (*Chapter 11 IV A 1-4*) It is imperative to know the Rh type of the infant's father (assuming no false paternity). An Rh-negative man cannot father an Rh-positive infant. In this situation, both Rh immune globulin and further testing can be avoided. Screening for a collagen vascular disorder is inappropriate in the clinical situation described in the question since there is concern about the red blood cell surface antibodies, not antinuclear antibodies. The reticulocyte count and bilirubin level in the mother should be *normal* since she is *not* anemic.

**105. The answer is A (1, 2, 3).** (*Chapter 33 IV F 2*) The luteal phase defect implies a deficiency in progesterone secretion from the corpus luteum, which is treated by either exogenous progesterone or increased endogenous progesterone. Progesterone suppositories during the postovulatory part of the cycle (the luteal phase) can help, and human chorionic gonadotropin given once or twice during the luteal phase can stimulate the corpus luteum to produce more endogenous progesterone. Clomiphene citrate stimulates the entire ovarian cycle, hopefully creating a more functional ovulation with a better quality corpus luteum. Low-dose estrogen is used in the preovulatory period to stimulate the production of good cervical mucus.

**106. The answer is A (1, 2, 3).** (*Chapter 24 IV B 4; V C 1-8; VII A 1 a*) Ectopic pregnancy should always be considered in a young woman who presents with abdominal pain, particularly if she is sexually active and late for a period. The pain of an ectopic pregnancy results from tubal distension and intra-abdominal blood. The Fitz-Hugh–Curtis syndrome is a pelvic inflammatory infection that spreads to the upper abdomen; it presents as perihepatitis and right upper quadrant abdominal pain. Appendicitis should always be part of the differential diagnosis of lower abdominal pain when there is evidence of an infectious process. Pelvic tuberculosis *rarely* causes acute abdominal pain; it is usually suspected as a causative agent when there is a history of pulmonary TB, a pelvic mass, long-standing infertility, and pelvic adhesions.

**107. The answer is A (1, 2, 3).** (*Chapter 22 II A 1-3, D 2, 3*) Condoms, sponges, diaphragms, and spermicides are good forms of birth control when used faithfully. However, even with regular use they are not 100% effective. The barrier forms of contraception do protect against sexually transmitted diseases and are relatively free from side effects, but they do not prevent cervical cancer; however, it is thought that because they protect the cervix from sexually transmitted diseases that promote cervical neoplasia, they may indirectly decrease the incidence of cervical cancer.

**108. The answer is E (all).** (*Chapter 9 III A 3 a*) Heavy cigarette smoking is associated with many pregnancy complications due to abnormalities of early placental development and supply. This is manifested by bleeding complications (i.e., placenta previa and abruptio), abnormalities of fetal growth and development because of low oxygen and elevated carbon monoxide levels, and the worst case scenario, fetal or neonatal death. Premature rupture of membranes is an additional associated risk.

**109. The answer is C (2, 4).** (*Chapter 26 II B 1; V B 3, C 3*) *Gardnerella vaginalis* vaginitis is characterized by a malodorous watery vaginal discharge. Vaginal pH in the normal vagina is approximately 4.5. In *G. vaginalis*, the vaginal pH is between 5.0–5.5. Wet mount preparations in *G. vaginalis* always have "clue" cells and a "clean" background, unlike Trichomonas infections in which there is an intense inflammatory reaction. The "clue" cell, a squamous cell whose sharp borders and cytoplasm are obscured by coccobacillary bacteria, are diagnostic of *G. vaginalis*. Application of a 10% potassium hydroxide to the wet mount specimen produces a fishy odor and is called a positive "whiff" test.

**110. The answer is D (4).** [*Chapter 23 V D; VI A 1 b (1)–(5)*] The patient described in the question presents a diagnostic dilemma. On the one hand, she has some reproductive complaints and an associated low human chorionic gonadotropin (HCG) β-subunit level. On the other hand, she has an enlarged uterus and a history of irregular menses. The differential diagnosis includes a normal intrauterine gestation with an associated corpus luteal cyst, causing right lower quadrant pain, a threatened abortion, or an ectopic pregnancy. The woman should be informed of these possibilities and told to notify the physician if the pain increases. The optimum management at this point is to repeat the HCG β-subunit measurement in 24 hours to see if it is rising appropriately. Pelvic ultrasound would not be helpful since its sensitivity begins at 5800 mIU/ml. Diagnostic laparoscopy and culdocentesis are invasive procedures that should be used only after more information is obtained.

**111. The answer is E (all).** (*Chapter 33 III A 1, 2*) A poor postcoital test means that there is some problem with the semen, the deposition of the semen onto the cervix, or the cervical mucus. If the sperm count is too low, the sperm motility is poor, or the coital technique does not provide adequate insemination of the cervix, few motile sperm will be seen in the cervical mucus. Hostile cervical mucus may not support the sperm so that none of them move in the mucus.

**112. The answer is B (1, 3).** (*Chapter 20 V E 1, 2; VI C 2, 6 a, c*) Hypo- or hypergonadotropic syndromes are characterized by the absence of secondary sex characteristics, infantile but normal genitalia, and streak-like gonads that are devoid of germ cells. The presence of the Y chromosome in some of these syndromes dictates early removal of the gonads because of their propensity for malignancy. Androgen insensitivity syndrome, or testicular feminization, has XY gonads, which are removed after puberty. Mixed gonadal dysgenesis has a mosaic chromosomal pattern that contains a Y chromosome (45,X/46,XY), and these gonads are removed before puberty. There is no Y chromosome in either Turner's syndrome or Kallmann's syndrome.

**113. The answer is C (2, 4).** [*Chapter 23 VI B 1 b (1) (a), (b)*] The young woman described in the question complains of primary dysmenorrhea. Because her menses were 5 weeks ago, a physician is obligated to rule out pregnancy. This can be done by measuring human chorionic gonadotropin β-subunit levels or by having the patient call within the next few weeks if the menses have not occurred. However, irregular menses within several years after menarche are not unusual. After her next menses, this young woman should begin oral contraceptives or antiprostaglandins to limit the prostaglandin effects on the uterus, which are responsible for her symptoms. However, pregnancy must be ruled out before contraceptives or antiprostaglandins are begun.

**114. The answer is A (1, 2, 3).** (*Chapter 28 III C 2, D 1, 2; IV B*) Motile spermatozoa found on a wet mount preparation indicate the deposition of those spermatozoa within the previous 24 hours. The Wood's light fluoresces in the presence of histone, which is found in semen. Acid phosphatase is suggestive of semen, especially if found in large concentrations. A pregnancy test (high human chorionic gonadotropin levels) could not possibly be positive at the time of the rape unless a woman was already pregnant.

**115. The answer is E (all).** (*Chapter 31 I B 2 a–d*) Observations of the appearance, growth, and disappearance of myomas during the life of a woman suggest that they are dependent on estrogen for growth. At times of low or absent estrogen secretion, such as the premenarchal or postmenopausal years, myomas are either absent or stop growing. At times of high estrogen secretion, such as pregnancy or conditions that lead to endometrial hyperplasia, myomas may grow rapidly or appear for the first time.

**116. The answer is B (1, 3).** [*Chapter 16 II E 1 b (1)–(3)*] Meconium plugs cause segmental airway obstruction of portions of lung tissue, resulting in atelectasis with associated ventilation perfusion irregularities. There is also pulmonary arteriolar spasm, resulting in pulmonary arterial hypertension. This encourages persistent flow of blood from right to left through the ductus arteriosus. Meconium is not laden with bacteria. It is a sterile substance that can act as a culture medium for bacteria, but it does not carry bacteria normally in utero. Meconium does not pass into the general circulation through the lung. Its particulate matter is trapped in the lung parenchyma.

**117. The answer is B (1, 3).** (*Chapter 25 IV B 2, 4; V A–C*) The woman described in the question has mild endometriosis. Because she is a young single college student, she has no immediate plans for pregnancy; therefore, using gonadotropin-releasing hormone agonists or danazol is not appropriate therapy at this time. Although danazol can be used prophylactically, it is not helpful in a woman with minimal disease. The nonsteroidal anti-inflammatory agents are helpful with the dysmenorrhea, and the birth control pills help to prevent further growth of the endometriosis by minimizing the amount of monthly endometrial shedding.

**118. The answer is C (2, 4).** (*Chapter 34 II A 2 a, b*) Transverse vaginal septum and imperforate hymen in a teenager with pelvic pain demand immediate attention. In both cases, the vaginal outlet, which is blocked, must be opened to prevent the continued reflux of menstrual material into the pelvis with the potential for endometriosis and impaired fertility. Although vaginal agenesis and uterine anomalies result in reproductive failure, they are not medical emergencies.

**119. The answer is C (2, 4).** (*Chapter 24 VI C 1 a–c*) The combination of medications used to treat acute pelvic inflammatory disease must cover a variety of pathogens—aerobes, anaerobes, gonorrhea, and *Chlamydia.* Therefore, the two anaerobic agents (clindamycin/metronidazole) or two aerobic agents (ampicillin/cefoxitin) would not be effective. The other two possibilities represent combinations that adequately cover all the potential pathogens.

**120. The answer is A (1, 2, 3).** (*Chapter 31 IV A–D*) A 49-year-old woman with significant myomas can be treated in a number of ways. Because she is near menopause at which time the myomas should regress spontaneously, she can be observed. However, it is hard to palpate the ovaries with a big uterus; thus, the ovaries should be followed with periodic pelvic ultrasound examinations. Gonadotropin-releasing hormone agonists can be administered to make the myomas regress, although the regression is temporary (i.e., while the medication is given); however, menopause may intervene before the myomas regain their former size. A hysterectomy would be appropriate for a woman with a myomatous uterus of the size described in the question. Myomectomy should not be performed on a woman so close to menopause as it is a temporizing operation, and there is no need to preserve a myomatous uterus in a 49-year-old woman.

**121. The answer is B (1, 3).** (*Chapter 18 IV C 1 a–d, 3 a–e*) A potential problem with obstetric anesthesia is aspiration of gastric contents during anesthesia, especially general anesthesia. Gastric emptying is significantly delayed during labor, even if the last meal was many hours prior to the need for anesthesia. Prior to and during intubation during general anesthesia, the gastric contents are neutralized with antacids, and cricoid pressure is used to compress the esophagus. Corticosteroids and antibiotics are used after aspiration has occurred in an effort to control the often severe pneumonitis that develops.

**122. The answer is A (1, 2, 3).** (*Chapter 37 III C 2 b*) In this case, endometrial hyperplasia was discovered in a young woman who wants children. First, the hyperplasia must be treated; this is done with cyclic combination estrogen/progestin therapy for three cycles followed by an endometrial biopsy at the conclusion of the therapy to see if the lesion has cleared. Since she wants to become pregnant and since her cycles are irregular, and probably anovulatory with unopposed estrogen, ovulation induction is appropriate. Cyclic progestin therapy would make her have regular cycles, but she would not ovulate without stimulation with clomiphine.

**123. The answer is B (1, 3).** (*Chapter 6 III A 1; IV*) The workup of a woman who presents with prolonged pregnancy should begin with measurement of urinary and plasma estriol but only after a nonstress test of fetal well-being has been performed. More than one test of fetal well-being should always be performed in assessing fetal risk and making the decision to deliver or await spontaneous labor. Therefore, the least invasive yet reliable tests are performed first.

**124. The answer is C (2, 4).** (*Chapter 29 II A 1, D*) Chronic salpingitis leads to fibrosis and scarring of intraluminal structures so that a fertilized egg can become trapped in the tube. When pregnancies occur with an intrauterine device in situ, a disproportionate number of these pregnancies are ectopic. Endometritis and adenomyosis have no association with ectopic pregnancies as neither has an effect on the tube.

**125. The answer is A (1, 2, 3).** (*Chapter 11 IV A 1–3; V B 1–10, C*) The clinical presentation of a woman who is sensitized during a pregnancy, despite administration of $Rh_o$ (anti-D) immune globulin illustrates several points. Immune globulin only protects against D-positive cells. If there was a fetomaternal hemorrhage of fetal red cells that have other foreign antigens on the surface, then sensitization could occur. In 1% of pregnancies, sensitization occurs from a fetomaternal hemorrhage during the pregnancy. If there was a significant transplacental bleeding before 28 weeks gestation when the $Rh_o$ (anti-D) immune globulin was given, then the immune globulin would not have been fully protective against maternal sensitization. Moreover, if an inadequate dose of immune globulin was given, such as after a massive fetomaternal hemorrhage sometimes found in placental abruption or after traumatic amniocentesis, then sensitization would have occurred despite the $Rh_o$ (anti-D) immune globulin. The fact that this woman's first child might have been Rh negative would not explain erythroblastosis fetalis in the second and third pregnancies. It is possible to have an Rh-negative infant and in future pregnancies become sensitized against non-Rh antigens. It is impossible for the first infant to be D negative or Du negative and develop anti-D antibodies.

**126. The answer is D (4).** (*Chapter 35 II E 1–3*) Excessive hair growth depends on an increased availability of the biologically active androgen, testosterone. This can occur either by increasing the amount of free testosterone or by increasing the conversion of testosterone to dihydrotestosterone within the cell. Thus, an increased amount of free, not bound, testosterone, an increase, not a decrease, in $5\alpha$-reductase activity, or a decrease in the concentration of sex hormone binding globulin with the consequent increase in free testosterone are expected in hirsutism. Increased levels of luteinizing hormone may suggest polycystic ovary disease but are not directly involved in the pathophysiology of hirsutism.

**127 and 128. The answers are: 127-C, 128-A.** (*Chapter 37 II E 1 a–c; III A 2 b, B, D 5 a*) In any woman whose colposcopically directed biopsy demonstrates microinvasion, an adequate histologic assessment in the form of a conization must be done to evaluate the extent of disease and the lymphovascular space involvement. The definition of microinvasive carcinoma of the cervix includes a depth of invasion of less than 3 mm, no lymphovascular space involvement, and no confluent tongues of tissue. If these conditions are met, appropriate treatment is total abdominal hysterectomy.

The woman with a normal appearing cervix with normal cytology is unlikely to have a cervical malignancy. The pattern of bleeding, that is, every 3–6 months, suggests anovulatory cycles, which places this patient at risk for endometrial hyperplasia or carcinoma. Thus, this woman with abnormal uterine bleeding must be evaluated by fractional dilatation and curettage to exclude endometrial cancer prior to initiating any therapy.

**129–131. The answers are: 129-B, 130-A, 131-C.** (*Chapter 15 Table 15-2*) Valproic acid taken during pregnancy has been associated with several congenital malformations, including renal defects, neural tube defects, and cleft palate. This drug should be avoided, if possible, by women during their reproductive years.

The warfarin embryopathy includes multiple structural and developmental defects. This is one situation in which in utero exposure to a medication (i.e., warfarin) produces a syndrome with features typical of an inherited disorder, Conradi's syndrome (chondrodysplasia punctata).

The administration of tetracycline to a pregnant woman before the tenth week postconception has no known deleterious effect since bone development occurs later. If given beyond the fourth month, there may be yellowing of the deciduous teeth and difficulty with enamel development, but there are no adverse effects on permanent teeth.

**132–134. The answers are: 132-B, 133-D, 134-B.** (*Chapter 35 IV B 1–4; V B 3–5, C 1–2, D*) In congenital adrenal hyperplasia, there is an enzyme deficiency along the steroid pathway to cortisol. As a result of the deficiency, there is a deficiency of the end product, cortisol, and a buildup of the intermediate metabolites in front of the enzyme block. Increased levels of both dihydroepiandrosterone sulfate and 17-hydroxyprogesterone reflect the results of the enzyme deficiencies seen in congenital adrenal hyperplasia. In each case, therefore, the appropriate therapy would be the administration of a corticosteroid, such as prednisone. The provision of the end product would cause a decrease in adrenocorticotropic hormone secretion, a reduced stimulus of adrenal steroidogenesis, and a decreased secretion of the androgenic substances along the steroid pathway.

Since elevated levels of androstenedione point to an ovarian source of the androgens, the birth control pills would be appropriate therapy to shut down ovarian steroidogenesis, thereby reducing the androgenic substance (androstenedione) coming from the ovary.

**135–138. The answers are: 135-B, 136-D, 137-C, 138-D.** (*Chapter 27 I C, D; III B 1–4; IV A 1 a–c, B 1, C 1, D 1, E 1–4*) The 30-year-old woman with menorrhagia is experiencing excessive bleeding during her menses. Her cycles are regular, and she has no ongoing, heavy bleeding that must be controlled. Because her cycles are regular, she is probably ovulating and needs no diagnostic curettage. Because her bleeding is neither irregular nor continuous, she does not need estrogen or progestin therapy. Nonsteroidal anti-inflammatory drugs, however, can help to lessen menstrual blood loss in women who ovulate.

The 35-year-old woman who presents with continuous uterine bleeding despite hormonal therapy should not have been given any kind of medication without a diagnosis. Because of the continuous bleeding, it is inappropriate to add either estrogen or progestin alone without first sampling the endometrium to rule out an endometrial pathology. The only management at this point is to perform a dilatation and curettage, which should have been done before starting her on the oral contraceptives.

The 23-year-old woman who presents with a 3-week history of bleeding needs something immediately to control her uterine bleeding because of the amount of blood she has lost. She is most probably not ovulating as indicated by the long periods between her bleeding episodes. She is experiencing estrogen breakthrough bleeding associated with long periods of unopposed estrogen. The therapy that will most quickly control this type of bleeding is oral conjugated estrogen, which provides rapid growth of endometrial tissue. A progestin should be added after the bleeding is under control. A progestin or oral contraceptives are not as effective in controlling dysfunctional uterine bleeding as estrogen alone.

The 47-year-old woman with menometrorrhagia is a classic presentation of perimenopausal bleeding, which is characterized by irregular anovulatory bleeding and sustained unopposed estrogen. The combination of obesity, hypertension, and irregular perimenopausal bleeding demands that an endometrial neoplasia be ruled out. Because of the possibility of endometrial hyperplasia or carcinoma in women of this age, no hormonal therapy can be given until there is a diagnosis, which can be achieved with a dilatation and curettage to sample the endometrium.

**139–142. The answers are: 139-C, 140-A, 141-D, 142-B.** (*Chapter 38 VII A 1, 2, B 1 a, b; VIII A–C*) In cases involving wrongful conception, namely parents seeking compensation for a normal child resulting from a failed sterilization, willingness to compensate has been low. In cases where the resulting child was abnormal, medical expenses for the care of the infant have been granted. Important to the determination of wrongful conception is documentation of whether the mother was informed of the possibility of failure of the sterilization procedure.

Wrongful life actions brought by a child, alleging that no life would have been better than life with congenital defects, have generally been unsuccessful. Compensation prior to these injuries, however, may be granted on negligence theory.

Wrongful birth actions are brought by the parents of a child with a congenital defect, alleging that a physician was remiss in genetic counseling, and because of this, a defective child was allowed to be born. Generally, these cases have been successful, especially in cases where testing would have been easy, such as in parental testing for Tay-Sachs disease.

Birth injury cases are one of the most common cases plaguing obstetricians. A child may bring suit until the age of majority, after which time he or she is dependent upon the statute of limitations of the particular jurisdiction. Since children represent a great emotional investment, a child who is abnormal in any way may be suspected of being a victim of an obstetrician's malpractice. Although the association of cerebral palsy and fetal heart monitoring is a weak one, these cases often prevail on public sentiment and liability based on a preponderance of evidence.

Negligence in obtaining informed consent is illustrated by the claim by the woman who stated that she was not informed of the possibility of a cesarean section. The physician's defense would be based on the fact that most pregnant women should know that cesarean section is always a possibility in the delivery process. This type of case is not a strong one.

**143–145. The answers are: 143-D, 144-A, 145-C.** (*Chapter 2 IV E 1 b, c, 2*) With the prolonged rupture of membranes and the *Streptococcus* organism in the vagina, a postpartum uterine infection must be suspected. The high temperature early in the puerperium and the lower abdominal tenderness, probably myometrial and parametrial in origin, are highly suggestive of a parametritis.

A spiking fever that does not respond to antibiotic therapy 5 days after cesarean section is classic for pelvic thrombophlebitis, especially after pelvic surgery. There are no pelvic physical findings.

The diagnosis is confirmed when heparin therapy is instituted, and the temperature curve gradually returns to normal.

Conduction anesthesia may result in postpartum urinary retention because of the effect of the anesthesia on the autonomic nervous system. When this happens a woman often needs to be catheterized once or twice before the anesthesia has completely worn off. The catheterization can easily introduce bacteria and seed a urinary tract infection. The chills and back (kidney) pain are common symptoms of a urinary tract infection.

**146–148. The answers are: 146-D, 147-B, 148-B.** (*Chapter 19 II B 2, C 5, E 1–4; III D 2 c*) The human chorionic gonadotropin level is unimportant in the woman who presents with vaginal bleeding and no fetal heart sounds. Because the most recent pregnancy was 4 months ago, this pregnancy could only be 12 weeks at the most; however, the top of the uterus is at the level of the umbilicus (i.e., the size of a 20-week pregnancy) and, thus, is larger than expected. With a positive pregnancy test, bleeding, a large uterus, and no fetal heart tones, a molar pregnancy must be ruled out. Thus, a pelvic ultrasound examination is indicated to determine the uterine contents.

It appears that the woman with metastatic gestational trophoblastic neoplasia (GTN) has been successfully treated. The follow-up protocol suggests monthly HCG titers for a year after they have been negative every 2 weeks for 3 months. A chest x-ray is required every 3 months. This woman had her previous chest x-ray 3 months ago and needs another one to compliment the monthly HCG levels in order to follow the course of the disease appropriately.

Following evacuation of a molar pregnancy, the HCG titer is expected to become negative over 2–3 months; this must be carefully followed with weekly and then monthly HCG titers. If the HCG values plateau or rise, there is a possibility of metastatic or nonmetastic GTN. Therefore, a chest x-ray is indicated to rule out pulmonary metastasis.

**149–151. The answers are: 149-A, 150-B, 151-D.** (*Chapter 14 I E 1; II D 1–3*) The picture of painless vaginal bleeding in the third trimester strongly suggests placenta previa. One of the associated findings of placenta previa is fetal malpresentation, such as a breech. In this case, the placenta occupies the lower uterine segment, which causes the fetal head to occupy the roomier fundus of the uterus.

The clinical picture of a woman at term who presents with a tender uterus, tense abdomen, and no audible fetal heart tones is one of severe abruptio placentae, in which there has been almost complete separation of the placenta, leading to a rigid uterus and a dead fetus. One of the complications of severe abruptio placentae is renal failure, the chief sign of which is oliguria.

The clinical picture of a woman with moderate vaginal bleeding and uterine contractions and a uterus that does not entirely relax between contractions is one of mild to moderate abruptio placentae in which there is usually fetal distress but not to the extent that the fetus is immediately in jeopardy.

**152–154. The answers are: 152-B, 153-D, 154-A.** (*Chapter 4 II C 2; IV B 4, 5 b*) The woman who is having frequent painful contractions that have no gradient and cause no change in the cervical dilation is experiencing hypertonic uterine dysfunction. Oxytocin will not help this situation, but strong sedation, such as morphine, will relieve the pain, relax the patient, and result in a normal labor pattern.

The woman who is experiencing progressive cervical dilation is in the active phase of labor. Contractions are at intervals of 2–4 minutes and are strong enough to effect the cervical change.

The woman with contractions at irregular intervals is in false labor. The intensity of the contractions remains the same and is chiefly in the lower abdomen. There is no change in the cervical dilation. The contractions are usually relieved and often stopped with a sedative.

**155–157. The answers are: 155-A, 156-D, 157-B.** (*Chapter 36 IV A–C; V A 1 a, b, 2 a, b, C 3 b, 4*) The 55-year-old woman described in the question has vasomotor instability and needs hormonal replacement. She has two contraindications to estrogen replacement, migraine headaches and a family history of breast cancer. Even though it is less effective than estrogen in offering symptomatic relief, medroxyprogesterone (Depo-Provera) is useful when estrogen cannot be used.

The 49-year-old woman described in the question presents with signs of unopposed estrogen stimulation of the endometrium—namely, irregular cycles and intermenstrual bleeding. Even though she is symptomatic at this point, additional estrogen is contraindicated without first sampling the endometrium to rule out endometrial neoplasia. Therefore, an endometrial biopsy is indicated as treatment for this patient.

The 54-year-old woman described in the question presents with symptoms of atrophic changes of the vagina—an atrophic vaginitis with dyspareunia. Discharge and spotting are common with atrophic vaginitis. Because she had an endometrial biopsy 3 months previously, it is unlikely that the spotting is a result of endometrial pathology. Because the vagina is the chief source of discomfort, local vaginal estrogen cream should correct the problem.

**158–160. The answers are: 158-E, 159-A, 160-B.** (*Chapter 34 I A 1 a; II A 1 a, b; IV C, D*) Hypogonadotropic, or secondary, amenorrhea is the most common amenorrhea. It occurs after a menstrual pattern has been established and has a number of etiologies, including emotional stress, drugs, nutritional deficiencies, abnormalities of the hypothalamic–pituitary axis, and excessive exercise. There is an inadequate amount of pituitary gonadotropin secretion, which, in turn, does not stimulate the ovary, leading to low estrogen output and absent menses. Marathon runners are known to occasionally experience this reversible type of amenorrhea.

A key feature of eugonadotropic amenorrhea is functional ovaries, which secrete estrogen, ovulate, and have the normal feedback on the pituitary, resulting in normal gonadotropin levels. This type of amenorrhea is characterized by an abnormality of the outflow tract, either congenital or acquired, which prevents the egress of menstrual blood. In this particular case, the young woman has normal ovaries and a congenital defect of the müllerian ducts, resulting in an absent uterus and a vestigial vagina, as seen in the Rokitansky-Küster-Hauser syndrome.

Physiologic amenorrhea occurs quite normally at certain times in the lives of most women, such as after menarche, during pregnancy and lactation, and after menopause. The gonadotropin levels may be in the normal range (after menarche), low (during pregnancy and lactation), or high (after menopause). The adolescent described in the question had a normal menarche with three subsequent menses before becoming amenorrheic. It is not uncommon to have intervals of amenorrhea lasting 2–12 months within the first 2 years after menarche.

**161–164. The answers are: 161-B, 162-A, 163-D, 164-E.** (*Chapter 20 II C; IV A 2; V A; VI B 1, E 1 a*) Bleeding in adolescent girls that lasts more than 7 days accompanied by anemia is known as dysfunctional uterine bleeding. The usual etiology is an immature hypothalamic–pituitary axis. Other causes include psychogenic factors, juvenile hypothyroidism, and coagulation disorders (von Willebrand's disease).

Heterosexual precocious puberty is the development of secondary sex characteristics that are *inconsistent* with genetic sex. The etiology may include tumors (arrhenoblastomas), congenital adrenal hyperplasia, or chronic ingestion of androgenic preparations.

Congenital adrenal hyperplasia results when enzymatic regulation of the biosynthesis of cortisol and aldosterone is impaired. The 21-hydroxylase defect is the most common cause of distinct virilization of the female newborn.

Clear cell adenocarcinoma, which is found at or after puberty, is a malignancy associated with maternal ingestion of diethylstilbestrol (DES).

Epidermolysis bullosum is a skin disease that is often responsible for labial agglutination (adhesion of the labia minor in the midline).

**165–166. The answers are: 165-D, 166-A.** (*Chapter 33 III E 2 b, F 3 a–b*) With persistent short luteal phases determined by a basal body temperature graph and a very early spontaneous abortion, the suspected diagnosis must be an inadequate corpus luteum. The best test for an inadequate corpus luteum is the endometrial biopsy, which shows whether or not the endometrial histology or the preparation of the endometrium is in phase with the luteal phase as seen on the temperature graph.

With apparently good sperm and mucus interaction and regular ovulations, but infertility for 3 years, the woman's tubal function must be evaluated. The acute appendicitis and appendectomy may be precursors of pelvic adhesions and poor tubal function. Direct observation of the pelvic structures via the laparoscope would be the indicated procedure at this point.

**167–169. The answers are: 167-A, 168-D, 169-B.** (*Chapter 12 III C 2 a, b, D 2 a; IV*) The maternal serum α-fetoprotein is the best screening test for a neural tube defect, such as meningomyelocele. If the α-fetoprotein is 2.5 times normal or above, an amniocentesis is then performed to see if the α-fetoprotein is also elevated in the amniotic fluid.

The diagnosis of Tay-Sachs disease is made by demonstrating a deficiency of the enzyme β-D-hexosaminidase A in cultured amniotic fluid cells. These fetal cells are obtained via amniocentesis.

The chorionic villi sampling supplies trophoblastic tissue for evaluation. Chromosomal abnormalities and hemoglobinopathies, such as sickle cell disease, can be identified with this procedure.

**170–171. The answers are: 170-B, 171-D.** (*Chapter 7 II E 1 b, 2 a–d*) The patient who demonstrated on two occasions that she has mild preeclampsia with a blood pressure of 140/90 and 1+ proteinuria is no longer a candidate for outpatient management. She must be hospitalized and confined to bed with fetal monitoring. She does not yet need to be delivered, and she is not sick enough to begin magnesium sulfate.

The patient with blood pressure of 160/110 and 3+ proteinuria is a very sick woman with severe preeclampsia. The pregnancy should be terminated to control the disease process, but she is too sick at this point to think of either induction of labor or cesarean section. She may have a seizure at any moment and, therefore, must be stabilized before any attempt at delivery is undertaken. She must be started on magnesium sulfate, an anticonvulsant, and stabilized before the pregnancy is terminated.

**172–175. The answers are: 172-A, 173-D, 174-B, 175-C.** (*Chapter 17 I D 4 b (1)–(3); IV B 1, 2*) The eponyms Kerr, Sellheim, and Sanger are derived from physicians who pioneered procedures for cesarean section; McDonald and Shirodkar pioneered procedures for cervical cerclage. The differences in the procedures for cesarean section are in type and location of the incision into the uterus: the Kerr incision is low and transverse, the Sellheim incision is low and vertical, extending into the corpus, and the Sanger incision is a longitudinal incision that is entirely in the corpus.

The cerclage procedures differ in the technique of placing the encircling suture about the cervix. Because the Shirodkar suture is buried beneath the cervical mucosa, it can be left in place for a subsequent pregnancy if a cesarean section is performed. The McDonald suture is the simplest procedure that incurs less trauma to the cervix and less blood loss than the Shirodkar suture. It is a simple purse-string suture of the cervix.

**176–178. The answers are: 176-C, 177-D, 178-A.** (*Chapter 18 II C; III C 1, E 1, 2*) The woman who is still only 2 cm dilated after 3 hours of labor is experiencing a nonprogressive, painful type of labor; she is having regular contractions but showing no cervical change. Her labor is hypertonic and dysfunctional. Potent sedation with morphine is indicated at this point to give the patient a much needed rest. Often the labor will become progressive when the morphine wears off.

The woman who is 5–6 cm dilated is in active, progressive labor. The vertex is well within the pelvis. A vaginal delivery can be anticipated in such a patient. Since she needs pain relief, an epidural is appropriate because she needs an anesthetic that will last for an indefinite period of time and that will provide anesthesia for the delivery. Neither a spinal nor a pudendal block will accomplish both purposes; a spinal lasts for a finite period of time, and a pudendal block will not block the pain of labor.

The woman who is in the delivery room needs a quick, simple anesthesia for the perineum. A spinal and an epidural would take too long to take effect and involve much more anesthesia than is needed. An analgesic like meperidine would have a depressant effect on the infant and would do nothing for the perineal pain. The pudendal block is a quick, easy method of applying last minute anesthesia of the perineum so that an episiotomy can be cut and repaired.

**179–180. The answers are: 179-A, 180-A.** (*Chapter 28 IV B 1, 2*) Because the patient presented to the office more than 72 hours after the assault, pregnancy prevention, such as diethylstilbestrol (DES) therapy or intrauterine device (IUD) insertion, would be ineffective. A dilatation and evacuation (D and E) is not indicated in the absence of a pregnancy diagnosis. Therefore, the physician should follow the patient with weekly human chorionic gonadotropin (HCG) tests until a period occurs or until the test becomes positive. If the latter occurs, a D and E would be indicated.

Once again, it would be important to follow the patient with HCG tests. Because she presented within 72 hours of the assault, she is a candidate for DES therapy; however, DES preventive therapy is not 100% successful. Because of the teratogenic effects of DES on the fetus, it would be important to determine whether or not the patient becomes pregnant as termination of the pregnancy is recommended because of the teratogenicity.

# Index

Note: Page numbers in italics denote illustrations; those followed by t denote tables; those followed by Q denote questions; and those followed by E denote explanations.

## A

Abdominal examination, circumference measurement, 129–130
enlargement as a sign of pregnancy, 13
myomata uteri, 297
for pelvic pain, 231, 238Q, 240E
role in antepartum care, 24
Abdominal incision for cesarean section, 178
Abdominal pain, diagnosis, 230
Abnormal fetal development and ultrasonography, 128–129
Abnormal labor, risk assessment for, 87–88, 88t–89t
Abnormal menstruation, myomata uteri, 296
Abnormal uterine bleeding, in ectopic pregnancy, 280
myomata uteri, 299Q, 301E
ABO antigen as rape evidence, 274
ABO incompatibility in Rh disease, 103
Abortion, 180–181
federal financing, 362
induced, 181, 182Q, 183E, 233
intrauterine vs. ectopic pregnancy, 280
medicolegal considerations, 361–362
pelvic pain with, 232–233
risk assessment and, 77
state restrictions, 362
surgical intervention and, 182Q, 183E
trimester model for, 361–362, 368Q, 370E
ultrasound diagnosis, 128
Abruptio placentae, see Placental abruption
Acetylcholine, prostaglandin inhibition, 69
Acetysalicylic acid, prostaglandin synthesiis inhibition, 32
Acid phosphatase as rape evidence, 274
Actin, role in labor, 31
Actinomyces israelii, 242, 247Q, 248Q, 249E
Actinomycosis, 246
Active management of postdatism, 50
Active phase of labor, 34, 43Q, 45E
Acute pyelonephritis, 58–59
Acute tubular necrosis, with placental abruption, 142
placenta previa and, 140
Acyclovir, 263
Addiction (drug and alcohol), fetal morbidity and mortality, 76
Adenocarcinoma, clear cell, 399Q, 418E
of endometrium, 356Q, 358E
Adenomatous hyperplasia, 347
Adenomyosis, 84, 236
uterine rupture and, 90Q, 91E
Adhesive disease, 236
Adnexal torsion, with ectopic pregnancy, 280
Adolescent gynecology, 207–209
Adrenal disorders, high-risk pregnancy and, 80

hirsutism and, 329
ovulatory defects and, 314
Adrenal hyperplasia, 206
characteristics of, 210Q, 212E
congenital, 399Q, 418E
Adrenal tumors, 206
Adrenocorticotropic hormone and postdatism, 47
Aerobic metabolism, fetal, 69–70
Afterpains, 16
Age, endometrial cancer and, 347
high-risk patient assessment, 80
maternal mortality risk assessment, 75–76, 90Q, 91E
Agglutination test, HCG levels, 4
Alcohol, see also Fetal alcohol syndrome
and fetal morbidity and mortality, 77
pregnancy and, 27
teratogenicity, 152
threshold dose, 152
Aldosterone, progesterone stimulation of, 7
Aldridge sling procedure, stress urinary incontinence (SUI) therapy, 306
"All or none" teratogenicity, 149
Alpha-fetoprotein levels, Down's syndrome, 384Q, 401E
meningomyelocele, 400Q, 418E
high-risk pregnancy assessment, 85–86
neural tube defect diagnosis, 117–118, 118t
role in antepartum care, 25
Trisomy 21 diagnosis, 121Q, 123E
Ambulatory care, contraindications for, 56
Amenorrhea in adolescent, 207–208, 390Q, 408E
with contraceptive methods, 226Q, 227E
defined, 319
eugonadotropic, 319–320, 399Q, 418E
evaluation of, 321–322
hirsutism and, 332Q, 334E
hypergonadotropic, 320
hypogonadotropic, 320–321, 399Q, 418E
in menopausal women, 336
physiologic, 399Q, 418E
postpill, 223, 226Q, 227E
transverse vaginal septum and imperforate hymen, 395Q, 414E
treatment, 322–323
Amniocentesis, chromosomal abnormality and congenital malformation, 117–118
complications, 116–117
genetic counseling and, 368Q, 370E
inborn errors of metabolism, 118, 118t–119t
indications for, 116
lecithin/sphingomyelin (L/S) ratio, 118, 120
Liley curve, 388Q, 406E
malpractice and, 361
phosphatidylglycerol detection, 120
in postdates pregnancy, 49, 51Q, 53E
prenatal diagnosis with, 116–120
procedures for, 121Q, 123E
Rh disease and, 106–107, 118, 385Q, 403E

Tay-Sachs disease diagnosis, 400Q, 418E
ultrasound visualization, 127
Amnioinfusion, 42
Amniotic fluid, congenital anomalies and, 137Q, 138E
excessive, 131
hormone levels in, 4
lecithin in, 15
phosphatidylglycerol in, 15
postdatism evaluation, 49
prolactin levels, 6
ultrasound measurements, 131
Amniotomy, 34
A-mode ultrasound, 125
Analgesia, see Anesthesia and analgesia
Androgen antagonists, hirsutism therapy, 330
Androgen excess, amenorrhea and, 320
free testosterone, 388Q, 406E
hirsutism and, 327–328, 332Q, 334E
ovarian sources of excess, 332Q, 334E
Androgen ingestion, 206
Androgen insensitivity, 206, 395Q, 413E
characteristics, 387Q, 405E
hypergonadotropic amenorrhea, 320
Androstenedione and hirsutism, 330
Anechoic tissue, 115, 125
Anemia, see also specific types
high-risk pregnancy assessment, 85
in pregnancy, 57–58
Anencephaly, 131
postdatism, 47
prolonged gestation, 384Q, 401E
Anesthesia and analgesia, for active labor, 392Q, 410E
aspiration prevention, 395Q, 414E
cardiac patients, 60
for cesarean section, 178
conduction anesthesis, 186–188
fetal distress, 42, 43Q, 45E
hazards of, 185
iatrogenic fetal distress, 42
for induced abortion, 181
general anesthesia, 188–189
during labor, 186
as mortality risk factor, 75
physiologic factors, 185–186
Animal research on teratogenicity, 151–152
Anorexia nervosa, amenorrhea and, 207
Anovulation, diagnosis, 271Q, 272E
dysfunctional uterine bleeding, 267–268, 270Q, 272E
endometrial cancer and, 347
postmenarchal bleeding, 270Q, 272E
Antepartum bleeding, placenta previa, 139–140
Antepartum care, components, 23
risk assessment in, 86–87, 86t
Antibiotics, general anesthesia aspiration, 189
neonatal infection, 169
for puerperal infections, 17
veneral disease in rape victims, 275

Antibody screening, high-risk pregnancy assessment, 84
  for pelvic pain evaluation, 232
  Rh disease, 105–106
Anticholinergics, stress urinary incontinence (SUI) therapy, 305
Anticholinesterase drugs, stress urinary incontinence (SUI) therapy, 305
Antihypertensives, for pregnancy-induced hypertension, 57
Antiprostaglandin agents, 32
Antral follicle, 215
Apgar scores, interpretation, 161–162, 172Q, 174E
  moderately depressed infants, 162
  preeclampsia and, 173Q, 174E
  risk assessment for low scores, 87–88, 88t–89t
  severely depressed infants, 162–163
Apnea, 166–167
Appendicitis, 235
  signs and symptoms, 248Q, 249E, 394Q, 412E
Arachidonic acid, characterists of, 36Q, 38E
  labor initiation, 95
  as prostaglandins precursor, 31
Arrhenoblastoma, heterosexual precocious puberty, 208, 399Q, 418E
  teratogenicity, 155
Arterial blood pressure, 69
Artery, Doppler waveforms, 134, *136*
Artificial insemination, 362, 369Q, 370E
Aschheim-Zondek rat test, HCG levels, 4
Asherman's syndrome, 320
  causes, 324Q, 326E
  endometrial potential, 322
  treatment, 322
Asphyxia, fetal, in postdatism, 48
Aspiration, during general anesthesia, 189
  pneumonitis, 190Q, 192E
Aspirin, use during pregnancy, 27
Asymptomatic bacteriuria, 59
Atrophic vaginitis, 261–262, 264Q, 266E, 337, 399Q, 417E
Attenuating tissue, 125
Autonomic nervous system, 229
Autosomal dominant disorders, malpractice and, 361
Autosomoal recessive disorders, malpractice and, 361

**B**

Bacteria, pelvic inflammatory disease (PID), 242
  sexually transmitted disease (STD), 287–288
*Bacteroides fragilis*, 248Q, 249E
Barium enema, ovarian cancer detection, 352
Barium-impregnated plastic IUD, 221
Barrier contraception methods, 220–221, 394Q, 412E
  pelvic inflammatory disease (PID), 242
  side effects of, 226Q, 227E
Basal body temperature, fertility cycle, 220, 390Q, 408E
  infertility and, 387Q, 405E
Baseline fetal heart rate (FHR), 40
Beat-to-beat variability in fetal heart rate (FHR), 40
Bed rest, for pregnancy-induced hypertension, 56
Beta$_2$-Adrenergic agents, preterm labor therapy, 97
Bilirubin, aspirin use and, 27
  in Rh disease, 104, 105
Binding proteins, elevations during pregnancy, 3

Biologic assays, HCG levels, 4
Biophysical profile, postdatism, 49, 51Q, 53E, 385Q, 403E
  pregnant diabetics, 54
  ultrasound, 133–134
Biparietal diameter, 129
Birth asphyxia, 39
Birth canal lacerations, 35
Birth control, *see also* Contraception
  in puerperium, 16, 19Q, 21E
Birth injury, 364, 397Q, 416E
Bladder neck elevation test (Marchetti test) stress urinary incontinence diagnosis, 304
Bladder, urethra and, 391Q, 409E
Bleeding, *see also* Hemorrhage; *specific types of bleeding*
  in first trimester, 14
  in puerperium, 16
  in second trimester, 14
  in third trimester, 15
Blood count, high-risk pregnancy assessment, 85
  for pelvic pain evaluation, 232
  as rape evidence, 274
Blood (fetal), oxygenated, nutriment-bearing blood, 67–68
Blood flow, Doppler evaluation, 134–136, *135*
Blood pressure, artery waveforms, 134, *136*
  fetal arterial, 69
  fetal blood flow, 69
  preeclampsia, 56
  in pregnancy-induced hypertension, 55–56
Blood studies, dysfunctional uterine bleeding, 268
  hirsutism, 329–330
  pelvic inflammatory disease (PID), 244
Blood sugar monitoring, high-risk pregnancy assessment, 85
Blood-type testing, 84
  for pelvic pain evaluation, 232
  Rh disease, 105
Bloody show, 143
B-mode ultrasound, 125
Bohr effect, oxygen-carrying capacity of fetal blood, 69
Bones in postmenopausal women, 336
Bowel habits during pregnancy, 27
Bradycardia, fetal, 40
Bradykinin in neonate, 70
Braxton-Hicks contractions, in second trimester, 14
  in third trimester, 14–15
Breach of duty, 359
Breast cancer, estrogen replacement therapy and, 338
  with oral contraceptives, 224
Breast changes, during lactation, 17
  in postmenopausal women, 336
  as sign of pregnancy, 13
Breech presentation, 16, 19Q, 21E
  complete, 16
  placenta previa and, 398Q, 417E
Bronchopulmonary dysplasia, 165
Bronchoscopy, general anesthesia, 189
Brow presentation of fetus, 15
"Buddha" stance of fetus, in Rh disease, 109–110
Burch procedure, stress urinary incontinence (SUI) therapy, 306, 308Q, 309E
Butorphanol, 186

**C**

CA 125 antigen, 352
Calcium, agonists, preterm labor, 97
  requirements in pregnancy, 26
Calcium ion (Ca$^{2+}$), labor and, 32, 37Q, 38E

Calories, requirements in pregnancy, 26
*Calymmato-bacterium granulomatis*, 288
Cancer chemotherapy teratogenicity, 154
*Candida* vaginitis, 260–261
  diagnosis, 264Q, 266E
Cardiac care, cesarean delivery contraindicated, 386Q, 404E
  heart disease and pregnancy, 60
Cardiac failure, pregnant cardiac patients, 60–61
Cardiac output, anesthesia and, 185
  fetal heart, 68–69, 386Q, 404E
Cardinal movements in labor and delivery, 33–34
Cardiopulmonary disorders, prenatal ultrasound diagnosis, 116
Cardiovascular disease, high-risk patient assessment, 78
  in menopausal women, 338
Catecholamine and prostaglandin inhibition, 69
CAT scan, abdominopelvic, 352
  prolactin level evaluation, 322
Caudal anesthesia, 188
Causation, malpractice and, 359
Cefoxitin, 248Q, 249E
Central nervous system abnormalities, 131–132
  alcohol teratogenicity and, 152
  amenorrhea and, 207
  postdatism, 47
  precocious puberty, 208–209
  prenatal ultrasound diagnosis, 116
Cephalic forceps, 179
Cephalic presentation of fetus, 15
Cervical cancer, 220
  abnormal Pap smear evaluation, 344–345
  diagnosis, 356Q, 358E
  epidemiology, 343, 357Q, 358E
  follow-up evaluation, 346
  hemorrhage and, 145Q, 146E
  with HSV-2, 263
  incidence, 343
  invasive and microinvasive, 345–347
    stages, 388Q, 406E
    therapy, 346
  with oral contraceptives, 224
  risk of recurrence, 346–347
  transformation zone (TZ), 343–344
  viral transmission, 343
Cervical cerclage, 95–96, 180
Cervical cultures, for pelvic pain evaluation, 232
Cervical dilation, pain during, 187
  uterine contractions, 187
Cervical Gram stain, pelvic inflammatory disease (PID), 244
Cervical incompetence, cervical cerclage, 180
  high-risk pregnancy assessment, 84
  premature labor, 95–96
  in second trimester, 14
  signs and symptoms, 393Q, 412E
Cervical infertility factor, 312–313, 316Q, 318E
  correction of, 315
Cervical mucus, with oral contraceptives, 223
Cervical transformation zone (TZ), 343–344
Cervix, abnormalities of, 312
  carcinoma, 220
  changes in
    during fertility cycle, 219–220
    in labor, 33
    as sign of pregnancy, 13
  conization of, 396Q, 415E
  dilation in pregnant cardiac patients, 61
  mucus, quality of, 312–313
Cesarean section, complications, 177
  contraindicated in cardiac patients, 61, 386Q, 404E

fetal scalp blood sampling, 44Q, 45E
history of, 177
incidence, 177, 182Q, 183E
incisions for, 389Q, 407E
indications for, 182Q, 183E, 388Q, 406E
inhalation anesthesia for, 178
intrauterine resuscitation and, 384Q, 402E
as mortality risk factor, 75
in pregnant diabetics, 55
techniques, 178-179
types of, 178
vaginal delivery after, 78
Chadwick's sign, 13
Chassar moir sling, stress urinary
  incontinence (SUI) therapy, 306
Chest x-ray, for GTN followup, 398Q, 417E
  ovarian cancer, 352
  teratogenicity, 390Q, 408E
*Chlamydia trachomatis*, 289
  pelvic inflammatory disease (PID), 242
Choanal atresia, 164
Cholesterol, estriol synthesis, 7
Cholinergic drugs, stress urinary
  incontinence (SUI) therapy, 305
Chorioamnionitis, ultrasound diagnosis, 134
Choriocarcinoma, 351
  HCG production, 4
Chorion and trophoblast sampling, *see*
  Chorionic villi sampling
Chorionic villi sampling, prenatal diagnosis
  with, 120
  for sickle cell anemia, 400Q, 418E
  ultrasound visualization, 127
Chorioretinis, 156
Chromosomal abnormalities, amniocentesis
  diagnosis of, 116-118
  hypergonadotropic amenorrhea, 320
  maternal age and, 76, 122Q, 123E
Chronic hypertension, 55
  high-risk patient assessment, 78
Chronic vascular disease, placental abrup-
  tion and, 141
Class A diabetes, management of pregnan-
  cy, 54
Clear cell adenocarcinoma, 205, 399Q,
  418E
  vaginal cancer, 353
Clindamycin, pelvic inflammatory disease
  (PID), 245
Clomiphene citrate, 315, 317Q, 318E
  ovulation stimulation, 324Q, 326E
Coagulation defects, 16
  placental abruption, 393Q, 411E
Cocaine as teratogen, 154
Coital infertility factor, 312
  correction of, 314
Coitus, during pregnancy, 27
  pelvic inflammatory disease (PID), 242
Collagen vascular disease, high-risk patient
  assessment, 79
Colposcopy, abnormal cervical Pap smears,
  344
Colpotomy, pelvic inflammatory disease
  (PID), 245
Complete abortion, 232
Conception, physiology of, 311
Condoms, characteristics of, 220-221
Conduction in fetal-neonatal transition, 162
Conduction anethesia, 186-188
Confirmation of pregnancy, with ultra-
  sound diagnosis, 126
Congenital adrenal hyperplasia, 206
  therapy for, 211Q, 212E
Congenital anemia, 58
Congenital anomalies, amniocentesis
  diagnosis of, 117
  amniotic fuild levels and, 137Q, 138E
  drug teratogenicity and, 151, 159Q, 160E
  ectopic ureter with vaginal terminus, 205
  ionizing radiation, 150-151
  IUD-related, 222

neonatal resuscitation, 163-164
precocious puberty, 208-209
risk assessment for pregnancy, 78, 86t, 87
teratogenic causes, 149, 149t
ultrasound diagnosis, 131-133
uterine pathology, 236
vaginal atresia, 205
vaginal ectopic anus, 205
viral infections, 159Q, 160E
Congestive heart failure, with hydrops
  fetalis, 104
  pregnant cardiac patients, 60
Consanguinity, high-risk patient assessment, 79-80
Constipation during pregnancy, 27
Consumption coagulopathy with placental
  abruption, 142
Contraception, for adolescents, 209
  barrier methods, 220-221
  hormonal, 224-225
  oral, 222-224
  pelvic inflammatory disease (PID),
    241-242, 384Q, 402E
  postcoital, 225
Contractions, *see also* Dysfunctional labor
  patterns
  character in labor, 32-33
  during coitus, 27
Contraction stress test (CST), 51Q, 53E
  in postdates pregnancy, 49
  pregnant diabetics, 54
Convection, fetal-neonatal transition, 162
Coombs' tests, interpretation of, 113Q, 114E
  Rh disease and, 104-106
Copper-bearing Copper 7 and Copper T
  IUDs, 221
Cordocentesis, Rh disease evaluation,
  385Q, 403E
Corpus luteum, bleeding in, 280, 284Q,
  285E
  fertilized ovum in, with ectopic pregnancy,
    278-280
  functional defects, 316Q, 318E
  oogenesis, 215
  progesterone production, 314
  transmigration, 283Q, 285E
Corticosteroids, general anesthesia aspira-
  tion, 189
  preterm labor therapy, 98
Cortisol and hirsutism, 330
Cortisol-binding globulin (CBG), elevations
  during pregnancy, 3
Couvelaire uterus, with placental abruption,
  142
Coxsackie B virus teratogenicity, 158
Crab louse, 292
Crowning, 35
Crown-rump length, 128, *128*
Culdocentesis, ectopic pregnancy diagnosis,
  281
  pelvic inflammatory disease (PID), 244
  for pelvic pain evaluation, 232
Curretage, *see also* Dilatation and curettage;
  Endocervical curettage
  postpartum hemorrhage, 17
Cyclic adenosine monophosphate (cAMP)
  and gonadotropin-releasing hormone,
  213-214
Cyclopropane, 188-189
Cystic myoma degeneration, 296
Cystitis, 58
  in menopausal women, 337
Cystometry, stress urinary incontinence
  (SUI) diagnosis, 304, 307Q, 309E
Cystoscopy, stress urinary incontinence
  (SUI) diagnosis, 304, 307Q, 309E
Cytologic screening, abnormal cervical Pap
  smears, 344
Cytomegalovirus, high-risk pregnancy
  and, 80
  teratogenicity, 155-156

**D**

Dalkon shield, 360
Damages, malpractice and, 359
Danazole treatment, endometriosis
  therapy, 253-254, 255Q, 257E
Dandy-Walker syndrome, 131
Danger signs in pregnancy, 25-26, 28Q,
  30E
Decreased variability, fetal heart rate
  (FHR), 40, 43Q, 45E
Dehydroepiandrosterone sulfate
  (DHEASO$_4$), dynamic testing of, 4
  estriol synthesis, 7-8
  prolactin levels and, 322
17beta-ol-Dehydrogenase, 339
Delayed puberty, 208
Delivery, in cesarean section, 178
  of high-risk fetus, 386Q, 404E
  indications for, insulin-dependent
    diabetics, 62Q, 64E
  Rh disease, 110
  with placental abruption, 142
  in pregnancy-induced hypertension, 57
  pregnant cardiac patients, 60
  risk assessment for spontaneous preterm,
    87
  timing in pregnant diabetics, 55
Delta OD$_{450}$, interpretation, 112Q, 114E
  Rh disease and, 106-108, *107-108*
Depo-Provera, 224
Depression, in rape victims, 275
Descent, 33
Detrusor instability, 305
  stress urinary incontinence (SUI)
    diagnosis, 307Q, 309E
Dextran instillation for endometriosis, 253
DHEA sulfate and hirsutism, 329
Diabetes, classification in pregnant women,
  53
  control in pregnancy, 53-54
  fetal-neonatal complication, 62Q, 64E,
    392Q, 410E
  gestational, 173Q, 174E
  high-risk patient assessment, 79
  pregnancy and, 53-55
  risk assessment for, 86t, 87
  teratogenicity, 155
  term infant disorders, 167-168
Diagnostic radiation, teratogenicity, 151,
  151t
Diaphragmatic hernia, 163, 173Q, 174E
  head:body ratio, 137Q, 138E
Diaphragms, 221
Diet in pregnancy, 383Q, 401E
Diethylstilbesterol (DES), adenocarcinoma,
  353
  high-risk pregnancy assessment, 84
  neoplasms associated with, 210Q, 212E
  prematurity and, 93
  for rape victims, 275, 400Q, 419E
  vaginal tumors, 205
Dilatation and curettage (D & C),
  dysfunctional uterine bleeding and,
    270Q, 272E
  ectopic pregnancy, 283Q, 285E
  fractional, for endometrial carcinoma
    diagnosis, 348
  myomata uteri, 300Q, 301E
Dilation of cervix, 33
Discharge summary, 366
Disclosure rules, informed consent, 364
Disphosphoglycerate, oxygen delivery to
  fetal tissue, 69
Diuretics, contraindicated in pregnancy, 26,
  56, 390Q, 408E
Doppler flow evaluation, ultrasound and,
  134-136, *135*
Down's syndrome, *see also* Trisomy 21
  alpha-fetoprotein levels, 384Q, 401E
  amniocentesis diagnosis of, 117

Doxycycline, 248Q, 249E
  pelvic inflammatory disease (PID), 245
Drug-depressed infants, 172Q, 174E
  resuscitation, 164
Drugs, congenital malformations, 159Q,
  160E
  fetal morbidity and mortality, 76–77
  hypogonadotropic amenorrhea and, 321
  perinatal effects, 82t
  "recreational," 152
  teratogenicity, 151–154, 153t–154t, 396Q,
  415E
Drug therapy, preterm labor, 97
Duchenne muscular dystrophy, 118
Ductus arteriosus, 68
  dynamics of, 69
  fetal hypoxia, 72Q, 73E
Ductus venosus, 67
  contraction, 71Q, 73E
Duty, negligence of, 359
Dynamic tests of hormone levels, 4
Dysfunctional labor patterns, 34
Dysfunctional uterine bleeding (DUB), in
  adolescent, 207
  characteristics of, 211Q, 212E
  defined, 267
  diagnosis, 268, 270Q, 272E
  estrogen replacement therapy, 268, 339
  etiology of, 267
  hypothalmic-pituitary axis, 399Q, 418E
  management, 268–269
  nonsteroidal anti-inflammatory drugs
  (NSAIDs), 269
  oral contraceptive therapy, 269
  in preimenopause, 337, 340Q, 341E
  physiology, 267–268
  therapy for, 211Q, 212E
Dysgerminoma, 351
  treatment, 352–353
Dysmaturity syndrome, 47–48
Dysmenorrhea, in adolescent, 207
  clinical aspects and management, 216
  endometriosis and, 252
  pelvic pain with, 235–236
  primary, 395Q, 413E
  therapy for, 211Q, 212E, 395Q, 414E
Dyspareunia, and endometriosis, 252,
  255Q, 257E

**E**

Early decelerations in fetal heart rate (FHR),
  40
Echogenic tissue, 115, 125
Eclampsia, defined, 55
  recurrent risk of, 78
  risk assessment for, 86, 86t
  seizures in pregnancy-induced hypertension, 55–57
  weight gain and, 26
Ectoparasites, 292
Ectopic pregnancy, defined, 279
  diagnosis, 280–281, 283Q, 285E, 387Q,
  405E, 390Q, 408E
  etiology, 279–280, 283Q, 285E
  human chorionic gonadotropin
  levels, 4, 271Q, 272E, 391Q, 409E
  incidence, 279, 396Q, 415E
  IUD-related, 222
  laboratory evaluation, 233
  pelvic inflammatory disease (PID), 241
  pelvic pain with, 233
  prognosis, 282, 283Q, 285E
  Rh disease, 105
  risk of recurrence, 78
  significance, 279
  signs and symptoms, 233, 239Q, 240E,
  248Q, 249E, 280, 394Q, 412E
  treatment, 281–282, 389Q, 407E
  ultrasound diagnosis, 126, 128

  unruptured, 281
Ectopic ureter with vaginal terminus, 205
Education of patient, liability and, 366
Effacement of cervix, 33
Embryo freezing, 363
Embryonal carcinoma, 351
Employment-risks in pregnancy, 77
Encephaloceles, 132, 164
Endocervical canal curettage (ECC),
  abnormal cervical Pap smears, 344
  cervical cancer diagnosis, 356Q, 358E
  for endometrial carcinoma, 349
Endocrine changes in pregnancy, 3
  for pregnancy diagnosis, 13–14
Endodermal sinus tumor, 351
Endogenous estrogen levels, 337, 340Q,
  341E
Endometrial biopsy, 399Q, 417E
  ectopic pregnancy diagnosis, 281
  infertility and, 399Q, 418E
Endometrial carcinoma, 348–349
  treatments for, 385Q, 403E
Endometrial hyperplasia, 347–348, 357Q,
  358E
  therapy, 395Q, 414E
Endometrial cancer, 347–349
  epidemiology, 347
  estrogen replacement therapy and,
  338–339
  incidence, 347
  pathophysiology, 347
  in perimenopause, 337
Endometrial sampling, 322
  dysfunctional uterine bleeding, 268
Endometrioid tumors, ovarian cancer, 350
Endometriosis, defined, 251
  diagnosis, 252, 255Q, 257E, 386Q, 404E
  management of, 252–253
  pathogenesis, 255Q, 257E
  pathogenic theories, 251
  radical surgery for, 388Q, 406E
  signs and symptoms, 236, 383Q, 401E
  therapy, 256Q, 257E, 387Q, 395Q, 405E,
  414E
Endometritis, 17
  signs and symptoms, 234
Endometrium, changes during menstrual
  cycle, 217Q, 218E
  dysfunctional uterine bleeding and, 268
  evaluation for infertility, 316Q, 318E
  with oral contraceptives, 224
  in postmenopausal women, 336
  progesterone and nidation in, 7
Endosalpingitis, 243
Endovaginal probe, 128
Enflurane, 189
Engagement, 33, 36Q, 38E
  in pregnant cardiac patients, 61
Enteroviruses (coxsackie B), teratogenicity,
  158
Enzymes, fetal metabolism and, 70
Epidural anesthesia, 188
  for active labor, 400Q, 419E
  complications, 190Q, 192E
  fetal distress, 42, 43Q, 45E
Epigastric pain, diagnosis, 230
  preeclampsia, 56
Epilepsy, teratogenicity, 155
Episiotomy, 35
  advantages of, 182Q, 183E
  mediolateral, 385Q, 403E
  pudendal block anesthesia, 400Q, 419E
  techniques of, 179
Epithelial tumors, ovarian cancer, 350
  treatment, 352
Ergonovine, postpartum hemorrhage, 17
Erythroblastosis fetalis, pathophysiology,
  104–105
  Rh disease, 112Q, 114E, 396Q, 415E
Erythrocyte sedimentation rate, for pelvic
  pain evaluation, 232
*Escherichia coli* in urinary infections, 59

Esophageal atresia, 163–164
Estimated date of confinement (EDC), 14,
  387Q, 405E
  calculation, 23
  high-risk pregnancy assessment, 83
  in pregnant diabetics, 54
Estradiol, chemical structure, 7
  estrogen replacement therapy, 339
  maternal, in postdatism, 48–49, *48*
Estriol, chemical structure, 7
  functions, 8
  levels of, 8
  maternal, in postdatism, 48–49, 48
  postdatism levels, 51Q, 53E
  production in pregnancy, 7, 10Q, 11E
  Ph isoimmunization and, 386Q, 404E
  smoking and intrauterine growth retardation, 383Q, 401E
  synthesis of, 7–8
Estrogen, chemical structure, 7
  elevations during pregnancy, 3
  excess endogenous, 337
  exogenous unopposed, endometrial
  cancer and, 347
  for labial agglutination, 204
  maternal, in postdatism, 48–49, *48*
  myoma dependence on, 395Q, 414E
  oral conjugated, 397Q, 416E
  in postmenopausal women, 335
  sequalae of menopausal reduction,
  336–337
  sources, 7–8
  target organ response, 335–336
  two-cell production hypothesis, 214,
  217Q, 218E
  vaginal discharge and, 204
Estrogen:progesterone ratio, labor
  initiation, 94
Estrogen replacement therapy, 337–339
  benefits of, 337, 340Q, 341E
  breast cancer, 338
  contraindications, 337, 340Q, 341E
  dysfunctional uterine bleeding, 268
  endometrial cancer, 338–339, 347–348
  indications for, 339
  limits of, 340Q, 341E
  metabolic contraindications, 338
  precocious puberty, 209
  relative contraindications, 338
Estrone, chemical structure, 7
  endometrial cancer and, 347
  in postmenopausal women, 335
Ethanol, preterm labor therapy, 97
Ethnicity and high-risk patient assessment,
  80
Eugonadotropic amenorrhea, 319–320,
  399Q, 418E
  treatment of, 322
Evaporation, fetal–neonatal transition, 162
Everard Williams procedure, stress urinary
  incontinence (SUI) therapy, 305
Exchange transfusion, kernicterus, 166
Exercise, during pregnancy, 27
  hypogonadotropic amenorrhea, 321
Expectant management of postdatism, 50
Expulsion, 34
Extension, 34
External rotation, 34
Extraperitoneal cesarean section, 178,
  182Q, 183E

**F**

Facial presentation of fetus, 15
Fallopian tube, narrowing of, 279
  reproductive functions, 313
  segmental resection of, 281
False labor, 32, 398Q, 417E
Family history, *see also* Genetic counseling;
  Patient history

high-risk pregnancy and, 82–83
role in antepartum care, 24
Faradism, stress urinary incontinence (SUI), 305
Fatigue as sign of pregnancy, 13
Feguson reflex, cervical incompetence, 95
Ferning technique for ruptured membranes, 34
Fertility cycle, 219
ovarian hyperstimulation syndrome, 234
Fetal abdomen, abnormalities, 132, *132*
ultrasound measurements, 130
Fetal activity test (FAT), in postdates pregnancy, 49
Fetal alcohol syndrome, 27, 77, 152, 154
Fetal blood sampling, prenatal diagnosis with, 120
procedures for, 121Q, 123E
Fetal bradycardia, 40
Fetal cardiac output defined, 386Q, 404E
Fetal circulation, conversion to adult circulation, 161
persistent, 395Q, 414E
physiology, 161
Fetal cortisol infusion during labor, 31
Fetal development, teratogens and, 149, 149t, *150*
Fetal distress, acid–base characteristics, 392Q, 410E
defined, 39
iatrogenic causes, 41–42
in pregnant diabetics, 54
transient, 398Q, 417E
Fetal head:abdominal circumference ratio, 130
ultrasound diagnosis, 137Q, 138E
Fetal head position and umbilical blood flow, 42
Fetal heart beat, as positive sign of pregnancy, 14
Fetal heart rate (FHR), abnormal changes, 40
decelerations, 40–41
conditions for, 388Q, 407E
head compression and, 393Q, 412E
elements of pattern, 40
fetal distress and, 39
maternal hypertension, 44Q, 45E
monitoring, 40–41, 43Q, 45E
in postdates pregnancy, 49
Fetal hormones, progesterone and, 7
Fetal hyposia, pathophysiology, 39–40
respiratory distress syndrome, 164
significance of, 39
umbilical circulation, 69, 72Q, 73E
Fetal limbs, ultrasound measurements, 130
Fetal lung maturity, 15
early onset of, 15
measurement of, 392Q, 410E
normal signs of, 19Q, 21E
Fetal metabolism, 70
Fetal monitoring, 39
during first stage of labor, 34
Fetal mortality, and maternal age (over 35), 76
prematurity and, 94
Fetal movement, in second trimester, 14
as sign of pregnancy, 13
ultrasound, 134
Fetal–neonatal transition to life, 162–163, 173Q, 174E
Fetal oxygenation, intrauterine resuscitation, 42
Fetal oxytocin, labor initiation, 94
Fetal pH, 39, 389Q, 407E
scalp blood sampling, 41, 44Q, 45E
Fetal scalp blood sampling, 41, 44Q, 45E
capillary sampling, 39
Fetal spine, ultrasound visualization, 131
Fetal tachycardia, 40
Fetal thorax, abnormalities, 132, *132*
ultrasound measurements, 130

Fetal viability, in second trimester, 14
Fetal weight, diameter, 129–130
Fetal well-being, anesthesia and, 185–186
ultrasound visualization, 127
Fetoplacental access, teratogenic agents, 151
Fetoscopy, benefits of, 121Q, 123E
prenatal diagnosis with, 120
Fetus, endocrine structures, 3
enzyme capabilities, 7
goiter development in, 59
growth and development, 15
hemodynamics of, 67–69, *68*
office evaluation of, 25
postdates risk assessment, 48–49
in pregnant diabetics, 54
protein and growth of, 26
status of, 15–16
in third trimester, 15
Fibroids, high-risk pregnancy assessment, 84
First stage of labor, 36Q, 38E
First trimester, defined, 14
signs and symptoms, 14
ultrasound visualization in, 127–129
miscarriage and maternal age (over 35), 76
Fitz-Hugh-Curtis syndrome, 243
signs and symptoms, 394Q, 412E
Flexion, 33
Folic acid, requirements in pregnancy, 27
Folic acid deficiency, 28Q, 30E
megaloblastic anemia, 57
Follicle-stimulating hormone (FSH), amenorrhea evaluation, 322
functions of, 217Q, 218E, 386Q, 404E
in menstrual cycle, 214, 319
Footling breech presentation, 16
Foramen ovale, 67
functions, 71Q, 73E
Forceps, 179–180
indications for, 386Q, 404E
prerequisites for use, 182Q, 183E
Foreign body salpingitis, 246
Frank breech presentation, 16
Free testosterone, 328
Friedman rabbit test, HCG levels, 4


**G**

Gamete intrafallopian transfer (GIFT), 315
Gap junctions, formation, 37Q, 38E
role in labor, 32
*Gardnerella vaginalis*, epidemiology, diagnosis and treatment, 288
vaginitis, 261
signs and symptoms, 394Q, 413E
Gas anesthetics, 188–189
Gastrointestinal disorders, prenatal ultrasound diagnosis, 116
Gastroschisis, 164
General anesthesia, indications for, 191Q, 192E
Genetic abnormalities in fetus and maternal age (over 35), 76
Genetic counseling, amniocentesis and, 368Q, 370E
defined, 115
high-risk patient assessment, 79–80
indications for, 391Q, 409E
medicolegal considerations, 360–361
patient history and, 90Q, 91E
routine, 360
Genetic lethals, amniocentesis diagnosis of, 118
Genetic susceptibility, endometriosis and, 251
to teratogens, 149
Genitalia, developmental defects, 206–207
Genital tract trauma, 16
Genitourinary abnormalities, 132–133

Germ-cell ovarian tumors, 206
ovarian cancer, 351
treatment, 352–353
Gestational age, ultrasound diagnosis, 126
Gestational hypertension, 55
Gestational sac mean diameter, 127
Gestational trophoblastic disease (GTD), clinical classification, 193
incidence, 193
Gestational trophoblastic neoplasia (GTN), characteristics, 193, 194
chemotherapy and, 198Q, 199E
chest x-ray for, 398Q, 417E
diagnosis, 195
good prognosis GTN, 195, 384Q, 402E
hysterectomy and, 198Q, 199E
incidence, 193
metastatic, 195–196
nonmetastatic, 195
recurrence rates, 196
signs and symptoms, 389Q, 407E
Glomerular filtration, in pregnancy-induced hypertension, 56
Glucose, elevation in infants of diabetic mothers, 167–168
presence in fetal blood, 70
Glucose control, in pregnant diabetics, 54
testing, 53
Glucose tolerance test, diagnosis with, 392Q, 410E
Glucosuira, diabetes in pregnancy, 53
interpretation of, 62Q, 64E
Goebell-Stoeckel-Frangenheim procedure, stress urinary incontinence (SUI) therapy, 306, 308Q, 309E
Gonadal dysgenesis, 324Q, 326E
amenorrhea and, 207–208
estrogen replacement therapy, 339
hirsutism and, 329
hypergonadotropic amenorrhea, 320
mixed, 395Q, 413E
Gonadoblastoma, 351
Gonadotropin levels, amenorrhea evaluation, 325Q, 326E
Gonadotropin-releasing hormone, agonists, for endometriosis, 253
myoma therapy, 297
amenorrhea in adolescence, 207
hirsutism, 330
menstruation and, 213–214, 319
control of, 217Q, 218E
synthesis, 3
Gonorrhea culture, high-risk pregnancy assessment, 84–85
Good variability in fetal heart rate (FHR), 40
"Grandmother theory" of Rh disease, 105
Granulomatous salpingitis, 245–246
Granulosa cell tumors, ovarian cancer, 350
Granulosa-theca cell tumors, endometrial cancer and, 347
ovarian cancer, 350
Gravida, 23
Grey scale, 115


**H**

Habitual abortion, 181
causes of, 182Q, 183E
Hair, *see also* Hirsutism
in postmenopausal women, 336
Halothane, 189
Hamartomas, 329
Headache, preeclampsia, 56
Heart disease, *see also* Cardiovascular disease
classification, 60
diagnosis, 60
incidence of in pregnancy, 60
as mortality risk factor, 76
in pregnancy, 60–61
Heat loss mechanisms, 162, 172Q, 174E

Hegar's sign, 13
Hematologic disorders, high-risk patient assessment, 79
Hematopoiesis, 70
Hemoconcentration, in pregnancy-induced hypertension, 56
Hemoglobin, content in fetal blood, 69
in fetal development, 70, 71Q, 73E
Hemoglobin A$_{1c}$ determination, in pregnant diabetics, 54
Hemoglobin (fetal), anesthesia and, 186
Hemoglobinopathies, congenital anemias, 58
Hemolytic disease of fetus and newborn, 104
Hemolytic streptococcal vaginitis, 204
*Hemophilus ducreyi*, diagnostic test, 293Q, 294E
epidemiology, diagnosis and treatment, 287–288
lesions associated with, 293Q, 294E
Hemorrhage, *see also types of bleeding*
concealed with placental abruption, 141
external, with placental abruption, 141
maternal mortality from uterine, 75
postpartum, 16–17, 19Q, 21E
prematurity and, 94, 99Q, 101E
recurrent risk of, 78
Heparin, for puerperal infections, 17
Hepatitis B, righ-risk pregnancy and, 80–81
Hepatocellular damage with hydrops fetalis, 104
Hepatosplenomegaly with hydrops fetalis, 104
Hermaphroditism, 206
Heroin teratogenicity, 154
Herpes simplex genitalis, 262–263
Herpes simplex virus (HSV), high-risk pregnancy and, 80
sexually transmitted disease, 290–291
Herpes simplex virus type 2 (HSV-2), cervical cancer and, 343
teratogenicity, 156
vaginitis, 262–263
Herpes zoster teratogenicity, 157
Heterosexual precocious puberty, 208
High forceps operation, 179
Hirsutism, classification, 328–329
defined, 327
diagnosis, 329–330
ethnic background and, 332Q, 334E
etiology, 327–328
laboratory testing for, 397Q, 415E
pathophysiology of, 328, 396Q, 415E
treatment, 330–331
Histamine, prostaglandin inhibition, 69
Holoprosencephaly, 131
Hormones, amenorrhea evaluation, 325Q, 326E
chemical nature of, 3
elevation during pregnancy, 393Q, 411E
function of, 4
gestational abnormality markers, 4
hypogonadotropic amenorrhea and, 321
levels during pregnancy, 3–4
placental, 9Q, 11E
sources of, 3
therapy endometriosis, 254
hypergonadotropic amenorrhea, 322–323
stress urinary incontinence (SUI), 305
Hospitalization, for preeclampsia, 400Q, 418E
for pregnancy-induced hypertension, 56
pregnant cardiac patients, 60
Hot flashes, 336, 385Q, 403E
Human chorionic gonadotropin (HCG), beta-subunit measurements, 394Q, 413E
in ectopic pregnancy, 280–281
changes in pregnancy, 3
chemical structure, 4

ectopic pregnancy diagnosis, 391Q, 409E
functions, 5, 9Q, 11E
gestational trophoblastic neoplasia diagnosis, 195
hydatidiform mole, 194
levels of, 4–5, 9Q, 11E
amenorrhea evaluation and, 321
normal patterns, 5
pelvic inflammatory disease (PID), 244
pregnancy diagnosis, 391Q, 409E
rape therapy and, 400Q, 419E
serum pregnancy tests, 13
source, 4
trophoblastic neoplastic disease and, 4
urine pregnancy tests, 13
vaginal ultrasound and, 386Q, 404E
Human immunodeficiency virus (HIV), 291
high-risk pregnancy and, 81–82, 82t
perinatal effects of drugs, 82t
risk factors, 293Q, 294E
Human ovum fertilization test, 312
Human papillomavirus (HPV), 290
cervical cancer and, 343
characteristics, 391Q, 409E
lesions associated with, 293Q, 294E
vulvar carcinomas, 354
Human placental lactogen (HPL), changes in pregnancy, 3
chemical structure, 5
functions, 5, 9Q, 11E
in postdatism, 49
source of, 5
Hyaline myoma degeneration, 296
Hydatidiform mole, characteristics, of, 193, 383Q, 401E
diagnosis, 194
human chorionic gonadotropin levels, 4, 271Q, 272E
incidence, 193
management, 194
follow-up, 197Q, 199E
signs and symptoms, 193–194, 197Q, 199E, 390Q, 408E
treatment modalities, 197Q, 199E
ultrasound diagnosis, 129
Hydramnios, 131
in pregnant diabetics, 54
Hydranencephaly, 131
Hydrocephaly, 131
head:body ratio, 137Q, 138E
Hydronephrosis in myomata uteri, 298
Hydrops fetalis pathophysiology, 104–105
16-alpha-Hydroxy DHEASO$_4$, estriol synthesis, 8
21-Hydroxylase defect, 399Q, 418E
17-Hydroxyprogesterone, hirsutism and, 330
3-beta-Hydroxysteroid dehydrogenase, inhibition during pregnancy, 3
progesterone and, 7
Hymen, imperforate, *see* Imperforate hymen
Hyperbilirubinemia, 166
clinical manifestations, 170
etiology, 170
incidence, 170
infants of diabetic mothers, 168
pathophysiology, 170–171
treatment, 171
Hypergonadotropic amenorrhea, 320
treatment of, 322–323
Hypermagnesemia, 393Q, 411E
Hypertension, defined, 55
fetal heart rate monitoring and, 44Q, 45E
maternal mortality and, 75
with oral contraceptives, 223, 384Q, 402E
placental abruption and, 141
pregnancy and, 55–57
Hyperthermia teratogenicity, 155
Hyperthyroidism, hydatidiform mole, 194
pregnancy and, 59
Hypertonic uterine dysfunction, 34

Hypertrichosis, 327
Hyperviscosity syndrome, infants of diabetic mothers, 168
Hypoestrogenic rate, endometriosis therapy, 255Q, 257E
Hypogastric artery ligation, postpartum, hemorrhage, 17
Hypogastric or suprapubic pain, 230
Hypoglycemia, neonatal, 168
Hypogonadropic amenorrhea, 320–321, 323, 399Q, 418E
Hypogonadotropic hypogonadism, 207
Hypotension with spinal anesthesia, 187
Hypothalamic disease, prolactin levels, 6
Hypothalamic-pituitary disorders, ovulatory defects and, 314
Hypothyroidism, 59
prolactin levels and, 6
Hypotonic uterine dysfunction, 34
Hypovolemia, placental abruption and, 142
Hypoxia, *see* Fetal hypocia
Hysterectomy, for endometrial carcinoma, 348–349, 385Q, 403E
for hydatidiform mole, 194
myomata uteri, 298, 300Q, 301E
Hysterosalpingography, fallopian tube function, 313
premature labor, 95
Hysteroscopy, dysfunctional uterine bleeding, 268

**I**

Iatrogenic events, fetal distress, 41–42
pelvic inflammatory disease (PID), 242
Imidazoles for *Candida* vaginitis, 260–261
Immunoglobulin M (IgM) in Rh disease, 103
Immunoglobulins in Rh disease, 112Q, 114E
Immunologic assays and HCG levels, 4–5
Immunology, endometriosis and, 251
Imperforate hymen, 204, 319, 395Q, 414E
treatment, 322
Inborn errors of metabolism 118–120, 118t–119t
Incompetent cervix, *see* Cervical incompetence
Incomplete abortion, 181, 232
Indomethicin, prostaglandin synthesis inhibition, 32, 69
Induced abortion, 181, 182Q, 183E, 233
Induced labor, in postdates pregnancy, 50
in pregnant diabetics, 54
Inevitable abortion, 181, 232
Infection, congenital malformations, 159Q, 160E
high-risk pregnancy and, 80–82
IUD-related, 222
neonatal, 168–169
pelvic inflammatory disease (PID), 234
routine screening for, 90Q, 91E
teratogenicity, 155–158
vaginal discharge with, 204
Infertility, cervical factor, 312
coital factor, 312
defined, 311
endometriosis, 251–252, 255Q, 257E
incidence of, 311
laboratory testing for, 399Q, 418E
male factor, 312
ovarian factor, 313–314
pelvic inflammatory disease (PID), 241
therapy, 314–315
tubal factor, 313
unexplained, 315
uterine factor, 313
Informed consent, 364
Inhalational anesthesia for cesarean section, 178
Injectable contraception, 224

Insulin requirements in pregnant diabetics, 54
Intermenstrual bleeding, 267
Internal rotation, 33
Intestinal tract, pelvic pain and, 236
Intracranial hemorrhage, 131
Intramural myomas, 295
Intramuscular spectinomycin, 248Q, 249E
Intraperitoneal blood transfusion, Rh disease, 108–109, *109*
Intrapartum risk assessment, 87–88, 88t–89t
Intrauterine device (IUD), 221–222
  complications in pregnancy, 387Q, 406E
  malpractice and, 360, 368Q, 370E
  pelvic inflammatory disease (PID), 242
  side effects of, 226Q, 227E
  ultrasound visualization, 129
  use in rape victims, 275
Intrauterine growth retardation, 99Q, 101E, 169–170
  asymmetric, 169
  complications, 172Q, 174E
  etiology, 169
  head:body ratio, 137Q, 138E
  incidence, 169
  long-term sequelae, 170
  management, 169–170
  risk assessment for, 86, 86t
  smoking and, 383Q, 401E
  symmetric, 169
  ultrasound diagnosis, 137Q, 138E
  umbilical artery waveforms, 136
Intrauterine infection, prematurity and, 94
Intrauterine resuscitation, 42
  with cesarean section, 384Q, 402E
Intrauterine transfusion, Rh disease, 108–109, *109*
  ultrasound visualization, 127
Intravascular transfusion for Rh disease, 109
Intravenous pylogram, amenorrhea evaluation, 321
Intraventricular hemorrhage, 167
Introitus and clitoris, pediatric patients, 203
Invasive cervical carcinoma, 345–347
Inversion, indications for amniocentesis, 116
In vitro fertilization, 263–363
  indications for, 317Q, 318E
In vitro fertilization/embryo transfer (IVF/ET), 315
Ionizing radiation, 150–151
  teratogenicity, 391Q, 409E
Iron requirements during pregnancy, 26, 57
Iron deficiency anemia, 57
  signs and symptoms, 63Q, 64E
Isosexual precocious puberty, 208
  therapy for, 211Q, 212E


J

Jaundice, *see* Hyperbilirubinemia


K

Kallmann's syndrome, 207, 320
Kernicterus, 166, 171
Kerr incision for cesarean section, 178, 400Q, 419E
Kielland and Barton forceps, 180


L

Labial agglutination, 203–204
Labor, active phase, 398Q, 417E
  analgesia and sedation during, 186
  cardinal movements, 36Q, 38E
  causes of, 31
  cervical changes, 33
  character of uterine contractions, 32–33
  conduct of, 34–35
  current model of, 94–95
  definition and characteristics, 31–33
  false, 398Q, 417E
  first stage, 34
  initiation mechanisms, 94–95, 100Q, 101E
  maternal age (over 35) and, 76
  occiput presentation, 33–34
  pregnant cardiac patients, 60
  second stage, 34–35
  stages of, 32
  true vs. false, 32
Laboratory testing, antepartum care, 28Q, 30E
  for amenorrhea, 321–322
  high-risk pregnancy assessment, 84–86
  pelvic inflammatory disease (PID), 235
  for pelvic pain, 231–232
  of rape victims, 274
  role in antepartum care, 24–25
Lacerations of birth canal, 35
Lactation, 17–18
  menstruation during, 16
*Laminaria* for induced abortion, 181
Laparoscopy, amenorrhea evaluation, 322
  ectopic pregnancy diagnosis, 281
  for endometriosis, 252
  fallopian tube function, 313
  infertility and, 399Q, 418E
  pelvic inflammatory disease (PID), 244, 248Q, 249E
  for pelvic pain evaluation, 232
Laparotomy, pelvic inflammatory disease (PID), 245
Large-for-dates uterus, 126
Large infant, *see also* Macrosomia
  risk assessment for pregnancy, 77
Late decelerations, fetal heart rate (FHR), 40–41, 43Q, 45E
Latent phase of labor, 34
Lecithin in amniotic fluid, 15
Lecithin/sphingomyelin (L/S) ratio, 55
Left occiput posterior (LOP), 33
Left occiput transverse (LOT), 33
Leiomyomas, 236
Leperous salpingitis, 246
Lesch-Nyhan syndrome, amniocentesis diagnosis of, 118
Leukemia, ionizing radiation, 150–151
  varicella zoster virus, 157
Leukocytosis, high-risk pregnancy assessment, 85
  in puerperium, 16
Leydig cells, HCG stimulation, 5
Lie of fetus, 15
Lightening in third trimester, 15
Liley curve, percutaneous umbilical blood sampling (PUBS), 107–108
  Rh disease and, 107, *108*
Lithium teratogenicity, 159Q, 160E
Liver disease, high-risk pregnancy and, 80
Liver tumors, with oral contraceptives, 223
Live-virus vaccines, teratogenicity, 152
Lochia, 16
Lochia alba, 16
Lochia rubra, 16
Lochia serosa, 16
Longitudinal lie of fetus, 15
Longitudinal scans of fetal spine, 131
"Lost" tampons, 262
Low forceps operation, 179
Lumbar epidural anesthesia, 188, 392Q, 410E
Luteal phase deficiency, correction of, 315
  etiology, 267
  signs of, 270Q, 272E
  therapy for, 394Q, 412E
Luteinizing hormone (LH), functions of, 217Q, 218E, 391Q, 409E
  levels with hirsutism, 333Q, 334E
  menstruation and, 214, 319
  urine pregnancy tests, 13
Luteinizing hormone-releasing hormone analogues, 224–225
Lymphogranuloma venereum, characteristics of, 393Q, 412E
Lymphatic supply to vulva, 259


M

Macrosomia, head:body ratio, 137Q, 138E
  infants of diabetic mothers, 168
Magnesium sulfate, 393Q, 411E
  functions, 37Q, 38E
  preeclampsia therapy, 400Q, 418E
  for pregnancy-induced hypertension, 57
  role in labor, 32, 97
Male frog, HCG levels, 4
Male infertility factor, 312
  correction of, 314
Male pseudohermaphroditism, hirsutism and, 207
Malignant teratoma, 351
Malmstrom vacuum extraction, 180
Malpractice issues, birth injury, 363–364
  defined, 359
  genetic counseling, 360–361
  informed consent, 364–365
  preconception issues, 359–360
  pregnancy termination, 361–362
  reproductive technology, 362–363
  risk management, 365–366
  terminology, 397Q, 416E
Mammary glands, prolactin stimulation, 6
Marijuana, teratogenicity, 154
Marshall-Marchetti-Krantz procedure, stress urinary incontinence (SUI) therapy, 305, 308Q, 309E
Mastitis, symptoms and etiology, 17–18
Maternal age, indications for amniocentesis, 116
  placenta previa and, 139
  prematurity and, 93, 99Q, 101E
Maternal blood studies, hormone levels, 4
Maternal disease, prematurity and, 94
Maternal hydration, intrauterine resuscitation, 42
Maternal mortality, 75–76
  causes of, 75, 391Q, 409E
  in cesarean section, 178–179
  defined, 90Q, 91E
  maternal age (over 35), 76
  risk factors, 76
Maternal physiology during pregnancy, 190Q, 192E
Maternal position, iatrogenic fetal distress, 41–42
  intrauterine resuscitation, 42
  umbilical blood flow, 42
Maternal size, high-risk pregnancy assessment, 83
Maternal virilizing tumor, 207
McDonald cervical cerclage, 180, 400Q, 419E
Meconium, in amniotic fluid, 389Q, 407E
  aspiration, 163
  persistent fetal circulation, 395Q, 414E
  risk assessment for, 88, 88t
Median episiotomy, 35, 179
Medical history, *see also* Family history; Patient history
  high-risk patient assessment, 78–82
Medical records, liability and, 366
Medication, *see also* Drugs
  during pregnancy, 27
  list of known teratogens, 152, 153t–154t

teratogenicity, 151–154, 159Q, 160E
Mediolateral episiotomy, 35, 179
Medroxyprogesterone (Depo-Provera), 224, 339Q, 417E
  as alternative to estrogen replacement therapy, 339
  endometrial cancer and, 347–348
  hirsutism, 330
Megaloblastic anemia, 57
Membranes, premature rupture of, 14–15
Meningitis, with spinal anesthesia, 188
Meningomyelocele (spina bifida cystica), 132
  alpha-fetoprotein levels, 400Q, 418E
Menometrorrhagia, 267
  dilatation and curettage, 397Q, 416E
Menopausal syndrome, 337
Menopause, atrophic changes, 337
  endometrial cancer and, 347
  estrogen replacement therapy, 337–339
  health hazards of, 336–337, 340Q, 341E
  physiology, 335
  premature, 335
  signs and symptoms, 336–337, 340Q, 341E, 390Q, 408E
Menorrhagia, 267, 385Q, 403E
  therapy for, 397Q, 416E
Menstrual cycle, amenorrhea and, 319
  average length, 217Q, 218E
  cessation of, 335
    as sign of pregnancy, 13
  characteristics, 215–216
  clinical problems, 216
  cycle integration, 213
  establishment and maintenance, 217Q, 218E
  factors influencing, 324Q, 326E
  fertile phase, 226Q, 227E
  follicular or proliferative phase, 213, 392Q 411E
  hirsutism and, 329
  length of, 213
  luteal or secretory phase, 213, 389Q, 407E
  menopausal changes in, 335
  natural family planning and, 390Q, 408E
  pelvic inflammatory disease (PID), 242, 247Q, 249E
  physiologic characteristics of, 393Q, 412E
  in puerperium, 16
Mental status evaluation, of rape victims, 274
Meperidine, indications for, 186, 191Q, 192E
Metabolism, fetal and neonatal, 70
Metallic bead chain cystourethrography, stress urinary incontinence (SUI) diagnosis, 304, *304*, 307Q, 309E
Methoxyflurane, 189
Methylene blue lavage, fallopian tube function, 313
Metrorrhagia, 267
Microcephaly, 131
Microinvasive carcinoma, cervical cancer, 345
  vulvar cancer, 354
Midforceps operation, 179
Midline incision for cesarean section, 178
Midstream urine specimen, 304
Mineral requirements in pregnancy, 26
Missed abortion, 181
Mittelschmerz, 233, 235, 238Q, 240E
Mixed gonadal dysgenesis, 208
M-mode ultrasound, 125
Model Uniform Parentage Act, 362
Molar pregnancy, *see* Hydatidiform mole
Molluscum contagiosum (MC), 291
  lesions associated with, 293Q, 294E
Monilial vaginitis, 204
Monodermal tumors, 351
Morphine for labor, 186, 400Q, 419E
Mortality rates, maternal, 75–76
Mucinous tumors, ovarian cancer, 350

Mucoid myoma degeneration, 296
Mucus (cervical), changes during fertility cycles, 219–220
Müllerian anomalies, 207
  amenorrhea evaluation, 322
Multigravida, 23, 392Q, 410E
Multiparity, 23
  placenta previa and, 139
  pregnancy-induced hypertension, 55
  risk assessment and, 77
Multiple births, Doppler waveform evaluation, 136
  HCG levels, 5
  maternal age (over 35), 76
  placenta previa and, 139
  prematurity and, 93
  ultrasound visualization, 129
Multiple of the mean (MOM), alpha-fetoprotein levels, 25
Mumps, teratogenicity, 158
Muscle relaxants, stress urinary incontinence (SUI) therapy, 305
*Mycoplasma hominis,* 242
Myeloceles, amniocentesis diagnosis for, 117
Myelomeningocele, 164
Myomas, *see also* Uterine leiomyomas
  clinical conditions for, 392Q, 410E
  degenerative changes, 296
  estrogen dependence on, 395Q, 414E
  gross and microscopic features, 299Q, 301E
  high-risk pregnancy assessment, 84
  intramural, 295
  subcerous, 295
  submucous, 295
  treatment, 297–298
Myomata uteri
  characteristics of, 295–296, 389Q, 408E
  signs and symptoms, 296–297, 299Q, 301E
  surgery indications, 386Q, 405E
  therapy, 300Q, 301E, 395Q, 414E
Myomectomy, 298, 300Q, 301E
Myometrium, hypotonic, 16
  physiology, 31–32
  in postmenopausal women, 336
  relaxation by progesterone, 7
Myosin, role in labor, 31
Myosin light-chain kinase, 32

**N**

Näegele's rule for pregnancy dating, 23, 28Q, 30E
  postdatism, 47
Nalbuphine, 186
Narcotic antagonists for labor, 186
Natural family planning, 219–220
  menstrual cycle and, 390Q, 408E
Nausea as sign of pregnancy, 13
Necrosis, myoma degeneration, 296
Necrotizing enterocolitis, 165–166
  susceptibility factors, 172Q, 174E
Negligence, elements of, 359
  home births and, 368Q, 370E
  theory of consent, 364
*Neisseria gonorrhoeae,* 248Q, 249E
  epidemiology, diagnosis and treatment, 287
  pelvic inflammatory disease (PID), 242
  signs and symptoms, 384Q, 402E
Neonatal infection, 168–169
  signs of, 172Q, 174E
Neonatal metabolism, 70
Neonatal mortality, decline in, 161
Neoplasia, *see also specific types*
  lesions, high-risk pregnancy assessment, 84
  with oral contraceptives, 224
  ovarian tumors, 205–206

vaginal, 205
  vaginitis with, 262
Nervous system, 259
Neural tube defects, amniocentesis diagnosis for, 116–117
  incidence of, 122Q, 123E
  risk of recurrence, 122Q, 123E
  ultrasound diagnosis, 131
Neurologic abnormalities, fetal hypoxia and, 39
Neurologic disorders, high-risk pregnancy and, 80
Neurologic examination of rape victims, 274
Neurophysiology of pelvic pain, 229–230
Nevi, 329
Newborn resuscitation, 161–164
Nipple stimulation test , in postdates pregnancy, 49, 51Q, 53E
Nitrazine test for ruptured membranes, 34
Nitrous oxide, 188
Non-germ cell ovarian tumors, 205
Nongranulomatous salpingitis, 246
Nonlactating women, menstruation in, 16
Nonmetastatic gestational trophoblastic neoplasia, 195
  treatment modalities, 197Q, 199E
Nonsteroidal anti-inflammatory drugs (NSAIDS), dysfunctional uterine bleeding therapy, 269
  for endometriosis, 254
  menorrhagia and, 397Q, 416E
Nonstressed fetus, 39
Nonstress test (NST), in postdates pregnancy, 49, 51Q, 53E
  pregnant diabetics, 54
Nulligravida, 23
Nulliparity, 23
  pregnancy-induced hypertension, 55
  risks for, 77
Nursing, 17
Nutrition, during pregnancy, 26, 28Q, 30E
  hypogonadotropic amenorrhea, 321

**O**

Obesity, endometrial cancer and, 347
  high-risk pregnancy assessment, 83
  postmenopausal estrogen levels, 335
Oblique lie of fetus, 15
Obstetric history, *see also* Patient history
  prematurity and, 93
  Rh disease and, 105–106
  risk assessment and, 77–78
Occiput anterior (OA), 33
Occiput posterior (OP), 33
Occiput presentation, 33–34
Occiput transverse (OT), 33
Office visits, first visit in early pregnancy, 47
  first visit in late pregnancy, 47
  frequency of, 25
  monitoring during, 25
Oligomenorrhea, in adolescent, 207
  defined, 213, 267
  in menopausal women, 336
Oliguria, with placental abruption, 398Q, 417E
Omphalocele, 164
One-hour glucose tolerance test, 53
Oogenesis, 214–215
Oophoritis, 243
Operative notes, liability and, 366
Oral contraceptives, 222–224
  cancer prevention and, 392Q, 411E
  dysfunctional uterine bleeding therapy, 269
  endometriosis therapy, 252–254, 255Q, 257E
  hirsutism therapy, 332Q, 334E, 397Q, 415E
  hypertension and, 384Q, 402E

malpractice and, 359–360
  pelvic inflammatory disease (PID), 242
  physiologic changes, 226Q, 227E
  side effects of, 226Q, 227E
  uterine bleeding and, 397Q, 416E
Organogenesis, critical period, 388Q, 406E
Orthopedic conditions, pelvic pain and, 236
O station, 33
Osteitis pubis, with stress urinary incontinence (SUI), 305
Osteoporosis, in postmenopausal women, 336–337, 383Q, 401E
  therapy for, 336–338, 340Q, 341E
Ovarian abnormalities, ultrasound diagnosis, 129
Ovarian accidents, 233–234
Ovarian cancer, diagnosis, 352
  dysgerminoma, 357Q, 358E
  epithelial tumors, 350
  germ cell tumors, 351
  incidence, 349
  pathophysiology, 349–350
  sex cord stromal neoplasms, 350–351
  staging, 350, 357Q, 358E
  treatment, 352–353
  tumor categories, 350–351
Ovarian disorders, hirsutism and, 329
Ovarian hyperstimulation syndrome, 234
Ovarian infertility factor, 313–314
  correction of, 315
Ovarian tumors, 205–206
  characteristics of, 210Q, 212E
  with oral contraceptives, 224
Ovaries, in female fetus, 3
  functions, 314, 317Q, 318E
  torsion of, 392Q, 410E
Overt insulin-dependent diabetes, 54–55
Ovulation, *see also* Anovulation
  documentation of, 314
  expected date of confinement (EDC) and, 23
  evidence of, 316Q, 318E
  induction of, 315
  with HCG, 5
  oogenesis and, 215
  predictors of, 217Q, 218E
  processes in, 390Q, 409E
  progesterone and, 6–7
Ovum, transmigration of fertilized, 279
Oxygen and general anesthesia aspiration, 189
Oxygenation of fetal blood, 70, 71Q, 73E
Oxygen delivery to fetal tissue, 69–70
Oxygen dissociation curve, fetal hemoglobin, 69–70
Oxytocin, iatrogenic fetal distress, 42
  infusion in postdates pregnancy, 50
  labor initiation, 94
  in lactation, 17
  postpartum hemorrhage, 17
  in puerperium, 20Q, 21E
  stimulation during labor, 31


**P**


Pain, myomata uteri, 296
  obstetric neuropathways, 186–187
  in third trimester, 15
  uterine contractions, 187
Painless vaginal bleeding, 144Q, 146E
Pap smear, classification of abnormal, 344–345
  evaluation of abnormal, 344
  high-risk pregnancy assessment, 85
Paracervical block anesthesia, 187, 392Q, 410E
Paramedian incision for cesarean section, 178
Parametritis, 17

signs and symptoms, 398Q, 416E
Parasympathetic nerves, 229, 238Q, 240E
Parathyroid disorders, high-risk pregnancy and, 80
Parity, 23
  risk assessment and, 77
Particulate matter, ultrasound visualization, 131
Parturition, pain with, 187
  uterine contractions, 187
Parvovirus infection, high-risk pregnancy and, 81
Past pregnancy history, 23–24
Paternity, surrogate motherhood, 363
Pathologic amenorrhea, 319
Patient education, during antepartum care, 23
  liability and, 366
Patient history, amenorrhea, 321
  antepartum care, 23–24
  for endometriosis, 252
  hirsutism, 329
  maternal mortality risk assessment, 76
  pelvic pain diagnosis, 230–231
  prematurity and, 99Q, 101E
  rape victims, 273–274, 277Q, 278E
  stress urinary incontinence (SUI) diagnosis, 303
Pco₂, in stressed fetus, 39
"Pediatric cocktail," 203
Pediculosis pubis, 292
Pelvic congestion syndrome, 236
Pelvic examination, for endometriosis, 252
  high-risk pregnancy assessment, 83
  myomata uteri, 297
  pediatric patients, 203
  for pelvic pain, 231
  placenta previa, 388Q, 406E
  of rape victims, 274
  role in antepartum care, 24
  vulvovaginitis, 260
Pelvic exercises, stress urinary incontinence (SUI), 305
Pelvic forceps, 179
Pelvic inflammatory disease (PID), 234–235
  acute, 241
  bacteriology, 242
  causative diseases, 247Q, 249E
  chronic, 241
  contraceptive methods and, 384Q, 402E
  defined, 241
  diagnosis, 243–244, 243t, 244t, 247Q, 249E
  ectopic pregnancy and, 279
  epidemiology, 241–242
  incidence, 241
  IUDs and bacterial infection, 383Q, 401E
  pathophysiology, 242–243, 387Q, 405E
  risk factors, 247Q, 249E
  sequelae, 243, 247Q, 249E
  signs and symptoms, 243, 243t, 244t, 248Q, 249E
  therapy, 245, 395Q, 414E
Pelvic pain, acute, 232–235
  acyclic, 236–237
  associated symptoms, 231
  characteristics, 238Q, 240E
  differential diagnosis, 232–237
  duration and recurrence, 230–231
  endometriosis and, 252
  neurophysiology, 229–230
  pregnancy-related, 232–234
  psychogenic etiology, 237–238Q, 240E
  quality of, 230
  unrelated to pregnancy, 233–234
Pelvic thrombophlebitis, signs and symptoms, 398Q, 417E
Penicillin therapy, pelvic inflammatory disease (PID), 245
Percutaneous umbilical blood sampling (PUBS), prenatal diagnosis with, 120
  Rh disease, 107–108

ultrasound visualization, 127
Pereya procedure, stress urinary incontinence (SUI) therapy, 306
Perforation with IUDs, 222
Peridural anesthesia, iatrogenic fetal distress, 42
Perihepatitis, 243
Perimenopause, characteristics of, 340Q, 341E
  endometrial cancer and, 347
  estrogen replacement therapy, 339
  physiology, 335
Perinatal mortality (stillborn or neonatal), cesarean birth, 177
  factors associated with, 90Q, 91E
  postdatism, 47, 51Q, 53E
  risk assessment for pregnancy, 77
Perineal anesthesia, 190Q, 192E
Periodic patient assessment, during antepartum care, 23
Peritonitis, 243
Periumbilical pain, 230
Pfannensteil incision for cesarean section, 178
pH, fetal, 39, 41, 389Q, 407E
Pharmacologic agents, preterm labor, 97
Phencyclidine (PCP) [angel dust] teratogenicity, 154
Phenolsulfonphthalein (PSP) testing, fallopian tube function, 313
Phenothiazines, prolactin levels and, 6
Phenylketonuria, teratogenicity, 155
Phosphatidylglycerol, in amniotic fluid, 15
Phospholipase A₂, labor initiation, 95
Phototherapy for kernicterus, 166
*Phthirus pubis*, 292
  characteristics of, 293Q, 294E
Physical examination, *see also* Pelvic examination
  in antepartum care, 24
  for amenorrhea, 321
  of high-risk pregnancy patient, 83–84
  of rape victims, 273–274
  stress urinary incontinence (SUI) diagnosis, 303–304
Physiologic amenorrhea, 319
Piper forceps, 180
Pituitary disorders, high-risk pregnancy and, 80
  hirsutism and, 329
  hypogonadotropic amenorrhea and, 321
Pituitary tumors, Cushing's syndrome, 324Q, 326E
  prolactin levels, 6
Placenta, endocrine changes and, 3
  enzyme capabilities, 7
  fetal surface, 67
  functions, 67
  hormones in, 3, 9Q, 11E
  maternal surface, 67
  penetration with amniocentesis, 116
  physiology, 67
  signs of separation in delivery, 35
  ultrasound visualization of, 133
Placenta accreta, 140
Placental abruption, clinical presentation, 141
  cocaine and, 154
  complications, 142
  defined, 141
  diagnosis, 142, 393Q, 411E
  etiology, 141, 144Q, 146E
  incidence, 141
  management, 142, 144Q, 146E
  oliguria, 398Q, 417E
  prematurity and, 94
  signs and symptoms, 144Q, 146E
  transient fetal distress, 398Q, 417E
  ultrasound diagnosis, 133
Placental insufficiency, postdatism, 48
Placental localization, 133
Placental membrane, 67

Placental perfusion, in pregnancy-induced hypertension, 56
Placental steroids, labor initiation, 94
Placental sulfatase deficiency, dynamic testing, 4
Placental thickness, 133
Placental tissue, retention of, 16
Placental transfer of anesthesia, 186, 190Q, 192E
Placenta previa, breech presentation and, 398Q, 417E
  clinical presentation, 139–140, 144Q, 146E
  complications, 140
  defined, 139
  etiology, 139
  incidence, 139
  pelvic examination and, 388Q, 406E
  signs and symptoms, 144Q, 146E
  ultrasound diagnosis, 133
Plasma fibrinogen in puerperium, 16
Plasma testosterone, 322
Plastic cup vacuum extraction, 180
Pneumonitis and general anesthesia, 189
  with aspiration, 190Q, 192E
Pneumosalpingography, fallopian tube function, 313
Po2, in intervillous blood, 69
  nonstressed fetus, 39, 43Q, 45E
Polycythemia, infants of diabetic mothers, 168
Polyhydramnios, 131, 389Q, 407E
Polymenorrhea, 213
  defined, 267
Poor prognosis metastatic gestational trophoblastic neoplasia characteristics, 195–196, 197Q, 199E
Postcoital contraception, 225
Postcoital testing, 317Q, 318E, 394Q, 413E
  male infertility factor, 312
Postdatism
  anencephaly and, 384Q, 401E
  management, 50
  risk assessment, 51Q, 52E
Posterior fossa cysts, 131
Postpartum care, pregnancy-induced hypertension, 57
  pregnant cardiac patients, 61
Postpartum hemorrhage, 16–17, 19Q, 21E
  anesthetic agents and, 190Q, 192E
  complications, 144Q, 146E
  management of, 17
  placenta previa and, 140
Postpill amenorrhea, 223
Preantral follicle, 215
Precocious puberty, 208–209
  characteristics of, 210Q, 212E
Preconception care, and malpractice, 359–360
  of pregnant diabetics, 54
Predifferentiation, teratogenicity and, 149
Prednisone, hirsutism and, 397Q, 415E
Preeclampsia, Apgar scores and, 173Q, 174E
  clinical aspects, 56
  defined, 55
  in diabetics, 54
  etiology, 56
  hospitalization for, 400Q, 418E
  magnesium sulfate and, 400Q, 418E
  multiparity and, 62Q, 64E
  pathophysiology, 62Q, 64E
  recurrent risk of, 78
  risk assessment for, 86, 86t
  weight gain and, 26
Pregnancy, abnormal signs, 28Q, 30E
  average duration of, 47
  clinical evidence, 13–14
  dating, 14
    *See also* Postdatism
  diagnosis of, 13–14
  effect on diabetes, 53–54
  endocrine changes, 3

HCG levels and, 391Q, 409E
  with herpes simplex virus-2 (HSV-2), 263
  IUD-related complications, 222
  positive signs of, 14
  presumptive symptoms, 13
  of rape victims, 275
  signs of, 19Q, 21E, 29Q, 30E
  termination of, 361–362
Pregnancy-induced hypertension, 55–56, 56–57
  classification, 55–56
  placental abruption and, 141
  recurrent risk of, 78
Pregnancy rates, condoms, 220
  diaphragms, 221
  with endometriosis, 254
  spermicides, 220
  vaginal sponges, 221
Pregnancy testing, HCG as marker, 5
  for pelvic pain evaluation, 232
Pregnanediol, progesterone levels, 6
Pregnenolone, estriol synthesis, 7
Preinvasive cervical disease, management of, 344–345
Premalignant lesions, 84
Premature labor, drug inhibition of, 385Q, 402E
  early signs of, 384Q, 402E
  recurrence risk, 385Q, 404E
  in second trimester, 14
Premature menopause, 335
Premature rupture of membranes in second trimester, 14
Prematurity, defined, 93
  disorders of, 164–167
  epidemiology, 93
  IUD-related, 222
  recurrence of, 99Q, 101E
  risk factors, 86, 86t, 93–94, 100Q, 101E
Premenstrual syndrome (PMS), 216
Prenatal care, duties of, 361
  of pregnant cardiac patients, 60
  of pregnant diabetics, 54
Prenatal diagnosis, defined, 115
Preovulatory follicle, 215
Preponderance of evidence, malpractice and, 359
Presentation of fetus, 15
Pressure, myomata uteri, 297
Preterm labor, early detection, 95
  management of, 95–97, 96
  risk factors, 77, 87, 99Q, 101E
  therapy for, 97–98, 100Q, 102E
Primary dysmenorrhea, 235, 395Q, 413E
Primary hysterectomy, for hydatidiform mole, 194
Primary infertility, 311
Primigravida, 23
Primipara, 23, 392Q, 410E
Primordial follicle, 214
Privacy, artificial insemination and, 362
  malpractice and, 359
Probenecid, pelvic inflammatory disease (PID), 245
Prochlorperazine, 275
Progesterone, chemical structure, 6, 9Q, 11E
  deficiency and corpus luteum insufficiency, 4
  functions, 6–7, 7, 37Q, 38E, 384Q, 402E
  HCG role in production, 5
  levels of, 6
  source, 6, 9Q, 11E
  withdrawal during labor, 31
Progesterone challenge test, amenorrhea evaluation and, 321
Progesterone-releasing T-shaped device, 221
Progestin, for endometriosis, 253
  estrogen replacement therapy and, 338–339
  impact on female fetus, 206
  properties of, 270Q, 272E

therapy for dysfunctional uterine bleeding, 268–269
Prolactin, amenorrhea evaluation, 322, 325Q, 326E
  chemical structure, 5
  functions, 6
  in lactation, 17
  levels, 6, 9Q, 11E
  sources, 5–6
Prolonged corpus luteum activity, 267
Promethazine, for labor, 186
Prophylactic chemotherapy, for hydatidiform mole, 194
Prophylaxis for aspiration, 189
Propylthiouracil, hyperthyroidism in pregnancy, 59
Prostaglandins, ductus arteriosus dynamics and, 69
  functions, 37Q, 38E
  gel, for postdates pregnancy, 50
  labor initiation, 94–95
  postpartum hemorrhage, 17
  presence during coitus, 27
  release during labor, 31
  synthesis, 36Q, 38E
    inhibition, 97, 99Q, 101E
  uterine contractions with dysmenorrhea, 216
Prostaglandins synthetase inhibitors, 216
Protein requirements in pregnancy, 26
Proteinuria and preeclampsia, 56
Protozoal sexually transmitted disease, 291–292
Psychogenic pelvic pain, 237, 238Q, 240E
Psychological stress, prematurity and, 93
  in rape victims, 275, 277Q, 278E
Psychosocial support during antepartum care, 23
Pudendal block anesthesia, 187
  for episiotomy, 400Q, 419E
Puerperal infection, 17
Puerperium, characteristics of, 16–17
Pulmonary disease, high-risk patient assessment, 78
Pulmonary embolism, maternal mortality and, 75
Pulmonary function, anesthesia and, 185
Pure gonadal dysgenesis, 208

**Q**

Q-tip or straight catheter test, stress urinary incontinence (SUI) diagnosis, 304
Quickening, *see* Fetal movement

**R**

Race, as mortality risk factor, 75
Radiation, *see also* Ionizing radiation
  fetal-neonatal transition, 162
Radiation therapy, dysgerminoma, 356Q, 358E
Radioactive iodine teratogenicity, 151
Radioimmunoassay, estriol levels, 8
  HCG levels, 4
Radioreceptor assay and HCG levels, 5
Radiotherapy, for endometrial carcinoma, 348–349
  invasive cervical cancer, 346
Rape, characteristics of, 277Q, 278E
  defined, 273
  epidemiology, 273
  evidence of ejaculation, 395Q, 413E
  HCG levels and, 400Q, 419E
  psychological sequelae, 275, 277Q, 278E
  support evidence for, 277Q, 278E
  vulvovaginal lesions, pediatric patients, 203–204

Reactivity in fetal heart rate (FHR), 40
Real-time B-scanner, 125
Real-time ultrasound, equipment, 115
  intraventricular hemorrhage, 167
  prenatal diagnosis, 121Q, 123E
  Rh disease, 106, 109–110
Recovery, malpractice and, 359
Recreational drugs, 152, 154
5α-Reductase enzyme, 328
Redundant hymen in pediatric patients, 203
Referred pelvic pain, 229
Regional anesthesia for cesarean section, 178
Renal disease, high-risk patient assessment, 79
  prenatal ultrasound diagnosis, 116
Renal failure, with placental abruption, 142
  placenta previa and, 140
Renal perfusion, in pregnancy-induced hypertension, 56
Renal vascular thrombosis, infants of diabetic mothers, 168
Reproductive disorders, myomata uteri, 297, 299Q, 301E
Reproductive technology, 362–363
Reproductive tract abnormalities, 83–84
Residual urine volume, stress urinary incontinence (SUI) diagnosis, 304
Resistant ovary syndrome, 320
Respiratory distress syndrome, 164–165
  infants of diabetic mothers, 168
  risk assessment for, 87t, 88, 88t
  signs and symptoms, 385Q, 403E
Resuscitation, newborn, 161–164
  respiratory distress syndrome, 164
Retrograde menstrual flow in endometriosis, 251
Retroplacental clot, placental abruption diagnosis, 142
Retropubic urethropexy, stress urinary incontinence (SUI) therapy, 305
Rheumatic disease, high-risk patient assessment, 79
Rh immune globulin, administration, 111
  dosage, 389Q, 407E
  failure of, 111
  prevention of Rh disease, 110–111
Rh isoimmunization, amniocentesis diagnosis of, 118
  clinical management, 105–110
  defined, 103
  degree of sensitization, 103
  epidemiology, 103
  erythroblastosis fetalis, 396Q, 415E
  estriol levels, 386Q, 404E
  evaluation of, 385Q, 403E
  immunology of, 103
  incidence, 103, 112Q, 114E
  laboratory testing, 28Q, 30E, 394Q, 412E
  prevention, 110–111
  risk factors, 105t, 112Q, 114E
Rifenstein syndrome, 206
Right occiput posterior (ROP), 33
Right occiput transverse (ROT), 33
Ringer's solution, postpartum hemorrhage, 17
Risk factors, characteristics of, 90Q, 91E
  drugs and medications in pregnancy, 152
  high-risk pregnancy, 86–89, 86t, 87t, 88t–89t
  identification in antepartum care, 24
Risk management, liability and, 365–366
Ritodrine, preterm labor therapy, 97
Rokitansky-Kuster-Hauser syndrome, 319
Roll-over test for preeclampsia and eclampsia, 86
Round ligament pain, in second trimester, 14
Rubella virus (German measles), high-risk pregnancy assessment, 85
  teratogenicity, 155
Ruptured membranes, 36Q, 38E

detection of, 34
  in third trimester, 15
Ruptured uterus, 143
  in cesarean section, 178–179
  risk factors, 84, 90Q, 91E
  signs of, 145Q, 146E
Ruptured vasa previa, 143

**S**

Salpingectomy, ectopic pregnancy treatment, 281
Salpingitis, ectopic pregnancy and, 279, 283Q, 285E
Salpingo-oophorectomy, ectopic pregnancy treatment, 281
  for endometrial cancer, 385Q, 403E
Salpingo-oophoritis, chronology of, 242–243
  signs and symptoms, 234
Salpingostomy, ectopic pregnancy treatment, 281
Sanger incision for cesarean section, 178
Sarcoidosis, 246
Sarcoma botryoides, 205
Sarcomatous myoma degeneration, 296
*Sarcoptes scabiei*, 292
Scabies, 292
Schiller-Duval body, 351
Schistosomiasis, 246
Secondary dysmenorrhea, 235–236
Secondary infertility, 311
Second trimester, complications, 14, 19Q, 21E
  signs and symptoms, 14, 19Q, 21E
  ultrasound visualization, 129–131
Sellheim incision for cesarean section, 178, 400Q, 419E
Semen analysis, male infertility factor, 312
  as rape evidence, 274
Sepsis, in abortion, 233
  maternal mortality from, 75
  prematurity and, 94
Serous tumors in ovarian cancer, 350
Sertoli-Leydig cell tumors, ovarian cancer, 351
Serum pregnancy tests, 13
Sex chromosomal abnormalities, indications for amniocentesis, 116
  and maternal age (over 35), 76
Sex cord stromal neoplasms, 350–351
  treatment, 352
Sex hormone binding globulin (SHBG), 328, 332Q, 334E
Sexual abuse of children, 275–276, 387Q, 405E
  incidence, 275
  risk factors, 276, 277Q, 278E
Sexual ambiguity, conditions for, 207–208, 210Q, 212E
Sexual assault, defined, 277Q, 278E
Sexually transmitted disease (STD), bacterial, 287–288
  chlamydial, 289
  condoms and, 220
  ectoparasites, 292
  protozoal, 291–292
  spirochetal, 288–289
  viral, 290–291
Sheehan's syndrome, 140
Shirodkar cervical cerclage, 180, 400Q, 419E
Shoulder dystocia, prevalence in pregnant diabetics, 55
Shoulder pain, 230
Shunts, insertion of, ultrasound visualization, 127
Sickle-cell anemia-hemoglobin SS disease, 58
Sickle-cell-β-thalassemia disease, 58
Sickle cell-hemoglobin SC disease, 58
  signs and symptoms, 63Q, 64E

Sickle cell trait, 58
  asymptomatic bacteriuria, 59
  chorionic villi sampling, 400Q, 418E
  risk factors for, 62Q, 64E
  signs and symptoms, 63Q, 64E
Simpson's forceps, 180
Skeletal abnormalities, 133
Skin changes in menopausal women, 338
Sling procedures in stress urinary incontinence (SUI) therapy, 306
Smoking, cervical cancer and, 343–344
  fetal morbidity and mortality, 76
  pregnancy and, 27
  effects on fetus, 391Q, 394Q, 409Q, 413E
  prematurity and, 93
Socioeconomic status, maternal mortality risk assessment, 76
  prematurity and, 93
Sodium requirements in pregnancy, 26
Sonolucent tissue, 125
Spectinomycin, pelvic inflammatory disease (PID), 245
  venereal disease in rape victims, 275
Spectrophotometry, Rh disease and, 106–108, *107–108*
Speculum examination, of rape victims, 274
  vulvovaginitis, 260
Sperm acrosin measurement, 312
Sperm antibodies, male infertility factor, 312
Spermatozoa, fertilizing capacity testing, 312
  motile, 386Q, 404E
  as rape evidence, 274
Spermicides, mechanism of action, 226Q, 227E
  mode of action and pregnancy rates, 220–221
Spina bifida occulta, amniocentesis diagnosis for, 117
Spinal anesthesia, 187
Spinal blockade and respiratory paralysis, 187–188
Spinal headache, causes of, 190Q, 192E
  with spinal anesthesia, 187–188
Spirochetal sexually transmitted disease, 288–289
Splanchnic pelvic pain, 229
Spontaneous abortion, 180
  IUD-related, 222
  pelvic pain with, 232–233
Squamous cell carcinoma, vaginal cancer, 353
  vulvar cancer, 355
Standard glucose tolerance test, 53
Static B-scanner, 125
Statutory rape, 277Q, 278E
  defined, 273
Sterilization, defined, 360
  involuntary, 360
  medicolegal considerations, 360
  tubal ligation, 368Q, 370E
  voluntary, 360
Steroids and labor initiation, 94
"Strawberry spots" in vulvovaginitis, 260
*Streptococcus vaginitis*, 383Q, 401E
Stress and hypogonadotropic amenorrhea, 321
Stressed fetus, 39–40
Stress urinary incontinence (SUI), defined, 303
  diagnosis, 303–305
  epidemiology, 303
  etiology, anatomical abnormalities, 304, 307Q, 309E
  management, 305–306
  pathophysiology, 303
  type I, 304
  type II, 305
  urethra and bladder in, 391Q, 409E
Subcerous myomas, 295
  etiology, 299Q, 301E
Submucous myomas, 295
  signs and symptoms, 299Q, 301E

Suction, general anesthesia aspiration, 189
Suction curettage, for hydatidiform mole, 194
Superior mesenteric plexus, 230
Supine pressor test for preeclampsia or eclampsia, 86
Support groups for rape victims, 275
Surgical techniques, for cesarean section, 178
  for endometriosis, 253–254
  invasive cervical cancer, 346
Surrogate motherhood, 363
Swyer's syndrome, amenorrhea in, 208
Sympathomimetics, stress urinary incontinence (SUI) therapy, 305
Syphilis, characteristics of, 288–289, 293Q, 294E
  teratogenicity, 156–157
Systemic illness, amenorrhea, 207

**T**

Tachycardia, fetal, 40
Tay-Sachs disease, amniocentesis diagnosis of, 118, 400Q, 418E
Teratogenic agents, defined, 149
  drugs and medications, 151–154
  embryonic and fetal sensitivity, 149, *150*
  hyperthermia (fever), 155
  infections, 155–158
  ionizing radiation, 150–151
  malpractice and, 361
  maternal medical disorders, 155
  range of effects, 159Q, 160E
  route and length of administration, 150
Teratogenicity, criteria for, 150
Terbutaline, preterm labor therapy, 97
Testes, male fetus, 3
Testicular feminization, 206, 320
  characteristics, 324Q, 326E
Testosterone, free, 328
  androgen activity and, 388Q, 406E
Tetracycline, and pelvic inflammatory disease (PID), 245
  teratogenicity, 396Q, 415E
Therapeutic abortion, in class III cardiac patients, 61
  in rape victims, 275
Third stage of labor, 35
Third trimester, hemorrhage in, 143
  laboratory testing in, 25
  in pregnant diabetics, 54–55
  symptoms, 14–15
  ultrasound visualization, 129–131
Thoracolumbar sympathetic nerves, 229–230
Threatened abortion, 181, 232
  ectopic pregnancy and, 284Q, 285E
  signs of, 248Q, 249E
Thromboembolism with oral contraceptives, 223, 226Q, 227E
Thrombophlebitis, 17
Thyroid-binding globulin (TBG), elevations during pregnancy, 3
Thyroid disease, high-risk patient assessment, 79
  ovulatory defects and, 314
  pregnancy and, 59
Thyroidectomy during pregnancy, 59
Thyroid hormone, hyperthyroidism in pregnancy, 59
  for hypothyroidism, 59
Thyroid indices (T$_3$ and T$_4$), 322
Thyroid-stimulating hormone (TSH), amenorrhea evaluation, 325Q, 326E
  hydatidiform mole, 194
  prolactin levels and, 322
Time position scan, 115
Tissue, acoustic properties, 125
Tobacco, *see* Smoking

Tocolysis, preterm labor, 97–98, 100Q, 102E
  prostaglandin synthesis, 99Q, 101E
Torsion, ovarian, 233–234
Toxoplasmosis, high-risk pregnancy and, 81
  teratogenicity, 156
Tracheoesophageal fistula, 163–164
Traction, for shoulder delivery, 35
Transient tachypnea, risk assessment for, 88, 88t
Translocation, indications for amniocentesis, 116
Transperitoneal cesarean section, 178
Transverse lie of fetus, 15
Transverse scans of fetal spine, 131
Transverse vaginal septum, 319–320, 395Q, 414E
  treatment, 322
Trauma, placental abruption and, 141
  vulvovaginal lesions in pediatric patients, 203–204
Traumatic vaginitis, 262
Travel during pregnancy, 27
*Treponema pallidum,* 288–289
*Trichomonas vaginalis,* 261, 291–292
Trimester model of abortion, 361
Trisomy 21, alpha-fetoprotein diagnosis of, 121Q, 123E
  amniocentesis diagnosis of, 117
  head:body ratio, 137Q, 138E
  maternal age and, 76, 122Q, 123E
  recurrence risk, 383Q, 401E
  risk factors for, 117t
Trophoblastic neoplasia, HCG as marker, 5
True labor, 32
Tubal infertility factor, 313
  correction of, 315
Tubal insufflation, fallopian tube function, 313
Tuberculosis, amenorrhea and, 320
Tuberculous salpingitis, 245–246
Tubo-ovarian abscess, pathogenesis, 247Q, 249E
  signs and symptoms, 234–235
Tucker-McLean forceps, 180
Turner's syndrome, amenorrhea in, 208
Two-cell hypothesis of estrogen production, 214, 217Q, 218E

**U**

Ultrasound, 125
  amniotic fluid in postdates pregnancy, 49
  biophysical profile, 133–134
  B-scan, 115
  clinical indications for, 115–116
  congenital anomaly diagnosis, 131–133
  Doppler flow evaluation, 134–136
  ectopic pregnancy diagnosis, 281
  erythroblastosis fetalis, 112Q, 114E
  functions, 126–127
  for molar pregnancy, 398Q, 417E
  ovarian cancer, 352
  pelvic inflammatory disease (PID), 244
  for pelvic pain evaluation, 232
  physics of, 125
  placental development, 133
  pregnancy dating with, 384Q, 402E
  prenatal diagnosis with, 115–116
  recognition of fetus, 14
  Rh disease and, 106
  scattered echoes, 115
  specular echoes, 115
  types of, 125
Umbilical artery waveforms, 134, 136, *136*
Umbilical circulation, 69
  fetal hypoxia and, 72Q, 73E
  intrauterine resuscitation, 42
  vein, 387Q, 405E
Umbilical cord, structure, 67

ultrasound visualization, 131
Umbilical structures, 67
Underweight patients, high-risk pregnancy assessment, 83
Urethra, and bladder, 391Q, 409E
  prolapsed, 204
Urinary tract infection, causes, 62Q, 64E
  characteristics of, 90Q, 91E
  pelvic pain and, 236
  during pregnancy, 58–59
  prematurity and, 94
  in puerperium, 16–17
  signs and symptoms, 398Q, 417E
Urination, changes in as sign of pregnancy, 13
Urine analysis, estriol levels, 8
  high-risk pregnancy assessment, 85
  hormone levels, 4
  for pelvic pain evaluation, 232
  pregnancy testing, 13
  role in antepartum care, 25
Uterine abnormalities, ultrasound diagnosis, 129
Uterine bleeding, with IUDs, 222
  labor initiation, 94
  ultrasound diagnosis, 126–127
Uterine contractions, active-phase characteristics, 386Q, 405E
  agents for, 17
  dysmenorrhea and, 216
  hypertonic, 398Q, 417E
  pain of, 187
  progesterone prevention of, 7
Uterine cramping, in first trimester, 14
Uterine decompression, placental abruption and, 141
Uterine evaluation, high-risk pregnancy assessment, 83
Uterine hemostasis, 35
Uterine incisions, for cesarean section, 389Q, 407E
Uterine infertility factor, 313
  correction of, 315
Uterine leiomyomas, defined, 295
  degenerative changes, 296
  etiology, 295
  pathology, 296
  signs and symptoms, 296–297
  treatment, 297–298
  types of, 295–296
Uterine relaxation, intrauterine resuscitation, 42
Uterine rupture, *see* Ruptured uterus
Uteroplacental insufficiency, 39, 43Q, 45E
Uterus, changes in as sign of pregnancy, 13
  gestational age and, 126
  involution of, 16
  overdistension, 94
  pathology, 236
  in postmenopausal women, 336
  reproductive role of, 313
  structural abnormalities, 83–84

**V**

Vacuum extractor, 179–180
  prerequisites for use, 182Q, 183E
Vagina, anatomy and physiology, 259
  health factors for, 264Q, 266E
  in postmenopausal women, 335–336
  visualization in pediatric patient, 203
Vaginal adenosis, 205–206
Vaginal atresia, 205, 207
Vaginal cancer, 353
Vaginal cystourethropexy, stress urinary incontinence (SUI) therapy, 305, 389Q, 407E
Vaginal delivery, after cesarean section, 179
  with placental abruption, 142
  spontaneous, 35

Vaginal discharge, causes of, 210Q, 212E
  in newborn females, 210Q, 212E
  vulvovaginal lesions, 204
  vulvovaginitis, 260
Vaginal ectopic anus, 205
Vaginal sonography, 128
Vaginal sponge, 221
Vaginal tumors, 205
Vaginitis, 261–263
  atrophic, 399Q, 417E
  therapy for, 265Q, 266E
    for both partners, 264Q, 266E
Valproic acid teratogenicity, 396Q, 415E
Variable decelerations in fetal heart rate
    (FHR), 41, 43Q, 45E
Varicella zoster, teratogenicity, 157
VDRL, assessment of fetal risk in maternal
    syphilis, 84
Vegetarian diets during pregnancy, 28Q,
    30E
Venereal disease in rape victims, 274–275
Venous thromboembolic disorders,
    high-risk pregnancy and, 80
Ventilation, general anesthesia aspiration,
    189
Vertex presentation of fetus, 15
Vessel spread in endometriosis, 251
Viral sexually transmitted disease, 290–291
Virilization, 327
Virilizing tumors, maternal, 207
  teratogenicity, 155
Visual disturbance in preeclampsia, 56
Vitamin A isomer, teratogenicity, 152,
    153t–154t
Vitamin B₁₂ requirements in pregnancy, 27
Vitamin C requirements in pregnancy, 27
Vitamin requirements in pregnancy, 27
Volatile anesthetics, 189
Von Willebrand's disease, platelet function
    test, 271Q, 272E
Vulva, physiology and anatomy, 259, 394Q,
    412E
Vulvaginitis, defined, 259
  diagnosis, 259–260
Vulvar cancer staging, 354, 356Q, 358E
  carcinoma, 353–355
Vulvar dystrophies, 262, 264Q, 266E
Vulvovagina, anatomy and physiology, 259

  atrophic vaginitis, 261–262
  *Candida* vaginitis, 260–261
  conditions, 260–263
  *Gardnerella vaginalis* vaginitis, 261
  herpes simplex genitalis, 262–263
  lesions in pediatric patients, 261
  neoplasia, 262
  traumatic vaginitis, 262
  *Trichomonas vaginalis* vaginitis, 261
  vulvar dystrophies, 262

**W**

Warfarin teratogenicity, 396Q, 415E
Weight gain in pregnancy, 26–27
  preeclampsia, 56
Weight reduction and stress urinary incon-
    tinence (SUI), 305
Work environment, prematurity and, 93
Wrongful birth and life, 363–364, 397Q,
    416E
Wrongful conception, 363, 397Q, 416E

**X**

X-chromosome abnormalities, 208
  malpractice and, 361
X-rays, for pelvic pain evaluation, 232
  teratogenicity, 390Q, 408E

**Y**

Yeast infections, patient history and,
    260–261, 266Q, 266E

**Z**

Zona-free hamster ovary penetration test,
    312